THALLIUM-BASED HIGH-TEMPERATURE SUPERCONDUCTORS

APPLIED PHYSICS

A Series of Professional Reference Books

Series Editor

ALLEN M. HERMANN

University of Colorado at Boulder
Boulder, Colorado

1. Hydrogenated Amorphous Silicon Alloy Deposition Processes, *Werner Luft and Y. Simon Tsuo*
2. Thallium-Based High-Temperature Superconductors, *edited by Allen M. Hermann and J. V. Yakhmi*
3. Composite Superconductors, *edited by Kozo Osamura*

Additional Volumes in Preparation

THALLIUM-BASED HIGH-TEMPERATURE SUPERCONDUCTORS

EDITED **BY**

ALLEN M. HERMANN
University of Colorado at Boulder
Boulder, Colorado

J. V. YAKHMI
Bhabha Atomic Research Centre
Bombay, India

CRC Press
Taylor & Francis Group
Boca Raton London New York

CRC Press is an imprint of the
Taylor & Francis Group, an **informa** business

CRC Press
Taylor & Francis Group
6000 Broken Sound Parkway NW, Suite 300
Boca Raton, FL 33487-2742

First issued in paperback 2019

© 1994 by Taylor & Francis Group, LLC
CRC Press is an imprint of Taylor & Francis Group, an Informa business

ISBN-13: 978-0-8247-9114-8 (hbk)
ISBN-13: 978-0-367-40227-3 (pbk)

Library of Congress Cataloging-in-Publication Data

Thallium-based high-temperature superconductors / edited by Allen M.
    Hermann, J.V. Yakhmi.
        p.   cm.—(Applied physics ; 2)
    Includes bibliographical references and index.
    ISBN 0-8247-9114-2 (alk. paper)
    1. High temperature superconductors.   2. Thallium compounds—
Electric properties.   I. Hermann, Allen M.   II. Yakhmi,
J. V.   III. Series: Applied physics (Marcel Dekker, Inc.)
; 2
QC611.98.H54T47   1994
537.6'223—dc20   93-26660
                                                           CIP

Visit the Taylor & Francis Web site at
http://www.taylorandfrancis.com

and the CRC Press Web site at
http://www.crcpress.com

Series Introduction

The Applied Physics series represents a commitment by Marcel Dekker, Inc., to develop a book series that provides up-to-date information in the new and exciting areas of physics emanating from the explosion of discoveries in new materials and new processing techniques. The advances in amorphous materials, layered copper oxide high-temperature superconductors, organic and inorganic cage structures, organic conductors, and a host of new thin-film deposition and nanofabrication processes have led to a host of new applications. It is the intent of the series editor to invite experts in the most important of these areas to assemble book-length manuscripts describing the latest developments. Many volumes will take the form of comprehensive multi-authored compendia that are linked logically by judicious choice of subject matter and editing.

The Applied Physics series is designed to bring the non-specialist scientist/ engineer to the forefront of each area. Researchers wishing to enter a particular area will find these volumes of significant benefit as a detailed introduction to the field. The books in the series are also expected to provide important summaries and to serve as reference guides to the experts in each field.

The scope of the series is broad. We will encourage volumes relating to new or improved sensors, memories, processing units, displays, energy conversion, storage and delivery, and to semiconducting and superconducting devices generating technologies in areas ranging from communications and computers to medical physics.

We sincerely hope that through this series we can help today's discoveries rapidly become tomorrow's technology.

Allen M. Hermann

Preface

Thallium-based cuprate superconductors occupy a special place in the ever-increasing family of high-temperature superconducting oxides because, as of this writing, the highest reproducible zero-resistance temperature record of 127 K is held by the superconductor $Tl_2Ba_2Ca_2Cu_3O_{10}$. Discovered in early 1988, the Tl-based superconductors encompass the broad homologous series $Tl_1Ba_2Ca_{n-1}Cu_nO_{2n+3+\delta}$ ($n = 1-5$), $Tl_2Ba_2Ca_{n-1}Cu_nO_{2n+4+\delta}$ ($n = 1-4$), $(Tl,Pb)Sr_2Ca_{n-1}Cu_nO_{2n+3}$ ($n = 1-3$), and various substituted phases, and constitute the largest single class of high-temperature superconducting cuprates. In fact the largest number of high-T_c superconductors with $T_c > 77$ K belong to this class.

Thallium-based cuprate conductors provide an excellent example of how the stacking arrangements of two structurally dissimilar but dimensionally compatible building blocks, the perovskite and rock-salt layers, can lead to a wide variety of superconducting phases. In addition, the richness and complexity of solid-state chemistry displayed by this class of oxide superconductors continue to provide new directions toward a better understanding of the high-T_c phenomenon. For instance, it is felt that Tl vacancies can play an important role in the creation of holes, at least in Tl double-layer series of superconductors. Although the mechanism of hole doping may turn out to be complex, Tl-based cuprates offer the possibility of two or more redox systems within a single superconducting compound, Tl(III)/Tl(I), Cu(III)/Cu(II), and Pb(IV)/Pb(II), thus providing a remarkable opportunity to adjust the internal redox mechanism (self-doping) via, say, $Tl^{III} + Cu^{II} \rightarrow Tl^{(III-\delta)}$. This also offers a versatile tool to provide a better understanding of the mechanism of hole doping and the relation of T_c with n_h, which may specifically establish a general methodology to suppress the metal–insulator transition in order to achieve superconductivity in many potential parent perovskite phases that exist as insulators. Thallium cuprate superconductors especially have proved to be a very fertile hunting ground for the synthetic chemist in view of the enormous possibilities they provide for continuous variation of the conduction behavior of a multi-cation parent material. Just to cite an example, $(Tl_5Pb_5)Sr_2(Ca_xY_{1-x})Cu_2O_2$ with $T_c(max) = 105$ K for $x = 0.8$ is an excellent system for demonstrating the fine control of the formal valence of copper in a wide range through variation of "x", so that one can observe a metal–insulator transition at $x = 0.4$, an underdoped state of $0.4 \leq x < 0.8$, and an overdoped state for $0.8 < x \leq 1.0$. In fact, the Tl-1212 superconductor, whose structure is similar to $YBa_2Cu_3O_{7-\delta}$, has prompted the development of a number of new superconductors, including the many thallium-free Pb-based ones, through systematic substitutions at different cation sites. In the applications arena, the low R_s values observed for Tl-2212 films are particularly encouraging and augur well for use in high-frequency devices. In addition, faster processes at higher temperatures and the higher T_cs (up to 127 K) make them more promising candidates for fabrication of high-speed, thin-film radiation detectors.

Despite a very large effort in the synthesis of Tl-based superconducting materials in bulk, thin films, and in single crystals and the emergence of an understanding of the physical, structural, and superconducting characteristics of the individual superconducting phases during the past five years, no attempt has been made to publish a comprehensive book (treatise) devoted exclusively to the thallium-based superconductors. Although a few review articles have been published on Tl–cuprate superconductors in certain conference proceedings or as chapters in books, they lie scattered and do not expose the reader to the whole spectrum of activity and the state of the existing knowledge regarding this special class of cuprates. Our effort to plan and edit the present book fills this gap. We have made an attempt to put together contributions from leaders in various aspects of the ongoing research and development of Tl-based cuprate superconductors. A look at the Table of Contents would convince the reader that we have tried to put together a "complete" book on the subject covering all the important aspects in a logical sequence. Thus the reader will find chapters providing up-to-date basic knowledge on the different categories of Tl-based superconducting materials, details of their crystallographic structure, their physical behavior as investigated by transport measurements, magnetization, specific heat, infrared, tunneling, NMR, XPS, etc., and details of different methods employed to synthesize Tl-based superconducting materials in bulk, single crystals, thin films and tapes, and their characterization. Separate coverage has been provided on thermal stability, site-selective substitution, redox mechanisms, auto-doping and band structure calculations. Care has been taken to include chapters on generally hard-to-get information on the chemistry of thallium, its handling and safety requirements, and its solid-state chemistry aspects. A chapter devoted exclusively to TBCCO electronics is included. A chapter on government-sponsored TBCCO research provides a perspective on the potential of Tl-based superconductors for use in devices.

Over the past four years, the state of existing knowledge of the superconducting characteristics of Tl-based superconducting materials and their general behavior has, we believe, attained a level of maturity, particularly as a result of the measurement of data on recently available single crystals. Therefore, we feel that this book will not only serve the high-T_c community in general but will also help to maintain and enliven the interest and enthusiasm of workers in the field by serving as a resource book for ready reference.

We wish to record our sincere thanks to the contributors for having agreed to write their respective chapters, in spite of their own tight schedules, and for meeting the deadlines for submission. We indeed appreciate the special efforts required on the part of all contributors to put together up-to-date material on their respective specializations, simultaneously taking care to avoid undue overlap with other contributions. We are very grateful to the publisher, Marcel Dekker, Inc., for inviting us to edit this book. We thank Gina Hill Mooney for the editorial support during the preparation of this book and last but not least, our wives, Leonora Christopher Hermann and Amar Upasana, for the cooperation and patience shown during the course of the preparation of this book.

Allen M. Hermann
J. V. Yakhmi

Contents

Contents

Contributors

Roland E. Allen Department of Physics, Texas A&M University, College Station, Texas

R. N. Bhattacharya Device Development Branch, National Renewable Energy Laboratory, Golden, Colorado

R. D. Blaugher Device Development Branch, National Renewable Energy Laboratory, Golden, Colorado

Hans B. Brom Department of Physics, Kamerlingh Onnes Laboratory, Leiden University, Leiden, The Netherlands

Ho Sou Chen Metallurgy and Ceramics Research, AT&T Bell Laboratories, Murray Hill, New Jersey

Timir Datta Institute for Superconductivity, University of South Carolina, Columbia, South Carolina

M. S. Davis Strategic Analysis Inc., Arlington, Virginia

L. Pierre de Rochemont Radiation Monitoring Devices, Inc., Watertown, Massachusetts

H. M. Duan Department of Physics, Superconductivity Laboratories, University of Colorado at Boulder, Boulder, Colorado

Peter P. Edwards School of Chemistry, The University of Birmingham, Birmingham, England

Robert A. Fisher Materials Sciences Division, Lawrence Berkeley Laboratory, University of California at Berkeley, Berkeley, California

Yasuo Fukuda Research Institute of Electronics, Shizuoka University, Hamamatsu, Japan

Patrick K. Gallagher Departments of Chemistry and Materials Science and Engineering, The Ohio State University, Columbus, Ohio

Ashok K. Ganguli* Central Research and Development, Experimental Station, E. I. du Pont de Nemours & Company, Inc., Wilmington, Delaware

David S. Ginley National Renewable Energy Laboratory, Golden, Colorado

John B. Goodenough Center for Materials Science and Engineering, The University of Texas at Austin, Austin, Texas

J. Gopalakrishnan Solid State and Structural Chemistry Unit, Indian Institute of Science, Bangalore, India

Joel E. Gordon Department of Physics, Amherst College, Amherst, Massachusetts and Lawrence Berkeley Laboratory, University of California at Berkeley, Berkeley, California

Allen M. Hermann Department of Physics, Superconductivity Laboratories, University of Colorado at Boulder, Boulder, Colorado

Maryvonne Hervieu Laboratoire de Cristallographie et Sciences des Matériaux, Caen, France

A. W. Hewat Institut Max von Laue–Paul Langevin, Grenoble, France

E. A. Hewat Centre d'Etudes Nucléaires, Grenoble, France

R. M. Iyer Chemistry Division, Bhabha Atomic Research Centre, Bombay, India

A. B. Kaiser Department of Physics, Victoria University of Wellington, Wellington, New Zealand

R. S. Liu University of Cambridge, Cambridge, England and Materials Research Laboratories, Industrial Technology Research Institute, Hsinchu, Taiwan

Arumugam Manthiram Center for Materials Science and Engineering, The University of Texas at Austin, Austin, Texas

J. S. Martens[†] Sandia National Laboratories, Albuquerque, New Mexico

Claude Michel Laboratoire de Cristallographie et Sciences des Matériaux, Caen, France

John Moreland Superconductor and Magnetic Measurements Group, National Institute of Standards and Technology, U.S. Department of Commerce, Boulder, Colorado

Norman E. Moulton Condensed Matter and Radiation Sciences Division, Naval Research Laboratory, Department of the Navy, Washington, D.C.

Present affiliations:

*Ames Laboratory, Institute for Physical Research and Technology, Iowa State University, Ames, Iowa

†Conductus, Sunnyvale, California

Gary E. Myers Industrial Hygienist, Naperville, Illinois

Masayasu Nagoshi Physical Analysis Laboratory, Applied Technology Research Center, NKK Corporation, Kanagawa, Japan

D. G. Naugle Department of Physics, Texas A&M University, College Station, Texas

Mariappan Paranthaman Department of Physics, Superconductivity Laboratories, University of Colorado at Boulder, Boulder, Colorado

G. M. Phatak Chemistry Division, Bhabha Atomic Research Centre, Bombay, India

Norman E. Phillips Lawrence Berkeley Laboratory and Department of Chemistry, University of California at Berkeley, Berkeley, California

C. N. R. Rao Solid State and Structural Chemistry Unit, CSIR Centre of Excellence in Chemistry, Indian Institute of Science, Bangalore, India

Bernard Raveau Laboratoire de Cristallographie et Sciences des Matériaux, Caen, France

Karl F. Renk Institut für Angewandte Physik, Universität Regensburg, Regensburg, Germany

Brent A. Richert Department of Physics, U.S. Air Force Academy, Colorado Springs, Colorado

Eli Ruckenstein Department of Chemical Engineering, State University of New York at Buffalo, Buffalo, New York

Ken H. Sandhage Department of Materials Science and Engineering, The Ohio State University, Columbus, Ohio

Yuichi Shimakawa* Fundamental Research Laboratories, NEC Corporation, Ibaraki, Japan

Earl F. Skelton Condensed Matter and Radiation Sciences Division, Naval Research Laboratory, Department of the Navy, Washington, D.C.

Michael R. Squillante Radiation Monitoring Devices, Inc., Watertown, Massachusetts

M. A. Subramanian Central Research and Development, Experimental Station, E. I. du Pont de Nemours & Company, Inc., Wilmington, Delaware

Teruo Suzuki Electronics Research Center, Applied Technology Research Center, NKK Corporation, Kanagawa, Japan

Koji Tada Basic High-Technology Laboratories, Sumitomo Electric Industries, Ltd., Osaka, Japan

Hiromi Takei Basic High-Technology Laboratories, Sumitomo Electric Industries, Ltd., Osaka, Japan

---

*Present affiliation: Materials Science Division, Argonne National Laboratory, Argonne, Illinois

Yasuko Torii   Basic High-Technology Laboratories, Sumitomo Electric Industries, Ltd., Osaka, Japan

S. A. Wolf   Naval Research Laboratory, Department of the Navy, Washington, D.C.

Nae-Lih Wu   Department of Chemical Engineering, National Taiwan University, Taipei, Taiwan

J. V. Yakhmi   Chemistry Division, Bhabha Atomic Research Centre, Bombay, India

1
Historical Perspective on the Discovery of Tl-Based HTSC Oxides

Allen M. Hermann
University of Colorado at Boulder
Boulder, Colorado

I. INTRODUCTION

The search for high-temperature superconductivity has proceeded from two rather separate sources of motivation. Empirical work on binary and ternary compound superconductors and oxides (for a comprehensive review, see Ref. 1) brought the critical temperature of superconductors to 23 K in the early 1970s, a critical temperature not to be exceeded until the discovery of superconductivity in the Cu−O system in 1986 by Bednorz and Müller [2].

A theoretical suggestion provided a separate source of motivation for searching for high-temperature superconductors. In 1964, Little [3] suggested the possibility of electronic (nonphonon) coupling of electrons to form Cooper pairs and a correspondingly high critical temperature. In 1968, Ginsburg [4] introduced a two-dimensional analog to the one-dimensional excitonic coupling in Little's calculation. In 1973, Allender et al. [5] suggested that excitonic-induced superconductivity might be found in systems in which a thin metallic film is in intimate contact with a narrow-gap semiconductor. This line of reasoning led this author to search experimentally for high-temperature superconductivity since the 1970s.

As part of a National Aeronautics and Space Administration (NASA)-funded team, we looked experimentally for excitonic coupling in thin metallic or superconducting films. Specifically, we deposited Al films onto (polar) polymeric substrates [6]. We subsequently studied Pb films on PbTe substrates, Pb films overcoated with PbTe, and coevaporated Pb/PbTe films [7]. In none of these experiments were we able to demonstrate excitonic coupling.

Additionally, we searched for evidence of excitonic coupling in other two-dimensional and one-dimensional systems. We studied the normal-state and superconducting properties of MoS_2 intercalated with alkali metals (K, Pb, Cs) [8].

We also investigated quasi-one-dimensional organic metals such as TIF-pseudo-halides [9] and $(TTT)_2I_3$ [10]. Again no evidence for excitonic coupling was uncovered.

In 1986, Bednorz and Müller [2] ushered in the period of high-temperature superconductivity with the Nobel Prize–winning discovery of superconductivity at 30 K in $(La,Ba)_2CuO_4$. This work was followed rapidly by the discovery of superconductivity in $YBa_2Cu_3O_7$ (90 K) [11] in Tl–Ba–Cu–O [12], in Bi–(Ca,Sr)–Cu–O (85 to 106 K) [13], and finally in Tl–Ca–Ba–Cu–O (105 to 125 K) [14]. Table 1 gives a sketch of the key materials discoveries in high-temperature superconductivity along with the rapid follow-up work. The remainder of this chapter provides a brief background sketch of my work at the University of Arkansas with my postdoctoral fellow Z. Z. Sheng, which culminated in the Tl–Ba–Ca–Cu–O discovery in 1988.

II. Tl–Ba–Ca–Cu–O SUPERCONDUCTIVITY DISCOVERY

Following the La–Ba–Cu–O and Y–Ba–Cu–O discoveries, our group at the University of Arkansas prepared a variety of rare earth (R) substitutions in the R–Ba–Cu–O system. It was during the early summer of 1987 that we completed the setup of rudimentary synthesis and characterization apparatus, and in pursuing the rare earth substitutions, discovered that Y–Ba–Cu–O could be formed using a melt reaction [15].

During the period May–October 1987, our substitution program was moving forward with great intensity. Early in this period, we discussed the Tl-incorporation idea as one of many substitutional studies to be attempted. Other substitutions being actively carried out in the systems $A_{III}–B_{II}–Cu–O$ included

rare earths $\rightarrow A_{III}$
Ca,Sr,Pb $\rightarrow B_{II}$

Additionally, we substituted Ag for Cu, Se or F for O, and so on. The rationale for Tl substitution as part of our program can be from a perusal of column IIIA, from which it is seen that Tl is an excellent choice in this column because of its ionic size in the $+3$ state (ca. 0.95 Å); that of the other IIIA elements is significantly

Table 1 Discoveries of High-T_c Superconductors

Material (T_c)	Discovery	Rapid follow-up
$(La,Ba)_2CuO_4$ (30 K)	IBM (Zurich)	University of Tokyo, University of Houston, AT&T, Bellcore
$YBa_2Ca_3O_7$ (90 K)	University of Alabama, University of Houston	Institute of Physics (Beijing), University of Tokyo, Bellcore–National Research Council Canada, AT&T, Argonne, IBM
Bi–(Ca,Sr)–Cu–O (85–106 K)	National Research Institute Metals, Tsukuba	University of Houston, Bellcore–National Research Council Canada, Du Pont, AT&T, IBM
Tl–Ca–Ba–Cu–O (105–125 K)	University of Arkansas	IBM, Du Pont, National Geophysics Laboratory, Johns Hopkins University, Sandia National Laboratories

Source: Adapted from *Physics Today*, November 1988.

smaller. The size consideration we felt to be the key because we wished to preserve the layered structure of the Cu–O and other planes, and consequently, we did not wish to deviate significantly from the ionic size of the rare earths ($\gtrsim 1$ Å), which was known to produce Jahn–Teller distortions and corresponding layered structures. Furthermore, not much was known about the A_{III} bonding in these crystal structures, and the various valence electron hybridizations present in the rare earths (which successfully produced layered high-temperature superconducting Cu–O structures) led us to believe that the p character of the valence electrons in the group IIIA elements might not be detrimental to the formation of these layered structures.

Early attempts at producing Tl–Ba–Cu–O superconductors in our laboratory were hindered by the high vapor pressure of Tl. It was decided that either sealed containers or rapid processing would be necessary for the reaction. We chose the latter for lack of apparatus to carry out reactions in sealed containers. These early attempts were also frustrated by the processing procedures attempted. Preparation in an oxygen atmosphere is now known to lead to overdoping of the 2201 phase (Tl–Ba–Ca–Cu–O), which in turn leads to metallic nonsuperconducting behavior. (See Chapters 3 and 5 for a discussion of the phases.)

In October 1987 a sample whose critical temperature was above that of the boiling point of liquid nitrogen was prepared in our laboratory. The initial synthesis route we followed is described as follows. We mixed $BaCO_3$ and CuO to form $BaCu_3O_4$ or $Ba_2Cu_3O_5$, which was then mixed and ground and heated for 24 h at 925°C. We then mixed and ground appropriate amounts of Tl_2O_3 and pressed this into a pellet of diameter 7 mm and thickness 2 mm. The pellet was then heated in flowing O_2 at 880 to 910°C for 2 to 5 min, removed from the furnace, and quenched to room temperature. Figure 1 shows the resistivity–temperature variation for two

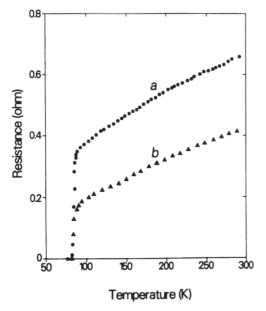

Figure 1 Variation of resistance with temperature for samples (a) $Tl_2Ba_2Cu_3O_x$ and (b) $TlBaCu_3O_x$. (From Ref. 12.)

early Tl–Ba–Cu–O superconducting samples (the compositions given are nominal).

Within the next 2 weeks, the magnetic measurements confirming the existence of superconductivity above 77 K in Tl–Ba–Cu–O were performed using an ac susceptibility apparatus at Phillips Petroleum in Bartlesville, Oklahoma. Patent invention disclosures were rapidly filed, and dc superconducting quantum interference device (SQUID) susceptibility measurements [16] were completed at the University of South Carolina. Figure 2 shows the susceptibility versus temperature (field cooled and zero-field cooled) for a 2201 sample. On January 19, 1988, papers on 90 K superconductivity in Tl–Ba–Cu–O were submitted to *Nature* [12] and *Physical Review Letters* [16]. A press conference was held on January 22, 1988, in Fayetteville, Arkansas, announcing this discovery, that of the first rare-earth-free high-T_c (above 77 K) superconductor. Coincidentally, in Japan on the same day, Maeda announced the discovery of a Bi-based Cu–O superconductor [13].

Within the next few weeks, substitution studies on the Tl–Ba–Cu–O system were carried out with great enthusiasm in our laboratory. One of the substitutions, a partial substitution of Ca for Ba, then led to the finding that Tl–Ba–Ca–Cu–O was a superconductor of T_c in excess of 100 K [14,17]. Figure 3 shows a resistivity temperature variation from two of the early Tl–Ba–Ca–Cu–O samples. The crystal structures (2201, 2212, and 2223) were worked out [17] in collaboration with the Geophysical Laboratory and Johns Hopkins University and the magnetic confirmation was made [18] in collaboration with the University of California at San Diego and San Diego State University. Figure 4 shows the susceptibility–temperature variation (field cooled and zero-field cooled).

One final comment should be made. The early resistivity–temperature variations we reported for Tl–Ba–Ca–Cu–O [14,17] suggested zero-resistance critical

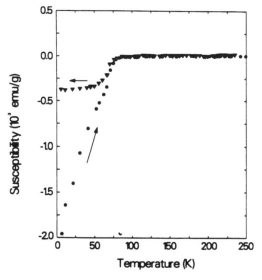

Figure 2 Susceptibility versus temperature for a TlBaCu$_3$O$_x$ sample. As indicated with arrows, the circles were obtained during warming in a field of 2 mT after cooling in zero field, and the triangles were obtained during cooling in the same field. (From Ref. 16.)

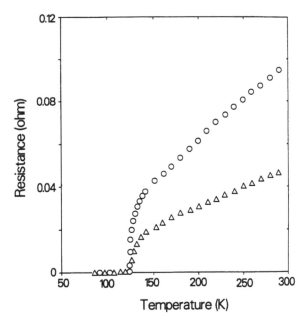

Figure 3 Resistance–temperature dependencies of samples $Tl_2Ca_2BaCu_3O_{9+x}$ (triangles) and $Tl_2Ca_4BaCu_3O_{11+x}$ (circles). (From Ref. 18.)

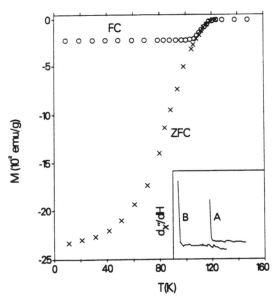

Figure 4 Field-cooled (FC) and zero-field-cooled (ZFC) magnetization as a function of temperature for a dc field of 1 mT for a sample of $Tl_2Ca_2BaCu_3O_{9+x}$. The two data traces in the inset illustrate the sharp onset of superconductivity observed by microwave technique. Sample A is $Tl_2Ca_2BaCu_3O_{9+x}$ with an onset temperature of 117.2 K, and for comparison, data for a sample of $EuBa_2Cu_3O_{7-x}$ (sample B) with an onset temperature of 94.4 K is plotted. Note that the difference in onset is 22.8 K. (From Ref. 18.)

temperatures below 110 K. In April 1988 we discovered a thermometry error in a partially calibrated thermocouple at the University of Arkansas. The thermocouple was providing voltages that corresponded to temperatures lower than those being sensed. The correct *RT* variation shown in Fig. 2 shows data taken on a sample prepared in February 1988; it shows a zero-resistance temperature of 121 K. The magnetization data [18] on early samples confirms the high critical temperatures.

REFERENCES

1. A. Sleight, *Phys. Today* 44, 24 (1991).
2. J. G. Bednorz and K. A. Müller, *Z. Phys. B* 64, 189 (1986).
3. W. A. Little, *Phys. Rev. A* 134, 1416 (1964).
4. V. L. Ginsburg, *Contemp. Phys.* 9, 355 (1968).
5. D. Allender, J. Bray, and J. Bardeen, *Phys. Rev. B* 37, 1020 (1973).
6. T. Russell, M. E. Jones, A. C. Johnson, R. J. Deck, and A. M. Hermann, *J. Appl. Phys.* 45, 1470 (1974).
7. A. C. Johnson, M. E. Jones, A. M. Hermann, and R. J. Deck, *Solid State Commun.* 16, 803 (1975).
8. A. M. Hermann, R. Somoano, V. Hadek, and A. Rembaum, *Solid State Commun.* 13, 1065 (1973).
9. R. B. Somoano, M. Jones, T. Datta, R. Deck, and A. M. Hermann, *Phys. Rev. B* 15, 595 (1977).
10. R. B. Somoano, S. P. S. Yen, V. Hadek, S. K. Khanna, M. Novotny, T. Datta, A. M. Hermann, and J. A. Woollam, *Phys. Rev. B* 17, 2853 (1978).
11. M. K. Wu, J. R. Ashburn, C. T. Torng, P. H. Hor, R. L. Meng, L. Gao, Z. J. Huang, Y. W. Wang, and C. W. Chu, *Phys. Rev. Lett.* 58, 908 (1987).
12. Z. Z. Sheng, and A. M. Hermann, *Nature* 332, 55 (1988).
13. H. Maeda, Y. Tanaka, M. Fuku-tani, and T. Asano, *Jpn. J. Appl. Phys. Lett.* 27, L209 (1988).
14. Z. Z. Sheng and A. M. Hermann, *Nature* 332, 138 (1988).
15. A. M. Hermann and Z. Z. Sheng, *Appl. Phys. Lett.* 51, 1854 (1987).
16. Z. Z. Sheng, A. M. Hermann, A. El Ali, C. Almasen, J. Estrada, T. Datta, and R. J. Matsen, *Phys. Rev. Lett.* 60, 937 (1988).
17. R. M. Hazen, L. W. Finger, R. J. Angel, C. T. Prewitt, N. L. Ross, C. G. Hadidiacos, P. J. Heany, D. R. Veblen, Z. Z. Sheng, A. El Ali, and A. M. Hermann, *Phys. Rev. Lett.* 60, 1657 (1988).
18. A. M. Hermann, Z. Z. Sheng, D. C. Vier, S. Shultz, and S. B. Oseroff, *Phys. Rev. B* 37, 9742.

2

Solid-State Chemistry of Thallium Oxides

J. Gopalakrishnan
Indian Institute of Science
Bangalore, India

I. INTRODUCTION

Thallium shares with its neighbors, lead and bismuth, certain unusual features that make their chemistry unique among the posttransition elements [1]. All three elements show a stable lower valency that is less by 2 units from the group valency. Thus Tl(I), Pb(II), and Bi(III), which are isoelectronic with the outer $6s^2$ electronic configuration, are stable oxidation states, in addition to the normal-group oxidation states, Tl(III), Pb(IV), and Bi(V), which are again isoelectronic ([Xe]$4f^{14}5d^{10}$). The occurrence of an oxidation state 2 units less than the group valence is known as the *inert-pair effect*, which refers to the reluctance of the pair of outer s electrons to ionize or participate in covalent bond formation. The inert pair effect becomes dominant with the heaviest posttransition elements (Hg, Tl, Pb, Bi), the underlying reason being that the outer s and p states are separated by a large difference in energy.

The inert (lone) pair of s^2 electrons often participates in bonding, conferring unsymmetrical stereochemistries on the metal atom in the solid compounds of Tl(I), Pb(II), and Bi(III). The influence of the inert pair on the structural chemistry of these elements was discussed in the early literature. Following early attempts by Sidgwick and Powell [2] and Gillespie and Nyholm [3], who introduced valence-shell electron-pair repulsion theory to explain the stereochemistries of lone-pair cations, Orgel [4] invoked hybridization of the outer s and p orbitals. Typically, in many compounds of Tl(I), Pb(II), and Bi(III), the $6s$ orbital mixes with the empty $6p$ orbitals, which are separated by about 6 eV, the lone pair taking up one of the hybridized $6s$–$6p$ orbitals and the anions bonded to the remaining orbitals. Together with the lone-pair orbital, the anions complete a distorted polyhedron around the metal ion. Andersson and Åstrom [5] have pointed out that the lone-

7

pair electrons indeed take up a volume comparable to that of an anion in solids where the lone pair is stereoactive.

Crystal structures of many Tl(I) compounds provide evidence for a stereoactive $6s^2$ lone pair [6]. TlF, for example, adopts a layered structure (Fig. 1a) very similar to that of yellow PbO. The lone-pair electrons on each thallium screen the sheets from one another. Tl_4O_3, which is actually $Tl(I)_3Tl(III)O_3$, adopts a monoclinic structure (Fig. 1b) where the trivalent cations are at the center of a distorted octahedron forming chains along the b axis. Tl(I) sticks out from the octahedral chains in trigonal pyramidal coordination, with the lone pairs screening the chains from one another. In general, low irregular coordination numbers around the metal are indicative of stereoactive lone pair, there typically being three or four short bonds all lying to one side of Tl(I). In structures where six or more bonds are distributed uniformly around the metal, the lone pair is not stereoactive. Tl(I) with a radius of 1.70 Å (twelve-coordination) behaves like K^+ or Rb^+ in such instances. Having a closed-shell electronic configuration and a radius of 0.98 Å (eight-coordination), Tl(III) exhibits more symmetric coordination geometries, the common geometries in oxides being cubic (distorted), octahedral, and tetrahedral. Although Tl(I), Pb(II), and Bi(III) are isoelectronic, their relative stability with respect to oxidation to the s^0 state varies significantly. Thus, Tl(I) is readily oxidized to Tl(III), and Bi(III) is quite stable to oxidation, rendering Bi(V) compounds both rare and highly oxidizing. Pb(IV) is moderately oxidizing, both Pb(II) and Pb(IV) being accessible in oxides under ambient conditions.

Another special feature of the chemistry of these elements is that they skip the valence state corresponding to the s^1 electronic configuration. This has been attributed to a negative intraatomic interaction $(-U)$ operating on the s^1 configuration of these elements [7]. In chemical terms [8], compounds of these elements where the metal atom has an apparent s^1 configuration exhibit a valence disproportionation to s^0 and s^2 configurations. Thus $TlBr_2$ is $Tl^I(Tl^{III}Br_4)$, TlS is $Tl^ITl^{III}S_2$, Pb_2O_3 is $Pb^{II}Pb^{IV}O_3$, and $BaBiO_3$ is $Ba_2Bi^{III}Bi^VO_6$. This tendency for valence disproportionation has important consequence for the electronic structure and

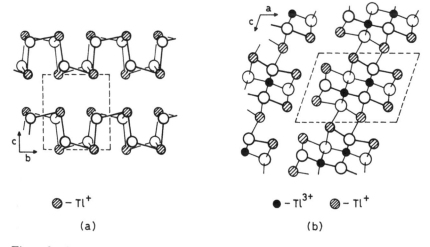

$\oslash - Tl^+$

(a)

$\bullet - Tl^{3+}$ $\oslash - Tl^+$

(b)

Figure 1   Structures of (a) TlF and (b) Tl_4O_3. (Adapted with permission from Ref. 6.)

properties of solids containing such cations. For example, valence disproportion-ation and superconductivity seem to be intimately related to each other in the $Ba(Pb,Bi)O_3$ and $(K,Ba)BiO_3$ systems.

The chemistry of thallium oxides, especially of thallium(III), has become important in the wake of the discovery of superconductivity in thallium cuprates [9]. A knowledge of thallium oxide chemistry is therefore essential for proper appreciation of the chemistry of superconducting thallium cuprates. In the following sections we describe in some detail the chemistry of representative thallium oxides and thallium cuprates.

II. THALLIUM OXIDES

There are four binary oxides of thallium [10]—Tl_2O, Tl_2O_3, Tl_4O_3, and TlO_2—of which the last is a peroxide. Tl_2O has a anti-CdI_2 (layered) structure with a short (3.52-Å) Tl–Tl distance in the sheets. Reportedly, Tl_2O is metallic, although it is not easily rationalized. Tl_2O_3 adopts the C–rare earth oxide structure where Tl(III) occurs in a distorted octahedral coordination. At high pressures, Tl_2O_3 transforms to the corundum structure [11]. Apparently, the size of Tl(III) does not permit close packing of oxygens at atmospheric pressure. The primary coordination of Tl(III) is not changed during this pressure transformation. Tl_2O_3 is black in both its modifications, showing a low electrical resistivity (10^{-3} to 10^{-4} $\Omega \cdot$ cm at room temperature) that is probably metal-like [12]. The electrical conductivity of Tl_2O_3 indicates an overlap of Tl(III):$6s°$ with the O:$2p$ states. Tl_2O_3 is unstable at elevated temperatures, decomposing to Tl_2O,

$$Tl_2O_3(s) \rightarrow Tl_2O(g) + O_2(g)$$

the enthalpy of vaporization at 1000 K being 90 kcal/mol.

TlOF is an interesting oxyfluoride crystallizing in a fluorite structure. Significantly, TlOF also exhibits a low electrical resistivity that shows a metal-like temperature dependence [12].

Thallium forms a number of ternary oxides exhibiting a rich variety of structures and electronic properties. In Table 1 we have listed representative of them, most of them being Tl(III) oxides. With alkali metals, Tl(III) forms brightly colored insulating oxides of the formula $ATlO_2$ (A = Na, K, etc.), where Tl(III) occurs in an octahedral coordination [10]. Obviously, the empty Tl($6s$) states in these compounds are well separated in energy from the O($2p$) states. In contrast to this, the alkaline earth metal oxides of thallium show an interesting diversity. $Ba_2Tl_2O_5$ is a unique oxide where Tl(III) occurs in alternating octahedral and tetrahedral coordination in a brownmillerite structure [13] (Fig. 2a). The material is a semiconductor [14] showing an unusual increase in activation energy for conduction with increasing temperature. $SrTl_2O_4$ (and probably $BaTl_2O_4$), possessing the $CaFe_2O_4$ structure [15] (Fig. 2b) wherein TlO_6 octahedra form edge-shared ribbons, exhibit a semimetallic behavior, the low resistivity being independent of temperature in the range 100 to 300 K. $Sr_4Tl_2O_7$ is another stable oxide [16] in the SrO–Tl_2O_3 system, crystallizing in a layered (defective K_2NiF_4) structure (Fig. 2c), wherein Tl(III) is present in an unusual (2 + 2) oxygen coordination. The golden yellow $Sr_4Tl_2O_7$ appears to be an insulator indicating considerable stabilization of O($2p$)

Table 1   Representative Ternary Oxides of Thallium

Oxide	Structural details	Properties	Refs.
$ATlO_2$ (A = alkali metal)	All, except $LiTlO_2$, adopt the α-$NaFeO_2$ structure; $LiFeO_2$ exists in three modifications; $Tl(III)$ is octahedrally coordinated; $NaTlO_2$: hexagonal, a = 3.35, c = 16.52 Å	Brightly colored (red, yellow, orange, etc.) insulators	10
$Ba_2Tl_2O_5$	Brownmillerite ($Ca_2Fe_2O_5$) structure	Black, semiconducting ρ (300 K) = $10^4 \, \Omega \cdot$ cm	13, 14
$SrTl_2O_4$	$CaFe_2O_4$ structure, orthorhombic, $Pnam$, a = 10.041, b = 11.697, c = 3.409 Å	Semimetallic, nearly temperature-independent resistivity	15
$Sr_4Tl_2O_7$	Tetragonal, $P4_2nm$, a = 5.006, c = 18.70 Å; structure related to K_2NiF_4; Tl in rectangular oxygen coordination	Insulating	16
$TlMO_3$ (M = Cr, Fe)	Perovskites: $GdFeO_3$ structure; a = 5.302, b = 5.405, c = 7.647 Å for $TlCrO_3$; high-pressure synthesis	—	17
$TlVO_3$	Orthorhombic, $Pbcm$, a = 5.16, b = 11.22, c = 5.73 Å; $Tl(I)$ phase isomorphous with KVO_3	—	18
$Tl_2M_2O_7$ (M = Ru, Rh, Os, Ir, Pt)	Pyrochlore structure, $Fd3m$, a = 10.10–10.30 Å	M = Rh, Os, Ir phases are metallic; M = Pt phase semiconducting; M = Ru phase shows metal–insulator transition	19, 20
$Tl_2Mo_2O_7$	Monoclinic, a = 8.426, b = 5.694, c = 7.934 Å; $Tl(I)$ compound	—	25
Ba_2TlTaO_6	Ordered perovskite, a = 8.42 Å	—	21
$Ba_2Bi_{1-x}Tl_xO_3$ (0.0 < x < 0.5)	Monoclinic-orthorhombic perovskites; $BaBi_{0.5}Tl_{0.5}O_3$: a = 6.076, b = 6.056, c = 8.549 Å	Superconductivity at 8 K for the composition $BaBi_{0.25}Tl_{0.25}Pb_{0.5}O_3$	22, 23
$BaPb_{1-x}Tl_xO_{3-x/2}$	Perovskite, pseudocubic, a = 4.3 Å	Electrical resistivity increases with x in the series	22
$LaTlO_3$	Hexagonal, a = 3.909, c = 12.365 Å	—	25
$TlMO_3$ (M = Nb, Ta, U)	Defect pyrochlores, a = 10.67–11.27 Å	—	24
$TlNbWO_6$	Defect pyrochlore, $Fd3m$, a = 10.36 Å	—	24
$TlVO_4$	Isostructural with $InVO_4$, orthorhombic, $Cmcm$, a = 5.839, b = 8.687, c = 6.80 Å	—	25
$TlReO_4$	Orthorhombic, scheelite-related structure	Undergoes pressure-induced phase transition corresponding to the valence change $Tl^+Re^{7+}O_4 \rightarrow Tl^{3+}Re^{5+}O_4$	26

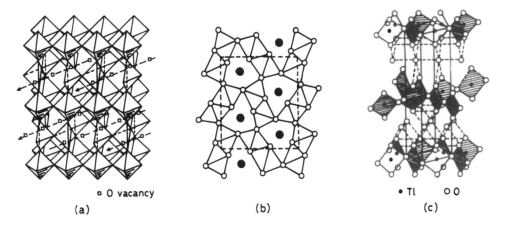

(a)   (b)   (c)

Figure 2 Structures of (a) $Ba_2Tl_2O_5$ (brownmillerite), (b) $SrTl_2O_4$, and (c) $Sr_4Tl_2O_7$. (Adapted with permission from Refs. 13, 15, and 16.)

states in the presence of strontium, leading to a finite gap in the density of states between Tl($6s$) and O($2p$).

Among the perovskite oxides containing Tl(III), mention must be made of $BaBi_{1-x}Tl_xO_3$ and $BaPb_{1-x}Tl_xO_{3-y}$, which have been investigated in recent times [22,23], the motivation being to understand the occurrence or otherwise of superconductivity in these systems in relation to the superconducting $BaPb_{1-x}Bi_xO_3$ and (K,Ba)BiO_3 systems [27,28]. For $BaPb_{1-x}Tl_xO_{3-y}$, a perovskite structure exists over a wide composition range. The increase in the pseudocubic cell parameter a with x is consistent with the larger size of Tl(III) relative to Pb(IV). Although electron count in the $6s$ band is not expected to change as a function of x in this system, the materials become insulating with increasing x, indicating that the $6s$–$2p$ separation increases with x. This can be understood because Tl is to the left of Pb in the periodic table and the samples are oxygen deficient. With $BaBi_{1-x}Tl_xO_3$, a perovskite phase forms over a wide range. Among these, $BaBi_{0.5}Tl_{0.5}O_3$, which is isoelectronic with $BaPbO_3$, is an orthorhombic perovskite ($a = 6.076$, $b = 6.056$, and $c = 8.549$ Å). Superconductivity occurs at 8 K when 50% of Bi and Tl are replaced by Pb, giving $BaBi_{0.25}Tl_{0.25}Pb_{0.5}O_3$. It has been suggested that the CDW instability of $BaBiO_3$ is suppressed by the substitution of Tl(III) ($6s^0$) for Bi(III) ($6s^2$), as revealed by the dramatic decrease of the resistivity in the series $BaBi_{1-x}Tl_xO_3$ ($0 < x < 0.5$). Insulating $BaTl_{0.5}Sb_{0.5}O_3$ crystallizing in an orthorhombic structure isostructural with $BaPbO_3$ has been reported [29]. We have prepared perovskite type $Ba_2Pb_{1-x}Tl_xCuO_5$, where superconductivity up to 70 K is observed (V. Manivannan, R. Mahindran, J. Gopalakrishnan, and C. N. R. Rao, personal communication).

Tl(III) can be incorporated at the eight-fold coordination site in pyrochlore and perovskite ($GdFeO_3$) structures when combined with suitable transition metal atoms. Thus $Tl_2M_2O_7$ (M = Ru, Rh, Ir, Os, Pt) all form cubic pyrochlores [19], of which M = Rh, Os, and Ir compounds are metallic, whereas $Tl_2Pt_2O_7$ is semiconducting. $Tl_2Ru_2O_7$ shows a metal–insulator transition [20] that seems to be related to the mixed valency of thallium and ruthenium, $Tl_{1-x}^{III}Tl_x^I Ru_{1-2x}^{IV} Ru_{2x}^V O_7$.

$TlReO_4$ is another interesting thallium oxide where pressure induces the valence change [26] $Tl^I Re^{VII}O_4 \rightarrow Tl^{III}Re^V O_4$. The valence change is obviously accompanied by drastic changes in the electrical and optical properties of the material. Pressure can stabilize Tl(III) at the A site of the perovskite structure as in orthorhombic, $GdFeO_3$-like $TlFeO_3$ and $TlCrO_3$ [17].

The foregoing brief survey of thallium oxides shows that unusual electronic properties result whenever there is a possibility of mixed valency for thallium. Thallium(I,III) mixed valency arises when it is coupled with an appropriate metal atom that can exhibit mixed valency [e.g., Tl(I,III) with Re(VII,V)]. Another factor that contributes to the interesting electrical properties of thallium oxides is the proximity of Tl(6s) states to the O(2p) states and the possibility of modulating the latter in the presence of electropositive countercations. Thus the semimetallic nature of Tl_2O_3 and $SrTl_2O_4$ stands in contrast to the insulating (semiconducting) nature of $Ba_2Tl_2O_5$ and $Sr_4Tl_2O_7$. These aspects of thallium oxide chemistry are of special relevance to the chemistry of the superconducting thallium cuprates, discussed in the following section.

III. THALLIUM CUPRATES

Superconductivity in thallium cuprates was first reported by Sheng and Hermann [9]. Following this discovery, intense research efforts by several groups [30–32] have not only yielded record-holding highest T_c superconductors but also revealed the occurrence of two homologous series of superconductors of formulas $Tl_2Ba_2Ca_{n-1}Cu_nO_{2n+4}$ and $TlBa_2Ca_{n-1}Cu_nO_{2n+3}$ and a number of their derivatives. Single-phase materials with n up to 3 are readily prepared in both series. Higher members have narrow composition and temperature regions of stability, often forming defective intergrowths with other members. The structures and properties of thallium cuprates have been reviewed by several workers [34–35]. In this section we capture the salient aspects of the chemistry of these materials.

More than 20 thallium cuprates belonging to the two homologous series and their derivatives have been well characterized. We list them in Table 2. The structures of these materials (Fig. 3) are related to those of the other superconducting cuprates [36,37] in the sense they all contain CuO_2 sheets comprising $CuO_{5/2}$ pyramids or $CuO_{6/2}$ octahedra. They provide finest examples of intergrowth structures [37], where two structurally different but dimensionally compatible units, derived from perovskite and rock salt structures, intergrow to give new homologous series of oxides. The entire thallium cuprate series may be represented by the formula $Tl_mBa_2Ca_{n-1}Cu_nO_{2n+m+2}$, where the subscript m defines the number of rock salt–like TlO layers and n refers to the number of adjacent CuO_2 sheets per formula unit. The thallium cuprates with $m = 2$ (double thallium layer) and the bismuth cuprates are structurally analogous, Tl replacing Bi and Ba replacing Sr in the series $Bi_2Sr_2Ca_{n-1}Cu_nO_{2n+4}$. The $m = 1$ (single thallium layer) cuprates are unique to the thallium cuprate family. The thallium cuprates are more symmetric (tetragonal) in structure and free of the structural modulation that plagues the bismuth cuprates [38]. Both the single- and double-thallium-layer materials contain perovskite-like units, $Ba_2Ca_{n-1}Cu_nO_{2n+2}$, that alternate with either single TlO sheet or double TlO sheets, the TlO sheets being formed by edge-sharing TlO_6 octahedra. In the $m = 1$ phases, the perovskite units are stacked exactly one above the other,

Table 2　Superconducting Thallium Cuprates

Oxide	Structural details	T_c (K)	Refs.
$Tl_2Ba_2CuO_6$	Orthorhombic, *A2aa*, $a = 5.468$, $b = 5.472$, $c = 23.238$ Å; tetragonal, *I4/mmm*, $a = 3.866$, $c = 23.239$ Å	0–90	39
$Tl_2Ba_2CaCu_2O_8$	Tetragonal, *I4/mmm*, $a = 3.855$, $c = 29.318$ Å	110	39
$Tl_2Ba_2Ca_2Cu_3O_{10}$	Tetragonal, *I4/mmm*, $a = 3.85$, $c = 35.88$ Å	115–125	39
$Tl_2Ba_2Cu_3Cu_4O_{12}$	Tetragonal, *I4/mmm*, $a = 3.854$, $c = 42.07$ Å	104–110	33
$TlBa_2CuO_5$	Tetragonal, *P4/mmm*, $a = 3.896$, $c = 9.694$ Å	Stoichiometric compound nonsuperconducting and metallic; superconductivity ($T_c \sim 10$ K) can be induced by oxygen removal or partial substitution of a lanthanide for barium	40
$TlBa_2CaCu_2O_7$	Tetragonal, *P4/mmm*, $a = 3.856$, $c = 12.754$ Å	90	39, 33
$TlBa_2Ca_2Cu_3O_9$	Tetragonal, *P4/mmm*, $a = 3.853$, $c = 15.913$ Å	110	39
$TlBa_2Ca_3Cu_4O_{11}$	Tetragonal, *P4/mmm*, $a = 3.85$, $c = 18.73$ Å	~120	33
$(Tl,Pb)Sr_2CuO_5$	Tetragonal, *P4/mmm*, $a = 3.736$, $c = 9.022$ Å	60	41
$(Tl,Pb)Sr_2CaCu_2O_7$	Tetragonal, *P4/mmm*, $a = 3.80$, $c = 12.05$ Å	85	42
$(Tl,Pb)Sr_2Ca_2Cu_3O_9$	Tetragonal, *P4/mmm*, $a = 3.81$, $c = 15.23$ Å	120	42
$(Tl,Bi)Sr_2CuO_5$	Tetragonal, *P4/mmm*, $a = 3.759$, $c = 9.01$ Å	45	43
$(Tl,Bi)Sr_2CaCu_2O_7$	Tetragonal, *P4/mmm*, $a = 3.796$, $c = 12.113$ Å	95	44
$(Tl,Bi)Sr_2Ca_2Cu_3O_9$	Tetragonal, *P4/mmm*, $a = 3.80$, $c = 15.0$ Å	115	45
$TlSr_2(Ca,Ln)Cu_2O_7$	Tetragonal, *P4/mmm*, $a = 3.80$, $c = 12.10$ Å	60–90	46
$TlBaSrCaCu_2O_7$	Tetragonal, isostructural with $TlBa_2CaCu_2O_7$	90	47
$TlBaSrCuO_5$	Tetragonal, isostructural with $TlBa_2CuO_5$	43	40
$TlSr_2CuO_5$	Tetragonal, isostructural with $TlBa_2CuO_5$ $a \sim 3.7$, $c \sim 9$ Å	Superconductivity is induced by substituting La/Nd for Sr	48
$TlSr_2CaCu_2O_7$	Tetragonal, *P4/mmm*, $a = 3.805$, $c = 12.14$ Å	~100	33
$TlBa_2(Ln_{1-x}Ce_x)_2$ Cu_2O_{10+y}	Tetragonal, *P4/mmm*, $a = 3.88$, $c = 17.28$ Å	No superconductivity	49

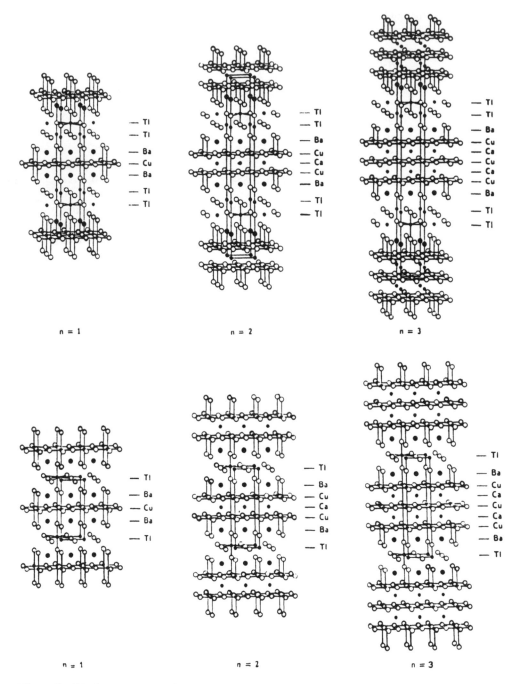

Figure 3   Ideal structures of $Tl_mBa_2Ca_{n-1}Cu_nO_{2n+m+2}$. Top: $Tl_2Ba_2Ca_{n-1}Cu_nO_{2n+4}$ ($n = 1$, 2, and 3). Bottom: $TlBa_2Ca_{n-1}Cu_nO_{2n+3}$ ($n = 1$, 2, and 3). Metal atoms are shaded and Cu–O bonds are shown. (From Ref. 30.)

whereas in the $m = 2$ phases, the perovskite units are displaced by the translation $(a_1 + a_2)/2$, resulting in a doubling of the c parameter. The T_c increases with n in both series up to $n = 3$; for higher values of n, the T_c decreases. The highest T_c, 125 K, is exhibited by the $m = 2$, $n = 3$ member [50], $Tl_2Ba_2Ca_2Cu_3O_{10}$, and the next-highest T_c, 122 K by $(Tl_{0.5}Pb_{0.5})Sr_2Ca_2Cu_3O_9$ [42]. $Tl_2Ba_2CuO_6$ is the most enigmatic of the thallium cuprates, exhibiting a T_c anywhere between 0 (no superconductivity) and 90 K, depending on the conditions of synthesis [51]. $Tl_2Ba_2CuO_6$ is also unusual in the sense that no other single CuO_2 sheet material shows such a high T_c (see later).

Several substitutions have been made in $Tl_mBa_2Ca_{n-1}Cu_nO_{2n+m+2}$, some of them yielding new superconductors. Substitution of strontium for barium in $Tl_2Ba_2CaCu_2O_8$ is possible [52] but T_c is lowered to 44 K. Substitution of K for Tl in the same material has been reported [53]. The strontium analogs of Tl-$Ba_2Ca_{n-1}Cu_nO_{2n+3}$ are not stable but have been stabilized by partial replacement of Tl by Pb, giving a new family of superconducting cuprates [42] of the formula $(Tl_{0.5}Pb_{0.5})Sr_2Ca_{n-1}Cu_nO_{2n+3}$ for $n = 1$, 2, and 3. In this series, the T_c reaches up to 122 K for the $n = 3$ material. Similar Bi-containing phases, $(Tl,Bi)Sr_2CaCu_2O_7$ and $(Tl,Bi)Sr_2Ca_2Cu_3O_9$ with T_cs of 95 and 120 K, respectively, have been reported [44,45]. Pb is in the IV oxidation state and Bi in the mixed III/V state in these materials [54]. Their role appears to be optimizing the hole concentration and the structural mismatch between rock salt and perovskite fragments. The strontium analog of $TlBa_2CaCu_2O_7$ has also been stabilized by partial replacement of Ca by rare earth, giving rise to new superconducting phases [46], $TlSr_2Ca_{1-x}Ln_xCu_2O_7$, where T_cs up to 90 K has been found in the composition range $0.25 < x < 0.75$. A septenary phase $(Tl_{0.5}Pb_{0.5})Sr_2(Ca_{1-y}Y_y)Cu_2O_7$, which is isostructural with $TlBa_2CaCu_2O_7$, has been found to show a maximum T_c of 108 K at $y = 0.2$ [55].

Thallium cuprates with $m = 1$ and $n = 1$, $TlBa_2CuO_5$ and $TlSr_2CuO_5$, where the formal oxidation state of copper is III, are generally nonsuperconducting. They become superconducting, however, when the oxygen stoichiometry is appropriately varied [40] or when part of Ba/Sr is replaced by a trivalent lanthanide element [56,57]. Thus when prepared by slow cooling in a nitrogen environment $TlBa_2CuO_5$ shows a T_c of 9.5 K. A similarly prepared $TlBaSrCuO_5$ sample shows a T_c of 43 K. Clearly, the parent materials are overdoped; lowering the oxygen content or substituting a trivalent lanthanide brings down the hole concentration to an optimal value where superconductivity sets in. Apparently, no solid solution region seems to exist for $(Tl_{0.5}M_{0.5})Sr_2Ca_{n-1}Cu_nO_{2n+3}$ (M = Pb, Bi) except for $Tl_{1-x}Bi_xSr_2CuO_5$, where $0 < x < 0.5$.

An interesting structural problem with the thallium cuprates is the presence of short-range ordering due to displacements of Tl and O atoms in the TlO layer from their ideal positions [58]. The anomalously large temperature factors for these atoms, as seen in the structure refinements from single-crystal x-ray and neutron powder diffraction data [59], gave first indications of a disorder in Tl–O layers. However, the disorder is not the same as the incommensurate modulated superstructure of the bismuth cuprates, because thallium cuprates do not show an obvious superstructure in the diffraction patterns. Pair distribution function (PDF) data obtained from pulsed neutron scattering have provided a model for the disorder in the TlO layers. The disorder involves a strongly correlated local displacement of both Tl and O atoms forming chains/pairs. However, the displacements remain short range and do not alter the average symmetry. A likely origin of the displacement and short-range ordering of Tl and O atoms is the mismatch between the atomic distance imposed by the lattice structure ($\sqrt{2}$ Cu–O bond distance, ~2.73 Å and the ideal Tl(III)–O distance given by the sum of the ionic radii, 2.28 Å). The disorder in TlO layers does not, however, involve insertion of extra oxygens.

Another problem with the thallium cuprates is the occurrence of widespread nonstoichiometry and associated defects. Microprobe chemical analysis shows significant thallium and calcium deficiencies in double-thallium-layer cuprates [60]. For example, deficiencies up to 12% Tl/30% Ca and 8.5% Tl/28% Ca have been reported for $Tl_2Ba_2CaCu_2O_8$ and $Tl_2Ba_2Ca_2Cu_3O_{10}$, respectively. Another complication is the partial occupation of Tl and Ca on each other's sites [50]. In addition to these point defects, extended defects that involve intergrowths of perovskite slabs of varying n values and of single-thallium-layer defects in double-thallium-layer cuprates, and vice versa, have been seen by high-resolution electron microscopy [61]. Indeed, it appears that all the superconducting thallium cuprates are metastable phases stabilized by entropy arising from defects and disorder [62].

Superconductivity in thallium cuprates arises from a hole concentration on CuO_2 sheets, just as in other superconducting cuprates. However, the mechanism of hole doping in the thallium cuprates is not straightforward. Thus, for stoichiometric $Tl_2Ba_2Ca_{n-1}Cu_nO_{2n+4}$, one would expect no hole concentration (n_h) because Tl is in the formal III oxidation state and Cu in the II state, and these materials should therefore be Mott–Hubbard (or charge-transfer) insulators just as are La_2CuO_4 or $YBa_2Cu_3O_6$. For the single-thallium-layer cuprates, the stoichiometric formula itself provides for a mixed-valent copper. The oxidation state of Cu in these materials, $TlBa_2Ca_{n-1}Cu_nO_{2n+3}$, would be $(2 + 1/n)$, assuming Tl(III). Accordingly, we would expect the hole concentration to decrease with increasing n and since T_c $\propto n_h$ to a first approximation, the T_c should decrease with increasing n in the series. But what we observe experimentally is exactly the opposite: $n = 1$ members of the $m = 1$ series are nonsuperconducting, $n = 2$ members show a T_c of 80 to 90 K, and $n = 3$ members show a T_c of 100 and above, clearly indicating that the origin of n_h and its relation to T_c are quite complex in these materials.

Jung et al. [63] were the first to point out a difference between single-thallium-layer and double-thallium-layer cuprates in their electronic structure. Their band-structure calculations, taking into account distortions in Tl–O layers, showed that the bottom of the Tl(6s) band lies significantly below the Fermi level for the double-thallium-layer cuprates (Fig. 4). On the other hand, for the single-thallium-layer cuprates, the bottom of the Tl(6s) lies well above the Fermi level (Fig. 4). The implication is that an internal redox mechanism (self-doping),

$$Tl^{III} + Cu^{II} \rightarrow Tl^{III-\delta} + Cu^{II+\delta}$$

might be operative in the double-thallium-layer cuprates, while a conventional doping mechanism might be operative in single-thallium-layer cuprates [64]. However, since the calculations do not take defects and nonstoichiometry into account, the real situation would be much more complex than band-structure calculations suggest.

The present understanding of hole doping in thallium cuprates may be summarized as follows: There are two different sources for a hole concentration in thallium cuprates [59]. One is the chemistry of the material that includes chemical

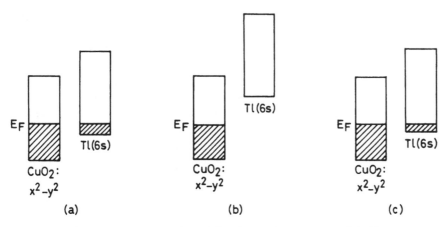

Figure 4 Schematic representation of the electronic structures of thallium cuprates showing relative energies of the conduction band of CuO_2 sheets and $Tl(6s)$ band: (a)$Tl_2Ba_2CuO_6$; (b) $TlBa_2Ca_{n-1}Cu_nO_{2n+3}$; (c) $TlSrLnCuO_5$ (Ln = La, Nd). (Adapted with permission from Ref. 64.)

composition, nonstoichiometry, defects, and so on, and the other is the overlap of $Tl(6s)$ with the conduction band of CuO_2 sheets at the Fermi level (Fig. 4). We designate the former as uncompensated holes (type I) and the latter as compensated holes (type II), because in type II an equivalent number of electrons would be present in the $Tl(6s)$ band. In a given thallium cuprate, the holes present may be type I, type II, or both. Band-structure calculations, as mentioned above, have shown that in general, the holes are type I in single-thallium-layer cuprates, whereas they are type II in double-thallium-layer cuprates. The presence of both types of holes in some thallium cuprates is a distinct possibility. Moreover, type II holes may exist in certain single-thallium-layer cuprates [64]: for example, $TlSrLnCuO_5$ (Ln = La, Nd) (Fig. 4c).

In view of the foregoing, an experimental (chemical) determination of hole concentration in thallium cuprates where type II holes are present is an impossible problem at present. Added to this is the problem due to Tl(III), which interferes with the determination of hole concentration by conventional wet-chemical titrations using I^- or Fe(II). Attempts have been made by Manthiram et al. [65] and by us [66] to determine the hole concentration in thallium cuprates using wet-chemical methods. The method of Manthiram et al. determines the total oxidation present in the material, above the formal valences of Tl(I), Cu(I), and O^{2-}, using iodometric titration. Combining the data with an independent determination of thallium content in the material, Manthiram et al. [65] compute what would be a formal oxidation state for copper assuming that all the thallium is present as III. A copper oxidation state less than II is taken to indicate the existence of $Tl(6s)$–CuO_2 band overlap (type II holes). In cases where the oxidation state of copper comes out more than II, the value in excess of II is taken as the hole concentration.

Our method [66] is based on selective oxidation of bromide ions by holes on copper. Thallium (III) does not interfere with the determination, because it does not oxidize Br^- ions under the experimental conditions. Accordingly, our method directly determines type I holes arising from the chemistry of the material. We have shown that in $TlBa_2Ca_{1-x}Y_xCu_2O_7$, where the holes appear to be exclusively of type I [no $Tl(6s)-CuO_2$ band overlap], the hole concentration determined by the bromide method correlates with the T_c in a manner analogous to other superconducting cuprates [67], showing a maximum T_c at an optimal hole concentration (Fig. 5). In cases where the holes are both type I and type II, our method determines only the type I holes. This appears to be the case with nominal $Tl_2Ba_2CaCu_2O_8$ and $Tl_2Ba_2Ca_2Cu_3O_{10}$ superconductors. When the holes are type II, the bromide method shows a dramatic negative result: that there are *no holes* although the material is superconducting! $Tl_2Ba_2CuO_6$ is the only cuprate known at present that shows this negative result. Thus $Tl_2Ba_2CuO_6$ closely approximates case (a) in Fig. 4, where the holes appear to be exclusively type II.

Before closing, we must mention thallium cuprate synthesis. It is now recognized that all the superconducting copper oxides, including the thallium cuprates, are metastable materials stabilized by entropy arising from defects and disorder [62]. It appears from experience that there is no combination of temperature, pressure, and composition where the thallium cuprates could be synthesized as equilibrium phases. These materials are generally made at temperatures close to the melting point of the mixture of constituent binary oxides. They have broad ranges of composition that often do not include the ideal (stoichiometric) composition. Since these phases decompose at the same temperatures at which they form, reaction time is a crucial parameter in the synthesis. For instance, synthesis of the thallium cuprate with a $T_c \sim 125$ K is best achieved by reaction in sealed gold tubes at 890°C for 1 h followed by quenching [50]. Synthesis of all the thallium cuprates is best done by carrying out the reaction in sealed gold or silver tubes around 880 to 910°C for periods ranging from 0.5 to 3 h. Instead of Tl_2O_3, which

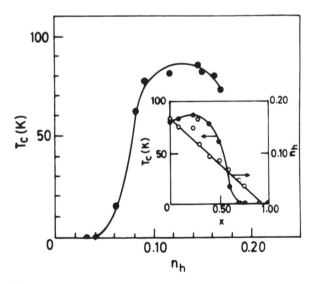

Figure 5 . Dependence of superconducting transition temperature T_c on the hole concentration n_h in the series $TlBa_2Ca_{1-x}Y_xCu_2O_7$. The variations of T_c and n_h with x for the same series of oxides are shown in the inset. The solid lines connecting the data points are guides for the eye. (From Ref. 66.)

decomposes and volatilizes at elevated temperatures. $Ba_2Tl_2O_5$, $Sr_4Tl_2O_7$ and similar Ca$-$Tl$-$O oxides could be employed as precursors in the synthesis.

IV. CONCLUDING REMARKS

The foregoing brief survey of thallium oxide chemistry would reveal the special role played by thallium in determining the structure and properties of thallium oxides, especially the superconducting property of thallium cuprates. The particular combination of Tl(III/I) with Cu(II/III) in oxides appears to be unique, producing novel metastable phases with an electronic structure that can support high-temperature superconductivity. The challenge for the future lies in finding another similar combination, possibly with thallium, a transition metal other than copper and an anion, not necessarily oxide, that can support still higher T_cs. The present writer is optimistic that the challenge will be met soon.

ACKNOWLEDGMENTS

The author would like to thank A. W. Sleight, who gave him an opportunity to work on superconducting cuprates at Du Pont during 1987–1989. His thanks are also due to C. N. R. Rao for encouragement and to the Department of Science and Technology, Government of India, for support. This work represents Contribution 842 from the Solid State and Structural Chemistry Unit of the Indian Institute of Science, Bangalore, India.

REFERENCES

1. A. F. Wells, *Structural Inorganic Chemistry*, 5th ed., Clarendon Press, Oxford, 1984.
2. N. V. Sidgwick and H. M. Powell, *Proc. R. Soc. London A* 176, 153 (1940).
3. R. J. Gillespie and R. S. Nyholm, *Q. Rev. Chem. Soc.* 11, 339 (1957).
4. L. E. Orgel, *J. Chem. Soc.* 3815 (1959).
5. S. Andersson and A. Åstrom, in *Solid State Chemistry*, NBS Special Publication 364, U.S. Dept. of Commerce, Washington, D.C., 1972.
6. F. Hulliger, *Structural Chemistry of Layer-Type Phases*, D. Reidel, Dordrecht, The Netherlands, 1976.
7. C. M. Varma, *Phys. Rev. Lett.* 61, 2713 (1988).
8. A.W. Sleight, *Proc. Robert A. Welch Foundation Conference on Valency*, 1988.
9. Z. Z. Sheng and A. M. Hermann, *Nature* 332, 55, 138 (1988).
10. A. G. Lee, *The Chemistry of Thallium*, Elsevier, Amsterdam, 1971.
11. C. T. Prewitt, R. D. Shannon, D. B. Rogers, and A. W. Sleight, *Inorg. Chem.* 8, 1985 (1969).
12. A. W. Sleight, J. L. Gillson, and B. L. Chamberland, *Mater. Res. Bull.* 5, 807 (1970).
13. R. von Schenck and Hk. Muller-Buschbaum, *Z. Anorg. Allg. Chem.* 405, 197 (1974).
14. M. Itoh, R. Liang, and T. Nakamura, *J. Solid State Chem.* 82, 172 (1989).
15. C. Michel, M. Hervieu, B. Raveau, S. Li, M. Greaney, S. Fine, J. Potenza, and M. Greenblatt, *Mater. Res. Bull.* 26, 123 (1991).
16. R. von Schenck and Hk. Müller-Buschbaum, *Z. Anorg. Allg. Chem.* 396, 113 (1973).
17. R. D. Shannon, *Inorg. Chem.* 6, 1474 (1967).
18. M. Ganne, Y. Piffard, and M. Tournoux, *Can. J. Chem.* 52, 3539 (1974).
19. A. W. Sleight and J. L. Gillson, *Mater. Res. Bull.* 6, 781 (1971).

20. H. S. Jarrett, A. W. Sleight, J. F. Weiher, J. L. Gillson, C. G. Frederick, G. A. Jones, R. S. Swingle, D. Swartzfager, J. E. Gulley, and P. C. Hoell in *Valence Instabilities and Related Narrow-Band Phenomena*, ed. R. D. Parks, Plenum Press, New York, 1977.
21. A. W. Sleight and R. Ward, *Inorg. Chem.* 3, 292 (1964).
22. R. Subramanian, M. A. Subramanian, and A. W. Sleight, *Mater. Res. Bull.* 24, 1413 (1989).
23. S. Li, K. V. Ramanujachary, and M. Greenblatt, *Physica C* 166, 535 (1990).
24. M. A. Subramanian, G. Aravamudan, and G. V. Subba Rao, *Prog. Solid State Chem.* 15, 55 (1983).
25. M. Touboul and D. Ingrain, *J. Less-Common Metals* 71, 55 (1980); M. Touboul and Toledano, *J. Appl. Crystallogr.* 8, 398 (1977); N. Jouini et al., *Rev. Chim. Miner.* 14, 435 (1977).
26. A. Jayaraman, G. A. Kourouklis, and L. G. van Uitert, *Phys. Rev. B* 36, 8547 (1987).
27. A. W. Sleight, J. L. Gillson, and P. E. Bierstedt, *Solid State Commun.* 17, 27 (1975).
28. R. J. Cava, B. Batlogg, J. J. Krejewski, R. Farrow, L. W. Rupp, A. E. White, K. Short, W. F. Peck, and T. Kometani, *Nature* 332, 814 (1988).
29. M. A. Subramanian, A. K. Ganguli, K. L. Willmer, and J. E. Greedan, *J. Solid State Chem.* 95, 447 (1991).
30. A. W. Sleight, *Science* 242, 1519 (1988); A. W. Sleight, J. Gopalakrishnan, C. C. Torardi, and M. A. Subramanian, *Phase Transitions* 19, 149 (1989).
31. S. S. P. Parkin, V. Y. Lee, A. I. Savoy, R. Beyers, and S. J. LaPlaca, *Phys. Rev. Lett.* 61, 750 (1988).
32. C. N. R. Rao and B. Raveau, *Acc. Chem. Res.* 22, 106 (1989).
33. M. Greenblatt, S. Li, L. E. H. McMills, and K. V. Ramanujachary, in *Studies of High-Temperature Superconductors*, ed. A. V. Narlikar, Vol. 5, Nova Science Publishers, New York, 1990, and the references given therein.
34. J. Gopalakrishnan, in *Chemistry of High-Temperature Superconductors*, ed. C. N. R. Rao, World Scientific, Singapore, 1991.
35. J. B. Goodenough and A. Manthiram, *J. Solid State Chem.* 88, 115 (1990).
36. R. J. Cava, *Science* 247, 656 (1990).
37. A. Santoro, F. Beech, M. Marezio, and R. J. Cava, *Physica C* 156, 693 (1988).
38. Y. Gao, P. Lee, P. Coppens, M. A. Subramanian, and A. W. Sleight, *Science* 241, 954 (1988).
39. For original references on the crystal structures of high T_c cuprates, see the review paper: K. Yvon and M. Francois, *Z. Phys. B* 76, 413 (1989).
40. I. K. Gopalakrishnan, J. V. Yakhmi, and R. M. Iyer, *Physica C* 175, 183 (1991).
41. C. Martin, D. Bourgault, C. Michel, J. Provost, M. Hervieu, and B. Raveau, *Eur. J. Solid State Inorg. Chem.* 26, 1 (1989); M.-H. Pan and M. Greenblatt, *Physica C* 176, 80 (1991).
42. M. A. Subramanian, C. C. Torardi, J. Gopalakrishnan, P. L. Gai, J. C. Calabrese, T. R. Askew, R. B. Flippen, and A. W. Sleight, *Science* 242, 249 (1988).
43. M.-H. Pan and M. Greenblatt, *Physica C* 184, 253 (1991).
44. S. Li and M. Greenblatt, *Physica C* 157, 365 (1989).
45. M. A. Subramanian, P. L. Gai, and A. W. Sleight, *Mater. Res. Bull.* 25, 101 (1990).
46. C. N. R. Rao, A. K.Ganguli, and R. Vijayaraghavan, *Phys. Rev. B* 40, 2565 (1989).
47. I. K. Gopalakrishnan, J. V. Yakhmi, and R. M. Iyer, *Physica C* 172, 450 (1991).
48. M. A. Subramanian, *Mater. Res. Bull.* 25, 191 (1990).
49. Y. Tokura, T. Arima, H. Takagi, S. Uchida, T. Ishigaki, H. Asano, R. Beyers, A. I. Nazzal, P. Lacorre, and J. B. Torrance, *Nature* 342, 890 (1989).

50. C. C. Torardi, M. A. Subramanian, J. C. Calabrese, J. Gopalakrishnan, K. J. Morrissey, T. R. Askew, R. B. Flippen, U. Chowdhry, and A. W. Sleight, *Science* 240, 631 (1988).

51. J. B. Parise, J. Gopalakrishnan, M. A. Subramanian, and A. W. Sleight, *J. Solid State Chem.* 76, 432 (1988); K. V. Ramanujachary, S. Li, and M. Greenblatt, *Physica C* 165, 377 (1990).

52. E. A. Hayri and M. Greenblatt, *Physica C* 156, 775 (1988).

53. A. Sequeira, H. Rajagopal, I. K. Gopalakrishnan, P. V. P. S. S. Sastry, G. M. Phatak, J. V. Yakhmi, and R. M. Iyer, *Physica C* 156, 599 (1988).

54. D. B. Kang, D. Jung, and M.-H. Whangbo, *Inorg. Chem.* 29, 257 (1990).

55. R. S. Liu, P. P. Edwards, Y. T. Huang, S. F. Wu, and P. T. Wu, *J. Solid State Chem.* 86, 334 (1990).

56. T. Manako, Y. Shimakawa, Y. Kubo, T. Satoh, and H. Igarashi, *Physica C* 158, 143 (1989).

57. A. K. Ganguli, V. Manivannan, A. K. Sood, and C. N. R. Rao, *Appl. Phys. Lett.* 55, 2664 (1989).

58. W. Dmowksi, B. H. Toby, T. Egami, M. A. Subramanian, J. Gopalakrishnan, and A. W. Sleight, *Phys. Rev. Lett.* 61, 2608 (1988).

59. D. E. Cox, C. C. Torardi, M. A. Subramanian, J. Gopalakrishnan, and A. W. Sleight, *Phys. Rev. B* 38, 6624 (1988).

60. S. J. Hibble, A. K. Cheetham, A. M. Chippindale, P. Day, and J. A. Hriljac, *Physica C* 156, 604 (1988).

61. M. Hervieu, B. Domenges, C. Michel, and B. Raveau, in *Chemistry of High-Temperature Superconductors*, ed. C. N. R. Rao, World Scientific, Singapore, 1991.

62. A. W. Sleight, *Phys. Today*, June 1991, p. 24.

63. D. Jung, M.-H. Whangbo, N. Herron, and C. C. Torardi, *Physica C* 160, 381 (1988).

64. M.-H. Whangbo and C. C. Torardi, *Acc. Chem. Res.* 24, 127 (1991).

65. A. Manthiram, M. Paranthaman, and J. B. Goodenough, *Physica C* 171, 135 (1990); M. Paranthaman, A. Manthiram, and J. B. Goodenough, *J. Solid State Chem.* 87, 479 (1990).

66. J. Gopalakrishnan, R. Vijayaraghavan, R. Nagarajan, and C. Shivakumara, *J. Solid State Chem.* 93, 272 (1991); 96, 468 (1992).

67. M. W. Shafer and T. Penney, *Eur. J. Solid State Inorg. Chem.* 27, 191 (1990); C. N. R. Rao, J. Gopalakrishnan, A. K. Santra, and V. Manivannan, *Physica C* 174, 11 (1991).

3

Crystallographic Aspects

Yuichi Shimakawa*
NEC Corporation
Ibaraki, Japan

I. INTRODUCTION

In this chapter we give an overview of the crystallographic aspects of Tl-based superconductors. Tl-based superconductors are one of the families of *layered perovskite compounds*, named for the perovskite-like layers. The family of Tl-based compounds shows various structural modifications. The ideal composition of these compounds is $Tl_mBa_2Ca_{n-1}Cu_nO_{2+m+2n}$ ($m = 1$ or 2, $n = 1,2,3, \ldots$), abbreviated as m:2:$n - 1$:n: for example, Tl-2201 or Tl-2212. The layer sequence of these compounds is $-(TlO)_m-(BaO)-(CuO_2)-Ca-(CuO_2)- \cdots -Ca-(CuO_2)-(BaO)-(TlO)_m-$. Figures 1 and 2 show schematic crystal structures for these compounds with TlO double layers ($m = 2$) and with a TlO monolayer ($m = 1$), respectively. These series of compounds include several two-dimensional CuO_2 layers between the TlO layers. The $n = 1$ compound has an octahedral CuO_6 block, the $n = 2$ compound has two pyramidal CuO_5 blocks, and the $n = 3$ compound has two pyramidal blocks and one square CuO_4 block. Single-phase samples with $n \geq 4$ are difficult to synthesize because they tend to include intergrowth structures with different numbers of CuO_2 layers. A few members with $n \geq 4$ are known [1–4], but their refined atomic coordinates are not yet available. These layered compounds also commonly contain a Ba site with ninefold oxygen coordinates, and a Ca site with nearly cubic eightfold oxygen coordinates. These structures might be thought of as the sequence of ionic charged layers of $(TlO)^+$, $(BaO)^0$, $(CuO_2)^{2-}$, and Ca^{2+}.

These Tl-based compounds are hole carrier (p-type) superconductors. Holes seem to be introduced into the octahedral and pyramidal type of CuO_2 layers, and these holes appear to contribute to superconductivity. Since the square type of

---

*Present affiliation: Argonne National Laboratory, Argonne, Illinois.

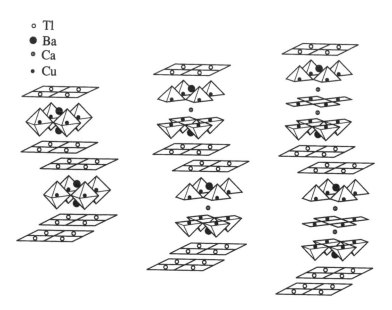

Figure 1 Crystal structures of TlO double-layer compounds ($m = 2$): left, $Tl_2Ba_2CuO_6$ (Tl-2201); center, $Tl_2Ba_2CaCu_2O_8$ (Tl-2212); right, $Tl_2Ba_2Ca_2Cu_3O_{10}$ (Tl-2223).

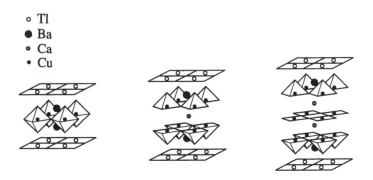

Figure 2 Crystal structures of TlO monolayer compounds ($m = 1$): left, $TlBa_2CuO_5$ (Tl-1201); center, $TlBa_2CaCu_2O_7$ (Tl-1212); right $TlBa_2Ca_2Cu_3O_9$ (Tl-1223).

CuO_2 layer, in contrast, lies between two Ca^{2+} layers, the electrostatic potential of the CuO_2 layer for holes is high, and holes tend not to be introduced into the layer. The inner square type of CuO_2 layer in $n \geq 3$ compounds may therefore not contribute to the appearance for superconductivity.

The structure of the Tl system includes both TlO monolayer and TlO double layers, whereas only BiO double-layer compounds are present in Bi-based super-conductors despite similar layered perovskite structures. This contrast is due to the different ionic characters of Tl^{3+} and Bi^{3+}; the electronic configurations for Tl^{3+} is $5d^{10}$, but the Bi^{3+} ion has lone-pair electrons of $6s^2$. In addition, Sr versions for TlO double-layer compounds of $Tl_2Sr_2Ca_{n-1}Cu_nO_{4+2n}$ are absent, although both Ba and Sr versions exist for TlO monolayer compounds. This fact is probably related to the difference in ionic radii with ninefold oxygen coordinates between Ba^{2+} (1.47 Å) and Sr^{2+} (1.31 Å) [5].

The first part of the chapter concerns the general structure of Tl-based compounds. Reference is made to crystal structures refined by single-crystal x-ray, powder neutron, and powder x-ray diffraction measurements. Nonstoichiometry and cation substitution are discussed in relation to hole carrier creation and structural stability. The results of studies using electron microscopy are also reported. In addition, as some Tl compounds show a wide range of T_c variations, there is a discussion of the relation between structural changes and variation in T_cs.

II. REFINED CRYSTAL STRUCTURES

A. $Tl_2Ba_2Ca_{n-1}Cu_nO_{4+2n}$ ($m = 2$) Series

Crystal structures in this series are refined using single-crystal x-ray, powder neutron, and powder x-ray diffraction data. Tl-2201 shows both tetragonal and orthorhombic structural symmetries, whereas Tl-2212 and Tl-2223 show only tetragonal symmetry. The tetragonal structures in this series are analyzed based on the space group of $I4/mmm$. The analysis of the orthorhombic Tl-2201 structure uses a face-centered or an A-centered space group. Both kinds of Tl-2201 are discussed in detail later in this chapter.

The refined crystallographic data at room temperature are listed in Tables 1 to 4, and the interatomic distances and angles are listed in Table 5 [6–9]. The anisotropic bond lengths around Cu atoms are worth noting. Given the ionic radii

Table 1 Crystallographic Data for Tetragonal Tl-2201 $[Tl_2Ba_2CuO_6]^a$

Atom	Site	x	y	z	B (Å^2)	g
Tl	4e	0.5	0.5	0.20265(2)	1.3(1)	1.0
Ba	4e	0.0	0.0	0.08301(3)	0.5(1)	1.0
Cu	2b	0.5	0.5	0.0	0.4(1)	1.0
O(1)	4c	0.0	0.5	0.0	0.7(1)	1.0
O(2)	4e	0.5	0.5	0.1168(4)	1.1(1)	1.0
O(3)	16n	0.595(5)	0.5	0.2889(5)	0.4(3)	1.0

Source: Ref. 6.
[a]Single-crystal x-ray diffraction, $T_c = 90$ K, $I4/mmm$, $a = 3.866(1)$ Å, and $c = 23.239(6)$ Å.

Table 2 Crystallographic Data for Orthorhombic Tl-2201 $[Tl_2Ba_2CuO_6]^a$

Atom	Site	x	y	z	B (Å^2)	g
Tl	8f	0.0	0.0	0.2018(1)	1.5(−)	1.0
Ba	8f	0.0	0.5	0.0822(2)	0.9(1)	1.0
Cu	4a	0.0	0.0	0.0	0.6(1)	1.0
O(1)	8e	0.75	0.25	0.0	0.2(1)	1.0
O(2)	8f	0.0	0.0	0.1161(2)	1.1(1)	1.0
O(3)	8f	0.045(3)	0.0	0.2887(3)	5.4(−)	1.0

Source: Ref. 7.
[a]Powder neutron diffraction, $T_c = 90$ K, $Abma$, $a = 5.4967(3)$ Å, $b = 5.4651(3)$ Å, and $c = 23.246(1)$ Å.

Table 3   Crystallographic Data for Tl-2212 [$Tl_2Ba_2CaCu_2O_8$][a]

Atom	Site	x	y	z	B (Å^2)	g
Tl	4e	0.5	0.5	0.21359(2)	1.7(1)	1.0[b]
Ba	4e	0.0	0.0	0.12179(3)	0.8(1)	1.0
Cu	4e	0.5	0.5	0.0540(1)	0.4(1)	1.0
Ca	2a	0.0	0.0	0.0	0.6(1)	1.0[b]
O(1)	8g	0.0	0.5	0.0531(2)	0.8(1)	1.0
O(2)	4e	0.5	0.5	0.1461(3)	1.5(2)	1.0
O(3)	16n	0.604(9)	0.5	0.2815(7)	3.9(11)	1.0

Source: Ref. 8.
[a]Single-crystal x-ray diffraction, T_c = 110 K, *I4/mmm*, *a* = 3.8550(6) Å, and *c* = 29.318(4) Å.
[b]Refined composition of $(Tl_{0.9}Ca_{0.1})_2Ba_2(Ca_{0.9}Tl_{0.1})Cu_2O_8$ is also suggested.

Table 4   Crystallographic Data for Tl-2223 [$Tl_2Ba_2Ca_2Cu_3O_{10}$][a]

Atom	Site	x	y	z	B (Å^2)	g
Tl	4e	0.5	0.5	0.2201(1)	2.1(1)	1.0[b]
Ba	4e	0.0	0.0	0.1448(1)	0.6(1)	1.0
Cu(1)	4e	0.5	0.5	0.0896(2)	0.7(1)	1.0
Ca	2a	0.0	0.0	0.0463(2)	1.2(1)	1.0[b]
Cu(2)	2b	0.5	0.5	0.0	0.3(1)	1.0
O(1)	8g	0.5	0.0	0.0875(6)	1.2(1)	1.0
O(2)	4e	0.5	0.5	0.1588(13)	2.3(9)	1.0
O(3)	4e	0.5	0.5	0.2719(14)	2.08(10)	1.0
O(4)	4c	0.5	0.0	0.0	0.9(9)	1.0

Source: Ref. 9.
[a]Single-crystal x-ray diffraction, T_c = 125 K, *I4/mmm*, *a* = 3.8503(6) Å, and *c* = 35.88(3) Å.
[b]Refined composition of $(Tl_{0.85}Ca_{0.15})_2Ba_2(Ca_{0.88}Tl_{0.12})_2Cu_3O_{10}$ is also suggested.

of Cu^{2+} (0.73 Å for sixfold, 0.65 Å for fivefold, and 0.57 Å for fourfold) and of O^{2-} (1.40 Å) [5], the ideal Cu–O bond lengths are 2.13 Å for an octahedron, 2.05 Å for a pyramid, and 1.97 Å for a square. The shorter bond lengths of about 1.9 Å in the CuO_2 layers for octahedrons and pyramids seem to result from a covalent character for these bondings. The bond lengths between Cu and apical oxygen, in contrast, are much longer than the ideal ionic bond length. Such large anisotropy in Cu–O bond lengths suggests a strongly two-dimensional nature of the CuO_2 layer. The Cu–O bond lengths in a square, on the other hand, are similar to the ideal ionic bond length, and this correspondence suggests an ionic character for these bondings.

The location of the oxygen atoms in the TlO layer deviates slightly from the ideal, and these local displacements must be indispensable for relaxation of the dimensional mismatch between the TlO and CuO_2 layers. Since the overall cell dimension is mainly controlled by the stiffer CuO_2 layer, the TlO layer must be "stretched," and the oxygen atoms in the TlO layer must deviate from their ideal positions in order to achieve a more desirable bond length. In fact, two intralayer

Table 5 Interatomic Distances (Å) and Angles (degrees) for the TlO Double Layer ($m = 2$) Compounds

n	1 (Tl-2201)				2(Tl-2212)		3(Tl-2223)	
	Tetragonal		Orthorhombic					
Cu(1)–O(1)	1.9330(5)	(× 4)	1.938	(× 4)	1.9277(3)	(× 4)	1.9252(3)	(× 4)
Cu(1)–O(2)	2.714(9)	(× 2)	2.698(5)	(× 2)	2.699(10)	(× 1)	2.48(5)	(× 1)
Cu(2)–O(4)					1.9275	(× 4)	1.927(1)	(× 4)
Tl–O(2)	1.995(9)	(× 1)	1.994(6)	(× 1)	1.978(10)	(× 1)	2.20(5)	(× 1)
Tl–O(3)	2.038(11)	(× 1)	2.034(7)	(× 1)	2.031(21)	(× 1)	1.92(3)	(× 1)
Tl–O(3)	2.496(12)	(× 2)	2.512(17)	(× 1)	2.462(22)	(× 2)	2.48(3)	(× 2)
Tl–O(3)	3.011(15)	(× 2)	3.002(17)	(× 1)	3.027(28)	(× 2)	3.02(4)	(× 2)
Tl–O(3)			2.752(2)	(× 2)				
Ba–O(1)	2.7309(7)	(× 4)	2.721[a]	(× 4)	2.788(4)	(× 4)	2.82(2)	(× 4)
Ba–O(2)	2.844(2)	(× 4)	2.844[a]	(× 2)	2.818(3)	(× 4)	2.768(9)	(× 4)
			2.859[a]	(× 2)				
Ba–O(3)	2.999(11)	(× 1)	3.011[a]	(× 1)	2.864(21)	(× 1)	2.99(3)	(× 1)
Ca–O(1)					2.478(4)	(× 8)	2.542(5)	(× 4)
Ca–O(4)							2.427(14)	(× 4)
O(1)–Cu(1)–O(1)	180.0	(× 2)	180.0	(× 2)	178.4(4)	(× 2)	175(1)	(× 2)
O(1)–Cu(1)–O(1)	90.0	(× 4)	90.3[a]	(× 2)	89.99(1)	(× 4)	89.91(6)	(× 4)
			89.7[a]	(× 2)				
O(1)–Cu(1)–O(2)	90.0	(× 4)	90.0	(× 4)	90.8(2)	(× 4)	92.3(7)	(× 4)
Cu(1)–O(1)–Cu(1)	180.0	(× 1)	180.0	(× 1)	178.4(4)	(× 1)	175(1)	(× 1)
O(4)–Cu(2)–O(4)							180.0	(× 2)
O(4)–Cu(2)–O(4)							90.0	(× 4)
Cu(2)–O(4)–Cu(2)							180.0	(× 1)

[a]Calculated from the refined structural data.

Tl–O bond lengths are about 2.5 Å, and the other two are about 3.0 Å. Such local displacements of the oxygen atoms in the TlO layer may be associated with weak modulations observed by transmission electron microscopy, which will be dealt with later in this chapter.

The structural changes in tetragonal Tl-2201 and Tl-2223 below room temperature are shown in Fig. 3 [10,11]. The temperature dependence of lattice parameters for each compound indicates no significant structural change or discontinuity in the refined parameters through the superconducting transition temperatures. The anisotropic nature of the thermal expansion coefficients is worth noting. The values of $\Delta a/a$ and $\Delta c/c$ are 5.9×10^{-6} K^{-1} and 8.3×10^{-6} K^{-1}, respectively, for a temperature change from 20 K to 294 K in Tl-2201, and 2.1×10^{-6} K^{-1} and 7.0×10^{-6} K^{-1} for a temperature change from 13 K to 150 K in Tl-2223. In addition, there is no indication in the profile fitting of any distortion from tetragonal symmetry because the refined values of half-width parameters, which would be affected by any broadening of the peaks, are essentially identical at each temperature. In contrast, local structural changes occurring around T_c are observed by extended x-

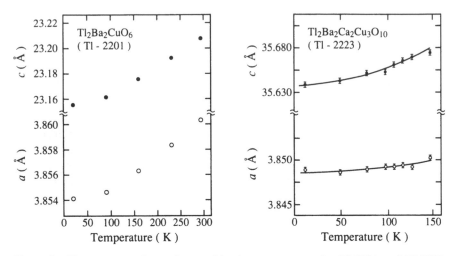

Figure 3   Temperature dependence of lattice parameters for Tl-2201 and Tl-2223.

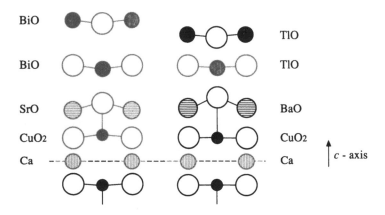

Figure 4   Schematic crystal structures for Bi-2212 (left) and Tl-2212 (right).

ray absorption fine structure (EXAFS) spectroscopy [12] and by pair-distribution function (PDF) analysis of pulsed neutron scattering [13], both of which suggest that the mechanism of superconductivity involves a electron–lattice interaction.

A comparison between Tl- and Bi-based compounds highlights some interesting bond lengths along the c axis. Schematic drawings of the crystal structure for Bi-2212 and Tl-2212 compounds are shown in Fig. 4. The TlO double-layer block along the c axis is more compact than the BiO double-layer block. Interlayer bond lengths of Tl–O are about 2.0 Å, while those of Bi–O are 3.2 Å for Bi-2201 [6] and 3.7 Å for Bi-2212 [14]. This structural difference is consistent with the fact that Tl-based compounds are less cleavable than Bi-based compounds. Moreover, bond lengths between Cu and apical oxygen for the Tl-based compounds are longer than those for the Bi-based compounds. A further difference is that the CuO_2 layers in the Tl compounds are much flatter than those in the Bi compounds, which may relate to the fact that the Tl-based superconductors show slightly higher T_cs than those of the isostructural Bi-based superconductors.

No hole carrier is created in the ideal TlO double-layer compositions if formal valences of Tl^{3+}, Ba^{2+}, Ca^{2+}, Cu^{2+}, and O^{2-} are assigned. Nonstoichiometries or cation substitutions or both therefore seem to play an important role in the creation of hole carriers. As shown in Fig. 5, for Tl-2201, the presence of excess oxygen atoms located at an interstitial site between double TlO layers is confirmed by high-resolution neutron diffraction [15,16]. A 9% excess oxygen in the TlO layer has also been reported for Tl-2212 [17]. The amounts of excess oxygen are estimated to be 0.1 to 0.2 per formula unit for fully oxygenated samples. Some refinements suggest Tl deficiencies of 3 to 13.5% [17,18] or TlO deficiencies of 12.5% [19]. Substitutions of Cu at about 5% for the Tl sites in Tl-2201 [16] and at about 10% of the Tl sites in Tl-2212 [20] have also been suggested. Substitutions of Ca for Tl, and vice versa, or mutual substitution of these atoms are seen in almost all samples for $n = 2$ and $n = 3$ compounds [8,9,17,18,20–23]. The electron probe microanalysis data also indicate these cation substitutions [22,24]. These nonstoichiometries and substitutions will create the hole carriers necessary for high-T_c superconductivity.

The creation of hole carriers by the self-doping mechanism can be examined by calculation. The calculation for minimum lattice energy based on the Madelung site potentials suggests charge transfer from Tl^{3+} to O^{2-} in the CuO_2 layer [25]. Tight-binding band calculation also suggests a relation between the modulated structure and hole creation [26,27]. In this case, when distortions such as ladder-, island-, and chainlike atomic arrangements in the TlO layers are taken into account in these calculations, the bottoms of the Tl-6s bands in the ideal compositions of $Tl_2Ba_2Ca_{n-1}Cu_nO_{4+2n}$ are significantly below the Fermi level, so that the compounds have large electron pockets to remove electrons from the CuO_2 layer. However, since nonstoichiometries and cation substitutions are outside the scope of these calculations, the electronic structures in the real compounds may differ from these calculated.

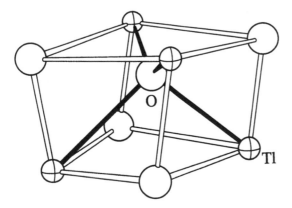

Figure 5   Excess oxygen atoms at an interstitial site between TlO double layers.

B. $TlBa_2Ca_{n-1}Cu_nO_{3+2n}$ ($m = 1$) Series

In this series of compounds, only $n = 2$ and $n = 3$ structures are refined using single-crystal x-ray diffraction data [28,29]. The unit cells of the TlO monolayer compounds are primitive tetragonal $P4/mmm$. Thus body-centered symmetry is absent, and the single TlO layer is perfectly flat. The results of structure refinement at room temperature for $TlBa_2CaCu_2O_7$ [28] and $TlBa_2Ca_2Cu_3O_9$ [29] are listed in Tables 6 and 7, and interatomic distances and angles are also listed in Table 8. Large anisotropies in Cu–O bond lengths similar to those in the TlO double-layer compounds are found in pyramids. In this system, too, pyramidal CuO_2 layers seem to show a covalent character. In contrast to the oxygen displacements in TlO double-layer compounds, the TlO monolayer system relaxes the dimensional mismatch by displacement of Tl atoms from their ideal positions.

Structural data for the $n = 1$ compound are not available because this compound is difficult to synthesize. In the ideal TlO monolayer compounds of Tl-

Table 6   Crystallographic Data for Tl-1212 [$TlBa_2(Ca_{0.83}Tl_{0.17})Cu_2O_{6.75}$][a]

Atom[b]	Site	x	y	z	B (Å²)	g
Tl	$1b$	0.0	0.0	0.5	1.10(12)	1.0
(Tl_D	$4m$	0.108(2)	0.0	0.5		0.125)
Ba	$2h$	0.5	0.5	0.2842(1)	0.79(5)	1.0
(Ba_D	$2h$	0.5	0.5	0.337(4)		1.0)
Cu	$2g$	0.0	0.0	0.1255(2)	0.46(7)	1.0
Ca	$2a$	0.5	0.5	0.0	0.68(15)	0.831(8)
Tl	$2a$	0.5	0.5	0.0	0.68(15)	0.169(8)
O(1)	$4i$	0.5	0.0	0.1210(8)	0.71(15)	1.0
O(2)	$2g$	0.0	0.0	0.342(1)	0.95(23)	1.0
O(3)	$1c$	0.5	0.5	0.5	2.6(6)	0.747(9)

Source: Ref. 28.
[a]Single-crystal x-ray diffraction, $T_c = 103$ K, $P4/mmm$, $a = 3.8566(4)$ Å, and $c = 12.754(2)$ Å.
[b]D denotes displaced atom positions.

Table 7   Crystallographic Data for Tl-1223 [$TlBa_{1.9}(Ca_{0.95}Tl_{0.05})_2Cu_3O_9$][a]

Atom	Site	x	y	z	B (Å²)	g
Tl	$4l$	0.085(2)	0.0	0.0	1.7(1)	1.0
Ba	$2h$	0.5	0.5	0.1729(1)	0.6(1)	0.94(1)
Cu(1)	$2g$	0.0	0.0	0.2911(2)	0.4(1)	1.0
Ca	$2h$	0.5	0.5	0.3953(3)	0.7(1)	0.95
Tl	$2h$	0.5	0.5	0.3953(3)	0.7(1)	0.05
Cu(2)	$1b$	0.0	0.0	0.5	0.4(1)	1.0
O(1)	$4i$	0.0	0.5	0.3019(7)	1.1(2)	1.0
O(2)	$2g$	0.0	0.0	0.1277(11)	0.6(3)	1.0
O(3)	$1c$	0.5	0.5	0.0	2.6(7)	1.0
O(4)	$2e$	0.0	0.5	0.5	0.9(3)	1.0

Source: Ref. 29.
[a]Single-crystal x-ray diffraction, $T_c = 110$ K, $P4/mmm$, $a = 3.853(1)$ Å, and $c = 15.913(4)$ Å.

Table 8 Interatomic Distances (Å) and Angles (degrees) for the TlO Monolayer ($m = 1$) Compounds

n	2 (Tl-1212)				3 (Tl-1223)	
	Idealized			Displaced		
Cu(1)–O(1)		1.929(1)	(× 4)		1.9270(6)	(× 4)
Cu(1)–O(2)		2.76(2)	(× 1)		2.73(2)	(× 1)
Cu(2)–O(4)					1.9265(5)	(× 4)
Tl–O(2)	2.01(4)	(× 2)		2.06(2)	(× 2) 2.06(2)	(× 2)
Tl–O(3)	2.727(1)	(× 4)		2.450[a]	(× 2) 2.503(4)	(× 2)
Tl–O(3)				3.036[a]	(× 2) 2.966(5)	(× 2)
Ba–O(1)	2.837(8)	(× 4)		3.36(4)	(× 4) 2.816(9)	(× 4)
Ba–O(2)	2.825(5)	(× 4)		2.728(2)	(× 4) 2.818(4)	(× 4)
Ba–O(3)	2.752(2)	(× 1)		2.079[a]	(× 1) 2.751(2)	(× 1)
Ca–O(1)		2.470(7)	(× 8)		2.433(8)	(× 4)
Ca–O(4)					2.547(2)	(× 4)
O(1)–Cu(1)–O(1)		176.9[a]	(× 2)		177.3(8)	(× 2)
O(1)–Cu(1)–O(1)		88.3(3)	(× 4)		89.97(2)	(× 4)
O(1)–Cu(1)–O(2)		91.7(3)	(× 4)		91.3(4)	(× 4)
Cu(1)–O(1)–Cu(1)		176.9[a]	(× 1)		177.3(8)	(× 1)
O(4)–Cu(2)–O(4)					180.0	(× 2)
O(4)–Cu(2)–O(4)					90.0	(× 4)
Cu(2)–O(4)–Cu(2)					180.0	(× 1)

[a]Calculated from the refined structural data.

$Ba_2Ca_{n-1}Cu_nO_{3+2n}$, the average Cu valence is $2 + 1/n$. The average Cu valence of an $n = 1$ compound of $TlBa_2CuO_5$ is $+3$, which is too high to be a Cu valence. This makes it difficult to obtain a single-phase sample unless Ba^{2+} ion, for example, are substituted by ions such as La^{3+}. Single-phase samples of the La-substituted Tl-1201, $Tl(Ba_{1-x}La_x)_2CuO_5$, have been synthesized [30]. Sr versions of the substituted Tl-1201 have also been reported: for example, $Tl(Sr,R)_2CuO_5$ (R = La, Nd) [31], $(Tl,Pr)(Sr,Pr)_2CuO_5$ [32], and $(Tl,Bi)(Sr,Ca)_2CuO_5$ [33,34]. The $n = 2$ and $n = 3$ compounds can be stabilized in the same way by substituting Y^{3+} for Ca^{2+} [35], or Pb^{4+} for Tl^{3+} [36–38], thus reducing the average formal Cu valences to $+2$. In addition, it is easy to introduce oxygen vacancies into the TlO layer of these TlO monolayer compounds. These oxygen vacancies also reduce the average Cu valences and thus stabilize the crystal structures, as a result of which the oxygen vacancies decrease as x increases in the series of $TlBa_2(Ca_{1-x}Y_x)Cu_2O_{7-\delta}$ [39].

The average Cu valence of $2 + 1/n$ in the ideal composition suggests the creation of hole carriers. In addition, nonstoichiometries and cation substitutions help control carrier concentration. Site occupancies for Ca of Ca/Tl = 0.83/0.17, and for oxygen in the TlO layer of 0.75, have been confirmed in the refined Tl-1212 compound above [28]. The presence of about 6% vacancy in the Ba sites and substitution of Tl at about 5% of the Ca sites are also observed in Tl-1223 [29]. Similar substitution of Tl for Ca sites and oxygen vacancies in the TlO layer have been confirmed using neutron diffraction data for the Sr version of Tl-1212, Tl-

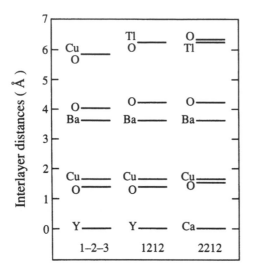

Figure 6 Interlayer distances for Y-123, Y-replaced Tl-1212, and Tl-2212.

$Sr_2CaCu_2O_7$ [40]. To achieve an appropriate carrier concentration for high T_c superconductivity in this TlO monolayer system, as well as to stabilize the crystal structures, nonstoichiometries and cation substitutions would be required.

The structure of Tl-1212 is very similar to that of $YBa_2Cu_3O_7$ (Y-123) as it is derived from $YBa_2Cu_3O_7$ by replacing Y by Ca and the liner Cu–O chains by square-planar TlO layers. The major difference is the location of oxygen atoms between the rock salt type of TlO layer and the CuO chain. In the Tl-1212 structure, oxygen atoms are located at $(\frac{1}{2},\frac{1}{2})$ in the a–b plane instead of at $(\frac{1}{2},0)$ in the Y-123 structure. Figure 6 shows interlayer distances for the Y-replaced Tl-1212 compound, together with those for Y-123 and Tl-2212 [35]. In Y-replaced Tl-1212, the Y–CuO_2 block is similar to that of Y-123, and the BaO–TlO block is similar to that of Tl-2212. These interlayer distances seem to be affected by the ionic radii and the valences of successive ionic charged layers.

III. HREM AND EDP INFORMATION

From the early stages of research on Tl-based superconductors, high-resolution electron microscopy (HREM) and electron diffraction patterns (EDP) have provided important information. Figure 7 shows a lattice image of the Tl-2201 compound [41]. The relatively heavy Tl and Ba atoms are seen as large dark blobs, while the lighter Cu atoms appear as smaller dark blobs. Oxygen atoms may not be imaged with detectable intensity because their scattering power for electrons is too small. The somewhat diffuse images in the Tl atom positions may be related to large thermal vibrations or atom displacement.

Intergrowth defects along the c axis are often observed in the Tl-based compounds [42–46], because the values of lattice parameters a are almost the same for all Tl-based compounds. Intergrowth bands with ordered structures of 100 to 1000 Å are sometimes seen in low-magnification lattice images, as are stacking faults in the matrix phase. More complex lattice images, which include several

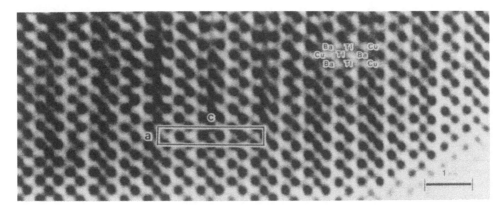

Figure 7   Lattice image taken from a Tl-2201 crystal oriented in the [010] direction.

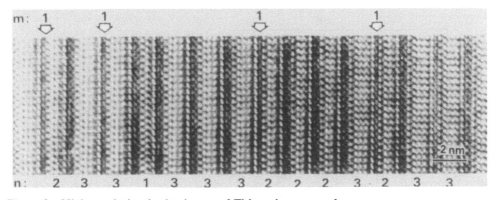

Figure 8   High-resolution lattice image of Tl-based compound.

kinds of layer sequence in a short scale, have also been reported. Figure 8 shows this type of intergrowth structure. Note the absence of c-glide symmetry, where there is a TlO monolayer block of $m = 1$. The presence of blocks with $n \geq 4$ has been confirmed. Lattice image by HREM of up to $n = 9$ block has been reported [4]. These intergrowth defects tend to disappear under appropriate compositional and annealing conditions.

Other defects, such as intercalation of additional layers, which can be either infinite or interrupted, have also been observed [47]. Amorphous areas such as bubble-coated crystals or crystallized microdomains are also seen [48]. Structural defects caused by substitution have been investigated by HREM to compare the experimental through-focus series with the images calculated for various crystal thicknesses. The suggested effects of substituting Tl for Ca sites in Tl-2212 and Tl-2223 [47,49] agree with the results of structural refinements using x-ray and neutron diffraction data.

EDP studies of incommensurate lattice modulations show them to be very weak in Tl compounds. Even with electron diffraction, the satellite reflections often smear out or disappear when samples are heated with an electron beam. It is

Table 9 Approximate Superlattice Wave Vectors for the
Tl-Based Compounds[a]

	Crystal structure	Wave vector
Tl-2201	b.c. tetragonal	$<-0.16, 0.08, 1>$
	f.c. or a.c. orthorhombic	$<0.24, \pm0.08, 1>$[b]
Tl-2212	b.c. tetragonal	$<0.17, 0, 1>$[c]
Tl-2223	b.c. tetragonal	$<0.17, 0, 1>$
Tl-1212	p. tetragonal	$<0.29, 0, 0.5>$
Tl-1223	p. tetragonal	$<0.29, 0, 0.5>$

Source: Ref. 50.
[a]No superlattice modulation is observed in Tl-1201.
[b]$<0.22, \pm0.07, 1>$ [51] and $<3/14, \pm1/14, 1>$ [52] are also reported.
[c]$<1/6, 0, 1>$ [17] is also reported.

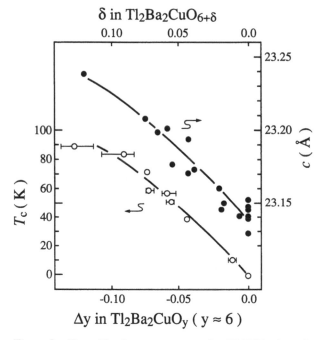

Figure 9 T_c and lattice parameter c for Tl-2201 plotted against oxygen content. The lower horizontal axis shows the relative change in oxygen content from that of the oxygen-annealed state. The upper horizontal axis shows δ based on the neutron diffraction data.

therefore thought that satellites cannot be detected by measuring x-ray diffraction. Such modulations are observed for almost all Tl compounds, but their wave vectors, listed in Table 9, are different. For Tl-2201, both tetragonal and orthorhombic structures show similar superlattice modulations with an approximate wave vector $\langle -0.16, 0.08, 1\rangle$ in the tetragonal cell ($\langle 0.24, \pm0.08, 1\rangle$ in the $a \sim \sqrt{2}\, a_{\text{tetra}}$ orthorhombic setting cell) [50–52]. The satellite reflections for the orthorhombic structure are spotty, whereas they are diffuse for the tetragonal structure. For Tl-

Table 10 Lattice Parameters (Å), Atom Coordinations, Occupation Factors, Oxygen Contents, and Interatomic Distances (Å) for Tl-2201 Samples with Various T_cs

		T_c (K)			
		0	48	58	73
a		3.86298(7)	3.86276(6)	3.86273(6)	3.86248(6)
c		23.1369(5)	23.1848(4)	23.1995(4)	23.2248(4)
$g(O(3))$		0.246(3)	0.250(3)	0.247(4)	0.247(2)
$g(O(4))$		0.028(3)	0.020(3)	0.017(4)	0.005(2)
$6 + \delta$		6.080	6.080	6.044	5.996
$z(Tl)$		0.20260(8)	0.20261(8)	0.2025(1)	0.20256(6)
$z(Ba)$		0.0842(1)	0.0834(1)	0.0833(1)	0.08299(8)
$z(O(2))$		0.1168(1)	0.1170(1)	0.1171(1)	0.11715(8)
$z(O(3))$		0.2098(2)	0.2105(2)	0.2106(2)	0.2109(1)
$y(O(3))$		0.096(4)	0.090(3)	0.091(3)	0.081(2)
Cu–O(1)	($\times$ 4)	1.93149(4)	1.93138(3)	1.93137(3)	1.93124(3)
Cu–O(2)	($\times$ 2)	2.703(3)	2.713(3)	2.716(3)	2.721(2)
Ba–O(1)	($\times$ 4)	2.743(2)	2.733(2)	2.733(2)	2.729(1)
Ba–O(2)	($\times$ 4)	2.8341(8)	2.8404(8)	2.8411(10)	2.8441(6)
Ba–O(3)	($\times$ 1)	2.929(5)	2.968(5)	2.972(6)	2.987(3)
Tl–O(2)	($\times$ 1)	1.985(3)	1.985(3)	1.983(4)	1.984(2)
Tl–O(3)	($\times$ 2)	2.490(10)	2.504(7)	2.501(6)	2.526(4)
Tl–O(3)	($\times$ 2)	3.009(11)	2.994(8)	2.997(7)	2.968(5)
Tl–O(3)	($\times$ 1)	2.061(5)	2.044(5)	2.047(5)	2.034(3)

2212 and Tl-2223 the wave vectors are $\langle 0.17, 0, 1 \rangle$ [50,17], while for Tl-1212 and Tl-1223 the wave vectors are $\langle 0.29, 0, 0.5 \rangle$ [50,53]. These superlattice wave vectors are distinctly different from those found in the Bi-based compounds, where wave vectors along the b axis cause strong satellite reflections [54]. These modulations may be caused by the defect structures or by the ordering of substituting atoms, but no clear relationship has yet been established between the wave vectors and the crystal structures.

IV. RELATIONSHIP BETWEEN STRUCTURAL CHANGE AND T_c VARIATION

Some Tl-based superconductors show a wide range of T_c variations, despite having almost identical crystal structures. In particular, the T_cs of Tl-2201 vary from 0 K (nonsuperconducting) to over 85 K [55]. In the tetragonal Tl-2201, there is a clear relationship among T_c, c-axis length, and oxygen content, as shown in Fig. 9 [56]. Both the Hall coefficient and resistivity measurements show this compound to be overdoped with hole carriers [57]. A decreased oxygen content of about 0.1 per formula unit increases the T_c to about 80 K and elongates the c axis by about 0.4%. The systematic structural changes that accompany changes in oxygen content have been clarified using powder x-ray and neutron diffraction data [16,58]. Table 10 lists important crystal data determined by measuring neutron diffraction for samples with different T_cs. [Note that O(4) is the interstitial oxygen site between TlO layers

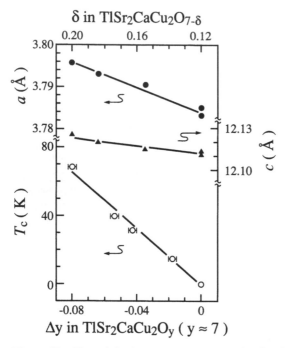

Figure 10 T_c and lattice parameters a and c for Tl-1212 plotted against oxygen content. Upper and lower axes as in Fig. 9.

in this compound.] The systematic change in the occupancy of the excess oxygen site O(4) suggests that incorporating and releasing this oxygen cause the change in T_cs. The occupancy of the O(4) site relates to the displacement of the O(3) atom. The degree to which the O(3) atom is displaced from its ideal site increases as the occupancy of the O(4) site increases. The interstitial oxygen atoms thus repel the neighboring O(3) atoms to increase distances between two O(3) atoms. In contrast, these O(3) atoms draw closer if a neighboring O(4) site is vacant. No structural discontinuity is observed in the change from a normal metal to a super-conductor, although changes in interatomic distances along the c axis are noted. The distance between Cu and apical oxygen increases by twice as much as the increase in c-axis length, but the Ba atom approaches the CuO_2 layer. These structural changes are explained by the release of oxygen atoms, which decreases the amount of hole carriers (positive charge) in the CuO_2 layer. Consequently, negatively charged apical O^{2-} ions are repulsed and leave the CuO_2 layer, whereas positively charged Ba^{2+} ions are attracted to it.

Similar variations in T_c and systematic changes in structure during the transition from a normal metal to a superconductor can also be seen in the Sr version of the Tl monolayer compound, $TlSr_2CaCu_2O_7$ (Tl-1212) [59], in which a decrease in oxygen content or about 0.08 per formula unit causes a T_c increase to 68 K. Figure 10 shows the relationship among T_c, lattice parameters, and oxygen content. The changes in oxygen content and the corresponding changes in hole concentration are confirmed to be caused by incorporating and releasing oxygen atoms in the TlO monolayer [40]. Changes in interatomic distances along the c axis suggest that the change in hole concentration is similar to that in the Tl-2201 compound. In the nonsuperconducting sample, confirmed to be more heavily doped with hole carriers than the superconducting sample, apical O^{2-} ions approach the CuO_2 layer and Sr^{2+} ions leave it. In contrast to what happens in Tl-2201, a decrease in the oxygen content of Tl-1212 primarily affects elongation in the *a*-axis length.

NOTE ADDED IN PROOF

Some interesting data have since been published. See [60] for temperature dependence of lattice parameters for tetragonal and orthorhombic Tl-2201 and Tl-1212. See [61] and [62] for pressure dependence and anomalous behavior of the pressure dependence of lattice parameters for Tl-2201.

REFERENCES

1. H. Ihara, R. Sugise, M. Hirabayashi, N. Terada, M. Jo, K. Hayashi, A. Negishi, M. Tokumoto, Y. Kimura, and T. Shimamura, *Nature* 334, 510 (1988).
2. M. Kikuchi, S. Nakajima, Y. Syono, K. Hiraga, T. Oku, D. Shindo, N. Kobayashi, H. Iwasaki, and Y. Muto, *Physica C* 158, 79 (1989).
3. S. Nakajima, M. Kikuchi, Y. Syono, T. Oku, D. Shindo, K. Hiraga, N. Kobayashi, H. Iwasaki, and Y. Muto, *Physica C* 158, 471 (1989).
4. Y. Syono, M. Kikuchi, S.Nakajima, T. Suzuki, T. Oku, K. Hiraga, N. Kobayashi, H. Iwasaki, and Y. Muto, *Mater. Res. Soc. Symp. Proc.* 156, 229, (1989).
5. R. D. Shannon, *Acta Crystallogr. A* 32, 751 (1976).
6. C. C. Torardi, M. A. Subramanian, J. C. Calabrese, J. Gopalakrishnan, E. M. McCarron, K. J. Morrissey, T. R. Askew, R. B. Flippen, U. Chowdhry, and A. W. Sleight, *Phys. Rev. B* 38, 225 (1988).
7. J. B. Parise, J. Gopalakrishnan, M. A. Subramanian, and A. W. Sleight, *J. Solid State Chem.* 76, 432 (1988).
8. M. A. Subramanian, J. C. Calabrese, C. C. Torardi, J. Gopalakrishnan, T. R. Askew, R. B. Flippen, K. J. Morrissey, U. Chowdhry, and A. W. Sleight, *Nature* 332, 420 (1988).
9. C. C. Torardi, M. A. Subramanian, J. C. Calabrese, J. Gopalakrishnan, K. J. Morrissey, T. R. Askew, R. B. Flippen, U. Chowdhry, and A. W. Sleight, *Science* 240, 631 (1988).
10. T. Kamiyama, F. Izumi, H. Asano, Y. Shimakawa, Y. Kubo, T. Manako, and H. Igarashi, *Physica C* 185–189, 881 (1991).
11. D. E. Cox, C. C. Torardi, M. A. Subramanian, J. Gopalakrishnan, and A. W. Sleight, *Phys. Rev. B* 38, 6624 (1988).
12. P. G. Allen, J. Mustre de Leon, S. D. Conradson, and A. R. Bishop, *Phys. Rev. B* 44, 9480 (1991).
13. B. H. Toby, T. Egami, J. D. Jorgensen, and M. A. Subramanian, *Phys. Rev. Lett.* 64, 2414 (1990).
14. P. Bordet, J. J. Capponi, C. Chaillout, J. Chenavas, A. W. Hewat, E. A. Hewat, J. L. Hodeau, M. Marezio, J. L. Tholence, and D. Tranqui, *Physica C* 156, 189 (1988).
15. J. B. Parise, C. C. Torardi, M. A. Subramanian, J. Gopalakrishnan, A. W. Sleight, and E. Prince, *Physica C* 159, 239 (1989).
16. Y. Shimakawa, Y. Kubo, T. Manako, H. Igarashi, F. Izumi, and H. Asano, *Phys. Rev. B* 42, 10165, (1990).
17. A. W. Hewat, E. A. Hewat, J. Brynestad, H. A. Mook, and E. D. Specht, *Physica C* 152, 438 (1988).

18. T. Zetterer, H. H. Otto, G. Lugert, and K. F. Renk, *Z. Phys. B* 73, 321 (1988).
19. A. W. Hewat, P. Bordet, J. J. Capponi, C. Chaillout, J. Chenavas, M. Godinho, E. A. Hewat, J. L. Hodeau, and M. Marezio, *Physica C* 156, 369 (1988).
20. M. Onoda, S. Kondoh, K. Fukuda, and M. Sato, *Jpn. J. Appl. Phys.* 27, L1234 (1988).
21. Y. Shimakawa, Y. Kubo, T. Manako, Y. Nakabayashi, and H. Igarashi, *Physica C* 156, 97 (1988).
22. M. Kikuchi, T. Kajitani, T. Suzuki, S. Nakajima, K. Hiraga, N. Kobayashi, H. Iwasaki, Y. Syono, and Y. Muto, *Jpn. J. Appl. Phys.* 28, L382 (1989).
23. T. Kajitani, K. Hiraga, S. Nakajima, M. Kikuchi, Y. Syono, and C. Kabuto, *Physica C* 161, 483 (1989).
24. S. S. P. Parkin, V. Y. Lee, A. I. Nazzal, R. Savoy, T. C. Huang, G. Gorman, and R. Beyers, *Phys. Rev. B* 38, 6531 (1988).
25. J. Kondo, *J. Phys. Soc. Jpn.* 58, 2884 (1989).
26. D. Jung, M.-H. Whangbo, N. Herron, and C. C. Torardi, *Physica C* 160, 381 (1989).
27. C. C. Torardi, D. Jung, D. B. Kang, J. Ren, and M.-H Whangbo, *Mater. Res. Soc. Symp. Proc.* 156, 295 (1989).
28. B. Morosin, D. S. Ginley, P. F. Hlava, M. J. Carr, R. J. Baughman, J. E. Schirber, E. L. Venturini, and J. F. Kwak, *Physica C* 152, 413 (1988).
29. M. A. Subramanian, J. B. Parise, J. C. Calabrese, C. C. Torardi, J. Gopalakrishnan, and A. W. Sleight, *J. Solid State Chem.* 77, 192 (1988).
30. T. Manako, Y. Shimakawa, Y. Kubo, T. Satoh, and H. Igarashi, *Physica C* 158, 143 (1989).
31. M. A. Subramanian, *Mater. Res. Bull.* 25, 191 (1990).
32. D. Bourgault, C. Martin, C. Michel, M. Hervieu, J. Provost, and B. Raveau, *J. Solid State Chem.* 78, 326 (1989).
33. P. Haldar, A. Roig-Janicki, S. Sridhar, and B. C. Giessen, *Mater. Lett.* 7, 1 (1988).
34. M.-H. Pan and M. Greenblatt, *Physica C* 184, 235 (1991).
35. T. Manako, Y. Shimakawa, Y. Kubo, T. Satoh, and H. Igarashi, *Physica C* 156, 315 (1988).
36. M. A. Subramanian, C. C. Torardi, J. Gopalakrishnan, P. L. Gai, J. C. Calabrese, T. R. Askew, R. B. Flippen, and A. W. Sleight, *Science* 242, 249 (1988).
37. H. H. Otto, T. Zetterer, and K. F. Renk, *Z. Phys. B* 75, 433 (1989).
38. J. B. Parise, P. L. Gai, M. A. Subramanian, J. Gopalakrishnan, and A. W. Sleight, *Physica C* 159, 245 (1989).
39. S. Nakajima, M. Kikuchi, Y. Syono, T. Oku, K. Nagase, N. Kobayashi, D. Shindo, and K. Hiraga, *Physica C* 182, 89 (1991).
40. F. Izumi, T. Kondo, Y. Shimakawa, T. Manako, Y. Kubo, H. Igarashi, and H. Asano, *Physica C* 185–189, 615 (1991).
41. S. Iijima, T. Ichihashi, Y. Shimakawa, T. Manako, and Y. Kubo, *Jpn. J. Appl. Phys.* 27, L1061 (1988).
42. S. Iijima, T. Ichihashi, and Y. Kubo, *Jpn. J. Appl. Phys.* 27, L817 (1988).
43. S. Iijima, T. Ichihashi, Y. Shimakawa, T. Manako, and Y. Kubo, *Jpn. J. Appl. Phys.* 27, L837 (1988).
44. S. Iijima, T. Ichihashi, Y. Shimakawa, T. Manako, and Y. Kubo, *Jpn. J. Appl. Phys.* 27, L1054 (1988).
45. H. W. Zandbergen, G. Van Tendeloo, J. Van Landuyt, and S. Amelinckx, *Appl. Phys. A* 46, 233 (1988).
46. T. Kotani, T. Nishikawa, H. Takei, and K. Tada, *Jpn. J. Appl. Phys.* 29, L902 (1990).
47. M. Hervieu, C. Martin, J. Provost, and B. Raveau, *J. Solid State Chem.* 76, 419 (1988).
48. B. Domenges, M. Hervieu, and B. Raveau, *Solid State Commun.* 69, 1085 (1989).
49. K. Hiraga, D. Shindo, M. Hirabayashi, M. Kikuchi, N. Kobayashi, and Y. Syono, *Jpn. J. Appl. Phys.* 27, L1848 (1988).

50. R. Beyers, S. S. P. Parkin, V. Y. Lee, A. I. Nazzal, R. Savoy, G. Gorman, T. C. Huang, and S. LaPlaca, *Appl. Phys. Lett.* 53, 432 (1988).
51. E. A. Hewat, P. Bordet, J. J. Capponi, C. Chaillout, J. Chenavas, M. Godinho, A. W. Hewat, J. L. Hodeau, and M. Marezio, *Physica C* 156, 375 (1988).
52. S. Iijima, in *Mechanisms of High Temperature Superconductivity*, ed. H. Kamimura and A. Oshiyama, Springer-Verlag, Berlin, 1989.
53. Y. Liu, Y. L. Zhag, J. K. Liang, and K. K. Fung, *J. Phys. C* 21, L1039 (1988).
54. H. W. Zandbergen, W. A. Groen, F. C. Mijlhoff, G. van Tendeloo, and S. Amelinckx, *Physica C* 156, 325 (1988).
55. Y. Shimakawa, Y. Kubo, T. Manako, T. Satoh, S. Iijima, T. Ichihashi, and H. Igarashi, *Physica C* 157, 279 (1989).
56. Y. Shimakawa, Y. Kubo, T. Manako, and H. Igarashi, *Phys. Rev. B* 40, 11400 (1989).
57. Y. Kubo, Y. Shimakawa, T. Manako, and H. Igarashi, *Phys. Rev. B* 43, 7875 (1991).
58. Y. Shimakawa, Y. Kubo, T. Manako, and H. Igarashi, *Mater. Res. Soc. Symp. Proc.* 169, 1061 (1990).
59. Y. Kubo, T. Kondo, Y. Shimakawa, T. Manako, and H. Igarashi, *Phys. Rev. B* 45, 5553 (1992).
60. Y. Shimakawa, *Physica C* 202, 199 (1992).
61. F. Izumi, J. D. Jorgensen, Y. Shimakawa, Y. Kubo, T. Manako, Shiyou Pei, T. Matsumoto, R. L. Hitterman, and Y. Kanke, *Physica C* 193, 426 (1992).
62. H. Takahashi, J. D. Jorgensen, B. A. Hunter, R. L. Hitterman, Shiyou Pei, F. Izumi, Y. Shimakawa, Y. Kubo, and T. Manako, *Physica C* 191, 248 (1992).

4

Extended Defects and Redox Mechanisms in Thallium Cuprates

Bernard Raveau, Maryvonne Hervieu, and Claude Michel

Laboratoire de Cristallographie et
Sciences des Matériaux
Caen, France

I. INTRODUCTION

The discovery of superconductivity at high temperatures in thallium cuprates [1–3] has opened the road to the exploration of a large family of oxides whose structures are closely related to each other and which can exhibit a wide range of critical temperatures, from several kelvin to 125 K. From a physical point of view, these materials appear as most promising since they are, up to now, the superconductors that exhibit the highest T_cs. Moreover, it has been shown that their critical temperature can be improved dramatically by adjusting the oxygen pressure during synthesis or by annealing in a reducing atmosphere. Thus there is no doubt that the superconducting properties of these oxides are very sensitive to the nonstoichiometry of oxygen. Another particularity of these phases deals with their difficult synthesis, which is due to the enormous volatility of thallium oxide at 600°C and above. Indeed, this property can have disastrous effects not only for the operator, who can be intoxicated by the thallium loss, but it can also introduce thallium and oxygen nonstoichiometry in the samples, thus drastically modifying the superconducting properties of these materials. The possible coexistence in the crystal of the redox systems Tl(III)/Tl(I) and Cu(III)/Cu(II) is an important feature that must be taken into account for an understanding of the solid-state chemistry of thallium cuprates and for an understanding of their superconducting properties. All these characteristics mean that in addition to the ideal structure of the various phases, one must consider all the defects that might affect superconductivity. In this chapter we focus on a study of various extended defects—intergrowths, order–disorder phenomena, cation nonstoichiometry, and oxygen nonstoichiometry—in connection with their influence on superconductivity.

41

II. GENERAL FEATURES ABOUT STRUCTURE AND CHEMICAL BONDING

More than 15 thallium cuprates have been discovered since 1988 [4–28]. The idealized crystal structure of these phases is based on the association of rock salt layers with oxygen-deficient layers, forming regular intergrowths. Two types of rock salt layers are involved (Fig. 1): double rock salt layers built up from thallium monolayers sandwiched between barium or strontium oxygen layers, and triple rock salt layers built up from thallium bilayers sandwiched between barium or strontium oxygen layers. The oxygen-deficient perovskite layers (Fig. 2) consist of single copper layers built up from CuO_6 octahedra, double copper layers built up from CuO_5 pyramids, or triple quadruple copper layers involving sheets of CuO_4 square-planar groups sandwiched between sheets of CuO_5 pyramids. Examples of

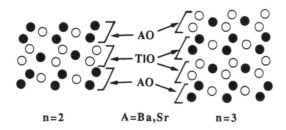

Figure 1   Idealized drawing of double and triple rock salt layers involved in thallium cuprates.

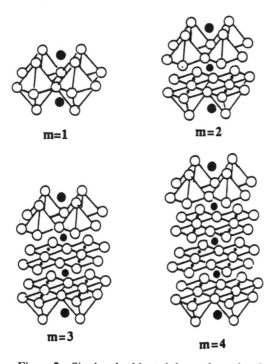

Figure 2   Single, double, triple, and quadruple oxygen-deficient perovskite layers.

the ideal structures of most of these phases are shown in Fig. 3. The two series can be represented by the ideal formulas $TlBa_2Ca_{m-1}Cu_mO_{2m+3}$ and $Tl_2Ba_2Ca_{m-1}Cu_mO_{2m+4}$. In this notation m represents the number of copper layers forming the perovskite slabs. Another notation is used alternatively to represent these formulas, in which only the numbers of cations are given [i.e., $12(m-1)m$ and $22(m-1)m$, corresponding to $TlBa_2Ca_{m-1}Cu_mO_{2m+3}$ and $Tl_2Ba_2Ca_{m-1}Cu_mO_{2m+4}$, respectively]. The various layered thallium cuprates so far isolated are summarized in Table 1.

In reality, we have represented here idealized structures as well as idealized formulas. Owing to its ability to take two oxidation states, Cu(II) and Cu(III), and to demonstrate the Jahn–Teller effect, copper can vary its coordination sphere so that anionic vacancies are susceptible to appearing in the octahedral layers of the $m = 1$ phases or on the apical sites of the pyramidal layers of compounds with $m \geq 2$. Tl(III) also exhibits a particular behavior since it is able to present a distorted

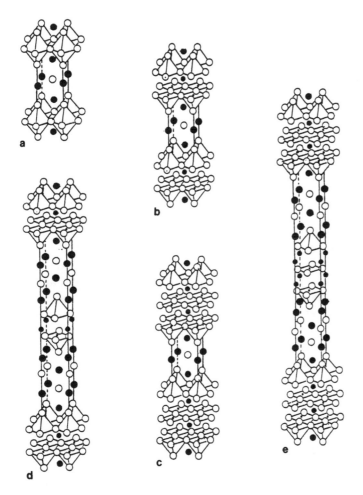

Figure 3 Schematic drawing of the structure of some superconducting thallium cuprates: (a) $TlBa_2CuO_5$; (b) $TlBa_2CaCu_2O_7$; (c) $TlBa_2Ca_2Cu_3O_9$; (d) $Tl_2Ba_2CaCu_2O_8$; (e) $Tl_2Ba_2Ca_2Cu_3O_{10}$.

Table 1 Thallium Cuprates with Bidimensional Structures Related to Perovskite (Ln = Pr, Y, Nd)

Copper oxide	T_c (K)	Copper oxide	T_c (K)
$TlSr_2CuO_{5-\delta}$	NS[a]	$TlBa_2CaCu_2O_7$	60–85
$Tl_{0.5}Pb_{0.5}Sr_2CuO_{5-\delta}$	NS	$TlBa_2LnCu_2O_7$	NS
$Tl_{1-x}Pr_xSr_{2-y}Pr_yCuO_{5-\delta}$	40	$TlBa_2Ln_2Cu_2O_9$	NS
($x = 0.2$, $y = 0.4$)		$Tl_2Ba_2CaCu_2O_8$	105
$TlSr_2CaCu_2O_7$	50	$Tl_{3(4x/3)}Ba_{1+x}LnCu_2O_8$	NS
$Tl_{0.5}Pb_{0.5}Sr_2CaCu_2O_7$	85	$TlBa_{2-x}La_{2+x}Cu_2O_9$	45
$Tl_{0.5}Pb_{0.5}Sr_2Ca_2Cu_3O_9$	115	$TlBa_2Ca_2Cu_3O_9$	110
$TlSr_2Ln_2Cu_2O_9$	NS	$Tl_2Ba_2Ca_2Cu_3O_{10}$	125
$Tl_2Ba_2CuO_6$	NS–90	$TlBa_2Ca_3Cu_4O_{11}$	108
$TlBa_2CuO_{5-\delta}$	NS	$Tl_2Ba_2Ca_3Cu_4O_{12}$	115
$TlBa_{1-x}La_{1+x}CuO_5$ ($x = 0.2$)	50	$TlBa_2Ca_4Cu_5O_{13}$	105

[a]NS, not a superconductor.

octahedral coordination, as in Tl_2O_3 [29], or a tetrahedral coordination, as in $Tl_2Ba_2O_5$ [30]. Consequently, the thallium oxygen layers are highly distorted, with two short apical Tl–O distances. As a result, the possibility of creation of anionic vacancies in the $[TlO]_\infty$ layers has to be considered. Another characteristic of thallium deals with its great mobility, as shown from the volatility of thallium oxide, so that a thallium deficiency is also possible. Thus these particular properties of copper and thallium mean that problems of nonstoichiometry related to thallium and oxygen are very complex in these structures, so that the formulas of these oxides should be written $Tl_{2-x}Ba_2Ca_{m-1}Cu_mO_{2m+4-\delta}$ or $Tl_{1-x}Ba_2Ca_{m-1}Cu_mO_{2m+3-\delta}$. Some recent studies have also suggested the possibility of intercalation of oxygen into the $[TlO]_\infty$ layers, leading to excess oxygen with respect to the ideal formulas as shown, for example, for $Tl_2Ba_2CuO_{6+\delta}$ [31].

The great mobility of thallium also means that the distribution of $[TlO]_\infty$ layers may vary according to the experimental conditions of synthesis, leading to extended defects, called intergrowth defects, which correspond to the replacement of thallium monolayers by thallium bilayers, or vice versa. The possibility of intercalation of additional layers in perovskite and rock salt–type layers must also be taken into account.

All of these complicated nonstoichiometry phenomena are important because they are capable of influencing the superconducting properties of these materials.

III. INTERGROWTH DEFECTS

As stated above, the layered thallium cuprates can be described as regular intergrowths of oxygen-deficient perovskite layers formed of m copper layers with n rock salt layers involving $n - 1$ $[TlO]_\infty$ layers, n being 2 or 3. Variations in layer stacking in relation to the ideal sequence are indeed the most probable event in intergrowth phases. We distinguish two types of variation: (1) the existence of defective slabs, characterized by m' or/and n' values, different from the m and n values of the nominal composition but built up of similar layers, and (2) the existence of defective slabs built up from the intercalation of layers of a different nature [26,32–40].

A.   Variation of the Layer Stacking

The appearance of m' defective layers in an $[m, n]$ member is a very classical defect in intergrowth phases built up from perovskite layers. In a general way, the number of defects increases with m and the $\Delta m = |m - m'|$ value is close to 1 for low m values and increases with m. This phenomenon is illustrated in Fig. 4. This high-resolution electron microscopy micrograph shows that in the matrix of the oxide $Tl_2Ba_2Ca_3Cu_4O_{12}$ [26] built up from quadruple copper layers ($m = 4$), large oxygen-deficient perovskite slabs form, built up of either five or seven layers of copper ($m' = 5$ or 7). Similar defects are observed in $Tl_2Ba_2Ca_2Cu_3O_{10}$ [38]. Figure 5 shows double and quadruple copper layers (black numbers) in an $m = 3$ matrix of the latter cuprate. Such extended defects, corresponding to a variation in the thickness of the perovskite layers, are observed systematically in phases built up from the intergrowth of perovskite or ReO_3 layers with another structural type, such as brownmillerite [41], intergrowth tungsten bronzes (ITBs) [42–44], and phosphate tungsten bronzes (PTBs) [45–47]. Their number generally increases with the thickness of the perovskite layer (i.e., with the m value).

On the other hand, when $[A^{II}]_\infty$ layers (A = Ca, Sr) are replaced by $[A^{III}]$ layers (A = Y, Nd, Pr, etc.) $m' = 3$ defects are never observed, whereas $m' = 1$ defects corresponding to single perovskite layers are sometimes detected, as shown, for example, in the HREM micrograph of $TlBa_2NdCu_2O_7$ in Fig. 6a. This feature could be related to the shape of the AO_8 cage, where these cations are located. The pseudocubic shape of the Y and Nd cages implies a displacement of the copper atoms to avoid short Cu–Cu distances, contrary to the elongated shape of the CaO_8 cages (Fig. 6b). In the $m = 2$ members, built up from pyramidal

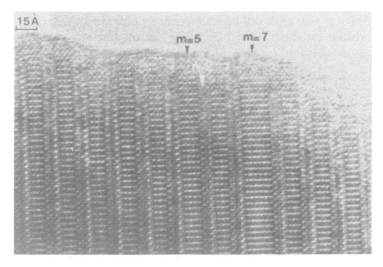

Figure 4 $Tl_2Ba_2Ca_3Cu_4O_{12}$ ($m = 4$, $n = 3$). Defective $m' = 7$ members ($\Delta m = 3$) and $n' = 2$ members (white arrow) are commonly observed.

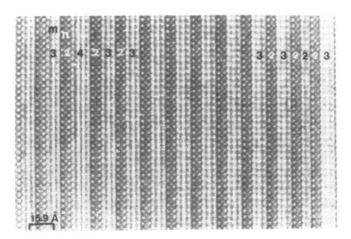

Figure 5 $Tl_2Ba_2Ca_2Cu_3O_{10}$ ($m = 3$, $n = 3$); (100) high-resolution image with two inter-growth defects in the perovskite layers. n and m values of the formula $(AO)_n(A'CuO_{3-\delta})_m$ are indicated.

layers, Cu atoms are displaced toward the apical oxygen atoms, but in the $m = 3$ members, the copper atoms, located in the median plane, cannot be displaced. The appearance of defective rock salt–type slabs obeys different rules. This behavior can be understood easily if one bears in mind that the existence of multiple rocks slabs ($n > 1$), interlayered with perovskite slabs is exceptional and observed for the first time in high-T_c superconducting cuprates.

Formation of thallium bilayers as extended defects (corresponding to triple rock salt layers, $n' = 3$) in a matrix built up of thallium monolayers (double rock salt layers, $n = 2$) is currently observed, as shown, for example, for the 1212 phase $TlBa_2Ca_{1-x}Nd_xCu_2O_7$ ($m = n = 2$) (Fig. 7a) [36]. On the contrary, $n' = 1$ defects, corresponding to the formation of single rock salt layers in a matrix formed of double ($n = 2$) or triple ($n = 3$) rock salt layers, are rarely observed. An example of such a defect is visible for the 1212 oxide in Fig. 7b. In the same way, the formation of thallium monolayers (double rock salt layers, $n' = 2$) in a matrix built up from thallium bilayers (triple rock salt layers, $n = 3$) is a common feature to all the $Tl_2Ba_2Ca_{m-1}Cu_mO_{2m+4}$ cuprates. An example of such a defect is shown in Fig. 8 for $Tl_2Ba_2CaCu_2O_8$.

In addition to the formation of such rock salt defects, which appear distributed in a disordered way in the structure, regular sequences can appear in some crystals of composition immediate between Tl_2 and Tl_1 cuprates. This is illustrated perfectly by the formation of polytypes, as shown by the oxide $Tl_{1.66}Ba_2CaCu_2O_8$ [48], for which a perfect sequence of two triple ($n = 3$) and one double ($n = 2$) rock salt slabs intergrown with double perovskite layers has been observed (Fig. 9).

B. Introduction of Additional Layers

Two types of extra layers can be intercalated in the original matrix of the thallium cuprates, at the levels of the rock salt layers and the calcium (or lanthanide) layers, respectively.

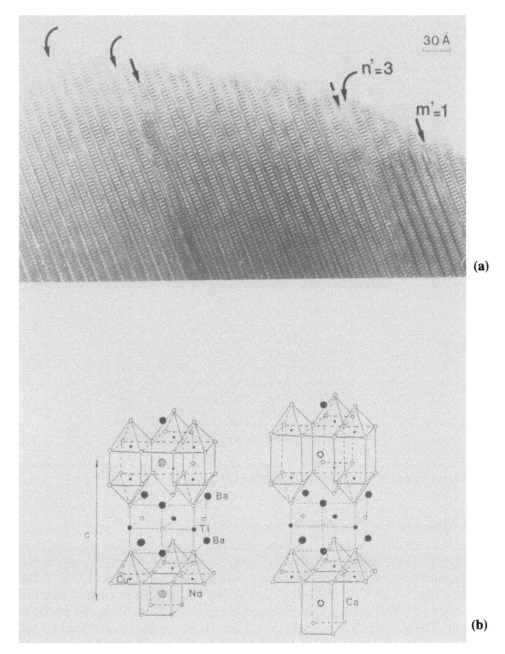

Figure 6   (a) In $TlBa_2LnCu_2O_7$ ($m = 2$, $n = 2$), only $m' = 1$ defective perovskite slabs have been observed (straight arrows). This feature is correlated with the shape of the AO_8 fluorite cage: elongated with Ca and pseudocubic with Nd and Y (b).

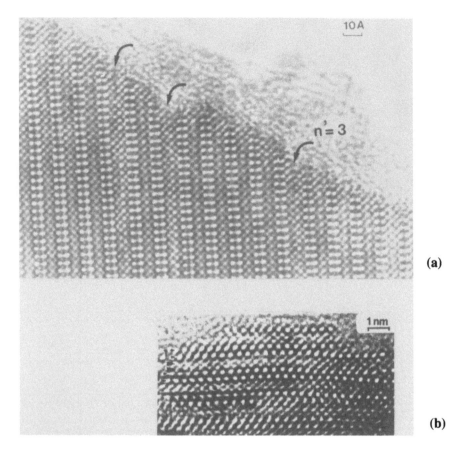

Figure 7 $TlBa_2(Ca_{1-x}Nd_x)Cu_2O_7$ ($m = 2$, $n = 2$). Defective $n' = 3$ members are shown by black arrows (a); $n' = 1$ is a very rare defect (b).

1. Addition of $[CaO]_\infty$ Layers in the Rock Salt Slabs

Thick rock salt layers with $n > 3$ have been observed in thallium cuprates only [32,39]. This is the case for $Tl_2Ba_2Ca_1Cu_2O_8$ ($n = 3$), for which two additional $[CaO]_\infty$ layers are intercalated between the two $[TlO]_\infty$ layers according to the sequence [BaO–TlO–CaO–TlO–BaO], as shown by the high-resolution electron microscopy (HREM) micrograph presented in Fig. 10a and the idealized model in Fig. 10b. The propagation of such extra layers is sometimes interrupted in the matrix by a "superdislocation" mechanism, where a row of polyhedra is stopped and the neighboring layers accommodate the defect. Such a superdislocation is illustrated in Fig. 11, which shows the disappearance of the $[CaO]_\infty$ layers when going from the right of the crystal to the left.

2. Introduction of Fluorite Layers

The intercalation of additional layers of a different nature is one way to obtain new compounds whose structure is related directly to the "parent structures." One example was observed in the $TlBa_2NdCu_2O_7$ oxide, where besides the classical intergrowth defects, original sequences in the layer stacking were characterized

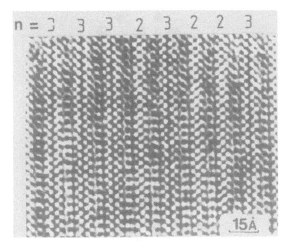

Figure 8 $Tl_2Ba_2CaCu_2O_8$ ($m = 2$, $n = 3$). Defect in rock salt–type layers: $[(BaO)_2(TlO)]$ layers ($n' = 2$) appear in a $[(BaO)_2(TlO)_2]$ matrix ($n' = 3$); they are noted 2 and 3, respectively.

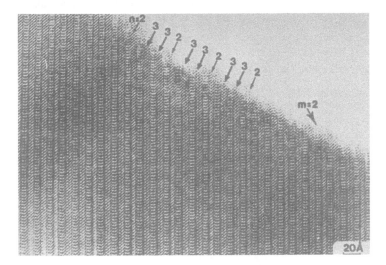

Figure 9 $Tl_{1.66}Ba_2CaCu_2O_8$; example of a regular intergrowth of $(2212)_2$ $(1212)_1$ members.

[36]. A HREM image of the defect (Fig. 12a) shows that at the level of the defect (labeled F), the single $[Nd]_\infty$ layer, intercalated between the $[CuO_2]_\infty$ layers, is replaced by a double layer. The shifting of the $[CuO_2]_\infty$ layers resulting from that intercalation, the layers thickness, and the contrast led us to propose a double fluorite layer $[Nd_2O_2]_\infty$ (Fig. 12b) as a new structural unit in the matrix.

The frequent appearance of this double fluorite layer in the form of defects suggested the existence of a new mechanism, confirmed by synthesis of the Tl-$Ba_2Ln_2Cu_2O_9$ phases [49]. The structure of the latter (Fig. 13) can be described as an intergrowth of double rock salt layers $[(AO)_2]_\infty$, double fluorite layers $[Ln_2O_4]_\infty$,

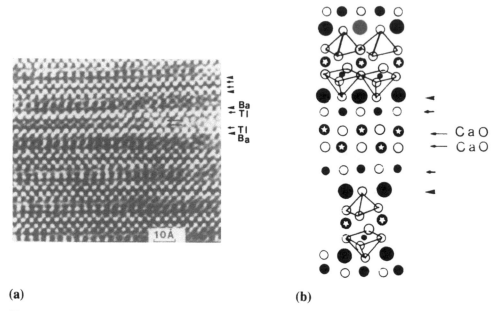

(a)   (b)

Figure 10 $Tl_2Ba_2Ca_1Cu_2O_8$ ($m = 2$, $n = 3$). Additional $[CaO]_\infty^{RS}$ layers are intercalated between the two $[TlO]_\infty^{RS}$ layers of the 2212 structure ($n' = 3$)($n' = 3$). (a) HREM image; (b) idealized model.

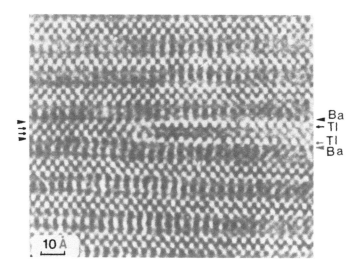

Figure 11 Stopping of the additional $[CaO]_\infty^{RS}$ layers in the 2212 matrix.

and oxygen-deficient perovskite layers $[ACu_2O_5]_\infty$, which are formed of single pyramidal layers. This study has opened the way to the synthesis of a large family of oxides, some containing thallium [49–55], whose structure is based on the following principle: In each oxide containing a double pyramidal copper layer, single fluorite layers, or more exactly the $[A]_\infty$ (A = Ca, Ln) layer sandwiched between

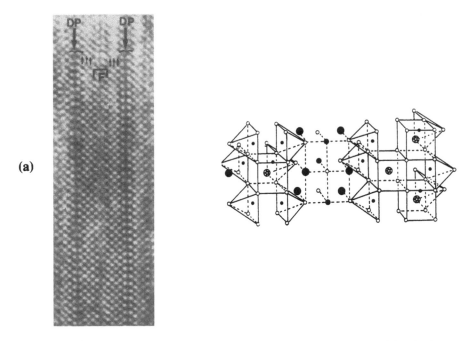

Figure 12 (a) Original defect observed in $TlBa_2NdCu_2O_7$, interpreted by the intercalation of an additional $[LnO_2]_\infty^F$ layer between the two copper layers; (b) idealized model of the defect.

the pyramids, can be substituted by a double fluorite layer, Ln_2O_4. An example of another member, $Tl_2Ba_2Ln_{2-x}Ce_xCu_2O_{10}$, is shown in Fig. 14. It can be seen that the structure of this phase derives from that of $Tl_2Ba_2CaCu_2O_8$ simply by replacing $[CaO_2]_\infty$ by $[Ln_2O_4]_\infty$ double fluorite layers.

The theoretical parameters of the resulting structures can easily be calculated from those of the parent structures, as shown in Table 2. This new mechanism can be applied not only to the thallium oxides but to every superconducting cuprate exhibiting a $[ACuO_{2.5}]_2$ unit, which can be ensured by a local variation of stoichiometry in defective slabs but which must be ensured by an appropriate cation balance in the pure phases.

An important point can be inferred from an analysis of these phenomena: The systematical appearance of a structural mechanism in the form of frequent defects can be considered as a signature of the existence of one phase built up from the implementation of this mechanism to the parent structure.

IV. ORDER–DISORDER PHENOMENA IN THE CATION AND OXYGEN LATTICE

A local cationic substitution such as a point defect cannot be detected in high-resolution images, except on the crystal edge if a strong disturbance of the neighboring ions is implied. This is why small deviations from the ideal composition cannot be observed by HREM.

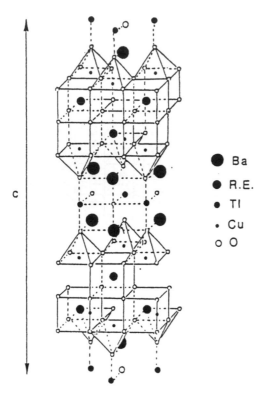

c

● Ba
● R.E.
• Tl
· Cu
○ O

Figure 13   Idealized structure of $TlA_2Pr_2Cu_2O_9$ with A = Tl, Sr, or Ba.

A.   Exchange Between Thallium and Calcium

In the simplest system, Tl–Ba–Ca–Cu–O, x-ray and neutron diffraction results give evidence of such small deviations in thallium and calcium ($[Tl_{2-x}Ca_x]_{Tl}$ and $[Ca_{1-x}Tl_x]_{Ca}$ with $x \approx 0.1$). The HREM images (Fig. 15) showed that locally, thallium ions can substitute for a complete row of calcium in the $m = 2$ members, involving the existence of $[Tl]_\infty$ layer intercalated between the two $[CuO_2]_\infty$ planes of the pyramidal layers [32]. On the other hand, no evidence of such a phenomenon has been observed in the $[TlO]_\infty$ layer, and we can thus conclude that the substitution, if it exists, occurs in a statistical way.

B.   Satellites and Modulations: Signature of Disordering

A large variety of order–disorder phenomena related to both the cation distribution and the anionic framework have been observed in thallium cuprates. They arise in the form of satellites and modulations extended either in the whole crystal or in more or less extended areas. It must first be pointed out that the nature of satellites in incommensurate positions in thallium cuprates is very different from that observed in bismuth cuprates. Unlike in bismuth cuprates, these satellites are not observed systematically, and their intensities vary from one crystal to the other in one sample and depend strongly on the thermal treatment. Moreover, they set up along different directions.

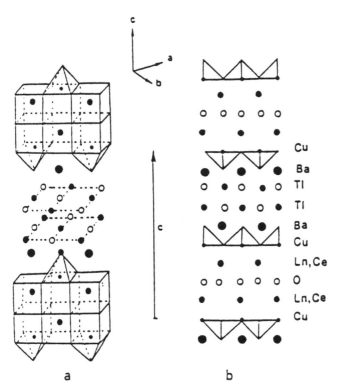

Figure 14 Crystal structure of the oxides $Tl_2Ba_2Ln_{2-x}Ce_xCu_2O_{10}$: (a) perspective view; (b) view projected onto (010).

Table 2 Relations Between Parent and Related Structures[a]

Parent structure	+D.F.	Related structure
Even n values		
a_p, b_p	—	a_p, b_p
c_n		$(c_n + f) \times 2$
Space group: P		SG: I
Odd n values		
a_p, b_p	—	a_p, b_p
c_n		$c_n/2 + f$
Space group: I		SG: P

[a]f, Thickness of a fluorite layer; c_n, c parameter, calculated from the thickness of the n rock salt layers and m perovskite layers.

In the system Tl–Ba–Ca–Cu–O, and especially for $Tl_2Ba_2CaCu_2O_8$, conflicting results were first published on the existence of satellites; most authors [13,23,32,35,38,56,57] observed satellite-free patterns for the various members (except 2201), whereas others reported the existence of weak satellites setting up along $[100]_p^*$, which disappear under beam heating [37,40,58]. In fact, it was shown that thermal treatments, especially oxygen pressure and actual composition (which can

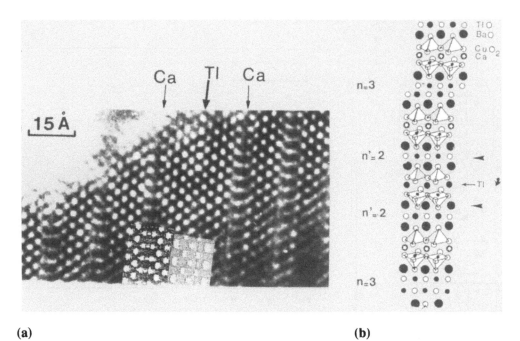

(a)   **(b)**

Figure 15   (a) Double defect in $Tl_2Ba_2CaCu_2O_8$: substitution of thallium for calcium between the CuO_5 pyramids (focus value -680 Å) and variation of the n value; (b) idealized drawing of the double defect imaged in (a).

differ from the nominal compositions according to the synthesis method), are essential factors for the appearance of satellites.

1.   Ba–Tl Substitution in the 2212 Structure: the Oxide $Tl^{III}_{2-x/3}Ba_{1+x}Tl^I_{1-x}LnCu_2O_8$

A series of oxides $Tl_{2-x}Ba_{1+x}LnCu_2O_8$ (Ln = Pr, Nd, Sm; $x \approx 0.25$) has been isolated [18] whose structure belongs to the 2212 type, $Tl_2Ba_2CaCu_2O_8$, but in which thallium has been substituted for barium and calcium is replaced by lanthanide. Taking into account the size of the cations and the charge balance, it has been proposed that the thallium ions distributed over the barium sites are univalent.

Electron diffraction and high-resolution electron microscopy studies of these oxides allow a comparison to be made with bismuth cuprates. Analysis of the HREM images of many crystals shows that the perovskite layers are highly ordered. It is indeed worth pointing out that no stacking defect was observed in the rock salt–type layers (Fig. 16), contrary to the Tl_2 cuprates, and especially to $Tl_2Ba_2CaCu_2O_8$, where thallium monolayers can easily replace thallium bilayers, forming extended defects [32]. In this respect these oxides are rather similar to the bismuth oxides, for which extended defects are never observed on the rock salt–type layers. Like the other thallium cuprates and contrary to the bismuth oxides, these compounds do not exhibit a mica-like morphology. Nevertheless, one observes a tendency to cleavage parallel to the layers (Fig. 17). Moreover, it appears that for those edges that are parallel to the layers (see the arrow, Fig. 17), the cleavage arises at the level of one of the $[(Ba,Tl^I)O]_\infty$ layers. The latter phenomenon

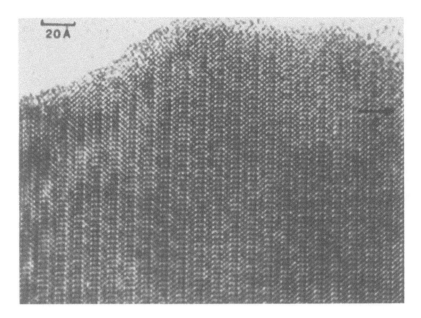

Figure 16 [010] HREM image. The absence of stacking defects in the rock salt–type layers as well in the perovskite-type layers is remarkable. Numerous crystals are perfectly ordered.

is in agreement with the stereoactivity of the $6s^2$ lone pair of Tl(I), a feature noted earlier for Bi(III) cuprates [59].

Electron diffraction (ED) study showed another similarity of these oxides to bismuch cuprates: The ED patterns of all the crystals exhibit satellites systematically. Zone patterns for [001] and [100] are reproduced in Fig. 18a and b. Along zones that deviate slightly from the ideal directions (Fig. 18c and d, respectively), the satellites are clearly visible. The reciprocal lattice can be reconstructed and the section of the reciprocal space along [001]*, [100]*, and [110]* is drawn in Fig. 19, where the large and small black dots represent basic spots and satellites at level 0, respectively, and the open circles correspond to level $\frac{1}{2}$.

These satellites are evidence of a bidimensional modulation that occurs in the structure. They can be characterized as follows:

1. The wave vectors are in (110)* planes, and the modulation wave pattern can be described in most crystals with a tetragonal symmetry. The satellites in (100)* planar section are not visible (Fig. 18b), but they can be observed in Fig. 18d because of the slight tilting of the crystal and the curvature of the Ewald sphere.
2. For all the crystals, only the first-order satellites are observed.
3. The wavelength of the modulation is incommensurate and close to six times the d_{110} distance (2.78 Å).

The existence of satellites in the copper superconductors was first reported in $Bi_2Sr_2CuO_6$ oxide [60], and afterward, this phenomenon appeared systematically [16, 61]. Contrary to the bismuth family, the thallium cuprates rarely exhibit such incommensurability. In the latter compounds their existence seems to depend on

Figure 17 [010] HREM image showing a crystal edge parallel to the perovskite and rock salt–type layers. The surface of the crystal (curved arrow on the right of the figure) is rigorously perpendicular to c. In this image the black dots are correlated with the cation positions and the four rows of darker spots to the four (AO) layers (four small arrows on the top of the micrograph): $-Ba/Tl^I - Tl^{III} - Ba/Tl^I-$ the cleavage appears at the level of the last $[(Ba,Tl)O]_\infty$ layer.

the thermal treatments and on the nature of the member belonging to this large family (see, e.g., Refs. 37 and 62 to 64). In $Tl^{III}_{2-x/3}Tl_{1-x}Ba_{1+x}PrCu_2O_8$ samples, the satellites run along a direction similar to those observed in the bismuth cuprates, but their periodicity as well as their symmetry are different. With the latter aspects, they are somewhat similar to the satellites observed by Zandbergen in $Tl_2Ba_2CaCu_2O_8$ [37]; however, contrary to these phenomena, they do not exhibit the same orientation and are stable; indeed, they do not disappear by prolonged exposure under the electron beam. As discussed earlier for bismuth and lead cuprates [58,65,66], the existence of such satellites is probably correlated to the stereoactivity of the $6s^2$ lone pair Tl(I), which involves a distortion of the $[(Ba,Tl^I)O]_\infty$ layers, leading to incommensurability phenomena.

2. *Thallium-Deficient 2212 Cuprates: $Tl_{2-x}Ba_2CaCu_2O_{8\pm\delta}$*

The study of thallium nonstoichiometry in the 2212 thallium cuprates has allowed a significant homogeneity range to be isolated, according to the formula $Tl_{2-x}Ba_2CaCu_2O_{8\pm\delta}$ with $0 \leq x \leq 0.40$. The systematic HREM observations performed on more than 200 crystals show a higher evenness of contrast at the level of the AO layers in most crystals and throughout large areas of the crystals (Fig. 20a). For the thallium-deficient oxide, this feature attests to the statistical distribution of thallium cations and vacancies in the rock salt layers. The experimental through-focus series give evidence of similar images in both stoichiometric and

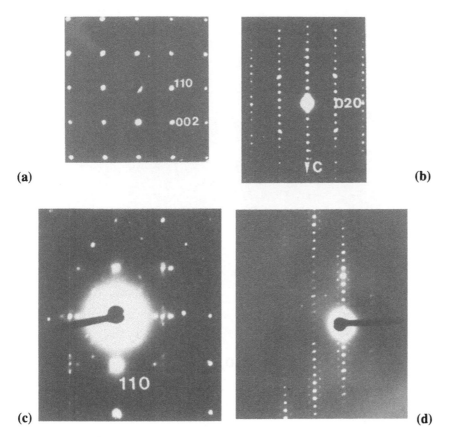

Figure 18   (a) [001] and (b) [100]: well-oriented electron diffraction patterns; (c) [001] and (d) [100]: slightly deviated patterns.

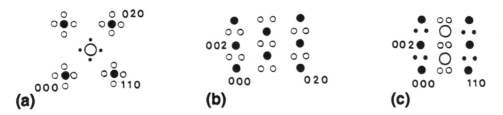

Figure 19   Reconstruction of the reciprocal lattice: (a) [001]*; (b) [100]*; (c) [110]*.

thallium-deficient oxides, in agreement with small differences in scattering factors in the [AO] layers. Figure 20b shows an example of where for focus values close to −700 Å, a direct analysis of the atomic planes allows the white dots to be correlated with cation positions.

The main difference between stoichiometric and the thallium-deficient phase deals with the existence of modulations of the structure. In the stoichiometric phase, one very rarely observes satellites in the ED patterns. On the other hand, weak satellites are frequently observed in the [100] ED patterns of the thallium-deficient

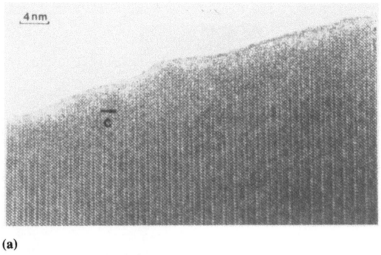

(a)

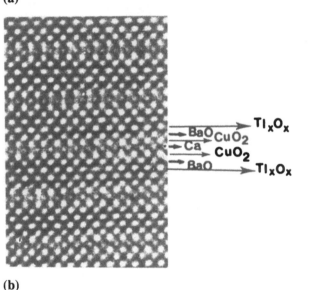

(b)

Figure 20   [100] HREM images of the thallium-deficient phase. (a) A high evenness of contrast is observed in most crystals. (b) Enlargement of an image (focus value $\Delta f \approx -700$ Å), showing the correlation between the rows of white dots and the atomic layers.

oxide. The incommensurate modulation with respect to the basic lattice is shown in Fig. 21; satellites lie along [106]* and the wavelength of the modulation along A is close to 5.4 times the d_{100} distance. This difference between stoichiometric and thallium-deficient samples may explain the disaccord regarding the existence of modulations that arose among the various authors at the beginning of studies of thallium cuprates [64,67,68].

The incommensurate satellites are systematically associated with the existence of more or less extended areas where the contrast is modulated. This suggests that the intensity of the satellites, which varies from one crystal to another, is directly correlated with the extent of the modulated zone. Through-focus series have been recorded for tiny modulated areas located on crystal edges. The interpretation of the contrast of such images does not allow a reliable model to be proposed since

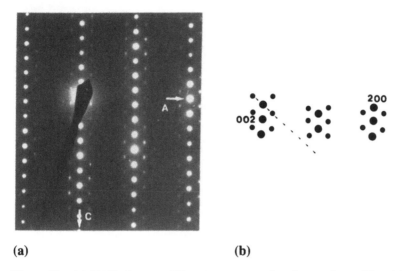

(a) **(b)**

Figure 21 (a) [010] electron diffraction pattern showing weak satellites (along [106]*); (b) schematic drawing.

the contrast variations can be correlated either with changes in the ion nature (and vacancies), or to ion displacements, or to both simultaneously. However, some important information can be obtained:

1. The direction of modulation is in agreement with the position of the satellites in ED patterns, and in the same way, the wavelength along the a axis is consistent.

2. Images with highlighted cations show that thallium and barium atoms in $[AO]_\infty$ layers are hit primarily by variations of contrast, whereas $[Ca]_\infty$ layers appear less disturbed. Images in which low-electronic-density zones are brightest show that oxygen sites are also involved in the modulation; an example is shown in Fig. 23. In that [010] image, recorded for a focus value close to -400 Å (close to Scherzer value), the white dots are correlated to oxygen atoms and oxygen vacancies; the image calculated for a nonmodulated zone (same focus value) is inserted for comparison, and the positions of cations are drawn as black dots and stars. It appears clear that the modulation of contrast also hit the anion (and vacancy) sites. Moreover, from these images, variations of the positions of the ions are also observable, leading to slightly wavy rock salt layers. These observations suggest that the modulations rise up from both variations in local composition and atomic positions. It is remarkable that the atomic planes affected by these variations are the $(111)_{RS}$ planes of the rock salt structure [i.e., the close-packed layers, shown in Fig. 22 (long arrow) and Fig. 23].

3. In some parts of the crystals, modulations exhibiting a different wavelength are observed. They correspond to similar variations in the contrast in the $(111)_{RS}$ planes. An example is shown in the right part of Fig. 24, where the q value is close to $2 \times d_{100}$; for this focus value (≈ -775 Å) the white dots are correlated with the cation positions and it appears clear that in the same way, the contrast variation is especially perceptible at the level of the rock salt layers (RS).

4. In other parts of the crystals, viewed along [100], strong undulations of the atomic planes are observed; the amplitude of the phenomenon is similar to that observed in the bismuth oxides. However, they differ by the direction $(100)_p$, the frequency (they are rare), and the nature of the atomic displacement; in the thallium

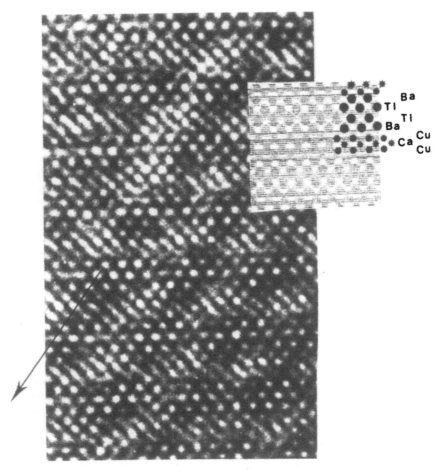

Figure 22 HREM [010] image of a tiny part of a crystal, showing modulations of the contrast. The focus value is close to -400 Å ($\approx$ Scherzer value). The corresponding calculated image is inserted and the cation positions are represented.

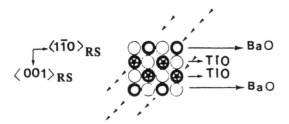

Figure 23 Idealized drawing of the triple rock salt slice built up from four AO layers; $\langle 110 \rangle_{RS}$ corresponds to $\langle 100 \rangle_p$ (i.e., $\langle 100 \rangle$ of the 2212 structure).

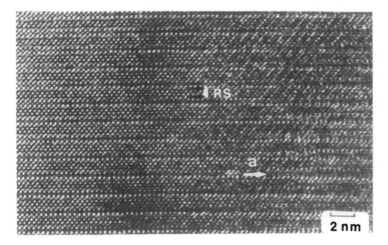

Figure 24 Another example of contrast variations in $(111)_{RS}$ planes. Modulation with $q = 2 \times d_{100}$ is observed in the right part of the image. The four AO layers of the rock salt slice are outlined (white).

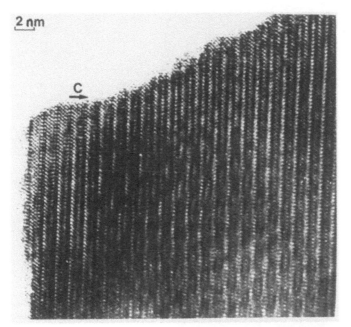

Figure 25 [100] image of a crystal, showing strong undulations in the atomic layers. The amplitude of the phenomenon is similar to that of the bismuth oxides, but the direction and frequency are different.

compound the contrast is locally so disturbed that no correlation can be made between the image and the structure projection (Fig. 25).

5. Along [001], modulations of the contrast are also observed, involving the existence of satellites on ED patterns. One example is given in Fig. 26. The ED pattern corresponding to the selected area (Fig. 26a) exhibits satellites lying along

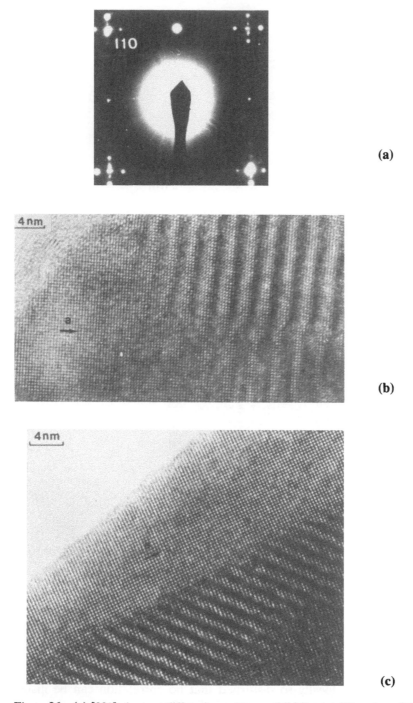

(a)

(b)

(c)

Figure 26 (a) [001] electron diffraction pattern exhibiting satellites along [100]* and [010]*; (b) [001] image showing a modulation of the contrast established along the *a* axis. In that part of the crystal, the periodicity is close to six times d_{100}; (c) [001] image showing a bidimensional modulation.

[100]* and [010]*, as a result of the establishment of the modulation along the two equivalent directions of the tetragonal cell. The examination of these satellites shows that three main systems of nodes are observed, corresponding to $q \approx 4.3$, $q \approx 5.4$, and $q \approx 12$; it should be noted that for such areas, the reflections corresponding to $h + k \neq 2n$ are sometimes observed, attesting to a local loss of symmetry. Figure 26b shows a zone where the modulation is established along a single direction but where the periodicity is variable along that direction. In the second zone of the crystal (Fig. 26c), the modulations set up along two orthogonal directions with two different periodicities ($5 \times d_{100}$ and $6 \times d_{010}$), leading to an apparent bidimensional modulation.

The fact that satellites, correlated with localized modulated domains in the crystals, are observed only in the thallium-deficient oxide suggests that they are correlated directly with the cation and anion deficiency. These modulations arise indeed in most crystals; however, their limited extents ($<< 10\%$) show that they do not correspond to a systematic phenomenon but correspond only to local orderings of Tl, O, and vacancies.

Study of the superconducting properties of these oxides shows that the critical temperature is not strongly affected by the thallium nonstoichiometry. The study of H_2–Ar annealing of these materials shows, in fact, that the predominant factor is the oxygen nonstoichiometry, as discussed in Section VI.

3. The 1212 Oxide $Tl_{0.5}Pb_{0.5}Sr_2CaCu_2O_7$: Particular Behavior with Respect to Pure Thallium Cuprates

Despite the fact that its structure is very similar to that of $TlSr_2CaCu_2O_7$, this oxide exhibits very different electron diffraction patterns [11]. The first striking feature deals with the highly lamellar character of the crystals, which is never observed in either barium thallium cuprates or $TlSr_2CaCu_2O_7$. Thus in this respect, $Tl_{0.5}Pb_{0.5}Sr_2CaCu_2O_7$ is very similar to bismuth cuprates [61, 69]. The second feature concerns the existence of numerous extra spots on the [001] electron diffraction patterns, contrary to $TlSr_2CaCu_2O_7$, which like barium cuprates, exhibits rather simple ED patterns. In that way, diffuse satellites are observed in some [100] ED patterns of $TlSr_2CaCu_2O_7$; these satellites are clearly visible when the crystal is slightly tilted, as shown in Fig. 27. It should be noted that these features are similar to those observed in the barium cuprate $TlBa_2CaCu_2O_7$ [9]. Some examples of the complex phenomena observed in $Tl_{0.5}Pb_{0.5}Sr_2CaCu_2O_7$ follow.

One observes, rather frequently, superstructures of the tetragonal cell described above. Besides the numerous $a \times a$ tetragonal crystals, a doubling of the a parameter accompanied by a slight orthorhombic disorder leading to a $2a \times 2b$ cell is often encountered (Fig. 28). Another example corresponds to a $a \times 4a$ periodicity (Fig. 29a); in the latter case the superstructure sets up along the two equivalent directions of the perovskite subcell. The corresponding bright-field image (Fig. 29b) shows perpendicular domains. The domain boundaries are mainly parallel to $\langle 110 \rangle$ and $\langle 1\bar{1}0 \rangle$ of the perovskite cell, and the 90° orientation of the $4a$ superstructure is clearly visible (Fig. 29c) through the $\approx$15-Å fringes. Another system of fringes, poorly resolved, is observed in these images, whose direction is roughly parallel to the a axis of the perovskite subcell, all over the crystal whatever the direction of the superstructure may be. It can be related to a fairly well established long period structure; the contrast in the corresponding high-resolution images is too disturbed to propose an interpretation.

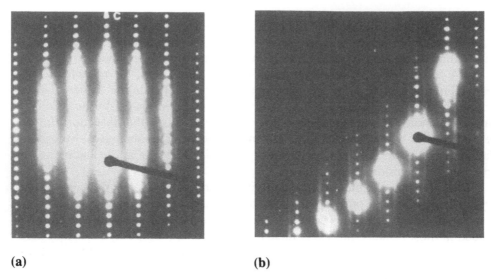

(a) **(b)**

Figure 27 $Tl_{0.5}Pb_{0.5}Sr_2CaCu_2O_7$ ($m = 2, n = 2$): (a) $TlSr_2CaCu_2O_7$ [100] electron diffraction pattern; (b) the satellites are clearly visible when the crystal is slightly tilted.

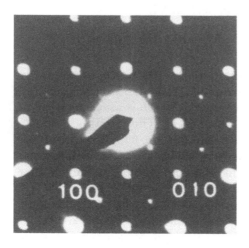

Figure 28 $Tl_{0.5}Pb_{0.5}Sr_2CaCu_2O_7$: [001] electron diffraction pattern with extra spots leading to an orthorhombic cell with $a' \approx 2a$ and $b' \approx 2b$.

Besides those simple superstructures, there exist other complex arrangements of extra spots. A typical example of such ED patterns is shown in Figs. 30 and 31, where many extra spots surround the mean positions of the basic reflections of the subcell. One possible interpretation of this feature is related to the occurrence of satellites whose direction does not correspond to a principal axis (i.e., $\langle 100 \rangle$ or $\langle 110 \rangle$ of the perovskite subcell) but is slightly disoriented. In fact, in the ED pattern (Fig. 30a), two systems of reflections appear as clearly brighter. This guides the interpretation toward the juxtaposition and/or the superimposition of misoriented areas in the crystal. In this image, two intense systems of spots give evidence of the respective orientations of both cells: They are shown in the schematic drawing

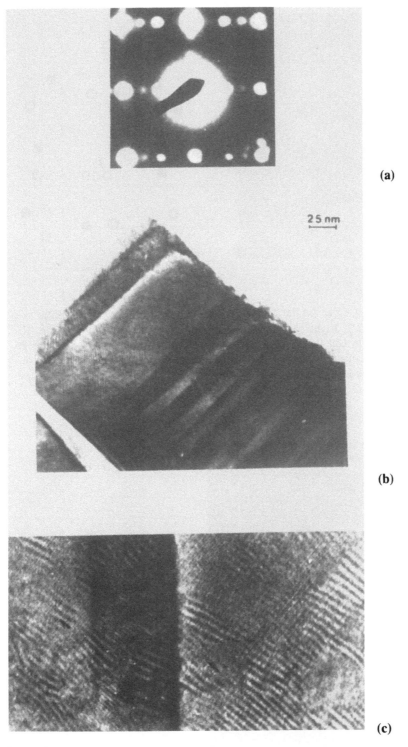

(a)

(b)

(c)

Figure 29 $Tl_{0.5}Pb_{0.5}Sr_2CaCu_2O_7$: (a) [001] electron diffraction pattern with $4a$ superstructure; (b) corresponding bright-field image; (c) enlargement showing the 15-Å fringes and their change of direction (arrows) through the domain boundaries.

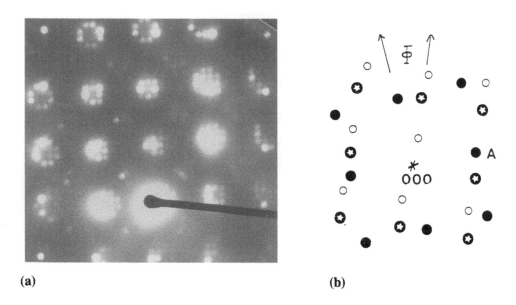

(a)   (b)

Figure 30 $Tl_{0.5}Pb_{0.5}Sr_2CaCu_2O_7$: (a) [001] electron diffraction pattern interpreted from two misoriented cells ($\Phi \approx 7°$). Light spots arise from multiple diffraction phenomena. (b) Schematic drawing of both cell (black spots and stars); the value of the angle Φ has been widened artificially to clarify presentation. Small circles are extra spots due to double diffraction with A as a secondary source.

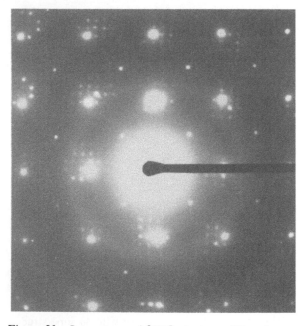

Figure 31   Same type of [001] electron diffraction pattern as in Fig. 30, but one of the components is less intense than the other.

(Fig. 30b). The other weak spots arise from multiple diffraction phenomena. Figure 30b shows both cells (black dots and stars, respectively) and the mechanism of double-diffraction-spot generation: Only one system of extra spots due to A as a secondary source is drawn (small points). The angle Φ between both axis systems is close to 7°. The second example (Fig. 31) is similar to the preceding one, but one of the components is less intense than the other. It results in lighter spots. The angle Φ between the two systems is again close to 7°.

The HREM images show contrast modulations related to cation displacements or local ordering. An example of very localized ordering is shown in the micrograph of Fig. 32. In zones labeled *a* we observe a periodicity of 5.4 Å ($\approx a\sqrt{2}$), which is in agreement with the doubling of the parameter, often observed. Complex superstructures are observed in areas labeled *b*, which are bidimensional. Sometimes, these features strongly affect the contrast, as in areas labeled *c*. Modulations of the contrast without any straightforward superstructures are observed in area *d*. A stacking defect is arrowed in the middle of the image.

The interpretation of the existence of such satellites is very difficult. It is now commonly admitted that lead exhibits the tetravalent state in this phase. The first type of additional spots, which corresponds to a multiplicity *a*, could be due to a local ordering of Pb(IV) and Tl(III), although these cations do not exhibit very different sizes. The second sort of extra spots, which correspond to disorientated area, might be due to the presence of small amounts of Pb(II). One hypothesis is the existence of isolated PbO bilayers as a stacking defect. These layers can indeed be described as distorted Aurivillius-type layers, with a monoclinic distortion of the layers, leading to $a = 5.48$ Å and $b \approx 5.87$ Å. Such a distortion can generate twinning domains, which would involve disorientations of the adjacent perovskite layers and, in that way, of the 1212 slabs. The bright-field image corresponding to

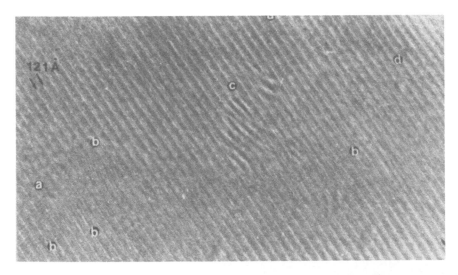

Figure 32 [110] image, witness of localized ordering. Areas correspond to: a, variations of the periodicity along ⟨110⟩; b, bidimensional superstructures; c, disturbed zones; d, modulations of the contrast.

such an ED pattern (Fig. 28) is shown in Fig. 33; the twinning domains here are clearly visible.

V.   COMBINATION OF DIFFERENT DEFECTS, ORIGIN OF WIDE NONSTOICHIOMETRY RANGE: THE 1201 PHASE OF THE Tl–Sr–La–Cu–O SYSTEM

When they are distributed statistically, the intergrowth defects, and the disordering phenomena of cations and oxygens, can be at the origin of wide ranges of non-stoichiometry. This is the case for the 1201-type superconductor isolated in the system Tl–Sr–La–Cu–O [70]. The 1201 structure (Fig. 3), which corresponds to the intergrowth of single perovskite layers with double rock salt layers (containing thallium monolayers), is represented by the ideal formula $TlSr_2CuO_5$. A study of the pseudoternary system Tl–Sr–La, considering the molar ratio (Tl + Sr + La)/ Cu = 3, shows that the domain of the 1201 phase is much wider than expected from the single replacement of Sr by La in the oxide $TlSr_2CuO_5$ (Fig. 34). Taking as a reference the line $TlSr_{2-y}La_yCuO_{5-\delta}$ ($0.4 \leq y \leq 1$), the domain of the 1201 oxide can be divided into two regions; the upper region (labeled A), which can be formulated $Tl_{1+x}Sr_{2-x-y}La_yCuO_{5-\delta}$, is characterized by a thallium excess with respect to the ideal composition, whereas the lower region (labeled B), which corresponds to the formula $Tl_{1-x}Sr_{2+x-y}La_yCuO_{5-\delta}$, exhibits a thallium deficiency.

Domain A, which is limited by the compositions $A_1(x = 0.15, y = 0.45)$, $A_2(x = 0.2, y = 0.6)$, $A_3(x = 0.3, y = 0.7)$, $A_4(x = 0.25, y = 0.85)$, $A_5(x = 0.1, y = 0.9)$, shows the ability of thallium to occupy the strontium sites located in the

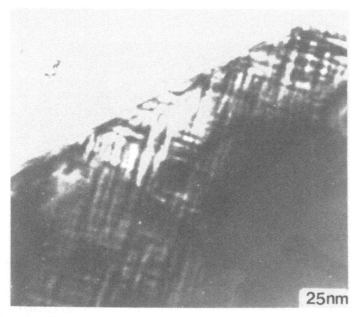

Figure 33   [001] bright-field image. The corresponding electron diffraction pattern is shown in Fig. 28.

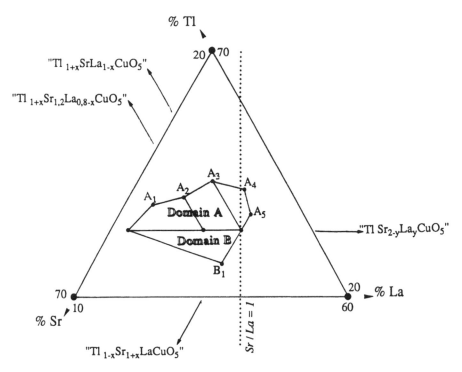

Figure 34 Pseudoternary diagram of Tl–Sr–La–Cu–O with (Tl + Sr + La)/Cu = 3.

planes of the apical oxygens of the CuO_6 octahedra (Fig. 3), despite the small size of Tl(III) compared to strontium (or lanthanum).

Domain B, whose most thallium-deficient composition corresponds to the formula $Tl_{0.8}Sr_{1.2}LaCuO_{5-\delta}(B_1)$, suggests the ability of strontium or lanthanum to occupy partially the thallium sites in the thallium monolayers (Fig. 3) despite its much greater size. This raises the question of the abnormally short Sr–O distance along the c axis for the mixed layer $[Tl_{1-x}Sr_xO]_\infty$. Nevertheless, it must be pointed out that the oxygen content of these phases is not known with accuracy, so that the existence of anionic vacancies in the rock salt layers may allow cations larger than thallium to be introduced. The x-ray diffraction patterns of different compositions of the wide A domain, as well as the ED study performed for more than 50 particles for each composition, did not allow any secondary phase to be detected, confirming the existence of a single domain.

The ED patterns of the compositions of the A domain, $Tl_{1+x}Sr_{2-x-y}La_yCuO_{5-\delta}$, are often characterized by diffuse streaks along the c axis. Two examples of typical [010] ED patterns are given in Fig. 35a and b, where streaks are absent or visible, respectively. The bright-field images give evidence of a strong difference between the corresponding types of crystal. The first one (Fig. 36) exhibits a very even contrast, whereas the second one exhibits a highly disturbed contrast (Fig. 37a). The enlargements of these areas show strong local contrast variations and local small variations of the c parameter's value (Fig. 37b); however, it can clearly be seen from these images that a systematic mechanism of intergrowth of single and double (TlO) layers cannot be implicated. This suggests the existence of extended

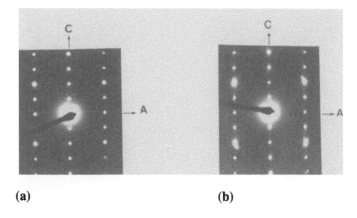

(a)   **(b)**

Figure 35 $Tl_{1.3}Sr_{0.7}LaCuO_{5-\delta}$: [010] electron diffraction pattern showing sharp spots (a) or diffuse streaks (b) along c.

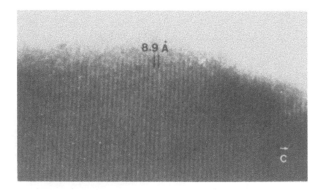

Figure 36 Bright-field image recorded for the electron diffraction pattern in Fig. 35a. An even contrast is observed.

defects, some 10 Å wide, resulting from inhomogeneities in the local composition which could correspond to thallium segregation.

Structure calculation performed from powder x-ray data did not allow the structure of the B domain, $Tl_{1-x}Sr_{2+x-y}La_yCuO_{5-\delta}$, to be refined. The electron microscopy investigation of this phase allows this result to be explained. Systematic investigations, performed on more than 100 crystals, showed intergrowth phenomena similar to those observed for the limit oxide $TlSrLaCuO_5$. However, contrary to the latter, these intergrowth phenomena become much more pronounced. Defect-free crystals are very rare, and various intergrowths of the La_2CuO_4, 0201-type structure with the $TlSr_2CuO_5$, 1201-type structure are observed as shown in Fig. 38. The first pattern (Fig. 38-1) corresponds to a defect-free crystal; on the second one, weak streaks are observed along the c axis (Fig. 38-2), whereas in the third (Fig. 38-3), the intensity of the diffuse streaks increases and a second system of reflections, corresponding to a 0201-type structure, is observed. In the fourth, Fig. 38-4, the superposition of both systems, 1201 and 0201, is clearly observed; in the two last patterns, the coexistence of the two systems of reflections indicates that contrary to what was observed in the $TlLaSrCuO_5$ oxide, several 0201-type

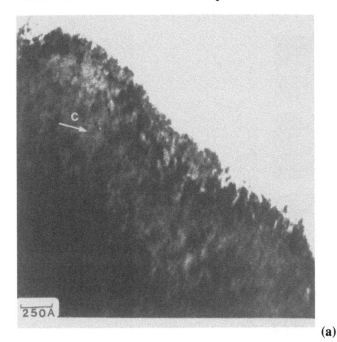

(a)

(b)

Figure 37 (a) Image recorded for the electron diffraction pattern in Fig. 35b exhibiting diffuse streaks along the *c* axis. A highly disturbed contrast is observed. (b) The enlargement shows local variations in the contrast and *c* value.

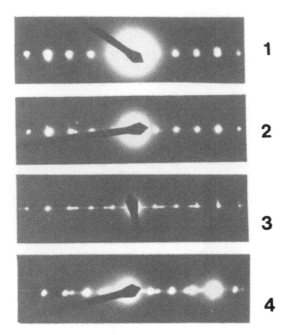

Figure 38   Examples of c axes observed in $Tl_{0.8}Sr_{1.2}LaCuO_{5-\delta}$ sample.

slices can be adjacent, leading to the formation of small 0201 domains. This is illustrated in Fig. 39, where the bright-field image, corresponding to the ED pattern of Fig. 38-3, gives evidence of the existence of both structural types and of their aleatory stacking. It should be noted that the [001] ED patterns are also sensitive to that coexistence since the intensity of the $h + k \neq 2n$ reflections decreases with the amount of 0201 members (Fig. 40). It must be emphasized that in that sample, we never observed a pure 0201-type crystal. Clearly, the main matrix consists of a regular intergrowth of double rock salt layers $[(TlO)(La_{1-x}Sr_xO)]_\infty$ with single perovskite layers, but from time to time some of the double rock salt layers are replaced statistically by single rock salt layers $[La_{1-x}Sr_xO]_\infty$.

The latter observation of intergrowth defects is at the basis of the discovery of a new superconductor $TlBa_{2-x}La_{2+x}Cu_2O_9$ [71], which exhibits a T_c of 42 K. The structure of this phase, deduced from the layer stacking observed by HREM and confirmed by XRD (Fig. 41), shows that it can be described as a regular intergrowth of the ideal 1201 ($TlBa_{2-x}La_xCuO_5$) and the 0201 ($La_{2-x}Ba_xCuO_4$) structures, which exhibit a similar T_c.

VI.   OXYGEN NONSTOICHIOMETRY, THE KEY FACTOR FOR SUPERCONDUCTIVITY

It is extremely difficult to evaluate the influence of the different defects on the superconducting properties of thallium cuprates. However, it is now certain that the presence of intergrowth defects does not significantly modify the superconducting properties of these materials. It is indeed now possible to prepare defect-free samples, and their comparison with samples exhibiting intergrowth defects

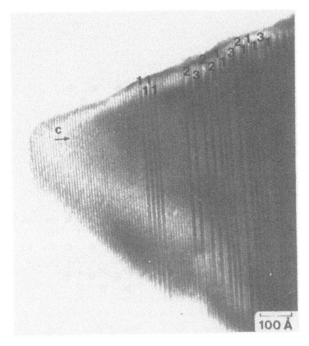

Figure 39 Bright-field image recorded for a crystal exhibiting the 38(3) electron diffraction pattern. We can observe 1201 and 0201 domains and areas where both are intergrown.

 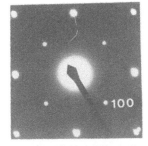

Figure 40 $Tl_{0.8}Sr_{1.2}LaCuO_{5-\delta}$: examples of two [001] electron diffraction patterns showing strong differences in the intensity of the $h + k \neq 2n$ reflections. This variation is correlated to the number of 0201 members intergrown with 1201 members.

shows that they exhibit the same critical temperature, provided that they are optimized by the same mode of thermal treatment, in particular at the same oxygen pressure, in order to obtain the same hole carrier density for both samples.

On the contrary, the intercalation of fluorite layers between pyramidal copper layers tends to destroy superconductivity. Indeed, both Tl_2 and Tl_1 cuprates, resulting from the intercalation of double fluorite layers [10,51], do not superconduct. It seems that this phenomenon can be generalized to most cuprates containing double fluorite layers. Some of them have been found to exhibit superconducting

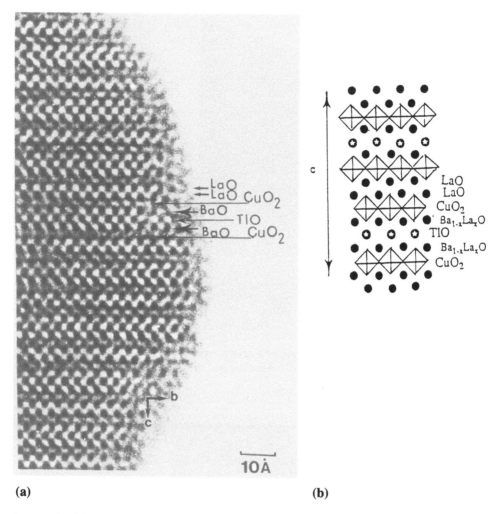

(a)   **(b)**

Figure 41 (a) High-resolution image and (b) structural model of the $[1201]_1[0201]_1$ super-conducting thallium oxide.

properties but either have not been reproduced by other authors or have been prepared under high pressure.

The disordering phenomenon of cations, and especially of oxygen within the layers, certainly has a greater influence on superconductivity than have intergrowth defects. The following two examples, which deal mainly with 2201, 2212, and 1212 cuprates, show clearly that a very small deviation from oxygen stoichiometry has a very dramatic effect on the critical temperature of these phases.

A. Optimization of Critical Temperature by Hydrogen Annealing: the 2201 and 2212 Thallium Cuprates

As early as 1988 we reported that the critical temperature of the oxide Tl-$Ba_2CaCu_2O_{7-\delta}$ could be increased by annealing under an argon flow [9]. This suggested that in the initial compound, the oxygen content (i.e., the hole carrier density) was too high, so that the creation of an oxygen deficiency allowed the T_c

to be increased. After these preliminary studies, many contradictory results were obtained by different authors. For instance, the 2201 oxide $Tl_2Ba_2CuO_6$ was found by different authors not to be superconducting, whereas others found it superconducting but with different temperatures, ranging from 40 to 80 K [1,8, 72–75].

To understand the role of oxygen nonstoichiometry, pure phases (i.e., free of intergrowth defects and of disordering) have been prepared. These oxides were then characterized by HREM and x-ray diffraction using magnetic measurements on the same sample. The same samples were then treated in very soft conditions (i.e., at low temperature and low oxygen pressure) and studied again. Study of the 2201 and 2212 thallium cuprates [76,77] shows the spectacular effect of oxygen nonstoichiometry on superconductivity. Both oxides, $Tl_2Ba_2CuO_{6\pm\delta}$ and $Tl_2Ba_2CaCu_2O_{8\pm\delta}$, were first synthesized at high oxygen pressure (i.e., at about 15 bar), starting with adequate mixtures of BaO_2, Tl_2O_3, and CuO heated in sealed quartz ampoules. In these conditions, a pellet of the 2201 phase does not superconduct, whereas that of the 2212 phase exhibits a critical temperature of 96 K (Fig. 42).

These pellets were then annealed under H_2–Ar flow (10% H_2) at low temperature (i.e., 290°C) to avoid thallium loss. The weight losses and gains were followed by microthermogravimetry. The samples were then systematically characterized with an ac susceptometer and studied by high resolution electron microscopy, after different annealing times. The spectacular effect of hydrogen an-

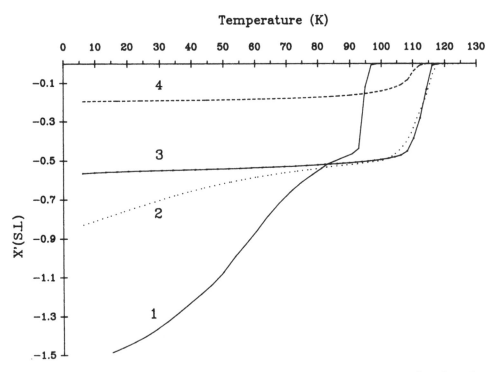

Figure 42   Ac susceptibility $\chi'(T)$ of the Tl-2212 sample for different annealing times in a gas mixture (argon 90% + hydrogen 10%): 1, untreated; 2, 15 min; 3, 120 min; 4, 720 min.

nealing is shown by the curves $\chi' = f(T)$ (Figs. 42 and 43). It can indeed be observed that the critical temperature of the 2212 phase increases from 96 to 118 K for 15 mn annealing and remains unaffected after 2 h of annealing (Fig. 42). It is also worth pointing out that for these experimental conditions, electron microscopy does not reveal any formation of extended defects: In both samples, 96-K and 118-K superconductors, the sequence stacking of $[CuO_2]_\infty$, $[BaO]_\infty$, and $[TlO]_\infty$ layers is absolutely regular and characteristic of the pure 2212 phase. Moreover, the weight loss, during hydrogen anealing, is only 0.4%, showing that the oxygen loss, in the form of H_2O, is weak.

The hydrogen annealing of $Tl_2Ba_2CuO_{6\pm\delta}$ is even more effective (Fig. 43). One observes a progressive increase in the critical temperature and of the superconductive volume fraction as the annealing time increases, reaching a critical temperature of 92 K after 480 mn. One also observes that the slopes of the intermediate curves $\chi'(T)$ indicate an inhomogeneous character of superconductivity that can be attributed to an inhomogeneous distribution of the oxygen vacancies during the reduction process. On the contrary, the final 92-K superconductor exhibits a sharp transition with a superconducting volume of 25%.

It must be emphasized that this phenomenon is perfectly reversible. Annealings of the hydrogen-reduced samples under an oxygen flow, or at higher oxygen pres-

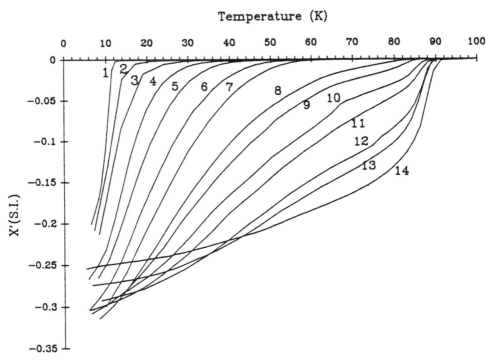

Figure 43   Real part $\chi'(T)$ of the magnetic ac susceptibility of the Tl-2201 sample, as synthesized 1, and after the succeeding and variable H_2–Ar annealing times $\Sigma\tau$: 2, $\Sigma\tau = 2$ min; 3, $\Sigma\tau = 6$ min; 4, $\Sigma\tau = 14$ min; 5, $\Sigma\tau = 26$ min; 6, $\Sigma\tau = 50$ min; 7, $\Sigma\tau = 74$ min; 8, $\Sigma\tau = 122$ min; 9, $\Sigma\tau = 146$ min; 10, $\Sigma\tau = 196$ min; 11, $\Sigma\tau = 242$ min; 12, $\Sigma\tau = 338$ min; 13, $\Sigma\tau = 430$ min; 14, $\Sigma\tau = 480$ min.

sures, at 400°C, induce a continuous decrease of the critical temperature and of the superconducting volume, leading finally to the nonsuperconducting state without any significant change in the structure according to x-ray and HREM observations. Note also that prolonged annealing in an H_2–Ar flow leads to a progressive disappearance of superconductivity (see Fig. 42), in agreement with the HREM observation, which shows the formation of amorphous zones, characteristic of the destruction of the structure.

These results demonstrate that the critical temperature of these phases goes through a maximum value for an optimal Cu(III) content in the sample (i.e., there is an optimal hole carrier concentration). Moreover, the weak weight variation, with regard to the spectacular variation in T_cs, suggests that the critical temperature is very sensitive to the hole carrier density. Although it seems most probable that the reduction results from the formation of H_2O, intercalation of protons leading to the formation of OH groups cannot be ruled out. Both phenomena lead, indeed, to the decrease in Cu(III) content, but so far there is no evidence of the introduction of protons in the structure. On the contrary, the formation of oxygen vacancies is evidenced from the weight loss. Thus it is most probable that the ordering of anionic vacancies and oxygen atoms plays a role in the optimization of T_cs, so that times and temperatures must be carefully controlled during annealing.

B. Oxygen Stoichiometry and Localization: Neutron Diffraction Study of the 1212 Thallium Cuprates $TlBa_2Ca_{1-x}Nd_xCu_2O_{7-\delta}$

Although the role of oxygen is of primary importance for superconductivity in thallium cuprates, the oxygen content in these oxides is usually determined by chemical methods alone and thus may be of questionable accuracy. Moreover, structure determination is not performed on samples studied for superconductivity. Consequently, it is not possible to draw conclusions regarding the influence of oxygen on superconducting properties. In this respect the solid solution $TlBa_2Ca_{1-x}Nd_xCu_2O_{7-\delta}$ exhibits interesting behavior since one observes a large variation in critical temperature, from 100 K for $x = 0.2$ to nonsuperconductor temperatures for $x = 1$ [9,10,73,78–80].

We present here a study of three compositions $x = 0.2$, $x = 0.5$, and $x = 1$ [81]. These three compositions were synthesized as pure phases in large amounts and were studied by neutron diffraction. $TlBa_2CaCu_2O_7$, which is also a good superconductor, is never absolutely pure and consequently was not studied.

To try to establish relationships between superconductivity and chemical bonding, the magnetic properties of these three samples were studied. The curves of magnetic susceptibility $\chi' = f(T)$ (Fig. 44) show that the oxide $TlBa_2Ca_{0.8}Nd_{0.2}Cu_2O_7$ ($x = 0.2$) exhibits a high critical temperature $T_c = 100$ K and a high superconducting volume (greater than 60%), whereas the intermediate composition, $TlBa_2Ca_{0.5}Nd_{0.5}Cu_2O_7$, is characterized by a much lower critical temperature $T_c = 40$ K and a weaker diamagnetic volume (30%). The limit phase, $TlBa_2NdCu_2O_7$, is not a superconductor.

Refinements of the structure of these phases from neutron diffraction data using profile Rietveld analysis confirm the structural principle described previously [10]: that the structure (Fig. 45) consists of double pyramidal copper layers intergrown with double rock salt–type layers, themselves formed of $[BaO]_\infty$ and $[TlO_y]_\infty$

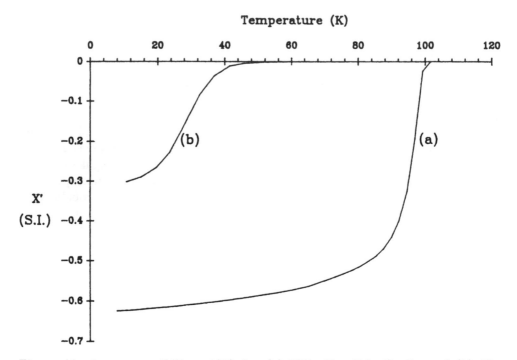

Figure 44 Ac susceptibility $\chi'(T)$ for (a) $TlBa_2Ca_{0.8}Nd_{0.2}Cu_2O_7$ and (b) Tl-$Ba_2Ca_{0.5}Nd_{0.5}Cu_2O_7$.

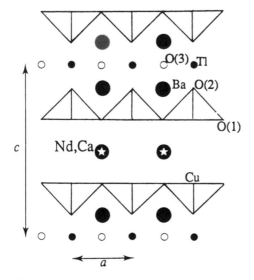

Figure 45   Idealized drawing of the 1212 structure.

layers. Moreover, they shed light on the nonstoichiometry phenomena, atomic distribution, and structural evolution in these phases in connection with their superconducting properties.

1. Nonstoichiometry and Atomic Distribution

Three features are of importance for superconductivity in these compounds:

1. One always observes a small thallium deficiency (i.e., 0.96 instead of 1 for the ideal formula whatever the composition is). This deviation from stoichiometry is in the limit of the experimental error; nevertheless, it is not unexpected if one takes into account the volatility of thallium oxide at high temperature.

2. The important point deals with oxygen stoichiometry, which appears either close to O_7 for $x = 1$ and to $O_{6.86}$ for $x = 0.2$ and 0.5. Thus no extra oxygen has been detected, contrary to what was found for $Tl_2Ba_2CuO_6$, for example [31,82]. Furthermore, one observes that the O(1) and O(2) oxygen sites, which belong to the basal plane and apical sites of the CuO_5 pyramids (Fig. 46), are fully occupied. In contrast, the oxygen deficiency on the O(3) sites of the $[TlO_y]_\infty$ layers, although rather small, is absolutely significant for the two superconducting phases $x = 0.2$ and $x = 0.50$, corresponding to $y \approx 0.86$ instead of 1.

Considering the chemical formulas deduced from the neutron refinements— $Tl_{0.96}Ba_2NdCu_2O_{6.96}$, $Tl_{0.95}Ba_2Nd_{0.5}Ca_{0.5}Cu_2O_{6.86}$, and $Tl_{0.96}Ba_2Nd_{0.2}Ca_{0.8}Cu_2O_{6.86}$— leads for copper to mean oxidations of $+2.02$, $+2.18$, and $+2.32$, respectively. This evolution is in perfect agreement with the increase in T_c as x decreases; for $x = 1$, one only obtains Cu(II) and consequently, the compound does not superconduct; for $x = 0.2$, which corresponds to the highest T_c, the mean oxidation state of copper of 2.32 is close to that obtained for the 92-K superconductor $YBa_2Cu_3O_7$, which is 2.33. These results suggest that the role of the hole reservoir of the $[TlO]_\infty$ layers is not so important, contrary to what is deduced from superconductivity in the oxides $Tl_2Ba_2Ca_{m-1}Cu_mO_{2m+4}$; they also confirm that Cu(III) content influences T_cs dramatically, in agreement with the annealings of these materials in H_2–Ar flow [76,77].

3. Contrary to other atoms of the structure, the thallium and O(3) oxygen atoms, which form the $[TlO]_\infty$ layers, exhibit a disordered arrangement since they are both split over four positions (Fig. 46). A similar splitting of thallium sites was found for the 1212 pure calcium phase: $Tl_{1.17}Ca_{0.83}Ba_2Cu_2O_{6.75}$ [78], the 2212 oxide:

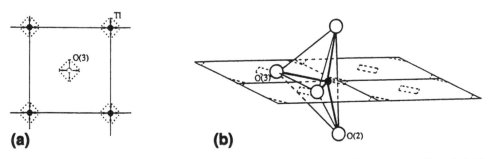

Figure 46 Schematic representation in the TlO plane of the splitting of the Tl and O(3) positions (a) leading to a pseudotetrahedral oxygen environment for thallium (b).

$Tl_{1.9}Ba_2Ca_{1.1}Cu_2O_8$ [17], and the 2201 compound: $Tl_{1.97}Ba_{2.03}CuO_{6.3}$ [31]. This particular distribution of thallium and oxygen in the rock salt layer can be compared with that observed for oxygen in $[PbO]_\infty$ layers of the single rock salt layers of the oxides $Pb_2Sr_2YCu_3O_8$ [83,84] and $PbBaSrNdCeCu_3O_9$ [85]. Such disordered displacements can be explained in terms of bond lengths as discussed later.

2. Structural Evolution and Superconductivity

Consideration of the interatomic distances (Table 3) allows the following points to be emphasized:

1. The TlO_6 octahedra are highly distorted; this disordered distribution of oxygen and of thallium in the $[TlO]_\infty$ layers comes from the very small size of Tl(III), so that Tl and O atoms move away from their ideal positions to realize shorter bonds (2.24 to 2.35 Å) in the basal plane, similar to those observed in pure thallium oxides such as Tl_2O_3 [29], $Sr_4Tl_2O_7$ [30], and $SrTl_2O_4$ [86,87]. Nevertheless, there remain very long Tl–O distances in the plane (2.78 to 2.94 Å) which are never observed in pure thallium oxides. The apical Tl–O distances parallel to the c axis are abnormally short (2.04 to 2.09 Å). Thus instead of a strongly distorted octahedral coordination, a pseudotetrahedral coordination can be proposed for thallium (Fig. 46b), with two short Tl–O distances along the c axis and two normal Tl–O distances in the $[TlO]_\infty$ layers, two other oxygens lying farther apart in the layer. The abnormally short apical Tl–O distances seem to be characteristic of the layered thallium cuprates since one observes Tl–O distances of 2.01 Å in a crystal of the 103-K cuperconductor $Tl_{1.17}Ca_{0.83}Ba_2Cu_2O_{6.75}$ [77], of 2.05 Å in the 81-K superconductor $TlSr_{1.3}Nd_{0.7}CaCu_2O_7$ [88], and distances ranging from 1.96 to 2.04 Å in cuprates involving thallium bilayers [17,80]. It is also worth pointing out that

Table 3 $TlBa_2Ca_{1-x}Nd_xCu_2O_{7-\delta}$: Calculated Interatomic Distances and Selected O–Cu–O Angles

		$x = 0.2$	$x = 0.5$	$x = 1.0$
		Distances (Å)		
Tl–O(2)	(× 2)	2.043(3)	2.056(3)	2.095(2)
Tl–O(3)[a]	(× 4)	2.353(9)–2.835(8)	2.335(7)–2.787(7)	2.239(5)–2.946(5)
Ba–O(1)	(× 4)	2.824(2)	2.840(2)	2.914(2)
Ba–O(2)	(× 4)	2.816(1)	2.818(1)	2.808(1)
Ba–O(3)	(× 1)	2.726(3)	2.703(3)	2.609(3)
Ca/Nd–O(1)	(× 8)	2.471(1)	2.474(1)	2.475(1)
Cu–O(1)	(× 4)	1.930(0)	1.941(0)	1.964(0)
Cu–O(2)	(× 1)	2.686(3)	2.626(3)	2.462(3)
O(1)–O(1)[b]		3.090(2)	3.075(2)	3.035(2)
Cu–Cu[b]		3.247(2)	3.287(2)	3.410(2)
		O–Cu–O angles (deg)		
O(1)–Cu–O(1)		175.34(6)	173.82(6)	168.91(6)
O(2)–Cu–O(2)		92.33(4)	93.09(4)	95.55(4)

[a]Shortest and largest in-plane Tl–O distances involved in the splitting of Tl and O(3) positions.
[b]Distances between two successive layers (along the c axis).

in the series $TlBa_2Ca_{1-x}Nd_xCu_2O_{7-\delta}$, the apical Tl–O distances increase progressively as x increases (i.e., as T_c decreases). In the same way, Tl–O equatorial distances decrease as x increases; nevertheless, this effect is too small for us to draw conclusions about its relationships with superconductivity.

2. The evolution of the geometry of the CuO_5 pyramids is interesting since it may be related to the superconducting properties of layered cuprates. One observes that the four equatorial Cu–O(1) distances are close to those generally observed for copper in square-pyramidal coordination in other layered cuprates whether or not they are superconductive. One also notices that these distances increase progressively as x increases (i.e., as the critical temperature decreases). But the most striking feature deals with the evolution of the Cu–O(2) apical distance: It slightly decreases as x increases but remains longer than 2.6 Å for superconductive phases ($x = 0.2, 0.5$) and then decreases abruptly to 2.46 Å for the nonsuperconducting oxide ($x = 1$). Thus copper is almost located in the basal plane of the CuO_5 pyramid in the superconducting cuprates, as shown from the O(1)–Cu–O(1) angles, which are close to 180° (Table 3), and is displaced toward the apical corner, leading to more puckered $[CuO_2]_\infty$ layers as superconductivity disappears, as shown from the O(1)–Cu–O(1) angle of 169° for the nonsuperconductive phase $TlBa_2NdCu_2O_7$. To determine whether this phenomenon can be generalized, we can analyze the geometry of the CuO_5 pyramids of layered cuprates, superconductive or not, involving double pyramidal copper layers whose structure has been determined either from single-crystal x-ray study or from neutron diffraction. The interatomic Cu–O distances and angles of these oxides with the corresponding T_cs are given in Table 4. From these data it appears that for the series of superconductive barium thallium cuprates formed of double layers of CuO_5 pyramids, one always observes long Cu–O apical distances ranging from 2.65 to 2.76 Å, leading to almost flat CuO_4 groups with O–Cu–O angles ranging from 176.6 to 178.4°, whether they contain single or double thallium layers. On the contrary, oxides containing strontium instead of barium exhibit much smaller apical Cu–O distances (whether or not they are superconductive), with or without thallium.

Clearly, the value of the Cu–O apical distance depends on the size of the cations of the rock salt layers (barium, strontium, or lanthanum). It decreases strongly as their size decreases, as shown by comparing barium cuprates with strontium cuprates. This does not rule out the fact that within one series, T_cs could be correlated with the Cu–O apical distance.

3. The Cu–Cu distance between two successive copper planes increases as x increases from 3.24 Å to 3.41 Å (Table 3). This is correlated with the variation in the Cu–O(2) apical distance, which is in the opposite direction. However, no general correlation between superconductivity and Cu–Cu distances (Table 4) can be seen.

4. The O(1)–O(1) distance parallel to the c axis and forming a fluorite-type cage, where the Ca or Nd cations are located, does not vary dramatically; nevertheless, one observes that it decreases slightly as x increases (i.e., as T_c decreases). However, the O(1)–O(1) distances cannot be considered as related to superconductivity, as shown in Table 4, where nonsuperconducting oxides such as $Nd_{1.8}Sr_{1.2}Cu_2O_6$ and $La_2SrCu_2O_6$ show longer O(1)–O(1) distances (3.21 to 3.37 Å) than superconductors.

Table 4   Cell Parameters, Superconducting Properties, Interatomic Distances, and Angles in Some Cuprates Involving Two Pyramidal Copper Layers

Phase	Ref.	Cell parameters (Å)	T_c (K)	Interatomic distance (Å)					Angle O–Cu–O[c] (deg)
				d(Cu–O) apical	d(Cu–O) equatorial	d(A–O) apical[a]	d(Cu–Cu)[b]	d(O–O)[b]	
$Tl_2Ba_2CaCu_2O_8$	17	3.8550 29.318	112	2.699	1.928	1.978 2.031	3.166	3.113	178.4
	89	3.8618 29.217	94	2.655	1.932	2.002 1.993	3.168	3.062	176.9
	89	3.8578 29.371	105	2.724	1.929	1.965 1.997	3.163	3.049	176.7
	89	3.8588 29.437	102	2.734	1.930	1.966 2.040	3.162	3.073	177.4
$TlBa_2CaCu_2O_7$	78	3.8566 12.754	103	2.762	1.929	2.01	3.201	3.086	176.6
$TlSr_{1.3}Nd_{0.7}CaCu_2O_7$	88	3.8282 11.9872	81	2.2852	1.917	2.053	3.345	3.133	173.7
$Tl_{0.5}Pb_{0.5}Sr_2CaCu_2O_7$	90	3.8023 12.107	85	2.443	1.903	1.996	3.238	3.155	177.5
$Nd_{1.8}Sr_{1.2}Cu_2O_6$	91	3.8365 19.952	—	2.152	1.928	2.342	3.600	3.210	168.7
$La_2SrCu_2O_6$	92	3.8647 19.941	—	2.210	1.938	2.363	3.664	3.370	171.3
$La_{1.9}Ca_{1.1}Cu_2O_6$	93	3.8245 19.420	—	2.306	1.913	2.339	3.306	3.191	176.6
$La_2CaCu_2O_6$	94	3.834 19.517	40 <1.5%	2.315	1.918	2.346	3.330	3.205	176.3

[a] A = Tl, Nd/Sr, La/Sr, La/Ca.
[b] Distances between two successive layers.
[c] Angle O–Cu–O in the basal plane.

VII.   CONCLUDING REMARKS

Although study of the relationships between structure, defects, redox mechanisms, and superconductivity in thallium cuprates is just beginning, several important points should be emphasized. Extended defects such as intergrowths should not dramatically influence the critical temperature of these superconductors. On the contrary, the oxygen nonstoichiometry plays a capital role for the T_cs, the latter being sensitive to very small deviations from oxygen stoichiometry. Clearly, the maximum critical temperature is obtained from an optimum hole carrier density per copper atom (i.e., for an optimum oxygen content) so that most of the time, the critical temperatures given by different authors do not correspond to the ideal compositions but to nonstoichiometric oxides, which can be formulated Tl-$A_2Ca_{m-1}Cu_mO_{2m+3\pm\delta}$ and $Tl_2A_2Ca_{m-1}Cu_{2m}O_{2m+4}O_{4\pm\delta}$. Besides the predominant role of the oxygen stoichiometry with respect to the number of copper layers forming the oxygen-deficient layers, there remains the important issue of the distribution of oxygen vacancies in the structure. The latter have been found to be located in the $[TlO]_\infty$ layers in the 1212-type structure, but systematic work remains to be done for other thallium cuprates, especially for those involving thallium bilayers. In any case it has been shown from x-ray absorption studies at the L_{III} edge of thallium [95] that a general shift in edge transitions toward low energies is observed for thallium superconductive cuprates with respect to other reference Tl(III) compounds. Such behavior, which indicates the presence of extra electrons in the $[TlO]_\infty$ layers, suggests that the $[TlO]_\infty$ layers play the role of hole reservoirs for the copper–oxygen layers accoding to the equilibrium

$$Tl(III) + Cu(II) \rightleftharpoons Tl(III) - \delta) + Cu(II + \delta)$$

It is most likely that these extra electrons, which are delocalized in a narrow band, induce semimetallic or metallic conductivity in the $[TlO]_\infty$ layers, whereas the origin of superconductivity takes place in the copper–oxygen layers. As a consequence, a model in which the thallium–oxygen layers become superconducting by proximity effects on the cell scale and play an indirect role in superconductivity has to be considered. An understanding of redox mechanisms, due to the presence in the same crystal of two redox couples, Tl(III)/Tl(I) and Cu(III)/Cu(II), is absolutely necessary to control the synthesis and to optimize the properties of thallium superconductors.

REFERENCES

1. Z. Z. Sheng and A. M. Hermann, *Nature* 332, 55 (1988).
2. Z. Z. Sheng and A. M. Hermann, *Nature* 332, 138 (1988).
3. Z. Z. Sheng, W. Kiehl, J. Benneth, E. El Ali, D. Marsh, G. D. Mooney, F. Arramash, J. Smith, D. Viar, and A. M. Hermann, *Appl. Phys. Lett.* 52, 1738 (1988).
4. R. Beyers, S. S. P. Parkin, V. Y. Lee, A. I. Nazzal, R. Savoy, G. Gorman, T. C. Huang, and S. La Placa, *Appl. Phys. Lett.* 53, 432 (1988).
5. C. Martin, D. Bourgault, C. Michel, J. Provost, M. Hervieu, and B. Raveau, *Eur. J. Inorg. Solid State Chem.* 26, 1 (1989).
6. D. Bourgault, C. Martin, C. Michel, M. Hervieu, J. Provost, and B. Raveau, *J. Solid State Chem.* 78, 326 (1989).
7. T. Manako, Y. Shimikawa, Y. Kubo, T. Satoh, and H. Igarashi, *Physica C* 158, 143 (1989).

8. C. C. Torardi, M. A. Subramanian, J. C. Calabrese, J. Gopalakrishnan, E. M. McCarron, K. J. Morrissey, T. R. Askew, R. B. Flippen, U. Chowdhry, and A. W. Sleight, *Phys. Rev. B* 38, 225 (1988).

9. M. Hervieu, A. Maignan, C. Martin, C. Michel, J. Provost, and B. Raveau, *J. Solid State Chem.* 75, 212 (1988).

10. C. Martin, D. Bourgault, C. Michel, M. Hervieu, and B. Raveau, *Mod. Phys. Lett. B* 3, 93 (1989).

11. C. Martin, J. Provost, D. Bourgault, B. Domenges, C. Michel, M. Hervieu, and B. Raveau, *Physica C* 157, 460 (1989).

12. M. A. Subramanian, C. C. Torardi, J. Gopalakrishnan, P. L. Gai, J. C. Calabrese, T. R. Askew, R. B. Flippen, and A. W. Sleight, *Science* 242, 249 (1988).

13. R. M. Hazen, D. W. Finger, R. J. Angel, C. T. Prewitt, N. L. Ross, C. G. Adidiacos, P. J. Heaney, D. R. Veblen, Z. Z. Sheng, A. El Ali, and A. M. Hermann, *Phys. Rev. Lett.* 60, 1657 (1988).

14. S. S. P. Parkin, V. Y. Lee, E. M. Engler, A. I. Nazzal, T. C. Huang, G. Gorman, R. Savoy, and R. Beyers, *Phys. Rev. Lett.* 60, 1539 (1988).

15. C. Politis and H. Luo, *Mod. Phys. Lett. B* 2, 793 (1988).

16. A. Maignan, C. Michel, M. Hervieu, C. Martin, D. Groult, and B. Raveau, *Mod. Phys. Lett. B* 2, 681 (1988).

17. M. A. Subramanian, J. C. Calabrese, C. C. Torardi, J. Gopalakrishnan, T. R. Askew, R. B. Flippen, K. J. Morrissey, U. Chowdhry, and A. W. Sleight, *Nature* 332, 420 (1988).

18. D. Bourgault, C. Martin, C. Michel, M. Hervieu, and B. Raveau, *Physica C* 158, 511 (1989).

19. C. Martin, C. Michel, A. Maignan, M. Hervieu, and B. Raveau, *C. R. Acad. Sci.*, 307, Ser. II, 27 (1988).

20. S. S. P. Parkin, V. Y. Lee, A. I. Nazzal, R. Savoy, R. Beyers, and S. La Placa, *Phys. Rev. Lett.* 61, 750 (1988).

21. A. Sulpice, B. Giordanengo, R. Tournier, M. Hervieu, A. Maignan, C. Martin, C. Michel, and J. Provost, *Physica C* 156, 243 (1988).

22. B. Domengès, M. Hervieu, and B. Raveau, *Solid State Commun.* 69, 1085 (1989).

23. C. C. Torardi, M. A. Subramanian, J. C. Calabrese, J. Gopalakrishnan, K. J. Morrissey, T. R. Askew, R. B. Flippen, U. Chowdhry, and A. W. Sleight, *Science* 240, 631 (1988).

24. M. Hervieu, C. Michel, A. Maignan, C. Martin, and B. Raveau, *J. Solid State Chem.* 74, 428 (1988).

25. H. Ihara, R. Sugise, M. Hirabayashi, N. Terada, M. Jo, K. Hayashi, A. Negishi, M. Tokumoto, Y. Kimura, and T. Shimomura, *Nature* 334, 511 (1988).

26. M. Hervieu, A. Maignan, C. Martin, C. Michel, J. Provost, and B. Raveau, *Mod. Phys. Lett. B* 2, 1103 (1988).

27. R. Sugise, M. Hirabayashi, N. Terada, M. Jo, T. Shimomura, and H. Ihara, *Physica C* 157, 131 (1989).

28. S. Nakajima, M. Kikuchi, Y. Syono, T. Oku, D. Shindo, K. Hiraga, N. Kobayashi, H. Iwasaki, and Y. Muto, *Physica C* 158, 471 (1989).

29. P. Pappamentellos, *Z. Kristallogr.* 126, 143 (1968).

30. R. V. Schenk and H. Müller-Busch Baum, *Z. Angew. Allg. Chem.* 405, 197 (1974).

31. J. B. Parise, C. C. Torardi, M. A. Subramanian, J. Gopalakrishnan, A. W. Sleight, and E. Prince, *Physica C* 159, 239 (1989).

32. M. Hervieu, C. Martin, J. Provost, and B. Raveau, *J. Solid State Chem.* 76, 419 (1988).

33. B. Domengès, M. Hervieu, and B. Raveau, *J. Solid State Commun.* 68, 303 (1988).

34. B. Domengès, M. Hervieu, and B. Raveau, *J. Solid State Commun.* 69, 1085 (1989).

35. M. Hervieu, B. Domengès, and B. Raveau, *J. Microsc. Spectrosc. Electron.* 13, 279 (1988).
36. B. Domengès, M. Hervieu, C. Martin, D. Bourgault, C. Michel, and B. Raveau, *Phase Transitions B* 19, 231 (1989).
37. H. W. Zandbergen, G. Van Tendeloo, J. Van Landuyt, and S. Amelinckx, *Appl. Phys. A* 46, 233 (1988).
38. J. P. Zhang, D. J. Li, H. Shibahara and L. D. Marks, *Supercond. Sci. Technol.* 1, 132 (1988).
39. M. Verwerft, G. Van Tendeloo, and S. Amelinckx, *Physica C* 156, 609 (1988).
40. S. Iijima, T. Ichihashi, Y. Shimakawa, T. Manako, and Y. Kubo, *Jpn J. Appl. Phys.* 27, L1054 (1988).
41. M. Hervieu, N. Nguyen, V. Caignaert, and B. Raveau, *Phys. Status Solidi A* 83, 473 (1984).
42. D. A. Jefferson, J. Gopalakrishnan, and A. Ramanan, *Mater. Res. Bull.* 17, 269 (1982).
43. D. A. Jefferson, M. K. Uppal, and D. J. Smith, *J. Solid State Chem.* 53, 101 (1984).
44. J. L. Hutchinson, G. R. Anstis, and R. J. D. Tilley, *International Congress on E. M.*, Hamburg, 1982.
45. M. Hervieu and B. Raveau, *Chem. Scr.* 22, 117 (1983).
46. M. Hervieu and B. Raveau, *J. Solid State Chem.* 43, 299 (1982).
47. B. Domengès, M. Hervieu, and B. Raveau, *Acta Crystallogr. B* 40, 249 (1984).
48. C. Michel, C. Martin, M. Hervieu, A. Maignan, J. Provost, M. Huvè, and B. Raveau, *J. Solid State Chem.*, 96, 271 (1992).
49. C. Martin, D. Bourgault, M. Hervieu, C. Michel, J. Provost, and B. Raveau, *Mod. Phys. Lett. B* 3, 993 (1989).
50. J. Akimitsu, S. Suzuki, M. Watanabe, and H. Sawa, *Jpn. J. Appl. Phys.* 27, L1859 (1988).
51. H. Sawa, S. Suzuki, M. Watanabe, J. Akimitsu, H. Matsubara, H. Watabe, S. Uchida, K. Kokusho, H. Asano, F. Izumi, and E. Takayama-Muromachi, *Nature* 337, 347 (1989).
52. Y. Tokura, T. Arima, H. Takagi, S. Uchida, T. Ishigaki, H. Asano, R. Beyers, A. I. Nazzal, P. Lacorre, and J. B. Torrance, *Nature* 342, 890 (1989).
53. T. Rouillon, D. Groult, M. Hervieu, C. Michel, and B. Raveau, *Physica C* 167, 107 (1990).
54. H. Sawa, K. Obara, J. Akimitsu, Y. Matsui, and S. Horiuchi, *J. Phys. Soc. Jpn.* 58, 2252 (1989).
55. T. Wada, A. Ichinose, Y. Yaegashi, H. Yamauchi, and S. Tanaka, *Phys. Rev. B* 41, 1984 (1990).
56. M. Kikuchi, N. Kobayashi, M. Iwasaki, D. Shindo, T. Oku, A. Tokiwa, T. Majitani, K. Hiraga, Y. Syono, and Y. Muto, *Jpn. J. Appl. Phys.* 27, L1050 (1988).
57. S. X. Dou, H. K. Lui, A. J. Bourduillon, N. X. Tan, J. P. Zhou, and C. C. Soviell, *Mod. Phys. Lett. B* 2, 875 (1988).
58. H. W. Zandbergen, W. A. Groen, F. C. Mijlhoff, G. Van Tendeloo, and S. Amelinckx, *Physica C* 156, 325 (1988).
59. B. Raveau, C. Michel, C. Martin, and M. Hervieu, *Mater. Sci. Eng. B* 3, 257 (1989).
60. C. Michel, M. Hervieu, M. M. Borel, A. Grandin, F. Deslandes, J. Provost, and B. Raveau, *Z. Phys. B* 68, 421 (1987).
61. M. Hervieu, C. Michel, B. Domengès, Y. Laligant, A. Lebail, G. Ferey, and B. Raveau, *Mod. Phys. Lett.* 2, 491 (1988).
62. E. A. Hewat, P. Bordet, J. J. Capponi, C. Chaillout, J. L. Hodeau, and M. Marezio, *Physica C* 156, 369 (1988).
63. T. Kijima, J. Tanaka, Y. Bando, M. Onoda, and F. Izumi, *Jpn. J. Appl. Phys.* 27, L369 (1988).

64. M. Hervieu, C. Michel, and B. Raveau, *J. Less Common Met.* 150, 59 (1989).
65. T. Rouillon, R. Retoux, D. Groult, C. Michel, M. Hervieu, J. Provost, and B. Raveau, *J. Solid State Chem.* 78, 322 (1989).
66. C. Michel, M. Hervieu, and B. Raveau, *Israel J. Technol.* 24, 65 (1988).
67. P. L. Gai, M. A. Subramanian, J. Gopalakrishnan, and E. D. Boyes, *Physica C* 159, 801 (1989).
68. R. Moret, P. Gougeon, M. Potel, J. C. Levet, and H. Noel, *Physica C* 168, 315 (1990).
69. J. M. Tarascon, Y. Lepage, P. Barboux, B. G. Bagley, L. H. Greene, W. R. McKinnon, G. W. Hull, M. Giroud, and D. M. Hwang, *Phys. Rev. B* 37, 9382 (1988).
70. M. Huvè, C. Michel, C. Martin, M. Hervieu, A. Maignan, J. Provost, and B. Raveau, *Physica C* 179, 214 (1991).
71. C. Martin, A. Maignan, M. Huvè, M. Hervieu, C. Michel, and B. Raveau, *Physica C* 179, 1 (1991).
72. S. Kondoh, Y. Ando, M. Onoda, M. Sato, and J. Akimitsu, *Solid State Commun.* 68, 1325 (1988).
73. S. S. P. Parkin, V. Y. Lee, A. I. Nazzal, R. Savoy, T. C. Huang, G. Gorman, and R. Beyers, *Phys. Rev. B* 38, 6531 (1988).
74. Y. Shimakawa, Y. Kubo, T. Manako, T. Satoh, S. Iijima, T. Ichihashi, and H. Igarashi, *Physica C* 157, 279 (1989); ibid., 162–164, 991 (1989).
75. K. V. Ramanujachary, S. Li, and M. Greenblatt, *Physica C* 167, 377 (1990).
76. C. Martin, A. Maignan, J. Provost, C. Michel, M. Hervieu, R. Tournier, and B. Raveau, *Physica C* 168, 8 (1990).
77. A. Maignan, C. Martin, M. Huvè, J. Provost, M. Hervieu, C. Michel, and B. Raveau, *Physica C* 170, 350 (1990).
78. B. Morosin, D. S. Ginley, P. F. Hlava, M. J. Carr, R. J. Baughman, J. E. Shirber, E. S. Venturini, and J. F. Kwak, *Physica C* 152, 413 (1988).
79. T. Manako, Y. Shimakawa, Y. Kubo, T. Satoh, and H. Igarashi, *Physica C* 156, 315 (1988).
80. S. Nakajima, M. Kikuchi, Y. Syono, N. Kobayashi, and Y. Muto, *Physica C* 168, 57 (1990).
81. C. Michel, E. Suard, V. Caignaert, C. Martin, A. Maignan, M. Hervieu, and B. Raveau, *Physica C* 178, 29 (1991).
82. D. M. de Leeuw, W. A. Groen, J. C. Jol, H. B. Brom, and H. W. Zandbergen, *Physica C* 166, 349 (1990).
83. R. J. Cava, M. Marezio, J. J. Krajewski, W. F. Peck, A. Santoro, and F. Beech, *Physica C* 157, 272 (1989).
84. W. T. Fu, H. W. Zandbergen, N. G. Haue, and J. L. de Jongh, *Physica C* 159, 210 (1989).
85. T. Rouillon, V. Caignaert, C. Michel, M. Hervieu, D. Groult, and B. Raveau, *J. Solid State Chem.*, 97, 56 (1992).
86. C. Michel, M. Hervieu, B. Raveau, S. Li, M. Greaney, S. Fine, J. Potenza, and M. Greenblatt, *Mater. Res. Bull.* 26, 123 (1991).
87. C. Michel, M. Hervieu, V. Caignaert, and B. Raveau, *Acta Crystallogr. C*, 48, 1747 (1992).
88. G. H. Kwei and M. A. Subramanian, *Physica C* 168, 521 (1990).
89. B. Morosin, D. S. Ginley, E. L. Venturini, R. J. Baughman, and C. P. Tigges, *Physica C* 172, 413 (1991).
90. J. B. Parise, P. L. Gai, M. A. Subramanian, J. Gopalakrishnan, and A. W. Sleight, *Physica C* 159, 245 (1989).
91. P. Labbé, M. Ledesert, V. Caignaert, and B. Raveau, *J. Solid State Chem.* 91, 362 (1991).
92. V. Caignaert, N. Nguyen, and B. Raveau, *Mater. Res. Bull.* 25, 199 (1990).

93. F. Izumi, E. Takayama-Muromachi, Y. Nakai, and H. Asano, *Physica C* 157, 89 (1989).

94. A. Fuertès, X. Obradors, J. M. Navarro, P. Gomez-Romero, N. Casan-Pastor, F. Perez, J. Fontcuberta, C. Miravitlles, J. Rodriguez-Carvajal, and B. Martinez, *Physica C* 170, 153 (1990).

95. F. Studer, N. Merrien, C. Martin, C. Michel, and B. Raveau, *Physica C* 178, 324 (1991).

5

Structures and Superstructures of Tl–Ba–Ca–Cu–O Superconductors

A. W. Hewat
Institut Max von Laue–Paul Langevin
Grenoble, France

E. A. Hewat
Centre d'Etudes Nucléaires, Grenoble, France

I. CRYSTAL STRUCTURES

A. Introduction: the Aurivillius Structure

As soon as it was realized that the structure of the 90-K superconductor $YBa_2Cu_3O_7$ consisted of layers of CuO-perovskite separated by CuO chains [1,2], much effort was expended in a search for other one- and two-dimensional copper oxide structures that might possibly be superconducting. The well-known Aurivillius phases [3], intergrowths of layers of perovskite and Bi_2O_2, are textbook examples of such structures, so it was not surprising that bismuth–copper oxide superconductors [4] were the first results of this search [5,6]. Sheng and Hermann [7,8] and Kondoh et al. [9] soon found that superconductivity was retained when Bi was replaced by Tl. They found that T_cs were higher in the new Tl compounds [10–13]. In fact, the structure of the BiO and TlO superconductors is not the Aurivillius[3]structure. The difference is important and is due to the different arrangement of oxygen atoms, which determines the oxidation state of the cations. These important details were only discovered with neutron diffraction, since the x-ray scattering power of the heavy bismuth and thallium atoms masks the weak scattering from oxygen.

Figure 1 shows how double bismuth layers are bound together by a common layer of oxygen in the Aurivillius phases. The bismuth atoms must be in oxidation state Bi^{3+}, and there is no space for extra oxygen. Figure 2 shows, for comparison, the structure of the simplest TlO superconductor, $(Sr,La)_2CuO_4 \cdot TlO$ [14,15], together with the structure of the original Bednorz and Müller [16,17] superconductor $(La,Sr)_2CuO_4$. Again there are layers of perovskite, but now the oxygen atoms have moved into the single plane of thallium. This TlO plane has stretched to fit

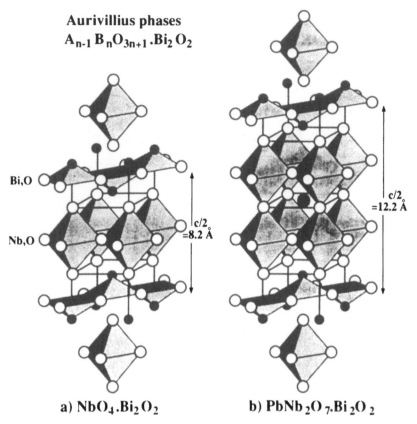

Figure 1 Aurivillius phases (a) $NbO_4 \cdot Bi_2O_2$ and (b) $PbNb_2O_7 \cdot Bi_2O_2$, showing perovskite slabs separated by layers of Bi_2O_2. The Aurivillius Bi_2O_2 layer is quite different from the double BiO layers in superconductors (e.g., Fig. 2b), since the oxygen is not in the Bi planes, but is between them in the Aurivillius phases.

the remaining perovskite structure, leaving too much space for a single oxygen atom at the center of the Tl square. To a first approximation, the single oxygen atom is therefore "disordered" over four equivalent positions, shown in Fig. 2 as a group of open circles corresponding to acceptable Tl–O distances. This mismatch between the ideal dimensions of two types of layers is common in composite materials; it leads to interesting possibilities for changing the stoichiometry of these layers, and therefore for changing the oxidation states of the cations and the superconducting properties. Although $(Sr,La)_2CuO_4 \cdot TlO$ looks very similar to the Bednorz and Müller superconductor $(La,Sr)_2CuO_4$ in Fig. 2, there are some important differences apart from the additional possibilities for changing the oxygen content of the TlO layers.

B.   Structures of Single- and Double-TlO-Layer Superconductors

The formulas for the single-TlO-layer series discovered by Parkin et al. [18] can be written $Ba_2Ca_{n-1}Cu_nO_{2n+2} \cdot TlO$ for $n = 1, 2, 3, \ldots$, where successive members are obtained by adding "neutral" $Ca^{2+}Cu^{2+}O_2^{2-}$. Just as the first member can be

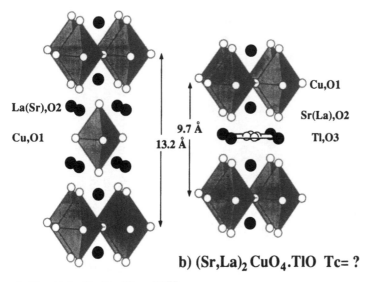

La(Sr),O2

Cu,O1

13.2 Å
9.7 Å

Cu,O1

Sr(La),O2

Tl,O3

b) (Sr,La)₂CuO₄.TlO Tc= ?

a) (La,Sr)₂CuO₄ Tc= 35K

Figure 2   The first member of the Tl₁ series can be written (Sr,La)₂CuO₄·TlO and is similar to the original Bednorz and Müller phase (La,Sr)₂CuO₄, with La³⁺ replaced by Sr²⁺ and single Tl³⁺O²⁻ layers intercalated between layers of CuO–octahedrae. The formal copper valence is Cu³⁺, which must be reduced by doping Sr²⁺ with La³⁺ to produce superconductivity.

compared to the original Bednorz and Müller material, the second member in the single-TlO-layer series, $Ca_1Ba_2Cu_2O_6$·TlO, can be compared to the structure of the original 90-K 123 superconductor $Y_1Ba_2Cu_3O_7$ if we write this formula as $Y_1Ba_2Cu_2O_6$·CuO. Y is replaced by Ca and Cu by Tl. Figure 3 shows that the structures are also similar, with the Y layer simply replaced by a layer of Ca, and the CuO chain by a layer of TlO. Figure 4 shows the first three members of the single-TlO-layer series.

Now suppose that we have two TlO layers instead of one, so that the formula may be written $Ba_2Ca_{n-1}Cu_nO_{2n+2}$·Tl_2O_2. For $n = 2$ this becomes $Ca_1Ba_2Cu_2O_6$·Tl_2O_2, and Fig. 5 shows that this structure is similar to the 124 superconductor $YBa_2Cu_4O_8$, especially if the latter formula is written $YBa_2Cu_2O_6$·Cu_2O_2. The single CuO chains in the 123 phase have been replaced by double Cu_2O_2 chains. Sera et al. [19], and in particular Hazen et al. [20], Subramanian et al. [21], Torardi et al. [21–23], and later Liang et al. [24], Kikuchi et al. [25], and Onoda et al. [26], characterized these double-layer Tl_2O_2 phases as being isomorphous with the Bi compounds, with general formulas $Ba_2Ca_{n-1}Cu_nO_{2n+2}$·Tl_2O_2. The structures of the first three members are shown in Fig. 6.

C.   Superconductivity and Mixed Cu^{2+}/Cu^{3+} Valence

Stoichiometric La_2CuO_4 is not a superconductor for what appears to be a very simple reason: If we assume that it contains $2 \times La^{3+}$ and $4 \times O^{2-}$, then for

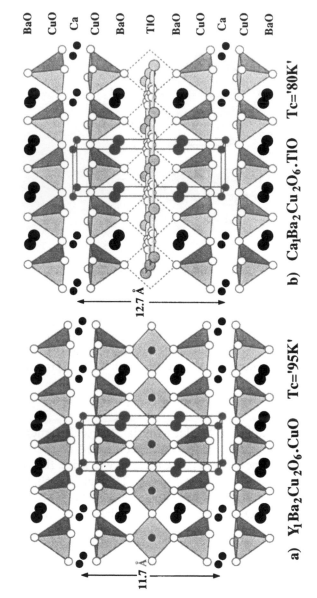

Figure 3 The second ($n = 2$) member of the Tl_1 series $CaBa_2Cu_2O_6 \cdot TlO$ can be compared to $YBa_2Cu_3O_7$ by writing the latter formula as $YBa_2Cu_2O_6 \cdot CuO$; Y^{3+} and the $Cu^{2+}O$ chains in the 123 superconductor are replaced by Ca^{2+} and a $Tl^{3+}O$ layer, so the valence electron counts are the same, and the $n = 2$ material should be naturally superconducting.

a) $Y_1Ba_2Cu_2O_6 \cdot CuO$ Tc='95K'

b) $Ca_1Ba_2Cu_2O_6 \cdot TlO$ Tc='80K'

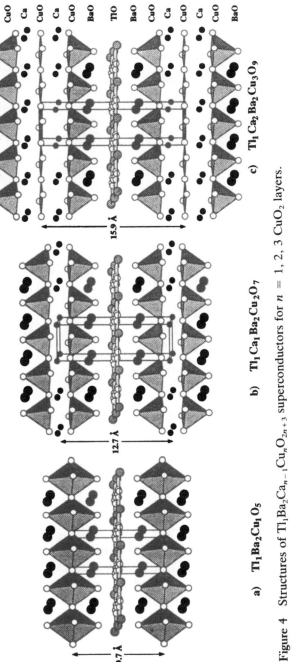

a) $Tl_1Ba_2Cu_1O_5$ b) $Tl_1Ca_1Ba_2Cu_2O_7$ c) $Tl_1Ca_2Ba_2Cu_3O_9$

Figure 4   Structures of $Tl_1Ba_2Ca_{n-1}Cu_nO_{2n+3}$ superconductors for $n = 1, 2, 3$ CuO_2 layers.

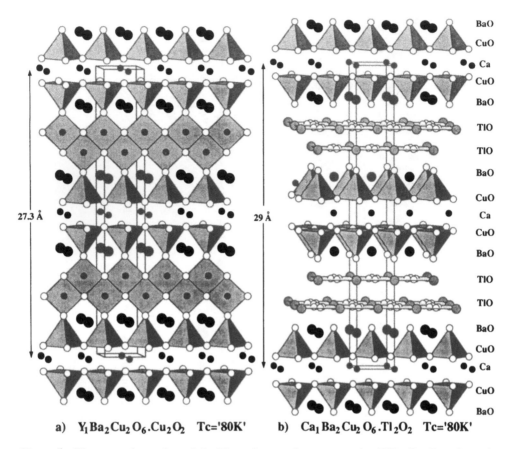

BaO
CuO
Ca
CuO
BaO
TlO
TlO
BaO
CuO
Ca
CuO
BaO
TlO
TlO
BaO
CuO
Ca
CuO
BaO

27.3 Å   29 Å

a) $Y_1Ba_2Cu_2O_6 \cdot Cu_2O_2$ Tc='80K'   b) $Ca_1Ba_2Cu_2O_6 \cdot Tl_2O_2$ Tc='80K'

Figure 5   The second member of the Tl_2 series may be compared to $YBa_2Cu_4O_8$, where the double CuO chains in $YBa_2Cu_2O_6 \cdot Cu_2O_2$ are replaced by double layers of TlO or BiO to produce $CaBa_2Cu_2O_6 \cdot Tl_2O_2$ (or $CaSr_2Cu_2O_6 \cdot Bi_2O_2$).

charge balance all copper must be Cu^{2+}. But high-temperature superconductivity apparently requires additional electron holes (formally, Cu^{3+}), and the T_c increases with increasing numbers of holes [27]. Then for La_2CuO_4 to become superconducting, either additional oxygen is needed, some La^{3+} must be replaced by Sr^{2+}, or some other means must be found to *oxidize* the material chemically.

Stoichiometric $Sr_2CuO_4 \cdot TlO$ is not a superconductor either. Counting the valences again we find $2 \times Sr^{2+}$, $1 \times Tl^{3+}$, and $5 \times O^{2-}$, implying that all copper is formally Cu^{3+}. Now we must either remove some oxygen, replace some Sr^{2+} by La^{3+}, or otherwise chemically *reduce* the material to make it a superconductor. However, the second member of the series, like $Y_1Ba_2Cu_3O_7$, is "naturally" superconducting, since successive members are obtained by adding Cu^{2+} as $Ca^{2+}Cu^{2+}O_2^{2-}$. $Y_1Ba_2Cu_3O_7$ was the first superconductor for which copper appears to have mixed valence without the need for doping. Assuming Y^{3+}, Ba^{2+}, and O^{2-}, we concluded [1] that in $Y_1Ba_2Cu_3O_7$, the copper formally exists as $Cu_2^{2+}Cu^{3+}$. When this material is reduced to $YBa_2Cu_3O_6$, by removing the oxygen from the CuO chains, superconductivity disappears. The valences of Y^{3+} and Ba^{2+}

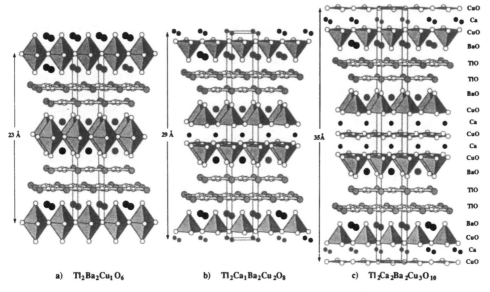

a) $Tl_2Ba_2Cu_1O_6$ b) $Tl_2Ca_1Ba_2Cu_2O_8$ c) $Tl_2Ca_2Ba_2Cu_3O_{10}$

Figure 6 Idealized structures of $Tl_2Ba_2Ca_{n-1}Cu_nO_{2n+4}$ superconductors for $n = 1, 2, 3$ CuO_2 layers. The oxygen within the TlO layers is shown disordered. Note that the formula of the first member of the series may be written $Ba_2CuO_4 \cdot Tl_2O_2$, and that its structure is again similar to the original $(La,Sr)_2CuO_4$ superconductor of Bednorz and Müller, with double TlO layers intercalated between the perovskite layers.

cannot be reduced, so we concluded [28] that copper in $YBa_2Cu_3O_6$ is formally reduced to $Cu_2^{2+}Cu^{1+}$, with loss of the $Cu^{2+}Cu^{3+}$ mixed valence necessary for superconductivity.

Similarly, the formula $Ca_1Ba_2Cu_2O_6 \cdot TlO$ implies, if we assume Ca^{2+}, $2 \times Ba^{2+}$, Tl^{3+}, and $7 \times O^{2-}$, that again copper should be "naturally" of mixed valence $Cu^{2+}Cu^{3+}$, and that again the material should normally be superconducting. Reduction, which is *necessary* for superconductivity in the first member of the single-TlO-layer series, might instead be expected to *reduce* superconductivity in the second member of the series. In general, the oxidation–reduction conditions of preparation will have a profound effect on the superconductivity of these materials [29].

Further increasing the value of n in the series will change the Cu^{2+}/Cu^{3+} ratio, and can be expected to change the superconducting properties. It was at first assumed that T_c might increase indefinitely with the number of CuO layers [30], but clearly as $n \Rightarrow \infty$ the effective valence of copper $\Rightarrow Cu^{2+}$, and superconductivity should disappear if mixed valence is a necessary condition. Indeed, Siegrist et al. [31] later showed that the $n = \infty$ compound $CaCuO_2$, stabilized as $(Ca_{0.86}Sr_{0.14})CuO_2$, was not superconducting.

Since the first and last members of the series, representing the two extremes of all Cu^{3+} and all Cu^{2+}, respectively, are not superconducting, the maximum T_c must be obtained for some intermediate member. However, consider the double-Tl-layer series with the formula $Ba_2Ca_{n-1}Cu_nO_{2n+2} \cdot Tl_2O_2$, where the second member has a structure similar to that of the 124 superconductor

$YBa_2Cu_2O_6 \cdot Cu_2O_2$. Remember that like the 123 phase, the 124 phase is naturally of mixed valence, since the combination Y^{3+}, $2 \times Ba^{2+}$, and $8 \times O^{2-}$ implies that copper is formally $Cu_3^{2+}Cu^{3+}$. For comparison, the formula $Ca_1Ba_2Cu_2O_6 \cdot Tl_2O_2$ implies that all copper should be Cu^{2+}, which is in fact true for *all* members of the double-Tl-layer series, since successive members are again obtained by adding Cu^{2+} as $Ca^{2+}Cu^{2+}O_2^{2-}$. It should then be necessary to oxidize the double-Tl-layer materials to obtain the mixed valence $Cu^{2+}Cu^{3+}$ necessary for superconductivity, or to dope them in some other way to obtain the necessary electron holes. In practice, the double-Tl-layer series are normally superconducting, so that either these materials do not conform to the stoichiometric formula above, or they must be naturally doped in some other way.

D.   Superconductivity Obtained or Destroyed by Doping or Nonstoichiometry

As for the original Bednorz and Müller superconductor, the natural copper valence can readily be changed by doping the divalent cation sites with trivalent cations. For example, the normally nonconducting $n = 1$ member becomes superconducting with composition $TlBa_{1.2}La_{0.8}CuO_5$ [32,33]. Conversely, the naturally superconducting second member of the series $Ca_1Ba_2Cu_2O_6 \cdot TlO$ is rendered nonconducting as $Y_1Ba_2Cu_2O_6 \cdot TlO$ [34]. Although this compound has much the same formula and structure as $YBa_2Cu_3O_7$, it cannot be superconducting because all copper is clearly Cu^{2+}. The replacement of divalent Ca^{2+} by trivalent Y^{3+} is a common method of changing electron hole content, and therefore superconducting properties.

Cava et al. [35] have argued that if the CuO planes in all these materials are the conducting layers, they become superconducting only when the effective valence of copper is mixed Cu^{2+}/Cu^{3+}, that is, when the copper oxide planes are doped by charge transfer from the other layers. In their original paper they demonstrated that oxidation of the CuO chains in $YBa_2Cu_3O_7$, called *charge transfer reservoirs*, also resulted in oxidation of the CuO planes by charge transfer, which was indicated by a shortening of certain Cu–O distances, in particular the apical Cu–O bond.

The Y^{3+}/Ca^{2+} planes are also potential charge transfer layers, and indeed, Parise and McCarron [36] have shown how nonsuperconducting $YBa_2Cu_3O_{6+\delta}$ can be made superconducting by doping Y^{3+} with Ca^{2+}. It is therefore not surprising that in thallium and bismuth superconductors, these other cation planes can play a similar role. Of course, the TlO and Tl_2O_2 planes are still the most interesting charge transfer reservoirs in these superconductors.

E.   Higher Members of the TlO and Tl_2O_2 Series

Iijima et al. [37–40], Zandbergen et al. [41], and Verwerft et al. [42] used high-resolution electron microscopy to show that intergrowths commonly occur between phases with different numbers of copper layers n, and observed phases up to $n = 6$. For the double-layer bismuth and thallium compounds, pure samples with at most $n = 3$ have been prepared [43], although small regions with larger numbers of copper oxide layers have been observed in the electron microscope [37,39,41]. However, the $n = 4, 5$, and 6 compounds have been prepared for the single-TlO-layer phases [30,44–46], and also in $Tl_2Ba_2Ca_3Cu_4O_y$ [47].

At the Interlaken meeting in March 1988, the IBM Almaden group announced the record T_c of 125 K for the $n = 3$ member of the double-Tl-layer series (see Parkin et al. [48] and also Itoh et al. [49]). In the single-layer series, T_c does increase for $n = 4$, and at 122 K is similar to that for the $n = 3$ double-TlO-layer compound, but then T_c decreases on adding more CuO layers for $n = 5$ [44], which we have seen is to be expected. Reports of onset T_cs of 155 K [50] and higher have not been confirmed. Eibschütz et al. [51] have reported superconductivity at 121 K in a thallium-depleted multiphase system, perhaps corresponding to single-layer Tl phases with every second TlO layer missing ($\frac{1}{2}$ 223).

T_c also depends on the precise conditions of preparation. Some of the reported variations are due to multiphase samples, but even in single-phase samples, important variations in T_c are observed [52–58]. The first members of the series, with $n = 1$, is an extreme case, since single-phase samples with apparently very similar stoichiometry can be either superconducting or not, depending on the precise heat treatment [59,60]. Nagashima et al. [61] and Sheng et al. [62] have found that superconductivity persists if Ba is replaced by Sr, but T_cs between 20 and 80 K are much lower. The substitution of other cations for bismuth or thallium has been successful in the case of lead. Lead doping has been found to increase T_c in some cases, to suppress it for larger dopings [63,64] and to modify the superstructure found in all these materials [4]. A new family of superconductors has been reported, based on $Pb_2Sr_2ACu_3O_8$, where A = Y, Ln, or Ln + (Sr or Ca), where Ln is a lanthanide [65–67]. For A $\sim$ $Ca_{0.5}Y_{0.5}$, T_cs near 70 K are obtained.

Since the first member of the Tl_2O_2 series should contain only Cu^{2+}, and the higher members are obtained by adding neutral $Ca^{2+}Cu^{2+}O_2^{2-}$, it should be necessary to oxidize all these phases before expecting superconductivity. In fact, samples are usually superconducting without the need for any special oxidation, and next we examine how this might occur.

F.   Structures of BiO-Layer Superconductors

The basic structures of all of these phases are well known and have been summarized by Beyers et al. [68] and Parkin et al. [69]. First we can compare the thallium oxide superconductor structures with those of bismuth, since there are instructive differences. The single-layer BiO phases corresponding to those of TlO in Fig. 4 do not exist. The double-layer Bi_2O_2 phases do exist, with structures that are similar but not identical to those of Tl_2O_2 in Fig. 6. The Bi_2O_2 materials are formed as lamella sheets, whereas the Tl_2O_2 materials are not. Atomic resolution microscope images of the surface of the Bi_2O_2 phases show that the double layer cleaves readily, leaving a single layer at the surface [70]. This was taken to mean that there are no strong bonds holding the pairs of the Bi_2O_2 layers together, and therefore that the structure is not that of the Aurivillius phases (Fig. 1), where oxygen binds the two Bi layers strongly. On the contrary, it was supposed that this ready cleavage implies bismuth Bi^{3+} with its lone-pair electrons accommodated between the Bi layers, rather than Bi^{5+} with no lone-pair electrons. The Bi^{3+} layers would then be held together only by very weak forces, unlike the Tl^{3+} layers, where there are no lone-pair electrons.

The need to accommodate lone-pair electrons between pairs of B^{3+} planes also explains why single-layer BiO phases do not exist, whereas single-layer TlO phases do. Unfortunately, the first x-ray determinations of the structure of the Bi_2O_2 phases could not locate the oxygen atoms in the presence of the strongly scattering bismuth. However, neutron diffraction [71], for which most atoms scatter almost equally well, confirmed that the oxygen atoms are within the Bi layers, not between them.

As in the Aurivillius phases (Fig. 1), the Bi (or Tl) atoms are centered below the apex oxygens of the perovskite (copper oxide) structure (Fig. 2). But there is then too much space at the center of the resulting square of Bi atoms for a single oxygen atom, and Bi–O distances would be too large for such a symmetric arrangement. The oxygen atom therefore moves off the center of the Bi square toward a pair of Bi atoms, as shown in Fig. 7. The neutron diffraction measurements [72] showed that these oxygen displacements were ordered, producing zigzag Bi–O–Bi–O chains (Fig. 8) and reducing the symmetry of the structure from tetragonal to orthorhombic. It is also clear that this resulting structure is no longer centrosymmetric; attempts to refine a centrosymmetric structure will inevitably result in apparent disorder or "split" positions for the oxygen atoms.

G.   Superstructures of the BiO-Layer Superconductors

Figure 8 shows only the average structure, with a relatively short repeat distance. From the beginning it was clear that this average structure was modulated. For example, the electron diffraction diagrams [70] show clear superlattice spots corresponding to a superstructure 4.75 times the average structure along the n axis.

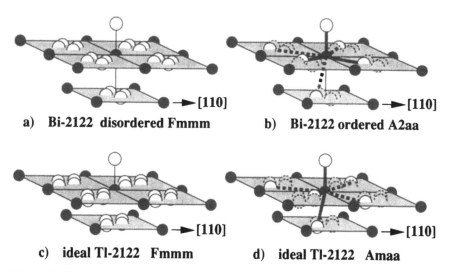

a)   **Bi-2122 disordered Fmmm**   b)   **Bi-2122 ordered A2aa**

c)   **ideal Tl-2122 Fmmm**   d)   **ideal Tl-2122 Amaa**

Figure 7   Disordered oxygen positions in (a) BiO layers and (c) TlO layers, producing an average tetragonal structure. This oxygen is probably ordered over at least short distances, as shown in (b) and (d). In BiO phases, this oxygen is displaced toward the line joining nearest Bi atoms; in TlO phases, Tl already has two short bonds to either side of the TlO planes, and oxygen is displaced mainly toward a single Tl atom. Longer-range order of oxygen implies orthorhombic symmetry, clearly seen for some samples, and associated with long-range order of the superstructure.

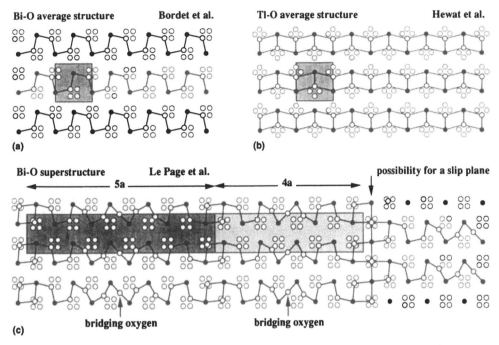

Figure 8 (a) and (b) Network of bonds within the BiO and TlO planes implied by Fig. 7. Zigzag metal-oxygen bonds make the x axis slightly shorter than the y axis. At higher temperatures, oxygen may be dynamically disordered over the four dotted sites. (c) BiO superstructure proposed by Le Page et al. by comparison with the structure of $Bi_{10}Sr_{15}Fe_{10}O_{46}$. Oxygen is displaced more and more toward the Bi–Bi midpoint, resulting in one extra oxygen every five subcells of (a). We have also indicated how $\times 4$ blocks might exist. Hewat et al. have observed series of 5, 5, 5, 4 blocks in high-resolution electron microscope images, resulting in an incommensurable $(5 + 5 + 5 + 4)/4 = 4.75a$ superstructure in $Bi_2Sr_2CaCu_2O_8$. The superstructure also consists of expansion around the short Bi–O–Bi bonds (not shown).

Since the period is not an integral multiple of the average unit cell, it is said to be an incommensurable modulation. Such modulations are known to be reflections of the electronic structure of some materials, and to correspond to charge density waves whose period has nothing to do with the period of the unit cell. Alternatively, incommensurable superlattice spots can mean that there are two repeat distances, for example $\times 4$ and $\times 5$. A pseudoperiodic sequence of blocks of 5, 5, 5, 4, 5, 5, 5, 4, . . . would produce sharp superlattice spots with the incommensurable period of $\times 4.75$ the basic unit, without the need for any incommensurable wave of that period. Such sequences are commonly found in mixed valence oxides, where they accommodate small changes in oxygen content.

It is difficult by diffraction measurements alone to distinguish between the incommensurable wave and the pseudoperiodic block models, since both give incommensurable superlattice spots. Atomic resolution microscopy should, of course, distinguish between the two, and in the original work the block model was favored [70]. However, for technical reasons associated with the problem of imaging periodic structures at the limit of instrument resolution, it is still not possible to be quite sure. If the block picture is correct, it would provide a natural explanation of the small changes in oxygen content, by analogy with other oxide block structures.

Further evidence that the incommensurable structure of Bi and Tl superconductors was associated with the oxygen stoichiometry of the Bi_2O_2 and Tl_2O_2 layers was provided by the Le Page et al. [73] structure of $Bi_{10}Sr_{15}Fe_{10}O_{46}$, a compound that is not itself superconducting, but which can be thought of as the Fe analog of the $n = 2$ Cu-superconductor phase $5 \times (Sr_2Ca_1Cu_2O_6 \cdot Bi_2O_{2+\partial})$. There is also a superstructure in this material, but it is a commensurable $\times 5$ multiple of the basic cell. This meant that it was much easier to determine the structure with precision.

It was found that the superstructure was due to a modulation of the displacement of the Bi_2O_2-plane oxygen atoms. As had already been found by neutron diffraction for the superconducting materials, these oxygen atoms were displaced from the centers of the Bi squares. But the displacement increased on going from one square to the next. In this way it was possible to accommodate one extra oxygen atom for every five squares, although all Bi–O distances remained reasonable, all Bi atoms had similar oxygen coordination, and it was not necessary to invoke oxygen "disorder" or split oxygen positions.

It is easy to see how this result can be applied to superconducting materials to modify the average Bi_2O_2 structure (Fig. 8a) to produce an incommensurable structure. In Fig. 8c we see that as well as $\times 5$ blocks, with one extra oxygen atom every five cells, we may also have (a smaller number of) $\times 4$ blocks, with one extra oxygen every four cells. Such different blocks can join together without large changes in the atomic positions. In Fig. 8c we have only shown the displacements of the oxygen atoms, but of course in the real structure the other atoms, especially the Bi or Tl atoms, would relax to accommodate the modulation in the oxygen content. The resulting modulation of the heavy atoms would, of course, give the main contribution to the electron and x-ray superlattice spots.

Thus we have a simple mechanism for adding extra oxygen to the Bi_2O_2 layers, as needed to produce the formal Cu^{2+}/Cu^{3+} mixed valence apparently necessary for superconductivity. The natural mismatch between the lengths of the CuO–perovskite and Bi_2O_2 layers results in a stretching of the latter to fit the former and makes it possible to accommodate one extra oxygen for every four or five unit cells by modulating the oxygen displacements within the Bi_2O_2 layer. The natural amount of extra oxygen must depend on the degree of mismatch, and it is fortunate that this corresponds with the degree of doping necessary to obtain a high T_c. The amount of oxygen can be changed by different conditions of sample preparation (oxidizing or reducing environments), and this produces changes in doping and therefore in T_cs. Unfortunately, this idealized picture is probably not sufficient, at least for the TlO superconductors [74].

H. Superstructures of TlO-Layer Superconductors

One of the first objectives of structure solution is to obtain the precise chemical composition. Of course, chemical, gravimetric, electron probe, and other methods of chemical analysis are also needed for this purpose, but these methods determine the *total* contents of a sample rather than the contents of a specific site in a specific phase. Also, the apparent valence of the various atoms can be calculated from precise measurements of the cation–oxygen bond distances, according to the empirical formulas of Zachariasen [75] and Brown-Shannon [76,77].

The earlier work showed that T_c cannot be a simple function of the number of electron holes. For example, as $Y_1Ba_2Cu_3O_{7-\delta}$ is reduced to $Y_1Ba_2Cu_3O_6$, T_c falls, but the fall is not linear: there is a plateau with $T_c \sim 60$ K near $\delta \sim 0.5$ [78], corresponding to an intermediate structure where every second CuO chain is absent [79]. T_c then depends on the details of the structure, not just on the electron-hole concentration, and for this reason, too, it is important to investigate structural details.

Neutron diffraction is needed to obtain these structural details, since x-ray scattering power of oxygen is very small compared to that of heavy elements such as thallium, and it is the oxygen stoichiometry and the distances to oxygen atoms that are of most interest. Neutron diffraction, however, requires large samples, often available only as powders; then only the average structure is determined, and the superstructure must be obtained by other means. Electron diffraction has played an important role in an understanding of the superstructures, as illustrated by the work of Iijima et al. [37–40,80], Beyers et al. [18,68,81,82], Domengès et al. [83,84], Hewat et al. [14,85,86], Hiraga et al. [87], Zandbergen et al. [41,88], and Fitzgerald et al. [89].

II. NEUTRON DIFFRACTION

A. $Tl_2Ba_2Ca_1Cu_2O_8$ and $Tl_2Ba_2Ca_2Cu_2O_{10}$

Cox et al. [90,91] and Hewat et al. [85] have analyzed neutron powder measurements on the $n = 2$ and $n = 3$ members of the Tl_2 series, using the tetragonal space group $I4/mmm$. They agree that for both materials, oxygen within the TlO plane appears disordered, with this oxygen moving toward a single thallium atom (Fig. 7c). They find two short Tl–O bonds (2.0 Å), two medium bonds (2.5 Å), and two long bonds (3.0 Å). Such oxygen displacements are in agreement with the short-range ordering observed by Dmowski et al. [92] and Toby et al. [93]. The other atoms appear well ordered. Toby et al. [94] reported an apparent structural change in $Tl_2Ba_2Ca_1Cu_2O_8$ at T_c, but this was not confirmed in the other neutron measurements [85,90]. The Zachariasen–Brown–Shannon valence for copper is 2.1, much lower than the 2.33 for $YBa_2Cu_3O_7$. On the other hand, Tl has an apparent valence considerably lower than the expected $3+$, although this calculation is rather sensitive to the precise model for the disordered oxygen.

These neutron measurements, as well as x-ray measurements by the Du Pont group, agree that the thallium site contains up to 5% vacancies in both phases, or else is substituted by up to 12% calcium. The calcium site may also contain up to 10% thallium in the 2122 phase, but appears fully occupied by calcium in the 2223 phase. Occupancy of the oxygen site within the TlO layer is subject to errors of a few percent because of correlation with its disorder, but within these limits it appears fully occupied in both phases.

In particular, there is no evidence of the additional oxygen, or cation vacancies [95,96], supposedly needed to increase the electron-hole concentration in the copper oxide layers to the values found in $YBa_2Cu_3O_7$. Indeed, the apparent copper valence of only 2.1 is a value that might be expected for a few percent Tl^{3+} vacancies. If it were true, substitution of the Ca^{2+} sites by Tl^{3+} would not help, since this would reduce the electron-hole concentration available for the copper oxide planes. The low values for the apparent valence of thallium, and the apparent oxygen disorder within the TlO plane, are the only remarkable aspects of otherwise well-characterized structures.

B. $Tl_2Ba_2Cu_1O_6$

In the Tl compounds the superstructures are slightly different, and may indeed be due to systematic Tl vacancies rather than O interstitials. In $Tl_2Ba_2Cu_1O_6$, Hewat et al. [59] found that as many as one in eight of the atoms in the TlO plane were apparently missing, and proposed that ordering of these vacancies may be responsible for the superstructure [86]. Removing $Tl^{3+}O^{2-}$ rather than adding O^{2-} is, of course, an alternative way of increasing the electron-hole concentration, and appears to be possible due to the relatively high vapor pressure of thallium oxide under the usual conditions of preparation. However, the higher members of the Tl series have higher T_cs, implying higher electron-hole concentrations, although the number of Tl vacancies is smaller and the superstructures less marked. Cox et al. [90] and Hewat et al. [85] found that the oxygen atoms within the TlO layers again appear disordered.

By slow cooling of the Tl-2021 phase, Hewat et al. [59] prepared an ordered phase, with pseudoorthorhombic (strictly monoclinic) symmetry, which was not superconducting as were the usual "tetragonal" phases. In such an ordered phase, the oxygen atoms are systematically displaced toward thallium atoms, occupying only one of the four positions equivalent in the tetragonal structure. Tl–O coordination is similar to Bi–O coordination, except that in the BiO phases oxygen is displaced toward *pairs* of Bi atoms, due to the different bonding requirements of bismuth and its lone-pair electrons [72].

Hewat et al. [59,86] examined several different samples of $Tl_2Ba_2Cu_1O_6$. As first prepared, the samples had a clear orthorhombic metric at relatively low temperature (815 K), but were not superconducting. Samples quenched at a slightly higher temperature (840 K) were superconducting and had an almost tetragonal metric, like the higher members of the series. These tetragonal samples showed the usual diffuse superlattice spots with electron diffraction, whereas the superlattice spots were quite sharp for the orthorhombic samples, indicating longer-range order. Since none of the well-ordered orthorhombic samples were superconducting, although their stoichiometry was apparently the same as for the tetragonal samples, Hewat et al. supposed that disorder might be necessary for superconductivity. However, other workers, such as Shimakawa et al. [60,97], have now produced superconducting samples of the orthorhombic form.

Since even the tetragonal samples were thought to have lower microscopic symmetry, all refinements of the neutron data were performed in the space group $A2aa$, a subgroup of $I4/mmm$. Clearly, the data for the tetragonal samples might also have been refined in $I4/mmm$ with disordered TlO–oxygen, as for the $n = 2, 3$ members of the series. However, all atoms appear ordered in the lower symmetry space group, and probably this is a better representation of the structure at the microscopic level. The apparent tetragonal metric was then assumed to be due to short-range disorder of the Tl–O bond network (and the superstructure).

As for the $n = 2, 3$ phases, oxygen within the TlO plane is displaced toward Tl atoms so that there are again two short Tl–O bonds, two medium and two long. Indeed, there is remarkably good agreement between the different measurements on the various samples, even though the earlier refinements assumed a disordered tetragonal cell.

III. ELECTRON MICROSCOPY

A. Superstructure

The microstructure of the Tl-based superconductors has been widely studied by electron microscopy and diffraction, and superstructures observed in all members of both the $Tl_2Ba_2Ca_{n-1}Cu_nO_{2n+4}$ and $Tl_1Ba_2Ca_{n-1}Cu_nO_{2n+3}$ families except in $TlBa_2CuO_5$. The superstructure in orthorhombic $Tl_2Ba_2CuO_6$ is well ordered, giving intense well-defined diffraction satellites of first and second order, but all the other superstructures give only very weak diffuse first-order satellites which vary in intensity from region to region in the same crystal and from preparation to preparation. These superstructures are therefore not very well ordered and they do not greatly perturb the basic structure.

In the $n = 2$ and $n = 3$ members of the $Tl_2Ba_2Ca_{n-1}Cu_nO_{2n+4}$ family, diffuse satellites are observed displaced from each Bragg reflection by $\langle \frac{1}{6}, 0, 1 \rangle$ (see Figs. 9a to c and 10). In the exact [001] orientation no satellites are visible, except at higher angles (see Fig. 9a). Tilting away from the exact [001] orientation reveals four distinct satellites, which in some specimens are joined by a circle of diffuse intensity about each Bragg reflection. (See the [1-1-2] diffraction pattern in Fig. 9b.) The diffuse circles indicate that the ordering is essentially short range in nature and that it is imperfectly aligned in the (001) plane. The superlattice modulation vectors are presumably rotated about [001] from one region to another. It is not, however, clear from electron microscopy and diffraction whether the superstructure modulation exists simultaneously on both (100) and (010) planes, or whether the modulation is unidirectional in any given region.

In high-resolution electron microscope (HREM) images, as shown in Fig. 11 for $Tl_2Ba_2Ca_2Cu_3O_{10}$, the superstructure is not visible. This is to be compared with the superstructure in $Bi_2Sr_2CaCu_2O_8$, which greatly perturbs the lattice and is seen clearly in HREM images. The superstructure can, however, be visualized in low-magnification images as seen in Fig. 12. The modulation is seen to be localized on the TlO planes, and the darker regions have a centered arrangement. Figure 11 also demonstrates the presence of intergrowths and stacking faults in a sample that was nominally $Tl_2Ba_2Ca_2Cu_3O_{10}$.

It will be noted that although the periodicity of the superstructures in the $Tl_2Ba_2Ca_{n-1}Cu_nO_{2n+4}$ and $Bi_2Sr_2Ca_{n-1}Cu_nO_{2n+4}$ families are similar (both are close to 2.5 nm), the direction of modulation differs by 45°. Figure 9b, c, and f

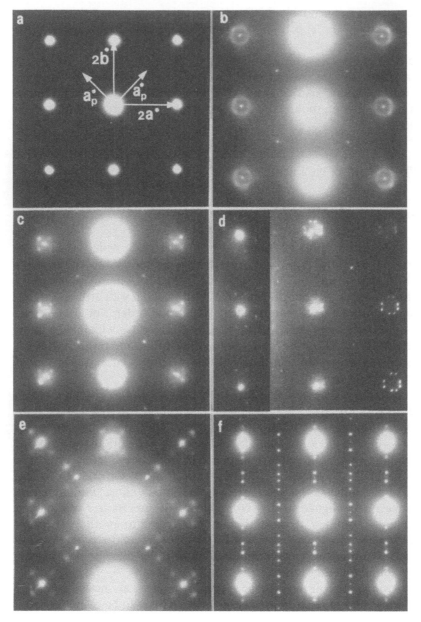

Figure 9   (a) [001] electron diffraction pattern (EDP) of $Tl_2Ba_2CaCu_2O_8$. The superlattice reflections are seen only at high angles. (b) [1-1-2] EDP of $Tl_2Ba_2CaCu_2O_8$ showing groups of four satellite reflections connected by a circle of diffuse intensity. (c) [1-1-2] EDP of $Tl_2Ba_2Ca_2Cu_3O_{10}$. (d) [001] EDP of orthorhombic $Tl_2BaCu_2O_6$. This is a montage showing the second-order diffraction satellites at low angle and the first-order satellites at high angles. The first-order satellites are given by the vector ±0.07, 0.22, 1. (e) EDP of $Tl_1Ba_2Ca_2Cu_3O_9$ tilted away from [001] to show the diffuse superstructure satellites at 0.29, 0, 0.5. (f) [001] EDP of $Bi_{1.6}Pb_{0.4}Sr_2CaCu_2O_8$. Note the unidirectional superstructure.

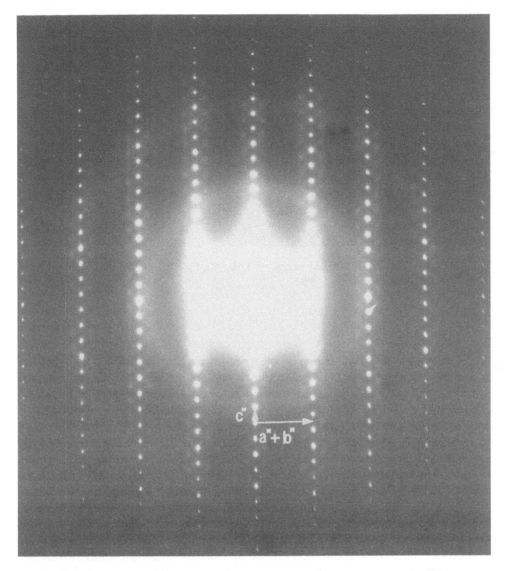

Figure 10 [100] electron diffraction pattern of $Tl_2Ba_2CaCu_2O_8$. Note the weak satellite reflections at $\langle \frac{1}{8}, 0, 1 \rangle$ relative to the main reflections.

demonstrates this point. The 45° difference in direction of modulation may be associated with the 45° difference in the direction of displacement of the oxygen atoms in the TlO layers compared to the BiO layers. While the superstructure in the $Bi_2Sr_2Ca_{n-1}Cu_nO_{2n+4}$ family is produced by oxygen interstitials in the BiO layer [73], it is unlikely that this is the case for the $Tl_2Ba_2Ca_{n-1}Cu_nO_{2n+4}$ family since the displacements of the cations are much smaller. The hypothesis of Tl vacancies (or interstitial Ca^{2+} or an ordering of Tl^{3+} and Tl^+) with associated displacements of the oxygen atoms in the Tl layers appears more plausible, but remains to be verified.

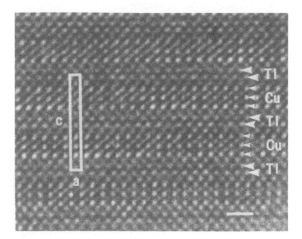

Figure 11 [100] projection HREM image of $Tl_2Ba_2Ca_2Cu_3O_{10}$. Heavy atom columns are white. The scale bar represents 1 nm.

Hewat et al. [86] have shown by electron microscopy and diffraction that the difference between the orthorhombic and tetragonal phases of $Tl_2Ba_2CuO_6$ is due to the presence of a superstructure that is well ordered in the orthorhombic phase and poorly ordered in the tetragonal phase. The superstructure modulation is unidirectional and is generated by the vectors $q = \pm 0.07, 0.22, 1$, which are not along principal directions of the sublattice (see Fig. 13). The local microscopic symmetry of each domain is therefore really monoclinic. Twinning of the basic structure and superstructure results in small domains with eight different orientations and an average symmetry that is orthorhombic. It was shown by neutron diffraction that the chemical compositions and apparent Cu valences of superconducting tetragonal and nonsuperconducting orthorhombic $Tl_2Ba_2CuO_6$ are surprisingly similar, with one in eight Tl and O atoms in the TlO planes missing. The geometry of the superstructure determined by electron microscopy and diffraction, and the number of vacancies determined by neutron powder diffraction, led Hewat et al. to propose that the superstructure in this case consists of a linear ordering of Tl and O vacancies as shown in Fig. 14.

Shimakawa et al. [97] have shown that it is possible to prepare superconducting and nonsuperconducting $Tl_2Ba_2CuO_6$ in both tetragonal and orthorhombic form. They find that annealing below 840 K gives the orthorhombic phase and annealing above 840 K gives the tetragonal phase, while superconductivity in both phases is determined by the cooling rate; fast quenching (or annealing in argon) induces superconductivity, and slow cooling gives a nonsuperconducting material. For both tetragonal and orthorhombic $Tl_2Ba_2CuO_6$, the superconducting material has a slightly larger c parameter, which Shimakawa et al. attribute to a slightly lower oxygen content. They believe that an extremely small oxygen decrease of $\partial = 0.037$ causes an increase of 40 K in the T_c and elongation of the c axis by 0.2%. Rao et al. [98] agree that the highest T_c is obtained for an "optimal" carrier concentration.

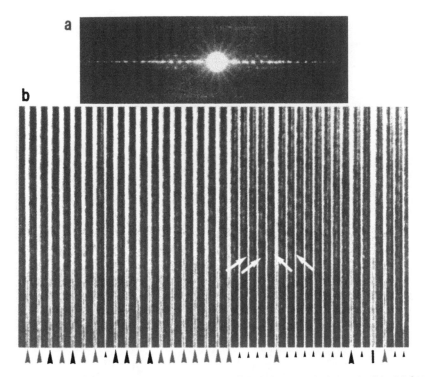

Figure 12   (a) Optical diffraction pattern of the micrograph shown in (b). (b) [100] projection image of $Tl_2Ba_2Ca_2Cu_3O_{10}$ with an intergrowth of $Tl_2Ba_2CaCu_2O_8$ and stacking faults. The superstructure is seen as a face-centered arrangement of weakly contrasted darker regions. The planes of Ca atoms are seen as white lines. Layers with two CuO_2 planes are indicated by small arrows, layers with three CuO_2 planes are indicated by large arrows, and a layer with four CuO_2 planes is indicated with a bar.

The superstructure in the $n = 2$ and $n = 3$ members of the $Tl_1Ba_2Ca_{n-1}Cu_nO_{2n+3}$ family is generated by the vector $q = 0.29, 0, 0.5$ [85] (see Fig. 9e). The superstructure viewed along the a-axis of the perovskite unit cell is centered as for the double-layered-Tl compounds, but the periodicity is roughly half as long. Again the origin of the superstructure is not established. Numerous experiments on partial substitution of Tl by Pb, Ba by Sr, Ca by Y, Cu by Fe, and so on, have been reported, but they are not discussed here (see, e.g., Subramanian et al. [99]). Hervieu et al. [100] report that some of the substituted compounds contain quite complicated superstructures that remain to be explained.

B.   Intergrowths and Stacking Faults

As mentioned earlier, intergrowths and stacking faults are a relatively common feature of Tl-based superconductors. The probability of finding these features increases with n, the number of CuO_2 planes per unit cell. Stacking faults involving varying numbers of CuO_2, and one or two TlO planes are the most common (see

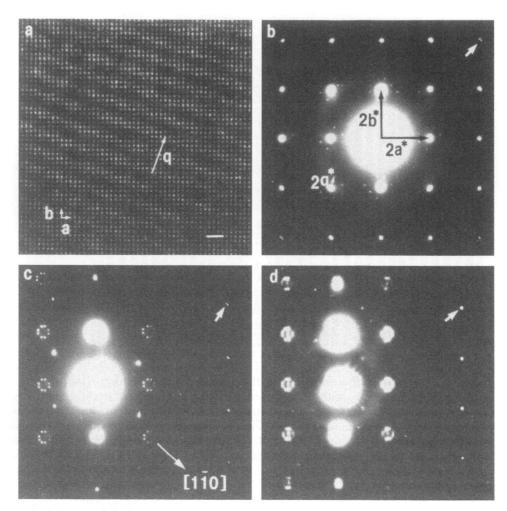

Figure 13 (a) [001] image of orthorhombic $Tl_2Ba_2Cu_1O_6$ showing the unidirectional super-structure in a $[-1,3,0]$ direction. The scale bar represents 1 nm. (b) Electron diffraction pattern corresponding to the crystal in (a). The area used for diffraction (approximately 1 μm^2) is so large that both superstructure variants from both twins are visible. Note the splitting of the fundamental spots (arrows). (c) Electron diffraction pattern of orthorhombic $Tl_2Ba_2Cu_1O_6$ tilted away from the [001] orientation about [100] to satisfy the first-order satellites of 020 at $l = 1$. Note the relative displacement of the pairs of satellites belonging to the two twinned crystals as one moves away from the $[1,-1,0]$ direction. (Some extraneous spots from an adjacent crystal are visible.) (d) Electron diffraction pattern of tetragonal $Tl_2Ba_2Cu_1O_6$ in the same orientation as for (c). Second-order diffraction satellites are not visible and the first-order satellites are very diffuse. (This diffraction pattern is overexposed.) The fundamental spots are not split.

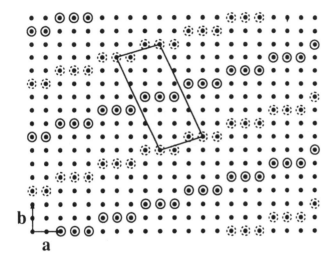

Figure 14 Schematic representation of the proposed model for the superstructure of the orthorhombic phase of $Tl_2Ba_2Cu_2O_6$. The solid points represent thallium atoms on one TlO bilayer, the open circles thallium vacancies on one TlO bilayer, and the dotted circles thallium vacancies on the succeeding bilayer. Compare with Fig. 13a. The oxygen atom vacancies are presumably associated with the Tl vacancies. It should be noted that in this model one in eight thallium atoms are missing, in agreement with the neutron diffraction results.

Fig. 12b), but stacking faults involving other variations in cation stacking have been reported. Isolated layers containing up to six CuO_2 planes have been reported by Iijima et al. [38] and up to seven CuO_2 planes by Raveau et al. [101].

Electron diffraction patterns have been particularly useful for rapid characterization of the structural quality of samples. This is demonstrated in Fig. 15 for $TlBa_2Ca_2Cu_3O_9$. Figure 15a shows the electron diffraction pattern for a well-ordered region of crystal. Figure 15b shows diffuse streaks along [001] passing through the Bragg diffraction spots. This is characteristic of stacking faults on (001) planes. The electron diffraction pattern in Fig. 15c shows the superposition of two diffraction patterns arising from regions with two different c axes, in this case $TlBa_2Ca_2Cu_3O_9$ and $TlBa_2Ca_3Cu_4O_{11}$. This is characteristic of a crystal with intergrowths. The corresponding HREM image is shown in Fig. 16. The case of an ordered long-period repeat of a stacking fault is shown in Fig. 15d. The fine spots indicate that the periodicity is six times the c-spacing of $TlBa_2Ca_2Cu_3O_9$. The corresponding low magnification image is shown in Fig. 17. HREM images (not shown) showed this superstructure to be due to the insertion of two CuO_2 planes instead of three after every five unit cells. This rather surprising type of long-range order has been reported by Verwerft et al. [42] in the $Tl_2Ba_2Ca_{n-1}Cu_nO_{2n+4}$ family, where very complicated series of stacking sequences up to 81 nm in length are observed. They propose that the most probable mechanism for their formation is spiral growth.

C. Conclusions

The thallium oxide superconductors are not fundamentally different from other copper oxide superconductors. The electron-hole content of the conducting CuO

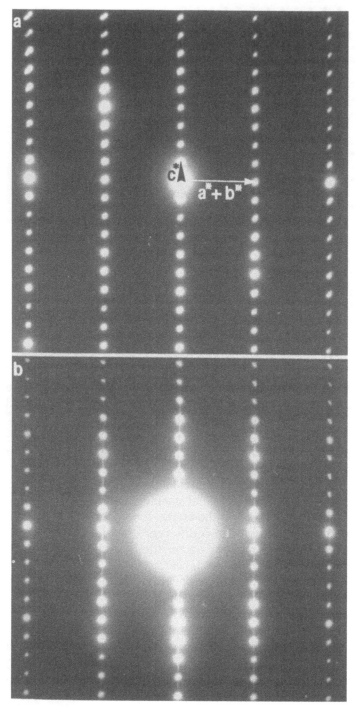

Figure 15 [110] EDP of (a) a perfect region of $TlBa_2Ca_2Cu_3O_9$, (b) a region of Tl-$Ba_2Ca_2Cu_3O_9$ with stacking faults on (001) planes, (c) a region of $TlBa_2Ca_2Cu_3O_9$ with an intergrowth of $TlBa_2Ca_3Cu_4O_{11}$, and (d) a region of $TlBa_2Ca_2Cu_3O_9$ with a long period repeat of six times c along the c axis.

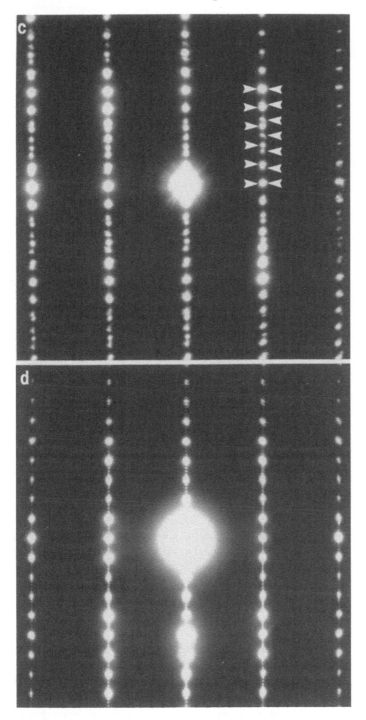

Figure 15 (*Continued*)

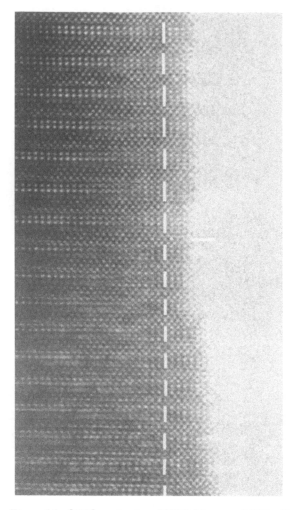

Figure 16   [110] projection HREM image of $TlBa_2Ca_2Cu_3O_9$ corresponding to the EDP in Fig. 15c. In the top half there are four CuO_2 layers, and in the bottom half there are three CuO_2 layers.

layers can be controlled by doping the charge reservoir layers of Cava et al. [35], which can be the Ba/Ca layers, but more interestingly, the TlO or Tl_2O_2 layers. The thallium oxide layers are not naturally commensurable with the copper oxide layers, and this results in superstructures which while not in themselves important for superconductivity, mean that dopants can be more easily accommodated. The relative volatility of thallium oxide also means that the materials are sensitive to the oxidation–reduction conditions of preparation. Apart from the fact that it is relatively easy to control electron–hole doping, there is no obvious reason why the thallium oxide materials should exhibit the highest T_cs. It would perhaps be more accurate to say that they exhibit the greatest *range* of T_cs.

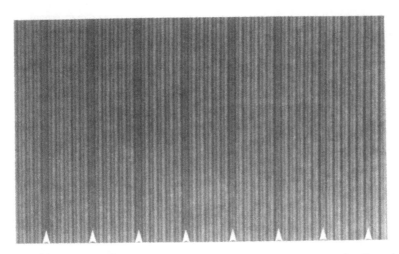

Figure 17   Low-magnification image of $TlBa_2Ca_2Cu_3O_9$ in the [110] projection with a long-period repeat of $6 \times c$ along the c axis. The corresponding EDP is shown in Fig. 15d. After every five unit cells of $TlBa_2Ca_2Cu_3O_9$ with three layers of CuO_2 there is one with two CuO_2 layers.

REFERENCES

1. J. J. Capponi, C. Chaillout, A. W. Hewat, P. Lejay, M. Marezio, N. Nguyen, B. Raveau, J. L. Soubeyroux, J. L. Tholence, and R. Tournier, *Europhy. Lett.* 3, 1301 (1987).
2. J. D. Jorgensen, M. A. Beno, D. G. Hinks, L. Soderholm, K. J. Volin, R. L. Hitterman, J. D. Grace, I. K. Schuller, C. U. Segre, K. Zhang, and M. S. Kleefisch, *Phys. Rev. B* 36, 3608 (1987).
3. B. Aurivillius, *Arki. Kemi* 2, 519 (1949).
4. P. Bordet, J. J. Capponi, C. Chaillout, J. Chenavas, A. W. Hewat, E. A. Hewat, J. L. Hodeau, M. Marezio, and J. L. Tholence, (1988) in *Studies of High Temperature Superconductors*, ed. A. C. Narlikar, Vol. 2, Nova Science, New York, p. 171.
5. C. Michel, M. Hervieu, M. M. Borel, A. Grandin, F. Deslandes, J. Provost, and B. Raveau, *Z. Phys. B* 68, 421 (1987).
6. H. Maeda, Y. Tanaka, M. Fukutomi, and T. Asano, *Jpn. J. Appl. Phys. Pt. 2* 27, L209 (1988).
7. Z. Z. Sheng and A. M. Hermann, *Nature* 332, 55 (1988).
8. L. Gao, Z. L. Huang, R. L. Meng, P. H. Hor, J. Bechtold, Y. Y. Sun, C. W. Chu, Z. Z. Sheng, and A. M. Herman, *Nature* 332, 623 (1988).
9. S. Kondoh, Y. Ando, M. Onoda, M. Sato, and J. Akimitsu, *Solid State Commun.* 65, 1329 (1988).
10. Z. Z. Sheng and A. M. Hermann, *Nature* 332, 138 (1988).
11. Z. Z. Sheng, A. M. Hermann, A. El Ali, C. Almasan, C. Estrada, T. Datta, and R. J. Matson, *Phys. Rev. Lett.* 60, 937 (1988).
12. Z. Z. Sheng, W. Kiehl, J. Bennett, A. A. El, D. Marsh, G. D. Mooney, F. Arammash, J. Smith, D. Viar, and A. M. Hermann, *Appl. Phys. Lett.* 52, 1738 (1988).
13. A. M. Hermann, Z. Z. Sheng, D. C. Vier, S. Schultz, and S. B. Oseroff, *Phys. Rev. B* 37, 9742 (1988).
14. D. Kovatcheva, A. W. Hewat, N. Rangavittal, V. Manivannan, T. N. Guru Row, and C. N. R. Rao, *Physica C* 173, 444 (1991).

15. J. S. Kim, J. S. Swinnea, and H. Steinfink, *J. Less Common Metals* 156, 347 (1989).
16. K. A. Müller and J. G. Bednorz, *Science* 237, 1133 (1987).
17. J. G. Bednorz, K. A. Müller, and M. Takashige, *Science* 236, 73 (1987).
18. S. S. P. Parkin, V. Y. Lee, A. I. Nazzal, R. Savoy, R. Beyers, and S. J. La Placa, *Phys. Rev. Lett.* 61, 750 (1988).
19. M. Sera, S. Kondoh, Y. Ando, K. Fukuda, S. Shamoto, M. Onoda, and M. Sato, *Solid State Commun.* 66, 707 (1988).
20. R. M. Hazen, L. W. Finger, R. J. Angel, C. T. Prewitt, N. L. Ross, C. G. Hadidiacos, P. J. Heaney, D. R. Veblen, Z. Z. Sheng, A. El Ali, and A. M. Hermann, *Phys. Rev. Lett.* 60, 1657 (1988).
21. M. A. Subramanian, J. C. Calabrese, C. C. Torardi, J. Gopalakrishnan, T. R. Askew, R. B. Flippen, K. J. Morrissey, U. Chowdhry, and A. W. Sleight, *Nature* 332, 420 (1988).
22. C. C. Torardi, M. A. Subramanian, J. C. Calabrese, J. Gopalakrishnan, E. M. McCarron, K. J. Morrissey, T. R. Askew, R. B. Flippen, U. Chowdhury, and A. W. Sleight, *Phys. Rev. B* 38, 225 (1988).
23. C. C. Torardi, M. A. Subramanian, J. C. Calabrese, J. Gopalakrishnan, K. J. Morrissey, T. R. Askew, R. B. Flippen, U. Chowdhry, and A. W. Sleight, *Science* 240, 631 (1988).
24. J. K. Liang, J. Q. Huang, G. H. Rao, S. S. Xie, Y. L. Zhang, G. C. Che, and X. R. Cheng, *J. Phys. D* 21, 1031 (1988).
25. M. Kikuchi, N. Kobayashi, H. Iwasaki, D. Shindo, T. Oku, A. Tokiwa, T. Kajitani, K. Hiraga, Y. Syono, and Y. Muto, *Jpn. J. Appl. Phys. Pt. 2* 27, L1050 (1988).
26. M. Onodo, S. Kondoh, K. Fukuda, and M. Sato, *Jpn. J. Appl. Phys. Pt. 2* 27, L1234 (1988).
27. M. W. Shafer, T. Penney, and B. L. Olson, *Phys. Rev. B* 36, 4047 (1987).
28. A. W. Hewat, J. J. Capponi, C. Chaillout, M. Marezio and E. A. Hewat, *Solid State Commun.* 64, 301 (1987).
29. C. Martin, A. Maignan, J. Provost, C. Michel, M. Hervieu, R. Tournier, and B. Raveau, *Physica C* 168, 8 (1990).
30. P. Haldar, K. Chen, Maheswaran, A. Roig-Janicki, N. K. Jaggi, R. S. Markiewicz, and B. C. Giessen, *Science* 241, 1198 (1988).
31. T. Siegrist, S. M. Zahurak, D. W. Murphy, and R. S. Roth, *Nature* 334, 231 (1988).
32. M. A. Subramanian, G. H. Kwei, J. B. Parise, J. A. Goldstone, and R. B. Von Dreele, *Physica C* 166, 19 (1990).
33. M. A. Subramanian, *Mater. Res. Bull.* 25, 191 (1990).
34. T. Manako, Y. Shimakawa, Y. Kubo, T. Satoh, and H. Igarashi, *Physica C* 156, 315 (1988).
35. R. J. Cava, A. W. Hewat, E. A. Hewat, B. Batlogg, M. Marezio, K. M. Rabe, J. J. Krajewski, W. F. Peck, and L. W. Rupp, *Physica C* 165, 419 (1990).
36. J. B. Parise and E. M. McCarron, *J. Solid State Chem.* 83, 188 (1989).
37. S. Iijima, T. Ichihashi, and Y. Kubo, *Jpn. J. Appl. Phys. Pt.* 27, L817 (1988).
38. S. Iijima, T. Ichihashi, Y. Shimakawa, T. Manako, and Y. Kubo, *Jpn. J. Appl. Phys. Pt. 2* 27, L837 (1988).
39. S. Iijima, T. Ichihashi, Y. Shimakawa, T. Manako, and Y. Kubo, *Jpn. J. Appl. Phys. Pt. 2* 27, L1054 (1988).
40. S. Iijima, T. Ichihashi, and Y. Kubo, *Jpn. J. Appl. Phys. Pt. 2* 27, L1168 (1988).
41. H. W. Zandbergen, G. Van Tendeloo, J. Van Landuyt, and S. Amelinckx, *Appl. Phys. A* 46, 233 (1988).
42. M. Verwerft, G. Van Tendeloo, and S. Amelinckx, *Physica C* 156, 607 (1988).
43. Y. T. Huang, S. F. Wu, C. K. Chiang, and W. H. Lee, *Appl. Phys. Lett.* 57, 2354 (1990).

44. H. Ihara, R. Sugise, M. Hirabayashi, N. M. J. Terada, K. Hayashi, A. Negishi, M. Tokumoto, Y. Kimura, and T. Shimomura, *Nature* 334, 510 (1988).
45. H. Ihara, R. Sugise, K. Hayashi, N. M. J. Terada, M. Hirabayashi, A. Negishi, N. Atoda, H. Oyanagi, T. Shimomura, and S. Ohashi, *Phys. Rev. B* 381, 1952 (1988).
46. H. Kusuhara, T. Kotani, H. Takei, and K. Tada, *Jpn. J. Appl. Phys. Pt. 2* L1772 (1989).
47. S. Hashiguchi, T. Sato, T. Endo, and M. Shimada, *J. Mater. Sci.* 25, 4886 (1990).
48. S. S. P. Parkin, V. Y. Lee, E. M. Engler, A. I. Nazzal, T. C. Huang, G. Gorman, R. Savoy, and R. Beyers, *Phys. Rev. Lett.* 60, 2539 (1988).
49. T. Itoh, H. Uchikawa, and H. Sakata, *Jpn. J. Appl. Phys. Pt. 2* 27, L559 (1988).
50. J. M. Liang, R. S. Liu, L. Chang, P. T. Wu, and L. J. Chen, *Appl. Phys. Lett.* 53, 1434 (1988).
51. M. Eibschütz, L. G. Van Uitert, G. S. Grader, E. M. Gyorgy, S. H. Glarum, W. H. Grodkiewicz, T. R. Kyle, A. E. White, K. T. Short, and G. J. Zydzik, *Appl. Phys. Lett.* 53, 911 (1988).
52. T. Usui, N. Sadakata, O. Kohno, and H. Osanai, *Jpn. J. Appl. Phys. Pt. 2* 27, L804 (1988).
53. I. K. Gopalakrishnan, P. V. P. S. Sastry, K. Gangadharan, G. M. Phatak, J. V. Yakhmi, and R. M. Iyer, *Appl. Phys. Lett.* 53, 414 (1988).
54. R. S. Lui, W. H. Lee, P. T. Wu, Y. C. Chen, and C. T. Chang, *Jpn. J. Appl. Phys. Pt. 2* 27, L1206 (1988).
55. A. K. Ganguli, K. S. Nanjunda Swamy, G. N. Subbanna, A. M. Umarji, S. V. Bhat, and C. N. R. Rao, *Solid State Commun.* 67, 39 (1988).
56. K. Takahashi, M. Nakao, D. R. Dietderich, H. Kumakura, and K. Togano, *Jpn. J. Appl. Phys. Pt. 2* 27, L1457 (1988).
57. R. Sugise, M. Hirabayashi, N. Terada, M. Jo, T. Shimomura, and H. Ihara, *Jpn. J. Appl. Phys. Pt. 2* 27, L1709 (1988).
58. M. Itoh, R. Liang, K. Urabe, and T. Nakamura, *Jpn. J. Appl. Phys. Pt. 2* 27, L1672 (1988).
59. A. W. Hewat, P. Bordet, J. J. Capponi, C. Chaillout, J. Chenavas, M. Godinho, E. A. Hewat, J. L. Hodeau, and M. Marezio, *Physica C* 156, 369 (1988).
60. Y. Kubo, Y. Shimakawa, T. Manako, T. Satoh, S. Iijima, T. Ichihashi, and H. Igarashi, *Physica C* 162, 991 (1989).
61. T. Nagashima, K. Watanabe, H. Saito, and Y. Fukai, *Jpn. J. Appl. Phys. Pt. 2* 27, L1077 (1988).
62. Z. Z. Sheng, A. M. Hermann, D. C. Vier, S. Schultz, S. B. Oseroff, D. J. George, and R. M. Hazen, *Phys. Rev. B* 38, 7074 (1988).
63. K. F. McCarty, D. S. Ginley, D. R. Boehme, R. J. Baughman, E. L. Venturini, and B. Morosin, *Physica C* 156, 119 (1988).
64. A. W. Sleight, J. Gopalakrishnan, C. C. Torardi, and M. A. Subramanian, *Phase Transitions* 19, 149 (1989).
65. R. J. Cava, B. Batlogg, J. J. Krajewski, L. W. Rupp, L. F. Schneemeyer, T. Siegrist, R. B. vanDover, P. Marsh, W. F. Peck, Jr., P. K. Gallagher, S. H. Glarum, J. H. Marshall, R. C. Farrow, J. V. Waszczak, R. Hull, and P. Trevor, *Nature* 336, 211 (1988).
66. M. Marezio, A. Santoro, J. J. Capponi, E. A. Hewat, R. J. Cava, and F. Beech, *Physica C* 169, 401 (1990).
67. C. Chaillout, P. Bordet, J. J. Capponi, R. J. Cava, J. Chenavas, E. A. Hewat, M. Marezio, A. Santoro, and B. Souletie, *J. Less Common Met.* 164, 816 (1990).
68. R. Beyers, S. S. P. Parkin, V. Y. Lee, A. I. Nazzal, R. Savoy, G. Gorman, T. C. Huang, and S. LaPlaca, *Appl. Phys. Lett.* 53, 432 (1988).

69. S. S. P. Parkin, V. Y. Lee, A. I. Nazzal, R. Savoy, T. C. Huang, G. Gorman, and R. Beyers, *Phys. Rev. B* 38, 6531 (1988).

70. E. A. Hewat, M. Dupuy, P. Bordet, J. J. Capponi, C. Chaillout, J. L. Hodeau, and M. Marezio, *Nature* 333, 53 (1988).

71. P. Bordet, J. J. Capponi, C. Chaillout, J. Chenavas, A. W. Hewat, E. A. Hewat, J. L. Hodeau, M. Marezio, J. L. Tholence, and D. Tranqui, *Physica C* 153, 623 (1988).

72. P. Bordet, J. J. Capponi, C. Chaillout, J. Chenavas, A. W. Hewat, E. A. Hewat, J. L. Hodeau, M. Marezio, J. L. Tholence, and D. Tranqui, *Physica C* 156, 189 (1988).

73. Y. Le Page, W. R. McKinnon, J. M. Tarascon, and P. Barboux, *Phys. Rev. B* 40 6810 (1989).

74. S. Kawashima, O. Inoue, and S. Adachi, *Jpn. J. Appl. Phys. Pt. 2 Lett.* 29, L900 (1990).

75. W. H. Zachariasen, *J. Less Common Met.* 62, 1 (1978).

76. I. D. Brown and D. Altermatt, *B* 41, 244 (1985).

77. I. D. Brown, *J. Solid State Chem.* 82, 122 (1989).

78. R. J. Cava, B. Batlogg, C. H. Chen, E. A. Rietman, S. M. Zahurak, and D. Werder, *Nature* 329, 423 (1987).

79. C. Chaillout, M. A. Alario-Franco, J. J. Capponi, J. Chenavas, P. Strobel, and M. Marezio, *Solid State Commun.* 65, 283 (1988).

80. S. Iijima, T. Ichihashi, Y. Shimakawa, T. Manako, and Y. Kubo, *Jpn. J. Appl. Phys. Pt. 2* 27, L1061 (1988).

81. R. B. Beyers, S. S. P. Parkin, V. Y. Lee, A. I. Nazzal, R. J. Savoy, G. L. Gorman, T. C. Huang, and S. J. La Placa, *IBM J.* 33, 228 (1989).

82. Y. Tokura, T. Arima, H. Takagi, S. Uchida, T. Ishigaki, H. Asano, R. Beyers, A. I. Nazzal, P. Lacorre, and J. B. Torrance, *Nature* 342, 890 (1989).

83. B. Domengès, M. Hervieu, and B. Raveau, *Solid State Commun.* 68, 303 (1988).

84. B. Domengès, M. Hervieu, and B. Raveau, *Solid State Commun.* 69, 1085 (1989).

85. A. W. Hewat, E. A. Hewat, J. Brynestad, H. A. Mook, and E. D. Specht, *Physica C* 152, 438 (1988).

86. E. A. Hewat, P. Bordet, J. J. Capponi, C. Chaillout, J. Chenavas, M. Godinho, A. W. Hewat, J. L. Hodeau, and M. Marezio, *Physica C* 156, 375 (1988).

87. K. Hiraga, D. Shindo, M. Hirabayashi, M. Kikuchi, N. Kobayashi, and Y. Syono, *Jpn. J. Appl. Phys.* 27, 1848 (1988).

88. H. W. Zandbergen, W. A. Groen, F. C. Mijlhoff, G. Van Tendeloo, and S. Amelinckx, *Physica C* 156, 325 (1988).

89. J. D. Fitzgerald, R. L. Withers, J. G. Thompson, L. R. Wallenberg, J. S. Anderson, and B. G. Hyde, *Phys. Rev. Lett.* 60, 2797 (1988).

90. D. E. Cox, C. C. Torardi, M. A. Subramanian, J. Gopalakrishnan, and A. W. Sleight, *Phys. Rev. B* 38, 6624 (1988).

91. M. Eibschutz, L. G. Vanuitert, W. H. Grodkiewicz, and D. E. Cox, *Physica C* 162, 530 (1989).

92. W. Dmowski, B. H. Toby, T. Egami, M. A. Subramanian, J. Gopalakrishnan, and A. W. Sleight, *Phys. Rev. Lett.* 61, 2608 (1988).

93. B. H. Toby, W. Dmowski, T. Egami, J. D. Jorgensen, M. A. Subramanian, J. Gopalakrishnan, A. W. Sleight, and J. B. Parise, *Physica C* 162, 101 (1989).

94. B. H. Toby, T. Egami, J. D. Jorgensen, and M. A. Subramanian, *Phys. Rev. Lett.* 64, 2414 (1990).

95. A. K. Cheetham, A. M. Chippingdale, S. J. Hibble, and C. J. Woodley, *Phase Transitions* 19, 223 (1989).

96. S. J. Hibble, A. K. Cheetham, A. M. Chippindale, P. D. and J. A. Hriljac, *Physica C* 156, 604 (1988).

97. Y. Shimakawa, Y. Kubo, T. Manako, and H. Igarashi, *Phys. Rev. B* 40, 11400 (1989).

98. P. Somasundaram, R. Vijayaraghavan, R. Nagarajan, R. Seshadri, A. M. Umarji, and C. N. R. Rao, *Appl. Phys. Lett.* 56, 487 (1990).
99. M. A. Subramanian, C. C. Torardi, J. Gopalakrishnan, and P. L. Gai, *Science* 242, 249 (1988).
100. M. Hervieu, V. Caignaert, R. Retoux, and B. Raveau, *Mater. Sci. Eng. B Solid State Mater. Adv. Technol.* 6, 211 (1990).
101. B. Raveau, C. Martin, M. Hervieu, D. Bourgault, C. Michel, and J. Provost, *Solid State Ionics* 39, 49 (1990).

6
Synthesis of Tl-Based High-T_c Superconductive Oxides

Eli Ruckenstein
State University of New York at Buffalo
Buffalo, New York

Nae-Lih Wu
National Taiwan University
Taipei, Taiwan

I. INTRODUCTION

The ability to synthesize powders of high phase purity is essential to many basic studies and practical applications of high-T_c superconductors. From the J_c (critical current density) point of view, the presence of impurity phases among superconductive grains impedes the transport of the electrical current, resulting in a low J_c. Even in the cases in which the presence of certain finely dispersed impurity phases has been thought to increase the flux pinning, the preparation processes in general employ single-phase superconductive powders as starting material and only subsequently, by following a well-defined heating procedure, does one generate the desired secondary phases in a controlled manner.

Like the other high-T_c superconductive oxides, the Tl-based superconductive oxides have most commonly been prepared by the solid-state reaction method. Shortly after identification of $Tl_2Ca_nBa_2Cu_{n+1}O_y$ superconductive oxides [1–5], many reports [6–15] in the literature called for the use of off-stoichiometric starting compositions to synthesize one particular phase of these oxides. For example, starting metal-cation ratios of Tl/Ca/Ba/Cu = 2:2:2:3 were found to give mainly $Tl_2CaBa_2Cu_2O_y$ (the 2122 phase), while a large percentage of $Tl_2Ca_2Ba_2Cu_3O_y$ (the 2223 phase) was obtained with an overall stoichiometry of Tl/Ca/Ba/Cu = 1:3:1:3. Simple mass-balance calculations immediately indicate that impurity phases, such as $BaCuO_2$, Ca_2CuO_3, and CuO, may actually be present in the powder in larger amounts on a molar basis than the intended superconductive phase. Powders prepared with off-stoichiometric starting compositions typically contain large superconductive crystals surrounded by nonsuperconductive impurity phases (Fig. 1).

Subsequently, methods have been developed to avoid the use of off-stoichiometric starting compositions, and many successful results have been reported [16–

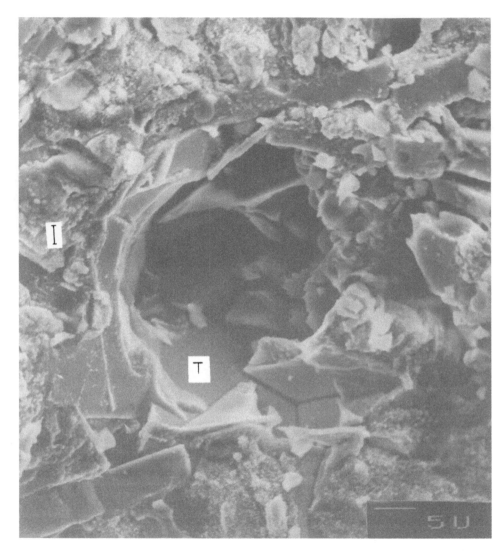

Figure 1   Tl–Ca–Ba–Cu–O sample prepared by using off-stoichiometric starting compositions which contain a large amount of impurity phases. T, 2122 crystals; I, impurity phases.

21]. Basically, these methods synthesize the desired superconductive phase either in a single step or through a series of calcination steps with well-defined intermediate compounds formed after each calcination. Until now, with proper selection of calcination conditions, Tl-based superconductive powders of high phase purity could be synthesized in short calcination times without resorting to complicated wet processes. However, some wet processes have been also attempted [22–25].

In this chapter, we discuss the major causes of the difficulties in phase control encountered during the synthesis of Tl–Ca–Ba–Cu superconductive compounds, review some of the synthesis methods, along with the reaction chemistries involved, that have been employed in the earlier studies, and examine the effects of post-treatments of the prepared materials on their superconducting properties.

II. DIFFICULTIES IN PHASE CONTROL

The need for using off-stoichiometric starting compositions emphasized in earlier reports can be attributed primarily to two causes: incongruent melting and Tl loss by vaporization during high-temperature calcination. Incongruent melting produces a solid phase with a stoichiometry different from that of the melt and hence from the overall composition. Consequently, the incongruent melting point (IMP) should be the upper temperature limit for synthesizing single-phase Tl-based superconductive oxides from stoichiometric starting compositions. Incongruent melting takes place over a wide range of Tl–Ca–Ba–Cu compositions, and as a result, slight differences in cation ratios could have a profound effect on the distribution of the final products [6–12]. To avoid incongruent melting, one should identify the IMP for the composition of the intended phase. Once the IMP is determined, the synthesis should be conducted at a temperature as far below the IMP as possible.

In studying the solidification of double-Tl-layer superconductive oxides, Kotani et al. [26] determined a phase diagram (Fig. 2) for cases in which the atomic Tl/Ba ratio was fixed at 1.0, as in $Tl_2Ca_nBa_2Cu_{n+1}O_y$, but the total content of Ca + Cu was varied. The IMP was found to decrease with increasing (Ca + Cu) content. For the stoichiometry Tl/Ca/Ba/Cu = 2:1:2:2, the IMP is about 905°C. Calcination of the powder with this stoichiometry above 905°C (e.g., point A in Fig. 2) produces the 2021 phase as the dominant superconductive phase. For the 2:2:2:3 stoichiometry, however, there are two melting points, including a congruent one at 905°C and an incongruent one at about 895°C. Between these two temperatures (e.g., point B in Fig. 2), the 2122 phase will be the predominant solid phase for the starting stoichiometry 2:2:2:3. On the other hand, as shown by point C in Fig. 2, the 2223 compound becomes dominant for 1:3:1:3 stoichiometry between 885 and 895°C. This explains the relation between the starting compositions and the resulting phases obtained in the earlier studies.

Melting involving only some oxides of the mixture can, however, also occur at temperatures lower than the IMP values presented in Fig. 2. Tl_2O_3 melts at about 715°C [27]. Cheung and Ruckenstein [28] reported melting for a Tl/Ba/Cu/O = 2:2:3:x composition at 865°C and observed a glassy phase, an indication of liquid formation at elevated temperatures, with this stoichiometry during synthesis of the 2223 oxide. Complex glassy phases with a wide range of Tl–Ca–Cu–O and Tl–Ba–Cu–O compositions have also been reported [29] at temperatures as low as 850°C in processes where the higher member phases of $Tl_2Ca_nBa_2Cu_{n+1}O_y$

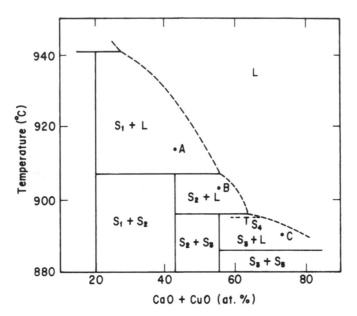

Figure 2   Phase diagram for $Tl_2Ca_nBa_2Cu_{n+1}O_x$. The symbols S_1, S_2, S_3, S_4, S_5, and L represent the 2021, 2122, 2223, and 2324 phases, the Ca–Cu–O mixture, and the liquid phase, respectively. A, B, and C denote the nominal compositions of Tl/Ca/Ba/Cu = 2:1:2:2, 2:2:2:3, and 1:3:1:3, respectively. (From Ref. 26.)

were synthesized from the lower member phases. Although these types of sub-system melting processes may not alter the thermodynamic stabilities of the $Tl_2Ca_nBa_2Cu_{n+1}O_y$ phases, such as those shown in Fig. 2, they do cause material segregation, which delays the formation of the thermodynamically stable phase.

Loss of T1 during high-temperature treatment is due primarily to the formation of Tl_2O vapor. Below the melting point of Tl_2O_3 (ca. 715°C) in air, Tl_2O vapor is formed directly via the decomposition of solid Tl_2O_3:

$$Tl_2O_3(s) = Tl_2O(g) + O_2(g) \tag{1}$$

The vapor pressure of Tl_2O is given by [30]

$$\log P = -19.63(10^3/T) + 18.60$$

where P is the vapor pressure of Tl_2O in atm and T is the temperature in K. Above the melting point, part of Tl_2O_3 decomposes into $Tl_2O(l)$ to form a Tl_2O_3–Tl_2O liquid mixture [31], whose Tl_2O content increases with temperature. Tl_2O vaporizes at these temperatures via

$$Tl_2O_3\text{–}Tl_2O(l) = Tl_2O(g) + xO_2(g) \tag{2}$$

Vapor pressure data for this reaction are not available. However, it is known that the liquid boils at about 895°C [27], which is close to the temperature range that was adopted in many earlier studies for calcination. Depletion of Tl changes the overall stoichiometry and hence the compositions of the resulted phases. Vapor-ization of Tl from the already formed Tl-based superconductive phase takes place at a slower rate than from the free Tl_2O_3, but still should not be overlooked. By heating their samples at 885°C in flowing oxygen, Sugise et al. [32,33] observed phase transformations among Tl-based superconductors in the sequence 2122 → 2223 → 1324 → 1425 → Ca–Ba–Cu–O oxide mixture caused by Tl depletion and reported that Tl was lost completely in less than 2 h. Measures to prevent Tl loss from the already formed superconductive phase are particularly necessary in a large-scale process, where a prolonged calcination is needed to achieve compositional homogeneity.

To reduce the loss of Tl, the synthesis of the Tl-based superconductive oxides have commonly been carried out in a closed vessel. The sample is either wrapped with a metal, mostly gold, foil, or placed inside a sealed quartz tube. In some cases a powder rich in Tl is placed next to the sample to serve as a source for supplying Tl$_2$O vapor to fill up the sealed vessel. An increase in Tl$_2$O partial pressure pushes reactions (1) and (2) to the left and hence helps to reduce Tl loss from the desired sample.

A high-oxygen pressure as an alternative way to reduce Tl loss should be used with caution, since the stability of each superconductive phase could change with oxygen pressure. In general, a higher member phase of Tl$_2$Ca$_n$Ba$_2$Cu$_{n+1}$O$_y$ is less stable than a lower member with increasing oxygen pressure [12,34]. For example, it has been found [34] that when a 2223-dominated powder is annealed below the IMP at an oxygen pressure above 1 atm, the amount of the 2223 compound decreases while that of the 2122 compound increases with time. For a fixed annealing time (2 h) at 870 and 880°C, the relative amount of the 2223 compound with respect to the sum of the 2122 and 2223 compounds drops initially rapidly with increasing oxygen pressure and then reaches a plateau with a molar amount of the 2122 compound twice as large as that of the 2223 compound above 25 atm (Fig. 3A). The T_c of the powder changes accordingly (Fig. 3B).

III. SYNTHESIS METHODS AND REACTION MECHANISMS

A. Solid-State Reaction Method

1. One-Step Method

The most straightforward synthesis method is to calcine the starting powder mixture containing the elemental oxides in one step to form the Tl-based superconductive oxides. Alkaline earth metal dioxides, such at BaO$_2$ and CaO$_2$, are often employed to replace the corresponding monoxides as the starting ingredients, because the dioxides are less hygroscopic and easier to handle. Torardi et al. [16] reported the formation of essentially single-phase 2223 powder, following this approach by calcining the reactant mixture at 890°C for 1 h. However, they also noted that calcinations at the same temperature for longer times or at temperatures higher than 900°C for shorter times resulted mainly in the 2122 compound. Presumably, either extensive incongruent melting or significant Tl loss has occurred.

More extensive experiments regarding the one-step method described above have been carried out by Ruckenstein and Cheung [20], who identified the sequential transformations that occur in the formation of various Tl–Ca–Ba–Cu phases and studied the kinetics of the processes. Table 1 lists the amounts of various compounds that appear during calcination for the starting compositions of Tl/Ca/Ba/Cu = 2:2:2:3, 1:1:2:2, 1:3:2:4, and 1:2:2:3 [20]. The calcination was conducted

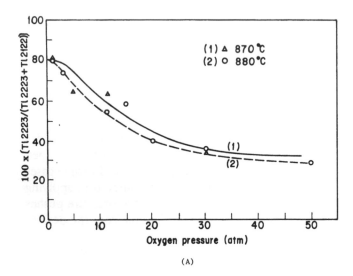

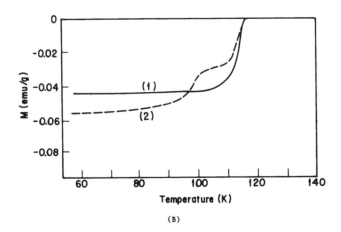

Figure 3   Effect of the oxygen pressure on the stability of the 2223 phase. (A) Molar ratio of 2223 with respect to the sum of 2122 and 2223 as a function of oxygen pressure and temperature. (B) Field-cooled magnetization curves for an initial 2223 sample (curve 1) and for the same sample after being annealed at 880°C and 12 atm O_2 for 2 h (curve 2). (From Ref. 34.)

at 865°C, and the elemental oxides Tl_2O_3, BaO_2, CaO, and CuO were used as the starting ingredients. After each calcination, the sample was quenched and analyzed by x-ray diffraction (XRD) for determining its composition. The amount of each compound detected is indicated in the tables by the normalized intensity of one of its XRD reflections. Despite different starting compositions, the data presented in these tables exhibit many common trends, which indicate that the reaction pathways for the formation of various Tl–Ca–Ba–Cu compounds are similar.

Table 1   Relative Amounts of Various Phases Determined by Powder X-Ray Diffraction Analysis for Tl–Ca–Ba–Cu–O Samples Prepared via the One-Step Method

(a) Starting composition: Tl/Ca/Ba/Cu = 2:2:2:3

Reaction time (min)	Compound							
	2021	2122	2233	$BaCuO_2$	Tl_2O_3	CaO	BaO_2	CuO
0	—	—	—	—	1543	122	406	176
2.5	26	—	—	—	96	46	15	35
5	36	17	—	—	36	28	—	32
10	—	79	34	—	—	23	—	26
20	—	50	103	10	—	13	—	17
30	—	50	86	14	—	25	—	25
60	—	42	104	13	—	21	—	21
120	—	18	115	9	—	12	—	12

(b) Starting composition: Tl/Ca/Ba/Cu = 1:1:2:2

Reaction time (min)	Compound								
	1020	2021	2122	1122	$BaCuO_2$	Tl_2O_3	CaO	BaO_2	CuO
0	—	—	—	—	—	716	70	448	119
2.5	47	—	—	—	68	—	63	12	39
5	—	33	—	—	93	—	56	—	37
10	—	23	43	15	98	—	35	—	20
20	—	12	35	12	105	—	23	—	35
30	—	—	35	18	88	—	20	—	31
60	—	—	24	29	88	—	27	—	49
120	—	—	—	46	100	—	32	—	76

(c) Starting composition: Tl/Ca/Ba/Cu = 1:3:2:4

Reaction time (min)	Compound								
	2021	2122	2223	1324	$BaCuO_2$	Tl_2O_3	CaO	BaO_2	CuO
0	—	—	—	—	—	487	106	248	124
2.5	38	—	—	—	21	—	103	15	68
5	—	44	56	—	38	—	79	—	62
10	—	17	97	—	53	—	77	—	60
20	—	—	116	—	38	—	65	—	49
30	—	—	116	—	44	—	63	—	63
60	—	—	46	24	36	—	44	—	41
120	—	—	17	103	46	—	37	—	40

(d) Starting composition: Tl/Ca/Ba/Cu = 1:2:2:3

Reaction time (min)	Compound									
	1021	2021	2122	2223	1223	$BaCuO_2$	Tl_2O_3	CaO	BaO_2	CuO
0	—	—	—	—	—	—	582	130	451	150
2.5	26	18	—	—	—	8	54	97	21	50
5	9	24	—	—	—	13	—	63	—	51
10	—	—	23	56	—	39	—	33	—	23
20	—	—	—	114	—	43	—	33	—	16
30	—	—	—	93	—	36	—	36	—	17
60	—	—	—	31	48	35	—	43	—	22
120	—	—	—	—	57	38	—	48	—	12

Source: Ref. 20.

First, the XRD intensities of Tl_2O_3, BaO_2, and CuO drop rapidly after the very first 2.5 min of calcination. During this time, only the 2021 and/or 1021 phases, which do not contain Ca, were formed. Second, the amount of CaO, which is consumed at a slower rate than other reactant species, decreases slowly as the dominant superconductive phase changes from the lower to the higher member phase of $Tl_mCa_nBa_2Cu_{n+1}O_y$. Third, the final products with the same cation ratios as those of the starting compositions are not formed in the one-step procedure from the starting elemental oxides. Rather, they are formed through sequential phase changes. The 2021, 2122, and 2223 constitute the major intermediate compounds, irrespective of the starting composition. In particular, as shown in Table 1c and d, the higher member phases of $TlCa_nBa_2Cu_{n+1}O_y$ are not formed at the expense of the lower member phase of the same group but at the expense of $Tl_2Ca_nBa_2Cu_{n+1}O_y$. Finally, there is a slight increase in the amount of $BaCuO_2$ during the later stage of calcination, suggesting that the decomposition of existing superconductive phases caused by T1 loss already takes place before the reaction is complete. It is also clear that since the formation of the final phases involves many intermediate compounds, it is possible to obtain a dominant superconductive phase whose stoichiometry bears no direct relation to the starting composition if the process is not completed.

Figure 4A and B show the scanning electron microscope (SEM) micrographs of one partially reacted sample having a Tl/Ca/Ba/Cu = 2:2:2:3 starting composition [28]. Energy dispersive x-ray analysis (EDX) identified a number of crystalline phases, such as unreacted CaO (labeled C), CuO (U), and a superconductive domain (T) containing primarily the 2122 phase. Large voids, filled by the mounting material (M) during the SEM sample preparation, are observed surrounding large CuO particles. In addition, a fairly large amount of glassy phase (labeled G in Fig. 4B) (an indication of liquid formation at the calcination temperature) is found. The glassy phase, as detected by EDX, contains variable proportions of Tl,Ca, Ba, and Cu. Tl_2O_3 melts at 715°C [27], while BaO_2 melts and decomposes at 450°C [35]. The glassy phases containing different Tl–Ca–Ba–Cu compositions as identified by EDX could be formed via interactions of the Tl–Ba–O liquid with CuO and CaO particles to extract Cu and Ca into the liquid. In an experiment where Tl_2O_3, BaO_2, and CuO were mixed in the ratio Tl/Ba/Cu = 2:2:n and calcined at 865°C, it was found [28] that the extent of powder melting increased as n increased. For $n = 1$, the powder melting is negligible and the 2021 phase is obtained. In contrast, extensive powder melting is observed for $n = 3$. It is possible that during the early stages of calcination of the samples involved in Table 1, Tl–Ba–O liquid is quickly formed and reacts with CuO particles to generate the 2021 phase near the surface of the CuO particles. The 2021 phase, however, occurs only momentarily and subsequently melts as more Cu diffuses into that region. The melting of Tl_2O_3 and BaO_2, the rapid consumption of CuO, and the formation of only the 2021 phase at the beginning of the reaction are consistent with the x-ray diffraction (XRD) data shown in Table 1. The assumption that the already formed 2021 phase is consumed via melting rather than by the solid-state reaction with CaO and CuO to generate the 2122 phase is supported by the following observations [29]. Cheung and Ruckenstein [29] investigated the interfacial interactions between $Tl_2Ca_nBa_2Cu_{n+1}O_x$ and various Ca–Cu–O oxide mixtures containing Ca_2CuO_3, CaO, and/or CuO. Figure 5A shows an example of the interfacial structure observed

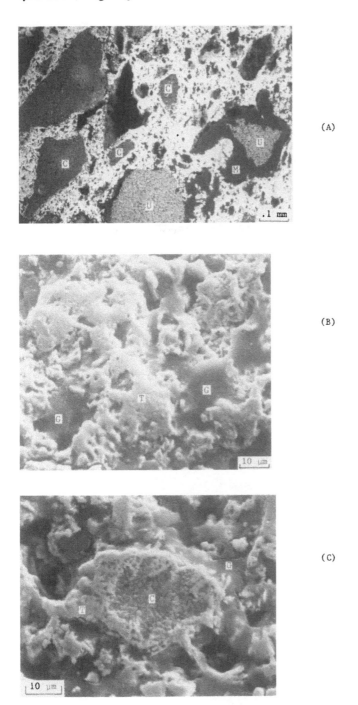

Figure 4 Micrographs of a polished cross section of a partially reacted sample with a starting composition of Tl/Ca/Ba/Cu = 2:2:2:3 calcined at 865°C. C, CaO; U, CuO; T, 2122 dominated superconductive mixture; G, glassy phase; M, polyester mounting material. (From Refs. 20 and 28.)

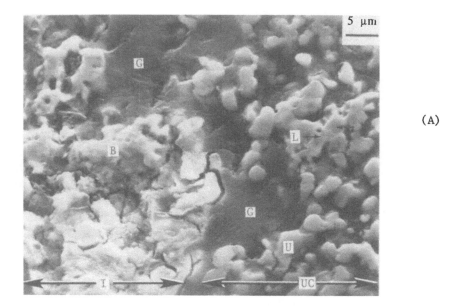

(A)

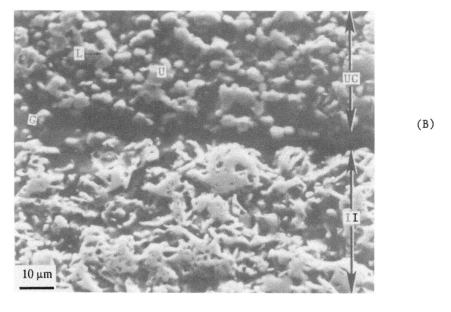

(B)

Figure 5 Micrographs of (A) 2021/Ca$_2$CuO$_3$ + CuO and (B) 2122/Ca$_2$CuO$_3$ + CuO interfaces after the composite pellets were calcined at 850°C. I, region initially occupied by the 2021 phase; II, region initially occupied by the 2122 phase; UC, region initially occupied by Ca$_2$CuO$_3$ + CuO mixture; U, Ca$_2$CuO$_3$; B, Ba–Cu-rich phase; L, Tl–Ca–Cu phase; G, glassy phase. (From Ref. 29.)

when two pellets that contain, respectively, the 2021 phase and Ca–Cu–O oxides were bonded by mechanical compression and then calcined at 850°C. It was found that prior to the occurrence of the 2122 phase, a thick layer of a glassy phase was first formed at the location of the original 2021/Ca–Cu–O interface. Melting of the 2021 phase resulted in a Ba–Cu-rich phase (labeled G in Fig. 5A) at the boundary of the glassy-phase layer near the 2021 side, while some Tl apparently diffused across the glassy phase to form a Tl–Ca–Cu phase (L) at the boundary near the Ca–Cu–O side. Furthermore, the amount of the interfacial glassy phase, and hence the extent of melting, was found to be larger when Cu was present in the starting powder in an uncombined form, namely CuO, than in a combined form, Ca_2CuO_3. When only Ca_2CuO_3 was available, extraction of Cu from the Ca_2CuO_3 phase was found to result in the formation of a Ca–O layer at the boundary of the glassy phase. These results suggested the requirement of the presence of excess Cu to induce the melting of the 2021 phase at the interface.

The phase (labeled T in Fig. 4A) that contains primarily the 2122 oxide is located more closely to the CaO particles than to the CuO ones. Layers of 2122 have also been detected in many cases on the peripheries of large CaO agglomerates (Fig. 4C) [20]. The 2122 surface layers were, in turn, found to be surrounded by the glassy phase. These observations, along with those of the interfacial studies described above, lead us to believe that the 2122 phase is formed via the reaction between the melt and CaO particles. It is also clear that the 2122 phase cannot be generated until the Cu content in the melt near the CaO particles reaches a certain level; this explains why the 2122 phase is not formed at the very beginning of the reaction, as shown in Table 1. The formation of the 2122 solid phase near CaO particles can induce a net transport of the melt away from CuO particles, leaving voids on the surfaces of the large CuO particles as observed in Fig. 4A.

Similar to what happens at the interface region between 2021 and Ca–Cu–O mixtures of various Ca/Cu ratios, it is shown [29] in Fig. 5B that when a 2122 powder is heated in contact with such a Ca–Cu–O mixture, a glassy phase is formed at the interface prior to the formation of the 2223 phase. These results indicate that the phase changes from 2122 to 2223 shown in Table 1 are likely to follow a mechanism that involves the melting of the formed 2122 phase and the reaction of the melt with CaO to generate the 2223 phase, similar to that leading to the phase change from 2021 to 2122, as discussed earlier.

2. Multistep Method

The multistep method is a process in which the synthesis of the desired superconductive phase is achieved through a series of calcination steps conducted at different temperatures. The required starting ingredients are either mixed all together at the very beginning of the first calcination step or introduced separately between consecutive calcination steps. Compared with the one-step method, the multistep method has numerous advantages, such as reduction in the extent of powder melting, less Tl loss, and better compositional homogeneity, to name a few.

For the synthesis of Tl-based superconductive oxides, the multistep approach was first developed in the processes in which alkaline-earth metal carbonates, rather than oxides, were employed as the starting ingredients. The carbonates are frequently employed in the synthesis of high-T_c superconductors because of their superior chemical stabilities, which allow easier handling and storage, to those of oxides. Decomposition of these carbonates, however, are slow even above 900°C [36]. To reduce the Tl loss caused by excessive heat treatment when carbonates are used, the complete solid reaction has often been divided into two stages: a precalcination of a mixture containing CuO and alkaline earth metal carbonates above 900°C in the first stage and a subsequent reaction of the resulted oxide mixture with Tl_2O_3 at a lower temperature in the second stage.

Use of a mixture containing $BaCO_3$ and excess CuO in precalcination, as employed in many earlier studies [1–5,12], resulted mainly in a mixture of $BaCuO_2$ and CuO [37]. Table 2 presents the XRD results obtained during the syntheses of various $Tl_mCa_nBa_2Cu_{n+1}O_y$ compounds when $BaCuO_2$ is used as the Ba source in the second-stage calcination, which is conducted at 865°C. The experimental method employed in these tables is similar to that employed for Table 1 except that different approaches have been adopted for normalizing the XRD intensities. Comparisons between Tables 1 and 2 corresponding to the same starting Tl/Ca/Ba/Cu composition show that the use of $BaCuO_2$ as the Ba source results in the same types of phase changes: namely, from the lower to the higher member phases of $Tl_2Ca_nBa_2Cu_{n+1}O_y$ during calcination, as when BaO_2 was employed. The fact that the 1324 phase is formed from the 2223 phase but not from the lower member phase of $TlCa_nBa_2Cu_{n+1}O_y$, as observed during the one-step method (Table 1c), remains true when $BaCuO_2$ is employed (Table 2c). However, the use of $BaCuO_2$ does accelerate the formation of the higher member phases of $Tl_2Ca_nBa_2Cu_{n+1}O_y$. For example, as shown in Table 1a to c, when BaO_2 is employed, the 1021 or 2021 phase constitutes the dominant product after 5 min of calcination. In contrast, in the case of $BaCuO_2$ (Table 2a to c), the 2122 compound is formed rapidly in less than 5 min. The phase change from either 1021 or 2021 to 2122 that occurs in Tables 1a to c is mostly skipped in Table 2. Furthermore, for 2–2–2–3 starting composition, the 2223 phase is not formed in the one-step process until after 10 min of calcination (Table 1a), while it becomes the dominant phase after only 5 min of calcination in the second stage of the two-step process (Table 2a). A compound (denoted as $Tl_xCa_yCu_zO_a$ in Table 2) that does not contain Ba has been detected when $BaCuO_2$ is employed, presumably due to insufficient free BaO to react with Tl_2O_3 in the early stages of reaction. Nevertheless, this Tl–Ca–Cu–O compound is formed only in a small amount and is not expected to have a significant effect on the reaction.

Combination [17–19,39,40] of $CaCO_3$, $BaCO_3$, and CuO as the starting ingredients in precalcination leads to $BaCuO_2$, Ca_2CuO_3 [41], and various Ca–Ba–Cu–O compounds [17,42]. Wu and Ruckenstein [17] compared the reaction rates and the superconductive properties of resulted powders for the syntheses of the 2122 phase when mixtures with and without Ca were used during precalcination. In one case, a mixture of $CaCO_3$, $BaCO_3$, and CuO in the ratios 1:2:1 was precalcined between 900 and 930°C. The precalcined powder was then reacted with Tl_2O_3 and additional CuO to produce the 2122 phase. In another experiment, $BaCuO_2$ was first prepared from $BaCO_3$ and CuO in precalcination and then reacted with additional Tl_2O_3, CaO, and CuO. As shown in Fig. 6, using the $CaBa_2CuO_4$ step, the single XRD-phase 2122 powder can be obtained between 800 and 850°C in less than 20 min. In contrast, reactions under the same conditions when the $BaCuO_2$ is employed are mostly incomplete. As a result, the 2122 powder prepared

Table 2 Relative Amounts of Various Phases Determined by Powder X-Ray Diffraction Analysis for Tl–Ca–Ba–Cu–O Samples Prepared via a Two-Step Method Using BaCuO$_2$ as the Ba Source in the Second-Step Calcination

(a) Starting composition: Tl/Ca/Ba/Cu = 2:2:2:3

Reaction time	Compound						
	2021	2122	2223	BaCuO$_2$	CaO	CuO	Tl$_x$Ca$_y$Cu$_z$O$_a$
0 min	0	0	0	41	10	8	0
2.5 min	6	72	8	5	4	4	9
5 min	0	39	55	0	3	0	0
10 min	0	19	67	0	0	0	0
20 min	0	9	73	0	0	0	0
40 min	0	0	87	10	2	1	0
2 h	0	0	37[a]	44	4	3	0

[a]Broadening of 002 peak observed, indicating the presence of the 1324 phase.

(b) Starting composition: Tl/Ca/Ba/Cu = 1:1:2:2

Reaction time	Compound					
	2021	2122	1122	BaCuO$_2$	CaO	Tl$_x$Ca$_y$Cu$_z$O$_a$
0 min	0	0	0	61	8	0
2.5 min	14	45	0	31	4	11
5 min	0	67	0	35	0	0
10 min	0	66	0	35	0	0
20 min	0	62	0	39	0	0
40 min	0	59	0	42	0	0
80 min	0	30	19	50	0	0
2 h	0	6	27	67	0	0

(c) Starting composition: Tl/Ca/Ba/Cu = 1:3:2:4

Reaction time	Compound					
	2122	2223	1324	BaCuO$_2$	CuO	CaO
0 min	0	0	0	45	17	16
2.5 min	29	14	0	27	10	11
5 min	5	37	0	28	8	9
10 min	0	45	0	27	8	9
20 min	0	43	0	28	8	9
40 min	0	38	0	32	9	7
80 min	0	12	30	38	7	6
10 h	0	0	10	48	11	11

Source: Ref. 38.

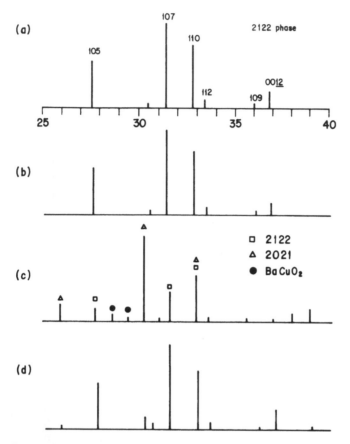

Figure 6 XRD data for the calcined pellets prepared by the two-step method. (a) Ca-Ba_2CuO_4 was used as the Ba source and calcination was conducted at 800°C for 20 min; (b) $CaBa_2CuO_4$, 850°C, 10 min; (c) $BaCuO_2$, 800°C, 20 min; (d) $BaCuO_2$, 850°C, 10 min. These samples correspond to samples A1, A2, B1, and B2 in Table 3, respectively.

by using $CaBa_2CuO_4$ exhibited a higher T_c and a larger diamagnetic signal, indicating better composition homogeneity, than those obtained via $BaCuO_2$ (Table 3).

The experiments described above consistently indicate that the higher member phases of $Tl_2Ca_nBa_2Cu_{n+1}O_x$ are formed more rapidly as more of the required alkaline-earth elements are combined with CuO before the second-stage calcination. As discussed earlier, the sequential phase changes among $Tl_2Ca_nBa_2Cu_{n+1}O_y$ oxides in the one-step process can largely be attributed to the presence of inhomogeneity in the local Cu and Ca contents. For example, the deficiency of Ca near CuO particles facilitates the formation of the 2021 phase, while the 2122 and 2223 phases cannot be formed until the Cu content near the CaO particles reaches a certain level. When $BaCuO_2$ is used as one of the starting ingredients, there is an additional Cu source other than CuO, and hence the Cu content is more homogeneously distributed within the reactant powder. On the other hand, use of powders, such as Ca_2CuO_3 and $CaBa_2CuO_4$, containing both Ca and Cu allows mixing

Table 3   Effect of Ba Source on Synthesis of 2122 Phase via the Two-Step Method

Sample[a]	Calcining conditions	T_c^b (K)	$-4\pi M^c$ ($\times$ 10^{-2} emu/g)	
			4.5 K	78 K
A1	800°C 20 min	108	7.16	1.77
A2	850°C 10 min	110	13.4	8.61
B1	800°C, 20 min	95	1.94	0.308
B2	850°C, 10 min	109	6.68	2.83

Source: Ref. 17.

[a]Samples A1 and A2 were prepared by using CaBa$_2$CuO$_4$ as the Ca and Ba sources in the starting powder; samples B1 and B2 were prepared by using CaO and BaCuO$_2$ as the Ca and Ba sources.

[b]T_c is taken as the temperature at which the magnetization becomes zero. The measurements were carried out at 100 G.

[c]The magnetization measurements were carried out under field-cooled conditions (Meissner effect) at 100 G.

between these two elements at an atomic scale. This also facilitates the formation of the higher member phases of Tl$_2$Ca$_n$Ba$_2$Cu$_{n+1}$O$_y$.

In the multistep methods described above, Tl$_2$O$_3$ is introduced into the reaction at the latest calcination stage presumably to reduce the calcination time experienced by Tl$_2$O$_3$ in order to minimize Tl loss. A completely reversed approach in which Tl$_2$O$_3$ is introduced in the early stage of a series of calcinations has also been proposed [18,19,21]. In the latter approach, a lower member phase of Tl$_2$Ca$_n$Ba$_2$Cu$_{n+1}$O$_y$ is first synthesized and the resulting phase powder is reacted with additional Ca and Cu-containing oxides at a higher temperature to convert the lower member phase to the higher one. For synthesizing either the 2122 or the 2223 compound, Ruckenstein and Narain [21] have first prepared the 2021 phase from powders with 2–1–2–2 or 2–2–2–3 stoichiometries at 775°C for 2 h and then calcined the resulted powder at 865°C to produce the single-phase 2122 or 2223 powder in less than 30 min. The loss of Tl at 775°C was reported to be small. Furthermore, the heating time at 865°C for complete reaction has been greatly shortened in comparison with those in Tables 1 and 2. As a result, while Tl$_2$O$_3$ was introduced early during synthesis, Tl loss along the complete synthesis process was actually suppressed effectively.

Wu et al. [18,19] demonstrated improved superconductive properties of the resulted 2223 sample by precalcination to produce the 2122 compound prior to formation of the 2223 compound. Table 4 compares the $T_{cs,zero}$ of the 2223 samples obtained either by directly heating the mixture of Tl$_2$O$_3$, CaBa$_2$CuO$_4$, CaO, and CuO above 870°C (samples 1 and 2) or by a two-step procedure with a precalcination conducted at 830°C, during which the 2122 compound is formed (samples 3 through 6). All the 2223 samples have been single phase according to the XRD analysis, but their T_cs varied. When the sample was first precalcined at 830°C (sample 3),

Table 4   Effect of Thermal History on the T_c of 2223
Samples Prepared Using $CaBa_2CuO_4$ as Reactant Precursor

Sample	Calcination condition	$T_{c,zero}$[a] (K)
1	870°C, 2 h	99
2	880°C, 2 h	98
3	830°C, 3 h + 870°C, 1 h	103
4[b]	870°C, 1 h	110
5[b]	870°C, 2 h	111
6[b]	870°C, 6 h	117

[a]Zero-resistivity temperature measured by the four-probe method for
a current density of 50 mA/cm^2.
[b]Precalcined at 830°C for 5 h to produce the 2122 phase before being
treated under the conditions as indicated.

it showed a higher T_c (103 K) than those of samples 1 (99 K) and 2 (98 K), even
though it was heated in the second step (870°C) for a shorter time than were samples
1 and 2. A slight increase in the precalcination time further increased the T_c (samples
3 and 4). Powder melting has been observed during calcination at 870°C for all the
samples shown in Table 4, but the extent of melting was greatly reduced once the
powder was precalcined. Furthermore, the process that employs a precalcination
step reduces the Tl loss and hence the superconductivity deterioration caused by
it. By using the usual gold-wrapping technique, a single-phase 2223 can be obtained
with a T_c of 117 K even after a long thermal treatment which consists of a precalcination at 830°C for 5 h followed by a second calcination at 870°C for 6 h
(sample 6). The allowance for long calcination makes this process particularly
suitable for large-scale processes.

3.   Synthesis of $TlCa_mBa_2Cu_{m+1}O_x$

Synthesis of single-phase $TlCa_mBa_2Cu_{m+1}O_x$ (the single Tl–O layer compounds)
has been less successful than that of $Tl_2Ca_nBa_2Cu_{n+1}O_y$ (the double Tl–O layer
compounds). The data presented in Tables 1b to d and 2b and c show that by either
the one-step or multistep methods, calcination of mixtures with stoichiometries Tl/
Ca/Ba/Cu = 1:n:2:n + 1 does not directly generate the single Tl–O layer compounds. Rather, the double-Tl–O-layer compounds are first generated and the
single-Tl–O-layer compounds are subsequently formed via phase changes from the
already formed double-Tl–O-layer compounds. For example, the 1122 phase is
formed at the expense of the 2122 phase (Tables 1b and 2b), while the 1223 and
the 1324 phases are formed from the 2223 phase (Tables 1c and d and 2c).

The phase changes from the double- to the single-Tl–O-layer compounds may
proceed via two reaction routes. First, the already formed double-layer compounds
react with Ca- and Cu-containing starting ingredients, such as CaO, CuO, and
$BaCuO_2$, to form the single-layer compounds. In this case the formation of the
single-layer compound should be accompanied by significant consumptions of both
Ca and Cu. This is, however, inconsistent with the data shown in Tables 1 and 2,

which show no decrease in the amounts of CaO, CuO, and $BaCuO_2$ when single-layer compounds are formed at the very late stage of calcination.

The second, and also the more plausible route for phase transformation is the depletion of Tl by vaporization from the already formed double-Tl-O-layer compounds. It has been reported [14,33] that Tl loss from $Tl_2Ca_nBa_2Cu_{n+1}O_x$ can lead to the formation of $TlCa_mBa_2Cu_{m+1}O_x$ (with m equal to or larger than n), and when these phase changes occur, $BaCuO_2$ is generated simultaneously as the dominant by-product. For example, by gradually depleting Tl by long heating at 890°C, Nakahima et al. [14] synthesized the 1122 phase from a 2122 powder and the $TlCa_mBa_2Cu_{m+1}O_x$ ($m = 2 - 4$) phases from the 2223 powders (Table 5). In all these cases, an appreciable amount of $BaCuO_2$, as identified by XRD, was formed along with the resulted single-Tl-O-layer compound. A significant increase in the amount of $BaCuO_2$ was observed in both Tables 1 and 2 whenever the phase change from the double- to the single-layer Tl-O compound takes place, suggesting that depletion of Tl by vaporization is the dominant mechanism leading to such phase changes.

The fact that for calcinations conducted at temperatures above 860°C, the single Tl-O layer compounds are not formed until significant Tl loss takes place suggests that these compounds may not be thermodynamically stable for the compositions corresponding to their stoichiometries but stable only under Tl-deficient situations. That is, the single-phase powders of these compounds cannot be synthesized from stoichiometric reactant powders at these temperatures. One possible cause for such an instability is the incongruent melting. Further studies to establish phase diagrams similar to Fig. 2 for the single-Tl-O-layer compounds under various overall compositions will be very useful for the identification of the real causes.

B. Gas-Solid Reaction Method

The loss of Tl leads to the decomposition of the Tl-based superconductive compounds into nonsuperconductive Ca-Ba-Cu-oxides. The reverse reaction (i.e., the reaction between Ba-Ca-Cu-oxides and Tl_2O vapor), on the other hand, has been employed, first by Sheng et al. [43] and Qui and Shih [44], as a method to synthesize the Tl-based superconductors in both bulk [32,43] and film [44-48] forms. For bulk processing [32,43], the Ca-Ba-Cu-oxide precursors were prepared by the solid-state reaction method, whereas for film processing they were predeposited onto the substrates by various thin-film techniques, including sput-

Table 5 Synthesis of $TlCa_mBa_2Cu_{m+1}O_x$ from $Tl_2Ca_nBa_2Cu_{n+1}O_y$

Starting phase	Heating temperature (°C)	Heating time (h)	Products	
			Main	Minor
2122	890	10	1122	$BaCuO_2$
2223	890	4	1223	$BaCuO_2$
2223	890	6	1324	$BaCuO_2$
2223	890	10	1425	$1324 + BaCuO_2$

Source: Ref. 14.

tering [44], evaporation [45], chemical vapor deposition (CVD) [46,47], and spraying [48]. The Tl_2O vapor was generated simply by heating Tl_2O_3 or Tl powder at
high temperatures. For calcination, the Ca–Ba–Cu–oxide precursors and the Tl
source can be heated either side by side at the same temperature [43–47] (the one-
temperature configuration) or separately in two temperature zones [48] (the two-
temperatures configuration) (Fig. 7). In the latter case, the rate of Tl vaporization
and that of gas–solid reaction can be controlled independently. To minimize the
loss of Tl_2O vapor from the reaction system, the calcination, using either the one-
or two-temperatures configuration, has often been carried out in a sealed reaction
chamber.

   Sheng et al. [43] prepared a layer consisting of 2223 and 2122 on the surface
of Ca–Ba–Cu–oxide pellets by using one-temperature calcination at 900°C for 3
min, while Qiu and Shih [44] obtained a Tl–Ca–Ba–Cu–O film which exhibited
a T_c of 75 K on ZrO_2 by reaction at 730°C for 1 h. It is noted that at the same
temperature, the Tl_2O vapor pressures in equilibrium with the Tl-based superconductors are in general lower than those with either Tl_2O_3 or pure Tl. As a result,
equilibrium cannot be simultaneously reached at the same temperature among the
Tl source, Tl_2O vapor, and the intended Tl-based superconductor. That is, in the
one-temperature configuration operation, the Tl-based superconductor will never
be an equilibrium product, but an intermediate one. The Tl content, and hence
the composition of the dominant Tl-containing phase, at the precursor side keeps
changing with time. Thus, in addition to the reaction temperature, calcination
duration is also an important process variable that determines the composition of
the resulting Tl-based superconductive phases in experiments using the one-temperature configuration.

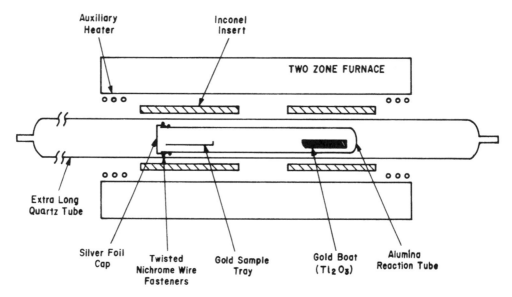

Figure 7   Two-temperature-zone suggested by DeLuca et al. [48] for the gas–solid reaction
method.

In the two-temperature process, however, one can employ a temperature at the Tl source side lower than that at the precursor side, so that the Tl$_2$O vapor inside the reaction chamber can be in equilibrium not only with the Tl source powder but also with the intended superconductive phase. In this case, the intended superconductive phase will be the equilibrium product at the precursor side. For preparing the 2223 films on Y-stabilized ZrO$_2$ by using the two-temperatures configuration, DeLuca et al. [48] reported an optimum Tl source temperature of about 805°C when the gas–solid reaction at the presursor side was carried out at about 895°C in oxygen. Films thus prepared contained a highly oriented 2223 phase with only a trace amount of 2122 [48].

Sugise et al. [32] reported that when the Ca–Ba–Cu–O presursor reacts with Tl$_2$O vapor, the dominant phases at the precursor side changes from the original starting ingredients to the single-Tl–O-layer compound and then to the double-Tl–O-layer compounds. These results indicate that the reaction mechanisms taking place during the gas–solid reaction method simply follow a pathway reverse to that which occurs during Tl loss from the formed Tl-based superconductive phases as discussed earlier.

C.   Wet-Process Method

The wet process, also known as the solution process, has drawn much attention (see, e.g., Refs. 22 to 25 and 49 to 52 and the references cited therein) as an alternative to the solid-state reaction method for fabricating the high-T$_c$ superconductive oxides. In the YBa$_2$Cu$_3$O$_x$ systems [49–52], this method has been shown in many cases, not only to lower the calcination temperature and/or shorten the calcination time but also to be a feasible technique for preparing thick films. However, much less research has been reported for using this method in fabricating the Tl-based superconductive oxides. In general, the wet-process method is less mature in the synthesis of the Tl-based superconductors than of other high-T$_c$ oxides.

Primarily two types of wet processes, coprecipitation [49–51] and sol-gel methods [24,52], have been used to prepare high-T$_c$ superconductors. In the coprecipitation method, water-soluble salts containing the constituent elements of the intended superconductive phase are first dissolved and mixed in water. Subsequently, either the solution is dried directly or a precipitating agent such as citric acid is introduced into the solution to form insoluble salts of the constituent elements, which coprecipitate from solution. In the latter case, the precipitates can be separated from the solution by filtration. Dried or filtrated precursor powder is then heated at intermediate temperatures to convert the salts into oxide mixtures and then finally calcined at higher temperatures to form high-T$_c$ superconductors. To prepare thick films, the solution is sprayed onto the surface of the substrate and then dried and calcined [22]. In the sol-gel technique, alkoxides of the constituent elements of the superconductive compound are dissolved in a water–alcohol mixture. Sol formation or gelation takes place on hydrolysis. The resulted sol solution or the gel is subsequently dried and calcined by procedures similar to those employed in the coprecipitation method. For making film samples, a surface layer of the solution can be deposited on the surface of the substrate by painting, dipping, or spinning coating.

Kordas and Teepe [24] have synthesized the 2122 powder by following the sol-gel technique. The alkoxides, including Tl-, Ba-, and Cu-methoxyethoxide and Cu ethoxide, were prepared by the following reactions:

$$Ba + 2CH_3OC_2H_4OH \rightarrow Ba(CH_3OC_2H_4O)_2 + H_2 \quad \text{in Ar}$$

$$Ca + 2CH_3OC_2H_4OH \rightarrow Ca(CH_3OC_2H_4O)_2 + H_2 \quad \text{in Ar}$$

$$Tl_2O + 2CH_3OC_2H_4OH \rightarrow 2Tl(CH_3OC_2H_4O) + H_2O$$

and

$$CuCl_2 + 2Li(C_2H_5O) \rightarrow Cu(C_2H_5O)_2 + 2LiCl$$

The 2122 powder was then prepared from these alkoxides by the procedures depicted in Fig. 8. Barboux et al. [23], on the other hand, prepared the Tl-based superconductive films by spraying an aqueous-glycerol solution containing Tl, Ca, Ba, and Cu onto a hot (200°C) substrate (SrTiO$_3$, MgO, or ZrO$_2$) and then baking and firing the sample at 500 and above 900°C, respectively. The use of a water–glycerol mixture, rather than pure water, as the carrier solution was reported to facilitate the formation of the high member phases of the Tl-based superconductive oxides.

In both examples described above, off-stoichiometric starting compositions were employed and the resulted samples were multiphased [23,24]. In particular, long calcination in flowing oxygen that are employed to decompose the precursor powders, which mostly contain organics and nitrates, before the final calcination was found to result in extensive Tl loss. The successful application of the wet-process method to the fabrication of the single-phase Tl-based superconductive powders requires much more effort in developing new solution–precipitation sys-

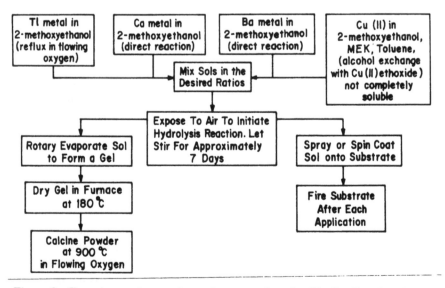

Figure 8   Experimental procedures for preparing the Tl–Ca–Ba–Cu–O via the sol-gel method as employed by Kordas and Teepe [24].

tems that will allow the precise control of the precipitate composition, as well as in seeking the optimum calcination procedure.

IV. POSTTREATMENTS

While the phase composition of a superconductive powder is determined mainly by the calcination method and conditions, the superconductive properties, including T_c and J_c, of the superconductive bulk prepared from the powder is heavily affected by posttreatment procedures, such as the shape-forming methods [53–57], sintering and annealing conditions [34,55,58,59], and oxygenation processes [34,60–62]. For example, for XRD single-phase polycrystalline 2223, a wide variation in T_c, ranging from about 100 to 130 K, can be obtained simply by varying the annealing and oxygenation procedures [34,58,59]. Furthermore, the employment of a high sintering temperature to cause partial decomposition of a 2223 bulk into a 1223 phase has been found to increase the flux pinning and hence improve J_c [55,56,63].

Wide ranges of structural defects, such as nonstoichiometries, site vacancies, cation substitutions, and intergrowth of $(CuO_2–CaO–Cu_2)_n$ slabs with different n values, have been reported by various groups [8,64–71], and these defects have been found [8,69,70] to play a significant role in determining the T_c of Tl-based superconductive oxides. In studying the microstructures and the T_cs of various single or double-Tl–O-layer phases, Parkin et al. [8] observed a correlation between the density of $(CuO_2–CaO–CuO_2)_n$ intergrowth and T_c. For the 2122, 2223, and 1223 phases, the T_c was found to decrease with an increase in intergrowth density. The causes leading to the extensive structural disorderings in the Tl-based superconductive oxides have not been completely understood. However, it is clear that structural disorderings are associated with higher entropies of the overall systems, and they are favored by high temperatures. Consequently, a prolonged annealing at temperatures lower than the calcination temperature can reduce the extent of disordering created during calcination, and hence increase the T_c. Kaneko et al. [58] observed improvement in the T_c of single-phase 2223 powder by annealing the powder at 750°C (Table 6). The T_c increased steadily, but rather slowly, from 118 K at the as-calcined state to 127 K after 250 h of annealing. Similar results were reported by Liu et al. [59].

Table 6 Effect on T_c of the 2223 Phase by Postannealing at 750°C

Sample	Annealing time (h)	$T_{c,zero}$ (K)
1	0	118
2	2	121
3	10	123
4	80	125
5	150	126
6	250	127

Source: Ref. 58.

A heavy dependence of T_c on oxygen content was first discovered for $YBa_2Cu_3O_{7-x}$ and is now known to be true for all high-T_c superconducting oxides. Wu and Chu [61] studied the effect of oxygenation on T_c of the 2122 oxide by consecutively postannealing a quenched (from 830°C) sample in oxygen at increasing temperatures between 250 and 800°C and monitoring the changes in both oxygen content and T_c taking place after each annealing. Figure 9A plots the cumulative change in the total oxygen content after consecutive oxygenations as a function of oxygenation temperature. An annealing time of 0.5 h was employed at each temperature selected. The cumulative change in the total oxygen content was taken as the difference between the sample weight measured after each oxygenation and the weight before any oxygenation and was expressed in terms of the number of

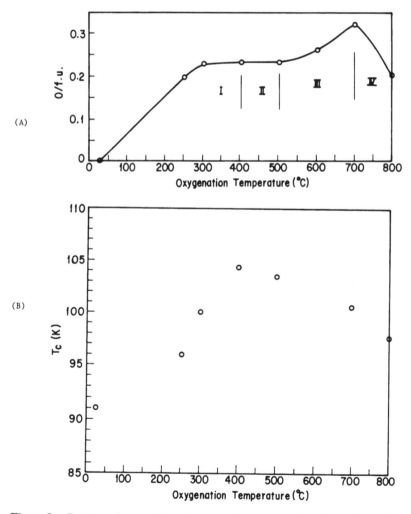

Figure 9   Oxygenation and its effect on the superconductivity of the 2122 phase. (A) Cumulative oxygen uptake versus oxygenation temperature. (B) $T_{c,mag}$ versus oxygenation temperature. (From Ref. 61.)

oxygen atoms adsorbed per formula unit of the superconductive phase (2122 or 2223) (O/f.u.). The T_cs (Fig. 9b) were obtained by magnetization measurements, which confirmed the homogeneous change in T_c throughout the oxide grains after each oxygenation.

As shown in Fig. 9A, four temperature regions, within which the total oxygen content changed following different trends with increasing temperatures, were identified: region I, between 250 and 400°C, in which the oxygen content increases; region II, with a plateau between 400 and 500°C; region III, between 500 and 700°C, in which a second increase occurs; and region IV, above 700°C, in which the oxygen content decreases. The T_c (Fig. 9B), on the other hand, first increases and then decreases with increasing oxygenation temperature. Since the oxygenation time (0.5 h) in these experiments is short, the changes in T_c shown in Fig. 9B cannot be attributed to the ordering–disordering effects but to changes in the amount of oxygen vacancies. By comparing Fig. 9A and B, it can be seen that there is no simple correlation between T_c and the total oxygen content except for oxygenation conducted below 400°C, for which the T_c increases monotonously with increasing oxygen content. The oxygenation behavior of the 2223 compound was found to be similar to that of the 2122 compound (Fig. 10A and B) [34]. For both

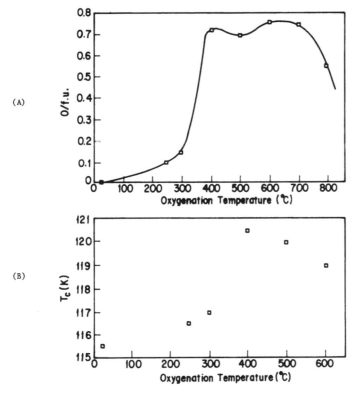

(A)

(B)

Figure 10 Oxygenation and its effect on the superconductivity of the 2223 phase. (A) Cumulative oxygen uptake versus oxygenation temperature. (B) $T_{c,\mathrm{mag}}$ versus oxygenation temperature. (From Ref. 34.)

the 2122 and 2223 compounds, the maximum T_cs were obtained after oxygenation at 400°C. Schilling et al. [60] also reported improvements in T_c for the 2223 compound with increasing oxygenation time at temperatures below 400°C.

In contrast to the 2122 and 2223 compounds, the 2021 compound showed a decrease in T_c after low-temperature oxygenation [62]. The 2021 phase can exist in either orthorhombic or tetragonal forms [62]. For both structural symmetries, Shimakawa etal. [62] reported that a T_c of about 70 K was obtained either by quenching the sample in liquid nitrogen at the end of calcination or by postannealing the calcined sample in argon. They also found that oxygenation of these samples in oxygen at 400°C completely destroyed the superconductivity. Kikuchi et al. [57] also observed an increase in the T_c of the 2021 compound due to oxygen loss caused by shock loading.

As for J_c, several significant improvements in raising the J_c of the poly-crystalline, Tl-based superconductive materials have been reported [53,55,56,72–74]. All these reports clearly indicate a great sensitivity of J_c on the processing protocols, but no general conclusion can be made regarding the optimum processing conditions for achieving high J_c values. Ginley et al. [72–74] reported the fabrication of polycrystalline 2223 and 2122 films, which were synthesized at 850°C, with J_c values greater than 10^5 A/cm^2 at 77 K and zero field. The J_c values are larger than that for any type of polycrystalline sample from the other families of high-T_c superconductive oxides. The correct Tl stoichiometry of the intended phase was suggested to be the key for strong intergranular links. The Tl-deficient films were normally weak linked. On the other hand, Goto et al. [55,56] prepared Tl–Ca–Ba–Cu–O superconductive filaments of about 250 μm diameter from pure 2223 powder by the suspension-spun method. J_c values greater than 10^4 A/cm^2 at 77 K and zero field were obtained for filaments sintered at 840°C for 20 min. A phase transformation of a fraction of the starting 2223 phase to the 1223 phase due to Tl loss took place during sintering. It has been suggested that defects created by the phase transformation increase the flux pinning in the filaments.

REFERENCES

1. Z. Z. Sheng and A. M. Hermann, *Nature (London)* 332, 55 (1988).
2. Z. Z. Sheng and A. M. Hermann, *Nature (London)* 332, 138 (1988).
3. Z. Z. Sheng, A. M. Hermann, A. El Ali, C. Almason, J. Estrada, T. Datta, and R. J. Matson, *Phys. Rev. Lett.* 60, 937 (1988).
4. A. M. Hermann, Z. Z. Sheng, D. C. Vier, S. Schutz, and S. B. Oseroff, *Phys. Rev. B* 37, 9742 (1988).
5. R. M. Hazen, L. W. Finger, R. J. Ahgel, C. T. Prewitt, N. L. Ross, C. G. Hadidiacos, P. J. Heaney, D. R. Veblen, Z. Z. Sheng, A. El Ali, and A. M. Hermann, *Phys. Rev. Lett.* 60, 1657 (1988).
6. S. S. Parkin, V. Y. Lee, E. M. Engler, A. I. Nazzal, T. C. Huang, G. Gorman, R. Savoy, and R. Beyers, *Phys. Rev. Lett.* 60, 2539 (1988).
7. S. S. Parkin, V. Y. Lee, A. I. Nazzal, R. Savoy, R. Beyer, and S. J. LaPlaca, *Phys. Rev. Lett.* 61, 750 (1988).

8. S. S. Parkin, V. Y. Lee, A. I. Nazzal, R. Savoy, T. C. Huang, G. Gorman, and R. Beyer, *Phys. Rev. B* 38, 6531 (1988).

9. M. A. Subramanian, J. C. Calabrese, C. C. Torardi, J. Gopalakrishnan, T. R. Askew, R. B. Flippen, K. J. Morrissey, U. Chowdhry, and A. W. Sleight, *Nature (London)* 332, 420 (1988).

10. M. Kikuchi, T. Kajitani, Y. Syono, and Y. Muto, *Jpn. J. Appl. Phys.* 28, L382 (1989).

11. F. Seidler, P. Bohm, H. Geus, W. Braunisch, E. Braun, W. Schnelle, Z. Drzazga, N. Wild, B. Roden, H. Schmidt, D. Wohlleben, I. Felner, and Y. Wolfus, *Physica C* 157, 375 (1989).

12. K. C. Goretta, J. G. Chen, N. Chen, M. C. Hash, and D. Shi, *Mater. Res. Bull.* 25, 791 (1990).

13. H. Ihara, R. Sugise, M. Hirabayashi, N. Terada, M. Jo, K. Hayashi, A. Negishi, M. Tokumoto, Y. Kimura, and T. Shimomura, *Nature (London)* 334, 510 (1988).

14. S. Nakajima, M. Kikuchi, Y. Syona, T. Oku, D. Shindo, K. Hiraga, N. Kobayashi, H. Iwasaki, and Y. Muto, *Physica C* 158, 471 (1989).

15. A. Sulpice, B. Giordanengo, R. Tournier, M. Hervieu, A. Maignan, C. Martin, C. Michel, and J. Provost, *Physica C* 156, 243 (1988).

16. C. C. Torardi, M. A. Subramanian, J. C. Calabrese, J. Gopolakrishnan, K. J. Morrissey, T. R. Askew, R. B. Flippen, U. Chowdhry, and A. W. Sleight, *Science* 240, 631 (1988).

17. N. L. Wu and E. Ruckenstein, *Mater. Lett.* 7, 169 (1988).

18. N. L. Wu, S. N. Lee, and Y. D. Yao, *Jpn. J. Appl. Phys.* 28, L1349 (1989).

19. N. L. Wu, Y. D. Yao, S. N. Lee, S. Y. Wong, and E. Ruckenstein, *Physica C* 161, 302 (1989).

20. E. Ruckenstein and C. T. Cheung, *J. Mater. Res.* 4, 1116 (1989).

21. E. Ruckenstein and S. Narain, *Mater. Lett.* 8, 421 (1989).

22. P. Barboux, J M. Tarascon, F. Shokoohi, B. J. Wilkens, and C. L. Schwartz, *J. Appl. Phys.* 64, 6382 (1988).

23. P. Barboux, J. M. Tarascon, L. H. Greene, G. W. Hull, and B. G. Bagley, *J. Appl. Phys.* 63, 2725 (1988).

24. G. Kordas and M. R. Teepe, *Appl. Phys. Lett.* 57, 146 (1990).

25. C. Chiang, C. Y. Shei, S. F. Wu, and Y. T. Huang, *Appl. Phys. Lett.* 58, 2435 (1991).

26. T. Kotani, T. Kaneko, H. Takei, and K. Tada, *Jpn. J. Appl. Phys.* 28, L1378 (1989).

27. K. Wade and A. J. Banister, in *Comprehensive Inorganic Chemistry*, ed. J. C. Bailar, H. J. Emeleus, R. Nyholm, and A. T. Trot-Dickenson, Vol. 1, Pergamon Press, London, 1973, p. 1146.

28. C. T. Cheung and E. Ruckenstein, *J. Mater. Res.* 5, 245 (1990).

29. C. T. Cheung and E. Ruckenstein, *J. Mater. Res.* 5, 1860 (1990).

30. D. Cubicciotti and F. J. Keneshea, *J. Phys. Chem.* 71, 808 (1967).

31. A. B. F. Duncan, *J. Am. Chem. Soc.* 51, 2697 (1929).

32. R. Sugise, M. Hirabayashi, N. Terada, M. Jo, M. Tokumoto, T. Shimomura, and H. Ihara, *Jpn. J. Appl. Phys.* 27, L2310 (1988).

33. R. Sugise, M. Hirabayashi, N. Terada, M. Jo, T. Shimomura, and H. Ihara, *Physica C* 157, 131 (1989).

34. H. T. Chu and N. L. Wu, *Bull. Coll. Eng. Nat. Taiwan Univ.* 53, 87 (1991).

35. R. C. Weast and M. J. Astle, eds. *Handbook of Chemistry and Physics*, 60th ed., CRC Press, Boca Raton, Fla., 1979, p. B134.

36. Y. A. Chang and N. Ahmed, *Thermodynamic Data on Metal Carbonates and Related Oxides*, The Metallurgical Society of AIME, New York, 1982.

37. R. S. Roth, K. L. Davis, and J. R. Dennis, *Adv. Ceram. Mater.* 2(3B), 303 (1987).

38. S. Narain and E. Ruckenstein, *Supercond. Sci. Technol.* 2, 236 (1989).
39. T. W. Huang, M. P. Hung, T. S. Chin, H. C. Ku, and S. E. Hsu, *Mod. Phys. Lett. B* 4, 885 (1990).
40. C. Wang, H. L. Chang, M. L. Chu, Y. S. Gou, and T. M. Uen, *Jpn. J. Appl. Phys.* 28, L587 (1989).
41. A. M. M. Gadalla and J. White, *Trans. Br. Ceram. Soc.* 65, 181 (1966).
42. K. A. Kubat-Martin, E. Garcia, and D. E. Peterson, *Physica C* 172, 75 (1990).
43. Z. Z. Sheng, L. Sheng, H. M. Su, and A. M. Hermann, *Appl. Phys. Lett.* 53, 2686 (1988).
44. C. X. Qui and I. Shih, *Appl. Phys. Lett.* 53, 1122 (1988).
45. K. K. Verma, G. D. Verma, R. S. Tiwari, and O. N. Srivastava, *Jpn. J. Appl. Phys.* 29, L880 (1990).
46. J. A. Ladd, B. T. Collins, J. R. Matsy, J. Zhao, and P. Norris, *Appl. Phys. Lett.* 59, 1368 (1991).
47. N. Hamaguchi, R. Gardiner, P. S. Lirlin, R. Dye, K. M. Hubbard, and R. E. Muenchausen, *Appl. Phys. Lett.* 57, 2136 (1990).
48. J. A. DeLuca, M. F. Garbauskas, R. B. Bolon, J. G. McMullin, W. E. Balz, and P. L. Karas, *J. Mater. Res.* 6, 1415 (1991).
49. I. W. Chen, S. Keating, X. Wu, J. Xu, P. E. Reyes-Morel, and T. Y. Tien, *Adv. Ceram. Mater.* 2(3B), 457 (1987).
50. C. T. Chu and B. Dunn, *Commun. Am. Ceram. Soc.* 70, C-375 (1987).
51. M. Kujiki, M. Hikita, and K. Sukegawa, *Jpn. J. Appl. Phys.* 26, L1159 (1987).
52. H. Zheng and J. D. Mackenzie, *Mater. Lett.* 7, 182 (1988).
53. M. Qi and J. H. Wang, *Physica C* 176, 38 (1991).
54. R. Navarro, F. Lera, A. Badia, C. Rillo, J. Bartolome, W. L. Lechter, and L. E. Toth, *Physica C* 183, 73 (1991).
55. T. Goto and C. Yamaoka, *Jpn. J. Appl. Phys.* 29, L1645 (1990).
56. T. Goto, *J. Mater. Res.* 5, 2759 (1990).
57. M. Kikuchi, Y. Syono, N. Kobayashi, T. Oku, E. Aoyagi, K. Hiraga, K. Kusaba, T. Atou, A. Tokiwa, and K. Fukuoka, *Appl. Phys. Lett.* 57, 813 (1990).
58. T. Kaneko, H. Yamanchi, and S. Tanaka, *Physica C* 178 (1991).
59. R. S. Liu, J. L. Tallon, and P. P. Edwards, *Physica C* 182, 119 (1991).
60. A. Schilling, H. R. Ott, and F. Hulliger, *Physica C* 157, 144 (1989).
61. N. L. Wu and H. T. Chu, *Physica C* 167, 267 (1990).
62. Y. Shimakawa, Y. Kubo, T. Manako, T. Satoh, S. Iijima, T. Ichikashi, and H. Igarashi, *Physica C* 157, 279 (1989).
63. T. Doi, M. Okada, A. Soeta, T. Yuasa, K. Aihara, T. Kamo, and S. P. Matsuda, *Physica C* 183, 67 (1991).
64. S. Iijima, T. Ichihashi, Y. Shimakawa, T. Manako, and Y. Kubo, *Jpn. J. Appl. Phys.* 27, L837 (1988).
65. S. Iijima, T. Ichikashi, and Y. Kubo, *Jpn. J. Appl. Phys.* 27, L817 (1988).
66. B. Morosin, R. J. Boughman, D. S. Ginley, J. E. Schirber, and E. L. Venturini, *Physica C* 161, 115 (1990).
67. J. Z. Liu, K. G. Vandervoort, H. Claus, G. W. Grabtree, and D. J. Lam, *Physica C* 156, 256 (1989).
68. H. Takei, T. Kontani, T. Kaneko, and K. Taka, *Jpn. J. Appl. Phys.* 27, L1378 (1989).
69. J. Chrzanowski, J. C. Irwin, B. Heinrich, A. E. Curzon, N. Fortier, and A. Cragg, *Physica C* 182,231 (1991).
70. A. K. Singh, M. A. Imam, K. Sadananda, S. B. Qadri, E. F. Skelton, M. S. Osofsky, V. Le Tourneau, and D. U. Gubser, *J. Mater. Res.* 5, 1620 (1990).
71. C. P. Tigges, E. L. Venturini, J. F. Kwak, B. Morosin, R. J. Baughman, and D. S. Ginley, *Appl. Phys. Lett.* 57, 517 (1990).

72. D. S. Ginley, J. F. Kwak, R. P. Hellmer, R. J. Baughman, E. L. Venturini, and B. Morosin, *Appl. Phys. Lett.* 53, 406 (1988).
73. D. S. Ginley, J. F. Kwak, R. P. Hellmer, R. J. Baughman, E. L. Venturini, M. A. Mitchell, and B. Morosin, *Physica C* 156, 592 (1988).
74. D. S. Ginley, J. F. Kwak, E. L. Venturini, B. Morosin, and R. J. Baughman, *Physica C* 160, 42 (1989).

7

Chemical Characterization of Thallium Cuprate Superconductors

Mariappan Paranthaman
University of Colorado at Boulder
Boulder, Colorado

Arumugam Manthiram and John B. Goodenough
The University of Texas at Austin
Austin, Texas

I. INTRODUCTION

Soon after the discovery of high-temperature superconductivity in the Tl–Ba–Ca–Cu–O system by Sheng and Hermann [1], two homologous series of phases were identified [2–12], a Tl_1 and a Tl_2 family with, respectively, the ideal chemical formulas

$$TlBa_2Ca_{n-1}Cu_nO_{2n+3} \quad 1 \leq n \leq 5 \tag{1}$$

$$Tl_2Ba_2Ca_{n-1}Cu_nO_{2n+4} \quad 1 \leq n \leq 5 \tag{2}$$

These two families have, ideally, the intergrowth structures illustrated in Fig. 1. They each consist of intergrowths of superconductively active $Ca_{n-1}Cu_nO_{2n}$ layers and inactive rock salt layers bounded by (001) planes of BaO. Partial substitution of Tl by Pb in the Tl_1 family has given a third family,

$$Tl_{1-z}Pb_zSr_2Ca_{n-1}Cu_nO_{2n+3} \quad 1 \leq n \leq 3 \tag{3}$$

Other substitutions are possible; we discuss below systems with La substituted for Ba and Y substituted for Ca.

The crystal architecture of Fig. 1 requires bond-length matching across the intergrowth interface. In the simpler La_2CuO_4 structure, which has a single CuO_2 sheet intergrown by two (001) rock salt planes of LaO, a measure of the bond-length mismatch across the interface is the tolerance factor

$$t = \frac{La-O}{\sqrt{2}\,(Cu-O)} \tag{4}$$

where La–O and Cu–O are the respective equilibrium bond lengths. A $t = 1$ value corresponds to a matching of the bond lengths; this condition can only be met at a particular temperature for a given pressure since the thermal expansion and the

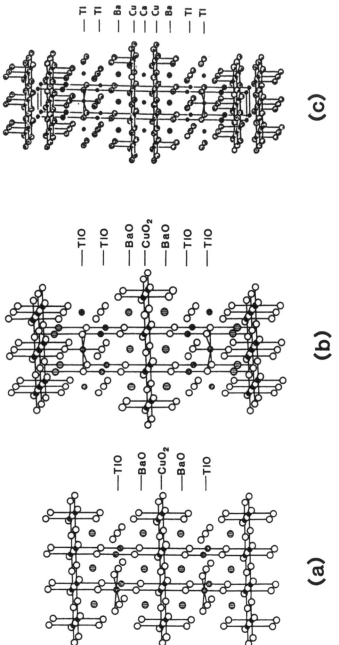

Figure 1   Ideal crystal structures of (a) TlBa$_2$CuO$_5$, (b) Tl$_2$Ba$_2$CuO$_6$, and (c) Tl$_2$Ba$_2$CaCu$_2$O$_8$. (Adapted from Ref. 4.)

compressibility are different for each bond. We have argued elsewhere [13–15] that since the La–O bond has a greater thermal expression than the Cu–O bond, a $t <$ 1 value in La_2CuO_4 decreases with decreasing temperature to place the CuO_2 planes under increasing compression as the temperature is reduced. The result is a large tetragonal ($c/a > 1$) distortion of the CuO_2 octahedra, a tendency to incorporate interstitial oxygen between the two LaO planes of a rock salt layer to form $La_2CuO_{4+\delta}$, a buckling of the Cu–O–Cu bonds from 180° via a cooperative tilting of the CuO_6 octahedra below an orthorhombic–tetragonal transition temperature T_t, and access to an oxidation of the CuO_2 sheets that relieves the internal stress. $La_2CuO_{4+\delta}$ and $La_{2-x}Sr_xCuO_4$ are, for example, p-type superconductors for a critical concentration p of holes in the antibonding x^2-y^2 bonds of the CuO_2 sheets. The situation is similar in the Tl_1 and Tl_2 families with $n = 1$; however, it is made more complex by the Tl–O layers between the BaO (001) rock salt "planes" that interface the CuO_2 sheets. In fact, the smaller Tl^{3+} ion relative to Ba^{2+} places the TlO rock salt layer under tension and the BaO "planes" under compression, and the Ba–O–Ba bonds within a BaO plane become buckled. Thus it is the Tl–O bond-length mismatch that is responsible for placing the Cu–O bonds within a CuO_2 sheet under compression.

Two necessary—but not sufficient—conditions for the appearance of high-temperature superconductivity in the intergrowth copper oxides such as the Tl_1 and Tl_2 families are (1) a similar oxygen coordination at each Cu atom of a given CuO_2 plane, and (2) mixed-valent CuO_2 planes having a mobile charge-carrier concentration within a narrow range [13–15]. The first condition ensures a nearly periodic potential within a superconductive CuO_2 sheet. The apparent need to fulfill this condition implies that the narrow-band charge carriers responsible for superconductivity have a momentum vector **k** that is a good quantum number, which is a prerequisite for Cooper pair formation in the superconductive state. The second condition is compatible with the suggestion [16] that the superconductive state is a distinguishable mixed-valent state appearing between the localized-electron state of the antiferromagnetic, semiconductive parent compound, which has $(CuO_2)^{2-}$ sheets, and the normal-metal state of more heavily oxidized (or reduced) CuO_2 sheets.

In the Tl_1 and Tl_2 families, where the CuO_2 sheets are under compression, the CuO_2 sheets may be oxidized, but not reduced [13]. Where the superconductive layers contain more than one CuO_2 sheet (i.e., $n > 1$), the Cu atoms are no longer coordinated by six oxygen atoms as in the $n = 1$ or La_2CuO_4 cases. For $n = 2$, each superconductive layer CuO_2–Ca–CuO_2 contains a plane of Ca atoms sandwiched between two CuO_2 sheets with fivefold oxygen coordination at each Cu atom (see Fig. 1). Moreover, the Cu–O–Cu bond angle is reduced from 180° by a movement of the bridging oxygen atoms away from the larger Ba^{2+} ions toward the smaller Ca^{2+} ions; these superconductive layers are analogous to the CuO_2–Y–CuO_2 layers of the p-type system $YBa_2Cu_3O_{6+x}$ with $T_c \approx 90$ K for $x > 0.9$. For $n > 2$, the two outer CuO_2 sheets of a superconductive layer (e.g., for $n = 3$, CuO_2–Ca–CuO_2–Ca–CuO_2) are like those for $n = 2$, but the inner ones are CuO_2 planes of Cu in the four-coplanar coordination with 180° Cu–O–Cu bonds. In this situation, each Cu atom of a given CuO_2 sheet has identical oxygen coordination, but the Cu atoms of the inner and outer sheets are distinguishable, and there is no reason to expect that for a given total oxidation state of the superconductive layers, the mobile holes are distributed equally between the two types of

CuO_2 sheets. In this chapter we restrict ourselves to the cases $n = 1$ or 2, so we do not need to address the problem of how the holes are distributed within a superconductive layer with $n > 2$. However, we point out here that a greater Madelung potential at the Cu in fivefold coordination can be expected to favor a greater hole concentration at the outer CuO_2 sheets, which is consistent with the observation that the maximum value of T_c is found for $n = 3$ (see Fig. 2).

Of more fundamental importance for the interpretation of the physical properties of the Tl_1 and Tl_2 families is a knowledge of the total oxidation state of the superconductive layers. This determination requires chemical analysis of both the thallium and oxygen contents of the product compounds since the concentrations of both elements vary greatly with the method of preparation and the sample thermal history. The determination is complicated further by the possibility of charge transfer between the superconductive and nonsuperconductive layers.

Lack of a suitable analytical procedure has plagued the early investigations and interpretations of the properties of the Tl_1 and Tl_2 families. For example, the ideal chemical formula $Tl_2Ba_2Ca_{n-1}Cu_nO_{2n+4}$ for the Tl_2 family should result in an oxidation state $(CuO_2)^{2-}$ for each CuO_2 sheet if all the thallium is present as Tl^{3+}. The observation of p-type superconductivity in these compounds indicates that the superconductive layer is more strongly oxidized, which has prompted the following suggestions as to the origin of the oxidation:

1. Thallium vacancies are responsible; their presence has been demonstrated by both electron microprobe analysis and neutron diffraction [7,9,18,19].

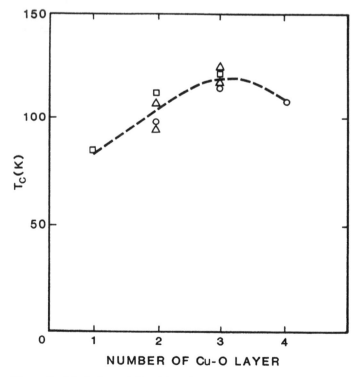

Figure 2   Variation of T_c with n for $Tl_2Ba_2Ca_{n-1}Cu_nO_{2n+4}$. (Adapted from Ref. 17.)

2. Excess oxygen is present in the Tl–O layer [20–22]; the analogous Bi_2 family $Bi_2Sr_2Ca_{n-1}Cu_nO_{2n+4+x}$ is known [23–26] to contain excess oxygen in the Bi–O layers where they give rise to a modulation of the structure along an orthorhombic *b* axis.

3. An internal oxidation–reduction reaction occurs as a result of electron transfer from the CuO_2 sheets to a Tl-6s band in the Tl–O layers [4,7,9]; this hypothesis is supported by photoelectron spectroscopy [27–29] and by band calculations [23,30,31].

4. Substitution of Ca^{2+} for Tl^{3+} occurs in samples with $n = 2$ and 3.

Without any quantitative chemical analysis for thallium and oxygen, it is not possible to know which of these alternatives or what mix of them is applicable for a given sample.

In this chapter we present a wet-chemical analytical procedure for determination of the thallium and oxygen content of a sample. We have applied this method to several systems, and we use our results to address not only the origin of the oxidation of the superconductive layers, but also the following additional questions: (1) control of the Tl/O content during synthesis and processing, (2) the origin of an orthorhombic distortion found in some superconductive samples and its influence on the critical temperature T_c, and (3) the nature of the superconductive phase.

II. WET-CHEMICAL PROCEDURE

Unlike other copper oxide superconductors, the thallium systems are not quantitatively soluble in a mixture of KI and HCl—the reagents used in the conventional iodometric determination of oxygen content [32]—due to the formation of a passive surface coating of TlI during the dissolution. The participation of Tl in the iodometric titration also makes impossible a determination of the oxygen content where the Tl content is unknown. We have overcome this problem by developing simple wet-chemical procedures for the separate determination of the thallium and oxygen contents [33–35].

The thallium content was determined by a volumetric method [33,36]. A known amount (ca. 100 mg) of the thallium cuprate was dissolved in 10 mL of 3.5 N HCl. The solution was diluted with water and perchloric acid to 30 mL such that the resulting solution contained 2% perchloric acid. An equal volume of 10% thiourea solution was then added, and the resulting thiourea perchlorate complex of thallium was stirred for 2 h on a magnetic stirrer. The precipitate was filtered on a sintered glass crucible, washed with a 5% thiourea solution containing about 0.5% perchloric acid, and dissolved in hot water. The solution was acidified with 3 mL of concentrated HCl and boiled with 1 mL of bromine until the color of the bromine disappeared. The solution was cooled and a few drops of bromine were added until the color turned yellow to ensure complete oxidation of all Tl to Tl^{3+}. The excess bromine was then removed by adding 5 mL of 5% phenol. The solution was next heated with 15 mL of 10% KI; the iodine liberated in accordance with the reaction

$$Tl^{3+} + 2I^- \rightarrow Tl^+ + I_2 \tag{5}$$

was titrated against sodium thiosulfate solution. The thallium content can be determined to an accuracy of 0.01 from the titer value. Alternatively, the Tl content in the solution having the precipitate of thallium–perchlorate–thiourea complex may be determined gravimetrically by precipitating Tl as Tl_2CrO_4 with K_2CrO_4 as precipitant in an ammoniacal medium [37].

The total oxidizing power of the system—and hence the oxygen content—was determined by a modified iodometric procedure [38]. A known amount (ca. 100 mg) of the thallium cuprate was dissolved in 10 mL of ice-cold 2 N HBr in a 250-mL Erlenmeyer flask fitted with a ground-glass stopper. The solution was shaken well and cooled in an ice bath; 15 mL of 10% KI was then added to this solution, which was shaken well and warmed to room temperature. The iodine liberated in accordance with the reactions

$$Tl^{(1+n)+} + nI^+ \rightarrow Tl^+ + (n/2)I_2 \tag{6}$$

$$Cu^{(1+n)+} + nI^- \rightarrow Cu^+ + (n/2)I_2 \tag{7}$$

was titrated against sodium thiosulfate. From the titer value, the total oxygen content could be determined.

With the procedures above we can determine the hole concentration in the CuO_2 sheets without any ambiguity in cases where we know that all Tl is present as Tl^{3+}. For example, in Tl_1 compounds $TlBa_2Ca_{n-1}Cu_nO_{2n+3}$, we can determine the hole concentration by this procedure since all Tl is present as Tl^{3+}. On the other hand, in cases where the $Tl^{3+/+}$ couple overlaps the $Cu^{3+/2+}$ couple (i.e., the Tl-6s band overlaps the Fermi energy E_F of the CuO_2 sheets) it is not possible to know the individual oxidations (hole concentrations) present in the CuO_2 sheets and Tl–O layers; in such cases we can only determine the total oxygen content, not the individual hole concentrations in the CuO_2 sheets and Tl–O layers. However, we could establish [33,34] with our procedure that (1) the Tl-6s band does not overlap E_F (i.e., all Tl is present as Tl^{3+}) in the Tl_1 compounds $TlBa_2Ca_{n-1}Cu_nO_{2n+4}$, and (2) the Tl-6$s$ band does overlap E_F (i.e., Tl has an average valence of $<3+$) in the Tl_2 compounds $Tl_{2-y}Ba_2Ca_{n-1}Cu_nO_{2n+4}$, the extent of overlap decreasing sensitively with increasing y (Tl vacancies).

Subsequently, Gopalakrishnan et al. [39] reported that they could determine the hole concentrations in the CuO_2 sheets by dissolving the thallium cuprates in a mixture of HBr and CCl_4 followed by extracting the liberated bromine into the CCl_4 layer and titrating against thiosulfate solution after treating with KI. Based on this procedure and a determination of the total Tl content by our own method, they agreed with our conclusions [33,34] that the oxidation of the CuO_2 sheets in $Tl_2Ba_2CuO_6$ is due to an overlap of the Tl-6s band with the conduction band of the CuO_2 sheets; however, they claim that in the two Tl_2 systems $n = 2, 3$ (e.g., $Tl_2Ba_2CaCu_2O_8$), it is due to Tl vacancies. We have addressed these issues in a recent comment [35]. The procedure of Gopalakrishnan et al. [39] can determine the hole concentration in excess of Tl^{3+} and Cu^{2+}, but not the actual hole concentration in the CuO_2 sheets, as they claim.

III. APPLICATION TO SYSTEMS WITH $n = 1$

A. $Tl_{2-y}Ba_2CuO_{6-x}$ System

1. Synthesis

Two synthetic procedures were used to prepare samples of $Tl_{2-y}Ba_2CuO_{6-x}$; in one, the reactants were wrapped in gold foil during firing, in the other they were sealed in gold tubes. In each procedure the reactants were an intimate mixture of the required quantities of Tl_2O_3, BaO_2, CaO, and CuO (and La_2O_3 or Y_2O_3, where

La is substituted for Ba or Y for Ca) to form a nominal starting composition; the intimate mixture was pelletized before being either wrapped in gold foil or sealed in a gold tube.

Reactants wrapped in gold foil were introduced for 4 min into a muffle furnace maintained at about 800°C and then quenched into liquid nitrogen. The product was ground, repelletized, rewrapped in gold foil, and introduced for 12 min into a muffle furnace maintained at 900 to 910°C followed by quenching into liquid nitrogen.

Reactants sealed in a gold tube were introduced for 2 h into a muffle furnace maintained at 880°C and then quenched into liquid nitrogen. The black product was ground, repelletized, resealed in a gold tube, and introduced for 4 h into a muffle furnace maintained at 860°C in air before quenching into liquid nitrogen.

2. Electron Transfer to Tl-6s Band

All samples were characterized by x-ray powder diffraction recorded with a Philips powder diffractometer. Figure 3a shows a typical pattern; no second phase is detected in samples prepared by our synthetic procedures. We found that quenching into liquid nitrogen is essential to obtain single-phase material; furnace cooling or air quenching results in the appearance of impurity phases such as Tl_2O_3 and $BaCuO_2$.

Table 1 presents our wet-chemical analytical data and the superconductive critical temperature T_c taken as the diamagnetic onset temperature. Sample 1, a product obtained from a $Tl_{2.1}$ starting composition wrapped in gold foil, is to be compared with samples 2 and 5 obtained from a Tl_2 starting composition that had been, respectively, wrapped in gold foil or sealed in a gold tube. The Tl concentrations of samples wrapped in gold foil could be tuned by varying either the temperature or the heating time in the second firing. However, it was not possible to get a reaction without some loss of Tl. The higher the temperature and the longer the firing time, the greater the loss of Tl.

All three samples showed a loss of oxygen as well as of thallium, which rules out the hypothesis that excess oxygen atoms are present; the Tl_2 family is not analogous to the Bi_2 family in this respect. Moreover, the loss of oxygen was $x >$ $3y/2$, which means that the Tl-vacancy concentration y is not great enough to account for an oxidation of the CuO_2 sheets even to the state $(CuO_2)^{2-}$. With the assumption that all the Tl are Tl^{3+} and the oxygen are O^{2-}, the formal valence $n+$ of the copper is calculated to be $n = 1.49, 1.58$, and 1.52, respectively, whereas the observation of p-type superconductivity demands an $n > 2.0$. This observation forces us to conclude that even with some Ca^{2+} substitution on Tl sites, the Tl cannot all be present as Tl^{3+}; the CuO_2 sheets can be oxidized beyond $(CuO_2)^{2-}$ only if there is an electron transfer from the CuO_2 sheets to Tl-6s states.

According to band calculations [23,30,31] for the ideal structures, the Tl-6s band is broad enough to overlap the top of the CuO_2-sheet valence band in the Tl_2 family, but not in the Tl_1 family. With a tight-binding bandwidth $W \approx 2zb$, where b is the energy-transfer integral for an overlap of nearest-neighbor Tl-6s orbitals, the bandwidth of the ideal Tl_2 family having $z = 8$ nearest Tl neighbors is twice that of the Tl_1 family having a $z = 4$. In fact, our Tl_{2-y} samples are not ideal; the introduction of Tl vacancies can be expected to introduce localized Tl-6s states—either at Tl^+ ions or within Tl_m clusters—at the bottom of the Tl-6s

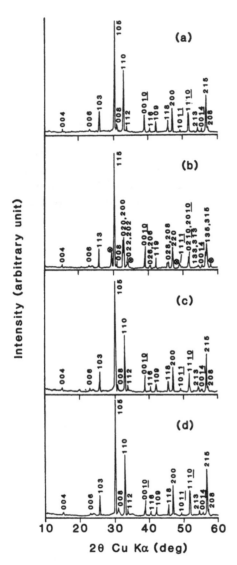

Figure 3   Room-temperature x-ray powder diffraction patterns of (a) $Tl_{1.82}Ba_2CuO_{5.52}$ (sample 2 in Table 1), (b) sample (a) after heating at 1°C/min in O_2 atm to 500°C in thermogravimetric analysis (TGA), (c) sample (b) after heating in O_2 atm at 850°C for 1 h, and (d) sample (c) after heating at 1°C/min in O_2 atm to 500°C in TGA. The symbol $\otimes$ in (b) refers to Tl_2O_3 reflections. The *hkl* values in (a), (c), and (d) are for tetragonal and in (b) for orthorhombic indexing.

band below a mobility edge E_μ [40]. As long as the Fermi energy is $E_F > E_\mu$, the Tl-6*s* band gives metallic conduction; but an $E_F < E_\mu$ would give semiconductive behavior of the Tl-6*s* electrons. The Tl-6*s* electrons give an *n*-type contribution to the transport properties; the holes in the CuO_2 sheets give a *p*-type contribution. This situation is illustrated schematically in Fig. 4a.

Table 1   Analytical Data of $Tl_{2-y}Ba_{2-z}La_zCuO_{6-x}$

| | | Analytical data[b] and T_c[c] | | | | | | | | |
| | | As quenched | | | | After O_2 annealing | | | | Tl_2O_3 |
Sample[a]	Nominal starting composition	Tl content	n of Cu^{n+}	O content $(6-x)$	T_c (K)	Tl content	n of Cu^{n+}	O content $(6-x)$	T_c (K)	extrusion upon oxidation
1	$Tl_{2.1}Ba_2CuO_6$	1.97	1.49	5.70	55	—	—	—	—	yes
2	$Tl_2Ba_2CuO_6$	1.82	1.58	5.52	63	—	—	—	—	yes
3	$Tl_{1.7}Ba_2CuO_6$	1.58	1.86	5.30	64	—	—	—	—	yes
4	$Tl_{1.65}Ba_2CuO_6$	1.50	1.94	5.22	65–50?	—	—	—	—	no
5	$Tl_2Ba_2CuO_6$	1.82	1.52	5.49	74	—	—	—	metal	yes
6	$Tl_2Ba_{1.95}La_{0.05}O_6$	1.80	1.63	5.54	67	—	—	—	15	yes
7	$Tl_2Ba_{1.90}La_{0.10}CuO_6$	1.74	1.68	5.50	54	—	—	—	20	yes
8	$Tl_2Ba_{1.85}La_{0.15}CuO_6$	1.69	1.56	5.39	55	—	—	—	22	yes
9	$Tl_2Ba_{1.80}La_{0.20}CuO_6$	1.60	1.56	5.28	55	—	—	—	38	yes
10	$Tl_{1.65}Ba_2CuO_{6-x}$	1.52	1.88	5.22	66	1.52	1.98	5.27	14	no
11	$Tl_{1.65}Ba_{1.9}La_{0.1}CuO_{6-x}$	1.48	1.96	5.25	64	1.48	2.00	5.27	26	no
12	$Tl_{1.65}Ba_{1.8}La_{0.2}CuO_{6-x}$	1.47	2.03	5.32	64	1.47	2.05	5.33	36	no

[a]Samples 1, 2, 3, and 4 were prepared in gold foils; samples 5–12 were prepared in sealed gold tubes.
[b]Cu^{n+} was obtained by assuming all Tl^{3+} and O^{2-} in the solid sample.
[c]T_cs are from diamagnetic onset except sample 4, which appears to show filamentary superconductivity in the resistance measurements.

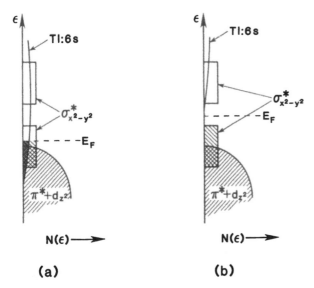

(a)   **(b)**

Figure 4   Energy density of states $N(E)$ versus energy E for nominal (a) $Tl_2Ba_2CuO_6$ and (b) $TlBaLaCuO_5$. (Adapted from Ref. 34.)

3. Formation of Tl_2O_3 and Orthorhombic Distortion

Heating sample 2, $Tl_{1.82}Ba_2CuO_{5.52}$, at 1°C/min in an atmosphere of flowing O_2 gave the thermogravimetric (TGA) curve of Fig. 5a. In the temperature interval $70 < T < 300$°C, the sample picks up oxygen rapidly. The room-temperature x-ray diffraction pattern of the product after the TGA run to 500°C is shown in Fig. 3b; it reveals two important changes from the pattern of Fig. 3a: (1) the appearance of Tl_2O_3 as a second phase, and (2) a splitting of reflections such as (110), (118), and (215) indicative of a distortion of the parent phase to orthorhombic symmetry.

The appearance of Tl_2O_3 with a retention of the parent phase indicates that Tl has been extruded to interact with adsorbed surface oxygen. To test this conjecture, we subjected the two-phase product to an O_2 anneal at 850°C and quenched into liquid nitrogen. The room-temperature x-ray diffraction pattern of this product is shown in Fig. 3c; the Tl_2O_3 has vanished—it apparently sublimed at 850°C—and the $Tl_{2-y}Ba_2CuO_{6-x}$ phase that remains is again tetragonal. A repeat of the

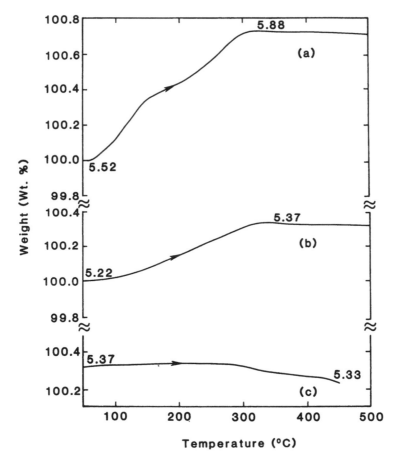

Figure 5 TGA plots at 1°C/min for (a) $Tl_{1.82}Ba_2CuO_{5.52}$ (sample 2 in Table 1) in O_2 atm, (b) $Tl_{1.5}Ba_2CuO_{5.22}$ (sample 4 in table 1) in O_2 atm, and (c) $Tl_{1.5}Ba_2CuO_{5.22}$ (sample 4 in Table 1) in N_2 atm after heating it in O_2 atm in TGA to 500°C [curve (b)] followed by cooling to ambient temperature in O_2 atm. The numbers refer to oxygen content.

TGA run to 500°C on this tetragonal samples showed oxygen uptake over the range $70 < T < 300°C$, but the room-temperature x-ray diffraction pattern of the product (Fig. 3d) shows no evidence of either Tl_2O_3 or a distortion from tetragonal symmetry. These observations suggest two conclusions: that (1) the Tl_2 structure allows topotactic extraction of thallium and/or oxygen, and (2) the orthorhombic distortion is associated with the extrusion of Tl at low temperature to react with adsorbed oxygen to form Tl_2O_3. To test these conclusions we investigated products formed from starting compositions having a lower thallium concentration.

4. Thallium-Deficient Nominal Compositions

The first question we asked was the range of y' in the starting oxide mix that will yield a single-phase product $Tl_{2-y}Ba_2CuO_{6-x}$, where loss of thallium during firing always yields $y < y'$. We found that single-phase samples can readily be obtained for starting compositions having $0 \leq y' < 0.5$, as illustrated by samples 3 and 4 in Table 1. Samples with $y > 0.5$ in the product gave impurity phases, such as $BaCuO_2$.

With our procedure, a starting mix with $y' = 0.35$ gave the product $Tl_{1.5}Ba_2CuO_{5.22}$ (sample 4 of Table 1) corresponding to $y = 0.5$ and $x = 0.78$. The room-temperature x-ray diffraction pattern gave a single-phase product with the tetragonal Tl_2 architecture (Fig. 6a). A TGA run in an O_2 atmosphere to 500°C of this product is shown in Fig. 5b. Considerable oxygen uptake occurs in the range $70 < T < 300°C$ without any segregation of Tl_2O_3, and the product remains tetragonal (Fig. 6b). On the other hand, compositions $Tl_{2-y}Ba_2CuO_{6-x}$ with $0 < y < 0.5$ showed the presence of Tl_2O_3 after a TGA run to 500°C in an O_2 atmosphere. In fact, they also showed Tl_2O_3 formation after an air anneal at 100°C. The amount of the Tl_2O_3 second phase decreased monotonically with increasing y until it disappeared near $y = 0.5$. A distortion to orthorhombic symmetry was always associated with the appearance of Tl_2O_3 as a second phase; moreover, the magnitude of the distortion was proportional to the amount of Tl_2O_3 formed. It should also be noted that the distortion to orthorhombic symmetry has not been observed in the $n = 2$ member of the Tl_2 family even where Tl_2O_3 is formed; our preliminary results indicate that it occurs in our heat treatments only where the Cu have sixfold oxygen coordination.

As illustrated by the system $La_2CuO_{4+\delta}$, distortions from tetragonal to orthorhombic symmetry below a transition temperature T_t may give similar line splittings on a powder x-ray diffraction pattern and yet have quite different origins [41]. In La_2CuO_4, a cooperative tilting of the CuO_6 octahedra [13] is induced by the compressive stress on the CuO_2 sheets; the tilting buckles the CuO_2 planes of the tetragonal structure to give CuO_2 sheets in the orthorhombic structure with a Cu–O–Cu bond angle reduced from 180°. In $La_2CuO_{4.08}$ the orthorhombic distortion is not due to a buckling of the CuO_2 sheets; it appears to reflect an ordering of the interstitial oxygen between the two LaO "planes" of a rock salt layer. The fact that the orthorhombic distortion is found only in association with the extrusion of Tl to form Tl_2O_3 clearly favors a distortion that is linked to the removal of Tl and/or O from the host matrix at temperatures below 500°C. We are thus led to propose that the distortion reflects an ordered removal of Tl from the matrix below 500°C.

The decrease in the amount of Tl_2O_3 formation with increasing y indicates that the driving force for the Tl extrusion is the presence of partially filled Tl-6s states, as illustrated in Fig. 4a. However, no Tl_2O_3 was formed during the TGA run in an

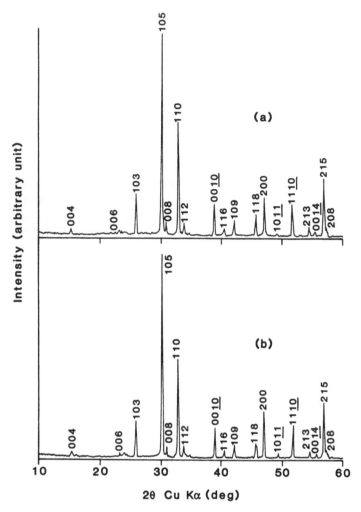

Figure   Room-temperature x-ray powder diffraction patterns of (a) $Tl_{1.5}Ba_2CuO_{5.22}$ (sample 4 in Table 1) and (b) sample (a) after heating at 1°C/min in O_2 atm to 500°C in TGA.

O_2 atmosphere to 500°C of sample 4 ($y = 0.5$) even though it was a superconductor—though perhaps only a filamentary one—before the run with $x > 3y/2$. Since superconductivity in the p-type copper oxides requires an oxidation of the CuO_2 sheets beyond $(CuO_2)^{2-}$, $x > 3y/2$ means that such an oxidation can occur only if there is some electron transfer from the CuO_2 sheets to Tl-6s states. This deduction indicates that the presence of partially filled Tl-6s states is a necessary but not a sufficient condition for the extrusion of Tl to form Tl_2O_3 on oxidation above 70°C.

To rationalize this observation, we note that Tl_2O_3 formation requires both a mobility of the Tl atoms within the $Tl_{2-y}O_{2-x}$ layer and the transfer of electrons from occupied Tl-6s states to the O_2/O^{2-} redox couple of the adsorbed surface oxygen. Above 70°C, at least the $(0.5 - y)$ Tl atoms in excess of $Tl_{1.5}$ are mobile within a $Tl_{2-y}O_{2-x}$ layer of $Tl_{2-y}Ba_2CuO_{6-x}$. In this respect, these excess Tl atoms act as guest atoms in a host. However, the oxidation of $Tl_{2-y}Ba_2CuO_{6-x}$ above

70°C shows that O atoms also act as guest species in a $Tl_{2-y}Ba_2CuO_{6-x}$ host. Moreover, oxidation of the compound requires electron transfer from the CuO_2 sheets to the O_2/O^{2-} redox energy of the adsorbed oxygen. Therefore, we must presume that the O_2/O^{2-} redox energy of the adsorbed oxygen species remains below E_F for all values of y in the range $0 \leq y \leq 0.5$. Since electron transfer between trap states at the bottom of the Tl-6s band would occur for $T > 70°C$ even were $E_F < E_\mu$, we seem forced to conclude that below 500°C only the $(0.5 - y)$ Tl atoms in excess of $Tl_{1.5}$ are mobile within a $Tl_{2-y}O_{2-x}$ layer of the host matrix. Removal of Tl below $Tl_{1.4}Ba_2CuO_{6-x}$ leads to destruction of the host. This conclusion is consistent with our deduction that an ordered removal of Tl is responsible for the orthorhombic distortion that accompanies Tl_2O_3 formation in the $n = 1$ system. On the other hand, $Tl_{1.5}$ samples quenched from 850°C are tetragonal (sample 4 of Table 1) and do not form Tl_2O_3 an oxidation, but are converted from superconductors to normal metals. Thus we conclude that a disordering of the Tl atoms within a $Tl_{1.5}O_{2-x}$ layer, at least on a macroscopic scale, does not render the Tl atoms of the host $Tl_{1.5}Ba_2CuO_{6-x}$ compound mobile below 500°C. We believe that there probably is a short-range ordering of the Tl/O vacancies in a quenched, macroscopically tetragonal $Tl_{1.5}Ba_2CuO_{6-x}$ compound.

Any assignment of the ideal ordered arrangement within a $Tl_{1.5}Ba_2CuO_{6-x}$ host must be speculative. The ideal metal–oxygen bonding arrangement within a nominal, tetragonal $Tl_2Ba_2CuO_6$ compound is illustrated in Fig. 7. An ordered removal of axial TlO units from alternate [110] axes of one of the Tl_2O_2 layers would, for example, yield an orthorhombic structure of the correct unit cell. Whether the resulting asymmetry in the compressive stress on the CuO_2 sheets would, in turn, induce a cooperative rotation of the CuO_6 octahedra in the $n = 1$ system needs to be explored.

5. Superconductivity

Single-phase samples quenched into liquid nitrogen from the reaction temperature were all superconductors (see Table 1). Oxidation of these superconductor

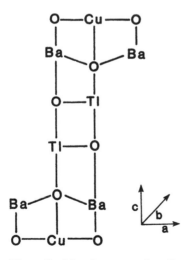

Figure 7 Metal–oxygen bonding arrangement in nominal $Tl_2Ba_2CuO_6$.

$Tl_{2-y}Ba_2CuO_{6-x}$ samples by an O_2 anneal to 500°C yielded, for all values of y in the range $0 \leq y \leq 0.5$, a normal metal having no superconductivity above 4 K whether or not Tl_2O_3 was formed. A subsequent anneal in an N_2 atmosphere at around 450°C caused a loss of oxygen, $\Delta x \approx 0.04$, as illustrated in Fig. 5c, and a reappearance of superconductivity with $T_c \approx 30$ K. Samples that had formed Tl_2O_3 on oxidation remained orthorhombic after the N_2 anneal; the extruded Tl was not reabsorbed into the host. Thus superconductivity is found for both tetragonal and orthorhombic structures, which should put to rest the controversy [19,22,42,43] over whether the superconductivity is dependent on the macroscopic tetragonal or orthorhombic symmetry. Moreover, these experiments clearly demonstrate that the transition from superconductivity to normal metallic behavior is associated with overoxidation of the CuO_2 planes, as originally pointed out by Shimakawa et al. [44], who found that annealing in argon or quenching from high temperature can increase T_c to 80 K, whereas annealing in O_2 gives a normal metallic behavior down to 4 K. The variation in T_c has been shown [43,44] to correlate with the room-temperature lattice c parameter (see Fig. 8); the mean Cu–O distance within the CuO_2 plane remains independent of T_c. The change in the c-parameter would normally be proportional to the change in the apical Cu–O bond length. In this case, from the sum of the Cu–O bond lengths, the c-axis should increase linearly with decreasing oxidation state of the CuO_2 planes for a fixed Cu–O bond length in the planes. Thus T_c is seen to increase linearly with decreasing hole concentration in the CuO_2 planes, reaching a value as high as 90 K [3]. The transition from the superconductive to the normal-metal state as a function of hole concentration in the CuO_2 planes appears to be a smooth but quite sharp function of the hole concentration.

B. $Tl_{2-y}Ba_{2-z}La_zCuO_{6-x}$ $(0 \leq z \leq 0.2)$ System

To understand better the variation of electronic properties with hole concentration, we [34,45] also investigated the system $Tl_{2-y}Ba_{2-z}La_zO_{6-x}$ prepared in sealed gold tubes. Single-phase samples are formed for $0 \leq z \leq 0.2$ as indicated by the lattice-parameter variation with z; this result is in agreement with Nakajima et al. [27]. The analytical data for samples with $0 \leq z \leq 0.2$ are summarized in Table 1, samples 5 to 12, for different starting compositions $y' = 0$ and $y' = 0.35$. All the as-quenched samples were tetragonal and superconductors.

After an O_2 anneal at 350°C, samples with $y \leq 0.4$ (samples 5 to 9 in Table 1) showed evidence of Tl_2O_3 and transformation from tetragonal to orthorhombic symmetry in the room-temperature x-ray diffraction patterns; samples with $y = 0.5 \pm 0.03$ remained tetragonal and showed no evidence of Tl_2O_3 formation. These results are consistent with our observations for the $z = 0$ system and with our interpretation of the orthorhombic distortion as driven by an ordered removal of Tl/O from the Tl–O layer.

Figure 9 shows the lattice-parameter variation with z for the as-quenched samples 5 to 9 of Table 1. In this system also, the Cu–O bond length remains essentially constant at 1.93 Å, which is comparable to the $a/2 = 1.90$ Å found in the superconductor $La_{1.85}Sr_{0.15}CuO_4$, where the in-plane Cu–O bond is known to be under a compressive stress. The cell volume, which decreases as expected for substitution of a smaller La^{3+} ion for Ba^{2+}, is therefore accommodated by a

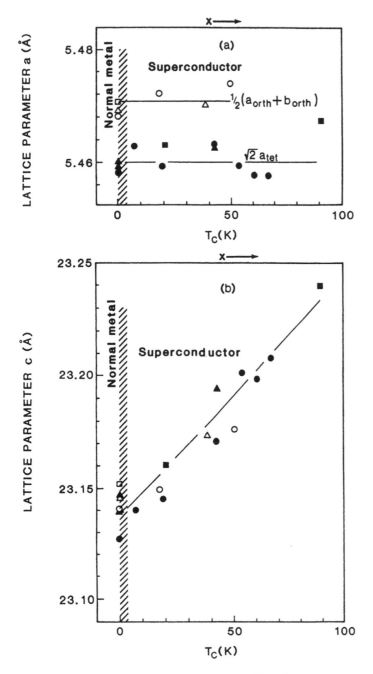

Figure 8 Variation of lattice parameters with T_c in nominal $Tl_2Ba_2CuO_6$. Closed and open symbols represent, respectively, tetragonal and orthorhombic structures; circles and triangles refer to data from Ref. 44, and squares to data from Refs. 3 and 43. (Adapted from Ref. 44.)

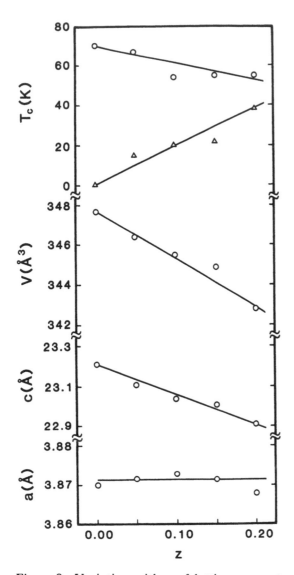

Figure 9 Variation with z of lattice parameters, volume, and T_c for as-quenched $Tl_{2-y}Ba_{2-z}La_zCuO_{6-x}$ with starting parameter $y' = 0.00$ (samples 5 to 9 in Table 1). Also shown is T_c versus z before (circles) and after (triangles) an O_2 anneal at 350°C. The upper error limits in the a and c parameters are, respectively, ± 0.002 and ± 0.01 Å.

decrease in the c parameter. To what extent the decrease in c also indicates a decrease in the apical Cu–O bond length was not determined. Substitution of La^{3+} for Ba^{2+} can be expected to decrease the hole concentration, which would normally translate to a longer apical Cu–O bond length (see the discussion of Fig. 8) despite a large decrease in the c parameter. Figure 9 also shows the variation in T_c with z after oxidation at 350°C for samples 5 to 9 of Table 1. Although oxidation suppresses T_c altogether in sample 5, the change in T_c decreases remarkably with increasing z. If a smaller hole concentration p decreases p to $p < p_s$, where p_s is the value of p at maximum T_c, a change $\Delta p > 0$ with oxidation would decrease the observed value of ΔT_c for a given Δp. Thus the data suggest that p decreases with z and that the as-quenched system $z = 0$ self-adjusts the charge transfer from the CuO_2 sheets to Tl-6s states to yield $p \approx p_s$, but $z > 0$ gives $p < p_s$.

Figure 10 shows the lattice-parameter, volume, and T_c variations with z for (a) the as-quenched and (b) the O_2-annealed (at 350°C) samples 10 to 12 of Table 1. The variations are qualitatively similar to those of Fig. 9 except that the T_c remains high in all the as-quenched samples, which suggests a somewhat smaller electron transfer from the CuO_2 sheets to the Tl-6s states for a smaller value of y.

To correlate the variation in T_c with the concentration p of mobile holes per Cu atom in the CuO_2 sheets, it is necessary to have an independent measure of the concentration n_s of Tl-6s electrons per formula unit since

$$p = n_s - n_d = n_s + n - 2 \tag{8}$$

where $n_d = 2 - n$ is obtained from the oxidation state n of Cu given in Table 1. Without a measure of n_s, we can only make qualitative arguments about the variation of T_c with p.

The as-quenched superconductive samples 10 to 12 of Table 1 have $n \leq 2.03$. From the $La_{2-x}Sr_xCuO_4$ system, it appears that p-type superconductivity requires

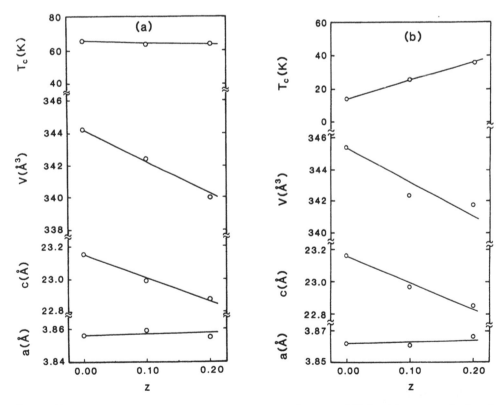

Figure 10 Variation with z of lattice parameters, volume, and T_c for (a) as-quenched and (b) O_2-annealed samples of $Tl_{2-y}Ba_{2-z}La_zCuO_{6-x}$ (samples 10 to 12 in Table 1) with starting parameters $y' = 0.35$.

that $p \geq 0.05$, and $p \geq 0.22$ is more probable for the transition with increasing p from a superconductor to a normal-metal phase. We are thus forced once again to conclude that Tl-6s electrons are still present within the $Tl_{1.5}O_{2-x}$ layers to give $n_s > 0$ even though no Tl is extruded from these samples during an O_2 anneal at 350°C. Moreover, since no Tl is extruded, we assume that n_s remains essentially unchanged not only with changing z, but also between the as-quenched and O_2-annealed samples. However, we find that $\Delta x = -0.05, -0.02$, and -0.01, after the O_2 anneal for, respectively, $z = 0.00, 0.10$, and 0.20 (see Table 1), corresponds to $\Delta T_c = -52, -38$, and -28°C. These data provide further evidence that the system is at the high oxidation side, where it undergoes a transition from superconductor to normal metal with increasing oxidation and that this transition is sharp but probably smooth. They also indicate a decreasing change with increasing z in the hole concentration of a CuO_2 sheet of the as-quenched and O_2-annealed samples.

If we assume that T_c varies with the hole concentration p, it is meaningful to compare quenched samples in Table 1 having similar values of T_c. For example, samples 6 and 10 each have $T_c \approx 66$ K. In one, $y = 0.20$ and $n_d = 0.37$; in the other, $y = 0.48$ and $n_d = 0.12$. If p is the same for each compound, $\Delta n_d = 0.25$ must compensate $\Delta n_s = -0.25$ on going from $y = 0.20$ to $y = 0.48$. This conclusion is reasonable since the concentration of Tl-6s states below E_F must decrease with increasing y. Similarly, a comparison of samples 7 and 8 with $T_c \approx 55$ K shows $n_d = 0.32$ for $y = 0.26$ and $n_d = 0.44$ for $y = 0.31$, which corresponds to $\Delta n_d = 0.12$ compensating for $\Delta n_s \approx -0.12$ with a change $\Delta y \approx 0.05$. On the other hand, comparison of as-quenched samples 8 and 9 shows that $\Delta n_d = 0$ for $\Delta y = 0.09$. We suspect that this anomaly is related to the fact that sample 9 is at the boundary of the solid-solubility limit for La^{3+}-ion substitution for Ba^{2+}. Apart from this anomaly, the data provide clear evidence that n_s decreases monotonically with increasing y, as anticipated from the model of Fig. 4a.

C. $Tl_{1-y}Ba_{1+z}La_{1-z}CuO_{5-x}$ System

Comparison of the Tl_2 family with corresponding members of the Tl_1 family is also instructive. The Tl-6s states of even an ideal Tl_1 compounds are calculated [23] to lie above E_F; we anticipate that the Tl are all present as Tl^{3+} ions. For $n = 1$, the ideal formula $TlBa_2CuO_5$ corresponds to a formal copper valence Cu^{3+} [i.e., to $(CuO_2)^-$]. This oxidation state is too high to be of interest for superconductivity. However, the ideal formula $TlBaLaCuO_5$ corresponds to a $(CuO_2)^{2-}$ plane, as in the parent compounds of all the copper oxide superconductors. Ideally, p-type doping can be achieved by having $z > 0$ in the $TlBa_{1+z}La_{1-z}CuO_5$ system.

The parent composition was prepared by both the gold-foil and sealed-gold-tube methods. The as-quenched samples, 1 and 3 of Table 2, were found to be tetragonal and semiconductive; they were also single phase for product compositions $Tl_{1-y}BaLaCuO_{5-x}$ having y in the range $0 < y \leq 0.2$. The measured oxidation for Cu is $n = 2.00 \pm 0.01$ for both samples given all Tl present as Tl^{3+}; the planes of all copper oxide parent compounds are semiconductive and antiferromagnetic. Any charge transfer to Tl-6s bands as in the Tl_2 family would give superconductor or normal-metal behavior, so we conclude that the Tl are all present as Tl^{3+}, as anticipated and illustrated in Fig. 4b.

Table 2   Analytical Data of $Tl_{1-y}Ba_{1+z}La_{1-z}CuO_{5-x}$ and $TlBa_2YCu_2O_7$

Sample[a]	Nominal starting composition[b]	As quenched				After O_2 annealing				Tl_2O_3 extrusion upon oxidation
		Tl content	n of Cu^{n+}	O content	T_c (K)	Tl content	n of Cu^{n+}	O content	T_c (K)	
1	$TlBaLaCuO_5$	0.83	2.01	4.75	semiconductor	—	—	—	—	no
2	$TlBa_2YCu_2O_7$	0.88	1.88	6.70	semiconductor	—	—	—	—	no
3	$TlBaLaCuO_5$	0.91	1.99	4.86	semiconductor	0.91	2.02	4.88	semiconductor	no
4	$TlBa_{1.1}La_{0.9}CuO_5$	0.86	1.95	4.72	semiconductor	0.86	2.05	4.77	semiconductor	no
5	$TlBa_{1.2}La_{0.8}CuO_5$	0.90	1.93	4.72	semiconductor	0.90	1.99	4.75	28	no
6	$TlBa_{1.3}La_{0.7}CuO_5$	0.86	1.92	4.60	38	0.86	1.98	4.63	38	no
7	$TlBa_{1.4}La_{0.6}CuO_5$	0.87	1.92	4.57	42	0.87	2.00	4.61	39	no
8	$TlBa_{1.5}La_{0.5}CuO_5$	0.87	2.06	4.59	40	0.87	2.09	4.61	36	no

[a]Samples 1 and 2 were prepared in gold foils; samples 3–8 were prepared in sealed gold tubes.
[b]Analytical data for $z = 0.6$ are not given as it is contaminated with small amounts of impurity phases.
[c]Cu^{n+} was calculated on the basis of all Tl^{3+} and O^{2-} in the solid sample.

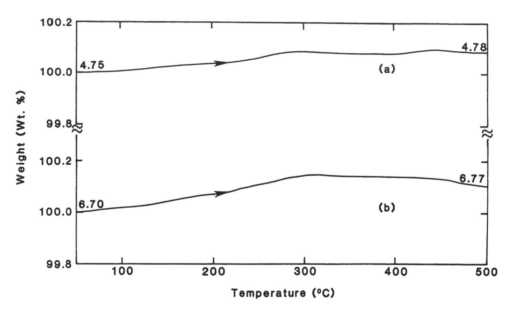

Figure 11 TGA plots at 1°C/min in an O_2 atm for (a) $Tl_{0.83}BaLaCuO_{4.75}$ (sample 1 in Table 2) and (b) $Tl_{0.88}Ba_2YCu_2O_{6.70}$ (sample 2 in Table 2). The numbers refer to oxygen content.

Figure 11a shows a TGA run at 1°C/min in flowing O_2 for sample 1 in Table 2; it picks up only a small amount ($\Delta x = -0.03$) of oxygen, x-ray diffraction shows no evidence of Tl_2O_3, and the fully oxidized sample was also found to be semiconductive. Similarly, a small amount of oxygen uptake was also found in the semiconductive $TlBa_2YCu_2O_7$ (see Fig. 11b).

Two groups have reported superconductivity in the $Tl_{1-y}Ba_{1+z}La_{1-z}CuO_{5-x}$ system, but each found it in a quite different range of z: Manako et al. [46] give $0.01 \leq z \leq 0.36$ and Ku et al. [47] $0.40 < z \leq 1.00$. Subramanian et al. [48] have reported $T_c = 52$ K for $z = 0.3$. All workers report an increase in the in-plane Cu–O bond length from its value in $TlBa_2CuO_5$ on substituting La^{3+} for Ba^{2+}, which is consistent with a reduction of the CuO_2 planes. In an attempt to clarify this situation, we used sealed gold tubes to prepare the system $Tl_{1-y}Ba_{1+z}La_{1-z}CuO_{5-x}$. We obtained single-phase products in the range $0 \leq z \leq 0.5$; a $z = 0.6$ sample contained a small amount of an impurity phase. The analytical data for the as-quenched and O_2-annealed samples (at 350°C) are given as samples 3 to 8 in Table 2. Superconductivity is found in the range $0.3 \leq z \leq 0.5$ for the as-quenched samples, in the range $0.2 \leq z \leq 0.5$ for the O_2-annealed samples.

Figure 12 shows the variation with z of the tetragonal lattice parameters and cell volume of the as-quenched samples. A remarkable change occurs on passing from the semiconductor $z = 0.2$ to the superconductor $z = 0.3$. From these preliminary data we conclude that there is a first-order transition between the antiferromagnetic semiconductor compositions and the superconductor compositions, probably with a two-phase region in the interval $0.2 < z < 0.4$. The increase in the c parameter and the volume V with increasing z are consistent with the substitution of a larger Ba^{2+} ion for La^{3+}. However, a discontinuous increase in c on changing from the semiconductor to the superconductor phase seems surprising

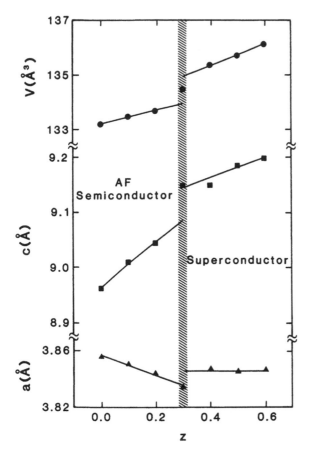

Figure 12 Variation of lattice parameters and volume with z for $Tl_{1-y}Ba_{1+z}La_{1-z}CuO_{5-x}$ (samples 3 to 8 in Table 2).

unless it is accompanied by a change in the position of the c-axis oxygen coordinating a Cu atom from above the Ba,La plane to below it. This conjecture needs to be checked with neutron diffraction.

Values of $n < 2.00$ for samples 4 to 7 are puzzling, particularly in the case of superconductor samples 6 and 7, since the Tl should all be present as Tl^{3+}. This observation suggests that there may be problems with chemical inhomogeneity, some $Tl_{1.5}$ regions intergrowing with the Tl_1 phase, for example. These types of difficulties may explain why different workers have reported superconductivity in such different ranges of z. Clearly, the work reported to date on this system must be considered preliminary.

IV. APPLICATIONS TO SYSTEMS WITH $n = 2$

A. $Tl_{2-y}Ba_2CaCu_2O_{8-x}$ System

Single-phase tetragonal samples of $Tl_{2-y}Ba_2CaCu_2O_{8-x}$ were obtained with y in the range $0 < y < 0.5$ by both the gold-foil and the sealed-gold-tube methods. The analytical data of Table 3 for the as-quenched samples indicate an electron transfer

Table 3   Analytical Data of $Tl_{2-y}Ba_2Ca_{1-z}Y_zCu_2O_{8-x}$

Sample[a]	Nominal starting composition	Tl content $(2 - y)$	n of Cu^{n+}	O content	T_c (K)	Tl_2O_3 extrusion upon oxidation
		Analytical data[b]				
1	$Tl_{2.1}Ba_2CaCu_2O_8$	1.90	1.78	7.63	77	yes
2	$Tl_2Ba_2CaCu_2O_8$	1.79	1.80	7.49	83	yes
3	$Tl_{1.6}Ba_2CaCu_2O_8$	1.51	2.05	7.32	80	no
4	$Tl_2Ba_2CaCu_2O_8$	1.79	—	7.59	94	yes
5	$Tl_2Ba_2Ca_{0.9}Y_{0.1}Cu_2O_8$	1.73	—	7.51	82	yes
6	$Tl_2Ba_2Ca_{0.8}Y_{0.2}Cu_2O_8$	1.70	—	7.49	72	yes
7	$Tl_2Ba_2Ca_{0.7}Y_{0.3}Cu_2O_8$	1.78	—	7.58	semiconductor	yes

[a]Samples 1, 2, and 3 were prepared in gold foils; samples 4, 5, 6, and 7 were prepared in sealed gold tubes.
[b]Total oxidizing power was calculated for all Tl^+, Cu^+, and O^{2-} at the end of the titration; Cu^{n+} was obtained on the basis of all Tl^{3+} and O^{2-} in the solid sample.

from CuO_2 sheets to Tl-6s states that decreases with increasing y just as in the $n = 1$ case of Table 1. Figure 13a shows a TGA run at 1°C/min in an O_2 atmosphere for sample 2 of Table 3, $Tl_{1.79}Ba_2CaCu_2O_{7.49}$; as in the $n = 1$ case, the sample rapidly picks up oxygen in the interval $70 < T < 300$°C, and the x-ray diffraction pattern after heating to 500°C showed the presence of Tl_2O_3. Moreover, on heating the oxidized product in N_2 to 375°C (Fig. 13b), the sample loses about $\Delta x \approx 0.06$ oxygen per formula unit. Thus this system behaves like the $n = 1$ analog as a $Tl_{1.5}Ba_2CaCu_2O_{8-x}$ host containing the excess $(0.5 - y)$ Tl and O atoms as guest species that are topotactically extracted from the $Tl_{2-y}O_{2-x}$ layer.

Despite this similarity, the $n = 2$ system differs from the $n = 1$ system in two significant respects: (1) the $n = 2$ system does not exhibit any macroscopic transformation to orthorhombic symmetry with the extrusion of Tl to form Tl_2O_3, and (2) the value of T_c remains high after annealing in O_2 or N_2. The lack of any orthorhombic distortion where the Cu have only one apical c axis oxygen near neighbor rather than two as in the $n = 1$ case would seem to suggest that a cooperative tilting of the CuO_6 octahedra in the $n = 1$ system may accompany an ordered removal of Tl to induce a long-range ordering. But this speculation needs to be checked out by neutron diffraction. The insensitivity of T_c to both the Tl and O content would seem to suggest that the charge transfer between CuO_2 sheets and Tl-6s states tends to be self-adjusted so as to maximize T_c. A similar trend can be noted for the as-quenched samples of Table 1. For all as-quenched samples, the short reaction times of the gold-foil syntheses appear to be more detrimental to T_c relative to the better equilibrated samples prepared in sealed gold tubes than do variations in y and x. This observation is consistent with the identification of superconductivity with a distinguishable thermodynamic state between the antiferromagnetic-semiconductor and normal-metal phases.

B. $Tl_{2-y}Ba_2Ca_{1-z}Y_zCu_2O_{8-x}$ System

To explore further the dependence of T_c on the hole concentration in the CuO_2 sheets of the $n = 2$ system, we [34,49] substituted Y^{3+} for Ca^{2+}. Nearly single-phase ($<5\%$ $BaCO_3$ impurity) $Tl_{2-y}Ba_2Ca_{1-z}Y_zCu_2O_{8-x}$ samples were obtained

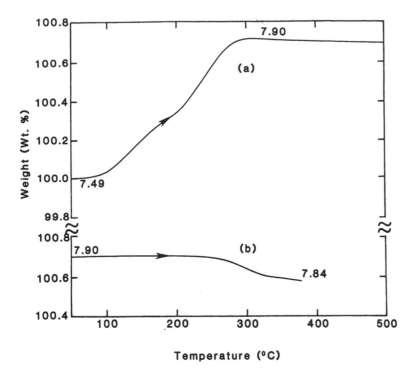

Figure 13 TGA plots at 1°C/min for (a) $Tl_{1.79}Ba_2CaCu_2O_{7.49}$ (sample 2 in Table 3) in O_2 atm and (b) $Tl_{1.79}Ba_2CaCu_2O_{7.49}$ (sample 2 in Table 3) in N_2 atm after heating it in O_2 atm in TGA to 500°C [curve (a)] followed by cooling to ambient temperature in O_2 atm. The numbers refer to oxygen content.

over the compositional range $0 \leq z \leq 0.3$ with the sealed-gold-tube synthesis. For $z = 0.4$, a small amount of $TlBa_2YCu_2O_7$ could be detected as an additional impurity. The analytical data for the as-quenched products are given as samples 4 to 7 in Table 3.

In Fig. 14 the variation of the lattice parameters with z of the as-quenched samples reflects the substitution of a smaller Y^{3+} ion for Ca^{2+} and reveals a first-order dilatation of the structure on passing from a superconductive phase for $z \leq 0.2$ to a semiconductive phase for $z \geq 0.3$. A first-order dilatation is expected for a transition from a superconductor phase of higher hole concentration in the CuO_2 sheets to an antiferromagnetic, semiconductor phase of small hole concentration in these sheets [50,51]. Comparison of the values of $(2 - y)$ and $(8 - x)$ for the superconductor and semiconductor samples 4 and 7, respectively, in Table 3 shows that the values of y and x are not responsible for the transition; it is the substitution of Y^{3+} for Ca^{2+} that depletes the CuO_2 sheets of holes and induces the transition. Clearly, the superconductor and semiconductor phases represent distinguishable thermodynamic states; a smooth transition between the superconductor and normal-metal phases makes their identification as distinguishable thermodynamic states less obvious.

Figure 14 also shows the variation of T_c with z. A linear decrease in T_c from 94 K for $z = 0.0$ to 72 K for $z = 0.2$ is characteristic of a linear decrease with z in the hole concentration in the CuO_2 sheets. Since the values of y and x do not

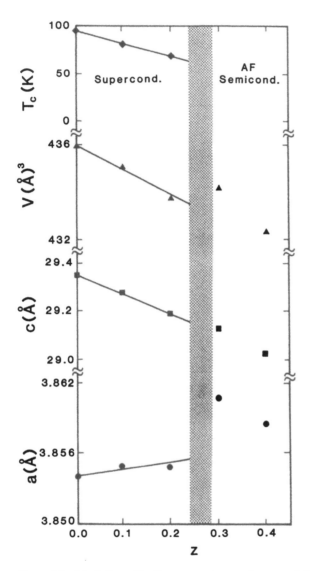

Figure 14   Variation of lattice parameters, volume and T_c with z for $Tl_{2-y}Ba_2Ca_{1-z}Y_zCu_2O_{8-x}$ (samples 4 to 7 in Table 3).

vary appreciably or systematically with z, it would appear that the decrease in hole concentration with z is due essentially to the substitution of Y^{3+} for Ca^{2+}, which leads to

$$\frac{dp}{dz} = -1 \tag{9}$$

The fact that the $z = 0.3$ sample is a semiconductor indicates not only an oxidation state close to $(CuO_2)^{2-}$ for the CuO_2 sheets, but also an $E_F < E_\mu$ in the $Tl_{2-y}O_{2-x}$ layers with $y \approx 0.22$. The fact that Tl is extruded to form Tl_2O_3 indicates that Tl-6s electrons are present even though the loss of Tl from the $Tl_{2-y}O_{2-x}$

layer has placed E_F below the mobility edge E_μ. We estimate the hole concentration in the CuO_2 sheets of the $z = 0.0$ semiconductor as follows. In the $La_{2-x}Sr_xCuO_4$ system, the CuO_2 sheets remain superconductive for a hole concentration $p \geq 0.04$ per Cu atom. If we therefore assume a hole concentration per Cu atom $p = 0.02 \pm 0.02$ for the semiconductor sample with $z = 0.3$, it follows from Table 3 that a total of 0.52 ± 0.04 electrons per formula unit (i.e., 0.26 ± 0.02 electrons per Cu atom) have been transferred from the CuO_2 sheets to the $Tl_{2-y}O_{2-x}$ layers to yield $p = 0.02 \pm 0.03$ holes per Cu atom. Since the values of y and x are nearly the same for $z = 0.0$ and $z = 0.3$, we assume that the charge transfer is the same for both samples. It follows from Table 3 that 0.10 electron must be transferred to make the CuO_2 sheets have the formal valence $(CuO_2)^{2-}$, which translates to $p = 0.16 \pm 0.02$ holes per Cu atom for a $T_c = 94$ K. This estimate agrees well with the $p = p_s$ value of maximum T_c for the other copper oxide superconductors, which is $0.15 \leq p_s \leq 0.20$. From the linear relationship between T_c and z, this value of T_c for $z = 0.0$ gives

$$T_c \approx (370p + 35 \pm 7) \text{ K} \tag{10}$$

Figure 15 shows the temperature dependence of the Seebeck coefficient for various values of z. By comparison with the $La_{2-x}Sr_xCuO_4$ system [50], the curves for $z = 0.0$ and 0.2 are typical for $p < p_s$ with p approaching p_s at $z = 0.0$, which means that $p_s \geq 0.16 \pm 0.02$. We choose

$$p_s \approx 0.18 \pm 0.02 \text{ per Cu atom} \tag{11}$$

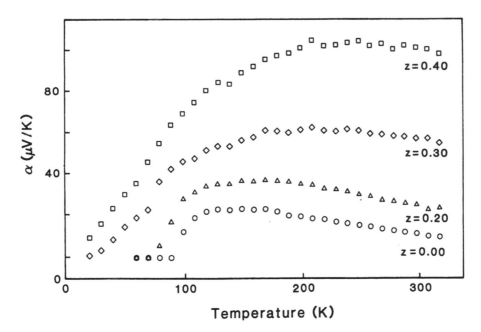

Figure 15 Variation of Seebeck coefficient with temperature for $Tl_{2-y}Ba_2Ca_{1-z}Y_zCu_2O_{8-x}$ (samples 4 to 7 in Table 3).

Finally, we note that the temperature dependence of the Seebeck coefficient for $z = 0.03$ is similar to that found [52] for sintered pellets of $La_{2-x}Sr_xCuO_4$ with $0.02 \leq x \leq 0.04$, where there is evidence for the onset of charge fluctuations below $T_\rho \approx 140$ K; we have interpreted the charge fluctuations to be a dynamic segregation into domains rich in holes and domains poor in holes via atomic displacements giving domains, respectively, of small and larger mean Cu–O bond length. We have taken these charge fluctuations to represent another signal of the distinguishable character of the semiconductor and superconductor phases.

V. CONCLUSIONS

In the Tl_1 systems, all the thallium is present as Tl^{3+}, and our wet-chemical analysis permits determination of the oxidation state of the CuO_2 sheets. Both thallium and oxygen are generally lost during formation of the $Tl_{2-y}Ba_2Ca_{n-1}Cu_nO_{2n+4-x}$ ($n = 1$ or 2) systems. The wet-chemical procedure presented here for determining the thallium and oxygen contents permits determination of the total oxidation state of a compound, but it does not allow determination of the oxidation state of the superconductive CuO_2 sheets without an independent determination of the electron transfer from the CuO_2 sheets to Tl-6s states in the $Tl_{2-y}O_{2-x}$ layers. Nevertheless, our preliminary experiments on systems with $n = 1$ and 2 have allowed us to determine the following:

1. The Tl_2 structural architecture of Fig. 1 is retained for compositions $Tl_{2-y}Ba_2Ca_{n-1}Cu_nO_{2n+4-x}$ with $0 \leq y \leq 0.5$. The systems act as though a $Tl_{1.5}Ba_2Ca_{n-1}Cu_nO_{2n+4-x}$ phase is the host matrix with O and $(0.5 - y)$Tl atoms as guest species that can be extracted from the $Tl_{2-y}O_{2-x}$ layer without disrupting the basic structural architecture. The guest O atoms act as acceptor O^{2-} anions; the guest Tl atoms act as donor cations of variable mean valence.

2. The CuO_2 sheets are under a compressive stress and may be doped p-type.

3. Analysis shows that p-type superconductors or normal metals generally have $x > 3y/2$, which means that p-type doping is achieved by electron transfer from the CuO_2 sheets to Tl-6s states in the $Tl_{2-y}O_{2-x}$ layers. The amount of electron transfer decreases with increasing y.

4. The Tl-6s band develops a tail of localized states at the bottom of the band below a mobility edge E_μ, and E_F falls below E_μ at larger values of y.

5. Retention of smaller y values can be accomplished by quenching from the reaction temperature into liquid nitrogen. Synthesis from starting compositions having $y > 0$ can yield single-phase samples with a final $y = 0.5$ in the as-ground sample; a final $y > 0.5$ results in impurity phases.

6. Heating the as-quenched samples in air or O_2 shows oxygen uptake in the temperature range $70 < T < 300°C$ and a distortion from tetragonal to orthorhombic symmetry in the $n = 1$ systems associated with the formation of Tl_2O_3. The magnitude of the distortion and the amount of Tl_2O_3 formed decreases with increasing y, disappearing as y approaches $y = 0.5$ in the as-quenched samples. Reduction of the $Tl_{2-y}O_{2-x}$ layer from all Tl^{3+} and Tl mobility within the layer above 70°C allows an extrusion of Tl to react with adsorbed O_2 to form Tl_2O_3. However, only $(0.5 - y)$Tl per layer is mobile below 500°C; removal of Tl at low temperature is ordered at long range to give a macroscopic distortion to orthorhombic symmetry in the $n = 1$ system.

7. The as-quenched samples are superconductors. The hole concentration per Cu atom in the CuO_2 sheets that corresponds to a maximum T_c is estimated to be $p = 0.18 \pm 0.02$. For the $n = 1$ system, $p > p_s$ is found in the as-quenched samples, and annealing in O_2 to 500°C transforms the superconductors to normal metals. However, substitution of La for Ba in the $n = 1$ system and Y for Ca in the $n = 2$ system creates $p < p_s$, and annealing in O_2 to increase p to $p > p_s$ does not increase p sufficiently to suppress the superconductivity. With such doping, it is possible to decrease p sufficiently that the system becomes a semiconductor. With increasing p, a first-order phase change separates the semiconductor and superconductor phases; the transition from superconductor to normal metal appears to be smooth, but sharp. The data are consistent with the hypothesis that the superconductors occur as a thermodynamically distinguishable phase between an antiferromagnetic semiconductor phase and a normal-metal phase.

ACKNOWLEDGMENT

Financial support by the National Science Foundation Grant DMR 90 15490 is gratefully acknowledged.

REFERENCES

1. Z. Z. Sheng and A. M. Hermann, *Nature* 332, 55 (1988); ibid., 332, 138 (1988).
2. R. M. Hazen, L. W. Finger, R. J. Angel, C. T. Prewitt, N. I. Ross, C. G. Hadidiacos, P. J. Heaney, D. R. Veblen, Z. Sheng, A. El Ali, and A. M. Hermann, *Phys. Rev. Lett.* 60, 1657 (1988).
3. C. C. Torardi, M. A. Subramanian, J. C. Calabrese, J. Gopalakrishnan, E. M. McCarron, K. J. Morrissey, T. R. Askew, R. B. Flippen, U. Chowdhry, and A. W. Sleight, *Phys. Rev B* 38, 225 (1988).
4. C. C. Torardi, M. A. Subramanian, J. C. Calabrese, J. Gopalakrishnan, K. J. Morrissey, T. R. Askew, R. B. Flippen, U. Chowdhry, and A. W. Sleight, *Science* 240, 631 (1988).
5. S. S. P. Parkin, V. Y. Lee, E. M. Engler, A. I. Nazzal, T. C. Huang, G. Gorman, R. Savoy, and R. Beyers, *Phys. Rev. Lett.* 60, 2539 (1988).
6. S. S. P. Parkin, V. Y. Lec, A. I. Nazzal, R. Savoy, R. Beyers, and S. J. La Placa, *Phys. Rev. Lett.* 61, 750 (1988).
7. D. E. Cox, C. C. Torardi, M. A. Subramanian, J. Gopalakrishnan, and A. W. Sleight, *Phys. Rev. B* 38, 6624 (1988).
8. H. Ihara, R. Sugise, M. Hirabayashi, N. Terada, M. Jo, K. Hayashi, A. Negishi, M. Tokumoto, Y. Kimura, and T. Shimomura, *Nature* 334, 510 (1988).
9. A. W. Sleight, *Science* 242, 1519 (1988) and references therein.
10. M. A. Subramanian, C. C. Torardi, J. Gopalakrishnan, P. L. Gai, J. C. Calabrese, T. R. Askew, R. B. Flippen, and A. W. Sleight, *Science* 242, 249 (1988).
11. M. Kikuchi, T. Kajitani, T. Suzuki, S. Nakajima, K. Hiraga, N. Kobayashi, H. Iwasaki, Y. Syono, and Y. Muto, *Jpn. J. Appl. Phys.* 28, L382 (1989).
12. M. Herview, A. Maignan, C. Martin, C. Michel, J. Provost, and B. Raveau, *J. Solid State Chem.* 75, 212 (1988).
13. J. B. Goodenough, *Supercond. Sci. Technol.* 3, 26 (1990).
14. A. Manthiram and J. B. Goodenough, in *Advances in the Synthesis and Reactivity of Solids*, Vol. 1, ed. T. E. Mallouk, JAI Press, Greenwich, Conn. 1991, p. 1.
15. J. B. Goodenough and A. Manthiram, *J. Solid State Chem.* 88, 115 (1990).

16. J. B. Goodenough, J.-S. Zhou, and J. Chan, in *Proceedings of Workshop on Lattice Effects in High-T_c Superconductors*, Sante Fe, N. Mex., Jan. 13–15, 1992. T. Egami, Y. Bar-Yam, J. Mastre-de Leon, and A. Bishop, eds. (World Scientific Publ. Co., Ltd, Singapore, 1993) p. 137.

17. M. Kikuchi, S. Nakajima, and Y. Syono, K. Miraga, T. Oku, D. Shindo, N. Kobayashi, H. Iwasaki, and Y. Muto, *Physica C*, 158, 79 (1989).

18. S. J. Hibble, A. K. Cheetham, A. M. Chippindale, P. Day, and J. A. Hsiljac, *Physica C* 156, 604 (1988).

19. A. W. Hewat, P. Bordet, J. J. Capponi, C. Chaillout, J. Chenavas, M. Godinho, E. A. Hewat, J. L. Hodeau, and M. Marezio, *Physica C* 156, 369 (1988); ibid., 156, 375 (1988).

20. J. B. Parise, C. C. Torardi, M. A. Subramanian, J. Gopalakrishnan, A. W. Sleight, and E. Prince, *Physica C* 159, 239 (1989).

21. J. B. Parise, N. Herron, M. K. Crawford, and P. L. Gai, *Physica C* 159, 255 (1989).

22. D. M. de Leeuw, W. A. Groen, J. C. Hol, H. B. Brom, and H. W. Zandbergen, *Physica C* 166, 349 (1990).

23. D. Jung, M.-H. Whangbo, N. Herron, and C. C. Torardi, *Physica C* 160, 381 (1989).

24. C. C. Torardi, J. B. Parise, M. A. Subramanian, J. Gopalakrishnan, and A. W. Sleight, *Physica C* 157, 115 (1989).

25. A. Manthiram and J. B. Goodenough, *Appl. Phys. Lett.* 53, 420 (1988); A. Manthiram and J. B. Goodenough, *Physica C* 162–164, 69 (1989).

26. C. C. Torardi, D. Jung, D. B. Kang, J. Ren, and M.-H. Whangbo, *Mater. Res. Soc. Symp. Proc.*, Vol. 156, ed. J. B. Torrance, K. Kitazawa, J. M. Tarascon, J. D. Jorgensen, and M. Thompson, Materials Research Society, Pittsburgh, Pa., 1989, p. 295.

27. S. Nakajima, M. Kikuchi, T. Oku, N. Kobayashi, T. Suzuki, N. Nagese, K. Hiraga, Y. Muto, and Y. Syono, *Physica C* 160, 458 (1989).

28. T. Suzuki, M. Nagoshi, Y. Fukuda, Y. Syono, M. Kikuchi, N. Kobayashi, and M. Tachiki, *Phys. Rev. B* 40, 5184 (1989).

29. H. M. Meyer III, T. J. Wagener, J. H. Weaver, and D. S. Ginley, *Phys. Rev. B* 39, 7343 (1989).

30. D. R. Hamann and L. F. Mattheiss, *Phys. Rev. B* 38, 5138 (1988).

31. P. Marksteiner, Jaejun Yu, S. Massidda, A. J. Freeman, J. Redinger, and P. Weinberger, *Phys. Rev. B* 39, 2894 (1989).

32. A. Manthiram, J. S. Swinnea, Z. T. Sui, H. Steinfink, and J. B. Goodenough, *J. Am. Chem. Soc.* 109, 6667 (1987).

33. M. Paranthaman, A. Manthiram, and J. B. Goodenough, *J. Solid State Chem.* 87, 479 (1990).

34. A. Manthiram, M. Paranthaman, and J. B. Goodenough, *Physica C* 171, 135 (1990).

35. A. Manthiram, M. Paranthaman, and J. B. Goodenough, *J. Solid State Chem.* 96, 464 (1992).

36. M. Kolthoff and R. Belcher, *Volumetric Analysis III*, Interscience, New York, 1957, p. 370.

37. C. Mahr and H. Ohle, *Z. Anal. Chem.* 115, 254 (1939).

38. E. H. Appelman, L. R. Morss, A. M. Kini, U. Geiser, A. Umezawa, G. W. Grabtree, and K. D. Carlson, *Inorg. Chem.* 26, 3237 (1987).

39. J. Gopalakrishnan, R. Vijayaraghavan, R. Nagarajan, and C. Shivakumara, *J. Solid State Chem.* 93, 272 (1991).

40. N. F. Mott and E. H. Davis, *Electronic Processes in Non-crystalline Materials*, Clarendon Press, Oxford, 1971.

41. J. D. Jorgensen, B. Dabrowski, Shiyou Pei, D. G. Hinks, and L. Soderholdm, *Phys. Rev. B* 38, 11337 (1988).

42. T. C. Huang, V. Y. Lee, R. Karimi, R. Beyers, and S. S. P. Parkin, *Mater Res. Bull.* 23, 1307 (1988).

43. K. V. Ramanujachary, S. Li, and M. Greenblatt, *Physica C* 165, 377 (1990).

44. Y. Shimakawa, Y. Kubo, T. Satoh, S. Iijima, T. Ichihashi, and H. Igarashi, *Physica C* 157, 279 (1989).

45. M. Paranthaman, A. Manthiram, and J. B. Goodenough, *J. Mater. Chem.*, 3, 317 (1992).

46. T. Manako, Y. Shimakawa, Y. Kubo, T. Satoh, and H. Igarashi, *Physica C* 158, 143 (1989).

47. H. C. Ku, M. F. Tai, J. B. Shi, M. J. Shieh, S. W. Hsu, G. H. Hwang, D. C. Ling, T. J. Watson-Yang, and T. Y. Lin, *Jpn. J. Appl. Phys.* 28, L923 (1989).

48. M. A. Subramanian, G. H. Kwei, J. B. Parise, J. A. Goldstone, and R. B. von Dreele, *Physica C* 166, 19 (1990).

49. M. Paranthaman, A. Manthiram, and J. B. Goodenough, *J. Solid State Chem.*, 98, 343 (1992).

50. J. B. Goodenough and A. Manthiram, *Physica C* 157, 439 (1989).

51. J.-S. Zhou, S. Sinha, and J. B. Goodenough, *Phys. Rev. B* 39, 12331 (1989).

52. J. B. Goodenough, J.-S. Zhou, and K. Allen, *J. Mater. Chem.* 1, 715 (1991).

8

Single-Crystal Growth and Characterization of Thallium Cuprate Superconductors: A Review

Mariappan Paranthaman, H. M. Duan, and Allen M. Hermann
University of Colorado at Boulder
Boulder, Colorado

I. INTRODUCTION

Research on single-crystal growth and characterization of thallium cuprate super-conductors has not yet been carried out thoroughly throughout the world even after 4 years of research on these systems. This could be due to the volatility of thallium during crystal growth and hence the uncertainty in Tl content, lack of phase stability, and inability to grow large crystals using standard flux methods. Many unusual normal-state properties have also been reported in the literature. To understand the possible superconducting mechanisms in thallium systems, one needs to know the following characteristics: (1) how in-plane (ab-plane) resistivities are compared with out-of-plane (c-axis) resistivities, (2) whether the resistivity versus temperature curve (in the normal-state region) in the c axis is metal-like or semiconductor-like, (3) interlayer interactions, (4) fluctuation effects, and (5) the dimensionality of these systems. Hence there is a need to grow large single crystals and to carry out all the transport property measurements on those crystals. Therefore, we have attempted to review all the single-crystal studies on thallium systems that are available at present. We report here single-crystal growth of thallium cuprate superconductors, anisotropic resistivity measurements, paraconductivity measurements, effects of annealing and corresponding magnetization, irradiation effects, and pressure dependence of T_c on different systems. In the text we represent the double-Tl–O compounds $Tl_2Ba_2CuO_6$, $Tl_2Ba_2CaCu_2O_8$, and $Tl_2Ba_2Ca_2Cu_3O_{10}$ as Tl-2201, Tl-2212, and Tl-2223, respectively. We represent the mono-Tl–O compounds $TlBa_2CuO_5$, $TlBa_2CaCu_2O_7$, and $TlBa_2Ca_2Cu_3O_9$ as Tl-1201, Tl-1212, and Tl-1223, respectively.

II. SINGLE-CRYSTAL GROWTH

Single crystals of T1 compounds have been grown primarily by the self-flux technique. Apart from the stoichiometric compositions, excess CaO and CuO were taken as the charge and they act like a flux to grow reasonably large crystals (on the order of several millimeters in size). To compensate for thallium loss during the crystal growth, excess Tl_2O_3 was also used. Gold crucibles [1–6], platinum crucibles [7–10], and alumina crucibles [11–14] were used as the containers for growth. Even though the melt is contaminated with Al_2O_3 [9,10], large crystals were obtained by using alumina crucibles [11–14]. The typical experimental setup used for the crystal growth is shown in Fig. 1. Either gold crucibles were sealed or alumina and platinum crucibles were covered with lids to protect Tl_2O_3 from direct evaporation. Sometimes even the lids were sealed to the alumina crucibles

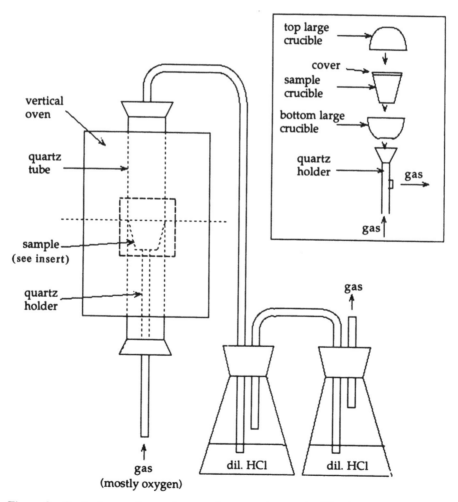

Figure 1 Typical experimental setup for crystal growth. The insert shows the enlarged picture of the crucible assembly inside the dashed rectangular portion. (Adapted from Ref. 14.)

by using ceramic adhesive (AREMCO, Model 569) [14]. Typically, 25 g of the sample was used as the charge [14]. Once the crucibles with charge are loaded in a vertical tube furnace, they are typically heated rapidly to 900 to 950°C and held for 1 to 3 h, then slowly cooled through the melt at the rate of 2 to 20°C/h to 750 to 780°C, and finally cooled down to room temperature. Instead of using one-step self-flux technique, some researchers have used a two-step process where large crystals are often obtained [14]. In the two-step process, the same starting compositions were used and the bulk Tl samples were made by heating at 920 to 950°C for 10 min and cooling fast, then grinding, pelletizing, and heating to the melting temperature and slowly cooling as in the above-mentioned one-step process. Care must be taken to keep the furnaces inside the hood and handling by latex gloves is imperative. The speculated phase diagram for the double–Tl–O compounds is given in Fig. 2 of Chapter 6. Table 1 summarizes the crystal growths of double- and mono-Tl–O compounds and their properties from different starting compositions (charge) in the self-flux technique. Oxygen was passed continuously through the tube during the crystal growth. Postannealing of the crystals has been done in different atmospheres to increase or decrease the T_c. We discuss this in detail in the following sections. Small (or no) crystals were obtained with the Tl concentration exceeding 4 and the Cu concentration exceeding 12 in the starting composition of 4:1:3:12 (Tl/Ba/Ca/Cu) [17]. This suggests that there is a Tl–O and Cu–O composition window in which various phases can be grown. Ginley et al. [7] appear to have produced small crystals predominantly Tl-2223 phase by using a Ca/Ba ratio of 3:1 in the charge (see also Table 1). Hence by using Table 1 and

Table 1 Properties of Various Thallium Cuprate Single Crystals Grown by the Self-Flux Technique

Starting composition Tl/Ba/Ca/Cu	Crystals obtained Tl/Ba/Ca/Cu	Lattice parameters (Å)		T_c (K)	Refs.
		a	c		
2202	2201	3.870	23.24	90	1
3202	2201	3.850	23.20	86	13, 14
2113	2212	3.855	29.32	112	2
2112	2212	—	—	114	9
2223	2212	3.850	29.30	110	6, 11, 15
2234	2223	3.850	35.90	125	3, 16
2223	2223	3.850	35.67	114	16
2266	2223	—	—	118	6
4136	2223	3.850	36.00	103	17
41310	2223	3.850	36.00	113	17
3148	2223	—	—	110	18
4148	2223	3.855	35.74	112	18
1112	1212	3.850	12.70	103	19
1234	1223	3.853	15.92	117	8
1115	1223	3.850	15.90	120	15
1133	1223	—	—	119	20
1135	1223	3.855	15.92	121	20

Fig. 2 of Chapter 6, one can conveniently choose the starting compositions and temperatures to grow the desired phase. Also, one needs precise temperature control to grow single crystals. Lee et al. [21] have grown epitaxial Tl-2223 thin films and made bulk samples in reduced O_2 pressure ($\cong$ 0.03 to 0.15 atm of O_2 and 830 to 860°C). The same treatment may also be applied for growing single crystals.

Recently, Manako et al. [22,23] have grown single crystals of Tl-2201 by using a KCl flux method. The starting compositions of [Tl-2201]/[KCl] = 1 to 10 wt % were used as the charge in a gold crucible was heated at 920°C for 3 h and cooled down to 750°C with a cooling rate of 2 to 10°C/h. The thin plates were grown in the *ab* plane with a typical size of 2 × 2 × 0.01 mm^3. Schneemeyer et al. [24] have already demonstrated the crystal growth of Bi-2212 by using alkali chloride flux. Large crystals (3 × 3 × 1.5 mm^3) of $YBa_2Cu_3O_7$ were grown using a K_2CO_3-flux method [25]. Based on this information, one may attempt growing large crystals of Tl compounds by either alkali chloride–flux or K_2CO_3-flux methods.

III.   TRANSPORT PROPERTY MEASUREMENTS

A. $Tl_2Ba_2CuO_6$ System

$Tl_2Ba_2CuO_6$ single crystals have been grown by the self-flux technique [13,14] and the KCl-flux methods [22,23]. The anisotropic resistivities, ρ_{ab} and ρ_c, are plotted against temperature in Fig. 2 for Tl-2201 single crystals annealed in O_2 at 350°C (metallic down to 4 K), in Ar at 300°C (T_c = 10 K), and in Ar at 400°C (T_c = 75 K) [22,23]. The anisotropic resistivity ρ_c is much greater than ρ_{ab} by a factor of 2 to 3 orders of magnitude. These results are similar to those obtained by Duan et al. [13]. The anisotropic resistivities were measured by the standard Montgomery method [26]. The temperature dependence of ρ_{ab} was fit to an expression of $\rho_{ab} = \rho_0 + AT^n$. The exponential factor n increased from 1.29 to 1.99 as T_c decreased. The room-temperature in-plane resistivity is on the order of $4 \times 10^{-4} \Omega \cdot cm$. The normal state of ρ_c was metallic for all the samples. The anisotropy of resistivity ρ_c/ρ_{ab} is plotted against temperature in Fig. 3. The anisotropy is about 600 at room temperature and increases as the temperature decreases for all the samples. It is interesting to note that considerable resistivity anisotropy remains even in the normal metallic samples. This could be due to the large separation of CuO_2 sheets in these systems.

To find the dimensionality of the superconductors, the temperature dependence of paraconductivity can be fit to [11]

$$\Delta\sigma(T) = A\left(\frac{T}{T_c} - 1\right)^{-\alpha} \tag{1}$$

where $\Delta\sigma$ is the excess conductivity, A is a constant, T_c is the critical temperature of the system, and α is 1 and $\frac{1}{2}$ for two- and three-dimensional superconductivity, respectively. This can be rewritten as [11]

$$\log\frac{-d(\Delta\sigma)}{dT} = \log\frac{\alpha A^{1/\alpha}}{T_c} + \left(1 + \frac{1}{\alpha}\right)\log\Delta\sigma \tag{2}$$

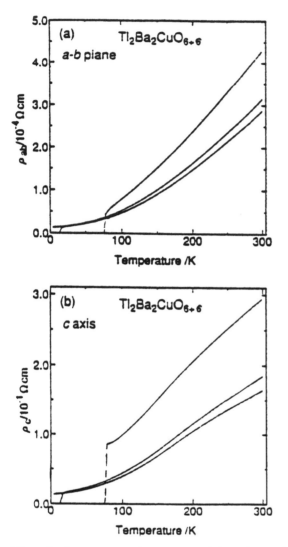

Figure 2   Temperature dependence of the anisotropic resistivities (a) ρ_{ab} and (b) ρ_c measured by the Montgomery method for Tl-2201 single crystals annealed at 350°C in O_2 (metallic), at 300°C in Ar (T_c = 10 K), and at 400°C in Ar (T_c = 75 K). (Adapted from Ref. 22.)

From Eq. (2), α, the dimensionality of the superconductors, can be determined from the slope of a $\log[-d(\Delta\sigma)/dT]$ versus $\log\Delta\sigma$ plot. Figure 4 is a plot of $\log[-d(\Delta\sigma)/dT]$ versus $\log\Delta\sigma$ for Tl-2201 crystals. The slope obtained was 2, indicating that Tl-2201 compounds are two-dimensional superconductors.

The temperature dependence of in-plane Hall coefficients for normal metallic T-2201 samples is shown in Fig. 5. The Hall coefficient is about $+1 \times 10^{-3}$ cm³/°C and shows a broad maximum around 100 K. The *ab*-plane single-crystal data are very similar to the data taken on ceramic samples. This suggests that the transport properties of ceramic samples mainly reflect the *ab*-plane nature.

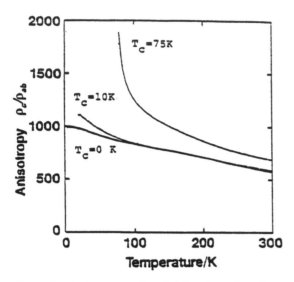

Figure 3 Anisotropy of resistivity plotted against temperature for Tl-2201 single crystals. The data points from Fig. 2 were replotted. (Adapted from Ref. 22.)

B. $Tl_2Ba_2CaCu_2O_8$ System

$Tl_2Ba_2CaCu_2O_8$ single crystals have been grown by the self-flux technique [2,6,9,11,15]. The anisotropic resistivities ρ_{ab} and ρ_c are plotted against temperature in Fig. 6. The out-of-plane resistivity, ρ_c, is greater than in-plane resistivity, ρ_{ab}, by about two orders of magnitude. This is similar to the results obtained for Tl-2201 single crystals [13,22]. Figure 7 shows the plot of ln $\Delta\sigma$ versus $\ln(T/T_c - 1)$. The paraconductivity is often discussed in such a plot according to Eq. (1). Two-dimensionality is ob-

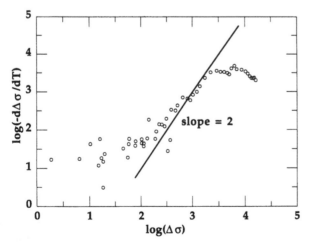

Figure 4 Plot of log $[-d(\Delta\sigma)/dT]$ versus log$\Delta\sigma$ for Tl-2201 single crystals. The slope of the line is 2, which corresponds to the two-dimensional superconductivity of the sample. (Adapted from Ref. 13.)

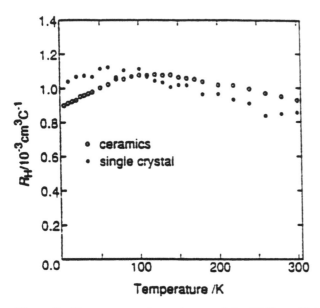

Figure 5 Temperature dependence of the Hall coefficients for normal metallic Tl-2201 samples. Single-crystal data in the *ab* plane (closed circles) and ceramic sample data (open circles) are compared. (Adapted from Ref. 23.)

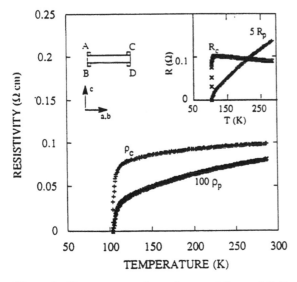

Figure 6 Temperature dependence of the resistivity for Tl-2212 single crystals. ρ_p is multiplied by 100. The insert shows the measured resistances by using the top-left configuration. (Adapted from Ref. 11.)

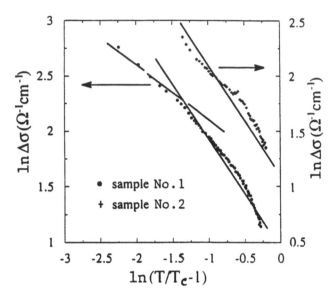

Figure 7 Plots of ln($\Delta\sigma$) versus ln($T/T_c - 1$) for Tl-2212 single crystals. The solid line and dashed line correspond to two- and three-dimensional superconductivity, respectively. In the high-temperature range, the superconductivity is two-dimensional. At low temperature, the superconductivity in some samples becomes three-dimensional. The T_cs used are the midpoints of the 10 to 90% resistance value. (Adapted from Ref. 11.)

served for all the samples in the high-temperature region. For some samples, the crossover from two-dimensional to three-dimensional occurs at a few degrees above T_c (for sample 1 in Fig. 7, the crossover occurs at 6 K above the T_c 101 K) [11]. The same behavior is observed in the plot of log[$= d(\Delta\sigma)/dT$] versus log$\Delta\sigma$ in Fig. 8. The crossover may be related to the formation of Josephson junctions between the Cu–O layers. The two-dimensional property in the region 105 to 155 K is consistent with results from Tl-2212 bulk samples and thin films [27,28]. No crossover was reported in these samples. The lack of crossover may also be due to the possible inhomogeneities of these polycrystalline samples.

The postannealing effects of Tl-2212 single crystals are shown in Fig. 9. The T_c (onset of Meissner signal) for the as-grown crystal is 104 K. The T_c decreased to 99 K after a vacuum anneal and increased to 106 K after annealing in O_2. This can be interpreted simple as an increase in the oxygen content and a corresponding increase in the number of holes. Similarly, when one removes oxygen by annealing in vacuum, the T_c decreases because of the decrease in the hole concentration. But Morosin et al. [19] observed that for some of the other crystals, O_2 annealing decreases the T_c and vacuum annealing increases it. This anomaly may be due to the possibility that the as-grown crystals may already be overdoped. Hence when one anneals in O_2, the T_c decreases. Morosin et al. [19] also suggested that strain, Tl content, and Tl/Ca site disorder might be as important as the oxygen content in determining T_c.

The anisotropic thermoelectric power of Tl-2212 single crystals has been measured [29]. The room temperature value of S_{ab} (*ab* plane) is about 14 μV/K and

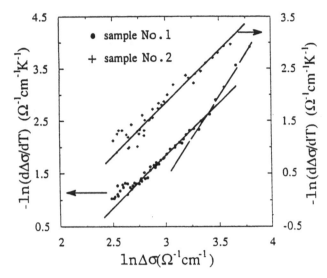

Figure 8 Plots of $\ln[d(\sim\Delta\sigma)/dT]$ versus $\ln\Delta\sigma$ for Tl-2212 single crystals. The solid and dashed lines correspond to two- and three-dimensional superconductivity, respectively. Superconductivity of sample 2 is two-dimensional. In the high-temperature range, the superconductivity of sample 1 is two-dimensional, and becomes three-dimensional in the low-temperature range. (Adapted from Ref. 11.)

S_c (along the c axis) is about 30 μV/K. The onset superconductivity determined by thermopower measurements shows that the transition measured along the c axis is always lower than that in the ab plane.

The scattering of light in Tl-2212 single crystals has been studied experimentally and analytically [10,30,31]. Besides probing photon effects, light-scattering spectroscopy yields information about electronic excitations and their symmetry. The tem-

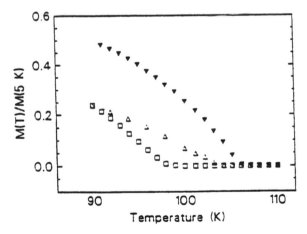

Figure 9 Meissner signal (normalized to its value at 5 K) versus temperature in 2.5 mT applied normal to a Tl-2212 single-crystal plate as-grown (open triangles), after a vacuum anneal (open squares), and then an oxygen anneal (solid triangles). (Adapted from Ref. 19.)

perature dependence of the electronic scattering intensity gives evidence for a pronounced anisotropy of the superconducting gap ($2\Delta_{min} < 50$ cm^{-1}, $2\Delta_{max} \approx 300$ cm^{-1}) [31]. The maximum value of the gap is estimated as $2\Delta_{max} \approx 300$ cm$^{-1} \approx 4k_bT_c$.

Novel flux motion has also been observed in Tl-2212 crystals [32]. Krasnov et al. [33] have extended the Bean critical state model for obtaining the bulk critical field, H_{c1}, and the magnetically determined critical current density, J_c, for Tl-2212 single crystals. This method is based on the original critical-state model and accounts for the shape of the sample, the change of demagnetization factor with flux penetration, the real distribution of the magnetic field, and the equilibrium flux density within the sample. They observed a small positive curvature of the $H_{c1}^{\perp}(T)$ dependence at $T < 50$ K; this curvature was suggested to be a consequence of two-dimensional superconductivity. The critical current density, J_c, has also been measured for Tl-2212 crystals from $I-V$ curves, and its temperature dependence is plotted in Fig. 10. The J_c value is observed to be 2.5×10^4 A/cm^2 at 77 K. Because of insufficient data, we can not comment on the dimensionality.

C. Tl$_2$Ba$_2$Ca$_2$Cu$_3$O$_{10}$ System

Tl$_2$Ba$_2$Ca$_2$Cu$_3$O$_{10}$ single crystals can be grown using a self-flux technique with different starting compositions [3,6,16–18]. Ginley et al. [7] have suggested that when one uses a Ca/Ba ratio 3:1 in the starting compositions, one obtains predominantly Tl-2223 crystals. The temperature dependence of anisotropic resistivities ρ_{ab} and ρ_c for Tl-2223 crystals are shown in Fig. 11. ρ_c is about two orders of magnitude greater than ρ_{ab}. This is similar to the results observed on Tl-2201 and Tl-2212 single crystals. The anisotropy (ρ_c/ρ_{ab}) is quite high and is on the order of 50 to 60 at 300 K. Tigges et al. [17] also observed from the resistivity data that the T_c and transition width for Tl-2223 crystal plates grown by two different melt

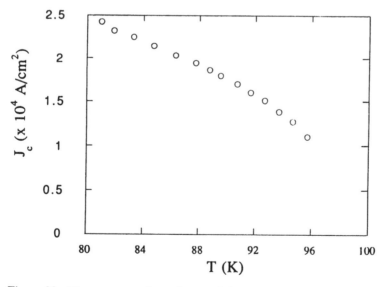

Figure 10   Temperature dependence of the transport critical current density J_c in the *ab* plane of a Tl-2212 single crystal. (Adapted from Ref. 34.)

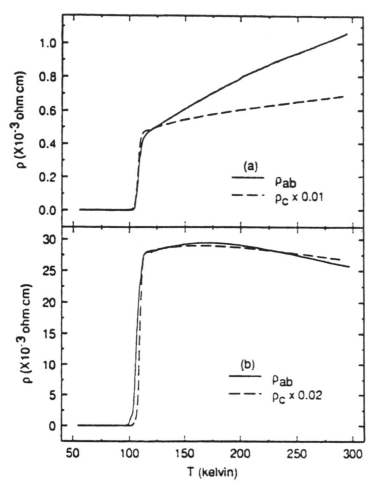

Figure 11 Temperature dependence of the anisotropic resistivities, ρ_{ab} and ρ_c for Tl-2223 single crystals. Crystal 1 (a) has a relatively sharp superconducting transition and high T_c. For crystal 2 (b), the resistivity does not increase monotonically with temperature in the normal state and is about 25 times greater than for crystal 1. (Adapted from Ref. 17.)

compositions were substantially different. A Tl–O and Cu–O-rich flux yielded approximately stoichiometric crystals with sharp transitions beginning near 113 K, while a Tl–O-rich flux produced crystals containing more Tl and less Ba with broad transitions starting near 103 K. These data demonstrate the extreme sensitivity of the superconductivity to cation site disorder in Tl systems [17].

Cox et al. [35] have refined the structure of Tl-2223 by neutron diffraction. The postannealing effects on Tl-2223 single crystals are reported by Morosin et al. [18]. They have suggested the possibility of the presence of small amounts of Tl^{1+} in the Tl–O layer of as-grown crystals—from single-crystal x-ray refinement data— which subsequently oxidizes to Tl^{3+} during mild oxygen anneals. The presence of mixed valency (Tl^{3+} and Tl^{1+} in a double-Tl–O system) has been reported from band structure calculations [36] and from wet-chemical analysis data [37–39]. By using the wet-chemical procedures that are available in the literature [37–41], one

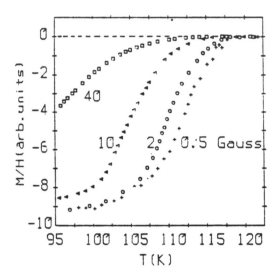

Figure 12 Shielding susceptibility M/H versus temperature for Tl-2223 single crystals. The numbers labeling the curves are the applied field in Gauss in the *c* direction. (Adapted from Ref. 42.)

can conveniently measure the Tl content, oxygen content, and hole concentrations for Tl single crystals.

The temperature dependence of the susceptibility with different applied magnetic fields for Tl-2223 crystals is shown in Fig. 12. The measured diamagnetic-onset temperature is reduced from 118 K to 113 K by the application of a magnetic field in the *c* direction in the range 0.5 to 40 G. Also, there is a systematic increase in the width of the transition with increasing field. The strong suppression of the diamagnetic-onset temperature with smaller magnetic fields could be due to the intrinsic property of the bulk material. Such an effect in the polycrystalline samples is attributed to weak coupling between grains [43]. Figure 13 shows shielding,

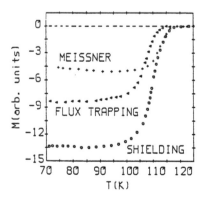

Figure 13 Flux trapping, Meissner, and shielding signals (in 2 G) versus temperature for Tl-2223 single crystals as in Fig. 12. One arbitrary unit corresponds to 0.031 G. For comparison, the positive flux trapping signal has been inverted. (Adapted from Ref. 42.)

Meissner, and flux trapping measurements on Tl-2223 single crystals in a field of 2 G along the *c* direction [42]. Just below T_c (above 100 K), the fraction of Meissner to shielding ratio increases to nearly 100%. Over the entire temperature range, the sum of the magnitudes of the Meissner and flux trapping signals equals the magnitude of the shielding signal, similar to the observations in $YBa_2Cu_3O_{7-\delta}$ [42].

D. TlBa$_2$Ca$_2$Cu$_3$O$_9$ System

$TlBa_2Ca_2Cu_3O_9$ single crystals have been grown by the self-flux technique [8,15,20]. The annealing effects on Tl-1223 crystals are shown in Fig. 14. The Meissner data show that the T_cs are 117 K after extended oxygen anneals and reach 121 K after nitrogen anneals. This suggests that there is only a small change in the hole con-

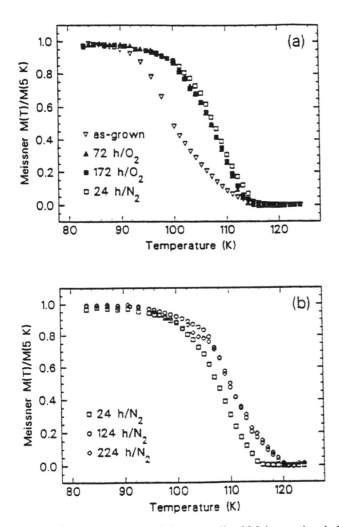

Figure 14 Comparison of the normalized Meissner signals for Tl-1223 single-crystal plates, (a) as-grown compared with initial oxygen (72 h), additional oxygen (172 h total) resulting in no change, and initial nitrogen (24 h) anneals, and (b) nitrogen for 24 h [same curve shown in (a)] and two additional 100-h nitrogen anneals. (Adapted from Ref. 20.)

centration in the CuO_2 planes. Because of the recent interest in introducing pinning centers and achieving high J_c values in different Tl-1223 phases [44–46], one should look forward to the growth of larger Tl-1223 single crystals by the self-flux technique or alkali halide–flux method and corresponding characterizations.

IV. IRRADIATION EFFECTS ON THALLIUM SYSTEMS

Venturini et al. [15] have studied the magnetic relaxation (flux creep) for Tl-2212 and Tl-2223 single crystals by recording the diamagnetic shielding signal versus time following a field change from 1 T to 50 mT applied normal to the crystal plate (along the c axis). The temperature dependence of flux pinning potentials for as-grown Tl-2223 crystals, Tl-2223 after proton irradiation, and Tl-2212 after neutron irradiation are shown in Fig. 15. Following irradiation with high-energy protons and neutrons [15], magnetic hysteresis at 50 K was observed above 1 T, and J_{cm} (magnetization critical current density) from the remanent moments increased by an order of magnitude, to 5 to 8 × 10^4 A/cm².

V. PRESSURE DEPENDENCE ON T_c

The pressure dependence for a single crystal of $Tl_5Ba_5Ca_2Cu_6O_y$ (composition close to Tl-2212) with a T_c of 106 K is shown in Fig. 16. The T_c seems to increase with hydrostatic pressure with a slope of $dT_c/dP = 0.23(4)$ K/kbar in a gaseous He pressure system. Very recently, Berkley et al. [48] have applied pressure on Tl-2223 single crystals and found a tremendous increase in T_c, from 118 K to 131.5 K under pressure. The pressure dependence of T_c for Tl-2212 single crystals is shown in Fig. 17. The T_c increases with pressure with a large initial positive slope of about 4 K/GPa at lower pressures. At higher pressures, the slope decreased and at the highest pressure, the T_c dropped from 122.7 K at 3.7 GPa to 121.3 at 4.8 GPa.

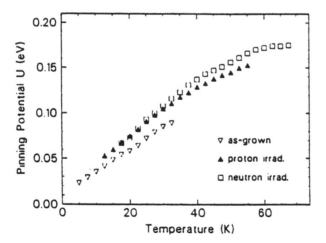

Figure 15 Comparison of flux pinning potential versus temperature determined from magnetization relaxation (flux creep) measurements in Tl-2223 single crystals as-grown (open triangles), Tl-2223 after proton irradiation (solid triangles), and Tl-2212 after neutron irradiation (open squares). (Adapted from Ref. 15.)

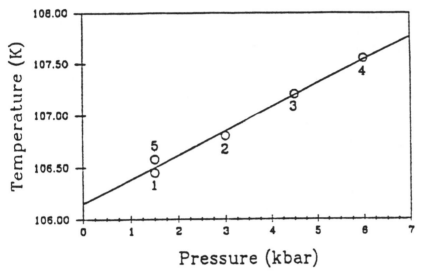

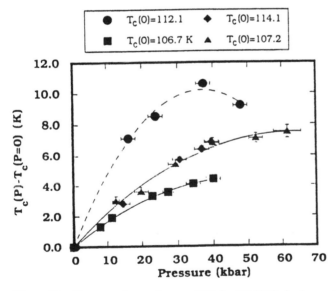

Figure 16 Pressure dependence of T_c for Tl-5526 single crystals, showing the onset of superconductivity near 106 K. The points are numbered in the sequence they were taken. (Adapted from Ref. 47.)

Figure 17 Pressure dependence of T_c for Tl-2212 single crystals. (Adapted from Ref. 49.)

VI. CONCLUSIONS

Our review of the present status of single-crystal growth and characterization of thallium cuprate superconductors has led us to make the following conclusions:

1. Reasonably large crystals (several millimeters) can be grown by self-flux and KCl-flux methods.

2. The anisotropic resistivity ρ_c (along the c axis) is greater than ρ_{ab} (along the ab plane) by two orders of magnitude for all double-Tl–O compounds. Hence these systems have high anisotropy values.

3. The paraconductivity data show that Tl-2201 is a two-dimensional super-conductor. Tl-2212 was also found to be two-dimensional, but some samples showed crossover from two-dimensional to three-dimensional a few degrees above the T_c.

4. The temperature dependence of the Hall coefficient for normal metallic Tl-2201 crystals (along the ab plane) is found to be similar to that of ceramic bulk samples. This suggests that the transport property measurements on ceramic samples reflect the properties of the ab plane.

5. A Tl–O- and Cu–O-rich flux yielded stoichiometric Tl-2223 crystals with sharp transitions around 113 K by using the self-flux technique. Also, a 3:1 Ca/Ba ratio in the starting composition was found to give predominantly Tl-2223 crystals.

6. Postannealing is necessary to alter the T_c of the as-grown crystals.

7. The strong suppression of the diamagnetic-onset temperature with small (0.5 to 40 G) applied magnetic fields for Tl-2223 crystals could be due to the weak coupling between grains.

8. The sum of the magnitudes of the Meissner and flux trapping signals equals the magnitude of the shielding signal over the entire temperature range below T_c for Tl-2223 single crystals. This behavior is similar to that observed for $YBa_2Cu_3O_{7-\delta}$.

9. Irradiation with high-energy protons and neutrons on Tl-2223 crystals increases the J_{cm} (magnetization critical current density) value deduced from the remanent moments by an order of magnitude to 5 to 8 $\times$ 10^4 A/cm^2 at 50 K.

10. The T_c seems to increase with the hydrostatic pressure with a slope of $dT_c/dP = 0.23(4)$ K/kbar in a gaseous He pressure system for a crystal of $Tl_5Ba_5Ca_2Cu_6O_y$ with a T_c of 106 K.

ACKNOWLEDGMENT

We gratefully acknowledge the support of Office of Naval Research under ONR grant N00014-90-J-1571.

REFERENCES

1. C. C. Torardi, M. A. Subramanian, J. C. Calabrese, J. Gopalakrishnan, E. M. McCarron, K. J. Morrissey, T. R. Askew, R. B. Flippen, U. Chowdhry, and A. W. Sleight, *Phys. Rev. B* 38, 225 (1988).

2. M. A. Subramanian, J. C. Calabrese, C. C. Torardi, J. Gopalakrishnan, T. R. Askew, R. B. Flippen, K. J. Morrissey, U. Chowdhry, and A. W. Sleight, *Nature* 332, 420 (1988).

3. C. C. Torardi, M. A. Subramanian, J. C. Calabrese, J. Gopalakrishnan, K. J. Morrissey, T. R. Askew, R. B. Flippen, U. Chowdhry, and A. W. Sleight, *Science* 240, 631 (1988).

4. A. W. Sleight, *Science* 242, 1519 (1988).

5. H. Takei, T. Kotani, T. Kaneko, and K. Tada, *Proc. First International Symposium on Superconductivity*, Nagoya, 1988, ed. K. Kitazawa and T. Ishiguro, Springer-Verlag, Tokyo, 1988, p. 229.

6. T. Kotani, T. Kaneko, H. Takei, and K. Tada, *Jpn. J. Appl. Phys.* 28, L1378 (1989).

7. D. S. Ginley, B. Morosin, R. J. Baughman, E. L. Venturini, J. E. Schirber, and J. F. Kwak, *J. Crystal Growth* 91, 456 (1988).

8. B. Morosin, D. S. Ginley, J. E. Schirber, and E. L. Venturini, *Physica C* 156, 587 (1988).

9. E. D. Bukowski and D. M. Ginsberg, *J. Low Temp. Phys.* 77, 285 (1989).

10. S. E. Stupp and D. M. Ginsberg, in *Physical Properties of High Temperature Superconductors*, Vol. III, ed. D. M. Ginsberg, World Scientific, Singapore, 1992, in press, and references therein.

11. H. M. Duan, W. Kiehl, C. Dong, A. W. Cordes, M. J. Saeed, D. L. Viar, and A. M. Hermann, *Phys. Rev. B* 43, 12925 (1991).

12. H. M. Duan, R. M. Yandrofski, T. S. Kaplan, B. Dlugosch, J. H. Wang, and A. M. Hermann, *Physica C* 185–189, 1283 (1991).

13. H. M. Duan, R. M. Yandrofski, T. S. Kaplan, B. Dlugosch, J. H. Wang, and A. M. Hermann, *Chin. J. Phys.* (1992) in press.

14. H. M. Duan, T. S. Kaplan, B. Dlugosch, A. M. Hermann, J. Swope, J. Drexler, and P. Boni, preprint.

15. E. L. Venturini, C. P. Tigges, R. J. Baughman, B. Morosin, J. C. Barbour, M. A. Mitchell, and D. S. Ginley, *J. Crystal Growth* 109, 441 (1991).

16. T. Kajitani, K. Hiraga, S. Nakajima, M. Kikuchi, Y. Syono, and C. Kabuto, *Physica C* 161, 483 (1989).

17. C. P. Tigges, E. L. Venturini, J. F. Kwak, B. Morosin, R. J. Baughman, and D. S. Ginley, *Appl. Phys. Lett.* 57, 517 (1990).

18. B. Morosin, E. L. Venturini, and D. S. Ginley, *Physica C* 175, 241 (1991).

19. B. Morosin, R. J. Baughman, D. S. Ginley, J. E. Schirber, and E. L. Venturini, *Physica C* 161, 115 (1990).

20. B. Morosin, E. L. Venturini, and D. S. Ginley, *Physica C* 183, 90 (1991).

21. W. Y. Lee, S. M. Garrison, M. Kawasaki, E. L. Venturini, B. T. Ahn, R. Beyers, J. Salem, R. Savoy, and J. Vasquez, *Appl. Phys. Lett.* 60, 772 (1992).

22. T. Manako, Y. Shimakawa, Y. Kubo, and H. Igarashi, *Physica C* 185–189, 1327 (1991).

23. T. Manako, Y. Shimakawa, Y. Kubo, and H. Igarashi, *Physica C* 190, 62 (1991).

24. L. F. Schneemeyer, R. B. van Dover, S. H. Glarum, S. A. Sunshine, R. M. Fleming, B. Batlogg, T. Siegrist, J. H. Marshall, J. V. Waszczak, and L. W. Rupp, *Nature* 332, 422 (1988).

25. P. Murugaraj, J. Maier, and A. Rabenau, *Solid State Commun.* 71, 167 (1989).

26. H. C. Montgomery, *J. Appl. Phys.* 42, 2971 (1971).

27. N. P. Ong, Z. Z. Wang, S. Hagen, T. W. Jing, and J. Hovarth, *Physica C* 153–155, 1072 (1988).

28. A. Poddar, P. Mandal, A. N. Das, B. Ghosh, and P. Choudhury, *Physica C* 159, 231 (1989).

29. Lin Shu-yuan, Lu Li, Zhang Dian-lin, H. M. Duan, and A. M. Hermann, *Europhys. Lett.* 12, 641 (1990).

30. V. B. Timofeev, A. A. Maksimov, O. V. Misochko, and I. I. Tartakovskii, *Physica C* 162–164, 1409 (1989), and references therein.

31. A. A. Maksimov, I. I. Tartakovskii, V. B. Timofeev, and L. A. Fal'kovskii, *Sov. Phys. JETP* 70, 588 (1990).

32. F. Zuo, M. B. Salamon, T. Dutta, K. Ghiron, H. M. Duan, and A. M. Hermann, *Physica C* 176, 541 (1991).

33. V. M. Krasnov, V. A. Larkin, and V. V. Ryazanov, *Physica C* 174, 440 (1991).

34. B. Dlugosch, H. M. Duan, T. S. Kaplan, and A. M. Hermann, preprint.

35. D. E. Cox, C. C. Torardi, M. A. Subramanian, J. Gopalakrishnan, and A. W. Sleight, *Phys. Rev. B* 38, 6624 (1988).

36. D. Jung, M.-H. Whangbo, N. Herron, and C. C. Torardi, *Physica C* 160, 381 (1989).

37. M. Paranthaman, A. Manthiram, and J. B. Goodenough, *J. Solid State Chem.* 87, 479 (1990).
38. A. Manthiram, M. Paranthaman, and J. B. Goodenough, *Physica C* 171, 135 (1990).
39. M. Paranthaman, M. Foldeaki, and A. M. Hermann, *Physica C*, in press.
40. J. Gopalakrishnan, R. Vijayaraghavan, R. Nagarajan, and C. Shivakumara, *J. Solid State Chem.* 93, 272 (1991).
41. A. Manthiram, M. Paranthaman, and J. B. Goodenough, *J. Solid State Chem.* 96, 464 (1992).
42. J. Z. Liu, K. G. Vandervoort, H. Claus, G. W. Crabtree, and D. J. Lam, *Physica C* 156, 256 (1988).
43. S. S. P. Parkin, V. Y. Lee, E. M. Engler, A. I. Nazzal, T. C. Huang, G. Gorman, R. Savoy, and R. Beyers, *Phys. Rev. Lett.* 60, 2539 (1988).
44. J. A. DeLuca, M. F. Garbauskas, R. B. Bolon, J. G. McMullen, W. E. Balz, and P. L. Karas, *J. Mater. Res.* 6, 1415 (1991).
45. T. Kamo, T. Doi, A. Soeta, T. Yuasa, N. Inoue, K. Aihara, and S. Matsuda, *Appl. Phys. Lett.* 59, 3186 (1991).
46. R. S. Liu, D. N. Zheng, J. W. Loram, K. A. Mirza, A. M. Campbell, and P. P. Edwards, *Appl. Phys. Lett.* 60 (1992).
47. B. Morosin, D. S. Ginley, E. L. Venturini, P. F. Hlava, R. J. Baughman, J. F. Kwak, and J. E. Schirber, *Physica C* 152, 223 (1988).
48. D. D. Berkley, E. F. Skelton, N. E. Moulton, and D. H. Liebenberg, paper presented at the *APS Meeting*, Indianapolis, Ind., Mar. 1992.
49. N. E. Moulton, S. A. Wolf, E. F. Skelton, D. H. Liebenberg, T. A. Vanderah, A. M. Hermann, and H. M. Duan, *Phys. Rev. B* 44, 12632 (1991).

9

Tl-Based HTSC Films for Microelectronics Applications

David S. Ginley
National Renewable Energy Laboratory
Golden, Colorado

I. INTRODUCTION

The discovery of high-temperature superconductors stimulated considerable speculation as to the potential applications benefits of these materials. Although large-scale magnetic levitation and power transmission gathered a majority of the attention, microelectronic applications will probably be the first to make a commercial impact. To minimize cooling power and system requirements it is desirable to operate at as high a temperature as possible. Nominally for $YBa_2Cu_3O_7$ (Y-123) materials, the operating point is liquid nitrogen temperature, 77 K, which is quite close to the T_c of 89 to 93 K. To this end the discovery of the Tl–Ba–Ca–Cu–O system [1–4] with transition temperatures to 127 K, the highest confirmed for any superconductor, appears to have real advantages for liquid nitrogen temperatures in terms of penetration depth (λ_1) and the stability of λ_1 and J_c with respect to temperature. Initial thin-film work demonstrated T_cs of >100 K and J_c values up to 10^5 A/cm^2 [5–9]. Early work on Tl films and bulk material indicated that the grain boundary chemistry of the Tl materials is considerably different from that of the Y-123 and Bi–Sr–Ca–Cu–O materials and that under the proper processing circumstances [7] the grain boundaries were not weak linked even for essentially randomly oriented material. Although this advantage has more implications for conductor application than for microelectronics, there too it may be important, as results [10] have shown that Tl step edge junctions have superior performance to similar Y-123 junctions, presumably due to different grain boundary chemistry. However, the complexity of processing Tl-based materials combined with perceived toxicity problems retarded the rapid development of improved films. Initially, there were only a few groups pursuing the development of thin films in this system; however, recent device results have shown considerable promise, and a large number of new laboratories have undertaken developmental programs in Tl films. Although much of this work is in industrial laboratories and has not been published in detail, in this chapter we summarize what is available. This work has demonstrated that a number of techniques allow high-quality films to be grown on wafers up to 2 in. in diameter. The critical temperatures of these films exceed those attainable with Y-123, the J_c values are somewhat less due to weaker pinning in the Tl–Ca–Ba–Cu–O system, and the surface resistances are comparable or superior. For many applications, especially those where flux flow is desirable, these films appear to be superior to the Y-123 and Bi–Sr–Ca–Cu–O materials overall. Only in applications where critical currents exceeding 10^6 A/cm^2 at 77 K are required is the Y-123 material clearly superior. Even in this area, recent results with the TlBa$_2$Ca$_2$Cu$_3$O$_9$ (1223) phase indicate that this limitation may not be as severe as once thought. One other area where the Y-123 materials are clearly superior is in the ability to do in situ growth combined with the lack of volatility of the constituents. This combination allows direct fabrication of multilayer structures, whereby interlayers of dielectrics and superconductor can be employed to fabricate complex vertical device structures. Their somewhat lower processing temperatures may also be useful for some applications. Although the Bi-based materials have T_cs greater than those for Y-123, the difficulties in synthesizing the high T_c phase compounded with even weaker pinning than the Tl system make these materials somewhat less attractive for development. In this chapter, wherever appropriate, we compare the properties of the Tl firms with those for typical in situ Y-123. We do not make comparisons with to the Bi system.

In this chapter we discuss Tl chemistry to provide a perspective on the processing difficulties and then discuss a variety of the current techniques for Tl film production. We then summarize and compare first the transport data and then magnetic data on J_c values in a field and pinning with and without material modification through implantation. We discuss briefly processing of the films compared to that for other high-temperature-superconductor (HTS) systems. Finally, we discuss future growth directions and the prospects for in situ growth.

II. TL CHEMISTRY

Although there are a variety of different processing schemes for the production of Tl films, the key determining factors are the chemistry of the Tl oxides and the variety of Tl superconducting phases. The nominal species in the processing regime normally employed are Tl, Tl$_2$O, and Tl$_2$O$_3$ [11,12]. During processing they are in a complex equilibrium as a function of the partial pressure of the Tl and O. Figure 1 illustrates a partial phase diagram for Tl–O as provided by William L. Holstein of Du Pont.

The primary volatile species is Tl$_2$O in the processing range used for film synthesis. If a suppression of volatility is desired, increased oxygen partial pressure will drive the equilibrium toward Tl$_2$O$_3$. While this may minimize the volatility of Tl from the film surface, it may be desirable to have excess Tl present to stimulate liquid-phase formation. The presence of a Tl–O-rich liquid phase appears to be critical for both superconducting phase formation and the beneficial grain boundary properties observed in the Tl–Ca–Ba–Cu–O system. Unlike the YBa$_2$Cu$_3$O$_7$ sys-

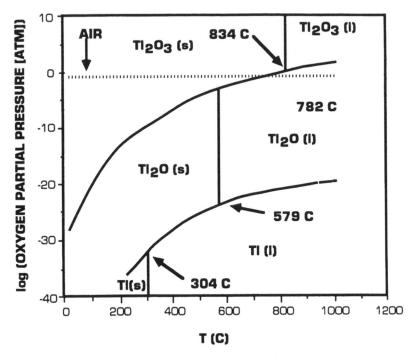

Figure 1   Phase diagram for Tl–O as a function of partial oxygen pressure and temperature.

tem, the Tl system does not have any chains and is composed primarily of planar Cu–O and Tl–O sheets. Bulk phases with a wide variety of combinations of stacked Tl–O (1–4) and Cu–O (1–3) sheets have been synthesized [13–15].

The in-plane lattice constants of all of these phases are essentially the same, allowing for epitaxial growth of each phase on any of the other phases. Thin films of a wide variety of Tl-based superconducting phases have been prepared, including $Tl_2Ba_2Ca_1Cu_2O_8$ (2212), $Tl_2Ba_2Ca_2Cu_3O_{10}$ (2223), $TlBa_2Ca_2Cu_3O_9$ (1223), $Tl-Ba_1Ca_2Cu_2O_7$ (1122), $Tl_2Ba_2Ca_0Cu_1O_6$ (2201), and a variety of complementary Tl–Pb–Sr–Ca–Cu–O phases that essentially substitute Tl/Pb for Tl and Sr for Ba. We discuss the Pb-doped materials later in the chapter. The main phases that are most attractive for thin-film applications have been 2212, 2223, and more recently, 1223. Thin films of the 2212 phase are the most common and have the smallest transition widths obtained to date, but do not have as high a T_c as the 2223 or 1223. Until recently, the most desirable phase was thought to be the 2223, because of its high T_c. It is, however, prone to a number of defects, including oxygen vacancies, cationic disorder, and interlaying of other phases [13–15]. Recently, the 1223 phase has been revisited predominately in unpublished work by Gray et al. at Argonne National Laboratories and Maley et al. at Los Alamos National Laboratories; it may be the most desirable phase because it has a high T_c and a single rather than double Tl–O interlayer. This may mean higher pinning because of a thinner layer–layer spacing for the Cu–O planes, resulting in a higher c-axis conductivity and less randomization of the pancake vortices. Both of those works and that of Nabatame et al. at Hitachi [16] indicate that the irreversibility line for the

1223 phase lies considerably closer to that for the more highly pinned Y-123 than does that for the 2223 phase. The latter reference found that H^* decreases exponentially with d, the spacing between the Cu–O layers (Y-123, 8.3 Å, 1223, 8.7 Å, 2223, 11.4 Å, and Bi22122, 12.1 Å).

While the complex chemistry may inherently produce some lateral defects and perhaps step-type faults, these may in fact help increase the pinning. In addition, this complex chemistry tends to build in a compositional tolerance. This results in producing interlayering on the single atomic layer to 100-Å block scales when the composition, especially that of the Tl, is off. Since most of the Tl phases are superconducting with comparable properties, this interlayering does not seem to affect adversely the transport, magnetic, and microwave properties of these materials. The fact that the grain boundary chemistry appears to be cleaner in these materials, resulting in no weak links in polycrystalline films, is probably the result of not forming grain boundary impurity phases.

III. SUBSTRATES

Although a wide variety of substrates have been investigated for Tl film growth, only a few have proven suitable and only three are of major current interest. The first films in the Tl system were predominately the 2212 phase and were grown on $SrTiO_3$ primarily because of the historical precedent from Y-123 films [17]. The lattice match to strontium titanate is reasonable and some of the highest-T_c films to date have been obtained on this substrate. For microwave applications, however, this substrate has too much loss to be of interest. Films have also been grown on MgO-[18] and Y_2O_3-stabilized ZrO_2 [19] with considerable success. However, the best substrate to date from both a lattice match and microwave loss viewpoint is $LaAlO_3$; it is still far from ideal and a substrate like sapphire would be preferable. The $LaAlO_3$ (0.5387 nm) has a much better match along the 100 surface to the 110 surface of the Tl compounds (0.545 nm) than does MgO (0.42 nm) or Y-ZrO_2 (0.526 nm) [20]. A number of other substrates have also been explored, including $LaGaO_3$, Al_2O_3, $NdGaO_3$, and $CaNdAlO_4$ [21,22]. With the exception of the Al_2O_3, all of the substrates suffer from interdiffusion problems. Sapphire does not lattice match as well as the other materials and may require buffer layers. Recently, Du Pont has demonstrated the growth of 2212 on sapphire with CeO_2 buffer layers [23]. The films grown at 870°C in air from Tl-free precursors demonstrated T_cs of 97 to 98 K and J_c values at 75 K of 2.8 × 10^5 A/cm^2. These results are very encouraging for microelectronic applications. Another interesting substrate is the growth on Ag [24] and stabilized silver alloys such as work at Los Alamos on Consil 995, where surface resistances were a few hundred mΩ at zero field and 18 GHz. The work on the silver and Consil (a stabilized silver alloy) have in general shown T_cs in the range 95 to 105 K and J_c values on the order of 10 to 60,000 A/cm^2, but the field dependence of J_c indicates that the materials are still weakly-linked. Whereas metallic substrates are not particularly suitable for microelectronic applications, for conductor applications and for large cavity applications they may be necessary. In general, $LaAlO_3$ has become the de facto industry standard and most of the growth reported below is on this substrate.

IV. FILM GROWTH

A. Precursor Production

Although there are a considerable number of variations on the processing of Tl films, from the deposition of precursors to ex situ sintering and annealing of the films, there are some common themes. For the deposition of precursors the primary techniques used to date are laser ablation, electron beam evaporation, metal organic chemical vapor deposition, and sputtering. Laser ablation involves a pulsed high-energy laser, typically a KrF [25] or a Nd-YAG [26], operating at relatively short wavelength to expel material rapidly from a target. Nominally, the plume produced has the stoichiometry of the target. When the plume strikes a heated substrate, in this case in an oxygen environment, a stoichiometric oxide precursor can be obtained. Because of volatility problems with the Tl, the most common use of laser ablation for the Tl system has been the deposition of Ba–Ca–Cu–O precursors. If the Tl flux problems can be overcome, perhaps through employing a second simultaneous ablation source, this technique may hold promise for in situ film production. In the Y-123 materials the most complex in situ structures grown to date have all been by laser ablation [27]. Other than the above, a major advantage is that targets may be readily replaced and new materials prototyped. The major disadvantages of this technique are the difficulty in making it into a scalable process and the fact that the ejection process results in some large particulates and a-axis inclusions, which adversely affect the surface smoothness and surface resistance.

In electron beam evaporation the constituent metals are typically evaporated in sequential order, producing a layered structure [28,29]. The process is normally done under a partial pressure of oxygen (10^{-5} torr) to prevent intermetallic phase formation. Although partially oxidized, the films are still metallic and are subject to further air oxidation. These films have been made both containing Tl [28] and Tl-free [30], with comparable results for the same type of processing. It appears that the major drawbacks of this technique are the macroscopic layering of the precursor films and the lack of oxidation. These can be partially rectified by coevaporation from multiple sources, which may be useful for in situ film development. Evaporation of the oxides or evaporation in a more oxidizing environment may help provide more stoichiometrically oxidized materials. An advantage of the technique is the typically long working distance between the source and substrate, which permits coating of large-area substrates, and 2-in. and larger wafers can be uniformly coated.

Ultimately, metal organic chemical vapor deposition may offer the best promise for a scalable production deposition technology for Tl and other HTS materials. In this technique volatile organometallic compounds of the constituent elements are carried to a heated substrate by a carrier gas. At the substrate, the compounds are decomposed to the correct elements, react with oxygen in the carrier gas, and deposit on the substrate. Clearly, the technique has some promise for in situ growth. The major disadvantage is that volatile precursors for the desired constituent are not readily available, oxygen is probably not the most desirable carrier gas because of reactions in the gas stream, and the organometallic compounds have potential environmental and safety problems. Nonetheless, there has been considerable progress recently in the preparation of precursors and films by metalloorganic chemical vapor deposition (MOCVD), particularly for $YBa_2Cu_3O_7$. The primary gaseous species employed have been metal β-diketonates or related carboxylic acid complexes [31]. BaCaCu precursors for Tl-film production have been deposited using fluorinated precursors [32]. The resulting films were processed ex situ and had reasonable surface resistances on both $LaAlO_3$ and Consil 995. The advantages are that the technique is scalable and has a high throughput. A possible variant of this technique employs a volatile organometallic Tl source with an oxygen carrier to process films ex situ. The controllability of the vapor pressure of the Tl and oxygen should lead to improved film processing.

The most common technique for precursor deposition is sputtering [33–40]. In this technique either a radio-frequency (RF) or dc plasma, typically formed from a mixture of argon or nitrogen and oxygen, produces high-energy ions that dislodge clusters of atoms from the target. These atoms or clusters of atoms can react with any gas or gases that are present during deposition on the substrate. Frequently, magnets are employed under the target material to confine and intensify the plasma, resulting in improved deposition rates. Materials can be deposited from a number of targets of the individual elements sequentially or simultaneously, or they can be deposited from a single mixed-metal or mixed-oxide target. The sequential method is similar to the sequential electron beam evaporation, and the simultaneous method suffers from the difficulties of trying to control four sources precisely at the same time. The best results to date and the most common technique is the deposition from a single or a pair of mixed-oxide targets. The normal alignment is with the axis of the target and substrate coincident (on-axis), but frequently this results in variations in sample stoichiometry with sputtering time unless the sample is carefully aged and power levels are controlled. More recently, considerable success in sputtering HTS materials has been obtained by having the target and substrate axis at 90° to one another (off-axis). When suitably burned-in such targets have produced stoichiometric materials for nearly the life of the target. The major disadvantage of off-axis sputtering is that the deposition rate is much less than that for on-axis sputtering and can be as low as a few angstroms per minute. Two opposed targets can be employed to enhance the rate, but care must be taken to ensure that the magnetic fields of the two sources do not interfere.

B.   Ex Situ Thermal Processing

Postdeposition processing of the films can be broken into two steps. First is the sintering of the correct phase, where the material is completely oxidized, growth of the correct crystalline phase occurs, and the morphology is established. Second is the annealing of the grown film to remove defects and ensure the proper degree of oxidation (which controls doping). The latter process is not as critical as in the case of the $YBa_2Cu_3O_7$ materials, because the Tl materials do not lose oxygen as readily and the oxygen vacancies are concentrated in the Tl–O planes, which do not appear to strongly affect the superconducting properties. Both processes are frequently combined by applying an appropriate cooling protocol after high-temperature sintering. As mentioned previously, here and in other chapters, there are a number of liquid Tl oxides at the normal processing temperatures and some of the superconducting phases may liquefy, especially in a Tl-rich environment. This liquid-phase formation promotes phase formation and grain growth and may be responsible for the grain boundary properties observed. In many ways the sintering process has similarities to liquid-phase epitaxy and may explain how epitaxial films can be produced with only a few minutes of thermal processing. While from both a crystallographic and a superconducting viewpoint, high-quality materials can be produced in one thermal cycle, there have been a number of studies [41,42], especially on single crystals, which show that long-term anneals in the range 500 to 700°C can substantially improve superconductor properties. This can be due to a number of effects, including removal of strain, altering the number of oxygen vacancies, and altering the cationic disorder, especially Ca on Tl sites. Controlled annealing is especially important for films of the 2201 phase, where small increases in the oxygen content can change the T_c from 93 K to 10 K.

Sintering is normally accomplished by one of three techniques: crucible anneals in air or oxygen, encapsulated annealing, and processes where the Tl content is actively controlled. We discuss these three approaches and, where appropriate, point out any annealing process. In the first two techniques the Tl and oxygen partial pressures are controlled in a passive fashion because the sintering environment is close to thermal equilibrium and the Tl pressure is defined by the initial composition of the Tl source. In the last technique there is the potential for in situ diagnostics and active feedback control to control Tl and oxygen partial pressures precisely.

C. Ex Situ Sintering

Table 1 summarizes most of the recent film work for precursors with Tl, precursors without Tl, encapsulated growth, and in situ. Some of the older work is summarized in Refs. 5 to 9. As can be seen, there is little uniformity in the results in terms of developing a unified process that will work for all situations. Rather, it appears that each precursor/process is unique. Recent results in our laboratory have shown that this is precisely the case when electron-beam-evaporated films were processed both with and without a preoxidation process, as described in Ref. 37. The resulting films did not show properties comparable to those processed with this precursor, and vice versa. There are some general observations that can be made.

The best films to date have been obtained by encapsulated growth where a completely sealed environment is employed to control the oxygen and thallium partial pressures. Normally, a sealed quartz tube is employed and the sample and the Tl source are wrapped in Au foil as illustrated in Fig. 2. Recent work at IBM [37,43] and elsewhere [44] has shown that encapsulated growth under reduced oxygen partial pressures (0.03 to 0.15 atm) in such an encapsulated environment can produce very high quality films. To date the best films in terms of T_c and J_c values have been made by this technique. The use of reduced oxygen pressure allows sintering at slightly lower temperatures, with some success obtained at 830 to 860°C. Part of this success may be a function of the very long sintering times allowed by the encapsulation (8 to 13 h). The films produced by this technique have established a figure of merit for the Tl system that clearly indicates that it can be competitive with the YBCO materials for nearly all applications. However, the main areas where it is not competitive are power handling and reproducibility.

Table 1 Examples of Process Variations as a Function of Precursor for a Number of Films in the Tl–Ca–Ba–Cu–O System[a]

Precursor	Substrate	Crucible	Thickness (μm)	Temperature and sintering conditions	Phase	Refs.
Tl-containing precursors						
Laser ablation, KrF, 10 Hz	LaAlO$_3$	?	0.9	830–900°C, 10 min	2212	25
RF sputtering, two targets	ZrO$_2$	Covered	1–2	850–900°C	?	33
Eximer laser 308 nm, 20-ns pulse, 10 Hz	MgO	Sealed alumina	1.3	840°C, 5–20 h	2223	26
Dc magnetron sputtering, two targets	LaAlO$_3$	Pt in tube furnace, flowing O$_2$	2	890°C, 2223, and 870°C, 2212, for 10 min at temperature then cool to 700°C and slow cool	2212 and 2223	35
Laser ablation, KrF, 10 Hz	CaNdAlO$_4$	?	0.9	830–900°C 10 min, some interdiffusion with substrate	2212	22
Magnetron sputtering	LaAlO$_3$	?	?	880–890°C for 15 min	2223	60
Electron beam	LaAlO$_3$	Pt covered	?	740°C 10 min, with electron-beam Tl film as Tl source	2212	61
Sequential electron beam	LaAlO$_3$ and others	Covered Pt	0.3–0.7	850°C 15 min, with bulk ceramic followed by a 750°C 10 min O$_2$ anneal with bulk ceramic	2223 2212 1223 1212	28

Method	Substrate	Container	Thickness	Conditions	Phase	Ref
Magnetron sputtering	Consil 995 metallic alloy; some films had BaF_2 buffer layers	Alumina	?	860–905°C for 2–20 min using a TlO_x source	Mixed phase	24
Dual dc magnetron sputtering	ZrO_2	Covered	1–2	850–900°C		59
Dual dc magnetron sputtering	ZrO_2	Covered quartz	1–2	850–900°C, slow 5-h cooling	2223	19
TL-free precursors Thick film	$Y-ZrO_2$	Au boats in two-zone furnace	?	Sample 865–905°C and Tl_2O source at 775–870°C in N_2/O_2 80:20	1223	45
Sputtering	$LaAlO_3$	Alumina sealed	0.2–1.2	800–900°C with Tl-2223 and Tl_2O_3 powder source	2212 2223 1223 1212	46
RF sputtering; stoichiometry corrected with BaF_2 or CaF_2 after preanneal	$LaAlO_3$		?	?	2212	40
RF magnetron sputtering	CeO_2 buffer on sapphire $LaAlO_3$ and Consil 995	Alumina sealed	0.7–0.8	870°C with Tl-2223 and Tl_2O_3 powder source	2212	23
MOCVD of BaCaCuOF precursor Encapsulated growth		Sealed	4 on Ag	Face down over Tl-2223 powder	2212	32

Table 1 (*Continued*)

Precursor	Substrate	Crucible	Thickness (μm)	Temperature and sintering conditions	Phase	Refs.
RF facing-target sputtering 2223	$LaAlO_3$	Au foil–wrapped film and bulk ceramic in sealed quartz tube	0.2–1.5	830–860°C with reduced O_2 (0.03–0.15 atm) pressure in sealed tube, annealed 1–8 h with a slow cool	2223	37, 38
Dc single-target sputtering	$LaAlO_3$	Au foil–wrapped film and bulk ceramic in sealed quartz tube	1	880°C for 15 min	2223	20
In situ growth						
RF magnetron; Ba, Ca, Cu, and evaporation of Tl	$LaAlO_3$ and $NdGaO_3$		0.05	490–650°C with deposition pressures of 50–400 mtorr	1212	52
Laser ablation	$LaAlO_3$				1212	62

[a]The superconducting properties for these films are given in Table 2.

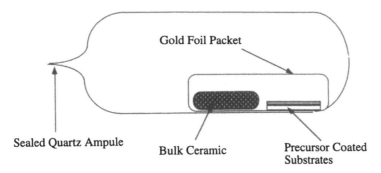

Figure 2   Geometry for encapsulated postprocessing of Tl precursors. The precursors may or may not contain Tl. The Au foil is required to prevent reaction of the Tl with the quartz of the tube walls.

The major challenge is to translate these results into a more manufacturable technology. Sample preparation for this technique is difficult, it is hard to scale, and it is time consuming.

More typical is the use of thalliated or nonthalliated precursors in a sealed or nearly sealed crucible to contain the precursor film and a bulk Tl source. Figure 3 illustrates a typical geometry employed in crucible anneals. Typically, the crucibles have been of Pt [28] or alumina [23]. The alumina crucibles are essentially expendable, can be sealed with a refractory cement, and can take large wafers. The precious metal crucibles are reusable but cannot readily be sealed and are quite expensive. Recently, a number of groups have switched to Au- or Au-coated crucibles because Tl does not react with Au as it can potentially with the Pt (Tl can form a $Pt_2Tl_2O_7$ phase), and it is easy to clean with dilute acids. A wide variety of Tl sources have been used. The foremost approaches have been to use bulk ceramic [28] of the same intended phase as that desired in the film or to use a powdered Tl oxide source such as Tl_2O_3 or Tl_2O or the oxides in combination with

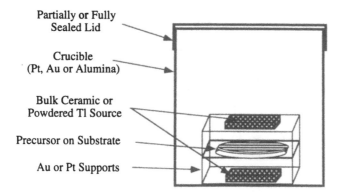

Figure 3   Typical process geometries for precursors with and without thallium, typically employing Pt or alumina crucibles. The figure shows an arrangement employing two pellets of bulk ceramic with the precursor film in between a similar arrangement can be employed using powdered oxide.

powdered bulk ceramic [23]. The most prominent Tl oxide species in the processing regime is the Tl_2O, so it is probably the preferable choice for a pure oxide source. With its high degree of volatility at normal processing temperatures, it lends itself to use of a two-zone furnace where the Tl source is kept at a lower temperature than the precursor film. This has recently been demonstrated with considerable success by GE [45] for thick-film production. The processing of the precursors can be broken down into two classes: those that contain Tl and those that are Tl-free. Initial results by Hermann et al. [30] indicated that it was more difficult to attain good grain boundary properties and surface morphologies with the latter precursors, but recent results, particularly those from Du Pont [23], indicate that high-quality films can be attained with either precursor. Somewhat surprisingly, Table 1 shows that the processing conditions for similarly prepared precursors with and without Tl are virtually identical. This may indicate that the partial pressure of the volatile Tl–O species and the chemistry of the liquidus Tl–O species dominate the chemistry to the extent that the precursor is not significant. Recent results in this laboratory have shown that 223 precursor can be turned into phase-pure 1223 with 1223 bulk ceramic pellets more readily than can 1223 precursor.

Key elements are the use of a bulk ceramic or Tl oxide source (Tl_2O_3 or Tl_2O) for a Tl source and orientation of the substrate face up or down with respect to the ceramic. In some cases two bulk ceramic pellets have been employed with the precursor film between them [28]. A number of investigators have also used powdered Tl ceramic (typically, 2223 phase) and Tl_2O_3 or Tl_2O powder [46]. Although the results are similar, minimizing contamination of the precursor surface by preventing particulate contamination from the ceramic is desirable for eliminating spurious nucleations. Although in this type of sintering the processing environment is further from equilibrium than the encapsulated growth, the processing times are appreciably shorter. Large wafers and two-sided wafers can be processed and turnaround can be very rapid. Typical processing conditions are 10 to 15 min at 850 to 890°C in air. This process is highly dependent on a large number of parameters, and consequently has been difficult to optimize. Recent results employing a two-zone furnace, as illustrated in Fig. 4, have been somewhat more successful and may indicate the way to a more manufacturable technology [45].

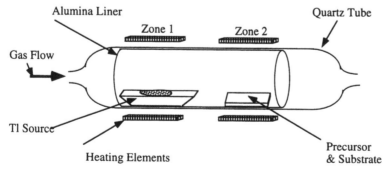

Figure 4 Possible configuration for a two-zone furnace for the processing of precursors that may or may not contain Tl.

Ideally, accurate control of the temperature, Tl, and oxygen partial pressures needs to be maintained throughout the processing cycle and across the surface of the precursor. This may actually call for a profile in the Tl and oxygen pressures to allow formation of the proper phase, the sintering and annealing of the film. In many ways, the growth process seems to resemble liquid-phase epitaxy wherein a liquidus Tl phase is responsible for film growth and, perhaps as important, the beneficial grain boundary chemistry.

D. Ex Situ Annealing

Although it is becoming more common to include a slow cool to anneal the films, there is some demonstrated benefit in annealing the Tl superconductors. Extensive annealing studies of single crystals have shown that either nitrogen or oxygen anneals can improve the T_c [41,42]. Beneficial effects have been observed for thin films up to 750°C under pure oxygen. At temperatures up to 500°C a Tl source is not required even for long-term (hundreds of hours) anneals. Above 600°C some initial Tl loss is observed and loss increases rapidly with temperature, requiring control of the Tl vapor pressure. Between 500 and 600°C some deterioration of the surface is seen if there is no external Tl source, which can be particularly detrimental to the microwave properties. These anneals do not seem to alter the grain boundary properties, the phase purity, or the morphology. They do effect the oxygen vacancies, particularly in the Tl–O plane, the cationic disorder between Tl and Ca, and the amount of strain in the films, all of which can substantially affect the superconducting properties [41,42,47,48]. Thus either a slow cool or a secondary anneal may be beneficial for obtaining optimum superconducting properties.

E. Process Variation Versus Precursor

Initially, it was thought that since the formation of liquid phases occurred during sintering, it did not make a difference what the initial precursor was since the kinetics of phase formation and grain growth dominated. Recent work in this laboratory and a comparison of the data in Tables 1 and 2 indicate that this may not be the case. Encapsulated growth of films from sputtered mixed-oxide targets and sequentially deposited electron-beam-evaporated metals gives distinctly different results even when the metals from the electron-beam process have undergone preoxidation. The sputtered precursors produced films with higher T_c and J_c values. This is probably because for the sequentially evaporated precursor, the various metal ions need to diffuse hundreds of angstroms to react to a stoichiometric 2223 composition. In the case of the sputtered precursor the ions essentially do not need to diffuse, and thus phase formation is relatively easy. The results with Tl-free precursors indicate that Tl transport through the film is not rate limiting, again supporting liquid-phase formation. The presence of Tl-rich liquid phases appears to promote Tl ion diffusion.

F. In Situ Films

We will discuss in situ films only briefly because of lack of success to date. In situ film growth for the Cu-O based high-temperature superconductors essentially en-

Table 2 Summary of the Superconducting Properties of the Films in Table 1

Precursor	Substrate	Phase	T_c (K)	R_s	J_c (A/cm^2)	Notes
Tl-containing precursors						
Laser ablation, KrF, 10 Hz	LaAlO$_3$	2212	100	0.2 mΩ at 9.55 GHz at 77 K	1.06×10^6 microbridge, 1 μV/mm	
RF sputtering two targets	ZrO$_2$	?	117		?	
Eximer laser 308 nm, 20-ns pulse, 10 Hz	MgO	2223	122		1.5×10^5	
Dc magnetron sputtering two targets	LaAlO$_3$	2212 and 2223	2212–98 2223–115		2212–1×10^5 2223–5.5×10^5	
Laser ablation, KrF, 10 Hz	CaNdAlO$_4$	2212	100.7		3.5×10^5	2.3-GHz microstrip with low-power Q of 2100
Magnetron sputtering	LaAlO$_3$	2223	116		5×10^5	
Electron beam	LaAlO$_3$	2212 2223	88 111		?	
Sequential electron beam	LaAlO$_3$ and others	2212 1223 1212		6 mW at 30 GHz, 77K	9.7×10^5	
Magnetron sputtering	Consil 995 metallic alloy; some films had BaF$_2$ buffer layers	Mixed phase	Check notes	14 mΩ at 18 GHz at 77 K		
Dual dc magnetron sputtering	ZrO$_2$		117			Dc SQUIDs[7] prepared amplitude of flux voltage 50 μV p-p

Process	Substrate	Phase	T_c (K)	Surface resistance	J_c	Comments
Dual dc magnetron sputtering	ZrO_2	2223	117		1.2×10^5 at 77 K	Dc SQUIDs with average noise power, 1.7×10^{-26} J/Hz, dc to 1 Hz
TL-free precursors Thick film Sputtering	$Y\text{-}ZrO_2$ $LaAlO_3$	1223 2212 2223 1223 1212	104–107 106, very sharp; 120, broadened; 110, rounded, poor transitions	<2 mΩ at 20 GHz and 80 K; 130 $\mu\Omega$ at 10 GHz and 77 K (Du Pont)	20–30,000 2.7×10^5	J_c value reduced $2300\times$ for a perpetual 5000-Oe field
RF sputtering stoichiometry corrected with BaF_2 or CaF_2 after preanneal	$LaAlO_3$	2212	100		1000 at 75 K, 0.7 T	
RF magnetron sputtering	CeO_2 buffer on sapphire	2212	97–98		2.8×10^5 at 75 K	
MOCVD of BaCaCuOF precursor	$LaAlO_3$ and Consil 995	2212	100	19.6 mΩ at 35 GHz and 88 K		
Encapsulated growth RF facing-target sputtering 2223	$LaAlO_3$	2223	120	3–4 mW at 30 GHz, 77 K	1.6×10^6 at 77 K, 0.38 μm, and 100 Oe	
Dc single-target sputtering	$LaAlO_3$	2223	110		1×10^7 A/cm^2 at 87 K	
In situ growth RF magnetron; Ba, Ca, CU, and evaporation of Tl	$LaAlO_3$ and $NdGaO_3$	1212	95 after 800°C ex situ anneal			
Laser ablation	$LaAlO_3$	1212	Phase synthesized but not superconducting			

tails deposition of the constituent elements, typically by laser ablation or sputtering, under sufficient oxygen partial pressure to eliminate intermetallic phases and produce a stoichiometric film. After deposition, the films are usually cooled under a controlled atmosphere. The resulting films in the case of the Y-123 materials produce high-quality films with T_cs of 88 to 91 K and critical currents in the middle range of 10^6 A/cm^2 [27]. These films do not have crystal quality as good as those of the best ex situ films [49], but they can be made routinely on up to 2-in. wafers. The main advantage of the in situ growth process for Y-123 materials is the ability to grow multiple-layer structures. These include Y-123/Pr-123/Y-123 SIS structures and a wide variety of growth on buffer materials, including MgO, CeO$_2$, LaAlO$_3$, Y-ZrO$_2$, and SrTiO$_3$, among others [50,51].

The difficulty in the Tl system arises primarily from the complex chemistry of the Tl oxides. Their volatility creates problems both during deposition and in loss from the heated substrate. Until recently, the best success was in unpublished results obtained by Olson et al. at STI. They used laser ablation to produce some of the 1212 phase, but the films were not superconducting. Unpublished work in this laboratory has shown that to try and get stoichiometric deposition of the 2223 phase, increasingly Tl-rich targets were employed to prevent excess Tl loss. To get close to the right stoichiometry, Tl excesses greater than 100:1 were required. At this concentration it was difficult to fabricate the targets and difficult to ablate with reproducible results. It is conceivable that with a separate Tl and O source the stoichiometry could be maintained. Unfortunately, to get synthesis of the correct phases requires elevated temperatures above 500°C and probably in the range 700 to 850°C. In this range the Tl oxide volatility is high enough to make maintenance of the correct stoichiometry very difficult. One possible solution is the use of metal organic chemical vapor deposition, where high Tl and/or oxygen overpressures may be possible. There have been some attempts to date but without great success other than in the production of precursor films.

Very recent results from Du Pont [52] have shown the production of phase-pure 1212 on LaAlO$_3$ and NdGaO$_3$ by sputtering Ba, Ca, and Cu from oxide targets in argon and oxygen and Tl$_2$O from a thermal evaporation source. Substrate temperatures were 490 to 650°C and deposition pressures from 50 to 400 mtorr. However, postdeposition annealing at 800°C was required to achieve T_cs to 95 K.

V. CHARACTERIZATION

Since the properties of Tl films are discussed in other portions of the book, here we discuss only those properties that are key diagnostics to film quality and/or applications readiness. Table 2 summarizes many of the key properties of the Tl films discussed earlier and in the same order as in Table 1. A comparison indicates that a key difficulty in the high-temperature superconductor field is a lack of standardization of the measurements used to report film quality. A key set of parameters would be the T_c, the J_c at 77 K−zero field and 1 T with a 1-μV/cm standard, and R_s at 77 K and a fixed frequency between 10 and 30 GHz. For microwave applications R_s is the crucial parameter. Despite measurement inconsistencies a number of salient observations can be made from Table 2. Most of the best films have similar properties, irrespective of initial precursor or process. For the 2223 phase the T_cs range from 110 to 120 K; the J_c values range from 5×10^5 to 1.5×10^6 A/cm^2 at 77 K and zero field. There have been reports of films processed by the encapsulated technique of J_c values to 1×10^7 A/cm^2 at 87 K [20]. Although samples are typically not entirely phase pure, T_c and J_c are not strongly affected. Figure 5 shows typical R versus T and J versus E curves for a Tl-2223 film.

The surface resistance varies more strongly as a function of film properties. It is very sensitive to the surface roughness and, in some cases, to film thickness, with thinner films showing a degradation in R_s. Figure 6 shows a typical R_s value for a 2223 film. Figure 7 shows the dependence of R_s on T for a film grown by the encapsulation technique.

In general to date, the 2212 phase has shown the narrowest transition width, 1 to 2 K, and the best crystalline purity (sharpest x-ray lines). The 1212 phase shows the poorest transitions, and in the case of in situ films was nonsuperconducting. The 1223 phase has shown transitions above 110 K, but they are more rounded than those for 2212. The 2223 phase has shown transitions to 122 K that are not as sharp as for 2212 (3 to 10 K), but the films are reasonably phase pure. The 2212 phase normally shows narrower x-ray line width than does the 2223 and other phases perhaps indicative of intergrowths or defects in the other phases. Films of the 2201 phase typically have depressed T_cs as grown (30 to 50 K), and nitrogen or vacuum annealing can raise this to >70 K. It has thus far been difficult to achieve the 93 to 94 K obtained in bulk material of this phase. It appears to be much more difficult to control the oxygen stoichiometry during growth, because slightly reducing conditions which produce the highest T_cs also result in the greatest Tl loss during growth.

As can be seen from Table 2, surface resistances in Tl materials vary widely but have been observed to be as low as 23 $\mu\Omega$ at 20 K and 10 GHz and 130 $\mu\Omega$ at 77 K and 10 GHz [23]. These values are best for any HTS material and are

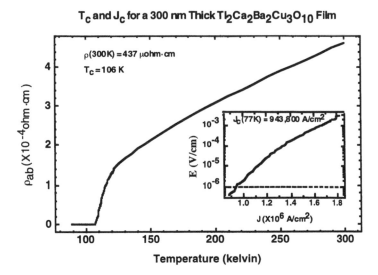

Figure 5 The larger portion of the figure illustrates the R versus T curve for a 0.3-μm Tl-2223 film. The film was part of a 2-in. wafer with similar characteristics processed as in Ref. 28. The inner portion of the curve illustrates the I versus V curve for the same film at 77 K. The measurements were done using dc currents and Ag contacts.

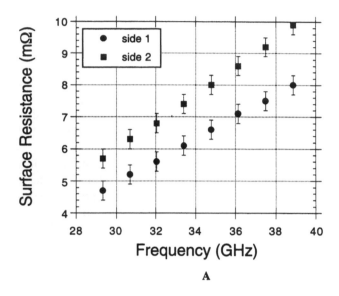

A

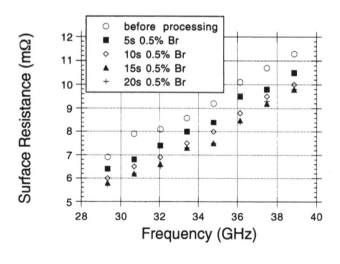

B

Figure 6 (A) Surface resistance of a 0.3-μm film similar to that in Fig. 5. The results shown are for a 2-in. wafer that had Tl-2223 film on both sides. (B) Effects of sequential Br–isopropanol etches on the surface resistance, indicating an improvement in R_s upon initial etching.

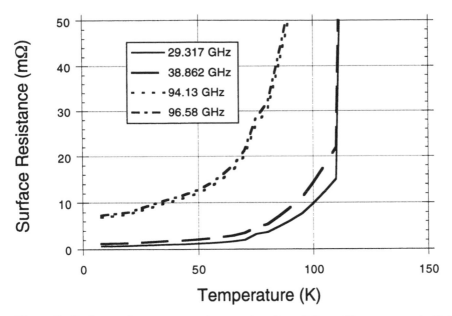

Figure 7 Surface resistance versus temperature for a 0.8-μm film grown as in Ref. 37.

better than those for 123 at all temperatures. The typical values of R_s at 77 K and 10 GHz are one or two orders of magnitude less than those for copper under the same conditions, and good values have been observed for film thicknesses from 0.1 to 2.0 μm. Unlike the 123 system, where films of thickness greater than 0.5 μm have reduced J_c and R_s values, the thicker films in the Tl system typically have better properties [37]. This may be due to the superior grain boundary properties in the Tl system and improved pinning in the thicker films, as discussed below.

Figure 8 shows a typical Meissner curve for a 1.1-μm-thick radio-frequency (RF) sputtered 2223 film. The films show onsets of >120 K and demagnetization factors indicating that the entire film is carrying superconducting current [29]. Figure 9 illustrates hysteresis loops at 5 K for 0.2- and 0.8-μm films as in Ref. 53. Note that the thinner film shows a much more collapsed hysteresis loop. Both films show a significant collapse of the hysteresis loop above 30 K due to decreasing pinning. The lack of pinning at higher temperatures is a significant drawback with respect to current-carrying applications for Tl materials. The deleterious effects are even more pronounced in a magnetic field, as shown in Fig. 10 for the same films as those in Fig. 9.

Note that the thinner film has essentially the same initial critical current as the thicker film, but J_c values fall off much more rapidly with increasing temperature [53]. For conductor development, this is a crucial result, indicating that tape and thick-film conductors may be feasible in this system. In addition, the increased pinning is thought to originate at the grain boundary structures, indicating that polycrystalline materials may even be desirable for conductor application. Results at GE [45] seem to support the thin-film results noted above. They have shown that thick 1223 films (3 μm) on $Y\text{-}ZrO_2$ have dc transport critical currents to 10^5 A/cm^2. More important, critical currents of 10^4 A/cm^2 have been obtained for these

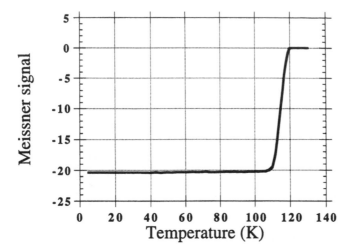

Figure 8 Typical static magnetization (Meissner) curve for a 1.1-μm Tl-2223 film as in Ref. 37 at 0.5 mT. Note that the onset is greater than 120 K, and the narrow transition.

polycrystalline films at 77 K and 1 T of field parallel to the *c* axis. Clearly, it would be desirable to be able to increase the pinning energy and the density of pinning centers. To date, attempts to introduce impurity phases (similar to what has been done with the 211 phase in 123 materials) or defects through ion or neutron irradiation have not been very successful [29]. While moderate J_c improvements are observed with field or at higher temperatures, as illustrated in Fig. 11 for an ion-irradiated sample, it appears that the density of the pinning centers is improved but that they are no stronger than those caused by the inherent defects in the materials.

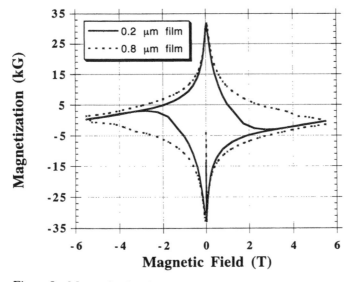

Figure 9 Magnetization hysteresis measured at 5 K for two Tl-2223 films as above with thicknesses of 0.2 and 0.8 μm.

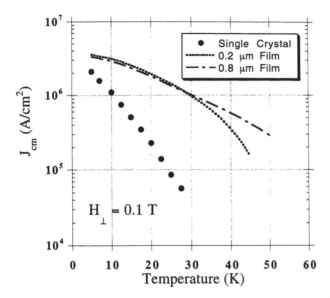

Figure 10 Critical current density J_{cm} versus temperature in a field of 0.1 T parallel to the *c* axis for the films in Fig. 9. The data were obtained from hysteresis curves as in Fig. 9. Results are also shown for a Tl-2223 single crystal illustrating even weaker pinning.

Although clearly not as strongly pinned as the 123 materials, recent thin-film results by IBM [37] and Du Pont [23] demonstrating greater than 10^6 A/cm² at 77 K indicate that some progress has been made in achieving better pinning. A number of laboratories are currently investigating the 1223 phase in hopes that it might have higher pinning than the 2223 phase. It has a comparable T_c and it is felt that

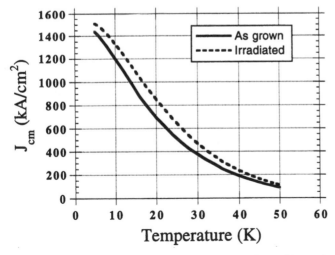

Figure 11 Critical current data versus field for a film as in Ref. 26 before and after implantation with 300-keV He⁺ ions, 2.5×10^{13} cm⁻² with a damage level of about 2×10^{-4} dPa.

the presence of a single rather than a double Tl–O layer between groups of Cu–O planes may allow better coupling and less wander of the flux lines. A coherent flux lattice is hence more favorable, and higher pinning energies are possible. Interlayer coupling appears to increase nearly linearly with decreased layer spacing based on recent work at Los Alamos National Laboratory and at Argonne National Laboratory [54,55]. Roughly, the spacing between Cu–O groups goes from 11.5 Å in 2223 to 9.5 Å in 1223. It is also thought that a substitution of Pb and/or Sr in these materials is beneficial by virtue of decreasing the layer spacing.

The weak pinning is not always a disadvantage. As discussed in Chapter 13, a number of new active-device structures have been based on the flux-flow properties of Tl thin films. These long-junction and flux-flow transistor devices have shown outstanding device properties, and circuits based on them exceed conventional electronics in some areas.

VI. CONTACTING/PATTERNING

Considerable progress has been made in processing technology for the utilization and characterization of thin films in the Tl system. The key areas are ohmic contacts and patterning. We also briefly discuss environmental stability and hazards. High-quality contacts have been obtained with thin-film Tl superconductors by a number of techniques. The most common is the evaporation or sputtering of Ag onto a fresh surface followed by a short (30 min or less) oxygen anneal between 500 and 600°C. A key element is the cleanliness of the surface. A surface precleaned by plasma etching, backsputtering, or chemical cleaning can produce high-quality contacts with no anneal. Detailed etching experiments indicate that frequently there is a thin, 10- to 100-Å layer on the surface that is highly defective and may not be superconducting. The best contact metals have been Ag and Au. Ag is preferred because of its ability to allow oxygen diffusion to the surface during anneals, maintaining the superconducting properties [56]. The best contact resistances obtained have been in the range 10^{-8} to 10^{-10} $\Omega \cdot cm^2$, adequate for all projected applications [56–58]. Recent results in this laboratory have shown that in some cases the thermal anneals can be replaced by current anneals (cycling the contacts to the normal state) to get good ohmic contacts. This may be important for heterostructures, where annealing a whole wafer may not be desirable.

Considerable progress has been made in the patterning of Tl films for testing and a variety of microwave devices (see Chapter 13). Both ion-beam patterning and wet-chemical etches have been employed. Ion-beam patterning, typically employing a Ar ion beam to erode material, can produce micrometer-sized features but is expensive, difficult, and costly. In addition, these materials are susceptible to peripheral ion-beam damage. Wet-chemical etching in acidic aqueous (HNO_3, HCl) solutions results in significant material degradation and increases surface resistance. More common has been standard photolithographic patterning employing a Br–alcohol solution. These require a negative photoresist, so feature size is limited to greater than 1 μm, but as shown in Fig. 6b, this etch can actually improve the surface resistance of these materials [57]. Other than feature size, the major drawbacks for this etch technology are the reactivity and chemical hazards of the etch itself. Since most Tl phases are compatible with a positive photoresist technology, if an appropriate etch could be developed, submicrometer wet-chemical lithography would be possible. Recent results in this laboratory have shown that a degassed, pH-controlled, aqueous EDTA-based etch can be utilized together with positive photoresist to obtain submicrometer feature sizes with little or no deterioration in the surface resistance [58]. This technique makes use of the co-ordination chemistry of the various ions to remove them from the etching surface and prevent the formation of carbonates.

As discussed in other chapters, Tl is a hazardous material. Particularly hazardous are solutions containing Tl ions. Great care must be taken to avoid exposure to etch and related solutions since the Tl ion can be transported readily through the skin. There is much less evidence that ready exposure is possible through direct contact with the superconductor. Results on films indicate that the superconducting properties of the 2212 and 2223 phases are stable over periods of at least a year and that although some surface reaction occurs on exposure to moisture, Tl loss is not observed and it appears that carbonate formation, particularly with the barium, occurs, creating a passivating layer.

VII. SUMMARY

A variety of techniques have been employed to grow applications-ready thin films in the Tl system. High-quality c-axis-oriented films have been grown in 1212, 2201, 2212, 1223, and 2223 on a variety of substrates, of which $LaAlO_3$, MgO, and $SrTiO_3$ are best. A typical T_c is greater than 100 K; J_c at 77 K is nearly 10^7 A/cm^2; and R_s is below 1 mΩ at 77 K and 10 GHz. Processing technologies have been developed suitable for obtaining submicrometer feature sizes with no surface degradation. To date, only ex situ processing has produced films with good superconducting properties, although recent results indicate that at least some of the correct phase can be formed in situ. It appears that the best precursors to date are those formed by either single-target or dual-facing-target radio-frequency or dc magnetron sputtering. The intimate mixing of cations appears to enhance phase formation and improve surface morphology. The best films to date have been obtained by encapsulating the films and a Tl source in Au foil. This does not represent a scalable technology. Recent results using alumina crucibles and two zone furnaces have produced results nearly as good as those for the encapsulated films. These techniques represent manufacturable approaches. However, until complete control of the oxygen and thallium partial pressures is obtained, perhaps by employing organometallic sources (MOCVD), a completely control process environment will not be possible.

Considerable recent work is directed at trying to improve pinning. A major thrust is in the area of the 1223 phase, where reduced c-axis spacing may improve coupling between the Cu–O sheets. Another new area is the partial substitution of Sr for Ba and Pb for Tl. Du Pont and others have recently shown that Tl,Pb–Sr–Ca–Cu–O materials do not form double-layer phases. They have obtained T_cs to 120 K in the 1223 phase, but it is difficult to make phase pure. The 1212 phase typically has a T_c of 95 K. There is some evidence that the Sr-based phases are more stable. A considerable drawback is that control of Tl, Pb, and O partial pressures must be maintained during processing. Another key observation is that thicker films (> 0.5 μm) show improved pinning and indicate that the Tl system may be the best for conductor applications if the weak pinning problems can be overcome.

Recent improvement in transport properties, the best surface resistance values, the highest T_cs, and the ability to do large-scale wafer production indicate that Tl-based thin films may be the leader among HTS systems for many microelectronic and conductor applications. The largest problem area is that of heterostructures where in situ growth is highly desirable.

ACKNOWLEDGMENTS

I would like to acknowledge assistance by J. S. Martens and E. L. Venturini in reviewing this manuscript. This work was supported partly by the Office of Basic Energy Sciences, Division of Materials Sciences, U.S. Department of Energy, and partly by Defense Programs, U.S. Department of Energy, under contract DE-AC04-76DP00789.

REFERENCES

1. R. M. Hazen, L. W. Finger, R. J. Angel, C. T. Prewitt, N. L. Ross, C. G. Haddicos, P. J. Heaney, D. R. Veblen, Z. Z. Sheng, A. E. Ali, and A. M. Hermann, *Phys. Rev. Lett.* 60, 1657 (1988).
2. S. S. Parkin, V. Y. Lee, A. I. Nazzal, R. Savoy, R. Beyers, and S. J. La Placa, *Phys. Rev. Lett.* 61, 750 (1988).
3. Z. Z. Sheng and A. M. Hermann, *Nature* 332, 55 and 138 (1988).
4. D. S. Ginley, E. L. Venturini, J. F. Kwak, R. J. Baughman, M. J. Carr, P. F. Hlava, J. E. Schirber, and B. Morosin, *Physica C* 152, 217 (1988).
5. D. S. Ginley, J. F. Kwak, R. P. Hellmer, R. J. Baughman, E. L. Venturini, and B. Morosin, *Appl. Phys. Lett.* 53, 406 (1988).
6. D. S. Ginley, J. F. Kwak, R. P. Hellmer, R. J. Baughman, E. L. Venturini, M. A. Mitchell, and B. Morosin, *Physica C* 156, 592 (1988).
7. J. F. Kwak, E. L. Venturini, R. J. Baughman, B. Morosin, and D. S. Ginley, *Physica C* 156, 103 (1988).
8. J. H. Kang, R. T. Kampwirth, and K. E. Gray, *Phys. Lett. A*, 131, 208 (1988).
9. W. Y. Lee, V. Y. Lee, J. Salem, T. C. Huang, R. Savoy, D. C. Bullock, and S. S. P. Parkin, *Appl. Phys. Lett.* 53, 329 (1988).
10. J. S. Martens, T. E. Zipperian, G. A. Vawter, D. S. Ginley, V. M. Hietala, and C. P. Tigges, *Appl. Phys. Lett.* 60, 1141 (1992).
11. D. Cubicciotti, *High Temp. Sci.* 2, 213 (1970).
12. R. H. Lamoreaux, D. L. Hildenbrand, and L. Brewer, *J. Phys. Chem. Ref. Data* 6, 419 (1987).
13. D. S. Ginley, B. Morosin, R. J. Baughman, E. L. Venturini, J. E. Schirber, and J. F. Kwak, *J. Crystal Growth* 91, 456 (1988).
14. C. T. Cheung and E. Ruckenstein, *J. Mater. Res.* 5, 1860 (1990).
15. Z. Z. Sheng and A. M. Hermann, *Nature* 332, 55 (1988).
16. T. Nabatame, J. Sato, Y. Saito, K. Aihara, T. Kamo, and S. Matsuda, *Physica C* 193, 390 (1992).
17. J. Talvacchio, J. R. Gavaler, J. Greggi, M. G. Forrester, and A. I. Braginski, *IEEE Trans. Magn.* 25, 2538 (1989).
18. T. Nabatame, Y. Saito, K. Aihara, T. Kamo, and S. Matsuda, *Jpn. J. Appl. Phys.* 29, L1813 (1990).

19. X. H. Zeng, S. G. Wang, J. Z. Li, J. Liu, Z. Z. Wang, Y. D. Dai, S. L. Yan, and L. Chen, *Cryogenics* 30, 915 (1990).
20. M. L. Chu, H. L. Chang, C. Wang, J. Y. Juang, T. M. Uen, and Y. S. Gou, *Appl. Phys. Lett.* 59, 1123 (1991).
21. D. S. Ginley, J. F. Kwak, R. P. Hellmer, R. J. Baughman, E. L. Venturini, M. A. Mitchell, and B. Morosin, *Physica C* 156, 592 (1988).
22. K. H. Young, D. Arney, E. J. Smith, and D. Strother, *Jpn. J. Appl. Phys.* L710 (1990).
23. W. L. Holstien, L. A. Parisi, D. W. Face, X. D. Wu, S. R. Foltyn, and R. E. Muenchausen, *Appl. Phys. Lett.* 61, 982 (1992).
24. D. W. Cooke, P. N. Arendt, E. R. Gray, and A. M. Portis, *IEEE Trans. Microwave Theory Tech.* 39, 1539 (1991).
25. R. B. Hammond, G. V. Negrete, L. C. Bourne, D. D. Strother, A. H. Cardona, and M. M. Eddy, *Appl. Phys. Lett.* 57, 825 (1990).
26. T. Nabatame, Y. Saito, K. Aihara, T. Kamo, and S. Matsuda, *Jpn. J. Appl. Phys.* 29, L1813 (1990).
27. K. Char, M. S. Colclough, S. M. Garrison, N. Newman, and G. Zaharchuk, *Appl. Phys. Lett.* 59, 733 (1991).
28. D. S. Ginley, E. L. Venturini, C. P. Tigges, T. E. Zipperian, R. J. Baughman, J. C. Barbour, and B. Morosin, *Physica C* 185, 2275 (1991).
29. C. P. Tigges, E. L. Venturini, D. S. Ginley, J. F. Kwak, B. Morosin, R. J. Baughman, J. C. Barbour, R. P. Hellmer, T. E. Zipperian, and J. S. Martens, in *Science and Technology of Thin Film Superconductors 2*, ed. R. D. McConnell and R. Noufi, Plenum Press, New York, 1990, p. 439.
30. A. M. Hermann, H. Duan, F. Arammash, R. J. Deck, D. Marsh, W. Kiehl, M. J. Saeed, J. Tang, D. L. Viar, P. S. Wang, and D. G. Naugle, in *Science and Technology of Thin Film Superconductors 2*, ed. R. D. McConnell and R. Noufi, Plenum Press, New York, 1990, p. 251.
31. K. Endo, Y. Moriyasu, S. Misawa, and S. Yoshida, *Physica C* 185–189, 1995 (1991); F. A. Kuznetsov, I. K. Igumenov, and V. S. Danilovich, *Physica C* 185–189, 1957 (1991).
32. N. Hamaguchi, R. Boerstler, R. Gardiner, and P. Kirlin, *Physica C* 185–189, 2023 (1991); N. Hamaguchi, R. A. Gardiner, and P. S. Kirlin, *Proc. SPIE: Processing of Films for High T_c Superconductor Electronics*, 1394 (1991), in press.
33. H. Yuhua, Y. Shaolin, H. Fengzhu, and X. Shouqing, *Solid State Commun.* 75, 991 (1990).
34. X. H. Zeng, S. G. Wang, J. Z. Li, J. Liu, Z. Z. Wang, Y. D. Dai, S. L. Yan, and L. Chen, *Cryogenics* 30, 915 (1990).
35. S. L. Yan, H. L. Cao, X. M. Yang, and X. D. Zhou, *J. Appl. Phys.* 70, 526 (1991).
36. M. L. Chu, H. L. Chang, C. Wang, J. Y. Juang, T. M. Uen, and Y. S. Gou, *Appl. Phys. Lett.* 59, 1124 (1991).
37. W. Y. Lee, S. Garrison, M. Kawasaki, E. Venturini, B. T. Ahn, R. Beyers, J. Salem, R. Savoy, and J. Vazquez, *Appl. Phys. Lett.* 60, 772 (1992).
38. W. Y. Lee, V. Y. Lee, J. Salem, T. C. Huang, and R. Savoy, *AIP Conf. Proc.* 199, 71 (1990).
39. W. Y. Lee, J. Salem, V. Lee, T. C. Huang, and R. Savoy, *Physica C* 162, 639 (1989).
40. P. Arendt, N. Elliott, R. Dye, K. Hubbard, M. Maley, J. Martin, Y. Coulter, and B. Bennett, *J. Elec. Mater.* 21, 499 (1992).
41. B. Morosin, E. L. Venturini, and D. S. Ginley, *Physica C* 185, 657 (1991).
42. B. Morosin, E. L. Venturini, and D. S. Ginley, *Physica C* 183, 90 (1991).
43. B. T. Ahn, W. Y. Lee, and R. Beyers, *Appl. Phys. Lett.* 60, 2150 (1992).

44. D. H. Kuo, R. J. Lin, and P. T. Wu, in *Science and Technology of Thin Film Superconductors 2*, ed. R. D. McConnell and R. Noufi, Plenum Press, New York, p. 291.

45. J. A. DeLuca, P. L. Karas, J. E. Tkaczyk, C. L. Briant, M. F. Garbauskas, and P. J. Bednarczyk, *Proc. 1992 Spring MRS Meeting*, San Francisco, Apr. 27 to May 1, 1992, Symposium S, *MRS Symp. Proc.* 275, in press (abs. S11.6).

46. W. L. Holstien, L. A. Parisi, C. Wilker, and R. B. Flippen, *Appl. Phys. Lett.* 60, 2014 (1992).

47. B. Morosin, E. L. Venturini, C. P. Tigges, D. S. Ginley, and S. R. Volk, *Proc. 1992 Spring MRS Meeting*, San Francisco, Apr. 27 to May 1, 1992, Symposium T, in press (abs. T4.8).

48. B. Morosin, M. G. Norton, C. B. Carter, E. L. Venturini, D. S. Ginley, and C. P. Tigges, *J. Mater. Sci.* 8 (1993).

49. M. P. Siegal, J. M. Phillips, Y. F. Hsieh, and J. H. Marshall, *Physica C* 172, 282 (1990).

50. Q. Li and T. Venkatesan, in *Science and Technology of Thin Film Superconductors 2*, ed. R. D. McConnell and R. Noufi, Plenum Press, New York, p. 153.

51. D. K. Fork, G. A. N. Connell, D. B. Fenner, J. B. Boyce, J. M. Phillips, and T. H. Geballe, in *Science and Technology of Thin Film Superconductors 2*, ed. R. D. McConnell and R. Noufi, Plenum Press, New York, p. 187.

52. D. W. Face and J. P. Nestlerode, *Appl. Phys. Lett.* 61, 1838 (1992).

53. E. L. Venturini, W. Y. Lee, B. Morosin, and D. S. Ginley, *Proc. 1992 Spring MRS Meeting*, San Francisco, Apr. 27 to May 1, 1992, Symposium S, *MRS Symp. Proc.* 275, in press.

54. M. P. Maley, P. H. Kes, G. J. Vogt, D. S. Phillips, and M. E. McHenry, *Bull. Am. Phys. Soc.* 36, 984 (1991).

55. D. H. Kim, K. E. Gray, R. T. Kampwirth, J. C. Smith, D. S. Richesson, T. J. Marks, J. H. Kang, J. Talvacchio, and M. Eddy, *Physica C* 177, 431 (1991) and K. E. Gray and D. H. Kim, *Physica C* 180, 139 (1991).

56. J. W. Ekin, T. M. Larson, N. F. Bergen, A. J. Nelson, A. B. Swartzlander, L. L. Kazmerski, A. J. Panson, and B. A. Blankenship, *Appl. Phys. Lett.* 52, 1819 (1988).

57. J. S. Martens, T. E. Zipperian, D. S. Ginley, V. M. Hietala, C. P. Tigges, and T. A. Plut, *J. Appl. Phys.* 69, 8268 (1991).

58. C. I. Ashby, J. S. Martens, T. A. Plut, and D. S. Ginley, *Appl. Phys. Lett.* 60, 2147 (1992).

59. H. Yuhau, Y. Shaoli, H. Frengzhu, and X. Shouqing, *Solid State Commun.* 75, 991 (1990).

60. D. J. Werder and S. H. Liou, *Physica C* 179, 430 (1991).

61. A. Naziripour, C. Dong, J. W. Droxler, A. B. Swartzlander, A. J. Nelson, and A. M. Hermann, *J. Appl. Phys.* 70, 6495 (1991).

62. J. Olsen, STI, private communication.

10
Spray Pyrolysis

L. Pierre de Rochemont and Michael R. Squillante
Radiation Monitoring Devices, Inc., Watertown, Massachusetts

I. INTRODUCTION

The discovery of superconductivity at high temperatures in the ceramic cuprates [1–6] has resulted in a rapid acceleration in superconductor materials research. Much of this research has been directed at producing and examining new compounds, motivated both to increase superconductor performance and to understand better the underlying phenomena. Most of the fabrication processes used to synthesize and study these superconductor materials are typically laboratory techniques limited to expensive, small-scale production that are not generally suited to the manufacture of very large structures. Thus the processes themselves, not the inherent properties of the superconductors, are restricting potential applications to small-scale devices. The commercial feasibility of any high-temperature ceramic superconductor application is enhanced by the availability of a fabrication process that can produce large amounts of high-quality material economically. One process that has the potential to meet the needs of superconductor production is spray pyrolysis. Spray pyrolysis is a technique that should enable production of high-quality films over areas of many square meters, including the mass production of continuous sheets. This technique has been used for many years to deposit good-quality films or coatings of selected electronic materials [7]. Spray pyrolysis is extremely attractive and relatively easy to commercialize because it is inherently low cost and simple, and it lends itself readily to flexible, high-throughput industrial processing.

In chemical spray pyrolysis, soluble metallic salts of the component metals of the ceramic are dissolved in solution where a high degree of molecular mixing is possible. Once thoroughly mixed, the salt solution is sprayed onto the surface of a hot substrate (Fig. 1). The solvent is driven off by evaporation, leaving the

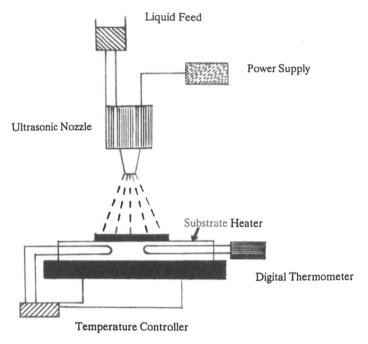

Figure 1   Spray process.

precursor salts to precipitate onto the substrate surface, where they are subsequently pyrolyzed into their component oxides. This sequence of events can be effected in one of two ways: either through a single step, if the substrate temperature is sufficiently hot [8–11], or by spray drying the salt residue onto the substrate at a temperature that volatilizes the solvent, followed by the calcination of the deposited ceramic formulation into its oxide form at a higher temperature [12–15]. The principal objective in adapting spray pyrolysis to the preparation of high-T_c superconducting ceramics is to adjust the process parameters to allow controlled deposition of a stoichiometric blend of precursor salts in a solid-state form that replicates the high degree of uniform molecular mixing available in the solution. A fundamental detraction in using spray pyrolysis to fabricate high-T_c superconducting ceramics is that initially, an inherently polycrystalline film is produced. As a result of this morphology, current transport through the material can be severely impeded by this granular structure if the grain boundaries are poorly aligned.

Of the families of high-T_c superconducting cuprates currently identified, the thallium-based superconductors appear to be the most forgiving when formed in polycrystalline structures. Even when deposited as polycrystalline films, they exhibit high-zero-resistance transition temperatures [16] and have registered critical current densities as high as 6×10^5 A/cm^2 at 77 K [17–19] in the absence of a magnetic load. Thallium-based high-temperature superconductors have been prepared via spray pyrolysis by dissolving all the precursor compounds into solution [13,14] and by diffusing vapor-transported thallium oxide into a spray-pyrolyzed Ba–Ca–Cu–O precursor film [15,20]. Recently, researchers at the General Electric Research and Development Center in Schenectady, New York, have demonstrated Tl diffusion on spray-pyrolyzed Ba–Ca–Cu–O films as a genuinely competitive materials processing technology in achieving critical current densities of 2×10^4 A/cm^2 at 77 K in a 1-T magnetic field, and 2×10^5 A/cm^2 at 4.2 K in a 10-T field [21] (Fig. 2).

In this chapter we describe experimental findings from our laboratory and the literature to identify key process parameters and how variations to these parameters can affect the quality of spray-pyrolyzed films. We also examine how different

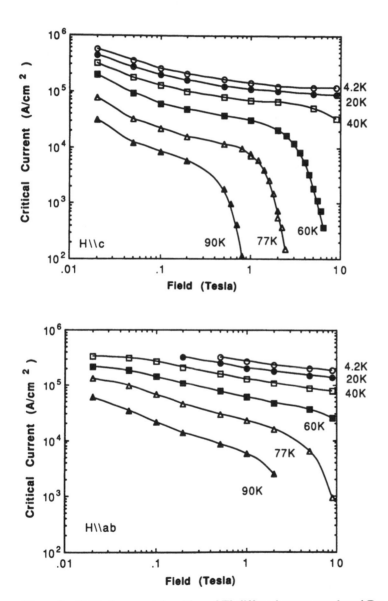

Figure 2 Critical current densities of Tl-diffused spray-pyrolyzed Ba–Ca–Cu–O films prepared by researchers at the General Electric Research and Development Center. (Reproduced with the permission of the General Electric Research and Development Center, Schenectady, N.Y.)

techniques that fit loosely into the general definition of spray pyrolysis could conceivably be used to improve on the process. The format of the chapter breaks the process into distinct synthesis elements. A brief preamble on the advantages of using solution processing is given to introduce the reader to the rationale for the process. This section is followed by short discussions on the influence of rheology and substrate temperature during the spray process, current issues regarding the calcination and sintering of the deposited films, the diffusion of thallium into spray-pyrolyzed Ba–Ca–Cu–O precursor films, and techniques that have shown some promise in densifying and texturing films that can be initially porous depositions. The chapter concludes with some thoughts on potential commercial applications for which spray-pyrolyzed films or coatings could conceivably be useful and a summary of processing parameters which, in our experience, deserve due consideration when implementing this processing technique.

II. ADVANTAGES AND POTENTIAL OF SOLUTION PROCESSES TO BULK SYNTHESIS

The high-T_c ceramic superconductors derive their unique physical properties from a very complex (in that they contain three or more metal-oxide components) and precise distribution of heavy-metal atoms that have been fused into a perovskite crystal structure. In using bulk or nonepitaxial methods, precursors of the individual heavy-metal oxide components found in the perovskite lattice are mixed together prior to fusing the crystal through solid-state reactions. The relative success or failure of any bulk synthesis method to produce these solids with reasonable crystal purity depends on the degree of intermixing that can be achieved among the individual precursors prior to crystal fusion. The conventional method used to prepare powder formulations of these ceramics requires calcination of a mechanically ground mixture of metal oxides and alkaline earth carbonates in definite proportions. Once components undesirable to the final product, such as carbon oxides, have been driven out of the solid mixture, the ceramic formulation can be sintered by thermally activated solid-state reactions.

The solid-state reaction that fuses the crystal lattice is a diffusion-controlled process. Interatomic diffusion mandates a high degree of intimacy between the various reacting species as well as a uniform distribution of each species prior to thermal processing in order to produce a thoroughly reacted and uniformly distributed product. Milling and grinding are normally used in this process to achieve a mixed state with very fine subdivision. A principal shortcoming to the bulk synthesis of ceramic materials is that abrasives used in preparing this finely subdivided state can introduce contaminants into the system. Furthermore, mechanically ground mixtures generally require prolonged calcination at high temperatures to break down carbonate species and to establish proper redox chemistry in the bulk. Protracted treatments at high temperatures can promote the growth of single-component or multicomponent crystallite phases that are undesirable in the synthesis of a phase-pure superconducting ceramic. Hence it is advantageous to use precursor compounds that decompose relatively easily at lower temperatures.

Solution processes offer an alternative approach to achieving a high degree of mixing and uniform distribution provided that all the precursors are mutually dissolved into the solvent. A number of parameters can influence how well the liquid concentrations are reproduced on the substrate surface. These include the selection of salts, the use of chemical binders in the solution, the choice of a solvent, the dynamics and length of the spray plume, and the substrate temperature.

Nitrate salts have been used almost exclusively in the development of spray-pyrolyzed Tl–Ba–Ca–Cu–O films [13,14,15,20,21], although other salt complexes, such as metal acetates [22] and organic acid salts [23–25], have also been used in solution processing other high-T_c ceramics. Nitrates of calcium, barium, copper, and thallium are all soluble in water. Although thallium can be sprayed directly with the solution [14] as a nitrate, it is often preferable to reduce health and safety hazards by diffusing the thallium into a precursor Ba–Ca–Cu–O film. Barium nitrate is weakly soluble in aqueous solution and will have a tendency to precipitate out if its solution concentration is too high. We have noted this problem in Ba–Ca–Cu aqueous nitrate solutions having molarities greater than 0.05 M. Another attractive feature of the nitrate salts is that they are not carbon based. Consequently, the risk of forming alkaline earth carbonates through the reaction of calcium and barium with carbon dioxide (CO_2) evolving from the lattice during calcination is avoided. A probable drawback to the use of nitrates is that they are not mutually soluble in the solid state.

Ideally, one would want to deposit a uniform amorphous film that reproduces the complete mixing achieved in solution. Figure 3 is an x-ray diffraction pattern (XRD) of the salt residue from an aqueous metal nitrate solution spray dried onto a magnesium oxide (MgO) substrate heated to 220°C, a temperature below the decomposition temperature of all three of the nitrate salts. The XRD pattern clearly reveals randomly oriented well-defined spectra from crystallite structures. Copper is present as copper hydroxynitrate salt $Cu_2(OH)_3 \cdot (NO_3)$ formed through its dis-

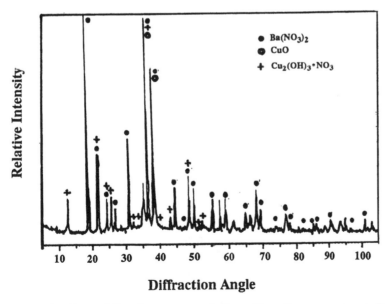

Diffraction Angle

Figure 3 X-ray diffraction pattern of salt residues from an aqueous $Ba(NO_3)_2$, $Ca(NO_3)_2$, $Cu(NO_3)_2$ sprayed at a substrate temperature below the decomposition temperatures of the precursor salts.

solution into the solvent. Both $Ca(NO_3)_2$ and $Ba(NO_3)_2$ retain their crystallinity in this deposit. While solutions can achieve a high degree of mixing, Fig. 3 shows that nitrate salt residues sprayed onto substrates clearly can contain phase-segregated crystallite structures. As shown below, the extent of the phase-pure domains of these crystallite structures can depend on the spray technique used to deposit the film.

By selecting appropriate salts and solvents, it is possible to modify the properties of the solution to enhance desired properties of the films. For example, a technique known as the Pechini process can be used to enhance the uniformity. This technique was originally developed to prepare high-quality large-area dielectric films of alkaline earth niobates ($BaNb_2O_6$, $SrNb_2O_6$, $CaNb_2O_6$, $MgNb_2O_6$, and $PbNb_2O_6$), titanates ($SrTiO_3$, $CaTiO_3$, $BaTiO_3$, $MgTiO_3$, and $PbTiO_3$), and zirconates ($CaZrO_3$, $SrZrO_3$, and $BaZrO_3$), useful to the fabrication of sheet capacitors, as well as more complicated mixed heavy oxides such as $Pb_{1-x}Sr_x(Zr_{1-y}Ti_y)O_3$, useful in piezoelectric devices [26]. As is the case with the ceramic superconductors, these dielectric materials are also complex oxides that have the perovskite crystal structure.

In its generalized form, the Pechini process takes advantage of the ability of certain α-hydroxycarboxylic acids, such as citric acid, lactic acid, glycolic acid, and gluconic acid, to form polybasic acid chelates with the heavy metals. These metal complexes can be formed by starting with alkoxides, hydroxides, or α-hydroxy-carboxylates of the metals and mixing them with two to eight times the molar ratio of citric acid in an excess of a polyhydroxy alcohol such as glycerol or polyethylene glycol. Alkaline earth metals (and lead) are added to the solution as oxides, hydroxides, carbonates, or alkoxides. The complexes thus formed can undergo polyesterification when heated with the polyhydroxy alcohol to form a network of cross-linked polymeric chains. When all the solvent is driven off, an intermediate resin is formed which, upon ignition, can be burned off to leave uniform and intimately mixed heavy-metal oxides. This resin can be thinned and sprayed, with care to safety procedures, onto hot substrates and pyrolyzed into a stoichiometric blend of oxides.

Chiang et al. report the deposition of a Tl-based superconductor using a "modified" Pechini process [14]. In their paper they describe using Tl_2O_3 and nitrate salts of bismuth, lead, strontium, calcium, and copper with oxalic acid to form $(Tl_{0.6}Pb_{0.2}Bi_{0.2})Sr_{1.6}Ca_{2.4}Cu_3O_y$ through a gel intermediate and prolonged sintering and calcination. While the oxalic acid can be used to chelate thallium ions and assist solubility with the other salts, the Pechini patent specifically states that compounds such as nitrates, chlorides, and acetates which are soluble in the liquid phase but insoluble in the solid phase are not useful to the process [26]. The same restriction is made for compounds such as oxalates, tartrates, and sulfates that form insoluble salts in the presence of alkaline earth cations.

The preparation of good-quality bismuth cuprate superconductors [25] and YBCO superconducting tapes using organic acid salts [24] has been reported. Since the Pechini process has been applied successfully to the fabrication of $YBa_2Cu_3O_{7-x}$ fibers, this could also prove to be a promising area for future research in the development of spray-pyrolyzed ceramic Tl–Ba–Ca–Cu–O films and coatings. However, these processes have such a heavy reliance on organic compounds that care must be taken during calcination to reduce the risk of forming alkaline earth carbonates. These carbonate phases severely impede transformation toward single-phase high-T_c superconducting ceramic coatings. Many researchers actually prefer sol-gel synthesis to the Pechini process as a solution process because of its ability to reduce the quantity of organic material used in the mixture.

III. INFLUENCE OF RHEOLOGY AND SUBSTRATE TEMPERATURE DURING THE SPRAY PROCESS

The predominant use of nitrate salts in spray pyrolysis has allowed researchers to infer the relative effects of rheological influences on the ultimate quality of spray-pyrolyzed films. As stated above, the key objective in using any solution process in the development of crystalline structures as complex as the high-T_c ceramics is to preserve the uniformity and intimacy of the mixed solution state. During the spray process the solution is nebulized into a mist which is then projected onto the hot substrate. Since the solution is subject to rapidly changing conditions during spray deposition, a number of factors can affect the preservation of its highly mixed state. Experimental results show that the choice of solvents, the length of the spray plume, and the substrate temperature can all affect stoichiometric control and the intimacy of precursor mixing in the deposited film.

Efforts to evaluate the influence of rheological factors related to the synthesis of high-T_c superconducting ceramics have focused on measuring the difference in film quality that results when chemical binders are added to the solution. The basic requirements for a chemical binder are that it must be soluble in the solution solvent and that it must have suitable functional groups along its molecular structure to bond loosely through intermolecular attractive forces with all of the precursor salts used in the solution. Polyhydroxyl alcohols become a very convenient choice as chemical binders for metal nitrate solutions (Fig. 4). If a solution is generated in conjunction with these chemical binders, high statistical probability exists for an ensemble average in which most of the microscopic states of the solution have the different precursor salts hydrogen bonded onto a polyhydroxyl organic chain. The enhanced intermolecular attraction with the binder molecules can help preserve this highly mixed state during the agitation of spray deposition. A slightly higher viscosity results from the enhanced intermolecular bonding in the solution, which can also affect the surface tension of liquid droplets nebulized into the solution mist to preserve the integrity of the mixed state.

In developing solution processes for bismuth cuprate and thallium cuprate superconductors, researchers at AT&T Bell Laboratories experimented with using glycerol as a chemical binder [13]. Glycerol has three hydroxyl groups in its chain onto which the different nitrate salts can loosely attach themselves through inter-molecular hydrogen bonding. In research at AT&T Bell Laboratories, films were prepared from pure nitrate aqueous solutions and nitrate aqueous solutions to which glycerol was added as a binder. In both instances the films produced were calcined and sintered with identical heat treatments to examine the rheology of the process. The films prepared with the glycerol solution had significantly sharper transitions to zero resistance than those prepared without a chemical binder (Fig. 5).

Another means to determine the effect of rheological influences is by studying the surface morphology of the sprayed films. The following series of micrographs

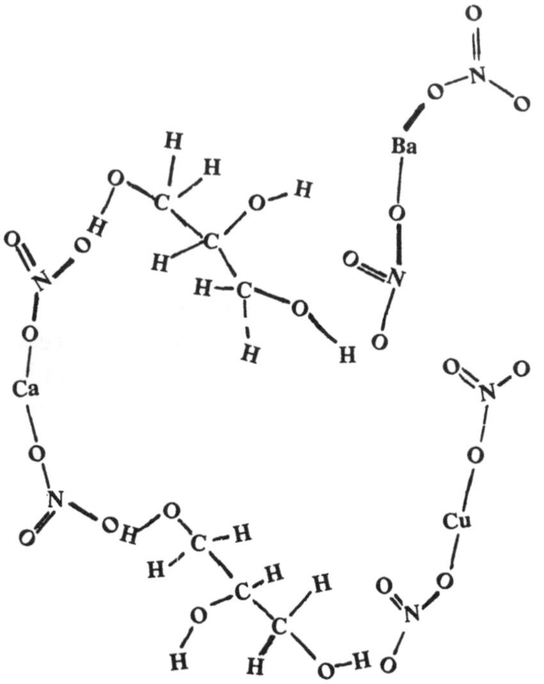

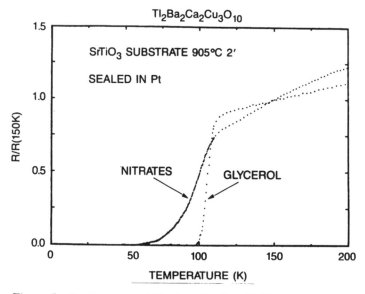

Figure 4 Molecular structure of the various nitrate salts allows them to bind loosely with hydroxyl groups through intermolecular hydrogen bonding. Polyhydroxyl alcohols have a number sites along their chain onto which the different nitrate salts can attach themselves.

$Tl_2Ba_2Ca_2Cu_3O_{10}$
$SrTiO_3$ SUBSTRATE 905°C 2′
SEALED IN Pt
NITRATES
GLYCEROL
R/R(150K)
1.5
1.0
0.5
0.0
0
50
100
150
200
TEMPERATURE (K)

Figure 5 $R-T$ curves of Tl–Ba–Ca–Cu–O films prepared at AT&T Bell Laboratories showing sharper zero-resistance transitions when glycerol is used. (From Ref. 13.)

demonstrate morphological features of spray-pyrolyzed films deposited on magnesium oxide (MgO) at substrate temperatures between 180 and 300°C by transporting chemically different solution mists over the depositing surface. All of these films were calcined and sintered using identical heat treatments of 500°C for 2 h in an air atmosphere and 850°C in oxygen for 4 h, allowing the furnace to cool down in both cases. For the films shown in Figs. 6 to 9, the solution was nebulized using an ultrasonic humidifier and the resulting mist was transported over a distance of 1 m to the hot substrate. As is evident from this series of micrographs, the best surface morphologies and greatest stoichiometric control observed in films prepared using these spray conditions were obtained with solutions containing glycerol binders rather than depositions from aqueous-only and aqueous ethanol solutions. Figure 6 shows the surface morphology of a Ba–Ca–Cu–O film prepared in our laboratory by transporting mist from an aqueous solution through a delivery chamber measuring 1 m in length before passing the vapor over a substrate heated to 265°C. The scanning electron micrograph (SEM) reveals relatively poor surface morphology and a rather nonuniform buildup of material across the face of the substrate. Robinson backscattered electron-imaged (RBEI) micrographs of this sample were highly spotted, indicating a high degree of single-species clustering in films prepared under conditions that allowed prolonged residence times of this salt solution in its vapor state.

Another factor that influences solution rheology and stoichiometric control is the ability of the solvent to maintain the individual precursors in their uniformly dissolved state throughout the deposition. Consequently, it is some times attractive to consider chemical modifications to the solution chemistry that might stabilize the less soluble solutes. The use of ethanol in barium nitrate solutions has been suggested to reduce $Ba(NO_3)_2$ precipitation [10]. The addition of ethanol in proportions of 20% by volume will substantially thin a water solution and noticeably

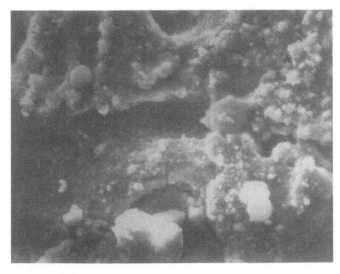

Figure 6 Scanning electron micrograph of a spray-pyrolyzed Ba–Ca–Cu–O film where the length of the spray plume was 1 m.

reduce the surface tension of droplets formed from the solution. However, $Ba(NO_3)_2$ does have a tendency to precipitate from pure acqueous solutions if the total metal ion molarity exceeds a solubility threshold. We have noticed precipitates of millimeter-sized crystals of barium nitrate form in pure aqueous nitrate solutions prepared with 2:2:3 Ba/Ca/Cu at molarities greater than 0.05 M. Ethanol can be used to raise that threshold [10]. Our experience in preparing films from nitrate salts in aqueous solutions containing 20% ethanol by volume has been characterized by a significant degradation in both the morphological texture and stoichiometric control when compared to results obtained with aqueous solutions. Although smooth surface textures were observed in these films (Fig. 7), they appeared merely as a surface crust supported by a void-riddled interior comprised of cylindrical or conically shaped deposits projecting upward from the substrate. These features, highly reminiscent of stalagmites, lent the impression of vigorous boiling on the substrate surface, causing the salt residue to take the form of the highly turbulent droplet. Films we have prepared from this mixture of solvents were appreciably depleted in calcium cations, thus suggesting that calcium may be transported from the rapidly drying residues as the solvents burn off.

Figure 8 clearly shows a dramatic improvement in the surface morphology of the films when they are deposited under these conditions from solutions containing 20% glycerol by volume. Compared to the other solvent chemistries, a much smoother surface texture is readily observable. Figure 9 shows an RBEI micrograph of the film. While some more lightly shaded and more darkly shaded areas in the film reveal some phase segregation in the deposit, the uniformity in these films is significantly better than those obtained without an organic binder in the solution. The impact on the homogeneity of spray-pyrolyzed films when the distance over which the solution must travel as a nebulized mist is shortened is demonstrated in Fig. 10.

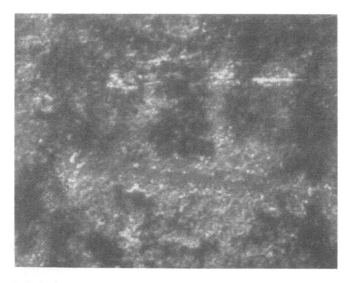

Figure 7   Scanning electron micrograph of a Ba–Ca–Cu–O film prepared from an aqueous solution containing 20% ethanol by volume.

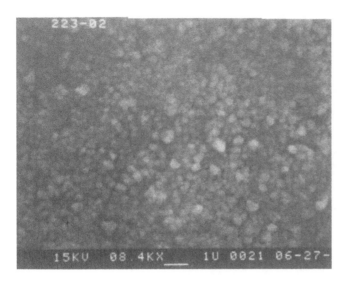

Figure 8 Scanning electron micrograph of a Ba–Ca–Cu–O film prepared from an aqueous solution containing 20% glycerol by volume.

Figure 10 shows an RBEI micrograph of a Ba–Ca–Cu–O film deposited using an identical solution chemistry to that used to prepare the film shown in Fig. 9, that is, Ba–Ca–Cu nitrate salts having 2:2:3 stoichiometric ratios with a total cationic molarity of 0.035 M dissolved into an aqueous solvent containing 20% glycerol by volume as an organic binder. The film examined in Fig. 10 was deposited using a Sono-Tek, Poughkeepsie, New York, 120 kHz microbore ultrasonic nozzle

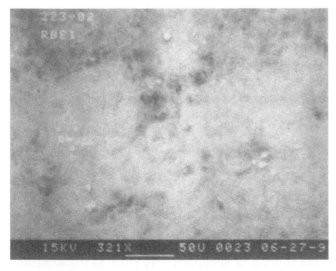

Figure 9 Robinson backscattered electron-imaged micrograph of a Ba–Ca–Cu–O film sprayed from a solution containing 20% glycerol by volume. The length of the spray plume was 1 m.

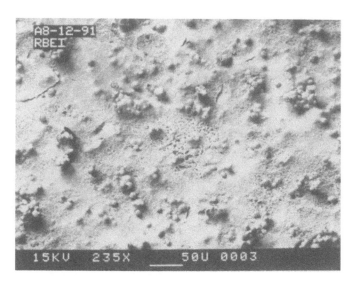

Figure 10 Robinson backscattered electron-imaged micrograph of a Ba–Ca–Cu–O film sprayed from a solution containing 20% glycerol by volume, where the length of the spray plume was 15 cm.

with the nebulizing head positioned roughly 15 cm from an MgO substrate heated to 265°C. In this configuration a quartz sleeve was used to confine the flow of the mist to the general area of the substrate. Use of the quartz sleeve can significantly improve deposition rates by reducing the volume mist driven away from the substrate by thermal convection currents.

The comparison of the micrographs shown in Figs. 9 and 10 appears to suggest that salt precursor phase segregation can be significantly diminished in spray-pyrolyzed superconducting films by reducing the lengths over which the spray solution must travel as a nebulized mist. The micrograph of the film, shown in Fig. 10, displays nearly perfect uniformity to within the resolution of the RBEI micrograph. Although we have not yet correlated an empirical relationship between salt precursor phase segregation and the amount of organic binder needed to achieve complete uniformity in films deposited with nebulizing source at given distances from the substrate surface, these results seem to suggest that the amount of binder needed to replicate the well-mixed state found in solution onto the film does seem to have a functional variance related to this parameter. It may be possible to use a lower volume fraction of organic binder in solution (i.e., < 20%) and still achieve exceptional homogeneity in the film when the solution is nebulized at a distance of 15 cm or less from the hot substrate. Researchers at the General Electric Research and Development Center Schenectady, New York, do not use any organic material in the preparation of Ba–Ca–Cu–O precursor films from aqueous nitrate salt solutions and have produced the best Tl–Ba–Ca–Cu–O films to date via spray pyrolysis. In the GE process the solution is nebulized very close to the substrate surface (J. A. DeLuca, private communication). Many researchers prefer to avoid introducing organic or carbonate material into the synthesis of high-T_c superconducting cuprates since carbon dioxide (CO_2) evolving from the lattice during calcination can react with alkaline earth oxides to form very stable carbonate material phases. The occurrence of carbonate phases in the material can obstruct the formation of the superconducting phases in the deposit if the kinetic transfer of CO_2 out of the lattice is not promoted efficiently. Consequently, in applying spray pyrolysis to the synthesis of high-T_c films or designing apparatus for this process, it is necessary to adjust the solution chemistry appropriately to the equipment configuration, and vice versa.

Substrate temperature can also affect both the flow dynamics of the solution mist and the quality of the film deposited. We have noticed that optimal substrate temperatures can be sensitive to the rheological properties of the solvent used in forming the precursor solution. For instance, if the solvent is thinned and droplets formed during nebulization have weaker surface tension, as is the case with aqueous solutions cut with 20% ethanol by volume, the flow dynamics and stoichiometric control are less stable with increasing substrate temperatures. Under these conditions very turbulent flow dynamics with apparently strong convection currents are observed in the deposition at substrate temperatures above 150°C. A great deal of material is transported and deposited away from the hot surfaces upon which the substrates are mounted. Furthermore, the salt precursor material is not uniformly transported under these conditions. We have noted that calcium nitrate appears to be preferentially lost from these turbulent mists, causing a severely reduced calcium incorporation in the deposit on the hot substrate surface.

Other interesting artifacts of the process can appear when using apparatus configured with a predisposition toward depositing phase segregated salts in dried films. In experimental configurations where the nebulized solution was transported over long distances (approximately 1 m) it may be possible to replicate the solution stoichiometry in the deposited film only over very specific substrate temperatures unless chemical binders are used. Empirical results tend to suggest that these observations are influenced by rheological factors in the solution. Figure 11 is a plot that compares metal cation ratios (2:1:2 Ba/Ca/Cu nitrates) dissolved in a 20% ethanol by volume aqueous solution to the ratios found in the deposited films versus substrate temperatures. The loss of calcium is clearly evident at all temperatures from these results. When these ratios are fitted to a curve detailing incorporation rates of the different metal cations an unusual result emerges (Fig. 12). In tabulating these data it was assumed that for substrate temperatures below 300°C the barium cations are 100% incorporated into the film. The fitted values for copper and calcium incorporation into the film show enhanced rates near a substrate temperature of 260°C. As shown in the following section, this temperature corresponds well with the decomposition temperature of the copper nitrate salt, possibly suggesting that the partial or whole decomposition of one of the salts may catalyze the incorporation of the other salts into the film. This speculation has been predicated on assuming 100% barium incorporation in the deposit. The validity of this assumption is consistent with results published by Kawai et al. [8], and the error in the fit is below 1% for calculated incorporation rates having substrate temperatures of 300°C or less (Fig. 13).

We have also experienced problems with stoichiometric control in pure aqueous solutions and observed a fairly strong sensitivity to substrate temperature. Hence if the laboratory objective is to use aqueous-only solutions, the design and construction of a hot-plate surface on which to place the substrate material that has

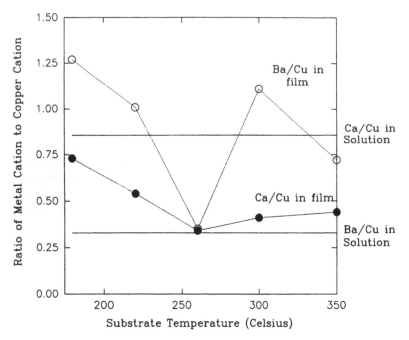

Figure 11 Ratios of the Ba and Ca cations to Cu cations versus substrate temperature in a film deposited using an aqueous solution containing 20% ethanol by volume. Straight lines represent the cation ratios in solution.

temperature control to within 1°C or better is strongly recommended. One should further check film morphology and cation incorporation in the deposited films and set the optimal substrate temperature on the basis of the morphological quality of the film. The loss of one or more cationic species by spraying at a given substrate temperature can always be compensated by increasing the fraction lost through nonstoichiometric additions to the solution. However, we have noticed that when glycerol is added as a chemical binder the solution stoichiometry is replicated in the film for substrate temperatures between 200 and 300°C. When substrate temperatures much above 300°C are used, convective forces have a strong tendency to propel the mist away from the hot substrate and cause it to condense on cooler surfaces nearby.

IV. CALCINATION AND SINTERING

Calcination is probably the single most important step in the process. During calcination the principal objectives are to decompose the salt precursor molecules into their irreducible oxide forms, to effectively burn off any organic material that might have been included in the solution without forming carbonate phases in the deposit, and to influence an oxidation-reduction chemistry among the irreducible oxides, which later facilitates the formation of phase-stable superconducting ceramic during sintering. In sintering the ceramic, it is desirable to identify a thermal pathway that activates the diffusive restructuring of the elemental oxides into the complex superconducting perovskite phases without actually melting or volatilizing

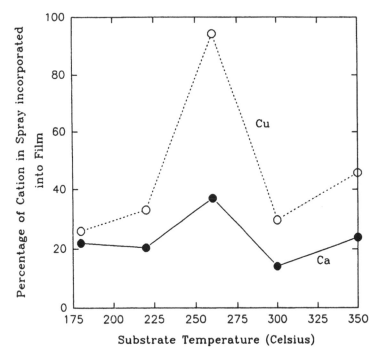

Figure 12 Percent incorporation rates for Ca and Cu cations versus substrate temperature in films sprayed from Ba–Ca–Cu–$(NO_3)_2$ solutions containing 20% ethanol by volume (calculated assuming 100% Ba incorporation over these substrate temperatures).

the material. Both processes are sensitive to atmospheric conditions, which in each case should be adjusted to play a specific role in the process.

To design a process effectively it is necessary to know the temperatures at which the individual precursor salts decompose, and at what temperatures and in which atmospheres the material is susceptible to volatilization. Certainly, rapid transport of by-product material, particularly the rapid removal of CO_2 from the lattice, will aid this objective. However, it is preferable to avoid the transport of volatile metal oxides, such as Tl_2O_3, out of the lattice during this processing step. Hence, knowledge of the temperatures over which the precursor materials decompose and the temperature (and atmosphere) above which the deflagrated material will sublime are necessary engineering parameters.

It has previously been reported that the nitrate salts decompose at temperatures around 500°C [13]. Although that is probably true for some of the nitrate salts, it is not specifically true for all the nitrate salts used to prepare thallium-based superconducting films. Thermal decomposition spectra of barium, calcium, and copper nitrate salts is provided below to assist the reader in forming or optimizing processing recipes. The denitration of the individual cation precursors was studied by drying $Ba(NO_3)_2$, $Ca(NO_3)_2$, and $Cu(NO_3)_2$ aqueous solutions on alumina substrates at 100°C. The dried material was scratched off the surface and analyzed using thermogravimetric analysis (TGA). These measurements were performed using a Perkin–Elmer Delta Series TGA7 thermal analyzer. All thermal analysis measurements presented here were done in oxygen atmospheres.

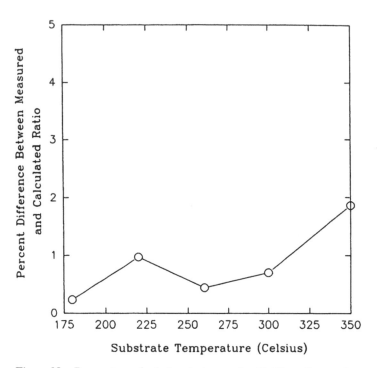

Figure 13 Percentage deviation between the Ca/Cu cation ratio measured in the films and the Ca/Cu ratios calculated from Ca and Cu loss estimates assuming 100% Ba cation film incorporation.

X-ray diffraction spectra on sprayed films show that $Cu(NO_3)_2$ in an aqueous environment is converted to $Cu_2(OH)_3 \cdot NO_3$ (see Fig. 3). TGA analysis of the dried material indicates that the temperature for maximal denitration/dehydration of $Cu_2(OH)_3 \cdot NO_3$ is 265°C, as shown in Fig. 14. It appears that between 10 and 210°C, residual water is driven from the film. The very sharp transition observed between 210 and 310°C is attributed to the denitration/dehydration of $Cu_2(OH)_3 \cdot NO_3$. The 33.7% percentage weight change between 210 and 1010°C agrees favorably with the formation of CuO from $Cu_2(OH)_3 \cdot NO_3$, on the basis of the following fractional weight percents:

fractional $Cu_2(OH)_3 \cdot NO_3 \rightarrow$ CuO   + oxides of nitrogen
wt %   100%   66.2%   33.8%

The chemical stability of CuO in oxygen at higher temperatures is evident.

Analysis on $Ca(NO_3)_2$ (Fig. 15) shows dehydration below 200°C and a percentage weight change of 66.1% between 400 and 750°C. Unlike the cuprous reaction, a series of weaker outgassing peaks are observed in the first derivative spectra prior to a very pronounced rate of weight loss at 733°C. These data agree quite favorably with the formation CaO through the following path:

fractional $Ca(NO_3)_2 \rightarrow$ CaO   + oxides of nitrogen
wt %   100%   34.16%   65.84%

CaO appears to be stable in oxygen above 740°C, after denitration.

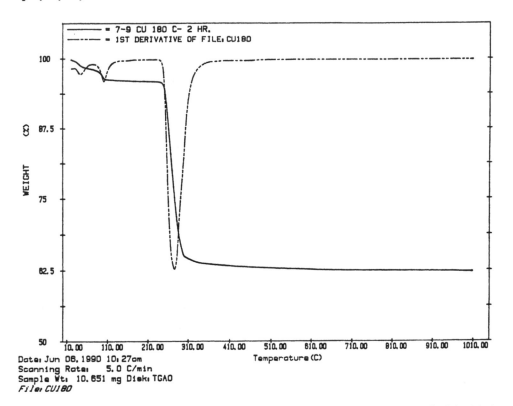

Figure 14 TGA spectrum of $Cu_2(OH)_3 \cdot NO_3$, the residue of aqueous solved $Cu(NO_3)_2$ dried at 170°C. The dashed line is the first derivative spectrum.

The barium nitrates, $Ba(NO_3)_2$, are the most stable of the starting compounds (Fig. 16). First derivative spectra indicates pyrolytic phenomena starts above 600°C. A stable regime is noted between 800 and 900°C. This stability breaks down above 900°C, but the first derivative spectrum shows the film tends toward stability again above 1010°C. The percentage weight change between room temperature and 1010°C is 40.4%, which is close to the fractional weight that would be lost in the formation of BaO:

fractional $Ba(NO_3)_2 \rightarrow$ BaO   + oxides of nitrogen
wt %   100%   58.65%   41.35%

The stable oxynitrate phase observed in the $Ba(NO_3)_2$ derivative between 800 and 900°C cannot be identified at this time. The $Ba(NO_3)_2$ TGA data suggest that unless the presence of the other nitrates [$Cu(NO_3)_2$ and its derivatives, or $Ca(NO_3)_2$] can accelerate conversion of $Ba(NO_3)_2$ to BaO at lower temperatures, annealing above 1000°C would be required to denitrate the BaCaCuO precursor films completely.

To study and suggest optimal solution chemistries and the effect of chemical binder activity, we have performed thermal analysis on 2:1:2 Ba/Ca/Cu solution residues precipitated from aqueous-only solutions, ethanol–(20% vol)/water solutions and glycerol–(20% vol)/water solutions. Each solution was dried on a hot (120 to 230°C, depending on solvent volatility) alumina substrate. TGA scans were

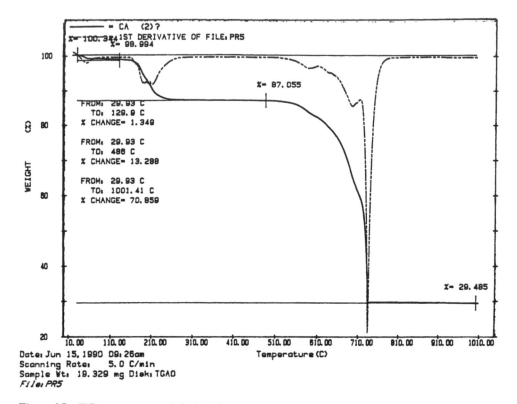

Figure 15   TGA spectrum of Ca(NO₃)₂ aqueous residue.

run to observe effects of the solvents and the other nitrates on chemical reactivity during denitration. Ethanol solutions were considered to reduce $Ba(NO_3)_2$ precipitation from solution. Glycerol solutions were considered to reduce single nitrate species agglomerations in solution and to enhance the chemical reactivity between the different nitrate compounds.

Figure 17 shows TGA spectra for the nitrates when dissolved in water only. Thermogravimetric changes below 200°C are being attributed to water absorbed in the nitrate film by exposure to ambient air. Copper denitration proceeds as before, unaffected by the presence of the other nitrate complexes. Its characteristic spectrum in the first derivative is once again observed at 265°C. The percentage weight change observed between 228 and 369°C was roughly 11% and is attributed to copper denitration. This is in general agreement with an 11.9% weight differential that would be expected using the fractional weight percentage change observed during $Cu_2(OH)_3 \cdot NO_3$ denitration/dehydration between 200 and 300°C.

In water solutions $Ca(NO_3)_2$ and $Ba(NO_3)_2$ denitration appears to be shifted to lower temperatures. Principal first derivative spectra for the denitration transitions are observed between 375 and 700°C. This is in rough agreement with the $Ca(NO_3)_2$ spectra but is contrary to the findings for $Ba(NO_3)_2$. It appears that $Ca(NO_3)_2$ and $Ba(NO_3)_2$ have a chemical affinity for one another and that their mutual denitration is accelerated by each other's presence or by the copper oxides in the films. Calculations of the fractional weight percent changes using the indi-

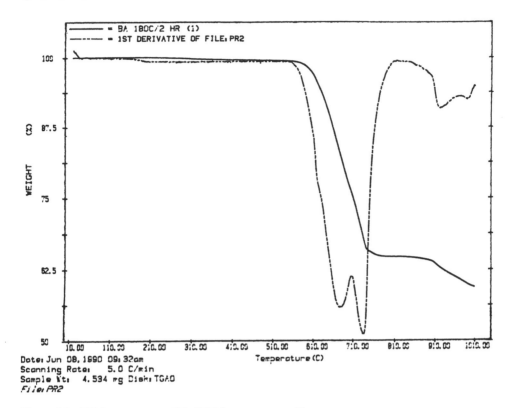

Figure 16   TGA spectrum of Ba(NO$_3$)$_2$ aqueous residue.

vidual denitration data predict a total percentage weight change of 57.56%, in rough agreement with the 54.35% weight change observed in the mixed films. It seems reasonable to assume that in water-only solution most denitration should be complete above temperatures of 700°C.

Figure 18 shows TGA spectra from the 20 vol % ethanol solution. Spectral signatures in the zeroth and first derivative spectra are identical, indicating Cu$_2$(OH)$_3$·NO$_3$ denitration again proceeds unaffected by the presence of either Ca(NO$_3$)$_2$ or Ba(NO$_3$)$_2$ in ethanol solvents. Once again, the thermal spectra suggest that the alkaline earth nitrates are complexed together.

Since glycerol is a chain molecule with many hydrogen-bonding sites, the individual nitrate species have a better chance of being physically complexed closer to each other if they bind at different sites along the chain. It is intended that the close physical proximity of the individual ntirate species provided by the glycerol chain would improve uniformity, inhibit single species clustering, and enhance the formation of Ba$_w$Ca$_x$Cu$_y$O$_z$ crystallites. Stoichiometric formation of Ba$_w$Ca$_x$Cu$_y$O$_z$ crystallites is preferred to CuO crystallites segregated from CaO/BaO polymorphous phases. This will help ensure chemical integrity during subsequent processing and single-phase formation during thallium diffusion.

Figure 19 shows that the qualitative character of glycerol-complexed BaCaCu denitration in oxygen changes substantially. TGA spectra below 182°C (the boiling point of glycerol) is due primarily to dehydration. Spectra between 182 and 250°C

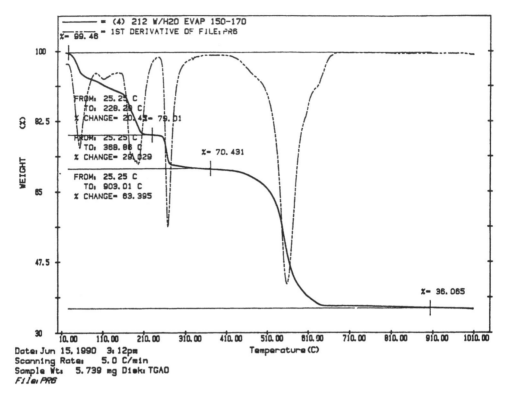

Figure 17   TGA spectrum of salt residue precipitated from a 2:1:2 Ba–Ca–Cu–$(NO_3)_2$ aqueous solution.

is attributed to unbound glycerol outgassing. Figure 20, which shows thermal spectra collected on the glycerol solvent used in these studies, clearly indicates that this is a reasonable assumption. In these solutions 20% glycerol by volume establishes a ratio of roughly 10.7 mol of glycerol per mole of 2:1:2 Ba/Ca/Cu–$(NO_3)_2$ precursor salts in the solution. It is difficult to assess how much glycerol remained in the residue during calcination since some most certainly could have lost while baking the material at 230°C to dry the precipitate.

The most notable feature in the spectra shown in Fig. 19 is that $Cu_2(OH)_3 \cdot NO_3$ no longer appears to denitrate by itself. Furthermore, the largest relative change in weight occurs below 500°C. The weight change between 250 and 500°C accounts for roughly 67% of all outgassing above 250°C. This suggests that some $Ca(NO_3)_2$ and $Ba(NO_3)_2$ denitration may actually be participating in the copper oxide conversions and that BaCaCu complexes are being formed in solution. Roughly one-third of the relative weight change occurs above 500°C, which can be solely attributed to $Ca(NO_3)_2$ and $Ba(NO_3)_2$ (complexed) denitration. Changes in the first derivative spectra above 900°C suggest that this precursor compound becomes volatile above this temperature in oxygen but can be safely processed below 900°C.

The high-T_c superconducting ceramics are now known to have thermodynamically unstable reactions with water (H_2O) and carbon dioxide (CO_2). They are also thermodynamically reactive with oxygen; and depending on the thermal and

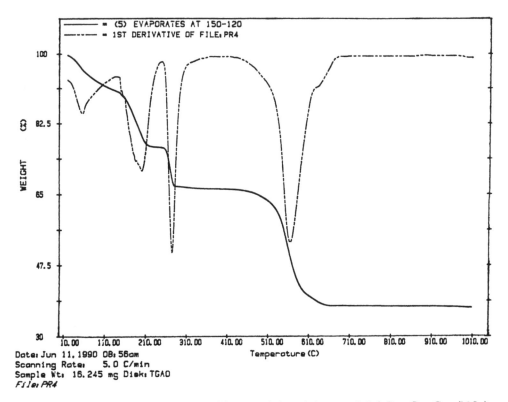

Date: Jun 11, 1990 08: 56am
Scanning Rate: 5. 0 C/min
Sample Wt: 16. 245 mg Disk: TGAO
File: PR4

Figure 18 TGA spectrum of salt residue precipitated from a 2:1:2 Ba–Ca–Cu–(NO₃)₂
aqueous/ethanol (20 vol %) solution.

atmospheric history of the material, this reactivity can be favorable or unfavorable
under given conditions. In preparing phase-stable films it is necessary to foster
conditions that alleviate the occurrence of unfavorable reactions. This most certainly means that every attempt must be made to shunt reactions between the
deposited material and H_2O or CO_2. These chemical reactivities make spray pyrolysis a somewhat vulnerable process. Since in its most conventional sense this
technique exposes the material to copious amounts of water, or with modifications
intended to improve upon the process, can incorporate organic material that when
heated will generate gaseous CO_2 and H_2O during firing. Furthermore, if the proper
chemical kinetics are not established during calcination and the material is exposed
to CO_2 at temperatures sufficiently elevated for it to become highly reactive, the
pure oxides will decompose in favor of forming alkaline earth carbonates and
insulating metal-oxide phases. Thus it is necessary to exercise particular caution
during calcination. It is extremely important that these materials are thoroughly
dry and free of carbon prior to sintering, and calcination should include steps that
specifically consider the removal of CO_2 and H_2O at temperatures over which these
chemical entities are not likely to create material phases harmful to high-quality
superconductor structures.

At present the thallium-based superconductors are probably the least characterized, in terms of fundamental chemical and thermodynamic relationships, of

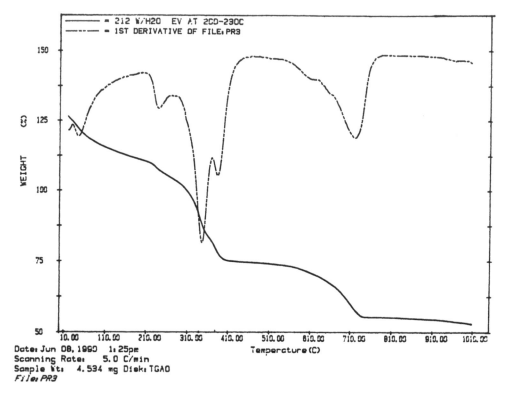

Figure 19 TGA spectrum of salt residue precipitated from a 2:1:2 Ba–Ca–Cu–$(NO_3)_2$ aqueous/glycerol (20 vol %) solution.

the cuprate superconductors. However, the alkaline earth ions are likely to be the most reactive chemical component in the ceramic matrix to the deleterious thermodynamic reactions initiated by water and carbon dioxide. While the kinetics of their reactivities are likely to differ, the chemistries of the alkaline earth components are probably similar within the different compositional families of cuprate superconductors. Some insight toward optimizing synthesis chemistry in spray-pyrolyzed thallium cuprates could conceivably be obtained by examining some of the better characterized superconducting oxides. Even though basic chemical behavior could indeed be similar, the rates at which these reactions occur are likely to be different.

$YBa_2Cu_3O_{7-y}$ is probably the best characterized of all the superconducting ceramic oxides. Experimental results have recently been reported and diffusion coefficients have been measured which indicate that there are possibly kinetic aspects to the hydration of this material. Hydration in $YBa_2Cu_3O_{7-y}$ is believed to be controlled by hydroxyl (OH^-) diffusion especially at low water vapor partial pressures [27]. A concern in having any hydroxyl anions present in the lattice is that they can easily solvate the alkaline earth cations and thus perturb their positions within the perovskite structure enough to alter the internal interaction(s) responsible for the occurrence of the superconductive phase. If allowed to occur in sufficient quantity, this process could totally decompose the material [28,29]. We have noticed that films prepared by diffusing thallium into a Ba–Ca–Cu–O precursor

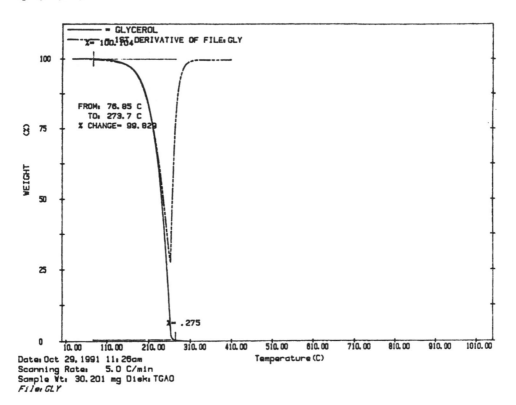

Figure 20 TGA spectrum of glycerol.

film spray pyrolyzed from aqueous-based solutions are phase unstable over time. That is, lower T_c knees began emerging in the $R-T$ characteristics of Tl–Ba–Ca–Cu–O films that initially exhibited zero-resistance temperatures as high as 111.5 K immediately after the Tl diffusion several months after synthesis. This phase instability was not a condition observed in Tl-diffused superconducting films prepared from radio-frequency-sputtered precursor films stored under similar conditions for comparable periods of time (A. M. Hermann, University of Colorado at Boulder, personal communication). Effects similar to this have prompted some laboratories to eliminate water completely from bulk processing of the high-T_c ceramics. For example, researchers at Argonne National Laboratory feel that the elimination of organic products from these materials is a more manageable process and now prefer to use nonaqueous solvents as dispersants in the extrusion of $YBa_2Cu_3O_{7-y}$ magnetic coils [30] (R. B. Poeppel, Science and Technology Center for Superconductivity, Argonne, Illinois, personal communication).

In studying 1:2:3 Y–Ba–Cu–O, Kabe et al. [27] report that the hydroxyl groups appear to diffuse into the material at rates similar to the rates known for oxygen diffusion in this material system. They also note that the rate of hydroxyl incorporation into Y–Ba–Cu–O at low equilibrium water vapor partial pressures ($p_{H_2O} = 270$ Pa) is considerably enhanced at temperatures between 100 and 160°C over rates observed in samples kept at room temperature. They further suggest the formation of a chemisorbed barrier that inhibits water from entering the lattice when the samples are heated above 160°C. These results imply that a chemically, and probably electronically insulating surface coverage results when these ceramic materials are exposed to water vapor at elevated temperature. Perhaps, this surface coverage is depleted in alkaline earth cations. This would lead directly to the fabrication of ceramic material that is superconducting within its grains but incapable of transporting current between them. Further work using high-resolution electron microscopy (HREM) of such surface coverages and correlating the magnetization and current transport properties of material heated and exposed to water vapor with material synthesized without this exposure is needed to confirm this hypothesis.

Although it may not be possible to map these results exactly to the thallium cuprates, some of these factors might also have a physical basis in the chemical dynamics of these materials. It would be exceedingly useful to compare films spray pyrolyzed from aqueous-based solutions to those prepared using nonaqueous solvents. The use of nonaqueous solvents invariably exposes the material to both CO_2 and H_2O at elevated temperature, as they are the result of an organic burn. However, more extensive studies on the reactive effects of CO_2 with the YBCO system appear to suggest that this is a manageable problem with these ceramics. In general, the stability and rates of formation of alkaline earth carbonates are increased with descending rank on the periodic table; that is, barium carbonate is more stable and forms more readily than calcium carbonate, while strontium carbonate assumes intermediate properties between the two. It has been shown that $YBa_2Cu_3O_{7-x}$ reacts strongly with CO_2 at temperatures ranging from 600 to 950°C and that the reaction products depend on the annealing temperature [31]. The reaction products all include $BaCO_3$ and insulating or semiconductive complex oxide phases. All of these adverse reactions occur at the grain boundaries, but rather than forming thin uniform surface coverages that coat the grain boundary, the deleterious material phases develop as island-type precipitates that, to some extent, depend on grain boundary structure.

Experimental results appear to suggest that a kinetic relationship underlies the reactivity of the ceramic with CO_2 and that the rate at which the organic combustion products are removed from the material can be carefully controlled by atmospheric flow conditions [32]. Firing the ceramic formulations under reduced total pressure (2 mmHg) and flowing oxygen ($\approx$1160 cm^3/min) can reduce the decomposition of YBCO observed to occur at ambient pressures. Since organics can generally be removed at lower temperatures (240 to 350°C) by thermal decomposition, calcination cycles for ceramic formulations containing organic material should have explicit lower-temperature processing steps. These steps should be designed to stimulate the complete removal of CO_2 and H_2O from the solid prior to subsequent calcination and sintering at elevated temperatures. It is at elevated temperatures (i.e., > 650°C) that exposure to organic decomposition products are quite harmful. It has been suggested that as little as 50 ppm CO_2 in the sintering atmosphere at elevated temperature can deteriorate transport superconductivity in these ceramics [33]. However, it has also been suggested that since this deterioration proceeds through the formation of island-type defects, tiny amounts of these secondary decomposition phases in ceramic having good texture and well-connected grain structures might actually serve as flux pinning centers and increase J_c [31].

In overview, the literature seems to recommend that optimal calcination of formulations containing organic material should include reduced total pressure (2 mmHg) heat treatments in flowing oxygen (1160 cm³/min) at ramping rates that prevent organic decomposition from proceeding too rapidly. It is further recommended to heat treat above 800°C but below volatilization temperatures of the material to convert back to oxides any carbonate phases that may have formed during binder removal. This is highly recommended when Ba–Ca–Cu–O precursor films are prepared for Tl diffusion. It also appears that flash treatments during a postanneal with 2 to 5% CO_2/O_2 in densified films might actually enhance critical current densities of the material.

After calcination, and prior to film densification and postannealing, the films should be sintered. During sintering the principal objective is to activate the solid-state reactions which organize uniformly distributed single-phase lattice structures throughout the ceramic's bulk. Ruckenstein and Cheung [34] have used powder x-ray diffraction analysis and scanning electron microscopy to study the reaction pathways leading to the formation of the Tl–Ba–Ca–Cu–O superconducting phases. Their studies were performed by sintering pellets formed from stoichiometric blends of the metal oxides at 867°C in air. This study shows that the 2223, 1223, 1234, and 1212 phases in the Tl–Ba–Ca–Cu–O system could be prepared from microcrystalline mixtures of the individual oxides having the same stoichiometry as the desired final phase. The study also shows that the evolution of these particular phases follows sequential transformations which appear to occur along fairly well-defined reaction pathways that depend on the sintering time. It should be noted that this study was conducted with the ceramic exposed to the atmosphere. These results may not be valid for ceramic material encapsulated by silver, which permits atmospheric oxygen diffusion. The formation of the 2223 phase was observed to proceed via the sequential transformations $2201 \rightarrow 2212 \rightarrow 2223$. The lower T_c 1223 and 1234 phases followed similar synthesis routes, forming $1201 \rightarrow 2201 \rightarrow 2212 \rightarrow 2223$ before emerging as 1223 or 1234, whereas the 1212 phase evolves through a shorter synthesis route: $1201 \rightarrow 2201 \rightarrow 1212$.

It is clear from these results that lower T_c phases can emerge from prolonged sintering of the higher T_c phases, suggesting that a point is reached after which continued sintering might actually deter the stabilization of the higher T_c phases. Ruckenstein and Cheung also note the appearance of a "glassy" intermediate phase, which they note as possible precursors in the formation of the Tl–Ba–Ca–Cu–O phases. The "glassy" phases were not confirmed to actually have amorphous structures, but were defined as such because they lacked distinct crystalline texture and were located between regions of the material that exhibited explicit crystallinity.

This analysis appears to suggest that during sintering, microcrystallites of the individual oxide precursors initially blend together possibly to form an amorphous phase in which the crystalline phases are generated first with structures that contain a single layer of Cu–O in the unit cell. Over time, additional Cu–O and structurally required Ca–O layers were successively incorporated into the structure to form the higher T_c phases. The 2212 and 2223 phases were observed to degenerate into the lower-rank thallium 1212 and 1223 compositions. The depletion of Tl was observed during extended sintering of 2223 material. Tl_2O_3 evaporation was suggested as a possible reaction pathway and may be applicable in both cases. Similar phase dynamics (i.e., the expulsion of Tl_2O_3 from the local crystal structure) may also be significant in the transformation of 2223 → 1234, suggesting that Tl_2O_3 may become more volatile as more layers of Cu−O accompanied by Ca−O are incorporated into the structure.

It is not yet clear what actually serves as the seed for these nucleations. Since these solid-state reactions are diffusion limited, it could be suggested that microstructural regions locally rich in thallium and barium oxide could conceivably get a "heat start" on forming the higher-T_c phases (2212 and 2223). These phases would form as calcium and copper oxides are incorporated into the local crystalline structure. Regions depleted in barium and thallium oxides would require the local influx of these precursors prior to initial nucleation. If sintering can begin with a uniformly mixed glassy precursor deposit that provides a high degree of intimacy between all the different precursor oxides, diffusion kinetics favor the formation of single-phase material. If the deposit does not contain a uniform glassy phase, it appears possible to obtain microstructural domains that may initially have been rich in barium and thallium, which will be further along the reaction pathway. These regions run the risk of degenerating into the lower T_c 1223 and 1234 phases, while intermediately mixed local regions are forming 2212 and 2223 phases. Similarly, other regions initially depleted thallium might end up nucleating the 1201 through 1212 phases. It is easy to see that successful sintering of phase-pure and phase-stabilized ceramic could be intimately linked to one's ability to engender glassy deposits having a very high degree of intimate and uniform mixing of the precursor oxides at the molecular level.

V. Tl−Ba−Ca−Cu−O FILMS PREPARED VIA Tl_2O_3 DIFFUSION INTO SPRAY-PYROLYZED FILMS

Except for the actual spray deposition of the initial, or "precursor" film, the preparation of thallium-based superconductors by spray pyrolysis is accomplished in essentially the same manner as with many other thin-film processes as described by Hermann et al. [16]. A precursor film containing the appropriate proportions of all the metals except Tl is deposited by spray pyrolysis, sintered, and then annealed. After these steps, the proper amount of Tl is then thermally diffused into the film [16,20]. The initial concept of diffusing Tl into an existing precursor film was reported by Shih et al. in 1988 [35]. A substrate coated with a film containing Ba, Ca, and Cu oxides was first prepared by vacuum evaporation. Tl was then vapor deposited directly onto the heated surface and the Tl allowed to diffuse into the film.

Hermann et al. used a Tl_2O_3 evaporation source in a single-zone furnace to accomplish the same effect in a single step by placing the Ba−Ca−Cu−O precursor film on top of a boat containing thallium oxide and flowing Ar−O_2 through the furnace [16,36]. Researchers at the Center for Superconductivity at the University of Colorado, Boulder have also investigated using crystals of TlBaCaCuO superconductor instead of thallium oxide as the source material with improved results. It is believed that this approach provides greater thermodynamic control over the process by allowing a 2212 source pellet to achieve equilibrium with a 223 Ba−Ca−Cu−O precursor film. Precursor films, prepared by spray deposition or other means, are placed in the furnace at about 840°C along on top of the boat containing the

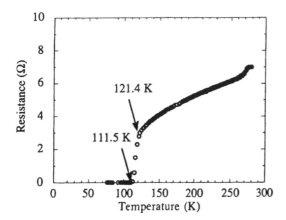

Figure 21 *R–T* curve of a Tl–Ba–Ca–Cu–O film prepared by diffusing Tl_2O_3 into a spray-pyrolyzed Ba–Ca–Cu–O precursor film.

Tl source. The process, which is carried out under flowing atmospheres, takes about 3 h and has resulted in films with good transition temperatures, as shown in Fig. 21.

This process has been further refined using a two-zone furnace that appears to offer better control of the deposition process. This process, developed by DeLuca et al. [20], offers better control of the vapor deposition process. A sketch of the furnace and reactor is shown in Fig. 22. The thallium reactor system consists of an open fused alumina tube that is 1.25 cm outside diameter and 25 cm long. As

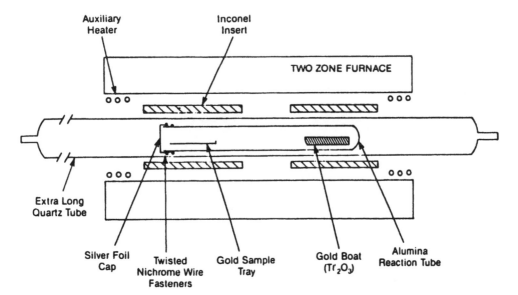

Figure 22 Two-zone thallium oxide vapor reactor. (Reprinted with the permission of John A. DeLuca, General Electric Research and Development Center, Schenectady, N.Y., and the *Journal of Materials Research*.)

shown in the figure, a boat containing Tl_2O_3 is placed in one of the controlled heater zones and the precursor sample tray in the other. The reaction takes place at atmospheric pressure under flowing oxygen or air. The superconducting transition temperature obtained depended strongly on the sample temperature. To obtain transition temperatures above 95 K, the sample temperature was held between 875 and 895°C. The best results were obtained with the sample at 885°C in an air ambient. The Tl_2O_3 boat was held at temperatures between 775 and 800°C during the process. Figure 2 shows the critical current curves obtained for a TlBaCaCuO film prepared in this manner.

VI. FILM DENSIFICATION

Films produced from any solution process are going to be much more porous than those deposited using other techniques. An example of the porosity one can encounter in spray-pyrolyzed films is shown in Fig. 23, which is an SEM micrograph of a Ba–Ca–Cu–O film prepared using a nitrate solution with 20% glycerol by volume as an organic binder. To achieve good critical current transport properties in the film, it is necessary to eliminate voids and to promote good grain boundary alignment to improve intergranular electrical connections. A number of techniques have been developed to overcome these deficiencies in solution processed ceramic superconductors. Some of these techniques are reviewed below.

In general, these films can be densified using three general techniques: cold-rolling (which requires that the films be deposited on a maleable substrate), flux sintering, and melt quenching or rapid thermal annealing. Hsu et al. have adapted flux sintering and melt quenching to the preparation of densified spray-pyrolyzed Bi–Sr–Ca–Cu–O ceramic exhibiting high degrees of crystallographic c-axis orientation perpendicular to the substrate [37]. While its chemistry is not identical,

Figure 23   SEM micrograph showing the porous nature of spray-pyrolyzed Ba–Ca–Cu–O precursor film.

the BiSCCO compositional system is likely to have the most similar crystal chemistry to the thallium family of superconductors. The flux sintering was achieved by adding an agent, such as sodium nitrate or potassium fluoride, to the spray solution with one-tenth the total metal ion concentration of the nitrate solution. Spray-deposited precursors are then put through a "normal" sintering process; that is, they are not heated to the point of melting.

Jin et al. have found cold pressing to be a necessary processing step in the preparation of spray-coated Bi–Pb–Sr–Ca–Cu–O ceramic superconducting silver tapes [38]. In this instance, fine superconducting ceramic powders were used as film precursors and mixed into a solution containing acetone and 10 vol % organic binder that was spray coated onto the silver substrate. These researchers consider that the cold-rolling step with sufficient pressure to densify the deposit to roughly 30 to 90% of its original cross-sectional area is essential in producing material with strong high-field critical current densities. After cold rolling, the films are partially melt treated at 880°C for an hour and subsequently sintered for 40 h, all under low-partial-pressure oxygen ($p_{O_2}/p_{N_2} = 0.08$). A critical aspect to this technology is that the ceramic must be brought into its superconducting phase prior to cold pressing. Hence if this technique is to be applied to spray-pyrolyzed films, it is recommended that it be done after all the necessary processing steps (spraying, calcination, diffusion, sintering) have been completed.

An interesting technique has recently been developed jointly by Daniel F. Ryder, with the Tufts University Laboratory for Materials and Interfaces, Medford, Massachusetts, and Michael J. Suscavage, with Rome Laboratory, Hanscom AFB, Massachusetts, which is in line with the cost-effective industrially compatible approach championed by spray pyrolysis. Ryder and Suscavage have developed a rapid thermal annealing process that is based on using a 6-kW bank of tungsten-halogen infrared lamps to flash heat solution processed films at various stages in their processing. Figure 24 shows a series of high-magnification ($\times 4000$) micrographs of Pb-modified (2223) Bi–Sr–Ca–Cu–O film prior to and after this rapid thermal processing (RTP) treatment. These films were prepared using a metal-organic-derived (MOD) solution process and sequentially deposited onto magnesium oxide substrates by spin coating. The micrograph on the left shows the initially porous microstructure of the film immediately after calcination in flowing oxygen to 450°C, a temperature below which all of the organic material is known to have thermally decomposed. The micrograph in the center shows the effect of a 10-s 6-kW RTP treatment, and the one on the right shows the very smooth dense microstructure produced after an 18-s exposure to the infrared light bank.

Figure 25 provides a lower-magnification ($\times 400$) perspective on the ability of RTP to promote grain growth in initially porous calcined films. The micrographs moving from left to right are of the same film viewed after 15-, 18-, and 20-s RTP treatments. This film was insulating prior to and after the RTP treatment but was transformed into its superconducting phase after a conventional sintering step. The ability of this technique to promote large grains can be observed in the series of micrographs presented in Fig. 26, which shows, from left to right, the grain structure of a 2212 Bi–Sr–Ca–Cu–O MOD processed film after a 3-h sintering step in a tube furnace at 865°C, subsequent RTP for 15 s, and RTP with a postanneal at 865°C for another 3 h. This film was observed to undergo phase transformations as a result of these processing steps, moving from the 2212 phase prior to RTP to

MOD-derived Oxide Film   RTP   RTP
(450 °C)   (6 kW, 10 s)   (6 kW, 18 s)

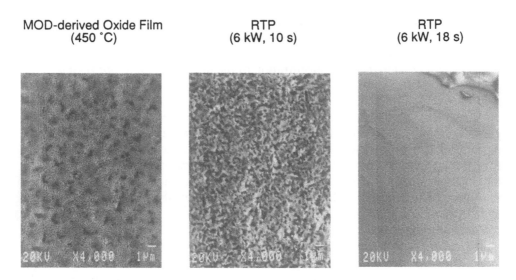

Figure 24   Effects of rapid thermal processing (RTP) on reducing porosity in calcined metal-organic-derived films.

predominantly 2201 after postannealing. A drawback to this processing technique is its lack of control. Infrared pyrometers are not likely to be useful in gauging the temperature of the film, since under these processing conditions, the film appears to be volatile. Consequently, the film's emissivity will change with its composition as its more volatile components vaporize. The success of RTP may yet be deter-

RTP   RTP   RTP
(6 kW, 15 s)   (6 kW, 18 s)   (6 kW, 20 s)

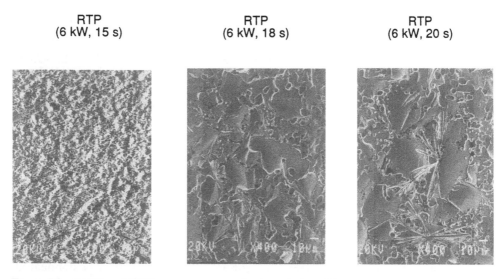

Figure 25   Ability of RTP to promote grain growth in a bismuth cuprate film after 15-, 18-, and 20-s treatments, respectively.

Tube Furnace Anneal
(865 ˚C, 3hrs)

RTP
(6 kW, 15 s)

RTP w/ Post-Anneal
(865 ˚C, 3 hrs.)

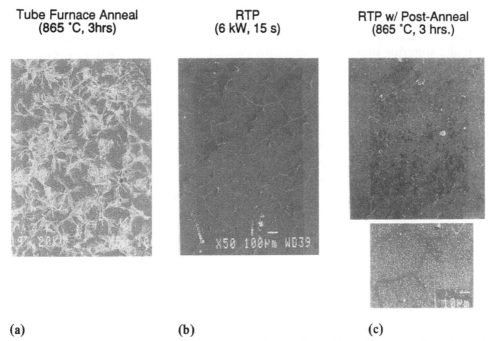

(a)   (b)   (c)

Figure 26 Promotion of large grains by RTP observed in a bismuth cuprate film sintered for 3 h at 865°C prior to RTP (a) immediately after RTP (b) and after a 3-h post-RTP annealing step (c).

mined by implementing it at the right stage in the process, for example, after low-temperature calcinations but prior to phase-stabilizing sintering.

VII. CONCLUSIONS AND SUMMARY

Recent experimental findings clearly indicate that one of the greatest applications potential for the high-T_c superconducting ceramics will be found in technologies that exploit their ability to sustain persistent critical current densities under the application of intense magnetic loads [39]. Progress toward developing and understanding methods to synthesize samples of these new materials with properties befitting these applications has been swift and rapid. A great deal of research has focused on optimizing current transport properties by developing single-crystal or epitaxial films of these ceramics. However, many of these synthesis techniques are not readily scalable for the economic production of high-quality large-area films. Spray pyrolysis represents a means toward developing a low-cost bulk synthesis method suitable for the manufacture of high-T_c superconducting components that require good-quality ceramic coatings applied over large areas or complex topological surfaces. In addition to potential commercial applications in high-field high-T_c superconducting magnets, spray-pyrolyzed films could also serve commercial needs in superconducting printed circuit board or as targets for laser ablation or thin-film sputtering.

In summary, in applying spray pyrolysis to the preparation of Tl-based superconductors, it is necessary to adapt the apparatus as well as the solution chemistry of the process to ensure the initial deposition of uniformly and intimately mixed precursors in the sprayed deposit. The use of organic material as a chemical binder can prove beneficial to the process, depending on the spray configuration. Calcination and sintering steps should explicitly consider the solution chemistry (organic or inorganic) and be adjusted accordingly. Health and safety hazards can be sharply reduced by diffusing thallium into a spray-pyrolyzed Ba–Ca–Cu–O precursor film. A double-zone furnace appears to provide the best means identified so far for vapor processing spray-pyrolyzed thallium superconductors. These films can initially be porous, and a number of techniques have been suggested as possible routes toward densifying the films. All these techniques show relative promise and have their trade-offs. Finally, in view of the ability for the thallium-based superconductors to sustain high critical current densities even in a polycrystalline state, it appears to be a material system that is very well suited to commercial bulk production using a process as simple and cost-effective as spray pyrolysis.

ACKNOWLEDGMENTS

We would like to thank John G. Zhang and Carlton E. Oakes for their assistance in our laboratory, as well as H. M. Duan and Allen M. Hermann, of the University of Colorado, Boulder for productive and fruitful collaborations in this field. The authors would also like to thank John DeLuca with GE Research and Development Center, Schenectady, New York; Frough Shokoohi, Bell Communications Research, Red Bank, New Jersey; Daniel F. Ryder, with the Tufts University Laboratory for Materials and Interfaces, Medford, Massachusetts; and Michael J. Suscavage, Rome Laboratory, Hanscom AFB, Massachusetts, for allowing us to improve the content of this chapter by incorporating data collected in their laboratories. We would also like to acknowledge Kristen Healey, Radiation Monitoring Devices, for her contributions in writing this chapter.

REFERENCES

1. G. Bednorz and K. A. Müller, *Z. Phys. B* 64, 189 (1986).
2. M. K. Wu, J. R. Ashburn, C. J. Torng, P. H. Hor, R. L. Meng, L. Gao, Z. J. Huang, Y. Q. Wang, and C. W. Chu, *Phys. Rev. Lett.* 58, 908 (1987).
3. R. J. Cava, B. Batlogg, R. B. van Dover, D. W. Murphy, S. Sunshine, T. Siegrist, J. P. Remeija, E. A. Reitman, S. Jahurak, and G. P. Espinosa, *Phys. Rev. Lett.* 58, 1676 (1987).
4. H. Maeda, Y. Tanaka, M. Fukutomi, and T. Asano, *Jpn. J. Appl. Phys.* 27, L209 (1988).
5. Z. Z. Sheng and A. M. Hermann, *Nature* 332, 55 (1988).
6. Z. Z. Sheng and A. M. Hermann, *Nature* 332, 138 (1988).
7. J. B. Mooney and S. B. Radding, *Annu. Rev. Mater. Sci.* 12, 81 (1982).
8. M. Kawai, T. Kawai, H. Masuhira, and M. Takahasi, *Jpn. J. Appl. Phys.* 26, L2740 (1987).
9. H. Nobumasa, K. Shimizu, Y. Kitano, M. Tanaka, and T. Kawai, *Jpn. J. Appl. Phys.* 27(8), 1544 (1988).
10. A. Gupta, G. Koren, E. A. Giess, N. R. Moore, E. J. M. O'Sullivan, and I. Cooper, *Appl. Phys. Lett.* 52, 163 (1988).
11. E. I. Cooper, E. A. Giess, and A. Gupta, *Mater. Lett.* 7, 5 (1988).

12. M. Awano, K. Kani, Y. Takao, and H. Takagi, *Jpn. J. Appl. Phys.* 30(5A), L806 (1991).

13. P. Barboux, J. M. Tarascon, F. Shokkohi, B. J. Wilkens, and C. L. Schwartz, *J. Appl. Phys.* 64(11), 6382 (1988).

14. C. Chiang, C. Y. Shei, S. F. Wu, and Y. T. Huang, *Appl. Phys. Lett.* 58(21), 2435 (1991).

15. L. P. de Rochemont, J. G. Zhang, M. R. Squillante, H. M. Duan, A. M. Hermann, R. J. Andrews, and W. C. Kelliher, *Proc. Aerospace Applications of Magnetic Suspension Technology*, NASA Langley Research Center, Hampton, VA, September 25–27, 1990, NASA Conf. Publication 10066, Part 2, pp. 607–627.

16. A. M. Hermann, H. M. Duan, H. Kuwahara, L. P. de Rochemont, and M. R. Squillante, *Proc. 2nd International Symposium on Superconductivity and Its Related Character in Layered Compounds*, Tokyo Institute of Technology, Yokohama, Japan, Dec. 1990.

17. D. S. Ginley, J. F. Kwak, R. P. Hellmer, R. J. Baughman, E. L. Venturini, M. A. Mitchell, and B. Morosin, *Physica C* 156, 592 (1988).

18. D. S. Ginley, J. F. Kwak, E. L. Venturini, B. Morosin, and R. J. Baughman, *Physica C* 160, 42 (1989).

19. D. S. Ginley, J. F. Kwak, R. P. Hellmer, R. J. Baughman, E. L. Venturini, and B. Morosin, *Appl. Phys. Lett.* 53(5), 406 (1988).

20. J. A. DeLuca, M. F. Garbauskas, R. B. Bolon, J. G. McMullen, W. E. Balz, and P. L. Karas, *J. Mater. Res.* 6(7), 1415 (1991).

21. General Electric World Wide Press Release, Dec. 6, 1991, General Electric Research and Development Center, Schenectedy, N.Y.

22. D. F. Vaslow, G. H. Dieckmann, D. Dawson, A. B. Ellis, D. S. Holmes, A. Lefkow, M. MacGregor, J. E. Nordman, M. F. Petras, and Y. Yang, *Appl. Phys. Lett.* 53(4), 324 (1988).

23. S. C. Zhang, G. L. Messing, W. Huebner, and M. M. Coleman, *J. Mater. Res.* 5(9), 1806 (1990).

24. I. Amemiya, H. Kobayasi, T. Nakamoto, and T. Hasegawa, *IEEE Trans. Magn.*, 27(2), 905 (1991).

25. M. Shoyama, H. Nasu, and K. Kamiya, *Jpn. J. Appl. Phys.* 30(5), 950 (1991).

26. M. P. Pechini, Method of preparing lead and alkaline earth titanates and niobates and coating method using the same to form a capacitor, U.S. Patent 3,330,697, July 11, 1967.

27. Y. Ikuma, M. Yoshimura, and S. Kabe, *J. Mater. Res.* 5(1), 17 (1990).

28. R. L. Barns and R. A. Laudise, *Appl. Phys. Lett.* 51, 532 (1987).

29. J. Wang, R. Stevens, and J. Bultitude, *J. Mater. Sci.* 23, 3393 (1988).

30. S. E. Dorris, J. T. Dusek, M. T. Lanagan, J. J. Picciolo, J. P. Singh, J. E. Creech, and R. B. Poeppel, *Bull. Am. Ceram. Soc.*, Apr. 1991.

31. Y. Gao, K. L. Merkle, C. Zhang, U. Balachdran, and R. B. Poeppel, *J. Mater. Res.* 5(7), 1363 (1990).

32. R. B. Poeppel, S. E. Dorris, J. J. Picciolo, U. Balanchandran, M. T. Lanagan, C. Z. Zhang, K. Merkle, Y. Gao, and J. T. Dusek, *Proc. 3rd International Symposium on Superconductivity*, Sendai, Miyagi Prefecture, Japan, Nov. 6–9, 1990.

33. Y. Gao, Y. Li, K. L. Merkle, J. N. Mundy, C. Zhang, U. Balachandran, and R. B. Poeppel, *Mater. Lett.* 9(10), 347–352 (1990).

34. E. Ruckenstein and C. T. Cheung, *J. Mater. Res.* 4(5), 1116 (1989).

35. Q. X. Qui and I. Shih, *Appl. Phys. Lett.* 53, 523 (1988).

36. Z. Z. Sheng, L. Sheng, H. M. Su, and A. M. Hermann, *Appl. Phys. Lett.* 53(26), 2686 (1988).

37. H. M. Hsu, I. Yee, J. DeLuca, C. Hilbert, R. F. Miracky, and L. N. Smith, *Appl. Phys. Lett.* 54(10), 957 (1989).
38. S. Jin, R. B. van Dover, T. H. Tiefel, and J. B. Graebner, *Appl. Phys. Lett.* 58(8), 868 (1991).
39. D. Larbalestier, *Phys. Today* 44(6), 74 (1991).

11

Liquid–Gas Solidification Method for Synthesis of Tl-Based HTSC Oxides

Ho Sou Chen
AT&T Bell Laboratories
Murray Hill, New Jersey

I. INTRODUCTION

The discoveries of bulk superconductor materials in a large family of compounds of prototype $YBa_2Cu_3O_7$ [1,2], $Bi_2Sr_2CaCu_2O_8$ [3], and $Tl_2Ba_2Ca_2Cu_3O_{10}$ [4] with resistive-superconducting transitions above 90 K opened a door to wider applications, because the structures can be cooled with liquid nitrogen, which is both cheaper and easier to handle than liquid helium. Liquid nitrogen is generally considered to be one of the most advantageous cryogenic refrigerants, and attainment of superconductivity at or above liquid nitrogen temperature was a long-sought goal which until recently appeared almost unreachable.

Some potential applications of superconductors are power transmission lines, rotating machinery, and superconducting magnets for fusion generators, magnetohydrodynamic (MHD) generators, particle accelerators, levitated vehicles, magnetic separation, and energy storage as well as junction devices and detectors. It is expected that many of the foregoing and other applications of superconductivity would materially benefit if high-T_c superconductive material could be used instead of the previously considered relatively low-T_c materials.

Two general approaches for forming superconductive oxide bodies are known. Thin films are formed by deposition of material on a substrate (e.g., by sputtering, evaporation, or decomposition of a solution), followed by heat treatment that produces the appropriate crystal structure and composition. On the other hand, bulk bodies and thick films are generally produced by synthesizing a powder of the appropriate composition (e.g., $YBa_2Cu_3O_x$, $x \sim 7$), forming the powder into the desired shape (e.g., by hot pressing, drawing, extrusion, or silk screening of a slurry) and heat treating the resulting body such that sintering occurs and such that the sintered material has the appropriate crystal structure and composition. An additional method involves melting the oxide powder and forming bulk bodies by solidification of the oxide melt [5].

The critical temperature T_c (i.e., the temperature at which a given body becomes superconductive) is an important parameter of a superconductor. Another important parameter is the maximum current density that can be supported by a body in the superconductive state. This "critical current density" J_c decreases with both increasing temperature and increasing magnetic field.

Work to date has shown that at least some thin films of high-T_c superconductors (e.g., $YBa_2Cu_3O_7$) can have high J_c (of order 10^6 A/cm² at 77 K) [6,7], with J_c being relatively weakly dependent on magnetic field. Although individual particles (crystallites) of superconductive oxides (e.g., $YBa_2Cu_3O_7$) can have large internal critical current density J_c (on the order of 10^6 A/cm²), the critical current density of bulk bodies produced by sintering of the particles is relatively small, generally of order 10^3 A/cm² in zero magnetic field ($H = 0$), and strongly dependent on magnetic field. This huge difference between the J_c value of a single particle and of an assembly of particles is generally attributed to the presence of weak links between adjacent particles. A critical current density on the order of 10^3 A/cm² at $H = 0$ is generally thought to be too small for most technologically important applications. The "metallurgical" processing technique of Jin et al. [5] can result in essentially 100% dense single-phase material in which the grains typically are of relatively large size and are nonrandomly oriented. Bulk bodies produced by this technique can have substantially high J_c values ($H = 0$, 77 K), about 5×10^5 A/cm², and significantly, J_c can decrease more slowly with increasing magnetic field than has been reported for sintered bodies.

In view of the immense economic potential of high-T_c superconductors, a simple, scalable processing method that is readily applicable to continuous processing and the formation of composite structures and which has the potential for producing improved bodies, especially material with improved compositional uniformity would be of great interest. Great progress has been made on forming Ag-sheathed superconducting tapes which exhibit desirable transport properties. (For an overview, see Chapter 14.)

In this chapter we overview the development of a novel, simple, scalable processing method: a liquid–gas-solidification (LGS) process which has the potential for producing improved materials and particularly promising for making high-T_c ceramic superconductors. One of the applications of this process consists of the oxidation and coating of metallic alloy on substrates. In Section II we discuss the principle and requirements of the LGS process. The distinguishing features and merits of the LGS process are described. In Section III, the growth of high-temperature superconductor (HTSC) films by the LGS method is exemplified for $YbBa_2Cu_3O_7$. The morphologies, microstructure, and transport properties are presented. Ag addition on the epitaxially grown $YbBa_2Cu_3O_7$ is shown to improve the transport properties of the resultant film. In Section IV we deal with the growth of Tl-based HTSC films. HTSC $TlBa_2Ca_2Cu_3O_9$ films have been grown successfully on MgO(100) and $SrTiO_3$(100) substrates in a single step. The as-grown films show a flat, uniform morphology, with a c axis perpendicular to the films. The films show a superconducting onset at 117 K and zero resistance at 103 K. A postanneal is seen to improve slightly the morphology and transport properties. J_c attains zero-field values of 5×10^4 A/cm² at 77 K and 5×10^5 A/cm² at 4 K, being comparable

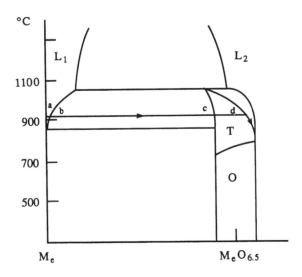

Figure 1 Schematic phase diagram of Me (e.g., YbBa$_2$Cu$_3$)–O.

to that of a single crystal. In Section V we address the growth mechanism of LGS process. Measurement of the activity, diffusivity, and solubility of oxygen in precursor alloy melts is described. Based on these thermodynamic data, the limiting growth mechanism of the film is presented. In Section VI, the formation and stability of bulk Tl-based HTSC by the LGS method are briefly presented. Finally, in Section VII we summarize the development and present knowledge of the LGS process.

II. PRINCIPLES AND REQUIREMENTS OF LIQUID–GAS-SOLIDIFICATION PROCESSING

This process consists of (1) forming a melt (the precursor melt, e.g., YbBa$_2$Cu$_3$) composed of the metallic components of the compound (e.g., the cuprate oxide) to be formed, and (2) introducing the missing element, oxygen, into the precursor melt to form the oxide. This process produces single phase materials with uniform composition. It is shown schematically in Fig. 1. The precursor alloy [e.g., Yb-Ba$_2$Cu$_3$($T_m \sim 870°C$)] is heated to 900°C in an inert atmosphere to form a uniform melt (a). Oxygen is then introduced into the melt. Initially, oxygen is absorbed in

Table 1 Melting Temperature of Alloys and Oxides and Operation Temperature

Cation composition	$T_{m_{oxide}}$ (°C)	$T_{operation}$ (°C)	$T_{m_{alloy}}$ (°C)
Yb$_1$Ba$_2$Cu$_3$	>1050	900	~870
Yb$_1$Ba$_2$Cu$_3$ + Ag$_{0.50}$	>1050	930	~910
Yb$_1$Ba$_2$Cu$_3$ + Ag$_1$	>1050	930	~920
Yb$_1$Ba$_2$Cu$_3$ + Ag$_3$	>1050	980	~930
Tl$_1$Ba$_2$Ca$_2$Cu$_3$	>950	870	~600

the melt until the oxygen concentration attains the solubility limit (b). With further introduction of oxygen, a tetragonal $YbBa_2Cu_3O_{7-\delta}$ phase (c) precipitates from the melt until the melt converts completely to a solid oxide. With further introduction of O_2, the tetragonal phase (c) transforms to the oxygen-rich *t*-phase (d). Upon cooling from 900°C to room temperature, the tetragonal phase transforms to an orthorhombic superconducting phase.

It is clear that the basic requirements of the LGS process are as follows:

1. The melting temperature of the precursor alloy is lower than that of the oxide compound to be formed, and the oxidation process is conducted between these two temperatures:

$$T_{m(\text{alloy})} < T_{\text{oxid.}} < T_{m(\text{oxide})}$$

The melting temperatures of high-T_c superconducting precursor alloys and their oxides are listed in Table 1. In all cases LGS processing is carried out at about 900°C.

2. The precursor alloy melts have to be structurally and compositionally homogeneous. The LGS process cannot be applied to $REBa_2Cu_3$ alloys, except for RE = Yb and Eu, which form immiscible melts of RE–Cu and Ba–Cu, where RE = Y, La, Nd, Sm, Gd, Tb, Dy, and Ho. It is not easily applicable for Bi-based superconductors, due to the formation of stable intermetallic compound Ca_3Bi_2, which exhibits a relatively high melting temperature of about 1200°C.

3. The melts have a limited solubility of oxygen ($C_0^s \sim 10^{-4}$) in order to attain reasonable growth of films, which will be discussed later.

The merits of LGS processing are:

1. The stoichiometry of superconducting films is directly controlled by that of the precursor alloy.
2. The entire process is conducted in a closed system that facilitates safety and handling of toxic and volatile component (e.g., Tl-based alloys).
3. Uniform distribution of the constituents in the melt leads to structurally homogeneous oxide film with improved transport properties.
4. Oxide films form rapidly. It takes only 1 to 10 min to grow a film of ~2 μm thickness.
5. It is scalable to continuous coating films and wires.

III. $YbBa_2Cu_3O_7$ EPITAXIAL FILMS GROWN BY THE LIQUID–GAS-SOLIDIFICATION PROCESS

A.   From a $YbBa_2Cu_3$ Precursor Alloy [8]

A ternary alloy of $YbBa_2Cu_3$ was prepared by induction melting of high-purity metals. The alloy was placed in a high-purity alumina crucible and was sealed together with $SrTiO_3$(100) substrate that was mounted on a translational stage shown in Fig. 2a. Due to the high reactivity of rare earth metals, Yb, Ba, and Ca, alloy preparation and assembly setup were carried out in a glove box with purified He. The assembly filled with inert gas was heated in a resistance furnace above the melting temperature T_m (= 870°C) of the alloy. (The heating processes are shown schematically in Fig. 2b.) The $SrTiO_3$ substrate was then immersed into and

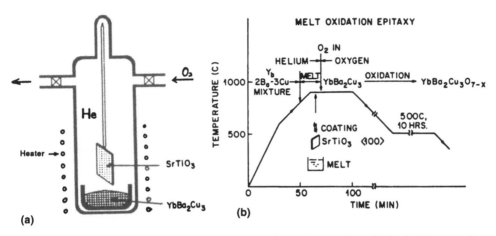

Figure 2   Schematic diagram of setup (a) and experimental procedure (b) for LGS processing.

pulled out of the melt. A melt layer about 2 μm thick coated the substrate. Oxidation of the melt coating was conducted in situ by introduction of oxygen into the tube. The duration of oxidation was typically 1 to 10 min. The resultant oxide film was annealed at 500°C in flowing oxygen.

Figure 3 illustrates the x-ray diffraction patterns of the oxidized materials: (a) a film grown on $SrTiO_3(100)$ and (b) a film grown on $SrTiO_3(110)$. As noted in Fig. 3a and b, films grown on an $SrTiO_3$ substrate show an epitaxially oriented structure. The film grown on $SrTiO_3(100)$ exhibits a growth habit with the c axis normal to the film plane, while the film grown on $SrTiO_3(110)$ has a predominant orientation with the (110) direction normal or the c axis parallel to the substrate plane. Clearly, the substrate orientation plays a decisive role in determining the

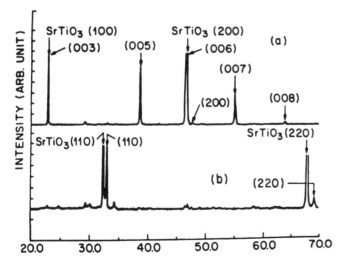

Figure 3   X-ray diffraction patterns of an epitaxial $YbBa_2Cu_3O_7$ film grown on (a) SrTiO_3(100) and (b) $SrTiO_3(110)$ substrates.

growth direction of the film. This strongly suggests that the formation of oxide is initiated at the substrate–liquid interface rather than at the liquid–gas interface. The lattice parameters of the orthorhombic $YbBa_2Cu_3O_7$ are $a = 3.808$ Å, $b = 3.887$ Å, and $c = 11.659$ Å. It is clear from Fig. 3 that the lattice parameters b and $c/3$ of the o phase have a nearly perfect match to the cubic lattice $a_0(3.89$ Å$)$ of $SrTiO_3$. The mismatch between the a parameter of the o-$YbBa_2Cu_3O_7$ and the a_0 of c-$SrTiO_3$, however, is small, being about 2% and 1% for the (100) and (110) substrates. The film grown on $SrTiO_3$(110) grows epitaxially with ($\bar{1}$10) and c axes in the plane [i.e., (110) axis normal to the plane]. It was also observed that the film grows epitaxially with (103) or (013) and b or a axes in the plane [i.e., ($\bar{1}$03) or (0$\bar{1}$3) axis normal to the plane].

One may expect that the film grown on $SrTiO_3$(100) would grow epitaxially with the b and c axes in the plane, because this is the orientation with the best lattice match. However, the growth rates of the o-$YbBa_2Cu_3O_7$ crystal are highly anisotropic, and it grows as a platelet with the c axis normal to the plane. The film thus also grows with a and b axes in the plane at the expense of small mismatch between a and a_0 of about 2%. Indeed, it has been reported that films grown on $SrTiO_3$(100) have a growth orientation with either the c or a axis normal to the plane [9].

Figure 4 shows a scanning electron micrograph (SEM) of a film with the c axis normal to the substrate plane ($c_\perp$) and a film with the (110) direction normal to the plane [(110)$_\perp$]. As shown in Fig. 4A, the $c_\perp$ film shows rectangular tile patterns with platelet dimensions of about 10 μm × 10 μm × 2 μm. The platelets are arranged such that the longer edges (i.e., the a and b axes) are parallel to the a_0 axis of the $SrTiO_3$(100) substrate. On the other hand, the (110)$_\perp$ film (Fig. 4B) shows an array of prisms with the longer edge along the c axis, which lies parallel to the (110) lattice plane of the $SrTiO_3$. Occasionally, platelets with the a or b axis parallel to the a_0 axis of the $SrTiO_3$, inclined about 45° from the $SrTiO_3$(110) plane, are seen. Closer examination near the film–substrate interface shows a continuous interconnection between each grain.

B. From Precursor Alloys, Ag–$YbBa_2Cu_3$

Effects of Ag addition on the epitaxially grown $YbBa_2Cu_3O_7$ superconductor on a $SrTiO_3$(100) substrate by the LGS process have been investigated [10]. Alloys of $Ag_xYb_1Ba_2Cu_3$ ($x = 0, 0.5, 1$, and 3) were prepared from high-purity metals. The differential thermal analysis (DTA) studies of these alloys found the liquidus temperature to be 870, 910, 920, and 930°C, respectively for $x = 0, 0.5, 1$, and 3. Due to the change in the melting temperature, the optimum growth temperatures were set at 900, 930, 930, and 980°C.

X-ray diffraction patterns of these films are similar. Strong peaks at (003), (005), (006), and (007) indicate an oriented structure with the c axis perpendicular to the film plane. In the x-ray diffraction data, no measurable peaks exist corresponding to either Ag or AgO, although the energy-dispersive x-ray analysis (EDX) revealed the Ag particles precipitated on film surfaces. This is explained by the too weak x-ray diffraction from powder Ag particles compared to strong diffraction from the epitaxial film.

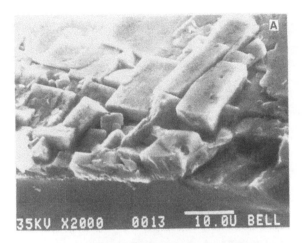

Figure 4 Scanning electron micrograph of film corresponding to Fig. 3.

The morphology of these films examined by SEM is shown in Fig. 5. The film exhibits irregular tile patterns with platelet dimensions of about 10 μm × 10 μm × 2 μm, which is similar to that of in $Yb_1Ba_2Cu_3$-grown film. The platelets are arranged in such a way that platelet edges (i.e., the *a* and *b* axes) are nearly parallel to the a_0 axis of the cubic $SrTiO_3$ substrate. The present film appears to have smoother interconnection between the platelets. The EDX spectrum taken from a large area of the film, shown in Fig. 6a, consists of Ag, Yb, Ba, and Cu spectra. Figure 6b and c show the spectra taken from bright particles and the background platelets, respectively. The former consists only of the Ag spectrum, while the latter is comprised of Yb, Ba, and Cu. It was clear that Ag did not corporate in the oxidation process and precipitated on the oxide surfaces. This finding is also consistent with the unchanging lattice parameters of the film with Ag concentration.

A conventional four-probe dc transport measurement was made using indium contacts. In all cases the resistance decreases linearly with temperature from 300 K to the transition. Extrapolation of the normal-state resistivity to $T = 0$ K yields

Figure 5   SEM of the $AgYbBa_2Cu_3$ oxide film.

a small value, approximately one-fifth of the normal resistance at 300 K. Typical values of $T_c(R = 0)$ of Ag-containing films are about 90 K, which may be compared with $T_c = 85$ K for the film containing no Ag (see Fig. 7).

J_c measurements were made on the samples ($x = 0.5$, 1, and 0). For J_c measurements, the films were shaped with a carbide tool to restrict current flow to a narrow path about 100 µm in width. Critical current density data are shown in Fig. 7. The critical current J_c was defined by a flux-flow resistance of $10^{-3}R_N$, where R_N is the normal resistance at T_c. J_c attains typical values of 3×10^4 A/cm^2 at 77 K zero field. The functional dependency of J_c on T appears similar and shifts uniformly with $T_c(R = 0)$. The addition of Ag improves the superconducting properties, probably as a result of a decrease in film–substrate interaction.

The uniform distribution of metallic constituents is a critical factor in producing improved superconducting materials. Superconducting cuprate oxides produced by the oxidation of precursor solid-phase alloys have been reported [11,12]. The resultant oxides show a multiphase structure and exhibit low critical current density $J_c < 10$ A/cm^2. Apparently, compositional and structural inhomogeneity result from the decomposition of solid-phase alloy during the oxidation process. The distinguishing feature and fundamental advantage of the LGS process is that the desired compound formation occurs at an accessible and kinetically favorable liquidus-phase boundary rather than the more limited solid–solid phase transformation path.

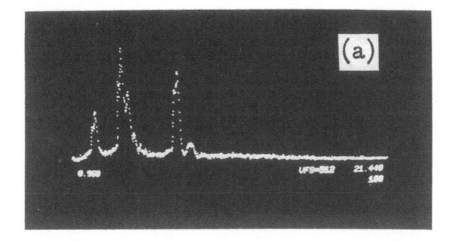

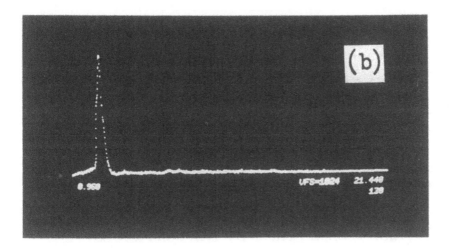

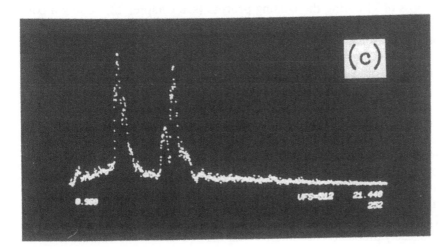

Figure 6   EDAX spectra of sample shown in Fig. 5.

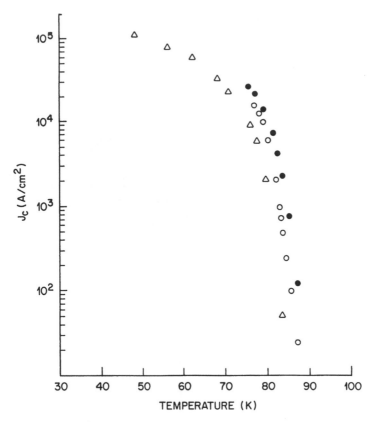

Figure 7   Critical current density J_c ($H = 0$) of $Ag_xYbBa_2Cu_3$ oxide films. Open circles, $x = 0.5$; closed circles, $x = 1.0$; triangles, $x = 0$.

IV.   GROWTH OF Tl-BASED FILMS BY THE LGS PROCESS

The major problem with and difficulty in coating a Tl-based superconducting film is the excess loss of Tl during the coating process (see Chapter 12). Using a conventional method it will take a long time to complete the coating process, and the immense loss of Tl results in nonstoichiometry of the film and nonsuperconducting. Therefore, postannealing is required to obtain the desired composition and structure. Alternatively, precursor alloys, Ba–Ca–Cu–O or Ba–Sr–Ca–Cu–O, are deposited on various substrates. To obtain the desired composition and structure, these predeposited films are sealed together with Tl_2O_3 or Tl-based compound in a Au, Pt, or Al_2O_3 container, and heated to about 900°C in a flow of oxygen to compensate for the Tl deficiency. In the LGS process, it takes only 5 min of oxidation to form the superconductor phase. This short processing time reduces the problem of Tl loss. In addition, with proper design of the oxidation chamber, containing the coated melt film in a near-equilibrium surrounding, the Tl loss can be greatly reduced. Finally, Tl oxide is very toxic and requires strict handling procedures. Since the LGS process is simple and can be operated in a complete enclosure, it eliminates the safety problems encountered with the conventional method.

Superconducting TlBa$_2$Ca$_2$Cu$_3$O$_y$ films have been grown on MgO⟨100⟩ and SrTiO$_3$⟨100⟩ substrates by the LGS process in a single step [13,14]. The films are grown in situ, as deposited. There are no postanneal or a compensation of Tl applied after the initial deposition. The as-grown film shows a flat and uniform morphology with a *c* axis perpendicular to MgO⟨100⟩ or SrTiO$_3$⟨100⟩.

A. TlBa$_2$Ca$_2$Cu$_3$O$_x$ Grown on MgO(100) Substrate

A typical SEM micrography of the film is shown in Fig. 8. The film shows a flat, uniform morphology of 0.5×1.00 mm^2 domain and a thickness of 1 μm with no obvious grain boundaries. The compositions of films are identified by EDX to be 1212, 1223, 2212 phases, which are consistent with the result of a single-crystal x-ray diffraction. Some impurity crystals are dispersed on the superconducting film surface and oriented randomly. These impurities are identified to be Ca–Cu–O and Ba–Cu–O by EDX and powder x-ray diffraction.

Single-crystal x-ray diffraction pattern reveals that the film consists of 1223 and 2212 with a trace of 1212 superconducting phases (Fig. 9). The lattice parameters of the *c* axis are $c \sim 12.2$, 15.91, and 14.4 Å for 1212, 1223, and 2212 phase, respectively. Predominant (00*l*) peaks indicate that the film is an epitaxial film with the *c* axis perpendicular to the MgO⟨100⟩ substrate. Very small amounts of impurities were detected because they were dispersed over the film and oriented randomly. The rocking curve of the (006) peak of the 1223 phase shows a strong center peak with a broadened tail (Fig. 10). The width of the strong center peak and the broadened tail are $\theta = 1.017°$ and $4.818°$, respectively, while the substrate has $\theta = 0.340°$. This suggests that most of the superconducting grains lie perfectly on the substrate and some are tilted within 5° of the substrate surface, which may be caused by strain distortion during quenching. It may be noted that x-ray intensities scan over five orders of magnitude on a dynamic scale such that even ⟨001⟩

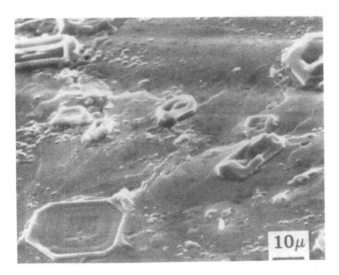

Figure 8 SEM micrograph of as-deposited TBCC film by the LGS process.

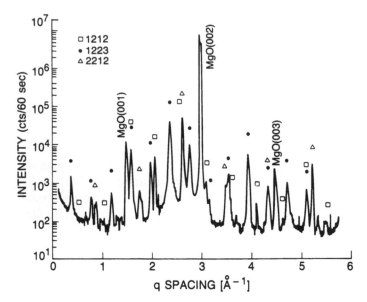

Figure 9   Single-crystal x-ray diffraction pattern of the film shown in Fig. 8.

and $\langle 003 \rangle$ peaks of MgO substrate are detected. The film exhibits a superconducting onset at 117 K and zero resistance at 103 K.

B. $Tl_1Ba_2Ca_2Cu_3O_y$ Film on $SrTiO_3$ $\langle 100 \rangle$ Substrate

To ensure an adequate Tl vapor pressure during the oxidation process, a minor modification was made. An alumina crucible was mounted upside down to cap on

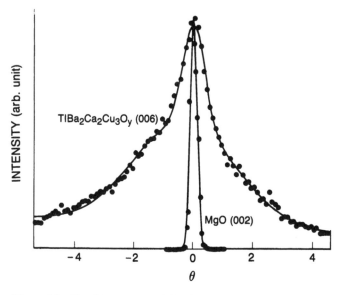

Figure 10   Rocking curve of (006) peak of the 1223 phase.

the sample holder. Tl vapor was in equilibrium with the liquid alloy inside the semiclosure space. $SrTiO_3$ substrate was then immersed into and pulled out of the melt. The composition of the melt coating on $SrTiO_3$ substrate remained unchanged due to the equilibrium of Tl vapor pressure in the semiclosure space. A purified Ar was filled with the assembly instead of He to reduce the diffusion of Tl vapor. Oxygen was introduced through a guiding tube near the cap with a flow rate from 0 to 100 cm^3/min in 30 s, while lifting the cap up 0.5 cm. After 1 min of oxidation, the assembly was pulled out from the furnace and air cooled.

A SEM micrography is shown in Fig. 11. The films show a two-dimensional layer growth with a 1×1 mm^2 domain and 1 μm thickness. The composition of the film is identified by EDX to be 1212 and 1223 phases, consistent with the result of a powder x-ray diffraction analysis. Some small CaO precipitates (white spots) at the edge of layers and pinholes (dark spots) are seen on the micrograph. The film is highly textured with its c axis perpendicular to the surface of $SrTiO_3$ substrate. The impurities surround the domain are identified to be $BaCuO_2$ and $CaCuO_2$. The resistivity superconducting transition was broad, due to the presence of a low-T_c 1212 phase, with transition onsets at 107 K and zero resistance at 95 K.

C. Postannealing Effect on a Film Grown on $SrTiO_3$ $\langle 100 \rangle$ Substrate

A postanneal was made following the conventional two-step procedure: (1) high-temperature anneal at 870°C for 15 min and then (2) low-temperature anneal at 700°C for 10 h followed by a furnace cool. A residual partially oxidized precursor alloy was used as a Tl source, which was used to growth the superconducting film previously. The film was placed together with the residual alloy in a high-purity alumina crucible. An Au foil was wrapped on the top of the crucible and was tightened by an alumel wire. This cell was placed in a quartz tube with an oxygen flow of 100 cm^3/min.

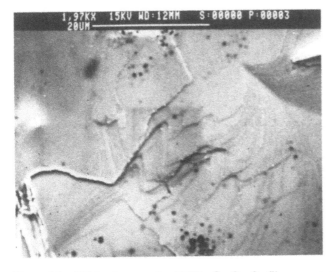

Figure 11 SEM micrograph of $TlBa_2Ca_2Cu_3O_y$ film grown on $SrTiO_3(100)$.

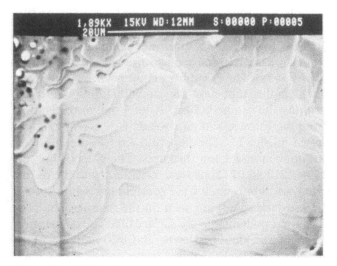

Figure 12   SEM micrograph of the postanneal film shown in Fig. 11.

A SEM micrography of the annealed film is shown in Fig. 12. The original layer structure remains unchanged after anneal. Ca–O precipitates are seen to disappear with the appearance of an additional 2212 phase. The film is smoother. As a result, the film exhibits improved transport properties with a superconducting transition onset at 117 K and zero resistance at 105 K (Fig. 13). To avoid burnout during measuring the critical transport current J_c, a pulse current source was used. Each pulse was set to 2 ms on and 8 ms off for heat dissipation. As shown in Fig. 14, J_c increases exponentially near T_c and then gradually below 70 K. Typically, J_c attains zero-field ($H = 0$) values of 5×10^4 A/cm² at 77 K and reaches 5×10^5 A/cm² at 4 K. These values are comparable to a single-crystalline Tl-based

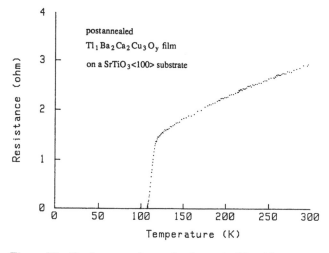

Figure 13   Resistance of sample shown in Fig. 12.

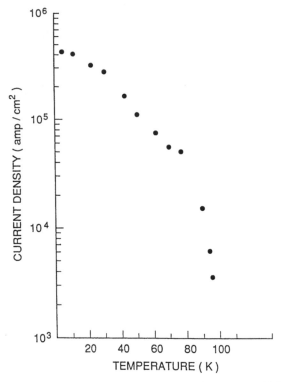

Figure 14 Critical current density of postanneal $TlBa_2Ca_2Cu_3O_y$ film.

superconductors. The film exhibits very weak flux pinning as expected for a structurally and compositionally homogeneous Tl-based superconductors.

V. GROWTH MECHANISM OF OXIDE FILMS BY THE LGS PROCESS

In the LGS process, the conversion of a liquid alloy to an oxide solid is governed by the thermodynamic and the transport properties of oxygen in the precursor liquid alloy. Oxygen is dissociated and adsorbed on the free surface of the liquid alloy. When the available sites on the free surface are nearly fully coveraged by the adsorbed oxygen, the present of a surface reactive solute can lower the rate of an interfacial reaction between the oxygen and the liquid alloy. The adsorbed oxygen then diffuses into the liquid alloy toward the liquid–substrate interface and reacts with metal atoms, forming a oxide film. The film growth is thus limited by the adsorption of oxygen at the oxygen–liquid interface or the diffusion of oxygen in the liquid. The nucleation and growth of films are a matter of thermodynamic and kinetic behavior of oxygen in the liquid alloys. It is therefore of scientific and technological importance to understand the interaction and the transport properties of oxygen in the precursor liquid alloy. In the following the experimental thermodynamics data of oxygen in precursor alloy melts are presented. Based on these data oxide film growth kinetics are discussed.

A.   Thermodynamics and Kinetics of Oxygen in Liquid Alloys

The activity, solubility, and diffusivity of oxygen in high-temperature superconducting precursor alloys. $YbBa_2Cu_3$, $Ag–YbBa_2Cu_3$, and $TlBa_2Ca_2Cu_3$ and related
binary Ba–Cu in a liquid state have been measured by modified coulometric titration method [15–17]. Detailed experimental procedures have been described in
the papers. In brief, a commercial-grade yttria-stabilized zirconia (YSZ) is used as
a solid electrolyte in a galvanic cell shown in Fig. 15 and described by

$$\text{liquid alloy} + O|ZrO_2 + (8\% \ Y_2O_3)|Pt,\text{air}$$
$$P'_{O_2} \qquad\qquad\qquad P_{O_2}$$

The electric potential across the electrolyte, E_1, of the cell is given by the
Nernst equation:

$$E_1 = \frac{RT}{4F} \ln \frac{P_{O_2}}{P'_{O_2}}$$

where R is the gas constant, T the temperature, F the Faraday constant, P_{O_2} the
oxygen partial pressure of the air, and P'_{O_2} the equilibrium oxygen partial pressure

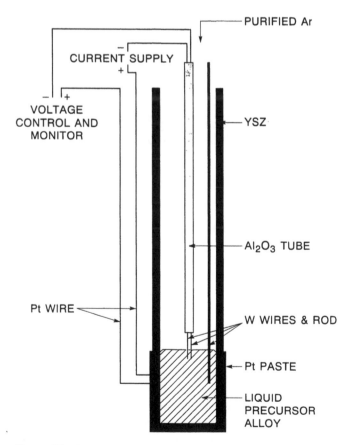

Figure 15   Schematic diagram of experimental galvanic cell.

of the liquid $Yb_1Ba_2Cu_3$. The equilibrium oxygen partial pressure inside the liquid can be monitored by measuring E_1 or can be preset to a new value by applying a different bias.

The quantity of electrical charge due to I_{ion} can be related to the concentration of dissolved oxygen in the liquid metal as [18]

$$C = 100Y\left[(Y + 1) + (Y - 1)\exp\left(-\frac{2\Delta EF}{RT}\right)\right]^{-1}$$

$$Y = \frac{Q_{ion}}{2F}\left(\sum_{i=1}^{n}\frac{W_i}{M_i}\right)^{-1}$$

where C is the oxygen concentration in the liquid metal in atomic percentage, Q_{ion} the quantity of pump-in electrical charge upon applied voltage change ΔE ($= E_2 - E_1$) and is calculated by integrating the ionic current with time, M_i and W_i the atomic weight and the weight of the liquid metal, and n the number of element in the alloy.

By assuming that Sievert's law is valid up to the saturation solubility, the saturation concentration $C^s = C(P^s_{O_2}/P'_{O_2})^{1/2}$ is determined from the equilibrium partial pressure of metal oxide $P^s_{O_2}$. To find the $P^s_{O_2}$ value, we pumped oxygen into the liquid and monitored the open-circuit voltage until a plateau was observed.

The standard Gibbs formation energy for the reaction,

$$\tfrac{1}{2}O_2(1 \text{ atm}) \rightarrow O(1 \text{ at } \%)$$

is given by

$$\Delta G = RT \ln \frac{\sqrt{P'_{O_2}}}{C}$$

Here the standard states for the reference gas is 1 atm, and for the oxygen atoms in the liquid alloy is 1 at %.

The diffusion coefficient of oxygen D and oxygen concentration C in a liquid alloy in a cylindrical tube of radius r is described by Fick's law:

$$\frac{\partial C}{\partial t} = \frac{1}{r}\frac{\partial}{\partial r}\left(rD\frac{\partial C}{\partial r}\right)$$

At a sufficiently long diffusion time, it can be approximated as [19]

$$\ln(I_{ion}) = \ln(B) - \frac{t}{\tau}$$

$$\tau = \frac{a^2}{D(2.405^2)}$$

where B is a constant. The diffusion coefficients are obtained from the slopes of $\ln(I_{ion})$ versus t curves.

The formation energies ΔG, the oxygen solubilities C^s, and the diffusion coefficients D are listed in Table 2. In all alloy systems investigated, the formation energy of oxygen dissolved in the liquid alloys is relatively high (ca. -255 kJ/g-atom) and is attributed to the present of rare earth elements of Ba, Ca, and Yb. The high affinity of oxygen with these elements results in a limited oxygen solubility

Table 2 ΔG Value of Alloys at 900°C

Alloys	ΔG (kJ/g-atom)	C^s (10^{-4})	D $(10^4$ cm²/s)
$Ba_{0.25}Cu_{0.75}$	−254	—	0.68
$Ba_{0.35}Cu_{0.75}$	−260	—	0.24
$Ag-Yb_1Ba_2Cu_3$	−257	9.1	1.52
$Yb_1Ba_2Cu_3$	−262	1.4	1.43
$Tl_1Ba_2Ca_2Cu_3$	−256	570	1.20

(10^{-4} M fraction except for Tl alloy for which $C^s \sim 5.7 \times 10^{-2}$); however, the diffusion coefficients of oxygen in the liquid alloys are high, on the order of 10^{-4} cm²/s, an order of magnitude higher than that ($D_s \sim 10^{-5}$ cm²/s) of self-diffusion of metal melts. These results suggest that oxygen atoms diffuse interstitially in the liquid alloys.

B. Growth Kinetics of Oxide Films

The kinetic of chemisorption of oxygen in a liquid alloy is very complex and barely understood. In the following we deal with the oxygen-diffusion-controlled growth of the oxide film. A simple model of the films grown by the liquid–gas-solidification process in an extreme condition is proposed. We assume that (1) there exists a negligible nucleation barrier for the oxide growth at the liquid–substrate interface because of a close match in chemical structure and lattice parameters between the oxide (e.g., $Yb_1Ba_2Cu_3O_7$ and the substrate $SrTiO_3$), and (2) oxygen atoms are absorbed on the liquid surface and supersaturated without forming oxide. The critical supersaturation is on the order of the homogeneous bulk nucleation such that $\Delta G \sim 0.1\ RT$ [20]. We thus arrive at $\ln(C_s/C_e) \sim 0.1$ or $C_s/C_e \sim 1.1$, where C_s and C_e are, respectively, the supersaturated concentration of oxygen at the free surface and the equilibrium saturation concentration of oxygen in liquid alloy. For steady-state one-dimensional growth, we obtain the growth velocity of oxide film,

$$V \approx \frac{D}{C_{\text{oxide}}} \frac{\partial C}{\partial x} \approx \frac{D}{h} \frac{0.1 C_e}{C_{\text{oxide}}}$$

where C_{oxide} (ca. 50 at %) is the oxygen concentration of the oxide $Yb_1Ba_2Cu_3O_7$ and h is the thickness of the melt coat. Taking $h \sim 2 \times 10^4$ cm and $C_e \approx 10^{-2}$ at % and $D \sim 10^{-4}$ cm²/s, it yields $V \approx 10^{-5}$ cm/s. Accordingly, an oxide film of ~2 μm thickness can be grown in 10 s by the LGS process, in harmony with the experimental results.

In $Yb_1Ba_2Cu_3O_7$ epitaxial films grown by an Ag-enhanced LGS process, Ag acts as flux and does not get incorporated in the oxide formation [10]. The excess Ag is rejected out of the film into the liquid in front of the interface. When the rejected Ag diffuses very fast compared with the film growth such that Ag is distributed uniformly throughout the melt, the amount of oxygen reaching the film–liquid melt interface determines the growth of the oxide film. By the same assumption as in $Yb_1Ba_2Cu_3$ precursor alloy, the growth velocity is related to the oxygen diffusion coefficient and the oxygen solubility in the precursor alloy Ag–$Yb_1Ba_2Cu_3$. With the diffusion coefficient of oxygen in Ag-doped $Yb_1Ba_2Cu_3$ (D

$\sim 1.52 \times 10^{-5}$ cm^2/s) and the oxygen solubility $C_e = 9 \times 10^{-4}$, a factor 5.5 higher than in Yb$_1$Ba$_2$Cu$_3$ alloy, the limiting growth rate is calculated to be

$$V_{\text{O control}} \approx 5.5 \times 10^{-5} \text{ cm/s}$$

On the other hand, if the Ag diffusion limits the film growth, the excess Ag rejected from the oxide film will build up in front of the oxide film–liquid interface. The amount of Ag diffused away from the oxide film–liquid interface determines the growth of the oxide film. We assume that the concentration of Ag decreases linearly from $C_{s(\text{Ag})} \approx 1$ at the oxide film–melt interface to $C_{e(\text{Ag})} = \frac{1}{4}$ at the free surface of the liquid melt. For one-dimensional growth, the Ag flux diffusing away from the interface is given by

$$J = D_{\text{Ag}} \frac{\partial C_{\text{Ag}}}{\partial x} \approx D_{\text{Ag}} \frac{C_{s(\text{Ag})} - C_{e(\text{Ag})}}{h}$$

where h is the diffusion layer thickness or the thickness of the melt layer, $\sim 2 \times 10^{-4}$ cm, and D_{Ag} is the diffusion coefficient of Ag in the liquid Ag–Yb$_1$Ba$_2$Cu$_3$ alloy, on the order of 10^{-5} cm^2/s. The number of Ag atoms diffused away into the liquid creates the same number of oxide unit cells. We then arrive at the growth velocity

$$V_{\text{Ag controlled}} \sim \frac{J}{C_{e(\text{Ag})}} \sim 3 \times 10^{-1} \text{ cm/s}$$

which is much greater than the $V = 5.5 \times 10^{-5}$ cm/s estimated for oxygen-diffusion-controlled growth. Thus the oxygen diffusion is seen to limit the growth of Yb$_1$Ba$_2$Cu$_3$O$_7$ film from precursor alloys of Ag–Yb$_1$Ba$_2$Cu$_3$ by the LGS process. The enhanced growth velocity simply reflects the higher oxygen solubility in the Ag-doped precursor liquid alloy, as the diffusion coefficients of oxygen in the precursor alloys are altered little by Ag doping. The oxygen solubility of the TlBa$_2$Ca$_3$Cu$_4$ melt, $C^s = 5.74$ at % at 900°C, is strikingly high. Since the oxygen diffusivities, $D \sim 10^{-4}$ cm^2/s, are nearly the same for all precursor alloy melts, we find the oxygen-controlled growth rate of Tl film to be rather high, $V \sim 5 \times 10^{-3}$ cm/s. It is probable that the surface absorption of oxygen will limit the growth of Tl film.

VI. Tl-BASED BULK SUPERCONDUCTORS

LGS processing has been used to fabricate Tl-based superconducting oxides by directly oxidizing a precursor liquid alloy Tl$_1$Ba$_2$Ca$_2$Cu$_3$ [21]. Oxidation of the melt was carried out in situ by introducing oxygen gas into the melt. The time duration of oxidation was typically 10 to 30 min.

X-ray diffraction revealed that the oxide contains mostly Tl$_1$Ba$_2$Ca$_1$Cu$_2$O$_7$, Tl$_2$Ba$_2$Ca$_1$Cu$_2$O$_8$, and Tl$_2$Ba$_2$Ca$_2$Cu$_3$O$_{10}$ phases. The morphology of the sample shows an array of flake-grained microstructure with grains about 2 to 5 μm in width and 1 to 2 μm in thickness. Often, the flake grains are assembled parallel to each other with voids between them. The electrical resistivity measurement shows an onset superconducting transition temperature of 120 K and zero resistance at 113 K. The critical current densities J_c obtained from the dynamic hysteresis loops at a sweep rate of 15.4 kOe/min and H_a = 9 kOe are about 10^6 A/cm^2 at 10 K and about 5×10^4 A/cm^2 at 60 K.

The LGS process has also applied to bulk superconductors of $(Tl_{0.64}Bi_{0.16}Pb_{0.2})Ba_{2-x}Sr_xCa_3Cu_4O_y$ to study the effect of Sr substitution [22]. The bulk has been synthesized directly from the corresponding precursor alloy. An x-ray diffraction pattern reveals the bulk oxides as being multiphase superconductors and the lattice parameter c decreasing with the Sr content. All samples consist of two phases: the 1223 and the 1234 in x = 0 and the 1212 and the 1223 in x =

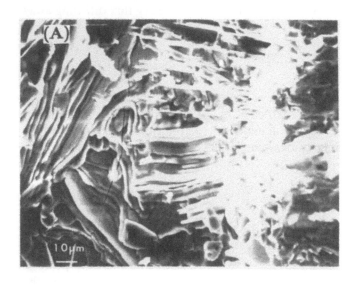

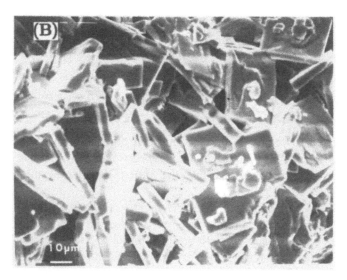

Figure 16   SEM micrographs of $(Tl_{0.64}Bi_{0.16}Pb_{0.2})Ba_{2-x}Sr_xCa_3Cu_4O_y$ for (A) x = 0 and (B) x = 2 samples.

1 and 2 samples. The substitution of Sr suppresses the formation of high Cu–O layered phase. The 1234 phase was formed only in the sample of $x = 0$, and the 1212 phase was always found in samples of $x = 1$ and 2. Oxidation time-dependent phase transformation from a Tl-rich phase to a Tl-poor phase was observed and was attributed to phase equilibration of kinetic origin. The resistivity measurements indicate percolation effects with a percolation limit just above 8% of the high-T_c phase. It suggests that during oxidation, the Tl-rich phase formed first and then transformed to the Tl-poor phase at the surface of a grain. The crystal growth is nonequilibrium and anisotropic in nature. The SEM micrograph shown in Fig. 16 reveals thin platelet crystals and a dendritic growth. Thin platelet crystals tend to pack up in pure Ba samples to form a tight and electrically well connected structure, but tend to form a loose and more isolated structure in a pure Sr sample. The bulk oxides typically exhibit a $T_c(R = 0)$ value of 110 K.

VII. SUMMARY

Superconducting films of $YbBa_2Cu_3O_7$ and $TlBa_2Ca_2Cu_3O_x$ about 1 μm thick have been grown on various substrates by the liquid–gas solidification (LGS) process. The $YbBa_2Cu_3O_7$ film grows epitaxially on $SrTiO_3(100)$ with the c axis normal to the film plane, while the film grown on $SrTiO_3(110)$ exhibits a predominant orientation with the (110) direction normal to the plane. The addition of Ag in the precursor alloy melt for the growth of $YbBa_2Cu_3O_7$ superconductor is found to improve the morphology and transport properties of the resultant film. The film shows a sharp superconducting transition at 90 K with a critical current $J_c(H = 0,$ 77 K) $\sim 3 \times 10^4$ A/cm², which is comparable to that of a single crystal. EDX examinations reveal that Ag acts as a flux and does not incorporate in the oxidation process. Because of strain relaxation, films grown on $SrTiO_3(100)$ exhibit irregular tile patterns with platelet dimensions of about 10 μm² $\times$ 2 μm, characteristic of lattice-mismatch thick epitaxial films. For a thinner film of about 1000 Å, a structurally homogeneous epitaxial film can be obtained.

$TlBa_2Ca_2Cu_3O_x$ film grown on $MgO(100)$ by the LGS process shows a flat, uniform morphology of 0.5×1.0 mm² domain and a thickness of 1 μm with a highly textured c axis normal to the film. The film consists of 1223 and 2212 phases with a trace of 1212 phase. There is no postannealing or Tl compensation after the initial deposition. The film exhibits a superconducting onset at 117 K and zero resistance at 107 K. On the other hand, the as-grown $TlBa_2Ca_2Cu_3O_x$ film on $SrTiO_3(100)$ shows a two-dimensional layer growth morphology with CaO precipitates appearing at the edge of layers. Upon anneal the film retaining the original layer structure becomes smoother and CaO precipitates disappear. The film consists of 1223 and 2212 phases with zero-field J_c of 5×10^4 A/cm² at 77 K and 5×10^5 A/cm² at 4 K. The film exhibits weak flux pinnings, as expected for a structurally and compositionally homogeneous Tl-based HTSC. To date, growth of a single-phase Tl-HTSC by the LGS process has been unsuccessful. All LGS-growth Tl films consist of multiphases due to the ease of intergrowth of CuO layers. The kinetics and thermodynamics of oxide formation by LGS process are not well understood. It appears that at present setup, the kinetics of oxygen–metal reactions govern the resultant HTSC phases, and the oxidation proceeds in quasi-thermodynamic equilibrium.

Progress has been made on understanding the thermodynamics and kinetics of the LGS process. In the LGS process, oxygen is dissociated and absorbed on the free surface of the liquid alloy. The absorbed oxygen then diffuses into the liquid alloy toward the liquid–substrate interface and reacts with metal atoms to form a oxide film. The nucleation and growth of films are a matter of the thermodynamic and kinetic behavior of oxygen in the precursor liquid alloys. Recently, the activity, solubility, and diffusivity of oxygen in a number of HTSC precursor alloys—$YbBa_2Cu_3$, $Ag–YbBa_2Cu_3$, and $TlBa_2Ca_2Cu_3$—have been measured by a coulometric titration method. In all alloy systems measured, the formation energy of oxygen dissolved in the liquid alloys is high (ca. -255 kJ/g-atom), which results in limited oxygen solubility (ca. 10^{-4} molar fraction). However, the diffusion coefficients of oxygen in the liquid alloys are high, on the order of 10^{-4} cm^2/s. Based on these thermodynamic data, the limiting growth rate of oxide films in the oxygen-diffusion-controlled mechanism is estimated to be 10^{-5} to 10^{-4} cm/s. Accordingly, an oxide film of about 2 μm thickness can be grown in seconds using the LGS process.

In principle, the LGS process is scalable to continuous coating films and wires, which has not yet been demonstrated. With a better understanding of the kinetics and thermodynamics of an oxidation process and improved coating design to control the oxygen partial pressure, reaction rate, and so on, uniform films or wires with improved superconducting properties may be attained using the LGS process.

ACKNOWLEDGMENTS

I would like to thank L. C. Kimerling, A. R. Kortan, F. A. Thiel, and H. Chou for their valuable collaborations.

REFERENCES

1. M. K. Wu, J. R. Ashburn, C. J. Throng, P. H. Hor, R. L. Meng, L. Gas, Z. J. Huang, Y. Q. Wang, and C. W. Chu, *Phys. Rev. Lett.* 58, 908 (1987).
2. R. J. Cava, B. Battlog, R. B. van Dover, D. W. Murphy, S. Sunshine, T. Siegrist, J. P. Remeika, E. A. Reitman, S. Zahurak, and G. P. Espinosa, *Phys. Rev. Lett.* 58, 1676 (1987).
3. H. Maeda, Y. Tanaka, M. Fukutomi, and T. Asano, *Jpn. J. Appl. Phys.* 27, L209 (1988).
4. Z. Z. Sheng, A. M. Herman, A. E. Ali, C. Almasan, J. Estrada, and T. Datta, *Phys. Rev. Lett.* 60, 937 (1988).
5. S. Jin, R. C. Sherwood, T. H. Thiel, R. B. van Dover, G. H. Kammlott, M. E. Davis, R. A. Fastnachi, S. Nakahara, M. F. Yan, and D. W. Johnson, Jr., *Phys. Rev.* B37, 9850 (1988).
6. P. M. Mankiewich, J. H. Scofield, W. J. Skocpol, R. E. Howard, A. H. Rayem, and E. Good, *Appl. Phys. Lett.* 51, 1753 (1987).
7. P. Chandhari, R. H. Kock, R. B. Laibowiz, T. R. McGuire, and R. J. Gambino, *Phys. Rev. Lett.* 58, 2684 (1987).
8. H. S. Chen, S. H. Liou, A. R. Kortan, and L. C. Kimerling, *Appl. Phys. Lett.* 53, 705 (1988).
9. M. Hong, J. Kwo, C. H. Chen, R. M. Fleming, S. H. Liou, M. E. Gross, B. A. Davidson, L. C. Kimerling, H. S. Chen, S. Nakahara, and T. Boone, *AIP Conf. Proc.* 165, 12 (1988).

10. H. S. Chen, A. R. Kortan, F. A. Thiel, and L. C. Kimerling, *Appl. Phys. Lett.* 55, 191 (1989).
11. R. Halder, Y. Z. Lu, and B. C. Giessen, *Appl. Phys. Lett.* 51, 538 (1987).
12. K. Matsuzaki, A. Inoue, H. Kimura, K. Aoki, and T. Masumoto, *Jpn. J. Appl. Phys.* 26, L1310 (1987).
13. H. Chou, H. S. Chen, A. R. Kortan, L. C. Kimerling, and F. A. Thiel, *Appl. Phys. Lett.* 58, 283 (1991).
14. H. Chou, Ph.D. thesis, Columbia University, 1992.
15. H. Chou, H. S. Chen, W. C. Fang, F. A. Thiel, and M. K. Wu, *Appl. Phys. Lett.* 60, 760 (1992).
16. H. Chou, H. S. Chen, W. C. Fang, and M. K. Wu, *J. Appl. Phys.* 72, 5635 (1992).
17. H. Chou, H. S. Chen, W. C. Fang, P. L. Trevor, and M. K. Wu, *J. Electrochem. Soc.* 139, 3545 (1992).
18. S. Otsuka, T. Sano, and Z. Kozuka, *Trans. Jpn. Inst. Met.* 22, 35 (1981).
19. K. E. Oberg, L. M. Friedman, W. M. Boorstein, and R. A. Rapp, *Met. Trans.* 4, 61 (1973).
20. D. Turnbull, *Solid State Phys.* 3, 225 (1956).
21. H. S. Chen, E. M. Gyorgy, L. C. Kimerling, A. R. Kortan, and F. A. Thiel, *Mod. Phys. Lett.* 3, 975 (1989).
22. H. Chou, H. S. Chen, E. M. Gyorgy, A. R. Kortan, and L. C. Kimerling, *Mod. Phys. Lett. B* 5, 1735 (1991).

12
Electrodeposited Tl–Ba–Ca–Cu–O Superconductors

R. N. Bhattacharya and R. D. Blaugher
National Renewable Energy Laboratory
Golden, Colorado

I. INTRODUCTION

Highly efficient energy conversion, storage, and transmission are required for the electric power industry. The introduction of superconducting devices into existing electrical generation and transmission grids would yield significant energy savings with respect to fuel costs and overall electrical losses. Consideration of superconducting power components would be more widely accepted if these devices could be operated at liquid nitrogen temperatures (77 K) rather than the present liquid helium temperature (4.2 K) required by conventional superconductors. The electrodeposited Tl–Ba–Ca–Cu–O (TBCCO) superconductors have considerable potential in the fabrication of long-length conductors such as wires or tapes operating at 77 K. The electrodeposition technique also offers a number of other advantages over conventional wire techniques: (1) they are simple, (2) they are inexpensive, (3) they have fast deposition rates, (4) they can be used in the fabrication of large nonplanar devices, and (5) they have high efficiency in raw material utilization. Electrodeposition has recently been used to fabricate high-temperature thin-film and thick-film superconductors [1–7]. The electrodeposited films provide a highly reactive mixture on an atomic scale, which markedly reduces the time and temperature compared to the usual oxide superconductor powder calcining approach. In principle, electrodeposition from both aqueous and nonaqueous solutions is possible—our preference being toward nonaqueous methods. We have electrodeposited TBCCO films from nonaqueous solution using dimethyl sulfoxide (DMSO) because in aqueous solution the reduction of water (-1.23 V versus NHE) is kinetically more favorable than the alkaline metals (< 2.0 V versus NHE), which limit the deposition of Tl, Ba, and Ca. The precursors of superconducting TBCCO films are obtained by codepositing constituent metals under constant-potential and pulsed-potential conditions, respectively. Pulse-potential deposition is generally found to improve the film properties (e.g., porosity, electrical conductivity, and roughness) and to increase the average deposition rate.

II. ELECTROCHEMICAL BACKGROUND

Electrodeposition is a process for depositing a coating (e.g., metals and oxides) with uniform coverage and properties on conducting substrates from a solution ("bath") containing the ions of interest (e.g., Tl^{1+}, Ba^{2+}, Ca^{2+}, Cu^{2+}). Almost always, one of the two phases contributing to an interface of interest in electro-deposition will be an electrolyte, a phase through which charge is carried by the movement of ions. The second phase at the boundary is the substrate through which charge is carried by electronic movement. In cathodic electrodeposition, when the potential of the substrate (electrode) is moved from its equilibrium value toward negative potentials, the cation that will be reduced first is the one with the least negative (or most positive) redox potential E^0.

In a solution containing Tl^{1+}, Ba^{2+}, Ca^{2+}, and Cu^{2+}, all four ions can be codeposited on the surface of the substrate when the potential is negative enough. To determine the deposition potential, a cyclic voltammogram experiment was performed. Figure 1 represents a cyclic voltammogram for a solution mixture of $TlNO_3$, $Ca(NO_3)_2 \cdot 4H_2O$, $Ba(NO_3)_2$, and $Cu(NO_3)_2$ dissolved in DMSO solvent. A set of anodic and cathodic waves were observed. The reduction peaks of the corresponding Tl, Ba, Ca, and Cu were clearly evident, with concomitant development of a black deposit on the electrode surface. This deposit was stripped from the electrode on the positive-going scan. This behavior was clearly due to the deposition of Tl–Ba–Ca–Cu precursor as described by the following reactions (the reduction potentials with respect to $Ag/AgNO_3$ are given in parentheses):

$$Tl^{1+}: \quad +1e \rightarrow Tl \quad (-1.36 \text{ V}) \tag{1}$$

$$Ba^{2+}: \quad +2e \rightarrow Ba \quad (-2.58 \text{ V}) \tag{2}$$

$$Ca^{2+}: \quad +2e \rightarrow Ca \quad (-2.58 \text{ V}) \tag{3}$$

$$Cu^{2+}: \quad +2e \rightarrow Cu \quad (-0.74 \text{ V}) \tag{4}$$

$$Cu^{1+}: \quad +1e \rightarrow Cu \quad (-0.55 \text{ V}) \tag{5}$$

The films can be codeposited by applying either constant potential or pulse potential. Pulse-potential deposition is used primarily to improve the film morphology. The morphology for the electrodeposited materials is a very important step in electrogrowth because it directly influences the structure of the annealed film and therefore its properties. In the electrodeposition process, the adatoms or adions are incorporated in the substrate. With time as the film thickness increases, the deposition continues either by the buildup on previously deposited material (old nucleation centers) or the formation and growth of new ones. These two processes are in competition and can be influenced by different factors. High surface diffusion rates, a low population of adatoms, and low overpotentials enhance the buildup of old nucleation centers, while conversely, low surface diffusion rates, a high population of adatoms, and high overpotentials on the surface enhance the creation of a new nucleation center. In pulse plating, the pulsed current density remains considerably higher with time than the corresponding direct-current (dc)

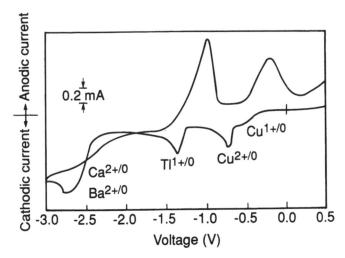

Figure 1 Cyclic voltammogram for a solution mixture of $TlNO_3$, $Ca(NO_3)_2 \cdot 4H_2O$, $Ba(NO_3)_2$, and $Cu(NO_3)_2$ dissolved in DMSO solvent.

density, which leads to a higher population of adatoms on the surface during pulse deposition than during dc deposition, resulting in an increased nucleation rate and therefore a finer-grained structure. Figure 2a and b represent the scanning electron micrographs (SEMs) of as-deposited TBCCO films deposited under constant-potential and pulse-potential conditions, respectively. As expected, the film deposited using pulse potential (Fig. 2b) has a finer-grained structure than that of the film deposited by applying constant potential (Fig. 2a). In addition, pulsed potential film is more uniform: High porosity is observed using constant-potential methods.

III. EXPERIMENTAL PROCEDURE

The constituents of the electrodeposition bath are adjusted empirically to produce the proper stoichiometric TBCCO superconductor thin films. All chemicals were of Analar or Puratronic grade purity and were used as received. A conventional three-electrode cell was employed, the reference electrode being $Ag/AgNO_3$ and the counterelectrode being platinum gauze. The electrodeposition system represented in Fig. 3 consists of a potentiostat/galvanostat (PAR 273A) interfaced with an IBM AT computer. This electrodeposition system allows sophisticated current/voltage profiles to be run automatically, gives precise control and reproducibility of deposition parameters, and facilitates monitoring of the processing parameters and archiving the numerous processing runs.

IV. EXPERIMENTAL RESULT

A. Deposition of the TBCCO Film at Constant Potential

The deposition bath consisted of 33 mM $TlNO_3$, 40 mM $Ca(NO_3)_2 \cdot 4H_2O$, 60 mM $Ba(NO_3)_2$, and 66 mM $Cu(NO_3)_2$ dissolved in DMSO solvent. The deposition potential was -4 V (versus $Ag/AgNO_3$), and the rate of deposition was roughly

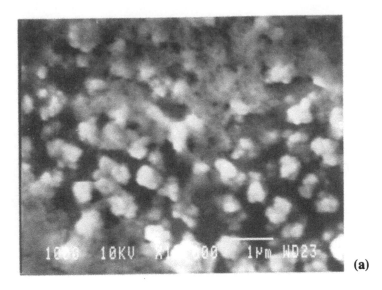

(a)

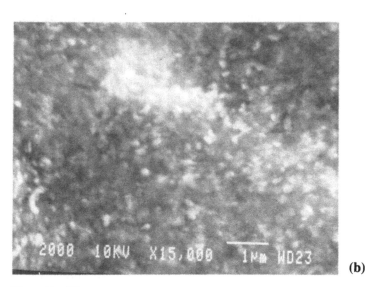

(b)

Figure 2   SEM of as-deposited TBCCO film deposited under (a) constant-potential and (b) pulse-potential conditions.

0.5 μm/min. The substrates used were silver-coated (about 500 Å) $SrTiO_3$ single crystals with [100] orientation. The electrodeposited TBCCO films were air-annealed for 10 min at 850°C in the presence of a TBCCO (2223) pellet. The films were subsequently annealed in oxygen (1 atm) at 750°C for 15 min, also in the presence of the 2223 pellet. A typical composition of the annealed films as measured by electron probe microanalysis (EPMA) was (Tl/Ba/Ca/Cu) 2:1.87:2.69:3.89. The

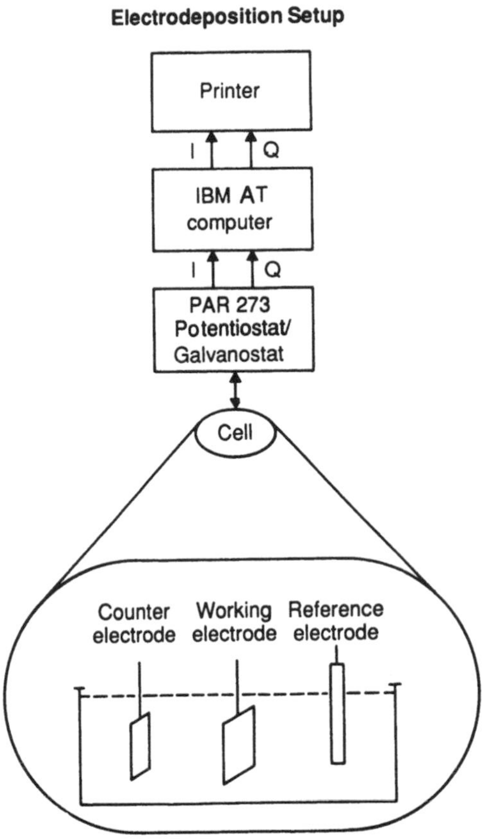

Electrodeposition Setup

Figure 3   Electrodeposition setup.

SEM micrograph of a representative film is shown in Fig. 4. The critical current density at 76 K of the oxygen-annealed film was 20,000 A/cm^2 in a zero magnetic field and 5000 A/cm^2 in a 1-T field parallel to the film plane (Fig. 5).

B.   Deposition of the TBCCO Film Under Pulse Potential Conditions

The deposition bath consisted of 20 mM TlNO$_3$, 38 mM Ca(NO$_3$)$_2$·4H$_2$O, 102 mM Ba(NO$_3$)$_2$, and 38 mM Cu(NO$_3$)$_2$ dissolved in DMSO solvent. The pulsed-potential precursors of superconducting TBCCO thin films were electrodeposited on silver-coated (500 Å) SrTiO$_3$ and on silver foil. The pulsed-potential cycle was 10 s at -4 V, followed by 10 s at -1 V (versus Ag/AgNO$_3$), and the rate of deposition was about 0.15 μm/min.

The electrodeposited TBCCO films deposited on silver-coated SrTiO$_3$ were annealed in oxygen (1 atm) for 10 min at 860°C in the presence of a TBCCO (mixture of 2223 and 2212) pellet. The thicknesses of the films were approximately 1 μm. The electrodeposited TBCCO films have exhibited zero resistivity as high as 112 K (Fig. 6). The I–V curve of the best film is shown in Fig. 7. The film width used for critical current density measurement is 4.5 mm. The critical current density of this film is 56,000 A/cm^2 at 76 K in a zero field. The results suggest that a

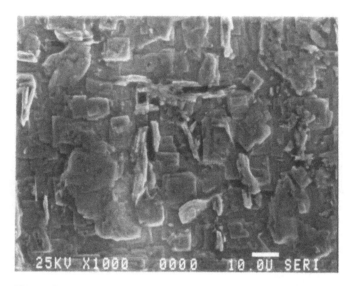

Figure 4   SEM of postannealed electrodeposited TBCCO film.

significant improvement in the critical current density of electrodeposited TBCCO superconductor film can be obtained when the films are deposited under pulsed-potential conditions as opposed to constant-potential conditions.

The electrodeposited TBCCO films deposited on silver foil were annealed in oxygen in the presence of thallium oxide. The films were processed for 10 min at 860°C with the thallium oxide source temperature held at 700°C. The temperature, $T_c(0)$, at which the sample dc resistance becomes zero (in a zero applied field) was about 100 K. The dependence of the critical current density, J_c, with applied field is shown in Figs. 8 and 9. The highest value of the zero-field J_c value at 77 K was about 10,000 A/cm^2. A strong weak-link behavior for electrodeposited TBCCO films fabricated on Ag foil is observed; this is in contrast to the electrodeposited TBCCO film on 500-Å Ag/SrTiO$_3$ substrates, which did not exhibit weak-link

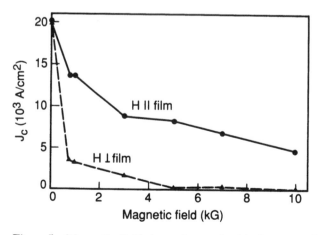

Figure 5   Magnetic field dependence of critical current density at 76 K for TBCCO film.

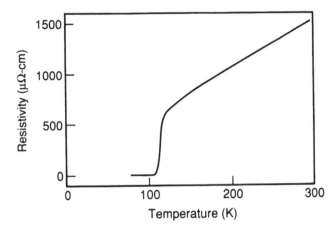

Figure 6   *R* versus *T* of an electrodeposited TBCCO film deposited on Ag/SrTiO₃.

characteristics. At present, we are investigating the effect of silver with respect to reaction conditions and nucleation and growth on TBCCO film. Our results to date indicate that Ag has a pronounced influence on the reaction of TBCCO films.

It is important to note that all of the electrodeposited films were highly polycrystalline. The absence of weak-link behavior in some of our polycrystalline films is noteworthy and demonstrates that under proper reaction conditions, "platelike" polycrystalline Tl-oxide films do not show evidence of weak-link behavior. This result is in contrast to the weak-link dominated grain boundary behavior observed for Y–Ba–Cu–O (YBCO, 123) material. It is suspected that the grain boundaries for the non-weak-link Tl films are compositionally correct, with no stoichiometric variation, and contain no extraneous phases within the grain boundaries that limit current transfer. Further studies on the Tl grain boundaries are urgently needed to define their character compared to YBCO (123).

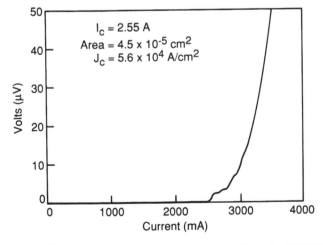

Figure 7   *I–V* curve of a pulsed-potential-deposited TBCCO film.

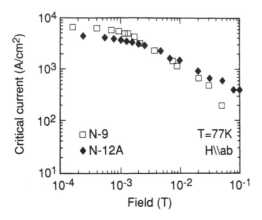

Figure 8 Magnetic field dependence of critical current density at 77 K for two TBCCO films deposited on Ag foil.

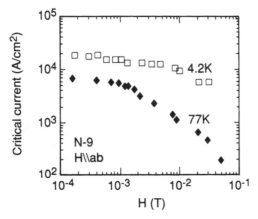

Figure 9 Magnetic field dependence of critical current density at 4.2 and 77 K, respectively, for TBCCO film deposited on Ag foil.

V. CONCLUSIONS

In conclusion, a cathodic electrodeposition method is described for fabricating TBCCO superconducting thin films. Optimization of fabrication and process conditions should offer higher critical current densities by improving the stoichiometries and the structural development. The simplicity and low cost of the electrodeposition process, as well as its utility for nonplanar and preengineered configurations, make it attractive for practical applications. Measurements of the films show relatively outstanding critical current densities in TBCCO samples. The outstanding critical current densities measured to date for our polycrystalline TBCCO films are as follows: (1) the critical current density at 77 K for a pulsed-potential TBCCO film on silver foil was 10,000 A/cm^2 in a zero field; (2) the critical current density at 76 K for a TBCCO film deposited at constant potential on a silver-coated SrTiO$_3$ substrate was 20,000 A/cm^2 in a zero magnetic field and 5000 A/cm^2 in a 1-T field parallel to the film plane, and (3) the critical current density of a pulsed-potential deposited TBCCO film on silver-coated $SrTiO_3$ was 56,000 A/cm^2 at 76 K in zero field. These values represent some of the better results to date in this field, and we are very encouraged that solutions to some of the major stumbling blocks in realizing an HTSC conductor can be overcome by this technique.

ACKNOWLEDGMENTS

The authors acknowledge Sandia National Laboratory (R. P. Hellmer and D. S. Ginley), Los Alamos National Laboratory (P. Arendt and N. Elliot), and General Electric (J. DeLuca) for collaboration. This work was performed at NREL under Contract DE-AC02-83CH10093 to the U.S. Department of Energy.

REFERENCES

1. M. Maxfield, H. Eckhardt, Z. Iqbal, F. Reidinger, and R. H. Baughman, *Appl. Phys. Lett.* 54, 1932 (1989).
2. R. N. Bhattacharya, R. Noufi, L. L. Roybal, R. K. Ahrenkiel, P. Parilla, A. Mason, and D. Albin, in *Science and Technology of Thin Film Superconductors 2*, ed. R. D. McConnell and R. Noufi, Plenum Press, New York, 1991, p. 243.
3. R. N. Bhattacharya, R. Noufi, L. L. Roybal, and R. K. Ahrenkiel, *J. Electrochem. Soc.* 138, 1643 (1991).
4. P. Slezak and A. Wieckowski, *J. Electrochem. Soc.* 138, 1038 (1991).
5. R. N. Bhattacharya, P. A. Parilla, L. L. Roybal, A. Mason, A. Swartzlander, R. K. Ahrenkiel, and R. Noufi, in *TMS Proc.*, *High Temperature Superconducting Compounds III*, 1991, p. 169.
6. R. N. Bhattacharya, P. A. Parilla, A. Mason, L. L. Roybal, R. K. Ahrenkiel, R. Noufi, R. P. Hellmer, J. F. Kwak, and D. S. Ginley, *J. Mater. Res.* 6(7), (1991).
7. R. N. Bhattacharya, P. A. Parilla, R. Noufi, P. Arendt, and E. Elliot, *J. Electrochem. Soc.* 139, 67 (1992).

13
Tl–Ca–Ba–Cu–O Electronics

J. S. Martens*
Sandia National Laboratories
Albuquerque, New Mexico

I. INTRODUCTION

Since the discovery of the Y–Ba–Cu–O, Bi–Sr–Ca–Cu–O, and Tl–Ca–Ba–Cu–O high-temperature superconducting systems, a great deal of progress has been made in the development of useful high-temperature superconducting (HTS) circuits. Much of the work has been done in Y–Ba–Cu–O. Because of this and because of the similarity between devices and circuits made from the various materials, devices made from various HTS materials are discussed in this chapter. Since this is a book on Tl–Ca–Ba–Cu–O, devices and circuits made from that material will be emphasized and only those devices that to this author's knowledge have been made successfully using Tl–Ca–Ba–Cu–O are discussed in any depth. A recurring theme in the chapter will be the differences and similarities in performance of the circuits made of the various materials. The main differences center around the ease of flux motion within Tl–Ca–Ba–Cu–O and the very different grain boundary structure from that seen in Y–Ba–Cu–O. The Tl–Ca–Ba–Cu–O circuits discussed have been made from both 2223 and 2122 phases. To date, no major inherent performance differences (aside from T_c) between these phases can be claimed, although little direct comparison has been done.

HTS electronics will be divided into three groups: passive electronics, Josephson-based electronics, and non-Josephson active devices. In each case a brief review of the relevant superconducting properties will be given along with demonstrated circuits and performance, advantages over conventional counterparts, and a materials discussion.

Present affiliation: Conductus, Sunnyvale, California

II. PASSIVE ELECTRONICS

Many passive components can be fabricated from HTS materials, but at frequencies below tens or hundreds of MHz, normal metals generally do well enough that it does not make sense to go to HTS. Hence much of the work on HTS passive electronics has centered on microwave or near-microwave circuits (above 100 MHz), including filters, delay lines, antennas, and related components. There are also a number of low-frequency applications, such as superconducting quantum interference device (SQUID) pickup coils [1] and lower-frequency transformers where the noise advantages of HTS materials warrant their use. These are not discussed here, for considerations of space.

All of the devices and circuits discussed are generally made by subtractive lithography of the HTS materials followed by the deposition and patterning of any other needed layers. The pattern of, for example, a filter is defined in photoresist and the unwanted HTS material is etched away. Various etches have been used, including many different acids, EDTA, ion milling, and bromine–alcohol solutions. These etches differ in how much they damage the remaining film, etch rate, and compatibility with different photoresists. To further complicate matters, the etch behavior is dependent not only on the type of superconductor but also on how it was deposited (different deposition or anneal techniques affect the morphology enough to change etch behavior). More details are given below, but we note here some general trends. Ion milling gives excellent definition but can possibly damage the films. Bromine–alcohol etches improve surface properties but are incompatible with positive resists, resulting in poor definition. Acid-based systems are compatible with positive resists and differ considerably in the amount of film damage observed. Finally, modified EDTA etches seem to cause little damage and are compatible with positive resists.

Central to the use of Tl–Ca–Ba–Cu–O and other HTS materials in the applications described above are advantages in loss, dispersion, and circuit size over conventional circuits. To better understand the motivation, in the next section we provide information on the behavior of superconducting transmission lines. Here transmission lines are generally defined as a group of one or more conductors used to propagate high-speed signals over some distance.

A. Background on Loss and Dispersion in Transmission Structures

It is important to understand that HTS transmission lines will probably be used because of their low loss and dispersion (frequency-dependent phase velocity). Obviously, in both normal and superconducting transmission lines there will be dielectric loss. (We are considering film transmission lines such as the microstrip and coplanar waveguide shown in Fig. 1.) The current substrates used for HTS growth are unfortunately not optimal. Direct growth on sapphire or quartz would be desirable since the dielectric constants are reasonable (about 10 and 4, respectively) and the anisotropy can be dealt with in the case of sapphire. While many groups are growing films on these substrates with the use of buffer layers [2], these layers (particularly $SrTiO_3$ and, to a much lesser extent, CeO_x) can lossy. Yttria-stabilized zirconia appears to be a better buffer dielectric and some recent work suggests that its added loss is not significant [3]. At any rate, the bulk of the electronics work that has been done to date has used films grown on $LaAlO_3$. It

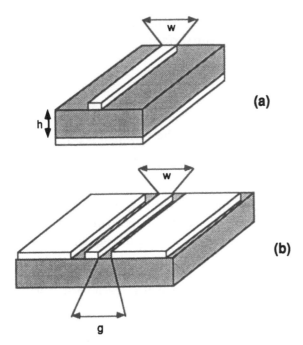

(a)

(b)

Figure 1   Schematics of the two most common high-speed transmission structures: (a) microstrip; (b) coplanar waveguide. The coplanar waveguide often uses a bottom ground plane in addition to the side ground planes. If the substrate is 20-mil LaAlO$_3$, the microstrip $w \approx$ 169 µm for a 50-Ω line. For a 50-Ω conductor-backed coplanar waveguide, possible dimensions are $w = 25$ µm and $g = 50$ µm.

has a dielectric constant of about 24.5 and a loss tangent [4] of about 0.0002 over microwave frequencies and at 77 K. While usable, both the dielectric constant and loss tangent are higher than desirable. A small loss tangent is needed to avoid loss in the transmission structure, and high dielectric constants are undesirable for circuit-design reasons. The widths of microstrip lines become smaller for the same impedance and the discontinuity models needed for circuit design are frequently ineffective for such materials. To get around this, many circuit designers have used more elaborate models or have gone to full-wave analysis of the structure [5]. A final problem with the present substrates is in twinning. Commonly, LaAlO$_3$ wafers are highly twinned. Unfortunately, the dielectric constant is anisotropic across these boundaries, which has caused problems with some high-Q filter designs [6].

The primary component of conductor loss can be expressed as surface resistance (real part of the ratio of electric to magnetic field amplitudes at the surface). For most common transmission lines such as microstrip and coplanar waveguide (shown in Fig. 1), the loss of the transmission line (per unit length) can be expressed as

$$\alpha_{\text{conductor}} = \frac{R_S}{Z_0} K \tag{1}$$

where R_S is the surface resistance, Z_0 the line characteristic impedance, and K is a factor dependent on geometry, frequency, and the dielectric properties. This neglects dielectric loss, surface wave losses, and so on, but for most normal metal lines it can be a reasonable approximation. The important point is that transmission loss (in dB/m or Np/m) commonly varies linearly with surface resistance. For normal metals, surface resistance varies as [7]

$$R_S = \sqrt{\frac{\omega\mu_0}{2\sigma}} \tag{2}$$

while for superconductors (using the two-fluid model) the variation is [8]

$$R_S = \frac{\omega^2\mu_0^2\lambda^3 n_n \sigma_n}{2n} \tag{3}$$

where λ is the penetration depth, n_n the density of normal electrons, n the total carrier density, and σ_n the normal state conductivity. The important point of the phenomenological superconductor model is the quadratic (or near-quadratic) frequency dependence. Thus while HTS materials may have enormous advantages over normal metals at lower frquencies (about an order of magnitude at 36 GHz and 77 K currently), eventually the two types of materials will reach parity (presently somewhere over 100 GHz).

For some applications, the dispersion a signal experiences along the transmission line is as important as, if not more so than, the loss. The phase velocity is inherently a function of frequency for microstrip and coplanar waveguides (and some other lines), and any loss present exacerbates the problem, so clearly HTS lines have a significant advantage in this area. Also, the inductance of a normal metal transmission line is a much stronger function of frequency, also increasing dispersion. The penetration depth (skin depth) in a normal metal goes as $[2/(\omega\mu_0\sigma)]^{1/2}$, so the volume the field sees (and hence the inductance) varies with frequency. The penetration depth in a superconductor (λ) is basically frequency independent [8] (this appears to hold at least up into the millimeter-wave range).

B. Interconnects and Transmission Lines

In the sub-50-GHz range, the differences in loss can be significant between HTS materials and cooled copper. For microstrip lines, this translates to the difference shown in Fig. 2 (both theoretical and measured results). Such improvements have been shown by many workers [9]. Several have pointed out that these gains do not justify the complexity of fabrication except in very specialized high-density circumstances, although this point of view appears to be changing. One such application may be high-speed interconnects between semiconducting computer chips in the form of multichip modules. The major advantage of HTS lines in these modules is that the linewidth can be reduced and the circuit density increased without performance degradation. However, the integration task is formidable, and this is the subject of current research in a number of laboratories.

Another application with somewhat smaller market size is delay lines for signal processing. Long delay lines, 25 to 50 ns or more, are needed for many analog signal-conditioning circuits, such as transversal filters [6]. Lines with 25 ns of delay and less than 0.2 dB/ns loss have recently been fabricated by Du Pont Using Tl–Ca–Ba–Cu–O on 2-inch wafers. Similar results have been reported using Y–Ba–Cu–O films [10,11].

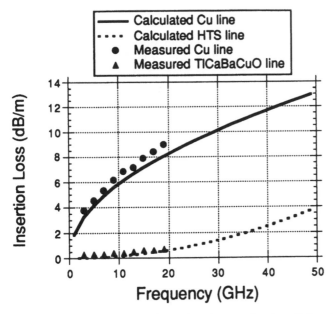

Figure 2 Conductor loss (insertion loss neglecting dielectric and other effects) per unit length in 50-Ω microstrip lines (on 20-mil LaAlO$_3$) as a function of frequency for Cu and HTS lines (both at 77 K). The frequencies shown here are low enough than anomalous surface resistance does not play too large a role.

The performance of transmission lines made of Tl–Ca–Ba–Cu–O does not differ much from that of their YBaCuO counterparts, at least at low power levels [12]. The available morphologies are similar, and the surface resistances reported by many workers are very similar [13,14]. The same processing sequences, including conventional photolithography and subtractive etching using acid systems, ion milling, solvent-based Br etches, and so on, all work to some degree with the various materials. The amount of damage, as measured by increases in surface resistance, that these processes inflict on the materials is dependent on the material being used and its deposition process, as discussed before. Some of these differences are obvious (more granular, weak-linked material being damaged more than more crystalline material), but some are not (more crystalline Y–Ba–Cu–O being more damaged than less crystalline Y–Ba–Cu–O or Tl–Ca–Ba–Cu–O).

Available J_c values are currently somewhat higher [15] in Y–Ba–Cu–O films, but typically by less than a factor of 3 at 77 K. The major difference, and one that is discussed often, is that of power-handling ability. The H_{c1} value of some Tl–Ca–Ba–Cu–O films is very low and it sometimes takes very little application of power in Tl–Ca–Ba–Cu–O materials before the loss starts to increase. According to the work of several groups [16], this does not appear to be a fundamental problem, so this difference may be resolved in the future (indeed over 100 mW has been used on some Tl–Ca–Ba–Cu–O delay lines with no degradation in performance). The poor power dependence also appears to be correlated with low pinning energies, suggesting that flux is nucleating on the edges of the conductor and creeping or oscillating in the film.

C. Resonators and Filters

Perhaps one of the most active areas of passive element research is in resonators and filters [17–19]. These devices are very attractive uses of HTS materials because the currents and concomitant I^2R losses get large in high-Q resonant circuits using normal conductors. The performance achievable with physically very small super-conducting resonators has exceeded that with normal resonators occupying much larger volumes: a critical result for space-based applications. Superconducting fil-ters are also of interest for stabilizing extremely low-phase-noise oscillators, which may use either semiconducting or superconducting devices as the active elements [20]. Many groups are working on this subject, including Superconducting Tech-nologies Incorporated, Du Pont (using Tl–Ca–Ba–Cu–O), and many others using Y–Ba–Cu–O.

Two main performance advantages are in passband loss and skirt steepness. The first is critical in power-budget calculations, which are so often pushed to their limit in space-based communications systems. The second refers to out-of-band rejection that is required in tightly packed communications bands. The loss of the material composing the filter is critical to both of these parameters and is why the low surface resistance of superconducting materials is important.

The structures most often investigated to date are microstrip or coplanar waveguide-based bandpass filters and narrow-band matching structures. Common microstrip examples are shown in Fig. 3. The matching structure may, as an ex-ample, be composed of several quarter-wave sections to accomplish a severe imped-ance transformation over a narrow frequency range (as needed, for example with some of the active devices discussed in later sections). Because large reactive currents can exist in such a transformation, the losses can be significant when using normal metal sections. Three section transformers (going from 50 Ω to 1 Ω) have been made from Au and Tl–Ca–Ba–Cu–O for use in amplifier circuits. When

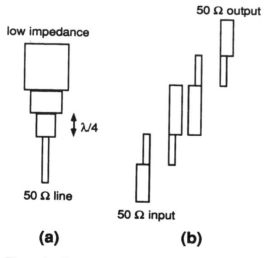

(a)   **(b)**

Figure 3 Examples of matching structures (a) and bandpass filters (b) that have been implemented in Tl–Ca–Ba–Cu–O and other HTS materials.

both transformers were tested at 77 K and 5 GHz, the Tl–Ca–Ba–Cu–O version showed less than 0.5 dB insertion loss, while the Au version showed a 3-dB insertion loss.

The passband filters cover a broader range of applications and many workers have produced excellent results. At the time of this writing, 2% passband Y–Ba–Cu–O filters have been achieved with a center frequency of bout 13 GHz and insertion losses of 0.2 dB by the Japanese consortium ISTEC (and perhaps other laboratories). One example of Tl–Ca–Ba–Cu–O filter performance is shown in Fig. 4. A comparable gold filter at 77 K had an insertion loss 2 dB higher and slightly broader skirts. Although the performance difference may not appear to be huge, the power budgets are tight enough in many satellite applications that 1 dB is very significant. While many of the filters demonstrated to date have used conventional Chebyshev designs, of perhaps greater interest are elliptical filters, with their much steeper skirts for use in channelized receivers. These designs are unfortunately difficult to implement using standard filter architectures. Work has been done by at least one group [19] using dual-mode microstrip elements to more easily implement elliptical transfer functions.

For some applications, such as in oscillator design, extremely narrow passbands are required. For these circuits, the extremely low loss of the HTS materials can lead to quality factors [Q, defined here as the (passband width)/(resonant frequency) where the width is measured between 3-dB points] of 10^7 or higher [21]. Some of the highest Q values reported to date have used been made by Du Pont using Tl–Ca–Ba–Cu–O in a loaded dielectric resonator configuration—inherently a very low loss structure.

Many results in both Tl–Ca–Ba–Cu–O and Y–Ba–Cu–O filters have been reported and there have not been wide variances in insertion loss or bandwidth performance among the various materials. Again the key test is in power-handling ability, and here we have again seen some advantages in the Y–Ba–Cu–O family. A number of laboratories have reported 1% bandwidth filters at the X-band that have third-order intercepts of around 1 W when made with Y–Ba–Cu–O. Most of the Tl–Ca–Ba–Cu–O results (until recently) have had power-handling capacities

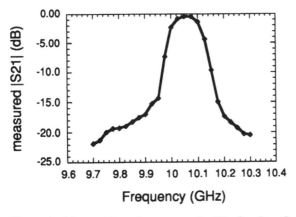

Figure 4 Measured performance of a Tl–Ca–Ba–Cu–O bandpass filter at 77 K designed for 1% bandwidth at 10 GHz. Note the low insertion loss; some groups have reported losses below 0.1 dB for similar structures.

one or two orders of magnitude lower. More heavily pinned Tl–Ca–Ba–Cu–O films grown recently (particularly 2122) have produced higher-powered filters.

D. Antennas and Related Components

As discussed previously, HTS components can perform useful functions with superior performance in many parts of a receiver front end. The next logical question involves the utility of HTS materials in the antenna and combiner circuitry itself. The major loss component in many normal metal microwave systems is in the matching networks [22,23]. In making the transition to very low impedance levels, very large reactive currents can exist. This is one reason why the efficiencies of conventional electrically small antennas are low. As discussed in the resonator section, superconducting matching networks are extremely useful in terms of low insertion loss. Several groups have done elaborate experiments on electrically small HTS antennas and matching networks with substantially improved efficiencies. Tl–Ca–Ba–Cu–O antennas have had efficiencies exceeding 40%, better than with any other material tested so far [22]. Power handling is still a problem, suggesting that it is less likely that the materials will be used in transmitting antennas.

III. JOSEPHSON JUNCTION-BASED ELECTRONICS

In the last few years, much progress has been made forming Josephson junctions from HTS materials. Most of the structures to date have used natural or engineered grain boundaries. The resulting junctions have very low capacitance (compared to the LTS tunnel junctions) and are hence nonhysteretic. Such junctions are very useful for applications such as SQUIDs and single flux quantum (SFQ) logic. In this section we begin with a brief review of junction properties before delving into HTS junction structure, performance, and applications.

A. Junction Basics

In this subsection we provide a little background information on Josephson junctions and their circuit properties. More complete coverage, including the physics of Josephson junctions, can be found in many books, such as those by Van Duzer and Turner [8] and Barone and Paterno [24]. The Josephson junction consists of two superconducting regions separated by a thin barrier, of which two forms are shown in Fig. 5. The barrier can take on many forms, including a thin dielectric layer, a constriction in a strip of superconductor, a semiconducting layer, or even a normal metal. Assuming that the distance between the superconducting regions is small enough, on the order of a coherence length, the electron wavefunctions in the two superconducting regions will be coupled. Depending on the barrier and its thickness, tunneling of electron pairs is allowed, and this results in the Josephson supercurrent. Normal particle tunneling is allowed under less severe restrictions. A pair of I–V curves are shown in Figs. 6 and 7. The curve in Fig. 6 is that of a hysteretic junction and is typical of LTS tunnel junctions. The supercurrent branch shown is that involving normal particle tunneling as well. At sufficiently high current, normal particle tunneling is all that is present. Important critical parameters are the critical current, the normal-state resistance, the subgap leakage resistance, and the gap voltage. The gap corresponds to that defined by the theory of Bardeen–

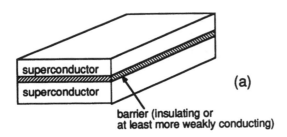

barrier (insulating or
at least more weakly conducting)

Weak link - either just very small or
poisoned in some way during growth
or processing

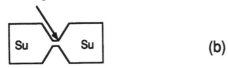

Figure 5 Common structurs of Josephson junctions: (a) a tunnel junction; (b) weak link. For the most part, the weak-link variety has been used to date in HTS materials. Since the coherence length is so small, some kind of additional weakening usually must take place either from grain boundaries or some kind of growth poisoning.

Cooper–Schrieffer (BCS) [8], and the observation of this kneelike structure in some HTS junctions suggests that BCS may apply in some context. Most HTS junctions to date have been of the nonhysteretic variety, as shown in Fig. 7. This is characteristic of junctions with low capacitance (e.g., the weak-link structures generally used for HTS junctions).

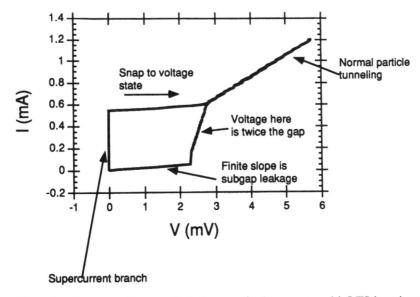

Figure 6 Theoretical hysteretic *I–V* curve (as is common with LTS junctions). The current drive is swept from 0 to 2 mA and back to 0, with the return path following the lower curve.

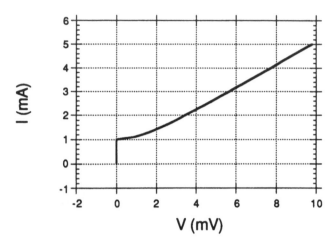

Figure 7   Theoretical nonhysteretic I–V curve which is similar to the experimental curves of many HTS grain boundary-based junctions.

A common equivalent circuit is the resistively shunted junction (RSJ) model shown in Fig. 8. If one allows the conductance to be nonlinear, one can do a reasonable job of modeling either type of junction. In the following equations, I and V are the current through and the voltage across the junction, ϕ the phase (wavefunction phase difference) across the junction, I_c the maximum critical current, C the junction capacitance, and G the junction conductance. The differential equation describing this circuit is

$$I = I_c \sin \phi + GV + C \frac{dV}{dt} \tag{4}$$

where the first Josephson relation, $I = I_c \sin \phi$ has been used for the current source and ϕ is the phase across the junction. Using the second Josephson relation [8],

$$\frac{\partial \phi}{\partial t} = \frac{2e}{\hbar} V \tag{5}$$

and a change of variable [$z = 2eI_ct/(\hbar G)$ and $\beta_c = 2eI_cC/(\hbar G^2)$], we get [8]

$$\frac{I}{I_c} = \beta_c \frac{d^2\phi}{dz^2} + \frac{d\phi}{dz} + \sin \phi \tag{6}$$

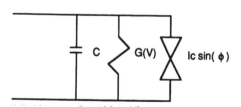

Figure 8   Resistively shunted junction (RSJ) equivalent circuit for a Josephson junction. Particularly for nonhysteretic junctions, the conductance must be considered a function of voltage for a reasonable fit.

In the limit of very low capacitance, β_c becomes small and the equation is directly soluble to give

$$V = \frac{I_c}{G} \sqrt{\left(\frac{I}{I_c}\right)^2 - 1}\ u(I - I_c) \tag{7}$$

which is the curve shown in Fig. 7 [$u(x)$ is the unit step function]. The RSJ model predicts other effects, including Shapiro steps. In Eq. (5) note that for any fixed voltage across the junction, there is a constant rate of change of the phase (i.e., oscillation frequency). Thus for any dc voltage across a junction, it is generating ac components. Hence if one applies a radio-frequency (RF) field of some frequency, it will match the Josephson frequency at some voltage and a corresponding dc voltage will be generated across the junction. If the applied frequency is f, then when a voltage $V = hf/(4\pi e)$ is across the junction, power exchange can take place. When the entire circuit model is included, steps occur in the I–V curve separated by this voltage. These are called Shapiro steps and are often used as an illustration that the Josephson effect is present. (This is not sufficient evidence: steps can arise from other effects [25].)

The effect of a static magnetic field on junction properties is also important. This dependence is related to the behavior observed in SQUIDs and is critical to understanding the long junction-based devices that are discussed later. It can be shown using the basic Josephson relations that if a magnetic field B_0 is applied in the y direction (current flow in the x direction) that the maximum critical current is

$$I_c(B_0) = \left| \int_{-\infty}^{\infty} J_c(z) e^{jSz}\ dz \right| \tag{8}$$

where $S = 4\pi e l B_0/h$ and l is an effective barrier thickness, which equals the physical barrier thickness plus twice the penetration depth. If J_c is assumed to be uniform, I_c will have a $\sin(x)/x$ dependence on applied B. The similarity between the measured dependence on B and the sine function is often used as a measure of junction uniformity. This relation is very handy in that given a measured B-field dependence, one can invert the transform to look at the distribution of current within the junction. In long junctions, the same types of effects occur, but complications arise. The junction current itself generates local magnetic fields, and any asymmetry in the applied field can have drastic effects that are actually useful for some device applications.

B. HTS Junction Structures and Performance

In the last few years, much progress has been made in generating useful HTS junctions [26–31]. Most of those studied to date are engineered grain boundaries capitalizing on the weak-link properties of the HTS grain boundaries. With few exceptions, HTS junctions are nonhysteretic because of the low capacitance in the structures used to date. At first this was seen as an unfortunate limitation, but recently the opinion has shifted to the view that this variety of junction is desired anyway for the majority of applications.

The first HTS junctions were patterned around existing grain boundaries. Although not very manufacturable, these were the first HTS "active devices." There have been an enormous variety of techniques for engineering the grain boundaries, and several are considered in this section. IBM's biepitaxial structures were among the first extensively studied and involved SrTiO₃ bicrystals [26]. The Y–Ba–Cu–O film grown on top of this then formed a matching grain boundary. Some extremely low-noise SQUIDs have been made with this technique. A similar but far more microelectronic technique developed by Conductus grows a misaligned MgO layer on part of the substrate [27]. Again, at the interface of misalignment, a grain boundary forms with high repeatability. This technique is sufficiently microelectronic in nature that Conductus has extended it to a 15-layer process flow. A third technique is that of step-edge junctions [28,29]. A pit is etched (usually through ion milling) into the substrate, typically LaAlO₃, prior to growth, and during growth, consistent junctions form at the step. These junctions have been made in both Y–Ba–Cu–O and Tl–Ca–Ba–Cu–O and have shown yields similar to those of the biepitaxy forms. Other techniques include weak-link formation by focused ion-beam etching and the selection of natural grain boundaries by pattern definition.

An extension of the step-edge technique has been used by Biomagnetic Technologies, Inc. and NIST to develop repeatable and uniform SNS Y–Ba–Cu–O junctions [30,31]. This structure purposely keeps the film much thinner than the step so that there is no connection between the two substrate mesas. A normal metal (Au, Ag, or sometimes alloys) is then deposited over the step to form the barrier.

Two example *I–V* curves of HTS junctions are shown in Fig. 9. The curves are of a Tl–Ca–Ba–Cu–O step-edge junction [29] at 77 K both with and without applied 93.6-GHz radiation. The first shows a reasonable RSJ fit and the other shows Shapiro steps at the appropriate $h\nu/2e$ positions. These results are material independent and illustrate that uniform junctions can be grown by a variety of methods. Further evidence is in the form of the field dependence of critical current

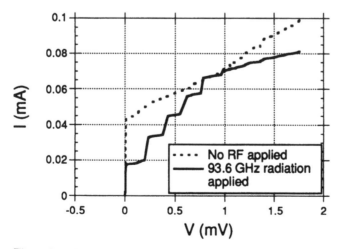

Figure 9 Measured *I–V* curves of a Tl–Ca–Ba–Cu–O step-edge junction with and without applied 93-GHz radiation.

is shown in Fig. 10. This Fraunhofer pattern is consistent with the result expected for a small-area uniform junction [8]. The dependence of Shapiro step height on radio-frequency (RF) applied field has also been examined by many researchers, with reproducible results consistent for RSJ small junctions [8]. This quality of magnetic field response (and better) has been observed with junctions made a variety of ways and with a variety of materials, suggesting that uniformity is less and less of a problem.

There appear to be a number of interesting differences between the grain boundaries in the various HTS materials. In point contact studies and junction attempts, the grain boundaries in Tl–Ca–Ba–Cu–O and in Bi–Sr–Ca–Cu–O have been observed to be more SIS-like than in Y–Ba–Cu–O. Also, the $I_c R_n$ products seen with Tl–Ca–Ba–Cu–O and Bi–Sr–Ca–Cu–O junctions have been considerably higher than in most Y–Ba–Cu–O junctions. The source of these differences is not yet completely understood but could have some very interesting application implications. Indeed, hysteretic junctions have been observed in both Bi–Sr–Ca–Cu–O and Tl–Ca–Ba–Cu–O systems but by very different techniques. In very elegant work by Varian, Bi–Sr–Ca–Cu–O/Bi–Sr–Cu–O/Bi–Sr–Ca–Cu–O trilayers are grown by MBE and produce true hysteretic junctions with a specific capacitance close to that calculated [32]. The care with which these structures could be grown is very important, and one hopes that it can be extended to all of the materials. In other work, capacitance has been added to Tl–Ca–Ba–Cu–O step-edge junctions through an added dielectric-normal metal cap to make them hysteretic [33].

C. Applications: SQUIDS

By far the most widespread Josephson application is the SQUID. This section is intended only to bring up the very basic group of applications and an estimate of the performance of Tl–Ca–Ba–Cu–O and other HTS SQUIDs to date. For more

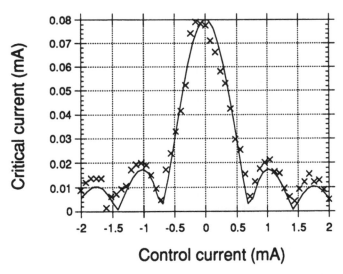

Figure 10 Measured magnetic field dependence of critical current data for a Tl–Ca–Ba–Cu–O step-edge junction at 77 K. The solid line is a best fit to the form $K \, |\sin(LI_{ctl})/(LI_{ctl})|$.

information, the interested reader is referred to one of many excellent treatments in the literature [8,34].

Because of their extreme sensitivity to small changes in magnetic field, SQUIDs are invaluable in a number of metrology applications, where their performance (sensitivity and noise primarily) is considerably better than that offered by conventional technology. Cardiography and encephalography are the most notable biomedical applications, and encephalography has one of the most stringent noise requirements among the applications. While Nb SQUIDs are used for both applications, HTS SQUIDs have achieved sensitivity sufficient for nondestructive evaluation and some cardiography applications [35]. There are literally hundreds of other applications, of which we mention only a few. Nondestructive evaluation of metals is of considerable importance (aging aircraft), and SQUID systems are currently being tested to map eddy current flow and to detect strain-induced anomalies [36]. Susceptometry for materials evaluation and the detection of seismomagnetism for geophysical applications are among other current areas of investigation.

The basic dc SQUID structure consists of two junctions in a superconducting loop with bias current split between them. As flux is admitted to the loop structure, the phases of the individual junctions adjust, thus changing the total critical current. For a SQUID with no inductance, the total critical current is given by

$$I_{\text{total critical}}(\Phi_{\text{applied}}) = \sqrt{(I_{c1} - I_{c2})^2 + 4I_{c1}I_{c2}\cos^2\left(\frac{\pi\Phi_{\text{applied}}}{\Phi_0}\right)} \qquad (9)$$

where I_{c1} and I_{c2} are the critical currents of the two junctions and Φ_{applied} is the applied flux. A more realistic dependence of critical current on flux is shown in Fig. 11. This plot assumes symmetric junctions but does include the effects of loop inductance. One of the most critical parameters for applications (because it helps determine sensitivity) is the depth of modulation available, which is about 30%

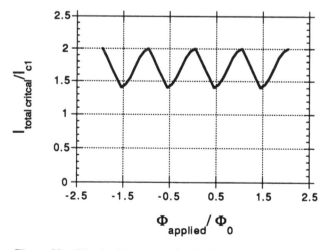

Figure 11   Classical I_c versus Φ SQUID response (symmetric junctions but including loop inductance).

for the example of Fig. 11. Practially, a SQUID has a larger pickup coil coupled to it, to increase sensitivity. A transformer, usually multilayer, is then used to match impedances between the SQUID loop and the pickup coil. Excellent multilayer Y–Ba–Cu–O structures incorporating SQUIDs, pickup loops, and so on, have already been obtained through laser ablation deposition [1].

Many HTS SQUIDs have been made to date, starting with junctions made by breaking bulk samples. To date, HTS SQUID noise (using any of the materials) has not been better than that of the best Nb SQUIDs, but there have been dramatic improvements [37] in 1990 and 1991. The lowest-noise HTS SQUID to date has been a Tl–Ca–Ba–Cu–O unit, but even this noise level is not yet adequate for encephalography. Strides are also being made in improving the noise performance of the pickup coils and other ancillary parts of the system [1].

D. Other Applications

As of this writing, relatively few HTS junction applications aside from SQUIDs have been demonstrated, but many laboratories are pursuing them. The difference between SQUID and non-SQUID applications is fine, so the distinction here is somewhat arbitrary. We separate some of the more obvious applications into three groups: single flux quantum and related logic families, microwave and millimeter-wave analog circuits, and long junction-based devices.

There has been much work in recent years on SFQ [38,39]. This logic is based on shuttling flux quanta about a circuit to perform the desired logic operations. Shift registers and analog-to-digital (A/D) converters are natural applications. Many high-speed circuits have already been built in Nb, including A/D converters operating [40] at over 40 GHz. While much development work remains to be done, probably in LTS, SFQ logic and its cousins are reasonably well suited to HTS materials because of the need for nonhysteretic junctions. Aside from uniformity concerns, another potential problem is in inductance. The inductance levels must be carefully controlled, within 20% or so [38], in these circuits. In HTS this will be made difficult by the high and sometimes variable penetration depth. Other logic families have already been demonstrated in HTS, including a more direct SQUID-based system [41]. In most cases the performance of these logic families vastly exceeds that of conventional technologies in the areas of power dissipation and speed.

There are many nonlinear high-frequency applications possible, including SIS mixers [42] and parametric amplifiers [43], both of which have been successful with LTS materials. Both have enormous noise advantages at LTS temperatures, which raised the question of their utility with HTS materials. The Josephson parametric amplifiers with HTS materials could still be interesting because of speed and noise advantages over conventional electronics at 77 K.

Among linear high-frequency applications are phase shifters, where the controllable inductance of a junction is used to alter the phase velocity of an artificial transmission line in which the junctions are embedded. A control line can provide a local magnetic field that alters the junction/SQUID inductance. Devices have been demonstrated showing reasonably controllable phase shift at frequencies of interest [44] in Y–Ba–Cu–O. Presumably, such structures are applicable to the other materials, although it would be interesting to see how the very different junctions produced in the Tl–Ca–Ba–Cu–O system would perform in such an application. The primary advantages of such circuits over their conventional counterparts are the low insertion loss and wide bandwidths that are possible.

One other category of applications is in controlled long junctions [45,46]. These devices, exploited in the LTS community, are potentially useful analog high-speed active devices. A long junction, one dimension larger than the penetration depth in the junction, is characterized by the ability to move vortices through the length of the junction. An external field or bias can strongly affect the generation of these vortices, hence providing a mechanism of control. The structure of such a device, made using Tl–Ca–Ba–Cu–O step-edge junctions and a control line, is shown in Fig. 12, along with a view of I–V curves as a function of control current. Treating the current suppression as an output variable, one can see real current gain is available. Like many other superconducting devices (see the next section), this one

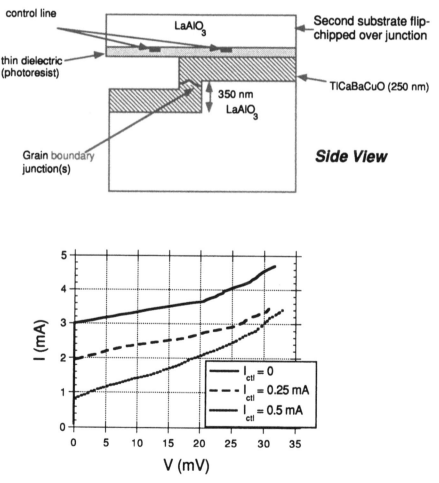

Figure 12 Structure of a long junction-based device and some I–V curves showing the effect of a control current. Note the real current gain available at small signal levels.

is plagued by a very low input resistance which makes matching difficult. With nearly perfect matching, however, real microwave power gain is available, albeit in a narrow bandwidth. Centered at 7 GHz with a bandwidth of about 100 MHz, 8 dB gain has been demonstrated. Potential significant advantages over conventional electronics are higher speed, low noise and impedance levels that are useful for specialized applications.

IV. NON-JOSEPHSON ACTIVE DEVICES

To date, relatively few non-Josephson active HTS devices have been attempted in the community. Three types of devices are considered in this section. One type of device that exploits the electric field dependence of the transport in a superconducting thin film, termed a SuFET [47], is considered only briefly because it has not, to my knowledge, been implemented in Tl–Ca–Ba–Cu–O, but it does promise to help in understanding film transport. The SuFET, which has been explored by workers at the University of Maryland and others, is pictured in Fig. 13. It consists of a thin, narrow superconducting bridge with a field electrode separated from it by a high-dielectric-constant insulator ($SrTiO_3$). The field applied by the "gate" electrode modifies the carrier density in the superconductor and hence its terminal characteristics. With such a dielectric requirement, high speed is unlikely, but small signal gain has been demonstrated. Of considerable interest is its ability to map density-of-states effects.

Another device type is a magnetic field–based device called the superconducting flux-flow transistor (SFFT). Although we consider one embodiment of that device, numerous workers have investigated the control of magnetoresistance. A third type of device worthy of a few sentences is the cryotron, a thermal switch. Although hardly new, it has some interesting HTS applications where great speed is not required (e.g., microwave signal flow control).

A. Flux-Based Three-Dimensional Devices

There is a class of devices analogous to the long junctions described earlier that have shown promise for some applications (major advantages possibly in speed range and signal levels). The SFFT consists of one or more weak links that are

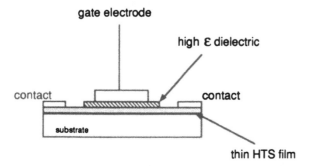

Figure 13 Structure of an electric-field-controlled device called the SuFET. An electric field modulates the density of state in the superconductor, altering its terminal behavior.

large relative to the coherence length but of the same scale as the effective penetration depth (see Fig. 14) [48,49]. The links are typically weak enough to allow for easy flux flow of Abrikosov vortices. Again an external magnetic field modulates this flux flow and hence the terminal characteristics. Since these devices are much larger than the coherence length, effects such as Shapiro steps are typically not observed, and critical current modulation does not follow Fraunhofer-like patterns. Useful characteristics are obtained and a sample $I-V$ curve is shown in Fig. 14. At low current drives, the links remain superconducting. Above the critical current, flux is admitted to the link system. By virtue of a Lorentz-type force, the Abrikosov vortices move through the link system and generate a voltage. At sufficiently high current drives, the links go entirely normal and the $I-V$ curve snaps to a resistive branch that, when extrapolated, passes through the origin.

The effect of an increase in control current is an increase in output voltage rather than a change in current (depending on loading, of course). A comparison of two equivalent circuits of the SFFT and the long junction device is shown in Fig. 15. Both devices show real gain and both have the very low input impedances of a near short circuit. The output impedances are similar when using HTS junctions for the long junction devices. As with the controlled long junctions, gain is available (over 15 dB demonstrated at X-band frequencies) with proper matching. The experiments are not yet conclusive as to the noise performance of these two devices, although it is suspected that the junction-based device will be less noisy since the Josephson vortex flow is more coherent. Other circuits, such as low-insertion-loss phase shifters [49], logic gates, and transimpedance amplifiers [50] have been demonstrated with this device. The transimpedance amplifiers are of some interest as a buffer between Josephson or other current-driven circuitry and conventional electronics. A junction can easily source on the order of 1 mA into the low-impedance control line of an SFFT (or long junction). The output impedance of the SFFT is at least several ohms, and about 200 mV can be supplied to the gate of a FET or similar device at high speed.

SFFTs have been made from at least Nb, Nb_3Sn, Tl–Ca–Ba–Cu–O, and Y–Ba–Cu–O. The key to high transresistance, large output voltage swing, and high speed is easy flux flow. Low H_{c1} values, low pinning, low viscosity, and good surface morphology all contribute to this. Most of the Y–Ba–Cu–O materials tested have somewhat higher pinning than the Tl–Ca–Ba–Cu–O materials, so one would expect slower device performance, and this was indeed observed. Devices made with weakly pinned Y–Ba–Cu–O performed very well, as is consistent. Good surface morphology helps keep pinning down as the bridges are thinned. It is anticipated that links below 0.25 μm can be used before the fundamental flux flow physics starts changing [46], and with these dimensions, operating frequencies near 150 to 200 GHz seem plausible.

B. HTS Cryotrons

The cryotron is a very old concept [51] and for reasons of speed and thermal control, was largely abandoned with the advent of the Josephson junction in the 1960s. There are some applications where those disadvantages are outweighted by the simplicity and isolation they offer. It has been seen earlier that HTS filters and delay lines offer performance considerably better than that available with conven-

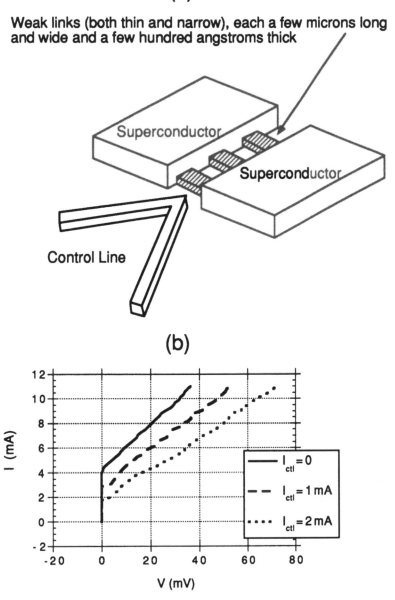

(a)

Weak links (both thin and narrow), each a few microns long and wide and a few hundred angstroms thick

Superconductor

Superconductor

Control Line

(b)

Figure 14 Structure (a) and some *I–V* curves (b) of the SFFT. The device consists of a parallel array of weak links operated in a non-Josephson flux flow mode (the links are much larger than the coherence length). The *I–V* curves show the effect of a control current. The control line need not be superconducting nor in the same plane as the links (overlay control lines increase sensitivity at the expense of crosstalk).

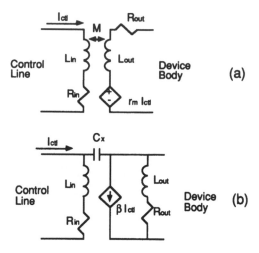

Figure 15 Comparison of the equivalent circuits of the SFFT (a) and the controlled long junction devices (b). Although somewhat arbitrary and dependent on loading, the circuits illustrate the predominant types of gain available and the relevant parasitics.

tional technology. High-performance switched delay lines and filter banks are needed, and for integration reasons it would be highly desirable to have HTS switches. The speed requirements are normally minimal, and cryotrons can have excellent insertion loss and isolation properties.

The object of the cryotron is to drive a section of superconductor normal through application of strong magnetic field or heat or both. Such a structure is shown in Fig. 16, although it is not unique [52]. Insertion loss in Tl–Ca–Ba–Cu–O

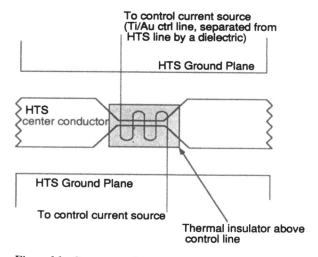

Figure 16 Structure of an HTS cryotron. The device relies on heat and magnetic field from the normal metal control line to drive the HTS bridge normal. While only fast enough for some signal control applications, the insertion loss and isolation properties are good.

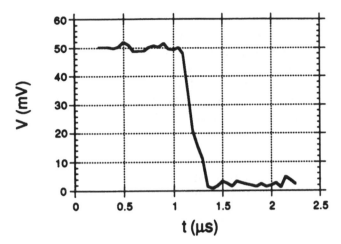

Figure 17 Transient response of a cryotron to a 10-mA step applied to the control line. A 50-mV dc supply is applied between one side of the switch and ground, and the signal between the other switch end and ground is measured. When the control current is removed, the return to the "on" state takes on the order of 1 μs.

cryotrons has been demonstrated to be <0.5 dB with isolation >35 dB at X-band frequencies. The transient response of the switch to a square-wave current on the control line is shown in Fig. 17. Without any significant thermal engineering, the response time is better than a few microseconds.

The usual differences between Tl–Ca–Ba–Cu–O and Y–Ba–Cu–O materials in applications surface again. With low pinning, Tl–Ca–Ba–Cu–O power handling is limited to about −5 dBm, while with well-pinned Y–Ba–Cu–O, the limit is closer to +10 dBm before self-switching starts to occur. As a trade-off, switching is a bit slower and more dc power is required for Y–Ba–Cu–O switches (at the same reduced temperatures) presumably because there is less help from the magnetic field in inducing the phase transition.

V. CONCLUSIONS

In this chapter we have covered considerable ground in the area of HTS/Tl–Ca–Ba–Cu–O electronics, which emphasized three main points: (1) HTS electronics is full of interesting ideas, work, and results that should offer advantages over conventional technologies (although the markets are now small in almost all cases); (2) differences between Tl–Ca–Ba–Cu–O and the other HTS materials as they relate to applications are not always understood (grain boundary and junction formations); and (3) apparent differences in performance may sometimes be more a result of process development than materials fundamentals (power handling in particular). Many devices and applications have had to be omitted for considerations of space, and presumably within the next few years many more applications and circuits will be developed.

ACKNOWLEDGMENTS

In preparing this chapter, the author is indebted to many persons, including workers at many laboratories who have discussed their accomplishments. For the work done at Sandia, I would like to thank my collaborators D. S. Ginley, V. M. Hietala, C. P. Tigges, T. E. Zipperian, and T. A. Plut. The work at Sandia has been supported in part by the Office of Naval Research, the Strategic Defense Initiative, and the U.S. Department of Energy under contract DE-AC04-76DP00789.

REFERENCES

1. P. C. Wellstood, J. J. Kingston, A. H. Milkich, M. J. Ferrari, J. Clarke, M.S. Colclough, K. Char,and G. Zaharchuk, *IC SQUID '91*, Berlin, June 18–21, 1991, to be published in proceedings by Springer-Verlag, Berlin, 1992.
2. D. K. Fork, D. B. Fenner, A. Barrera, J. M. Phillips, T. H. Geballe, G. A. N. Connell, and J. B. Boyce, *IEEE Trans. Appl. Supercond.* 1, 67 (1991).
3. J. Keßler, E. Dill, and P. Russer, *IEEE Microwave Guided Wave Lett.* 2, 6 (1992).
4. R. W. Simon, C. E. Platt, A. E. Lee, G. S. Lee, K. P. Daly, M. S. Wire, and J. A. Luine, *Appl. Phys. Lett* 53, 2677 (1988).
5. D. Nghiem, J. T. Williams, and D. R. Jackson, *IEEE Trans. Microwave Theory Tech.* 39, 1553 (1991).
6. W. G. Lyons, R. S. Withers, J. M. Hamm, A. C. Anderson, P. M. Mankiewich, M. L. O'Malley, R. E. Howard, and N. Newman, *IEEE 1991 Int. Microwave, Symp. Proc.* 3, 1227 (1991).
7. R. F. Harrington, *Time-Harmonic Electromagnetic Fields*, McGraw-Hill, New York, 1961, chaps. 2 and 7.
8. T. Van Duzer and C. W. Turner, *Principles of Superconductive Devices and Circuits*, Elsevier, New York, 1981), chaps. 2 to 5.
9. Z. Y. Shen, P. S. W. Pang, W. L. Holstein, C. Wilker, S. Dunn, D. W. Face, and D. B. Laubacher, *IEEE 1991 Int. Microwave Symp. Proc.* 3, 1235 (1991).
10. W. G. Lyons, R. R. Bonetti, A. E. Williams, P. M. Mankiewich, M. L. O'Malley, J. M. Hamm, A. C. Anderson, R. S. Withers, A. Meulenberg, and R. E. Howard, *IEEE Trans. Magn.* 27, 2537 (1991).
11. E. K. Track, G. K. G. Hohenwarter, L. Madhavrao, R. Patt, R. E. Drake, and M. Radparvar, *IEEE Trans. Magn.* 27, 2936 (1991).
12. L. C. Bournes, R. B. Hammond, M. Robinson, M. M. Eddy, W. L. Olson, and T. W. James, *Appl. Phys. Lett.* 56, 2333 (1990).
13. N. Newman, B. F. Cole, S. M. Garrison, K. Char, and R. C. Taber, *IEEE Trans. Magn.* 27, 1276 (1991).
14. J. S. Martens, V. M. Hietala, D. S. Ginley, T. E. Zipperian, and G. K. G. Hohenwarter, *Appl. Phys. Lett.* 58, 2543 (1991).
15. D. S. Ginley, J. F. Kwak, E. L. Venturini, B. Morosin, and R. J. Baughman, *Physica C* 160, 42 (1989).
16. M. L. Chu, H. L. Chang, C. Wang, J. Y. Juang, T.M. Uen, and Y. S. Gou, *Appl. Phys. Lett.* 59, 1123 (1991).
17. W. G. Lyons, R. R. Bonetti, A. E. Williams, P. M. Mankiewich, M. L. O'Malley, J. M. Hamm, A. C. Anderson, R. S. Withers, A. Meulenberg, and R. E. Howard, *IEEE Trans. Magn.* 27, 2537 (1991).
18. S. H. Talisa, M. A. Janocko, C. Moskowitz, J. Talvacchio, J. Billing, R. Brown, D. C. Buck, C. K. Jones, B. R. McAvoy, G. R. Wagner, and D. H. Watt, *IEEE Trans. Microwave Theory Tech.* 39, 1448 (1991).
19. J. A. Curtis and S. J. Fiedziuszko, *Appl. Microwave* 3(3), 86 (1991).

20. A. P. S. Khanna and M. Schmidt, *IEEE 1991 Int. Microwave Symp. Proc.* 3, 1239 (1991).
21. A. Lauder, *SC Global '92*, San Diego, Calif., Jan. 29–31, 1992.
22. R. J. Dinger, D. R. Bowling, and A. M. Martin, *IEEE Trans. Microwave Theory Tech.* 39, 1243 (1991).
23. H. Chaloupka, N. Klein, M. Peiniger, H. Piel, A. Pischke, and G. Splitt, *IEEE Trans. Microwave Theory Tech.* 39, 1513 (1991).
24. A. Barone and G. Paterno, *Physics and Applications of the Josephson Effect*, Wiley, New York, 1982.
25. K.K. Likharev, *Rev. Mod. Phys.* 51, 101 (1979).
26. D. Dimos, P. Chaudhari, J. Mannhart, and F. K. LeGoues, *Phys. Rev. Lett.* 61, 219 (1988).
27. K. Char, M. S. Colclough, S. M. Garrison, N. Newman, and G. Zaharchuk, *Appl. Phys. Lett.* 59, 733 (1991).
28. K. P. Daly, W. D. Dozier, J. F. Burch, S. B. Coons, R. Hu, C. E. Platt, and R. W. Simon, *Appl. Phys. Lett.* 58, 543 (1991).
29. J. S. Martens, T. E. Zipperian, G. A. Vawter, D. S. Ginley, V. M. Hietala, and C. P. Tigges, *Appl. Phys. Lett.* 60, Mar. 2, 1992.
30. M. S. DiIorio, S. Yoshizumi, K. Y. Yang, J. Zhang, and M. Maung, *Appl. Phys. Lett.* 58, 2552 (1991).
31. R. H. Ono, J. A. Beall, M. W. Cromar, T. E. Harvey, M. E. Johansson, C. D. Reintsema, and D. A. Rudman, *Appl. Phys. Lett.* 59, 1126 (1991).
32. J. N. Eckstein, I. Bozovoc, M. Klausmeier-Brown, and G. Virshup, *Fall 1991 MRS Meeting*, Boston, Dec. 1991.
33. J. S. Martens, V. M. Hietala, T. E. Zipperian, G. A. Vawter, D. S. Ginley, C. P. Tigges, T. A. Plut, and G. K. G. Hohenwarter, *Appl. Phys. Lett.* 60, Feb. 24, 1992.
34. Chapters by J. Clarke, G. L. Romani, and G. B. Donaldson, in *Superconducting Electronics*, ed. H. Weinstock and M. Nisenoff, Springer-Verlag, Berlin, 1989.
35. A. H. Miklich, J. J. Kingston, F. C. Wellstood, J. Clarke, M. S. Colclough, K. Char, and G. Zaharchuk, *Appl. Phys. Lett.* 59, 988 (1991).
36. H. Weinstock, R. B. Mignogna, R. S. Schechter, and K. E. Simmonds, *Proc. IC SQUID '91*, Springer-Verlag, Berlin, 1992.
37. A. Braginski, *Proc. IC SQUID '91*, Springer-Verlag, Berlin, 1992.
38. K. K. Likharev and V. K. Semenov, *IEEE Trans. Appl. Supercond.* 1, 3 (1991).
39. C. A. Hamilton and K. C. Gilbert, *IEEE Trans. Appl. Supercond.* 1, 157 (1991).
40. E. Stebbins, *Superconducting Digital Circuits and Systems*, Washington, D.C., Sept. 1991.
41. S. M. Schwarzbek, R. A. Davidheiser, G. R. Fisher, J. A. Luine, and N. J. Schneier, *Appl. Phys. Lett.* 59, 866 (1991).
42. C. A. Mears, Qing Hu, P. L. Richards, A. H. Worsham, D. E. Prober, and A. V. Raisanen, *IEEE Trans. Magn.* 27, 3363 (1991).
43. B. Yurke, P. G. Kaminsky, R. E. Miller, E. A. Whittaker, A. D. Smith, A. H. Silver, and R. W. Simon, *Phys. Rev. Lett.* 60, 764 (1988).
44. C. M. Jackson and D. J. Durand, *IEEE 1991 Int. Microwave Symp. Proc.* 1, 165 (1991).
45. B. J. Van Zeghbroeck, *Appl. Phys. Lett.* 42, 736 (1983).
46. K. K. Likharev, V. K. Semenov, O. V. Snigirev, and B. N. Todorov, *IEEE Trans. Magn.* 15, 420 (1979).
47. X. X. Xi, Q. Li, C. Doughty, C. Kwon, S. Bhattacharya, A. T. Findikoglu, and T. Venkatesan, *Appl. Phys. Lett.* 59, 3470 (1991).
48. J. S. Martens, D. S. Ginley, J. B. Beyer, J. E. Nordman, and G. K. G. Hohenwarter, *IEEE Trans. Appl. Supercond.* 1, 95 (1991).

49. J. S. Martens, V. M. Hietala, T. E. Zipperian, D. S. Ginley, C. P. Tigges, and J. M. Phillips, *IEEE Trans. Microwave Theory Tech.* 39, 2018 (1991).
50. J. S. Martens, D. S. Ginley, J. B. Beyer, J. E. Nordman, and G. K. G. Hohenwarter, *IEEE Trans. Magn.* 27, 3284 (1991).
51. V. L. Newhouse, *Applied Superconductivity*, Wiley, New York, 1964, pp. 155–169.
52. J. S. Martens, V. M. Hietala, T. E. Zipperian, D. S. Ginley, and C. P. Tigges, *IEEE Microwave Guided Wave Lett.* 1, 291 (1991).

14
TBCCO Ag-Sheathed Tapes

Koji Tada, Hiromi Takei, and Yasuko Torii
Sumitomo Electric Industries, Ltd., Osaka, Japan

I. INTRODUCTION

Since the discovery of superconductivity in the Tl–Ba–Ca–Cu–O system, a considerable number of superconducting phases have been reported in the Tl-based system. Single- and double-Tl–O-layered structures with the general formula $Tl_mBa_2Ca_{n-1}Cu_nO_{2n+m+2}$ ($m = 1, 2$; $n = 1$ to 6), where m and n are the number of Tl–O and Cu–O layers, respectively, have been obtained in the Tl–Ba–Ca–Cu–O system [1–5]. The critical temperature T_c increases with a rise in the number of Cu–O layers, reaching a maximum value of 122 K at $n = 4$ for the single-Tl–O-layer structure and of 127 K at $n = 3$ for the double-Tl–O-layer structure. In the Tl–Sr–Ca–Cu–O system, superconducting phases with the single T1–O layered structures $TlSr_2Ca_{n-1}Cu_nO_{2n+3}$ ($n = 1, 2,$ and 3) have been reported [6–8]. It is known that the addition of Pb and Bi stabilizes these superconducting phases, and a maximum T_c of 122 K is obtained at the phase of $n = 3$.

The superconductors of Tl-based systems have an excellent potential for practical applications based on their high critical temperature. For practical use in the fields of power transmission cables, high field magnets, and so on, a critical current density of above $10^4 A/cm^2$ is required. At the present time it is necessary to ensure that the current transportation property of the tape reaches this required level. While the J_c value at 77 K without an external magnetic field for short samples of Ag-sheathed tapes using high-T_c oxides of Tl–Bi–Sr–Ca–Cu–O, Tl–Ba–Sr–Ca–Cu–O, and Bi–Pb–Sr–Ca–Cu–O system exceeds $10^4 A/cm^2$, this density with a magnetic field is not sufficiently large [9–11].

In this chapter we report on the preparation of Pb-stabilized Tl–Ba–Ca–Cu–O superconductors with a single Tl–O layer structure and of Ag-sheathed tape using the oxide powder of $(Tl,Pb)Ba_2Ca_3Cu_4O_x$ phase.

II.   EXPERIMENTAL PROCEDURES

Bulk samples of the Tl–Pb–Ba–Ca–Cu–O system were prepared by the following method [4]. First, appropriate amounts of Tl_2O_3, PbO, BaO, CaO, and CuO powders were mixed well and formed into pellets under 100 kg_f/cm^2 pressure. These pellets were wrapped in Au foil, sintered at 1130 to 1170 K in an O_2 gas flow for 10 to 30 h and then cooled to room temperature.

The Ag-sheathed tapes are fabricated as follows [12]. The sintered bulk sample of the $(Tl,Pb)Ba_2Ca_3Cu_4O_x$ phase was ground into powder again. Then the superconducting powder was put into silver tubes (6 mm in diameter, 1 mm in thickness) and cold worked into tapes 3 mm in width and 0.15 mm in thickness. Specimens of the tapes were finally heat treated at around 1070 K.

The electrical resistivity of the bulk samples was measured by the standard four-probe method and dc magnetic susceptibility by a superconducting quantum interference device (SQUID) magnetometer. The transport critical current I_c of the tape samples was measured at 77 K by the four-probe method. The crystal structure was analyzed through the x-ray powder diffraction method. X-ray powder data were also refined using the Rietveld method. An energy-dispersive x-ray spectrometer (EDAX PV9900) was used to measure the atomic concentration of each phase found in the samples. The microstructure was observed using a scanning electron microscope (SEM) with an electron acceleration voltage of 25 kV and a transmission electron microscope (TEM) with a voltage of 200 kV.

III.   SYNTHESIS OF $(Tl,Pb)Ba_2Ca_{n-1}Cu_nO_{2n+3}$ SUPERCONDUCTING PHASES

The relationship between the critical temperature of zero resistivity T_{ci} and a nominal composition of Tl/Pb/Ba/Ca/Cu of $(1 - x):x:1:3:3$, where $0 \leq x \leq 1$, is

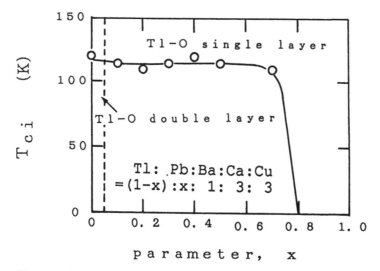

Figure 1   Relation between critical temperature T_{ci} and parameter x of nominal composition of Tl/Pb/Ba/Ca/Cu of $(1 - x):x:1:3:3$.

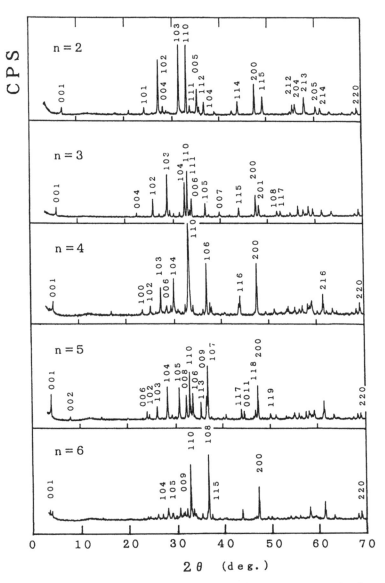

Figure 2 X-ray diffraction patterns of the superconducting tetragonal phases $(Tl,Pb)Ba_2Ca_{n-1}Cu_nO_x$ ($n = 2$ to 6).

shown in Fig. 1. Superconducting phases with T_{ci} higher than 110 K form until the parameter x is 0.7. From the results of x-ray diffraction, it is known that the double-Tl–O-layered structure of $Tl_2Ba_2Ca_2Cu_3O_x$ phase ($a = 3.85$ Å, $c = 35.9$ Å) forms in the sample without Pb ($x = 0$), and even a small substitution of Pb for Tl in the nominal composition makes the double-layered structure change to a single-layered structure. When almost single phases of $(Tl,Pb)Ba_2Ca_{n-1}Cu_nO_x$ ($n = 2$ to 6) are prepared by optimizing the synthetic conditions of the nominal composition and heat treatment, the number of Cu–O layers tends to increase with a small amount of additional Pb atoms and a higher sintering temperature.

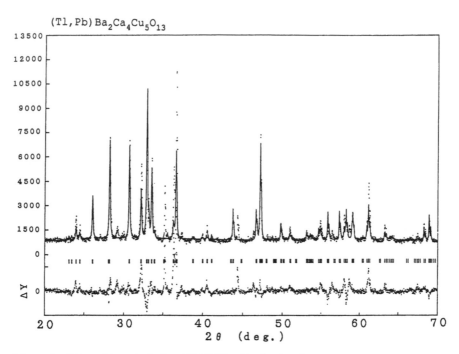

Figure 3   Rietveld analysis for $(Tl,Pb)Ba_2Ca_4Cu_5O_x$ phase.

X-ray diffraction patterns of the superconducting phases $(Tl,Pb)Ba_2Ca_{n-1}Cu_nO_x$ (n = 2 to 6) are shown in Fig. 2. The results of the Rietveld analysis for a $(Tl,Pb)Ba_2Ca_4Cu_5O_x$ phase using the initial structure model proposed by Ihara et al. [3] are shown in Fig. 3 and Table 1. As shown in Fig. 4, the simple tetragonal lattice structure with sixfold simple tetragonal subcells ($P4/mmm$), the four middle parts of which consist of Cu and Ca atoms, is confirmed between the R factors of R_{wp} = 0.19 and R_p = 0.13. Relatively high R factors may be due to intergrowths

Table 1   Structure Parameters of $(Tl,Pb)Ba_2Ca_4Cu_5O_x$ Phase[a]

Atom	Site	x	y	z	g
(Tl,Pb)	1a	0.0	0.0	0.0	0.78 (7)[b]
Ca(1)	2h	0.5	0.5	0.422(4)	1.0
Ca(2)	2h	0.5	0.5	0.281(4)	1.0
Ba	2h	0.5	0.5	0.124(1)	1.0
Cu(1)	2g	0.0	0.0	0.215(3)	1.0
Cu(2)	2g	0.0	0.0	0.358(3)	1.0
Cu(3)	1b	0.0	0.0	0.5	1.0
O(1)	1c	0.5	0.5	0.0	1.0
O(2)	2g	0.0	0.0	0.141(13)	1.0
O(3)	4i	0.0	0.5	0.226(10)	1.0
O(4)	4i	0.0	0.5	0.358(11)	1.0
O(5)	2e	0.0	0.5	0.5	1.0

[a]R_{wp} = 18.85, R_p = 13.42, R_E = 2.84, R_I = 20.45, R_F = 13.55.
[b]The occupation factor g of (Tl,Pb) is only refined.

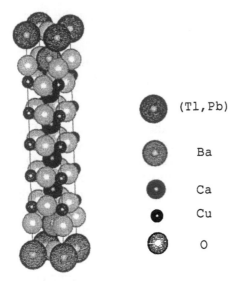

Figure 4 Crystal structure of $(Tl,Pb)Ba_2Ca_4Cu_5O_x$ phase.

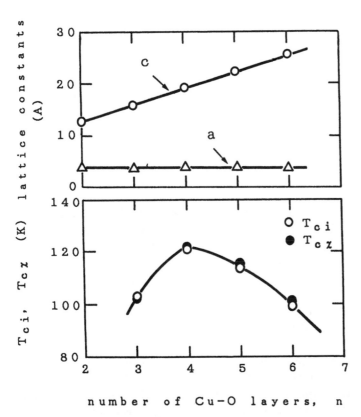

Figure 5 Critical temperatures of T_{ci} and T_{cx} and a,c-axis lattice constants as a function of the number of Cu–O layers n.

of other phases in the sample. The lattice constants, calculated using the least-squares method for the tetragonal phases of $(Tl,Pb)Ba_2Ca_{n-1}Cu_nO_x$, are shown as a function of the number of Cu–O layers n in Fig. 5. The c-axis lattice constant is in proportion to n, obeying an equation of $c = 6.38 + 3.19n$, because the tetragonal crystal structures are composed of two outer subcells and ($n = 1$) inner subcells.

Since the x-ray atomic scattering factors for Tl and Pb are very similar, it is not possible to differentiate between these two atoms in the crystal structure. The chemical composition of the observed phases was decided by EDX as being $Tl_{0.70}Pb_{0.21}Ba_{2.0}Ca_{1.9}Cu_{2.9}O_x$, $Tl_{0.5}Pb_{0.4}Ba_{2.0}Ca_{2.8}Cu_{3.9}O_x$, $Tl_{0.86}Pb_{0.27}Ba_{2.0}Ca_{4.1}Cu_{5.0}O_x$, and $Tl_{0.93}Pb_{0.26}Ba_{2.0}Ca_{4.9}Cu_{5.8}O_x$ for the tetragonal phases with three to six Cu–O layers. Atomic ratios of the sum of Tl and Pb to Ba, Ca, and Cu coincide with those of the stoichiometric compounds, so Pb possibly replaces Tl in the crystal lattice. The replacement of Tl with Pb is thought to stabilize the single-Tl–O-layered structure.

The temperature dependence of the electrical resistivity and the dc suscepti-bility of $(Tl,Pb)Ba_2Ca_{n-1}Cu_nO_x$ ($n = 3$ to 6) are shown in Fig. 6. The samples of $(Tl,Pb)Ba_2Ca_2Cu_3O_x$ and $(Tl,Pb)Ba_2Ca_5Cu_6O_x$ show two-step transitions, because these contain a small amount of $(Tl,Pb)Ba_2Ca_3Cu_4O_x$ and $(Tl,Pb)Ba_2Ca_4Cu_5O_x$, respectively. The critical temperature T_{ci} of zero resistivity and T_{cx} at the onset temperature, showing diamagnetism, are shown in Fig. 5 also as a function of the number of Cu–O layers n. With an increase in n, both critical temperatures T_{ci} and T_{cx} rise, reaching a maximum value of 121 K at $n = 4$ and then fall with any further increase in n. We adopted the $(Tl,Pb)Ba_2Ca_3Cu_4O_x$ phase for preparing the Ag-sheathed tapes because it has the highest critical temperature.

IV. PREPARATION OF $(Tl,Pb)Ba_2Ca_3Cu_4O_x$ Ag-SHEATHED TAPES

The external view of the Ag-sheathed tape using the superconducting oxide powder of $(Tl,Pb)Ba_2Ca_3Cu_4O_x$ phase is shown in Fig. 7. The highest J_c value of the tape at 77 K without an external magnetic field reaches 10,800 A/cm². The magnetic field dependence of J_c for the 10,000-A/cm²-class tape is shown in Fig. 8. Mea-surement of the transport critical current was carried out at 77 K between the external magnetic field range of zero to 2.5 T. The direction of the magnetic field applied perpendicularly to the transport current were both parallel and normal to the wide plane of the tapes. The critical current density is defined as current density at the point when an electrical resistivity of 10^{-10} Ω·cm is generated in the super-conductor. The J_c value of the tape decreases rapidly to 0.1 T; however, a decrease in J_c past 0.1 T is not as extreme, as shown in the figure. The direction of the magnetic field against the plane of the tape does not considerably influence the J_c values. The pinning force of the tape, F_p ($= J_c x B$), is shown in Fig. 9 as a function of the magnetic field. The value of F_p increases in proportion to the magnetic field up to 2.5 T.

A scanning electron micrograph of the superconducting core of the tape, after peeling the Ag sheath off, is shown in Fig. 10. A densificated structure of fine grains is seen at the surface of the core, and any defects such as microcracks are not observed. Figure 11 shows transmission electron micrographs of the tape spec-imen. The observed direction is perpendicular to the wide plane of the tape. The

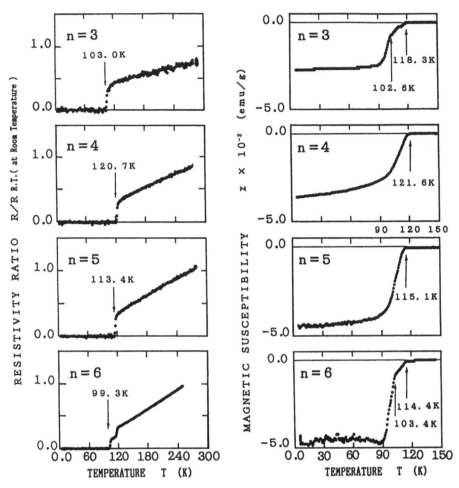

Figure 6 Temperature dependency of electrical resistivity and dc susceptibility of $(Tl,Pb)Ba_2Ca_{n-1}Cu_nO_x$ ($n = 3$ to 6) phases.

superconducting core is composed of randomly oriented, very fine grains about 0.5 μm in diameter. As shown in the magnified photograph, the grains of the core are a homogeneous $(Tl,Pb)Ba_2Ca_3Cu_4O_x$ phase that scarcely contains the intergrowth layers of other phases. This isotropic substructure causes a small influence of the magnetic field direction on the J_c, and the boundaries of the fine grains are thought to be one explanation for the pinning sites of magnetic flux lines.

V. SUMMARY

Almost single phases of $(Tl,Pb)Ba_2Ca_{n-1}Cu_nO_x$ ($n = 2, 3, 4, 5,$ and 6) were prepared. With an increase in the number of Cu–O layers, the critical temperature reached a maximum value of 121 K at $n = 4$ and declined with further increase in Cu–O layers. The J_c at 77 K without an external magnetic field for the Ag-sheathed tape that had been prepared using a superconducting oxide of $(Tl,Pb)Ba_2Ca_3Cu_4O_x$ phase was 10,800 A/cm². The pinning force of the tape was

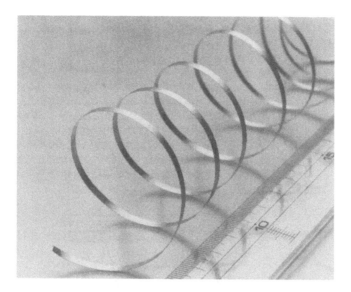

Figure 7   External view of the $(Tl,Pb)Ba_2Ca_3Cu_4O_x$ superconducting tape.

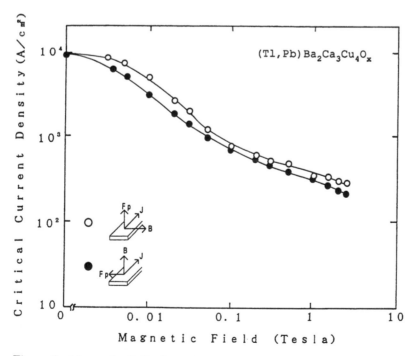

Figure 8   Magnetic field dependence of the critical current density J_c at 77 K for $(Tl,Pb)Ba_2Ca_3Cu_4O_x$ tape.

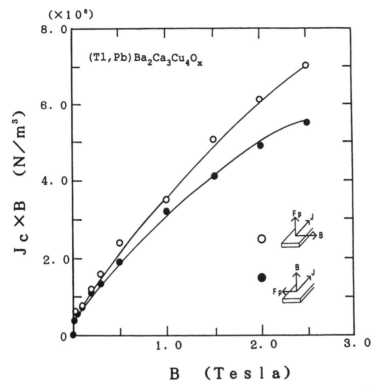

Figure 9 Magnetic field dependence of the pinning force F_p ($= J_c x B$) at 77 K for (Tl,Pb)Ba$_2$Ca$_3$Cu$_4$O$_x$ tape.

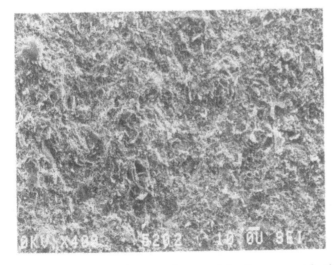

Figure 10 Scanning electron micrograph for the superconducting core of (Tl,Pb)Ba$_2$Ca$_3$Cu$_4$O$_x$ tape.

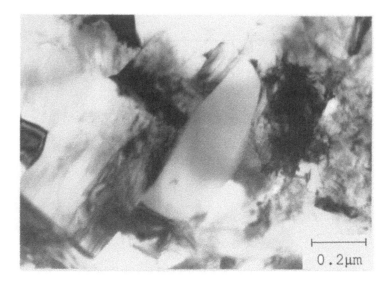

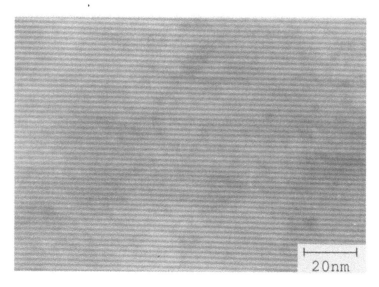

Figure 11 Transmission electron micrographs for the superconducting core of $(Tl,Pb)Ba_2Ca_3Cu_4O_x$ tape.

found to increase in proportion to the magnetic field up to 2.5 T. The direction of the external magnetic field, applied perpendicularly to the transport current, against the plane of the tapes does not greatly influence the values of J_c. The core of the tape is composed of randomly oriented fine grains of $(Tl,Pb)Ba_2Ca_3Cu_4O_x$ phase, which are thought to be the reason for a small influence in the magnetic field direction.

ACKNOWLEDGMENTS

The authors would like to thank Mr. Nishikawa and Mr. Hikata for the TEM measurement, and Ms. Yoshida for the computer graphics of the crystal structure.

REFERENCES

1. Z. Z. Sheng and A. M. Hermann, *Nature* 332, 138 (1988).
2. S. S. Parkin, V. Y. Lee, A. I. Nazzal, R. Savoy, T. C. Hung, G. G. Gorman, and R. Bayers, *Phys. Rev. B* 38, 6531 (1988).
3. H. Ihara, R. Sugise, K. Hayashi, N. Terada, M. Jo, M. Hirabayashi, A. Negishi, N. Atoda, H. Oyanagi, T. Shimomura, and S. Ohashi, *Phys. Rev. B* 38, 11952 (1988).
4. H. Kusuhara, T. Kotani, H. Takei, and K. Tada, *Jpn. J. Appl. Phys.* 28 L1772 (1989).
5. T. Kaneko, H. Yamauchi, and S. Tanaka, *Physica C* 178, 377 (1991).
6. S. Matsuda, S. Takeuchi, A. Soeta, T. Suzuki, K. Aihara, and T. Kamo, *Jpn. J. Appl. Phys.* 27, 2062 (1988).
7. M. A. Subramanian, C. C. Torardi, J. Gopalakrishnan, P. L. Gai, J. C. Calabrese, T. R. Askew, R. B. Flippen, and A. W. Sleight, *Science* 242, 249 (1988).
8. Y. Torii, H. Takei, and K. Tada, *Jpn. J. Appl. Phys.* 28, L2192 (1989).
9. Y. Torii, H. Kugai, H. Takei, and K. Tada, *Jpn. J. Appl. Phys.* 29, L952 (1990).
10. K. Aihara, M. Okada, T. Matsumoto, S. Matsuda, F. Hosono, and M. Seido, *IEEE Trans. Magn.* MAG-27, 894 (1991).
11. K. Sato, N. Shibuta, H. Mukai, T. Hikata, M. Ueyama, and T. Kato, *J. Appl. Phys.* 70, 6484 (1991).
12. H. Takei, Y. Torii, H. Kugai, T. Hikata, K. Sato, H. Hitotuyanagi, and K. Tada. *Proc. 2nd International Symposium on Superconductivity*, Springer-Verlag, Tokyo, 1990, p. 359.

15

Effects of Inherent (Auto) Doping and Oxygen Nonstoichiometry and Their Relationship to the Band Structure of Tl-Based Superconductors

R. S. Liu

University of Cambridge, Cambridge, England and
Industrial Technology Research Institute, Hsinchu, Taiwan

Peter P. Edwards

The University of Birmingham, Birmingham, England

I. INTRODUCTION

Following the discovery of high-temperature superconductivity in Tl-based cuprate material by Sheng and Hermann [1,2], the existence of a homogeneous superconducting series $Tl_mBa_2Ca_{n-1}Cu_nO_{2,+m+2}$ ($m = 1,2$) was proposed [3]. In such a series, one finds perovskite-type CuO_2 layers separated either by rock salt–type thallium double layers in, for example, $Tl_2Ba_2Ca_{n-1}Cu_nO_{2n+4}$ ($n = 1$ to 4), or rock salt–type thallium single layers, as in $TlBa_2Ca_{n-1}Cu_nO_{2n+3}$ ($n = 1$ to 6). The superconducting transition temperature, T_c, of the $Tl_2Ba_2Ca_{n-1}Cu_nO_{2n+4}$ and $TlBa_2Ca_{n-1}Cu_nO_{2n+3}$ phases increases with increasing number of Cu–O layers, up to $n = 3$ and $n = 4$, respectively, after which point T_c decreases.

Although both the single- and double-thallium-layer series possess similarities in both composition and structure, there are, nevertheless, remarkable differences in their chemical and physical properties. Such differences encompass synthesis conditions, phase stabilities, decomposition temperatures, and even the origins of hole concentrations within the CuO_2 (conducting) sheets. In fact, the hole concentration is of cardinal importance in dictating the details of the electronic structure of these materials, including the magnitude of T_c itself.

It is generally assumed that there are two sources for holes in the thallium cuprates. The first arises from the solid-state chemistry of the material (chemical compositions, defects, etc.), while the second, one might say, derives from the "physics" of the solid, involving considerations of the overlap of the Tl-6s band with the conduction band of the CuO_2 sheets at the Fermi level. From band structure calculations [4,5] it has been proposed that for the double-thallium-layer cuprates (e.g., $Tl_2Ba_2Ca_{n-1}Cu_nO_{2n+4}$) the origin of the holes appears to derive primarily from the spontaneous (internal) redox process that arises from the overlap of the Tl-6s and CuO$_2$ ($\sigma^*_{x^2-y^2}$) conduction bands.

However, for any thallium cuprate, the occurrence of both types cannot be ruled out unambiguously. In addition, band structure calculations do not take account of the presence of defects and the possibility of composition-induced modifications in the electronic structure, arising, for example, from changes in the (bandwidth) overlap of the Tl-6s band with the CuO$_2$ conduction band resulting from variation in the thallium content of the material. It is clear, therefore, that the precise origins and control of the hole concentration in the CuO$_2$ sheets in the thallium cuprates may originate from a variety of sources—some independent, some interrelated—and consideration must be given to *all* these effects.

In this review we outline a range of experimental studies in the thallium cuprates that we believe to be relevant to these issues. Overall, our aim is to reveal some of the richness and complexity of the solid-state chemistry of thallium cuprates with the hope that such studies may provide additional insights into the chemical control of superconductivity in these remarkable materials.

II. ORIGINS OF HOLE DOPING IN SINGLE- AND DOUBLE-LAYER THALLIUM CUPRATES

A. Band Structure Considerations

In the single-thallium-layer material of TlBa$_2$Ca$_{n-1}$Cu$_n$O$_{2n+3}$, each Tl has four near-neighbor Tl atoms, as compared to eight in the double-thallium-layer compound in Tl$_2$Ba$_2$Ca$_{n-1}$Cu$_n$O$_{2n+4}$. Enhanced Tl–Tl interactions in the double-thallium-layer materials are expected to increase the Tl-6s band width more than that in the single-thallium-layer series. Composition-induced broadening of the Tl-6s band may ultimately lower the bottom of this band to a stage where it overlaps the CuO$_2$ band. Such a proposal originated in tight-binding band calculations [4,5]. In Fig. 1 we show the energy density of states, $N(\varepsilon)$, as a function of energy ε for (a) Tl$_2$Ba$_2$Ca$_{n-1}$Cu$_n$O$_{2n+4}$ and (b) TlBa$_2$Ca$_{n-1}$Cu$_n$O$_{2n+3}$. The Tl-6s band of the double-thallium-layer compounds of Tl$_2$Ba$_2$Ca$_{n-1}$Cu$_n$O$_{2n+4}$ is sufficiently broad and lies significantly below the Fermi level and hence overlaps extensively with the CuO$_2$ ($\sigma^*_{x^2-y^2}$) band so that there exists a facile electron transfer from the ($\sigma^*_{x^2-y^2}$) bands of the CuO$_2$ layers to the thallium layers. This internal redox mechanism then gives rise to *both* holes (in the CuO$_2$ layers) *and* electrons (in the thallium 6s band) via

$$\text{Tl}^{\text{III}} + \text{Cu}^{\text{II}} \rightleftharpoons \text{Tl}^{\text{III}-\delta} + \text{Cu}^{\text{II}+\delta} \tag{1}$$

In contrast, the Tl-6s band of the single thallium layers of TlBa$_2$Ca$_{n-1}$Cu$_n$O$_{2n+3}$ was found to lie well above the Fermi level, so that this internal redox process cannot easily remove electrons from the ($\sigma^*_{x^2-y^2}$) bands of the CuO$_2$ layers. In single-thallium-layer cuprates, therefore, the origin of holes in the CuO$_2$ layers is generally assumed to arise from experimentally controlled thallium/oxygen defects in the structure. This proposal is consistent with the observation that stoichiometric TlBa$_2$Ca$_{n-1}$Cu$_n$O$_{2n+3}$ already has a high formal oxidation state for Cu, $(2n + 1/n)$, and hence further hole generation to produce superconductivity does not appear necessary.

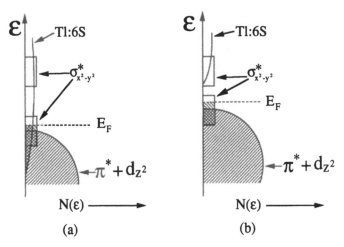

Figure 1   Energy density of states $N(\varepsilon)$ as a function of energy ε for (a) $Tl_2Ba_2Ca_{n-1}Cu_nO_{2n+4}$ and (b) $TlBa_2Ca_{n-1}Cu_nO_{2n+3}$.

The results proposed from a band structure simulations have been explored experimentally by Manthiram et al. [6], who suggested that the oxidation of the CuO_2 sheets (i.e., an increase in the hole concentration) in the $Tl_{2-y}Ba_2Ca_{n-1}Cu_nO_{2n+4-x}$ family (where y corresponding to the Tl vacancy and x corresponding to the oxygen deficiency) arises primarily from an overlap of the Tl-$6s$ band with the conduction band of the CuO_2 sheets (for small $y \sim 0$), but for larger values of y (≥ 0.5) it originates from thallium vacancies. On the intermediate region $0 < y < 0.5$, the Tl-$6s$ band overlap decreases sensitively with increasing y. This unusual situation of a partially filled Tl-$6s$ band is also assumed to provide a mechanism for the extrusion of Tl from the lattice as Tl_2O_3 upon exposure to air above 70°C, as observed experimentally [i.e., $Tl^{(3-x)+} + (3x/2)O_2 \rightarrow (1 - 2s)Tl^{3+} + xTl_2O_3$]. Extrusion of Tl appears to occur until the bottom of the Tl-$6s$ band moves above the Fermi energy, E_F, until all of the reduced Tl is oxidized. Elimination of Tl via this mechanism creates Tl vacancies in the Tl_2O_2 layers; the vacancies reduce the number of Tl–Tl near-neighbor interactions (from 8 at $y = 0$) and narrows the width of the Tl-$6s$ band. It appears that for $y \sim 0.5$, the Tl-$6s$ band becomes sufficiently narrow that its lowest energy lies just above E_F, so that no further Tl extrusion can take place. Therefore, any oxidation of the CuO_1 sheets beyond the formal oxidation state Cu^{2+} for $x < 3y/2$ in the $y = 0.5$ sample is expected to arise primarily from Tl vacancies.

These results are also consistent with the observation of charge transfer Tl^{3-t}–$Cu^{2+2t/n}$ in $Tl_2Ba_2CuO_6$ and $Tl_2Ba_2Ca_2Cu_3O_{10}$ by x-ray photoelectron spectroscopy (XPS) of the Tl-$4f$ core levels [7,8]. Such a charge transfer may also contribute to the stabilization of the thallium double layers via the internal electron transfer process. Therefore, Tl vacancies (y) may also play a very important role in the creation of the holes in the double thallium layers of $Tl_{2-y}Ba_2Ca_{n-1}Cu_nO_{2n+4-x}$. Routinely, a value of $y \sim 0.3$ is found for $Tl_{2-y}Ba_2Ca_{n-1}Cu_nO_{2n+4-x}$.

In contrast, the single-thallium-layer family $Tl_{1-y}Ba_2Ca_{n-1}Cu_nO_{2n+3-x}$ does not appear to have a partially filled Tl-$6s$ band which is stable to oxidation without

Tl extrusion. This observation is also consistent with the XPS of Tl $4f$ core levels in the single-thallium-layer system, which showed an ionic character for Tl^{3+} ions, in contrast to the reduced Tl state in the double-Tl-layer system [9]. Therefore, it is of great interest to discuss the relationship between the electronic band structure and the hole concentration in the single-thallium-layer system, and it is to such a system that we not turn.

B. Metal–Superconductor–Insulator Transition in $(Tl_{0.5}Pb_{0.5})Sr_2(Ca_{1-x}Y_x)Cu_2O_{7-\delta}$

An "overdoped" state in $TlBa_2Ca_{n-1}Cu_nO_{2n+3}$ is suggested by the high natural valence of Cu (2.5 to 2.25 +). One possible synthetic route to decrease the hole concentration is via the deliberate substitution of, for example, Y^{3+} for Ca^{2+} ions in the single-thallium-layer cuprates. Unfortunately, the fact that oxygen appears labile in systems such as $TlBa_2Ca_{n-1}Ca_nO_{2n+3}$ may lead to difficulties in trying to unravel the effects of inherent cation doping effects form those arising from changes in the oxygen nonstoichiometry.

Fortunately, we have discovered that the related system $(Tl_{0.5}Pb_{0.5})Sr_2$-$(Ca_{1-x}Y_x)Cu_2O_{7-\delta}$, structurally similar to $TlBa_2CaCu_2O_{7-\delta}$, has what appears to be a fixed oxygen stoichiometry across the solid solution series. The fixed stability of the oxygen content probably arises from the fact that substitution of high-valent Pb^{4+} for Tl^{3+} results in an attractive interaction for potentially labile oxygen (see Section III.D). Therefore, we have proposed that the series $(Tl_{0.5}Pb_{0.5})Sr_2$-$(Ca_{1-x}Y_x)Cu_2O_{7-\delta}$ represents a good contender for the experimental study of the intrinsic doping effects (in relation to band structure) by the substitution of high-valent cation (Y^{3+}) in low-valent cation (Ca^{2+}) sites [10]. Of course, our proposal for a fixed value of δ—across the homogeneity range—is an approximation. *Small* changes in δ will not seriously affect the following discussions.

The crystal structure of the phase $(Tl_{0.5}Pb_{0.5})Sr_2(Ca_{1-x}Y_x)Cu_2O_{7-\delta}$ can be described in terms of an intergrowth of double rock salt–type layers [({Tl/Pb}O)(SrO)] with double [Sr(Ca,Y)Cu$_2$O$_5$] oxygen-deficient perovskite layers, formed by sheets of corner-sharing CuO$_5$ pyramids interleaved with calcium and/or rare earth ions. The structure of $(Tl_{0.5}Pb_{0.5})(Ca_{1-x}Y_x)Sr_2Cu_2O_{7-\delta}$ resembles that of $YBa_2Cu_3O_7$: the (Tl,Pb)–O layers replacing the Cu–O chains, Sr cations replacing Ba cations, and Ca cations partial substituted rare earth atoms. A schematic comparison of the crystal structures of this system and that of $YBa_2Cu_3O_7$ is shown in Fig. 2.

The "parent" compound $TlSr_2CaCu_2O_7$ is itself a metal but exhibits no superconductivity at temperatures down to 4 K. The nominal Cu valency of this compound is 2.5 +, which indicates an excess of hole carriers present in the conducting CuO$_2$ layers, which gives rise to a so-called "over-doped" state. Such overdoping appears to be detrimental to the occurrence of high-temperature superconductivity, and it can be efficiently reduced by the stepwise substitution of Tl^{3+} by Pb^{4+} and/or by the substitution of Ca^{2+} by Y^{3+}, or, indeed, by the dual substitutions Tl^{3+}/Pb^{4+} and Ca^{2+}/Y^{3+}. Arising out of this strategy, we have observed a composition-induced metal–superconductor–insulator phase transition in the system $(Tl_{0.5}Pb_{0.5})Sr_2(Ca_{1-x}Y_x)Cu_2O_7$, and a representation of the entire electronic structure phase diagram is given in Fig. 3. In this system at least we appear

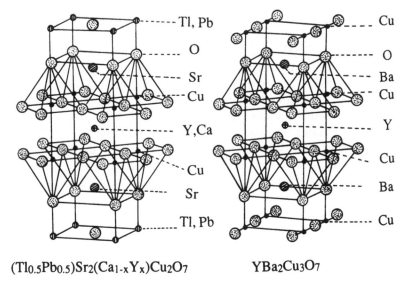

$(Tl_{0.5}Pb_{0.5})Sr_2(Ca_{1-x}Y_x)Cu_2O_7$ $YBa_2Cu_3O_7$

Figure 2 Presentation of the crystal structures $(Tl_{0.5}Pb_{0.5})Sr_2(Ca_{1-x}Y_x)Cu_2O_7$ and $YBa_2Cu_3O_7$.

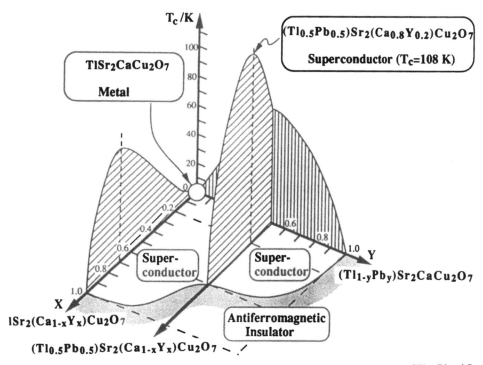

Figure 3 Metal-superconductor-insulator phase diagram for the system $(Tl_{0.5}Pb_{0.5})Sr_2$-$(Ca_{1-x}Y_x)Cu_2O_7$. (From Ref. 10.)

to be able to use chemical substitution to manipulate the properties of the material across the entire electronic phase diagram.

Both the $(Tl_{1-y}Pb_y)Sr_2CaCu_2O_7$ and $TlSr_2(Ca_{1-x}Y_x)Cu_2O_7$ systems have the highest T_c (ca. 80 K) for $y \sim 0.5$ or $x \sim 0.6$. However, the $(Tl_{0.5}Pb_{0.5})Sr_2$-$(Ca_{1-x}Y_x)Cu_2O_7$ system exhibits superconductivity over the homogeneity range $x = 0$ to 0.5, with the superconducting transition temperature showing a maximum of 108 K at $x = 0.2$. Across the homogeneity range $y = 0.6$ to 1.0, the material also undergoes a metal–insulator transition at temperatures about T_c (Fig. 3).

We have proposed [10] that substitution of Y^{3+} for Ca^{2+} in the system $(Tl_{0.5}Pb_{0.5})Sr_2(Ca_{1-x}Y_x)Cu_2O_7$ effectively reduces the hole concentration in the conducting CuO_2 sheets. Here we briefly summarize the experimental findings that support this proposal.

With increasing concentration Y^{3+} in the Ca^{2+} sites, the following effects have been observed: (1) an increase in the a lattice constant across the series, which is generally attributed to a decrease in the average copper oxidation state, leading to longer Cu–O distances within the electronically active copper oxygen plane; (2) an increase in the room-temperature resistivity; (3) a decrease in the effective number of holes in the conducting CuO_2 planes; and (4) an increase in the absolute value of the thermoelectric power. It is important to stress that no significant discontinuities or anomalies are observed in any of the physical measurements across the metal–insulator transition; there are clearly discontinuities across the metal–superconductor and superconductor–insulator boundaries (Fig. 3).

In Fig. 4 we demonstrate a possible band structure picture for both (a) before and (b) after the addition of excess electrons into the $(\sigma^*_{x^2-y^2})$ bands in the conducting CuO_2 planes via the substitution of high-valent cation (e.g., Y^{3+}) for low-valent cation (e.g., Ca^{2+}) in the single-thallium-layer cuprates. Note here that in contrast to Fig. 1, the width of the Tl-6s band is constant, and variations in chemical composition (Ca^{2+}/Y^{3+}) alter the position of the Fermi level in the $(\sigma^*_{x^2-y^2})$ band.

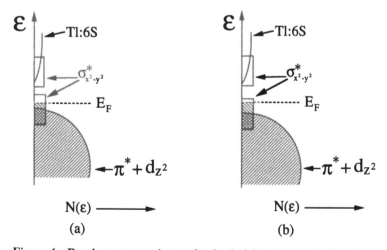

Figure 4 Band-structure picture for both (a) before and (b) after the addition of excess electrons into $(\sigma^*_{x^2-y^2})$ bands in the conducting CuO_2 planes via the substitution of high-valent cation of low-valent cation in single-thallium-layer cuprates.

As one can see, this is an efficient and highly controlled method for transforming the overdoped, metallic state into a high-temperature superconductor.

C. Inherent (Auto) Substitution

1. *(Cu/Tl) in Tl$_2$Ba$_2$CuO$_6$*

Due to the somewhat similar ionic radii of Cu^{2+} [0.73 Å for a coordination number (CN) of 6] and Tl^{3+} (0.885 Å for CN = 6) [11], it is likely that the substitution of Cu^{2+} can take place into the rock salt–type layers containing Tl^{3+} ions; this can lead to change in the hole concentration by "autosubstitution." Here we give a specific example of such inherent doping in single crystals of the double-thallium-layer cuprate Tl$_2$Ba$_2$CuO$_{6+\delta}$, to illustrate how one might determine experimentally the extent (and location) of such auto substitution effects [12].

High-quality single crystals of Tl$_2$Ba$_2$CuO$_{6+\delta}$ with T_c of 12.4 K were grown by a CuO self-flux method [12]. We carried out detailed electron probe microanalysis (EPMA) quantitative analyses which showed that three single crystals from the batch had the same compositional formula within a degree of experimental accuracy. The chemical composition of the as-grown superconducting single crystals were found to be Tl$_{1.85(4)}$Ba$_{1.94(4)}$Cu$_{1.15(2)}$O$_{6+\delta}$, and this observation suggests a substitution of about 7.5% Cu for Tl. The resulting chemical formula (Tl$_{0.925}$Cu$_{0.075}$)$_2$Ba$_2$CuO$_{6+\delta}$ shows a cation nonstoichiometry, also reported in polycrystalline sample [13], where 5% of Tl occupied sites were substituted with Cu.

Single-crystal x-ray diffraction experiments and structure refinement show that one crystal exhibits tetragonal symmetry (space group *I4/mmm*), with lattice parameters a = 3.8608(4) Å and c = 23.1332(9) Å. For this crystal we carried out site occupancy refinements which indicated completely occupied Ba and Cu sites within the structure but a deficiency in the apparent scattering power of the Tl site. The final structure refinement corresponds to the composition Tl$_{1.87(2)}$Ba$_2$Cu$_{1.13(2)}$O$_{6+\delta}$, which is consistent with EPMA results. The quality of the diffraction data allow Tl and O (which comprise the Tl$_2$O$_2$ layers) to be unambiguously assigned to split sites of the type *xOz*, as opposed to *xxz* sites. We have found no evidence for long-range ordering of the displacements of the atoms in the Tl$_2$O$_2$ layers of this crystal.

Thus for the highest T_cs (85 to 90 K) (with no excess oxygen) [13], one can propose that the hole carriers may be created by the inherent substitution of Cu^{2+} in the Tl^{3+} sites in the formula (Tl$_{0.925}$Cu$_{0.075}$)$_2$BaCuO$_{6+\delta}$ ($\delta \sim$ 0) is 0.15 per copper atom, which agrees with the typical value for the other cuprates superconductors. Therefore, inherent (auto) doping of the Cu^{2+} ions in the Tl^{3+} sites in Tl$_2$Ba$_2$CuO$_6$ may function as an effective oxidation route, and this may give rise to an increase in the hole concentration sufficient to induce superconductivity. However, we note that such substitutions are rarely observed in the higher Cu–O layer compounds, such as $n > 1$ in Tl$_2$Ba$_2$Ca$_{n-1}$Cu$_n$O$_{2n+4}$. Moreover, in this double-thallium-layer cuprate, we see that inherent (auto) doping may well compete with the often-proposed intrinsic redox process (1) to generate holes in the CuO$_2$ sheets.

2. *(Ca/Tl) in Tl$_2$Ba$_2$CaCu$_2$O$_8$ and Tl$_2$Ba$_2$Ca$_2$Cu$_3$O$_{10}$*

The ideal formula Tl$_2$Ba$_2$Ca$_{n-1}$Cu$_n$O$_{2n+4}$ (n = 1 to 4) would have all copper in the divalent state, all previously known superconductors based on cuprates having a mixed valency of Cu (ca. 2.1 to 2.3$^+$). Therefore, an increase in copper valence, corresponding to an increase in the hole concentration, is a major factor in promoting superconductivity. As outlined in Section II.A, Tl vacancies (y) and intrinsic redox process (1) in $Tl_{2-y}Ba_2Ca_{n-1}Cu_nO_{2n+4}$ could easily satisfy such a necessity. Moreover, Cox et al. [14] have found that mutual cation substitutions in the Tl-based superconductors, for example, substitution of Ca^{2+} for Tl^{3+}, gives rise to a true chemical composition of $(Tl_{1.8}Ca_{0.2})Ba_2(Ca_{0.9}Tl_{0.1})Cu_2O_8$ for $Tl_2Ba_2CaCu_2O_8$, and substitution between Ca^{2+} and Tl^{3+} results in a chemical composition of $(Tl_{1.76}Ca_{0.25})Ba_2Ca_2Cu_3O_{9.89}$ for $Tl_2Ba_2Ca_2Cu_3O_{10}$. Both of these substitutions would be effective as an inherent source of oxidation of the conducting CuO_2 sheets. We have found a similar substitutional chemistry for certain of the single-thallium-layer cuprates. For example, about 10% of Tl^{3+} substitution for Ca^{2+} in the series $(Tl_{0.5}Pb_{0.5})Sr_2(Ca_{1-x}Y_x)Cu_2O_7$ results in a chemical composition of $(Tl_{0.5}Pb_{0.5})Sr_2(Ca_{0.9-x}Tl_{0.1}Y_x)Cu_2O_7$ [15], which may be a source of the reduction of the overdoped state in these materials.

III. EFFECTS OF OXYGEN NONSTOICHIOMETRY

Oxygen nonstoichiometry is known to play a pivotal role in the geometric and electronic structure of high-T_c superconducting cuprates. The effect is perhaps most commonly associated with $YBa_2Cu_3O_{7-\delta}$, whereby oxygen is readily inserted into, or extracted from, the Cu(1)–O(1) chains [16]. Here the carrier concentration and superconducting properties are essentially controlled by the oxygen content. The significant effects of oxygen content on the physical properties of the $Bi_{2.1}(Ca_xSr_{1-x})_{n+1}Cu_nO_{2n+4+\delta}$ ($n = 1, 2, 3$) series of superconducting oxides have been investigated [17,18]. For the $n = 2$ material, T_c is raised to 90 K by annealing at high temperature (700 to 800°C) in a low partial pressure of oxygen, while for $n = 3$ system annealing the samples at low temperature (ca. 400°C) in 1 atm of oxygen-enhanced T_c to 105 K. Here we provide a brief overview of certain of the effects of oxygen nonstoichiometry in the superconducting Tl-containing cuprates.

A. Oxygen Nonstoichiometry in the Double Thallium Layer Cuprates

The relationship between T_c and oxygen loss (δ) for $Tl_2Ba_2CuO_{y-\delta}$, $Tl_2Ba_2CaCu_2O_{y-\delta}$, and $Tl_2Ba_2Ca_2Cu_3O_{y-\delta}$ has been studied in detail by Shimakawa et al. [19] and their results are shown in Fig. 5. $Tl_2Ba_2CuO_{y-\delta}$ and $Tl_2Ba_2CaCu_2O_{y-\delta}$ samples sintered in an oxygen atmosphere appear to have an excess of hole carriers [i.e., they are characterized by an overdoped state (normally, $y \sim 6.1$ and 8.1 for $Tl_2Ba_2CuO_{y-\delta}$ and $Tl_2Ba_2CaCu_2O_{y-\delta}$, respectively)] and a decrease in the number of excess oxygen atoms (leading to a reduction in the hole concentration in the conducting CuO_2 planes) results in an appropriate carrier concentration and an extremely high T_c. Parise et al. [20] and Shimakawa et al. [13] have confirmed by neutron diffraction that excess oxygen atoms are located at the interstitial site between the double thallium layers. In contrast, $Tl_2Ba_2Ca_2Cu_3O_{y-\delta}$ samples, synthesized in 1 atm of oxygen, appear to have a slightly reduced level of carriers [i.e., they are located in the slightly underdoped state (normally, $y = 9.9$ for $Tl_2Ba_2Ca_2Cu_3O_{y-\delta}$)], and a decrease in the oxygen content results in a lower carrier concentration than the optimum value, and T_c decreases. For the $Tl_2Ba_2Ca_3Cu_4O_{12}$ system, oxygen loading increases the T_cs from 112–113 K to 115–116 K, which reveals that the underdoped state probably exists in the as-synthesized samples [21,22].

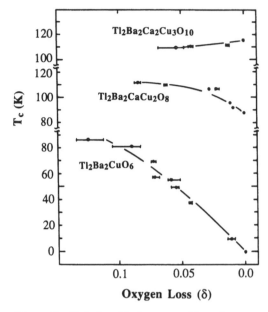

Figure 5 Relationship between T_c and oxygen loss (δ) for the systems of $Tl_2Ba_2CuO_{y-\delta}$, $Tl_2Ba_2CaCu_2O_{y-\delta}$, and $Tl_2Ba_2Ca_2Cu_3O_{y-\delta}$. (From Ref. 19.)

Presland et al. [23] have proposed a universal phase diagram for cuprate superconductors which showed a parabolic superconducting domain for $La_{2-x}Sr_xCuO_4$ and $(Tl,Bi)_2(Ba,Sr)_2Ca_{n-1}Cu_nCu_{2n+4+\delta}$ ($n = 1, 2, 3$). The parabolic curve can be fitted as

$$\frac{T_c}{T_{c(max)}} = 1 - 82.6(p - 0.16)^2 \qquad (2)$$

where p is the number of holes per Cu. Here we used previously published data for the variation of T_c with the oxygen nonstoichiometry for the system $Tl_2Ba_2Ca_{n-1}Cu_nO_{2n+4-\delta}$ ($n = 1, 2, 3, 4$) [19,21,22], where T_c varied from 0 K to $T_{c(max)} = 86$ K for $n = 1$, from 87 K to $T_{c(max)} = 112$ K for $n = 2$, from $T_{c(max)} = 116$ to 110 K for $n = 3$, and $T_{c(max)} = 116$ to 113 K for $n = 1$ 4, to fit the formula, and obtain $p = 0.270, 0.212. 0.135$, and 0.110 for $n = 1, 2, 3$, and 4, respectively. In Fig. 6 we show the variation of $T_c/T_{c(max)}$ as a function of the number of holes per Cu for $Tl_2Ba_2Ca_{n-1}Cu_nO_{2n+4}$ ($n = 1, 2, 3, 4,$). The good fit of the data to equation (2) leads us to propose that the $n = 1$ and 2 compounds lie on the "excess hole" (overdoped state) side of the T_c maximum, while the $n = 3$ and 4 compounds lie on the "deficient hole" (underdoped state) side of the maximum in T_c.

B. Important Effects of Intergranular Oxygen Nonstoichiometry in $Tl_2Ba_2Ca_2Cu_3O_{10-\delta}$

In granular high-T_cs cuprate superconductors, it is a common observation that the transport critical current density in a well-sintered bulk sample is very much smaller than the corresponding value deduced from magnetic hysteresis measurements.

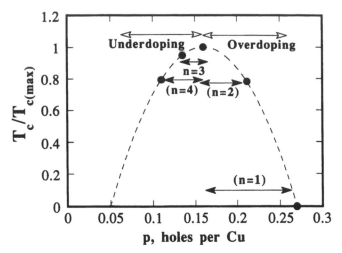

Figure 6 Variation of $T_c/T_{c(max)}$ as a function of the number of holes per Cu, p, for $Tl_2Ba_2Ca_{n-1}Cu_nO_{2n+4}$ ($n = 1, 2, 3, 4$).

This discrepancy is due to the fact that the two general experimental methods used to probe the existence of supercurrents (i.e., transport or magnetic measurements) correspond to the detection of the currents between, and within, the grains, respectively. With the multiconnected network consisting of weak links between grains, the intergrain supercurrent can relatively easily be destroyed under conditions of high transport currents or external magnetic fields. Thompson et al. [24] have reported the existence of a characteristic temperature, called the decoupling temperature (T_d), of granular $Tl_2Ba_2Ca_2Cu_3O_{10}$ at which the intergrain supercurrent, $J_{c(inter)}$, ceases to flow. These authors have also found that T_d moves to lower temperature on increasing an applied magnetic field. Although it is well established that variable oxygen nonstoichiometry in superconducting cuprates can have a dramatic effect on T_c [19,21–23], relatively little is known about the very important effects on inter- and intragrain supercurrents [25].

Here we review our systematic investigations of the effect of oxygen nonstoichiometry in the $Tl_2Ba_2Ca_2Cu_3O_{10-\delta}$ material; in particular, we have measured the effects of oxygen nonstoichiometry on the intergranular screening current, as monitored by the intergrain decoupling temperature, T_d.

Sample preparation details have been published previously [26]. X-ray diffraction (XRD) analysis confirmed that the as-sintered material is nearly single phase $Tl_2Ba_2Ca_2Cu_3O_{10}$. We have assumed that the as-sintered samples have the highest oxygen content, which correspond to complete oxygenation in $Tl_2Ba_2Ca_2Cu_3O_{10-\delta}$ ($\delta = 0$). Pellets of the compound were annealed over the temperature range between 300 and 600°C in flowing oxygen at 1 atm pressure for between 10 to 24 h. Following these heat treatments, the samples were quenched. This results in a decrease in oxygen content, which is subsequently determined by precise weight loss measurements. Although the initial oxygen content is not precisely known, our conclusions are based only on the effects of changes in the relative oxygen content, δ. A Lake Shore 7000 ac susceptometer [operating with a frequency (f)

of 333.3 Hz] was used to measure both the superconducting transition temperature and the decoupling temperature. The ac susceptibility was measured at various amplitude of h_0 (0.06 to 11 Oe) of the external magnetic field $h(t) = h_0 \sin 2ft$.

In Fig. 7 we show the dependence of h_0 on the real part of the ac susceptibility, χ', for the sample with $\delta = 0.009$. This shows clearly two successive transitions, at 116 K (superconducting transition temperature, T_c) and 111 to 100 K (decoupling temperature, T_d); the latter is strongly dependent on the amplitude of the driving ac field, h_0. Pulverization of the sample (typical grain dimension is ca. 5 μm) is effective in removing the second transition. This behavior is attributed to a bulk superconducting transition at 116 K and to a coupling by weak links across grain boundaries, which gives rise to a multiconnected Josephson network with a transition temperature corresponding to T_d (111 to 100 K).

In Fig. 8 we show complete data for the intragrain superconducting transition temperature (T_c) and intergrain decoupling temperature (T_d) of bulk $Tl_2Ba_2Ca_2Cu_3O_{10-\delta}$ as a function of the oxygen nonstoichiometry, δ. Note that the former characteristic temperature has a weaker dependence on δ than the latter, which suggests that the loss of oxygen within the interior of compound is slower than from the intergrain region.

In Fig. 9 we show the dependence of the decoupling temperature on h_0 for two samples of $Tl_2Ba_2Ca_2Cu_3O_{10-\delta}$ with $\delta = 0.009$ and 0.141. The data can be fit to simple polynominals: $T_d = 110.54 - 1.17h_0 + 0.026h_0^2$ for a $\delta = 0.009$ sample and $T_d = 98.09 - 2.25h_0 + 0.06h_0^2$ for a $\delta = 0.141$ sample. This indicates that samples with a higher oxygen content have a smaller magnetic field dependence than those of the oxygen-efficient materials.

In Fig. 10 we show the observed intergranular critical current, $J_{c(inter)}$, as a function of temperature for two samples of $Tl_2Ba_2Ca_2Cu_3O_{10-\delta}$ with $\delta = 0.009$ and 0.141. *This clearly demonstrates that higher-oxygen-content samples have a higher $J_{c(inter)}$.* This means that the fully oxygenated samples should have a higher intergrain critical current density than the oxygen-deficient sample. We point out that this information is very

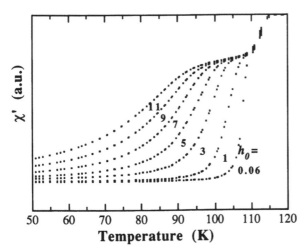

Figure 7 Dependence of the real part of ac susceptibility, χ', on h_0 (0.06 to 11 Oe) for $Tl_2Ba_2Ca_2Cu_3O_{10-\delta}$ with $\delta = 0.009$. (From Ref. 25.)

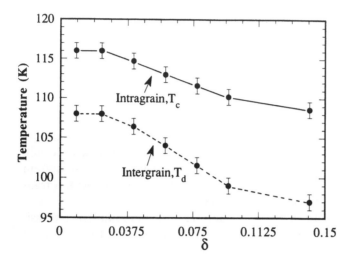

Figure 8 Intragrain superconducting transition temperature (T_c) and intergrain decoupling temperature (T_d) of bulk $Tl_2Ba_2Ca_2Cu_3O_{10-\delta}$ as a function of the oxygen nonstoichiometry, δ. (From Ref. 25.)

important when one considers the synthesis of thin films of $Tl_2Ba_2Ca_2Cu_3O_{10-\delta}$ for high-J_c applications in a magnetic field; clearly, to increase the critical current density of the film, a complete oxygenation is desirable, indeed essential.

C.   Oxygen Nonstoichiometry in Single-Thallium-Layer Cuprates

An oxygen vacancy, y, of about $y = 0.3$ has been found in oxygen-annealed compounds $TlBa_2Ca_{n-1}Cu_nO_{2n+3-y-\delta}$ ($n = 2, 3, 4$) which was determined by chemical iodometric titration [27]. Based on these samples, the relationship between T_c and oxygen loss (δ) for the single-thallium-layer cuprates

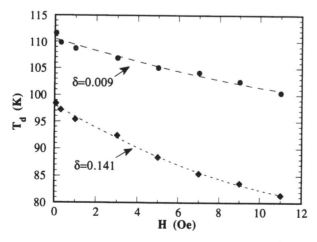

Figure 9 Dependence of the decoupling temperature (T_d) on h_0 for two samples $Tl_2Ba_2Ca_2Cu_3O_{10-\delta}$ with $\delta = 0.009$ and 0.141. (From Ref. 25.)

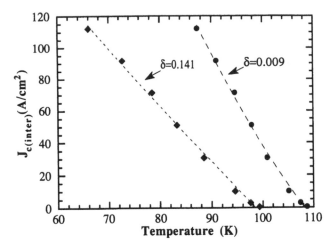

Figure 10 Intergranular critical current, $J_{c(inter)}$, as a function of temperature for two samples $Tl_2Ba_2Ca_2Cu_3O_{10-\delta}$ with $\delta = 0.009$ and 0.141.

$TlBa_2Ca_{n-1}Cu_nO_{2n+3-y-\delta}$ ($n = 2, 3, 4$) has been investigated by Nakajima et al. [27,28], and their results are shown in Fig. 11. The T_c of $TlBa_2 Ca_{n-1}Cu_nO_{2n+3-y-\delta}$ ($n = 2, 3, 4$) increased with increasing oxygen loss (δ). The maximum T_c of $TlBa_2CaCu_2O_{6.7-\delta}$, $TlBa_2Ca_2Cu_3O_{8.7-\delta}$, and $TlBa_2Ca_3Cu_4O_{10.7-\delta}$ was 110, 118, and 121 K, respectively, at about $\delta = 0.015$, which was independent of n. This

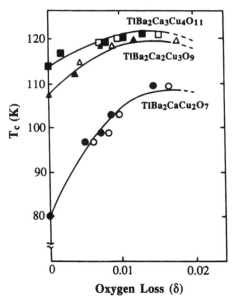

Figure 11 Relationship between T_c and oxygen loss (δ) for the single-thallium-layer system $TlBa_2Ca_{n-1}Cu_nO_{2n+3-y-\delta}$ ($n = 2, 3, 4$). (δ determined by thermogravimetric analysis and weight difference before and after quenching is marked by open and solid symbols, respectively.) (From Ref. 27.)

suggests that the oxygen-annealed samples were in the overdoped regime. Consequently, the loss of oxygen [by heating the samples at high temperatures (ca. 500°C) for 30 to 60 min in nitrogen and then quenching into liquid nitrogen] leads to a decrease in the hole concentrations within the CuO_2 planes, and one can then approach the maximum T_c inherent for that system. Recently, Izumi et al. [29] proposed that oxygen was released from the rock salt type of thallium–oxygen sheets in the single-thallium-layer cuprate superconductors.

Here we review the data derived from the previous publications for the variation of T_c with the oxygen nonstoichiometry for the system $TlBa_2Ca_{n-1}Cu_nO_{2n+3-y-\delta}$ ($n = 2, 3, 4$) [27,28], where T_c changed from 80 K to $T_{c(max)} = 110$ K for $n = 2$, from 108 K to $T_{c(max)} = 118$ K for $n = 3$, from 114 K to $T_{c(max)} = 121$ K for $n = 4$, to fit Eq. (2) $T_c/T_{c(max)} = 1 - 82.6(p - 0.16)^2$, and obtain the number of holes, p, per Cu of 0.217, 0.192, and 0.186 for $n = 2, 3,$ and 4, respectively. In Fig. 12 we show the variation of $T_c/T_{c(max)}$ as a function of p for $TlBa_2Ca_{n-1}Cu_nO_{2n+3}$ ($n = 2, 3, 4$) together with that for $Tl_2Ba_2Ca_{n-1}Cu_nO_{2n+4}$ ($n = 1, 2, 3, 4$). All the data can be fit to Eq. (2) for both $m = 1$ and $m = 2$ of $Tl_mBa_2Ca_{n-1}Cu_nO_{2n+m+2}$, which indicates that the parabolic shape of the curve of $T_c/T_{c(max)}$ versus hole concentration describes not only a particular cuprate system [with cation and/or oxygen doping (e.g., $La_{2-x}Sr_xCuO_4$ and $YBa_2Cu_3O_{7-\delta}$)], but also the range of Tl-based superconductors having different Cu–O layers and variable oxygen nonstoichiometry. This type of behavior appears to be a universal feature in the phase diagram of all the known cuprate superconductors. From Fig. 12 it is also worth pointing out that all the single-thallium-layer compounds $TlBa_2Ca_{n-1}Cu_nO_{2n+3}$ ($n = 2, 3, 4$) lie on the *excess*-holes site of the maximum, and these systems exhibit the overdoped state.

D. Dependence of Oxygen Nonstoichiometry in Single-Thallium-Layer Cuprates Simultaneous Cation Substitution

The variation of T_c with cation substitution (x) and oxygen nonstoichiometry (δ) of oxygen-annealed samples of $TlSr_2(Ca_{1-x}Lu_x)Cu_2O_{7-\delta}$ have been investigated

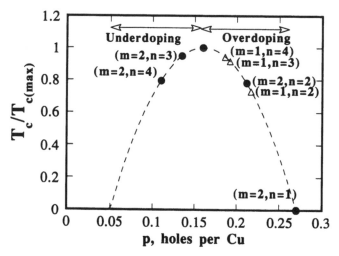

Figure 12 Variation of $T_c/T_{c(max)}$ as a function of holes per Cu, p, for $TlBa_2Ca_{n-1}Cu_nO_{2n+3}$ ($n = 2, 3, 4$) together with that for $Tl_2Ba_2Ca_{n-1}Cu_nO_{2n+4}$ ($n = 1, 2, 3, 4$).

by Kondo et al. [30] and their data are shown in Fig. 13. Due to the existence of the overdoped state for $TlSr_2CaCu_2O_7$ (as discussed in Section II.B), this material is itself a metal but not a superconductor (down to 4 K). Therefore, the substitution of high-valent Lu (3+) for the low-valent Ca^{2+} sites in $TlSr_2CaCu_2O_7$ gave rise to a decrease in the hole concentrations of CuO_2 planes and the appearance of superconductivity up to 68 K for $x = 0.6$. Subsequently, these oxygen-annealed samples were reduced in an argon atmosphere at 550°C for 5 h. It was found that the superconductivity in these materials was dramatically modified by oxygen nonstoichiometry. For the composition $x = 0$ to 0.6, T_c was raised from 0 to 80–85 K, respectively, by this type of chemical reduction. For $x > 0.6$, T_c exhibited a monotonic decrease. These authors [30] proposed that the samples having $x < 0.7$ were overdoped (by oxygen annealing) and their T_cs could be effectively increased by reduction. In contrast, the $x > 0.7$ samples were underdoped, the carrier concentrations in the materials were less than the optimum value and this resulted in a decrease in T_c. Moreover, it is interesting to point out that we have attempted to reduce the oxygen-annealed $x = 0$ and 0.2 samples in $(Tl_{0.5}Pb_{0.5})Sr_2(Ca_{1-x}Y_x)Cu_2O_{7-\delta}$ [31]. The crystal structure of $(Tl_{0.5}Pb_{0.5})Sr_2(Ca_{1-x}Y_x)Cu_2O_{7-\delta}$ (Fig. 2) is similar to that of $TlSr_2(Ca_{1-x}Lu_x)Cu_2O_{7-\delta}$ except for the substitution of Pb for Tl. The samples were reduced at 500 and 600°C in 2%, 0.2%, 0.009%, and 0% of oxygen in nitrogen at 1 atm. Following each of these reducing heat treatments, we found no significant shift in the diamagnetic onset temperature of about 80 and 110 K for the $x = 0$ and 0.2 sample, respectively. Izumi et al. [29] suggested that the release of oxygen atoms under reducing conditions was from the rock salt–type thallium sheets in $TlSr_2CaCu_2O_{7-\delta}$. Therefore, the labile oxygen derived from the single thallium layers in $TlSr_2(Ca_{1-x}Lu_x)CuO_{7-\delta}$ appear to act as an effective, and tunable, hole reservoir to control the carrier concentration in the conducting CuO_2 planes, and hence the superconducting transition temperature. However, the labile oxygen in the thallium–oxygen sheets may be attracted by the high valent Pb (4+) ions when substituted into the Tl sites in $(Tl_{0.5}Pb_{0.5})Sr_2)Ca_{1-x}Y_x)Cu_2O_{7-\delta}$. Therefore, we propose

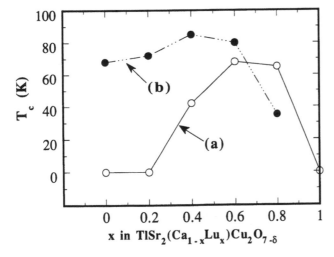

Figure 13 Variation of T_c and Lu content, x, for the (a) oxygen-annealed and (b) reduced $TlSr_2(Ca_{1-x}Lu_x)Cu_2O_{7-\delta}$ samples. (From Ref. 30.)

that the oxygen in rock salt–type layers of $(Tl_{0.5}Pb_{0.5})Sr_2(Ca_{1-x}Y_x)Cu_2O_{7-\delta}$ cannot easily be released. This behavior clearly contrasts with that found for $TlSr_2(Ca_{1-x}Lu_x)CuO_{7-\delta}$. Thus we have demonstrated that oxygen nonstoichiometry is highly dependent on the precise nature and location of cations in the rock salt–type sheets in single-thallium-layer cuprate systems.

IV.   TEST CASE: AN EFFICIENT AND REPRODUCIBLE APPROACH FOR ATTAINING SUPERCONDUCTIVITY AT 128 K IN $Tl_2Ba_2Ca_2Cu_3O_{10-\delta}$ VIA BOTH AUTODOPING AND OXYGEN NONSTOICHIOMETRY CONTROL

We have investigated an efficient and highly reproducible method for the preparation of almost-single-phase $Tl_2Ba_2Ca_2Cu_3O_{10-\delta}$ material via autodoping and oxygen nonstoichiometry, which is characterized by high-temperature superconductivity with $T_{c(midpoint)} = 128$ K and $T_{c(zero)} = 126$ K as measured by electrical resistivity, and a diamagnetic onset temperature $[T_{c(mag)}]$ of 128 K as measured by ac susceptibility [32].

This procedure draws on all the empirical observations outlined above; it involves synthesizing a material with nominal stoichiometry $Tl_{1.6}Ba_2Ca_{2.4}Cu_3O_{10-\delta}$ (i.e., autodoping for Ca for Tl) which was sintered at 910°C for 3 h, annealed at 750°C (10 days) in an evacuated quartz tube, and finally annealed at 600°C (2 h) in a 0.2% oxygen–nitrogen gas mixture. Based on these results, it appears that the optimization of the hole concentration in $Tl_2Ba_2Ca_2Cu_3O_{10-\delta}$, achieved here by precisely controlling the oxidizing–reducing power of the annealing atmosphere, appears to be crucially important in attaining such a high transition temperature.

High-purity powders of Tl_2O_3, BaO_2, CaO, and CuO were weighed in the appropriate proportions to form nominal compositions of $Tl_{2-x}Ba_2Ca_{2+x}Cu_3O_{10-\delta}$ ($x = 0$, 0.1, 0.2, 0.3, 0.4, 0.5, 0.6, 0.7, 0.8, and 1.0). The powders were then mixed using a mortar and pestle and pressed into a pellet (10 mm in diameter and 3 mm in thickness) under a pressure of 5 tons/cm². The pellets were wrapped in gold foil to prevent loss of thallium at elevated temperatures, then sintered at 910°C for 3 h in oxygen. After sintering, the furnace was cooled to room temperature at a rate of 5°C/min. The sample with $x = 0.4$ had the highest zero-resistance temperature (120.5 K) among the series. Consequently, we chose the $x = 0.4$ (i.e., $Tl_{1.6}Ba_2Ca_{2.4}Cu_3O_{10-\delta}$) sample for further investigations. The sample was wrapped in gold foil, encapsulated in an evacuated (ca. 10^{-4} torr) quartz tube and annealed at 750°C for 10 days. The resulting sample was then annealed in 0.2% and 2% oxygen–nitrogen atmosphere at 600°C for 2 h and rapidly quenched into liquid nitrogen.

In Fig. 14 we show the powder x-ray diffraction (XRD) pattern of the as-sintered $Tl_{1.6}Ba_2Ca_{2.4}Cu_3O_{10-\delta}$ sample. Most of the peaks of the XRD pattern of the sample can be indexed with the body-centered tetragonal (space group: *I4/mmm*) system having the lattics parameters of $a = 3.8522(2)$ Å and $c = 35.628(2)$ Å. Based on the structural model of $Tl_2Ba_2Ca_2Cu_3O_{10-\delta}$, positional parameters and isotropic thermal factors were successively refined except for oxygen atoms whose B were arbitrarily fixed at 1 Å². In the refinements, a splitting from the ideal position from 4e to 16n for the Tl/Ca and O(4) sites was introduced. In these conditions the different agreement factors were lowered to $R_p = 10.11\%$, $R_{wp} =$

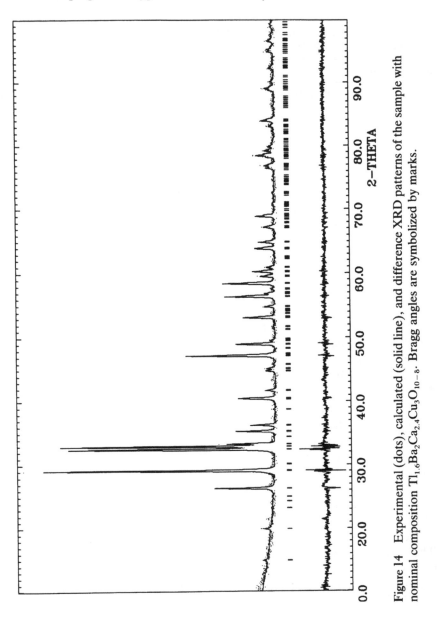

Figure 14 Experimental (dots), calculated (solid line), and difference XRD patterns of the sample with nominal composition $Tl_{1.6}Ba_2Ca_{2.4}Cu_3O_{10-\delta}$. Bragg angles are symbolized by marks.

Table 1 Results of Refinement for Sample with Nominal Composition
$Tl_{1.6}Ba_2Ca_{2.4}Cu_3O_{10}$ at 300 K[a]

Atom	Site	x	y	z	B (Å²)	g
Tl/Ca	$16n$[b]	0.041(6)	0	0.2798(2)	1.7(3)	0.24/0.01(1)
Ba	$4e$	0	0	0.1457(2)	0.4(2)	1
Ca/Tl	$4e$	0	0	0.0463(4)	0.4(6)	0.235/0.015(4)
Cu(1)	$2b$	0.5	0.5	0	0.9(5)	1
Cu(2)	$4e$	0.5	0.5	0.0893(3)	0.2(3)	1
O(1)	$4c$	0.5	0	0	1.0	1
O(2)	$8g$	0.5	0	0.090(1)	1.0	1
O(3)	$4e$	0.5	0.5	0.16(4)	1.0	1
O(4)	$16n$[c]	0.64(1)	0.5	0.280(2)	1.0	0.25

[a] R factors: R_{wp} = 12.99%, R_p = 10.11%, and R_i = 6.36%.
[b] Split into four pieces by shifting from ideal $4e$ site.
[c] Split into four pieces by shifting from ideal $4e$ site.

12.99.76%, and R_i = 6.36% for the different variables given in Table 1. Figure 14 also shows good agreement between experimental and calculated patterns. Based on the XRD refinements, the chemical composition of the sample with nominal composition $Tl_{1.6}Ba_2Ca_{2.4}Cu_3O_{10-\delta}$ can be determined as an ideal formula $(Tl_{1.92}Ca_{0.08})Ba_2 (Ca_{1.88}Tl_{0.12})Cu_3O_{10-\delta}$, which indicates the mutual autodoping between Tl and Ca in the $Tl_2Ba_2Ca_2Cu_3O_{10-\delta}$ phase. Such effects can be efficient to apply in preparing the monophasic $Tl_2Ba_2Ca_2Cu_3O_{10-\delta}$ compound. Following the vacuum and oxygen partial pressure heat treatments, the samples showed no dominant impurity phase in the XRD patterns.

In Fig. 15 we show the temperature dependence of the resistivity of the as-sintered (A) $Tl_{1.6}Ba_2Ca_{2.4}Cu_3O_{10-\delta}$ sample, that from the vacuum-annealed sample (B), that from the sample annealed in the 0.2% oxygen–nitrogen atmosphere (C), and finally, that from the sample annealed in the 2% oxygen–nitrogen atmosphere (D). The zero-resistance temperature, $T_{c(zero)}$, of the as-sintered sample was 120.5 K. The zero-resistance temperature could be enhanced to 124.5 K after vacuum annealing of the sample (see inset of Fig. 15, curves B). Believing this sample to be underdoped relative to maximum T_c, we then attempted to oxidize the vacuum-annealed sample in a partial oxygen pressure of 0.2% and 2% in nitrogen. When the vacuum-annealed sample was oxygenated in the 0.2% oxygen–nitrogen atmosphere, the $T_{c(zero)}$ increased from 124.5 K to 126.0 K (see curve C in inset of Fig. 15). This appears to be an optimally doped sample. When the same sample was further oxygenated in a 2% oxygen–nitrogen atmosphere, the $T_{c(zero)}$ was reduced to 125.7 K, compare curves C and D in inset of Fig. 15.

In Fig. 16 we show the temperature dependence of the normalized ac susceptibility [taking the limiting, low-temperature (5 K) values as −1] of the powdered $Tl_{1.6}Ba_2Ca_{2.4}Cu_3O_{10-\delta}$ sample after annealing in the 0.2% oxygen–nitrogen atmosphere. The onset of diamagnetism, $T_{c(mag)}$, appeared at 128 K (see inset, Fig. 16), and this was consistent with the $T_{c(midpoint)}$ = 128 K, measured from the resistivity curve C (Fig. 15). Under these synthetic conditions, the superconducting

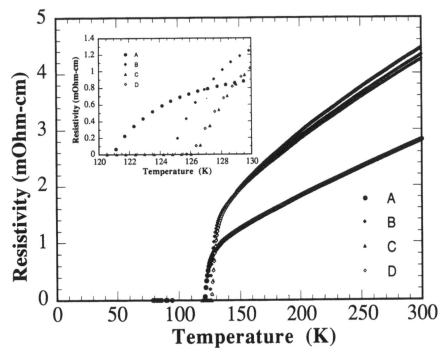

Figure 15 Temperature dependence of the resistivity of the as-sintered (A) $Tl_{1.6}Ba_2Ca_{2.4}Cu_3O_{10-\delta}$ sample, that from the vacuum annealed sample (B), that from the sample annealed in the 0.2% oxygen–nitrogen atmosphere (C), and finally that from the sample annealed in the 2% oxygen–nitrogen atmosphere (D). The inset gives greater detail of the temperature dependence of resistivity for samples A to D over the temperature range 120 to 130 K. (From Ref. 32.)

transition beginning at 128 K is very stable over a period of months, and easily reproducible.

As we discussed in Section III.A, the oxygen nonstoichiometry (δ) has a considerable effect on the superconducting transition temperature in $Tl_2Ba_2Ca_2Cu_3O_{10-\delta}$. For example, a decrease in δ arising from annealing the sample at low temperatures in oxygen results in a increase in T_c. These observations led us to propose that an increase in the hole concentration in the as-sintered samples may lead to possible improvements in T_c for the $Tl_2Ba_2Ca_2Cu_3O_{10-\delta}$ phase.

For the present systems, the weight loss between the as-sintered and low-temperature vacuum-annealed samples is about 5%. According to our previous studies [23,25], we found that the oxygen loss at 750°C for $Tl_2Ba_2Ca_2Cu_3O_{10-\delta}$ phase is less than 0.1%. Consequently, we propose that Tl loss may be the major contributing factor to the weight loss of 5%. The small amount of oxygen loss would lead to a decrease in the hole concentration in the $Tl_2Ba_2Ca_2Cu_3O_{10-\delta}$ phase. In contrast, any thallium loss (e.g., during low-temperature vacuum annealing) would create cation vacancies and subsequently increase the hole concentration in the $Tl_2Ba_2Ca_2Cu_3O_{10-\delta}$ phase. Some support for this proposal comes from the observation of a decrease in the normal-state thermopower values when one com-

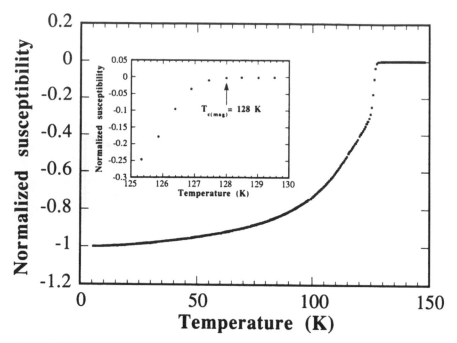

Figure 16 Temperature dependence of the normalized ac susceptibility [taking the limiting, low temperature (5 K) values as −1] of the powdered $Tl_{1.6}Ba_2Ca_{2.4}Cu_3O_{10-\delta}$ sample after annealing in the 0.2% oxygen–nitrogen atmosphere. The inset gives greater detail of the normalized ac susceptibility of the sample over the temperature range 125 to 130 K. (From Ref. 32.)

pares the values before and after vacuum annealing. The proposed increase in the hole concentration in the $Tl_2Ba_2Ca_2Cu_3O_{10-\delta}$ phase after the low-temperature vacuum annealing may be responsible for an increase in T_c. If this is the case, one still has the opportunity to further increase the hole concentration by annealing the compound in a partial oxygen pressure; this would naturally compensate for any oxygen loss occurring from the vacuum-annealed process. Based on this hypothesis, we have found that the optimizing oxygen partial pressure of 0.2% in nitrogen is effective in increasing the $T_{c(zero)}$ from 124.5 K to 126 K.

V. CONCLUDING REMARKS

It is now widely recognized that oxidation of the CuO_2 sheets above the formal valence of Cu^{2+} is one of the necessary prerequisites for inducing superconductivity in thallium cuprates. We have illustrated that such mixed valency can be created by a variety of sources involving consideration of both the geometric and electronic structure of these materials. These are (1) an internal, spontaneous redox reaction derived from strong overlapping of both oxidant (Tl-6s) and reductant (CuO_2 − $\sigma^*_{x^2-y^2}$) bands as in $Tl_{2-y}Ba_2Ca_{n-1}Cu_nO_{2n+4}$ for $0 < y < 0.5$ (Section II.A); (2) thallium vacancies as in $Tl_{2-y}Ba_2Ca_{n-1}Cu_nO_{2n+4}$ (Section II.A); (3) inherent (auto) substitution of constituent cations [e.g., Tl^{3+}/Cu^{2+} or Tl^{3+}/Ca^{2+} (Section II.C), which may also give rise to a redox process in the conducting sheets—if the substitution itself does not unduly perturb the crystal structure]; and (4) oxygen nonstoichiometry, previously known to be a crucial component in dictation of the electronic structure of $YBa_2Cu_3O_{7-\delta}$, also plays a substantial role in the occurrence and magnitude of superconductivity in thallium cuprates (Section III).

What seems clear is that many processes must be taken into consideration for the generation of excess holes within each specific class of thallium cuprates. The origin of these redox processes is thus complex, but arising out of such a combination of "crystal chemistry" and "crystal physics," considerable progress can be achieved in determining the type and extent of doping to maximize the superconducting properties. In Section IV we illustrated such effects for $Tl_2Ba_2Ca_2Cu_3O_{10-\delta}$, where consideration of both inherent (auto) doping (Tl/Ca) and oxygen nonstoichiometry can lead to an efficient and reproducible approach for attaining superconductivity at 128 K.

ACKNOWLEDGMENTS

We are grateful to J. L. Tallon (DSIR, New Zealand) for several useful discussions on oxygen nonstoichiometry effects and to C. Michel (ISMRa, France) for the XRD refinements of the $Tl_{1.6}Ba_2Ca_{2.4}Cu_3O_{10+\delta}$ sample. One of us (RSL) thanks the Materials Research Laboratories, Industrial Technology Research Institute, Taiwan, ROC, for support of my study in Cambridge. We also thank the BICC plc, Fellowship of Engineering, BP, and SERC for the support.

REFERENCES

1. Z. Z. Sheng and A. M. Hermann, *Nature* 332, 55 (1988).
2. Z. Z. Sheng and A. M. Hermann, *Nautre* 332, 138 (1988).
3. For reviews, see: (a) S. S. P. Parkin, V. Y. Lee, A. I. Nazzal, R. Savoy, T. C. Huang, G. Gorman, and R. Beyers, *Phys. Rev. B* 38, 6531 (1988); (b) A. W. Sleight, *Science* 242, 1519 (1988); (c) R. J. Cava, *Science* 247, 656 (1990); (d) B. Raveau, C. Michel, M. Hervieu, and D. Groult, *Crystal Chemistry of High-T_c Superconducting Copper Oxides*, Springer-Verlag, Berlin, 1991.
4. L. F. Mattheiss and D. R. Hamann, *Phys. Rev. B* 38, 5012 (1988).
5. D. Jung, M.-H. Whangbo, N. Herron, and C. C. Torardi, *Physica C* 160, 381 (1989).
6. A. Manthiram, M. Paranthaman, and J. B. Goodenough, *Physica C* 171, 135 (1990).
7. T. Suzuki, M. Nagoshi, Y. Fukuda, Y. Syono, M. Nikuchi, N. Kobayashi, and M. Tachiki, *Phys. Rev. B* 40, 5184 (1989).
8. S. Nakajima, M. Kikuchi, T. Oku, N. Kobayashi, T. Suzuki, K. Nagase, K. Hiraga, Y. Muto, and Y. Syono, *Physica C* 160, 458 (1989).
9. T. Suzuki, M. Nagoshi, Y. Fukuda, S. Nakajima, M. Kikuchi, Y. Syono, and M. Tachiki, *Physica C* 162–164, 1387 (1989).
10. R. S. Liu, P. P. Edwards, Y. T. Huang, S. F. Hu, and P. T. Wu, *J. Solid State Chem.* 86, 334 (1990), and the reference therein.
11. R. D. Shannon, *Acta Crystallogr. A* 32, 751 (1976).
12. R. S. Liu, S. D. Hughes, R. J. Angel, T. P. Hackwell, A. P. Mackenzie, and P. P. Edwards, *Physica C*, 198, 203 (1992).
13. Y. Shimakawa, Y. Kubo, T. Manako, H. Igarashi, F. Izumi, and H. Asano, *Phys. Rev. B* 42, 10165 (1990).
14. D. E. Cox, C. C. Toradi, M. A. Subramanian, J. Gopalakrishnan, and A. W. Sleight, *Phys. Rev. B* 38, 6624 (1988).

15. I. G. Gameson, R. S. Liu, H. S. Koo, S. F. Hu, P. T. Wu, W. I. F. David, S. Hull, and P. P. Edwards, *Conference of 1991 Condensed Matter and Materials Physics* (CMMP-91), Dec. 17–20, 1991, Birmingham, U.K.

16. J. D. Jorgensen, B. W. Veal, A. P. Paulikas, L. J. Nowiciki, G. W. Crabtree, H. Clause, and W. K. Kwok, *Phys. Rev. B* 41, 1863 (1990).

17. R. G. Buckley, J. L. Tallon, I. W. M. Brown, M. R. Presland, N. E. Flower, P. W. Gilberd, M. Bowden, and N. B. Milestone, *Physica C* 156, 629 (1988).

18. J. L. Tallon, R. G. Buckley, P. W. Gilberd, and M. R. Presland, *Physica C* 158, 247 (1989).

19. Y. Shimakawa, Y. Kubo, T. Manako, and H. Igarashi, *Phys. Rev.* 40, 11400 (1989).

20. J. B. Parise, C. C. Torardi, M. A. Subramanian, J. Gopalakrishnan, and A. W. Sleight, *Physica C* 159, 239 (1989).

21. M. R. Presland, J. L. Tallon, P. W. Gilberd, and R. S. Liu, *Physica C*, 191, 307 (1992).

22. T. Kaneko, T. Wada, K. Hamada, H. Yamauchi, and S. Tanaka, *J. Appl. Phys.* 71, 2347 (1992).

23. M. R. Presland, J. L. Tallon, R. G. Buckley, R. S. Liu, and N. E. Flower, *Physica C* 176, 95 (1991).

24. J. R. Thompson, J. Brynestad, D. M. Kroeger, Y. C. Kin, S. T. Sekula, D. E. Christen, and E. D. Specht, *Phys. Rev. B* 39, 6652 (1989).

25. R. S. Liu, S. Y. Lin, K. N. Tu, P. P. Edwards, and W. Y. Liang, *Physica C* 185–189, 1805 (1991).

26. R. S. Liu and P. P. Edwards, *Physica C* 179, 353 (1991).

27. S. Nakajima, M. Kikuchi, Y. Syono, T. Oku, K. Nagase, N. Kobayashi, D. Shindo, and K. Hiraga, *Physica C* 182, 89 (1991).

28. S. Nakajima, M. Kikuchi, Y. Syono, and N. Kobayashi, *Physica C* 185–189, 673 (1991).

29. F. Izumi, T. Kondo, Y. Shimakawa, T. Manako, Y. Kubo, H. Igarashi, and H. Asano, *Physica C* 185–189, 615 (1991).

30. T. Kondo, Y. Kubo, Shimakawa, T. Manako, and H. Igarashi, *Physica C* 185–189, 669 (1991).

31. J. L. Tallon, R. S. Liu, K. L. Wu, F. J. Blunt, and P. T. Wu, *Physica C* 161, 523 (1989).

32. R. S. Liu, J. L. Tallon, and P. P. Edwards, *Physica C* 182, 119 (1991).

16

Structural Origin of Holes in Thallium Cuprate-Based High-T_c Superconductors

M. A. Subramanian and Ashok K. Ganguli*

E. I. du Pont de Nemours & Company, Inc., Wilmington, Delaware

I. INTRODUCTION

The discovery of superconductivity in Tl–Ba–Ca–Cu–O systems [1] has not only provided compounds with the highest transition temperatures ($T_c \sim 125$ K) but has also revealed a structural pattern that is common to all superconducting copper oxides. Based on structural investigations of this system [2–8], it is now known that there exists a very large family of compounds of the type $Tl_mBa_2Ca_{n-1}Cu_nO_{2n+m+2}$ which contains copper–oxygen sheets interleaved with Tl–O sheets. The value of m may be 1 or 2. The number of consecutively stacked CuO_2 layers is indicated by n. So far, the maximum n value reached is 6 for Ba analog in the $m = 1$ series. Phases of the type $A_mM_2Ca_{n-1}Cu_nO_{2n+m+2}$ ($m = 1$ or 2) are readily prepared for M = Ba, but our attempts to prepare such phases with M = Sr have not been successful. However, it is now known that $TlSr_2Ca_{n-1}Cu_nO_{2n+3}$ ($m = 1$) phases can readily be stabilized by substitution of Bi (mixture of Bi^{3+}/Bi^{5+}) or Pb (Pb^{4+}) at the Tl site [9–13], and rare earth at the Sr or Ca site [13–17]. Although the detailed structural features of these compounds have been revealed by a combination of x-ray and neutron diffraction methods [6], the mechanisms giving rise to Cu(III) (or holes) and the role played by Tl–O layers in the creation of charge carriers are some of the issues that are not fully understood.

As examples of the $Tl_mBa_2Ca_{n-1}Cu_nO_{2n+m+2}$ family, crystal structures of the tetragonal $n = 1$ phases with $m = 1$ and 2 are shown in Fig. 1. The structures can be described as perovskite-like slabs alternating with double (Fig. 1a) or single (Fig. 1b) Tl–O rock salt–type layers. When n is greater than 1, the n number of sheets of corner-sharing square-planar CuO_4 groups are oriented parallel to the

---

*Present affiliation: Iowa State University, Ames, Iowa.

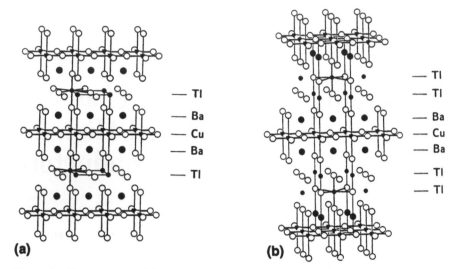

Figure 1 Structures of (a) $TlBa_2CuO_5$ and (b) $Tl_2Ba_2CuO_6$. Metal atoms are shaded and Cu–O bonds are shown.

(001) plane. Additional oxygen atoms, located above and below the consecutive Cu–O sheets, are bonded to the copper atoms of the outer two layers. There are no oxygen atoms between the Cu–O sheets. Barium ions are found above and below the Cu–O sheets in nine-coordination with oxygen and calcium atoms are located between Cu–O sheets in eightfold coordination. The structures of other members in the family are quite similar and differ from one another only by the number of Cu–O sheets.

Creation of holes in the Cu–O sheets for the $TlBa_2Ca_{n-1}Cu_nO_{2n+3}$ phases seems to be straightforward. For the series above, the formal oxidation state of Cu varies as $(2n + 1)/2$. For example the formal valence of state of Cu is 3.0 for $n = 1$, 2.5 for $n = 2$, and 2.33 when $n = 3$. In other words, in the materials above, chemistry forces the formal oxidation state of Cu to be above 2+ and hence holes in the Cu–O sheets. This is in contrast to stoichiometric $Tl_2Ba_2Ca_{n-1}Cu_nO_{2n+m+2}$ ($m = 2$ series) phases, where the chemical composition implies the formal oxidation state of Cu is always 2, irrespective of value of n. Several suggestions have been made as possible mechanisms for the introduction of holes in the Cu–O sheets for $Tl_2Ba_2Ca_{n-1}Cu_nO_{2n+4}$ phases. Tight-binding band electronic structural calculations for Tl–O double-rock-salt-layer phases have shown that the Tl-6s levels lie well below the Fermi level and remove electrons from the x^2-y^2 band of the CuO_2 sheets [18]. However, this is not the case for Tl–O single-rock-salt-layer phases, where the Tl-6s levels lie well above the Fermi level. Other possible mechanisms suggested are cation vacancies ($Tl_{2-x}Ba_2Ca_2Cu_3O_8$), cation substitutions ($Tl_{2-x}Ca_xBa_2CaCu_2O_8$), and oxygen nonstoichiometry [19].

To get insight into the mechanisms involved in the creation of holes in thallium-based superconductors, we have carried out many systematic studies involving substitutions of various cations in crystal structure of $Tl_mM_2Ca_{n-1}Cu_nO_{2n+m+2}$ and its implications on superconducting properties. In this chapter we discuss the results on the cation substitution studies in $Tl_2Ba_2CuO_6$ ($m = 2$, $n = 1$) and $TlBa_2CuO_5$-type ($m = 1$, $n = 1$) phases.

II. $Tl_2Ba_{2-x}Sr_xCuO_6$: A SYSTEM EXHIBITING COMPOSITIONALLY CONTROLLED SUPERCONDUCTOR–METAL TRANSITION

Perhaps the most enigmatic of $Tl_mBa_2Ca_{n-1}Cu_nO_{2n+m+2}$ phases is $Tl_2Ba_2CuO_6$ ($m = 2$, $n = 1$). This compound was first synthesized in our laboratory as a single phase in sealed gold tubes and shown to have T_cs close to 90 K [4]. However, this compound can be prepared as nonsuperconducting as well as superconducting with T_cs varying from 20 to 90 K [20]. This superconductor is unique in the sense that it has the highest T_c for any superconductor with one Cu–O sheet (Fig. 1a). The composition as written implies a formal oxidation state for Cu of 2+, and this composition, containing no excess holes, would not be expected to superconduct on the basis of comparison with other Cu–O-based superconductors. High-resolution neutron powder diffraction studies [21] of 90-K superconducting single-phase $Tl_2Ba_2CuO_6$ samples (prepared at Du Pont by heating the reactants in sealed gold tubes) have shown no significant oxygen excess or cation vacancies. As suggested by the tight-binding band-structure calculations, the creation of holes in this phase is due to the overlap of Tl-6s levels with Cu-3d levels at the Fermi level. As a part of our ongoing substitution studies in the thallium-based superconductors, we have studied the effect of Sr substitution on the superconducting properties of $Tl_2Ba_2CuO_6$.

Oxides of the type $Tl_2Ba_{2-x}Sr_xCuO_6$ were prepared by heating stoichiometric amounts of Tl_2O_3, BaO_2, SrO_2, and CuO in a sealed gold tubes at 875°C for 6 h. The samples were cooled to room temperature in the furnace. Single phases could be obtained until $x = 1.2$ and could be indexed on a body-centered tetragonal lattice (space group $I4/mmm$) as the parent $Tl_2Ba_2CuO_6$ oxide. Beyond $x = 1.2$, lines due to impurities started appearing and increased in intensity with further increase in Sr concentration. In Fig. 2 we show the variation of unit cell parameters with x. Both a and c parameters decrease smoothly as a function of x. This is easily explained from ionic size consideration since Sr is much smaller in size than Ba. Figure 3 shows the change in the superconducting transition temperature (T_c) as a function of x. Up to $x = 0.8$ there is a gradual change in T_c, after which T_c drops sharply and goes to zero at $x = 1.2$. The $x = 1.2$ composition is metallic down to 4.2 K. Similar behavior can be found in the $Tl_2Ba_{2-x}Sr_xCaCu_2O_8$ system, where T_c decreases from 105 K ($x = 0$) to 44 K ($x = 2.0$) [22].

Earlier, Jung et al. [18] carried out tight-binding band-structure calculations on the single ($m = 1$) and double ($m = 2$) Tl–O layers of $Tl_mBa_2Ca_{n-1}Cu_nO_{2n+m+2}$, taking into consideration the effect of local distortions. Their calculations revealed that in the case of double-Tl–O-layer compounds, the Tl-6s bands overlap at the Fermi level with the Cu-3d x^2-y^2 bands, whereas the Tl-6s bands are found well above the Fermi level in the case of single-Tl–O-layer compounds. In other words, creation of holes in the copper oxygen sheets is achieved by internal redox mechanism in the double Tl–O layer cuprates. Substitution of less electropositive Sr for Ba in $Tl_2Ba_2CuO_6$, should bring down the Tl-6s level and increase the overlap of Tl-6s with Cu-3d x^2-y^2, leading to an increase in the hole concentration. If this is true, the Cu–O distance should decrease as x increases. Since the Cu–O distance is roughly one-half of the a lattice parameter (this is true for thallium cuprates, as

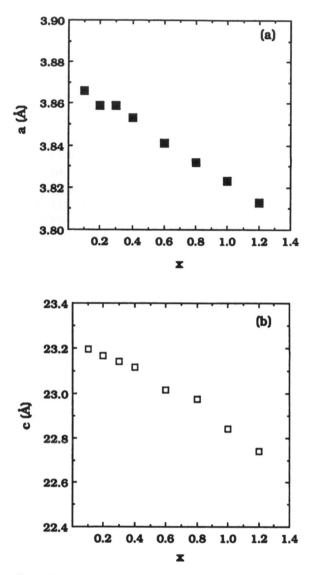

Figure 2   Variation of a and c lattice parameters as a function of x for $Tl_2Ba_{2-x}Sr_xCuO_6$.

the Cu–O sheets are flat), the a lattice parameter should decrease as x is increased. In the case of $Tl_2Ba_{2-x}Sr_xCuO_6$, the observed rapid decrease of a lattice parameter (Fig. 2a) is caused mainly by size effects, and any decrease due to increase in the covalency of Cu–O bonds should be negligibly small to distinguish. The decrease in T_c with increase in x is probably associated with the overdoping of Cu–O sheets, and when $x = 1.2$, the phase becomes a normal metal. Thus this transition is very similar to the one observed in $La_{2-x}Sr_xCuO_4$, where the superconductor–metal transition is brought about by an overdoping of holes through the substitution of Sr^{2+} for La^{3+} $(x > 0.3)$ [23].

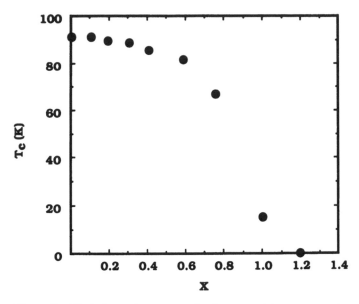

Figure 3 Variation of superconducting transition temperature as a function of x for $Tl_2Ba_{2-x}Sr_xCuO_6$.

Veal et al. [24] have examined the system $YBa_{2-x}Sr_xCu_3O_7$ and discussed the possible ways in which Sr substitution might lower T_c. In their view the depression of T_c is indirectly caused by a change in the structure around the Sr ion and due to additional oxygen vacancies at the O(1) sites (chain) in the Sr-doped samples. It is also argued that the electronic is not affected significantly by the substitution of Ba by Sr. The effect of Sr substitution on lattice parameter and the superconducting properties are similar in both $YBa_{2-x}Sr_xCu_3O_7$ and $Tl_2Ba_{2-x}Sr_xCuO_6$. However, in the case of $YBa_{2-x}Sr_xCu_3O_7$, the decrease in T_c may be related directly to the oxygen vacancies, whereas the changes in Cu–O bandwidth (resulting in an increase in overlap of Tl-6s at the Fermi level) due to Sr substitution may play an important role in the observed decrease of T_c in $Tl_2Ba_{2-x}Sr_xCuO_6$.

III. $TlBa_xSr_{1-x}LaCuO_5$: A SYSTEM EXHIBITING COMPOSITIONALLY CONTROLLED SUPERCONDUCTOR–SEMICONDUCTOR TRANSITION

The phase $TlBa_2CuO_5$ ($m = 1$ and $n = 1$ in $TlBa_2Ca_{n-1}Cu_nO_{2n+3}$, Fig. 1b) is reported to be a nonsuperconductor [25], presumably due to the presence of all the coppers in the $+3$ oxidation state. Our attempts to synthesize single-phase samples of this compound were not successful. However, TlM_2CuO_5 (M = Ba, Sr) can be stabilized by appropriate substitution: Pb or Bi at the Tl site [12], rare earth at the M site [15–17,26], or transition metal [Fe,Co] at the Cu site [27]. In the case of $TlBa_{2-x}La_xCuO_5$, single-phase samples could be prepared for $x = 0.5$ to 1.0. Superconductivity is observed for samples for $x = 0.5$ to 0.85. A plot of T_c versus x showed a maximum (57 K) for the sample with $x = 0.75$ (Fig. 4).

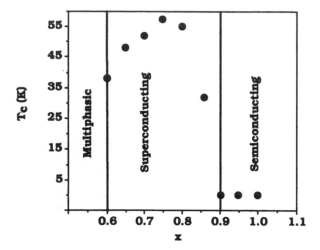

Figure 4 Variation of superconducting transition temperature as a function of x for $TlBa_{2-x}La_xCuO_5$.

Electrical resistivity and μSR studies (Y. J. Uemura and M. A. Subramanian, unpublished results) have shown that the end member, $TlBaLaCuO_5$ ($x = 1$) with all the coppers in the $+2$ formal oxidation state, is an antiferromagnetic semiconductor. Neutron diffraction studies [28] of single-phase $TlBaLaCuO_5$ and $TlBa_{1.2}La_{0.8}CuO_5$ have shown that compounds are stoichiometric with respect to cation and oxygen content. This shows clearly that the substitution of La for Ba in $TlBa_2CuO_5$ lowers the formal valence state of Cu below 3, thereby stabilizing the phases. Superconductivity is observed when the formal oxidation state lies between 2.15 and 2.5 and a maximum T_c for $TlBa_{1.25}La_{0.75}CuO_5$ with a formal oxidation state of 2.25.

Recently, we reported superconductivity in the cadmium-substituted thallium cuprates of the type $Tl_{1-x}Cd_xBaLaCuO_5$ [29]. A single-phase region exists for $x = 0$ to 0.4. Superconductivity is observed over the region $0.2 \le x \le 0.4$ with transition temperatures in the range 38 to 50 K. X-ray diffraction data suggest that divalent cadmium substitutes for trivalent thallium, thereby increasing the number of holes in the Cu–O $d(x^2-y^2)$ conduction band in $TlBaLaCuO_5$. A plot of T_c versus x showed a maximum when the formal oxidation state of copper is close to 2,3 as observed in the $TlBa_{2-x}La_xCuO_5$ series.

Similar investigations on $TlSr_{2-x}La_xCuO_5$ showed that the single-phase samples are formed for $x = 0.6$ to 1.0 [17]. Magnetic and electrical measurements showed that the compounds are superconducting for all the values of x. The end member $TlSrLaCuO_5$ with all the coppers in the $+2$ oxidation state is a superconductor, and this is in contrast to $TlBa_{2-x}La_xCuO_5$ phases, where the end member $TlBaLaCuO_5$ is a semiconductor. Recent neutron diffraction studies showed that the compound $TlSrLaCuO_5$ is stoichiometric and is similar to $TlBaLaCuO_5$ (G. H. Kwei and M. A. Subramanian, unpublished results). On the basis of tight-binding band structure calculations, we examined why $TlSrLaCuO_5$ is a superconductor, whereas its analog $TlBaLaCuO_5$ is not. Our study [30] strongly suggests that for a

Tl–O single-rock-salt-layer phase with a very short in-plane Cu–O distance such as TlSrRCuO$_5$, the CuO$_2$ layer x^2–y^2 band is raised high enough in energy for the Tl–O layer to create holes in the CuO$_2$ sheet. To test this hypothesis we synthesized solid solutions of the type TlBa$_x$Sr$_{1-x}$LaCuO$_5$. The compounds were prepared by heating stoichiometric amounts of Tl$_2$O$_3$, BaO$_2$, SrO$_2$, and CuO in sealed gold tubes at 860°C for 12 h. The samples were cooled to room temperature in the furnace. X-ray diffraction data showed that single-phase samples are formed for all values of x. Figure 5 shows the variation of a and c lattice parameters (tetragonal, space group: *P4/mmm*) with x. Substitution of larger Ba^{2+} for smaller Sr^{2+} gradually increased the a and c lattice parameters, as expected. The Cu–O bond distance is roughly one-half of the a lattice parameter. Superconductivity was observed for compounds in the range $0.3 \geq x \geq 0$. Compounds in the range $1.0 \geq x \geq 0.6$ are semiconducting. Compounds with x values of 0.4 and 0.5 showed temperature-independent electrical behavior (Table 1). So it is clear that there exists a correlation between the in-plane Cu–O distance and the occurrence of superconductivity in TlM$_2$CuO$_5$-type phases.

It is interesting to note that Tl$^{3+}_{0.5}$Pb$^{4+}_{0.5}$Sr$_2$CuO$_5$ is a metal and not a superconductor. Neutron diffraction studies (31) showed that the compound is stoichiometric, and hence the formal oxidation state of Cu is close to +2.5. Recently (32), we have successfully synthesized nearly single-phase TlSr$_2$CuO$_5$ (orthorhombic symmetry) and found it to exhibit metallic behavior. Both Tl$_{0.5}$Pb$_{0.5}$Sr$_2$CuO$_5$ and TlSr$_2$CuO$_{5-x}$ have very short in-plane Cu–O distance (1.865 and 1.867 Å, respectively), which is comparable to superconducting TlSrLaCuO$_5$ (1880 Å) with all the coppers in a +2 formal oxidation state. This suggests that the overlapping of Tl-6s (as well as Pb-6s for Tl$_{0.5}$Pb$_{0.5}$Sr$_2$CuO$_5$) bands with Cu-3$d(x^2$–$y^2)$ bands may actually increase the hole concentration either to an optimum level for super-

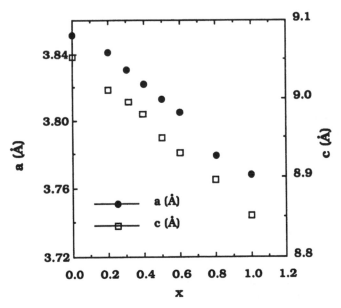

Figure 5 Variation of a- and c-lattice parameters as a function of x for TlBa$_{1-x}$Sr$_x$LaCuO$_5$.

Table 1 Electrical Data for $TlSr_{1-x}Ba_xLaCuO_5$ Phases

x	ρ at 300 K (Ω-cm)	ρ at 4.2 K (Ω-cm)	Comment
0.0	2.5×10^{-3}	—	$T_c \sim 42$ K
0.2	3.3×10^{-3}	—	$T_c \sim 44$ K
0.3	2.0×10^{-3}	—	$T_c \sim 37$ K
0.4	7.0×10^{-3}	6.5×10^{-3}	Semimetal
0.5	5.6×10^{-3}	6.0×10^{-3}	Semimetal
0.6	2×10^{-2}	1×10^{-1}	Semiconductor
0.8	4×10^{0}	2×10^{2a}	Semiconductor
1.0	1.3×10^{2}	4×10^{3a}	Semiconductor

[a]Measured at 77 K.

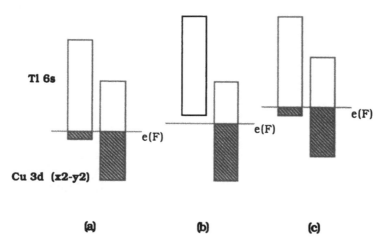

Figure 6 Schematic descriptions of the relative energies of the Tl-6s block bands and the Cu-$d(x^2-y^2)$ block bands in (a) $Tl_2Ba_2CuO_6$, (b) $TlBaLaCuO_5$, and (c) $TlSrLaCuO_5$. $e_{(F)}$ is the energy at the Fermi level.

conductivity (as in $TlSrLaCuO_5$ and $TlSr_{2-x}La_xCuO_5$, $0 \le x \le 0.2$) or to an overdoped region (as in $Tl_{0.5}Pb_{0.5}Sr_2CuO_5$ and $TlSr_2CuO_{5-x}$) giving rise to metallic behavior.

IV. SUMMARY

Based on the studies discussed above, it can be inferred that in $Tl_2Ba_2Cu_{n-}Cu_nO_{2n+4}$ (Tl–O double-rock-salt-layer) superconductors Tl-6s block bands overlap with Cu-3d bands at the Fermi level (i.e., the Tl–O layers create holes in the Cu–O sheet) (Fig. 6a). Incorporation of excess oxygen in Tl_2O_2 layers ($Tl_2Ba_2CuO_{6+x}$), cation substitutions ($Tl_{2-x}Ca_xBa_2CaCu_2O_8$), or cation vacancies ($Tl_{2-x}Ba_2Ca_2Cu_3O_{10}$) may play a role in terms of altering the number of charge carriers (holes) in the conduction band and hence superconducting properties. In many cases such defects seems to increase the number of holes beyond a certain optimum value (to an overdoped region), and this causes a decrease in the transition temperature or even destruction of superconductivity. In the case of $TlBa_2Ca_{n-2}Cu_nO_{2n+3}$ (Tl–O single-rock-salt-layer) superconductors, the metal chemistry forces the oxidation of Cu–O sheets. Successful synthesis of semiconducting $TlBaLaCuO_5$ with all the coppers in the $+2$ formal oxidation state (Tl–O single-rock-salt-layer analog of superconducting $Tl_2Ba_2CuO_6$) shows that the Tl-$6s$ block band lies well above the Fermi level and does not remove electrons from the Cu–O layer $d(x^2-y^2)$ band (Fig. 6b). Observation of superconductivity in $TlSrLaCuO_5$ and a superconductor–semiconductor transition in $TlSr_xBa_{2-x}LaCuO_5$ strongly suggests that for a single-Tl–O-rock-salt-layer phase with very short in-plane Cu–O distance, the Cu–O $d(x^2-y^2)$ band is raised high enough in energy for the Tl–O layer to create holes in the Cu–O sheet (Fig. 6c).

REFERENCES

1. Z. Z. Sheng and A. M. Hermann, *Nature (London)* 332, 138 (1988).
2. M. A. Subramanian, J. C. Calabrese, C. C. Torardi, J. Gopalakrishnan, T. R. Askew, R. B. Flippen, K. J. Morrissey, U. Chowdry, and A. W. Sleight, *Nature* 332, 420 (1988).
3. C. C. Torardi, M. A. Subramanian, J. C. Calabrese, J. Gopalakrishnan, K. J. Morrissey, T. R. Askew, R. B. Flippen, U. Chowdhry, and A. W. Sleight, *Science* 240, 631 (1988).
4. C. C. Torardi, M. A. Subramanian, J. C. Calabrese, J. Gopalakrishnan, E. M. McCarron, K. J. Morrissey, T. R. Askew, R. B. Flippen, U. Chowdhry, and A. W. Sleight, *Phys. Rev. B* 38, 225 (1988).
5. A. W. Sleight, *Science* 242, 1519 (1988).
6. A. W. Sleight, M. A. Subramanian, and C. C. Torardi, *MRS Bull.* XIV, 45 (1989).
7. C. N. R. Rao and B. Raveau, *Acc. Chem. Res.* 22, 106 (1989).
8. R. Beyers and T. M. Shaw, in *Solid State Physics*, Vol. 42, ed. H. Ehrenreich and D. Turnbull, Academic Press, New York, 1989, p. 135.
9. P. Halder, S. Sridhar, A. Roig-Janicki, W. Kennedy, D. H. Wu, C. Zahopoulos, and B. C. Giessen, *J. Supercond.* 1, 21 (1988).
10. M. A. Subramanian, C. C. Torardi, J. Gopalakrishnan, P. L. Gai, J. C. Calabrese, T. R. Askew, R. B. Flippen, and A. W. Sleight, *Science* 242, 249 (1988).
11. S. Li and M. Greenblatt, *Physica C* 157, 365 (1989).
12. C. Martin, D. Bourgault, C. Michel, J. Provost, M. Hervieu, and B. Raveau, *Eur. J. Solid State Inorg. Chem.* 26, 1 (1989).
13. M. A. Subramanian, P. L. Gai, and A. W. Sleight, *Mater. Res. Bull.* 25, 101 (1990).
14. M. A. Subramanian, P. L. Gai, and M.-H. Whangbo, *Mater. Res. Bull.* 25, 899 (1990).
15. D. Bourgault, C. Martin, C. Michel, M. Hervieu, J. Provost, and B. Raveau, *J. Solid State Chem.* 75, 212 (1988).
16. A. K. Ganguli, V. Manivannan, A. K. Sood, and C. N. R. Rao, *Appl. Phys. Lett.* 55, 2664 (1989).
17. M. A. Subramanian, *Mater. Res. Bull.* 25, 191 (1990).
18. D. Jung, M.-H. Whangbo, N. Herron, and C. C. Torardi, *Physica C* 160, 381 (1989).
19. M. A. Subramanian, *Ceram. Trans.* 13, 59 (1990).
20. Y. Shimakawa, Y. Kubo, T. Manako, and H. Igarashi, *Phys. Rev. B* 40, 11400 (1989).
21. J. B. Parise, C. C. Torardi, M. A. Subramanian, J. Gopalakrishnan, A. W. Sleight, and E. Prince, *Physica C* 159, 239 (1989).
22. E. A. Hayri and M. Greenblatt, *Physica C* 156, 775 (1988).

23. J. B. Torrance, Y. Tokura, A. I. Nazzal, A. Bezinge, T. C. Huang, and S. S. P. Parkin, *Phys. Rev. Lett.* 61, 1127 (1988).
24. B. W. Veal, W. K. Kwok, A. Umezawa, G. W. Crabtree, J. D. Jorgensen, J. W. Downey, L. J. Nowicki, A. W. Mitchell, A. P. Paulikas, and C. H. Sowers, *Appl. Phys. Lett.* 51, 279 (1987).
25. R. Beyers, S. S. P. Parkin, V. Y. Lee, A. I. Nazzal, R. Savoy, G. Gorman, T. C. Huang, and S. La Placa, *Appl. Phys. Lett.* 53, 432 (1988).
26. T. Manako, Y. Shimakawa, Y. Kubo, T. Satoh, and H. Igarashi, *Physica C* 158, 143 (1989).
27. A. K. Ganguli, M. A. Subramanian, and G. H. Kwei, *J. Solid State Chem.* 91, 397 (1991).
28. M. A. Subramanian, G. H. Kwei, J. B. Parise, J. A. Goldstone, and R. B. Von Dreele, *Physica C* 166, 19 (1990).
29. M. A. Subramanian and A. K. Ganguli, *Mater. Res. Bull.* 26, 91 (1991).
30. M.-H. Whangbo and M. A. Subramanian, *J. Solid State Chem.* 91, 403 (1991).
31. G. H. Kwei, J. B. Shi, and H. C. Ku, *Physica C* 174, 180 (1991).
32. A. K. Ganguli and M. A. Subramanian, *J. Solid State Chem.* 93, 250 (1991).

17

Dependence of T_c on the Hole Concentration in Superconducting Thallium Cuprates with Single Tl–O Layers

C. N. R. Rao

Indian Institute of Science, Bangalore, India

I. INTRODUCTION

Since the discovery of superconducting thallium cuprates [1] of the general formula $Tl_2Ca_{n-1}Ba_2Cu_nO_{2n+4}$, detailed investigations have been reported on a variety of thallium cuprates [2]. Some of the studies have been directed toward understanding the electronic structure and the origin of holes in thallium cuprates containing one or two Tl–O layers. Despite such efforts, the exact determination of hole concentration in the different thallium cuprates has remained somewhat problematic, making it difficult to describe the dependence of T_c on hole concentration in these materials. In cuprates of the $La_{2-x}M_xCuO_4$ (M = Ca, Sr, Ba), $YBa_2Cu_3O_{7-\delta}$, and $Bi_2Ca_{n-1}Sr_2Cu_nO_{2n+4}$ families, however, it is now fairly well established that in a given family, T_c attains a maximum value at an optimal hole concentration [3,4], as shown in Fig. 1. All the cuprates containing two CuO_2 layers, exhibit T_c maxima around a hole concentration of 0.2.

Band structure calculations [5] have indicated that in the double-Tl–O-layer cuprates, the bottom of the Tl(6s) band lies significantly below the Fermi level, while in the single-Tl–O-layer cuprates the Tl(6s) band lies well above the Fermi level. The metal chemistry of the single-Tl–O-layer cuprates is such as to create holes in the CuO_2 layers. In the two-Tl–O-layer cuprates, it has been suggested that an internal oxidation–reduction reaction mechanism could be operative. Other sources of holes are thallium vacancies, cation disorder, and excess oxygen in the Tl–O layer. Determination of hole concentration in the thallium cuprates directly by iodometric or redox titrations is difficult. Manthiram et al. [6] have described a useful wet-chemical method for determining the hole concentration in thallium cuprates. Their method involves the iodometric determination of the total oxidizing power (above the formal oxidation states of Tl^+, Cu^+, and O^{2-}) and an inde-

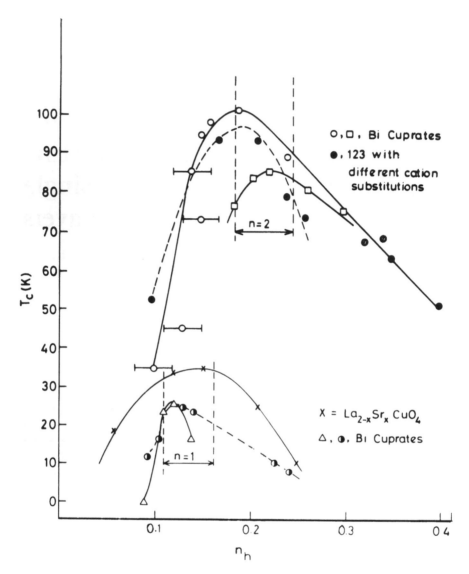

Figure 1 Variation of T_c with hole concentration n_h in superconducting cuprate families. (Adapted from Ref. 4.)

pendent determination of the thallium content. The oxidation state of copper is then computed; a value of less than $2+$ is taken to indicate $Tl(6s)$–CuO_2 band overlap, whereas a value of greater than $2+$ gives the hole concentration. The method has been applied successfully to determine concentration of holes in single-Tl–O-layer cuprates and in double-Tl–O-layer cuprates with certain assumptions [6]. Gopalakrishnan et al. [7] have proposed a method based on the selective oxidation of bromide ions by holes on copper; Tl(III) does not oxidize Br^- ions. The method gives satisfactory estimates of hole concentration in cuprates containing

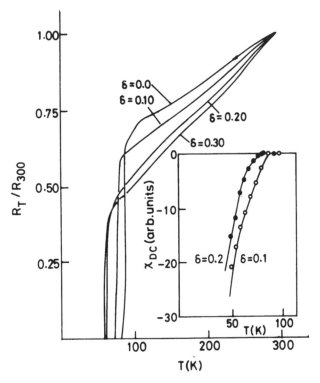

Figure 2 Temperature variation of resistivity of $Tl_{1-\delta}CaBa_2Cu_2O_7$. Inset shows dc susceptibility data. (Adapted from Ref. 8.)

uncompensated holes, especially in single-Tl–O-layer systems. Clearly, these wet-chemical procedures for the estimation of n_h are useful, although not universally applicable to both single- and double-Tl–O-layer cuprates, independent of the origin of the holes.

To describe the variation of T_c with hole concentration in single-Tl–O-layer cuprates, we can, in principle, prepare a set of derivatives with a range of n_h values and plot the experimentally determined T_cs against n_h determined by a wet-chemical procedure. Typical of such a system would be $TlY_{1-x}Ca_xSr_2(Ba_2)Cu_2O_7$, where the hole concentration would increase with increase in x, the $x = 0.0$ composition being an insulator. We have carried out an experimental study of T_c and n_h on $Tl_{1-\delta}CaBa_2Cu_2O_7$, $Tl_{1-\delta}CaBa_2Cu_2O_7$, and $TlLn_{1-x}Ca_xSr_2Cu_2O_7$. More important, we have systematically investigated the $Tl_{1-y}Pb_yY_{1-x}Ca_xSrCu_2O_7$ system, where substitution of trivalent Tl by quadrivalent Pb will have an effect opposite to that of the substitution of trivalent Y by divalent Ca. (Note that the composition $x = y = 0$ is an insulator.) If we assume that Pb^{4+} introduces electrons while Ca^{2+} introduces holes, $(x - y)$ will give the effective hole concentration. We show that T_c maximum occurs at the optimal $(x - y)$ value in this system. The present study of the $Tl_{1-y}Pb_yY_{1-x}Ca_xSr_2Cu_2O_7$ system therefore gives the variation of T_c with n_h ($\equiv x - y$) by a method independent of wet-chemical procedures. We also relate T_c to the in-plane Cu–O distance, which reflects the effect of hole doping.

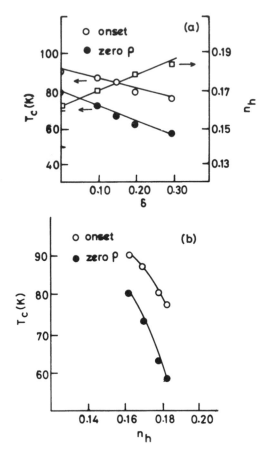

Figure 3 Variation of T_c with (a) δ and (b) n_h in $Tl_{1-\delta}CaBa_2Cu_2O_7$. (Adapted from Ref. 8.)

II. $Tl_{1-\delta}CaBa_2Cu_2O_7$

It has been shown [8] that some thallium deficiency is necessary to obtain mono-phasic compositions of the 1122 phase without any impurity of the corresponding 2122 phase. In Fig. 2 the resistivity data of a few members of this family is shown along with dc susceptibility data. We show the variation of T_c with δ in Fig. 3a and that with the experimentally determined n_h values [7] in Fig. 3b. The T_c decreases with increase in n_h and there appears to be a T_c maximum around $n_h \approx 0.16$. We do not, however, see the maximum because all the compositions belong to the overdoped regime. In the underdoped regime, T_c would have increased with n_h giving the left half of the curve in Fig. 3b.

III. $Tl_{1-\delta}Y_{1-x}Ca_xBa_2Cu_2O_7$

The parent member of this family with $x = 0.0$ is an insulator. In Fig. 4 we show resistivity data of a few members of this family. We have plotted the T_cs against the calcium content in Fig. 5. T_c shows a maximum at $x \approx 0.75$, suggesting that the hole concentration at this composition is optimal. In Fig. 6 we have plotted

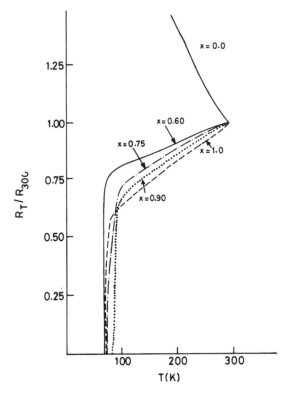

Figure 4   Resistivity data of $Tl_{0.9}Y_{1-x}Ca_xBa_2Cu_2O_7$. (Adapted from Ref. 8.)

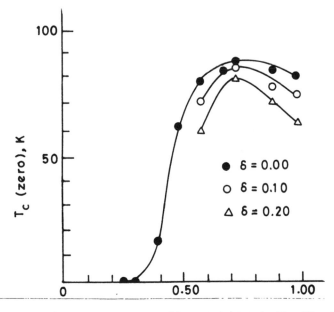

Figure 5   Variation of T_c with composition in $Tl_{1-\delta}Y_{1-x}Ca_xBa_2Cu_2O_7$. (Adapted from Ref. 8.)

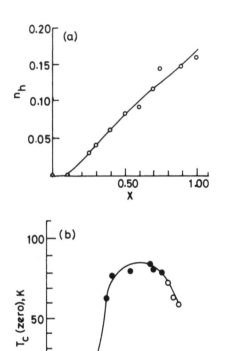

Figure 6 (a) Variation of n_h with x in $TlY_{1-x}Ca_xBa_2Cu_2O_7$; (b) variation of T_c with n_h in $TlY_{1-x}Ca_xBa_2Cu_2O_7$ (filled circles) and $Tl_{1-\delta}CaBa_2Cu_2O_7$ (open circles). (Adapted from Ref. 8.)

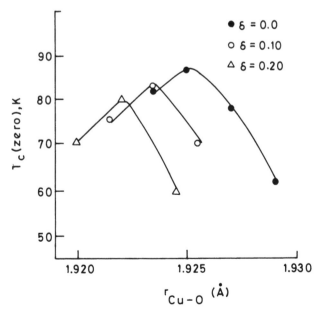

Figure 7 Variation of T_c with the in-plane Cu–O distance in the $Tl_{1-\delta}Y_{1-x}Ca_xBa_2Cu_2O_7$. (Adapted from Ref. 8.)

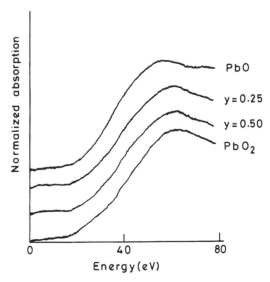

Figure 8 Pb-L$_3$ XANES of Tl$_{1-x}$Pb$_y$CaSr$_2$Cu$_2$O$_7$ along with that of PbO$_2$ and PbO. (Adapted from Ref. 13.)

the T_cs against the experimentally determined n_h values [7]. We observe the T_c maximum around an n_h value of 0.13.

Superconducting transition temperatures (T_c) of the cuprate superconductors have been correlated with a large number of structural parameters, such as in-plane Cu–O bond lengths, bond valence sums, Madelung potentials, and the electronegativity of the constituent ions. Among these, the in-plane Cu–O bond length (r_{Cu-O}) is an important structural parameter that reflects the extent of oxidation of the CuO$_2$ layer (i.e., the n_h value) on doping. Furthermore, unlike the r_{Cu-O}

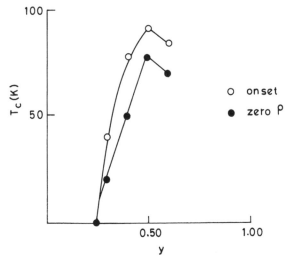

Figure 9 Variation of T_c with composition in Tl$_{1-y}$Pb$_y$CaSr$_2$Cu$_2$O$_7$. (Adapted from Ref. 13.)

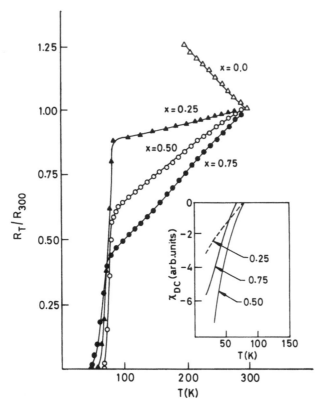

Figure 10 Variation of resistivity with temperature in $TlEu_{1-x}Ca_xSr_2Cu_2O_7$. Inset shows dc magnetic susceptibility data. (Unpublished results from this laboratory.)

values, accurate values of n_h are not known for all cuprate superconductors. The T_c versus r_{Cu-O} correlation therefore provides a simple and practical means of studying the T_c versus n_h relation. The r_{Cu-O} distance is obtained easily from the a parameter and is close to one-half of the a parameter in the thallium cuprates.

As n_h increases, the in-plane Cu–O bond length gets shortened, as expected, because oxidation of the CuO_2 layer removes electrons from the band which has antibonding character. In addition to this electronic factor, the r_{Cu-O} distance is affected by the steric factor caused by the cation located above and below the CuO_2 sheets at nine-coordinated sites. With the increasing size of the nine-coordinated site cation, the in-plane Cu–O bond gets lengthened to reduce the strain. For example, $TlBaLaCuO_5$ has a much larger r_{Cu-O} (1.925 Å) than $TlSrLaCuO_5$ (1.881 Å). It has been shown from tight-binding band structure calculations [9] that the very short Cu–O distance in $TlSrLaCuO_5$ ($T_c \sim 40$ K) raises the $x^2–y^2$ band of the CuO_2 layer sufficiently high in energy to cause an overlap between the CuO_2 layer and the Tl–O layer and creates holes in the CuO_2 layer. In the corresponding Ba compound, where the r_{Cu-O} is longer, the $x^2–y^2$ band of the CuO_2 layer lies well below the Tl-6s band and there is no overlap between the two. Accordingly, $TlBaLaCuO_5$ is not a superconductor. This shows that the in-plane Cu–O bond length is a relevant parameter to correlate the structure with properties.

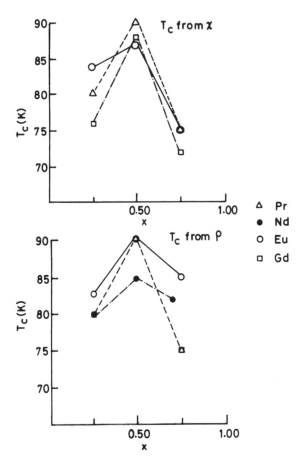

Figure 11 Variation of T_c with composition in $TlLn_{1-x}Ca_xSr_2Cu_2O_7$ (Ln = rare earth). (Unpublished result from this laboratory.)

Plots of T_c versus r_{Cu-O} distances for undoped and doped cuprate superconductors have been given in the literature [10] and they are grouped into three classes depending on the size of the nine-coordinated cation (La, Sr, and Ba classes). All three classes show dome-shaped curves similar to the T_c versus n_h curves. Each class shows a maximum in T_c at an optimal r_{Cu-O} distance. We considered it instructive to plot the T_cs against the r_{Cu-O} distance in the $Tl_{1-\delta}Y_{1-x}Ca_xBa_2Cu_2O_7$ system for different δ values. Such plots are shown in Fig. 7. It is evident from the figure that the T_c reaches a maximum at an optimal value of the Cu–O distance in the range 1.921 to 1.926 Å, depending on the value of δ. The value of maximum T_c and the optimal r_{Cu-O} value at the T_c maximum both increase with decrease in δ. The r_{Cu-O} values at maximum T_c found here are consistent with those of the Ba class of cuprates [10]. It should, however, be noted that the role of apical oxygen in relating the structure and properties in the cuprate superconductors should not be ignored.

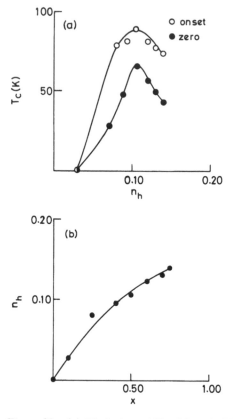

Figure 12   (a) Variation of T_c with n_h in $TlGd_{1-x}Ca_xSr_2Cu_2O_7$; (b) variation of n_h with x in the same series. (Unpublished results from this laboratory.)

IV. $Tl_{1-x}Pb_yCaSr_2Cu_2O_7$

It has been known for some time that substitution of Tl by Pb stabilizes the Tl–Ca–Sr–Cu–O system. XANES measurements (Fig. 8) show that Pb is in the 4+ state in such cuprates. Properties of $Tl_{1-y}Pb_yCaSr_2Cu_2O_7$ have been investigated [11] as a function of the Pb content. In Fig. 9 we show the variation of T_c with y. The T_c increases with increase in y up to $y = 0.5$ and then decreases, indicating that the hole concentration is optimal at this composition. It is not possible to obtain meaningful n_h values by chemical titrations in this system because of the presence of Pb. The T_c in this system also shows a maximum with the Cu–O distance around 1.905 Å corresponding to the range of the Sr class of thallium cuprates [10].

V. $TlLn_{1-x}Ca_xSr_2Cu_2O_7$

Just as the substitution of Tl^{3+} by Pb^{4+} stabilizes the Tl–Ca–Sr–Cu–O system, substitution of Ca^{2+} by a trivalent rare earth (Ln) also stabilizes the system [12]. In Fig. 10 we show the resistivity and magnetic susceptibility data of a series of cuprates of this family. In Fig. 11 we show the variation of T_c with x in several series of $TlLn_{1-x}Ca_xSr_2Cu_2O_7$ cuprates, all showing maxima around $x = 0.5$. We

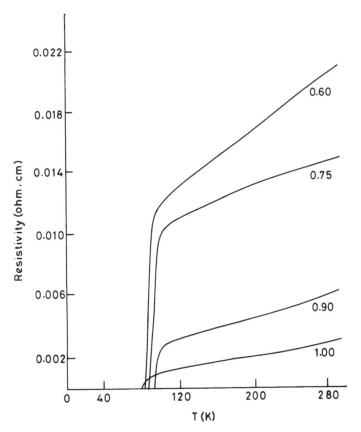

Figure 13 Temperature variation of resistivity in $Tl_{0.5}Pb_{0.5}Y_{1-x}Ca_xSr_2Cu_2O_7$. (Adapted from Ref. 13.)

have determined the hole concentration by the wet-chemical procedure [7] in the $TlGd_{1-x}Ca_xSr_2Cu_2O_7$ series and show the variation of T_c with n_h as well as of n_h with x in Fig. 12. The n_h value increases with Ca content as expected. The maximum T_c occurs at an n_h value of ~ 0.1, which is rather low compared to the other systems discussed earlier. It is possible that the n_h values at T_c maximum found by us are rather low because of the inherent limitation of the wet-chemical procedure, which does not take into account holes arising from $Tl(6s)$ band overlap; presence of any Tl^+ would also give rise to some errors.

VI. $Tl_{1-y}Pb_yY_{1-x}Ca_xSr_2Cu_2O_7$

In this system the parent cuprate $TlYSr_2Cu_2O_7$ corresponding to $x = y = 0$ is an insulator. The effect of varying x for different values of y has been investigated recently [13]. In Fig. 13 we show typical resistivity data in the case of $Tl_{0.5}Pb_{0.5}Y_{1-x}Ca_xSr_2Cu_2O_7$. In Fig. 14 we show the variation of T_c with x in this series ($y = 0.5$), along with similar plots for other y values. We see that T_c becomes maximum at a specific value of x (Ca content) for each values of y (Pb content). The value of x at maximum T_c shifts to higher values as y increases. This is under-

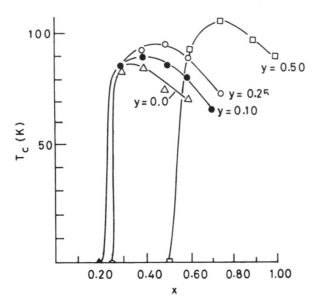

Figure 14 Plots of T_c against x in $Tl_{1-y}Pb_yY_{1-x}Ca_xSr_2Cu_2O_7$ for different values of y. (From Ref. 13.)

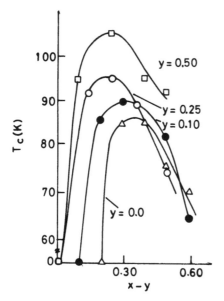

Figure 15 Plots of T_c against the effective hole concentration $(x-y)$ in $Tl_{1-y}Pb_yY_{1-x}Ca_xSr_2Cu_2O_7$. (From Ref. 13.)

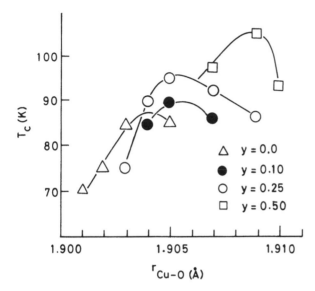

Figure 16 Variation of T_c with the Cu–O distance in $Tl_{1-y}Pb_yY_{1-x}Ca_xSr_2Cu_2O_7$. (From Ref. 13.)

standable since Pb in place of Tl introduces electrons, requiring a higher Ca content (number of holes) to give the maximum T_c. The value of maximum T_c itself increases with increasing y. Thus when $y = 0.5$, the maximum T_c is over 105 K (compared at 85 K when $y = 0.0$).

We would expect the difference between the concentrations of Ca and Pb $(x-y)$ in $Tl_{1-y}Pb_yY_{1-x}Ca_xSr_2Cu_2O_7$ to be a measure of the hole concentration. In Fig. 15 we have plotted the T_cs against $(x-y)$. The maximum T_c occurs at $(x-y)$ values of 0.35 and 0.30 when $y = 0.0$ and 0.10, respectively. The T_cs here are only in the range 85 to 90 K because of the high hole concentration. The value of $x-y$ at maximum T_c is, however, in the range 0.20 to 0.25 when $y = 0.25$ and 0.50. It is in this composition range that we see the highest T_c. This value of $x-y$ at maximum T_c is close to the hole concentration (0.20 to 0.25) found at maximum T_c in $YBa_2Cu_3O_{7-\delta}$ and several series of bismuth cuprates containing two CuO_2 sheets [4]. It therefore seems that $x-y$ in $Tl_{1-y}Pb_yY_{1-x}Ca_xSr_2Cu_2O_7$ represents the hole concentration. The value of $x-y$ found here at the T_c maximum value suggests that the n_h values from the wet-chemical procedure employed by us may be underestimates. On the other hand, the $x-y$ values at the T_c maximum value may be overestimated since we have not taken into account possible oxygen deficiency in these cuprates.

We discussed earlier some correlations between superconducting properties and the Cu–O distance. We show the plots of T_c against the in-plane Cu–O distance in $Tl_{1-y}Pb_yY_{1-x}Ca_xSr_2Cu_2O_7$ for different values of y in Fig. 16. We notice that T_c reaches a maximum at an optimal Cu–O distance. The maximum is found at slightly larger Cu–O distances as the Pb content (y value) increases.

ACKNOWLEDGMENTS

The author thanks his co-workers, especially R. Vijayaraghavan, for their unstinted cooperation and collaboration, and the National Superconductivity Board for support of the research.

REFERENCES

1. Z. Z. Sheng and A. M. Hermann, *Nature* 332, 55, 138 (1988).
2. C. N. R. Rao, ed., *Chemistry of High-Temperature Superconductors*, World Scientific, Singapore, 1992.
3. M. W. Shafer and T. Penney, *Eur. J. Solid State Inorg. Chem.* 27, 191 (1990).
4. C. N. R. Rao, J. Gopalakrishnan, A. K. Santra, and V. Manivannan, *Physica C* 174, 11 (1991).
5. D. Jung, M. H. Whangbo, N. Herron, and C. C. Torardi, *Physica C* 160, 381 (1988).
6. A. Manthiram, M. Paranthaman, and J. B. Goodenough, *Physica C* 171, 135 (1990); also *J. Solid State Chem.* 87, 479 (1990) and 98, 343 (1992).
7. J. Gopalakrishnan, R. Vijayaraghavan, R. Nagarajan, and C. Shivakumara, *J. Solid State Chem.* 93, 272 (1991).
8. R. Vijayaraghavan, J. Gopalakrishnan, and C. N. R. Rao, *J. Mater. Chem.* 2, 327 (1992).
9. M. H.Whangbo and M. A. Subramanian, *J. Solid State Chem.* 91, 403 (1991).
10. M. H. Whangbo and C. C. Torardi, *Acc. Chem. Res.* 24, 129 (1991).
11. R. Vijayaraghavan and C. N. R. Rao, *Solid State Commun.* 83, 73 (1992).
12. C. N. R. Rao, A. K. Ganguli, and R. Vijayaraghavan, *Phys. Rev. B* 40, 2565 (1989).
13. R. Vijayaraghavan, N. Rangavittal, G. U. Kulkarni, E. Gratscharova, T. N. Guru Row, and C. N. R. Rao, *Physica C* 179, 183 (1991).

18

Site-Selective Substitution in Tl Cuprates

R. M. Iyer and G. M. Phatak

Bhabha Atomic Research Centre, Bombay, India

I. INTRODUCTION

The thallium-based cuprates are represented by the general formula $Tl_m Ba_2$-$Ca_{n-1} Cu_n O_{2n+m+2}$ and can be classified into two families according to their structures. The members of the first family, $m = 2$, have the general formula $Tl_2 Ba_2 Ca_{n-1} Cu_n O_{2n+4}$, consisting of one double Tl–O layer and n-CuO (up to $n = 4$) layers per unit cell of the structure, whereas the second family, $m = 1$, is comprised of the phases with a single Tl–O layer and n-CuO (up to $n = 5$) layers, in their unit cell with a general formula $TlBa_2 Ca_{n-1} Cu_2 O_{2n+3}$. Similar to the Ba-containing single-layer-thallium cuprates, the Sr analog is also known (TSCCO).

Literature reports indicate that the thallium-based cuprates are thermally unstable phases and hence pose problems in their phase-pure synthesis. The double-Tl–O-layer phases are prepared more easily in pure-phase form as compared to mono-Tl–O-layer phases. Thus it is not impossible but is very difficult to prepare single-Tl–O-layer phases in pure form. Compared to rare earth–based cuprates and bismuth-based cuprates, the thallium-based cuprates are reported to be highly nonstoichiometric; the significant extent of disorder at Tl and Ca sites is probably responsible for superconductivity in these materials.

Most of the investigations on site-selective substitutions of cations at Tl/Ba/Sr/Ca sites in Tl superconducting oxides have focused on their effect on structural stability, ease of preparation of phases, effect on T_c, and optimization of hole concentration. Substitution at a particular cationic site may be inferred by various techniques, such as x-ray powder diffraction, neutron diffraction, high-resolution electron microscopy (HREM), and x-ray absorption near edge structure (XANES), although they all have inherent limitations. On the other hand, chemical analysis does not help in ascertaining the occupancy of a particular site by another cation, but gives the total content of the individual elements in the system.

In the present chapter (covering work reported in important international journals up to November 1991) we review the site-selective substitutions in thallium-based high-T_c cuprates. The chapter is not intended to be exhaustive. In the following sections we present the effects of substitution at Tl, Ba, Sr, and Ca sites on (1) improvement in the structural stability, (2) changes in T_cs, (3) ease in synthesis of pure phase products, and (4) hole doping for tuning superconducting properties.

II. THALLIUM–BARIUM–CALCIUM–COPPER OXIDE (TBCCO) SYSTEM

In general, Tl-based superconducting oxides can be considered to form oxygen-deficient perovskite-related structures [1]. The unit cell consists of alternating n Cu–O corner-sharing square-planar layers and single or double edge-sharing Tl–O octahedral layers. The edge-shared Tl–O octahedra are interconnected with Cu–O square planar/square pyramidal layers. The Ba cations are located in the interstices of Tl–O and Cu–O slabs and the Ca ions are between the Cu–O layers. The oxygen coordination number of Tl is 6, Cu is $\frac{4}{5}$, Ca is 8, and Ba(Sr) is 9 in double- and 12 in single-Tl–O-layered compounds. The valence state and ionic radii of different cations corresponding to different oxygen coordinations are given in Table 1. Tetragonal symmetry (space group $P4/mmm$ for single-Tl–O-layer and $I4/mmm$ for double-Tl–O-layer compounds) has been found common in all these compounds.

These phases are characterized by a large c parameter, and this parameter is found to increase with increase in n. For double-Tl–O-layer series it is 23.2461 Å for $n = 1$ [2], whereas it is 42.070 Å for $n = 4$ [3]. For single-Tl–O-layer series it is 9.694 Å [4,5] for $n = 1$ and 22.254 Å for $n = 5$ [6]. However, the a parameter does not show any appreciable change with increase in n.

In the stoichiometric (ideal) $Tl_2Ba_2Ca_{n-1}Cu_nO_{2n+4}$, the copper valence is estimated to be $2+$ for all the phases in the family. Since the existence of a copper oxidation state between $2+$ and $3+$ is believed to be a critical factor for super-

Table 1   Valence State, Coordination Number, and Ionic Radius of Thallium, Barium, Calcium, and Copper Forming $Tl_mA_2Ca_{n-1}Cu_nO_{2n+m+2}$[a]

Cation	Valence state	Existing coordination number	Ionic size (Å)
Tl	$+3$	6	0.885
Ba(Sr)	$+2$	9 ($m = 2$)	1.47 (1.31)
		12 ($m = 1$)	1.61 (1.44)
Ca	$+2$	8	1.12
Cu	$+2$	4	0.57
		5	0.65

Source: Ref. 10.
[a]High-T_c system ($m = 1$ or 2) ($n = 1$ to 5) A = Ba, Sr.

conductivity, this is brought about by compositional substoichiometry: that is, Tl and Ca deficiencies in the structure, partial occupancy of Tl and Ca on each other's site, and creation of holes in the CuO_2 layers via internal redox reaction $Tl^{3+} + Cu^{2+} \rightarrow Tl^{(3-\delta)+} + Cu^{(2+\delta)+}$. The foregoing factors are believed to be responsible for producing the higher copper oxidation state in these compounds. In the stoichiometric single-Tl-layer compounds $TlBa_2Ca_{n-1}Cu_nO_{2n+3}$, the copper cation would be in a high oxidation state $(2 + 1/n)^+$ (assuming Tl^{3+}, Ba^{2+}, Ca^{2+}, and O^{2-} and ideal stoichiometry) and would vary with n (i.e., number of Cu–O layers in the structure).

It is significant that XPS studies [7–9] show a valence state of Tl ions between Tl^{3+} and Tl^{1+} for double-Tl–O-layered compounds, whereas the lower valence state of the Tl ion is not observed in the single-Tl–O-layered phases and it is assumed that all the thallium is present at Tl^{3+} ion.

III. SUBSTITUTIONAL STUDIES: GENERAL CONSIDERATIONS

The thallium-based cuprates with Ba(Sr) possess tetragonal crystal structure with a large c parameter. Like other cuprate superconductors, Tl-based cuprates possess highly conducting two-dimensional Cu–O networks. High-temperature superconductivity is believed to be linked with these Cu–O networks. Essentially, therefore, the superconducting behavior influenced by substitutions is decided by the manner in which the Cu–O network would respond to substitution at various cationic sites in the compound, through modifications of the interlayer spacing and an electric field along the c axis.

The site occupancy in substitutional studies is controlled primarily by (1) ionic radii, (2) the valence state, and (3) the coordination number of the on-site cation. The valence state and ionic radii [10] corresponding to various coordinations for the dopants at the Tl and Ba/Sr/Ca sites are given in Tables 2 and 3, respectively.

A. Substitution at Tl Site

Tl^{3+}, Pb^{4+}, and Bi^{5+} have identical electronic configuration and somewhat similar chemistry. Their ionic radii for six-coordination are also close to each other and satisfy the criteria for substitution. Pb and Bi also exhibit multivalency (e.g., thallium)—namely, Pb^{4+}/Pb^{2+} and Bi^{5+}/Bi^{3+}—and would affect the copper valence on substitution and also the Tl^{3+}/Tl^+ ratio. However, during synthesis, the separation of $BaBiO_3$ and $BaPbO_3$, the most stable high-temperature phases at high levels of doping, has restricted the scope of studies in the TBCCO system. The TSCCO system does not pose such serious problems, and most studies on Bi and Pb substitution for thallium are reported in this system. Potassium is another important cation which resembles Tl^+ in chemical properties [11] and hence is expected to replace Tl^+ in double-Tl–O-layer cuprate.

B. Substitution at Ba/Ca Sites

Ba, Sr, Ca, and rare earths and yttrium exhibit high coordination numbers and the ionic radii for eight-, nine-, and twelve-coordination are also close to each other and thus satisfy the criteria for substitution. Ba(Sr) show more closeness in ionic radii to that of lighter lanthanides (La-Nd), whereas Ca and Y ionic radii for eight-

Table 2 Valence-State Coordination Number and Ionic Radius for
Dopant Used at Tl Site

Cation	Valence state	Existing coordination number	Ionic size (Å)
Cd	+2	4	0.78
		5	0.87
		6	0.95
		7	1.03
		8	1.10
		12	1.31
Pb	+2	4py	0.98
		6	1.19
		7	1.23
		8	1.29
		9	1.35
		10	1.4
		11	1.45
		12	1.49
	+4	4	0.65
		5	0.73
		6	0.775
		8	0.94
Bi	+3	5	0.96
		6	1.03
		8	1.17
	+5	6	0.76
K	+1	4	1.37
		6	1.38
		7	1.46
		8	1.51
		9	1.55
		10	1.59
		12	1.64
Sb	+3	4py	0.76
		5	0.80
		6	0.76
	+5	6	0.60

coordination are very close to each other. In view of this, it appears that lanthanides and yttrium can be substituted totally for Ba and Ca in these compounds.

Ample experimental evidence is available in the literature which indicates that the optimum hole doping corresponding to oxidation state of Cu between 2.2+ and 2.4+ leads to the highest T_c within a given structure. Hole doping can be achieved by substitution of a lower-valent cation (e.g., Cd^{2+}) for a higher-valent ion (e.g., Tl^{3+}) (which presumably leads to electron removal). Substitution of the higher-valent ion Pb^{4+} for the lower-valent ion Tl^{3+} would lead to electron addition. Substitution of trivalent rare earths/yttrium for Ba, Sr, and Ca would lead to electron doping. Thus electron coping can also be utilized to optimize the hole

Table 3 Valence State, Coordination Number, and Ionic Radius of Cations Substituted at Ba/Sr/Ca Site

Cation	Valence state	Existing coordination number	Ionic size (Å)
La	+3	6	1.032
		7	1.10
		8	1.16
		9	1.216
		10	1.27
		12	1.36
Nd	+3	6	0.983
		8	1.109
		9	1.163
		12	1.270
Pr	+3	6	0.99
		8	1.126
		10	1.179
	+4	6	0.85
		8	0.96
Tl	+1	6	1.50
		8	1.59
		12	1.70
Sr	+2	6	1.18
		7	1.21
		8	1.26
		9	1.31
		10	1.36
		12	1.44
Y	+3	6	0.90
		7	0.96
		8	1.019
		9	1.075

concentration where the copper oxidation state is already high: namely, $\sim 3+$ in the parent compound. Thus hole doping and electron doping substituents or a combination of these would help in controlling the formal oxidation state of copper between $2+$ and $3+$.

The substitutional studies in the Tl-based cuprates have been used effectively to stabilize the structure in the single-Tl–O-layer phases and has helped in the easy synthesis of these phases in bulk. It has been also helpful not only in stabilizing the phases but has helped in inducing/enhancing T_c by controlling the copper oxidation state to an optimum value (i.e., hole doping). Details regarding each of these aspects are dealt with in the following section.

IV. SUBSTITUTION AT Tl SITE

The cations that have been chosen for substitution at the Tl site as reported in the literature are cadmium, lead, bismuth, antimony, and potassium. Some of the results have been summarized in Table 4. The valence state and ionic radii corresponding to different coordinations are given in Table 2.

A. Mono-Tl–O-Layer Compounds

1. TBCCO System

Substitution of Pb or Bi stabilizes single-Tl–O-layer phases [12–15]. Pb and Bi substitution invariably results in lowering T_c [13–15]. Apparently, total replacement of Tl by Bi or Pb without affecting T_c is not possible. More than 50% Pb or Bi substitution for Tl is reported to cause precipitation of $BaPbO_3$ or $BaBiO_3$, the most stable phases in the system, and the superconductivity is destroyed [13–15].

Substitution of Cd for Tl in $TlBaLaCuO_5$ is reported to induce superconductivity in the parent compound, which is a semiconductor [16]. The decrease in the *a* parameter on Cd substitution is believed to be due to an increase in the copper oxidation state of the CuO_2 sheet. From the observed increase in the *c* parameter and on ionic size considerations (Table 2), it is believed that bigger Cd^{2+} essentially

Table 4 Substitution at Tl Site

No.	Composition	Lattice parameters (Å)		T_c (K)		Ref.
		a	*c*	Onset	R = 0	
1	$Tl_{1-x}Cd_xBaLaCuO_5$					
	$0.2 < x < 0.4$	~3.8	~9	38–48		16
2	$Tl_{1.6}Bi_{0.4}Sr_2Ca_2Cu_3O_9$	3.813	15.266		113	21
3	$(Tl_{1-x}Bi_x)Sr_2Ca_2Cu_3O_9$					
	$x = 0.1–0.5$	~3.8	~15	115–120	105–110	23
4	$(TlBi)Sr_2CaCu_2O_y$	3.796	12.113	100	90	24
5	$Tl_{1-x}Pb_xSr_2CaCu_2O_7$					
	$x = 0.25$	3.802	12.174	70		25
	$x = 0.50$	3.806	12.147	90		
6	$(TlPb)Sr_2Ca_{n-1}Cu_nO_{2n+3}$					
	$n = 1$	3.730	9.008		20	26
	$n = 2$	3.789	12.030		80	
7	$(TlPb)Sr_2CaCu_2O_7$	3.8023	12.107		85	27
	$(TlPb)Sr_2Ca_2Cu_3O_9$	3.8206	15.294		122	
8	$Tl_{0.5}Pb_{0.5}Sr_2CaCu_2O_7$	3.795	12.094		80	28
	$TlSr_2CaCu_2O_7$	3.794	12.133		47	
9	$Tl_{1-x}Bi_xSr_2CuO_5$					
	$x = 0.5$	3.7527	9.0278		[a]	31
10	$Tl_{2-x}Cd_xBa_2CuO_6$					
	$x = 0.25$	3.851	23.313		90	32
11	$SbPbBa_2CaCu_2O_x$	$a = 5.110$	18.512	NSC		33
		$b = 6.321$				
12	$Tl_2Ba_2CaCu_2O_x$	3.852	29.21		108	34
	$Tl_1K_1Ba_2CaCu_2O_x$	3.847	29.22		103	

[a]Not significant.

substitutes for the smaller Tl^{3+} ion (CN = 6), since substitution at the Ba site would have resulted in a decrease in the c parameter.

Results of pressure studies on T_cs of $TlPb_xBaCa_3Cu_3O_y$ compounds (i.e., Tl-1223) [15] are reported to be similar to other oxide cuprate superconductors—namely, La–Sr–Cu–O, Bi–Sr–Ca–Cu–O, and Tl–Ba–Ca–Cu–O systems—but different from Y–Ba–Cu–O (whose pressure dependence is negative) and the T_c is found to increase with pressure at a higher rate than in the Tl–Ba–Ca–Cu–O system. The effect is reported to be more pronounced at $x = 0.5$. The higher rate is attributed to the structural changes induced on Pb substitution, which might affect the structure of CuO planes. However, the reported in situ precipitation of the stable $BaPbO_3$ at high Pb concentration and thus the assumed phase purity in the samples used in this study is not unambiguous.

2. TSCCO System

Superconductivity in multiphase compounds in the Tl–Pb(Bi)–Sr–Ca–Cu–O system has been reported [17–20]. Preparation details of the pure-phase compound in the $TlBp(Bi)Sr_2Ca_{n-1}Cu_nO_{2n+3}$ ($n = 1, 2$, and 3) system in the form of polycrystalline powders [21–25] and $(TlPb)Sr_2Ca_{n-1}Cu_nO_{2n+3}$ in both powder and single-crystal form and structural studies have been reported [22,26–28]. The maximum stoichiometry of Tl/Bi(Pb) per formula unit is reported to be close to 1 [23,24,26]. A solid solution region has been reported in the nonsuperconducting $Tl_{1-x}Bi_xSr_2CuO_5$ ($0 < x \leq 0.50$) system only [31]. The oxidation state of Bi in $Tl_{1-x}Bi_xSr_2CuO_5$ and $Tl_{0.5}Bi_{0.5}Sr_2CaCu_2O_x$ has been reported to be $Bi^{(3+\delta)+}$ (i.e., mixed-valent state Bi^{3+}/Bi^{5+}) [29], while Pb is essentially in the 4+ state in $(TlPb)Sr_2CaCu_2O_7$ [27] and $(Tl_{0.5}Pb_{0.5})Sr_2CaCu_2O_7$ [30] phases. The existence of Bi as $Bi^{(3+\delta)+}$ in $(Tl_{1-x}Bi_x)Sr_2Ca_2Cu_3O_9$ has been deduced by comparison of the lattice parameters of the compound with that of the $(Tl_{0.5}Pb_{0.5})Sr_2Ca_2Cu_3O_9$ phase [22,23]. While the a parameter was found to be almost identical for the two compounds, the c parameter for the former was slightly higher than that of the latter. From ionic radii and c-parameter considerations the authors have deduced that the oxidation state of Bi is between 3+ and 4+. The T_c was found to increase with increased Pb doping [25].

B. Double-Tl–O-Layer Compounds

Substitution of Cd for Tl in $Tl_2Ba_2CuO_{6+\delta}$ [32] has shown that high Cd doping destroys superconductivity and that the oxidation state of copper is presumably a function of oxygen excess or cation doping. Decrease in the a parameter has been attributed to an increase in the copper oxidation state, while an increase in the c parameter has been attributed to substitution of bigger Cd^{2+} ion for smaller Tl^{3+}. Preparation of single-phase $SbPbBa_2CaCu_2O_x$, a nonsuperconducting oxide, by total substitution of Tl by Sb(Pb) in $Tl_2Ba_2CaCu_2O_x$ has also been reported [33].

Substitution of potassium in $Tl_2Ba_2CaCu_2O_8$ (Tl-2212) has been attempted successfully [34]. Chemical analysis of the compound confirmed that potassium was not lost during the synthesis. Neutron diffraction patterns did not show any appreciable change in the cell parameters on K substitution. Substitution of K^+ in Tl sites manifests itself in complete elimination of the mismatch of Ba sheets and

O(2) sheets along the c axis due to creation of a deficit of net positive charge in the Tl double layers consequent upon K substitution.

V. SUBSTITUTION AT Ba AND Sr SITES

Studies have been reported on substitution of lanthanum, strontium, and thallium (1+) for Ba in the TBCCO system and lanthanum, neodymium, and praseodymium for Sr in the TSCCO system. Some of the results are summarized in Table 5.

A. TBCCO System: Mono-Tl–O-Layer Compounds

The effect on the T_c of substitution of La^{3+} for Ba^{2+} in $TlBa_2CuO_5$ appears to depend on the formal oxidation state of copper in the parent compound. It is reported that $TlBa_2CuO_5$ is an insulator/semiconductor/low-T_c material [35–37], and substitution of La for Ba stabilizes the structure and induces T_c, presumably by reduction in the formal oxidation state of copper from +3, and monophasic $TlBa_{1+x}La_{1-x}CuO_5$ samples could be synthesized easily [35,36,38]. On the other hand, substitution of La for Ba in the oxygen-deficient compound $TlBa_2CuO_{5-y}$ lowers the T_c [39]. In the former case the maximum T_c has been variously reported to be 40 K [36] and 52 K [38] for 40% substitution, while in the latter case for the same extent of substitution the product is nonsuperconducting, regaining superconductivity at about 40 K on oxygen annealing. The apparent contradiction in these observations appears to be related to the reduction in formal copper valence from +3 to an optimum value in $TlBa_{1.2}La_{0.8}CuO_5$ (estimated to be ca. +2.2), whereas the substitution of La for Ba in $TlBa_2CuO_{5-y}$ reduces the copper valence much below the optimum value, probably to about +2, with increased La doping, and as expected, the compound is a nonsuperconductor. Thus it appears that doping a metal cation or adjusting the oxygen stoichiometry can induce superconductivity.

Neutron diffraction studies of $TlBa_{1.2}La_{0.8}CuO_5$ (52 K) [38] have revealed an increase in the in-plane Cu–O bond distance over that of $TlBa_2CuO_5$, attributed to lowering the formal oxidation state of copper from +3. Total substitution of Sr for Ba has been studied [40–42]. Synthesis of $TlSrBaCaCu_2O_x$ (Tl-1212 structure) in polycrystalline single-phase form with a T_c of 102 K [43] or 94 K [44] has also been reported.

Synthesis and crystal structure by x-ray powder diffraction of nonsuperconducting single-phase $TlSr_2CaCu_2O_7$ [45] has been reported and the structure is similar to that of $Tl_{0.5}Pb_{0.5}Sr_2CaCu_2O_7$ (85 K). It is generally believed that the higher copper valence of +2.50 in $TlSr_2CaCu_2O_7$ may be responsible for the nonsuperconducting nature of the compound [24]. More recently, superconductivity has been reported in $TlSr_2CaCu_2O_{7-\delta}$ (T_c = 68 K) by inducing oxygen deficiency [46].

B. Double-Tl–O-Layer Compounds

Partial substitution of Ba by Sr in $Tl_2Ba_2CaCu_2O_8$ and $Tl_2Ba_2Ca_2Cu_3O_{10}$ systems has been reported [47,48], resulting in a decrease in T_c. The a and c lattice parameters were found to decrease with increased Sr doping. It appears that partial substitution of Ba by Sr in $Tl_2Ba_2Ca_2Cu_3O_{10}$ (2223 phase) stabilizes the 2223 phase, and it is claimed that the 2223 phase can be synthesized more readily in pure form

Table 5 Substitution at Ba Site

No.	Composition	Lattice parameters (Å)		T_c (K)		Ref.
		a	c	Onset	$R = 0$	
1	$TlBa_{1+x}La_{1-x}CuO_5$					
	$x = 0.0-0.4$	3.84	9.01-9.23		40	36
2	$TlBa_{1.2}La_{0.8}CuO_5$	3.8479	9.0909		52	38
3	$TlBaSrCaCu_2O_x$	3.850	12.309		102	43
		3.827	12.27		94	44
4	$TlSr_2CaCu_2O_7$	3.7859	12.104			45
5	$Tl^{III}_{2-x/3}Tl^{I}_{1-x}Ba_{1-x}LnCu_2O_8$					
	$x = 0.25$ Sm	3.899	28.691		a	49
	Nd	3.911	28.749			
	Pr	3.920	28.787			
6	$Tl_2Ba_{2-x}Sr_xCaCu_2O_8$					
	$x = 0.0$	3.8530	29.342	116	106	47
	0.2	3.8449	29.224	113	94	
	0.6	3.8334	29.018	108	54	
	1.0	3.8232	28.852	90	82	
	1.4	3.8090	28.590	72	42	
	2.0	3.7930	28.280	61	28	
7	$TlSr_{2-x}R_xCuO_5$					
	$R = (La, Nd)$	~3.7	~9	35-46	24-38	51
	$x = 0.5-1.0$					
8	$Tl(La,Sr)_2CuO_5$					52
	$Tl_{1-x}LaSrCu_{1+x}O_5$					
	$x = 0.0$	3.7796	8.8060	38	30	
	0.2	3.7731	8.8268	42	32	
	0.4	3.7707	8.8528	40	26	
	0.6	3.7714	8.8558	38	22	
9	$Tl_{1.8}Sr_{1.6}Pr_{0.6}CuO_5$	3.741	8.875	40		50
	$Tl_{0.7}Sr_2Pr_{0.3}CuO_5$	3.733	8.951	a		
10	$TlSr_{2-y}La_yCuO_{5-\delta}$					
	$y = 0.4-1.0$	3.7507-3.7796	8.9400-8.8466	30-40		56
	Domain A					
	$Tl_{1+x}SrLa_{1-x}CuO_{5-\delta}$					
	$x = 0.0-0.3$	3.7796-3.7713	8.8466-8.8676			
	$Tl_{1+x}Sr_{1.2}La_{0.8-x}CuO_{5-\delta}$					
	$x = 0.0-0.2$	3.7704-3.7544	8.8676-8.9011			
	Domain B					
	$Tl_{1-x}Sr_{1+x}LaCuO_{5-\delta}$					
	$x = 0.0-0.2$	3.7796-3.7641	8.8466-8.8951			
	$Tl_{1-x}Sr_{2+x-y}La_yCuO_{5-\delta}$					
	$x = 0.0-0.2$	3.7507-3.7641	8.9400-8.8951			
	$y = 0.4-1.0$					

[a]Not significant.
[b]Ac susceptibility method.

than can the unsubstituted 2223. At high Sr doping values, crystallographic transformation from 2223 phase to 1223 has also been reported [48].

Synthesis of nonsuperconducting $Tl^{III}_{2-x/3}Tl^{I}_{1-x}Ba_{1+x}LnCu_2O_8$ (Ln = Pr and Nd)
corresponding to $x = 0.25$ has been reported by Bourgault et al. [49]. Based on
x-ray powder diffraction data the authors have concluded that the compound possesses 2212 structure with random distribution of Tl^+ and Ba^{2+} ions at the Ba site.
It is claimed that the occupancy of Tl^+ at the Ba site provides additional nonstoichiometry in addition to the existing defects at the Tl and Ca sites.

C. TSCCO System

Substitution of Sr^{2+} by La, Pr, and Nd in $TlSr_2CuO_5$ has been studied. Pr exhibits
multivalency (i.e., Pr^{3+} and Pr^{4+}), and it is reported that it substitutes at the Tl
site or Tl and Sr sites [50]. This can probably be attributed to the similarity of ionic
radii of Tl^{3+} and Pr^{4+} and Sr^{2+} and Pr^{3+}. Although $TlSr_2CuO_5$ is nonsuperconducting [51], substitution of La, Pr, and Nd induces superconductivity in the
$TlSr_{2-x}Ln_xCuO_5$ system and stabilizes the 1201 phase. From the observed decrease
in the c lattice parameter on increase in x and ionic size considerations of Sr^{2+},
Nd^{3+}, and La^{3+} in ninefold coordination, the authors have concluded that La and
Nd are substituted at the Sr site since substitution at the Tl site would have resulted
in an increase in the c parameter on La and Nd doping. Substitution of the smaller
Nd ion for Sr should have resulted in a decrease in the a parameter, but actually,
a slight increase in the a parameter has been observed. It is argued that substitution
of La and Nd results in a decrease in the formal oxidation state of copper and an
increase in the in-plane Cu–O bond length (i.e., the a lattice parameter and slight
increase in a parameter observed are the net result of the two opposing factors
above). A detailed phase diagram of 1201 (Tl–Sr–La–Cu–O system) has been
studied by Huve et al. using XRD, electron microscopy, and electron diffraction
techniques, particularly for thallium-rich and thallium-deficient samples [56].

It is interesting that $Tl(SrLn)CuO_5$ is superconducting [54], whereas the Ba
analog $Tl(BaLn)CuO_5$ (Ln = La, Nd) is a nonsuperconductor [53]. The superconductivity in the former is believed to be due to shortening of the in-plane Cu–O
bond on substitution of the smaller Sr^{2+} ion for Ba^{2+} and consequent creation of
holes in the CuO_2 layers [55]. The synthesis of pure-phase $(TlCu)(La,Sr)_2CuO_5$
has also been claimed by substitution of Cu for Tl from x-ray diffraction data [52].
Superconductivity in multiphasic samples in Tl–Pb–Ln–Sr–Cu–O and Tl–Ln–
Sr–Cu–O has also been reported [56–61].

VI. SUBSTITUTION AT Ca SITE

Yttrium and trivalent rare earths have been substituted for Ca in TBCCO and
TSCCO systems. Some of the data are summarized in Table 6.

A. Mono-Tl–O-Layer Compounds

Substitution of a trivalent ion (Y,Nd,Yb) for Ca^{2+} apparently reduces the copper
valence and stabilization of mono-Tl–O-layer phases, resulting in TBCCO and

Table 6 Substitution at Ca Site

No.	Composition	Lattice parameters (Å)		T_c $(R = 0)$ (K)	Ref.
		a	c		
1	$TlBa_2Ca_{1-x}Nd_xCu_2O_7$				66
	$x = 0.2$	3.85638	12.6533	100[a]	
	0.5	3.87677	12.6045	40	
	1.0	3.91021	12.4916	[b]	
2	$TlSr_2Ca_{1-x}L_xCu_2O_y$		~12	60–90	71
	L = Y and rare earths				
3	$Tl_{0.5}Pb_{0.5}Sr_2Ca_{0.8}A_{0.2}Cu_2O_x$	~3.81	~12	105–108	73
	A = Y and rare earths			(except Ce)	
4	$Tl_{1-y}Pb_ySr_2Y_{1-x}Ca_xCu_2O_7$				76
	$y = 0.25, x = 0.00$	3.830	12.031		
	0.25	3.826	12.040		
	0.50	3.814	12.066		
	0.75	3.807	12.124		
	1.00	3.804	12.165		
	$x = 0.5, \ y = 0.10$	3.813	12.058		
	0.25	3.814	12.066		
	0.50	3.824	12.060		
5	$Tl_{0.5}Pb_{0.5}Sr_2Ca_{0.9}Ce_{0.1}Cu_2O_x$	3.80	11.95	95	79

[a]Ac susceptibility method.
[b]Not significant.

TSCCO systems. This helps in the easy isolation of pure-phase mono-Tl–O-layer compounds.

1. TBCCO System

Substitution of yttrium [62,65,67] and neodymium [63,66] for Ca in $TlBa_2CaCu_2O_7$ (1212 structure) is reported to stabilize $TlBa_2Ca_{1-x}Y_xCu_2O_{7+\delta}$ and $TlBa_2Ca_{1-x}Nd_x$-$Cu_2O_{7-\delta}$ phases, and pure-phase compounds in bulk without traces of $Tl_2Ba_2CaCu_2O_8$ (2212 phase) could be prepared over the entire range of $0 \leq x \leq 1$. The substitution of Y and Nd is associated with decrease in T_c with increase in doping level. Complete substitution of calcium by Y and Nd destroys superconductivity without any apparent change in structure. The decrease in T_c with increase in x has been attributed to decrease in the formal oxidation state of Cu. The smooth variation of lattice parameters observed (i.e., increase in the a parameter, decrease in the c parameter, and increase in unit cell volume progressive with x) has been attributed by the authors to (1) the relative ionic sizes of Y and Ca and (2) the oxygen stoichiometry for different values of x [62]. Increase in the a parameter could be attributed partly to increase in the in-plane and Cu–O bond with decrease in the formal oxidation state of copper in the parent compound $TlBa_2CaCu_2O_7$ of Y substitution [64].

The insulator version of $TlBa_2YCu_2O_7$ [69] has been reported to show three-dimensional antiferromagnetic order similar to that of $La_2CuO_{4-\delta}$, $YBa_2Cu_3O_6$, and $Bi_2Sr_2YCu_2O_x$, with a Neel temperature of 350 K. Substitution of Ca and Ce for Y in $TlBa_2YCu_2O_7$, inducing superconductivity [67,70], and Yb substitution for Ca in 1212, with no adverse effect on T_c [68], have also been claimed.

B. TSCCO System

It is reported that total replacement of Ca by Y or rare earths destroys superconductivity, yielding semiconducting compounds [71,72,76]. Synthesis of monophasic $TlSr_2Ca_{1-x}L_xCu_2O_y$ (L = Y and rare earths) has been reported. A metal-to-insulator transition with a variation in x has been attributed to the decreased copper valence (ca. +2) at higher dopings [71]. Substitution of Ca by Y in the $(Tl_{0.5}Bi_{0.5})$ · $(Ca_{1-x}Y_x)Sr_2Cu_2O_y$ system was found to enhance the T_c from 83 K to 102 K (x = 0.2) [74,75]. Although the authors did not explain the variation in T_c with x, it can be understood if one considers that the Bi oxidation state in these compounds is reported to be $Bi^{(3+\delta)+}$ [23,29] and that Y^{3+} substitution for calcium is expected to introduce electrons and thereby reduce the formal oxidation state of Cu in the system. The estimated copper valence for the compounds will be slightly less than +2.5, +2.4, +2.25, and +2.15 for x = 0, 0.2, 0.5, and 0.7. The maximum T_c is obtained at x = 0.2, at which value the copper valence will be slightly less than +2.4, which is closer to the copper valence for maximum T_c in other cuprate superconductors: namely, Y-123 and Tl cuprates with a double copper layer. Thus the variation in T_c can be correlated with the change in hole concentration on Y doping for Ca. The slight expansion for the a parameter suggests a lower copper valence in the doped compound than in the parent compound. The observed contraction in the c parameter with increase in x could not be accounted for as being due to the difference in the sizes of the Y^{3+} and Ca^{2+} ions. However, the additional contraction in the c parameter has been attributed to the higher formal valence and different coordination number for Y.

Substitution of Pb^{4+} for Tl^{3+} and Ca^{2+} for Y^{3+} in $TlSr_2YCu_2O_7$, an insulator, resulted in superconducting cuprates with the general formula $Tl_{1-y}Pb_yY_{1-x}Ca_xSr_2Cu_2O_7$ (y = 0 to 0.50) [76]. By substituting Pb^{4+} for Tl^{3+} and Ca^{2+} for Y^{3+}, the authors could control electron doping and hole doping simultaneously in the parent compound. The observed decrease in the a parameter with increase in x and expansion of the a parameter with increase in y is evidence of an increased oxidation state for Cu (i.e., hole doping and decrease in copper oxidation state on electron doping, respectively). The increase in the c parameter is accounted for by the substitution of a larger calcium ion for a smaller yttrium ion. The difference ($x - y$) represents the net hole concentration of the system and could be correlated with the observed T_c variation with changes in x and y. The optimum hole concentration appears equal to 0.20 to 0.25 for maximum T_c in the system, which is in general agreement with what is observed in Y-123 and Bi cuprates.

Synthesis of single-phase $Tl_{0.5}Pb_{0.5}Sr_2Ca_{0.8}A_{0.2}Cu_2O_x$ (A = Y and rare earths) has been reported with a T_c above 100 K [73,79]. Superconductivity has also been reported in multiphase $TlSr_2Ca_{0.5}L_{0.5}CaCu_2O_{6.75}$ (L = Sm, Er) [77,78].

VII. EFFECTS OF NONSTOICHIOMETRY

The parent superconducting compounds in both the TBCCO and TSCCO systems, which are synthesized by a variety of methods employing a range of temperature of synthesis and environment (open/closed chambers), invariably produce a product having a range of cation concentration, particularly Tl and to a lesser extent, Ca. For example, Tl-rich starting composition does not necessarily produce a better product either in terms of phase purity or T_c, whereas Tl-deficient starting compositions invariably give good-quality products. In fact, our detailed studies on optimum concentrations of Tl and Ca in the TBCCO system (2212 and 2223) showed that Ca-rich compositions invariably gave a better-quality product in terms of both phase purity and T_c [80]. Tl-deficient, Ca-rich compositions are thus quite capable of forming the required product. The extraordinary flexibility of the structure, especially of the rock salt layers, to accommodate cations of various sizes brings into focus whether Ba-deficient composition can also stabilize the Tl 2212 structure, and this has recently been addressed by us (unpublished work). These studies show that over the composition range $1.25 \leq Ba \leq 2.0$, maintaining a Tl/Ca ratio of 2:1, the 2212 structure could be produced, indicating perhaps that cross-site occupancy among Tl, Ca, and Ba is responsible for structural stability. The T_c of such a product appears to be affected by these intrasite occupancies. Earlier studies [34] on K substitution have indicated K preference for Tl sites. Even in the Ba-deficient composition of Tl 2212, we have recently found that K can be substituted. Exact site occupancy data are not known.

VIII. GENERAL COMMENTS

Double- and single-copper-layer Tl and Bi cuprates are known to be amenable to Y/rare earth substitution at Ba/Sr/Ca sites. But for triple-Cu-layer systems in Tl and Bi cuprates [i.e., Tl2223 and Bi(Pb)2223], such substitutions have not been reported. In particular, our elaborate attempt to substitute Y at the Ca site in Bi(Pb)2223 have been a failure, resulting in the formation of the substituted Bi 2212 phase only. No parallel studies on Y/RE substitution in 2223/1223 systems in TBCCO/TSCCO have been reported so far.

After looking at the present status of substitution studies reported in the literature and reviewed briefly here, a word of caution on the reported results would not be out of place. As mentioned at the beginning of the chapter, there are at present no unequivocal methods available to confirm substitution at the chosen sites in these defect oxides. Even the extent of crystallinity of Tl cuprates is a debatable point. Even within the nominal composition range of these compounds (e.g., Tl-1212 or Tl-2212), the actual composition varies over a wide range and T_c is also known to vary over a wide range, unlike the case of the Y-123 system. For example, even though in the Tl-2223 compound the T_c is known to be as high as 125 K, no specific recipe has been successful in giving a reproducible $T_c = 125$ K when several samples of the same composition are prepared. The range of T_c could be as low as 100 K and as high as 130 K.

Since Tl cuprates can accommodate a very large degree of substoichiometry, particularly at Tl and Ca sites, and since experimental evidence further indicates scrambling of site occupancies among Tl, Ba, and Ca, the particular synthesis protocol adopted by the researcher may not ensure uniformity/reproducibility of stoichiometry/site scrambling when one is attempting to incorporate new cations into the system. Hence results of comparison studies such as the effect of cation substitution on T_c, cell parameter, and so on, will not be a true indicator of the effect of substitution. Nevertheless, a trend in behavior could provide qualitative clues to the subtle effect of chemical substitution on the complex phenomenon of superconductivity control in Tl cuprates.

REFERENCES

1. S. S. P. Parkin, V. Y. Lee, A. I. Nazzal, R. Savoy, T. C. Huang, G. Gorman, and R. Beyers, *Phys. Rev.* 38, 6531 (1988).
2. J. B. Parise, J. Gopalkrishnan, M. A. Subramanian, and A. W. Sleight, *J. Solid State Chem.* 76, 432 (1988).
3. Y. Tang, B. Lin, D. Zhou, W. Zhu, F. Chen, N. Li, K. Chen, and G. Lu, *Mod. Phys. Lett. B* 3, 853 (1989).
4. R. Beyers, S. S. P. Parkin, V. Y. Lee, A. I. Nazzal, R. Savoy, G. Gorman, T. C. Huang, and S. Laplaca, *Appl. Phys. Lett.* 53, 432 (1988).
5. S. S. P. Parkin, V. Y. Lee, A. I. Nazzal, R. Savoy, R. Beyers, and S. J. Laplaca, *Phys. Rev. Lett.* 61, 750 (1988).
6. S. Nakajima, M. Kikuchi, Y. Syono, T. Oku, D. Shindo, K. Hiraga, N. Kobayashi, H. Iwasaki, and Y. Muto, *Physica C* 158, 471 (1989).
7. H. M. Meyer III, T. J. Wagener, J. H. Weaver, and D. S. Ginley, *Phys. Rev. B* 39, 7343 (1989).
8. P. Marksteiner, J. Yu, S. Massidda, A. J. Freeman, J. Redinger, and P. Weinberger, *Phys. Rev. B* 39, 2894 (1989).
9. T. Suzuki, M. Nagoshi, Y. Fukuda, S. Nakajima, M. Kikuchi, Y. Syono, and M. Tachiki, *Physica C*, 162–164, 1387 (1989).
10. R. D. Shanon, *Acta Crystallogr. A* 32, 751 (1976).
11. A. G. Lee, *Chemistry of Thallium*, Elsevier, New York, 1971, and references therein.
12. H. Kusuhara, T. Kotani, H. Takei, and K. Tada, *Jpn. J. Appl. Phys.* 28, L1772 (1989).
13. Y. Torii, T. Kotani, H. Takei, and K. Tada, *Jpn. J. Appl. Phys.* 28, L1190 (1989).
14. I. Z. Kostadinov, M. D. Mateev, J. Tihov, V. Skumriev, E. Tsakin, O. Petrov, E. Dinolova, and V. Kovachev, *Physica C* 162–164, 995 (1989).
15. L. Mingzhu, T. Weihua, M. Xianren, L. Zhensin, H. Wei, T. Qingyun, R. Yanru, and L. Zhenxing, *Phys. Rev. B* 41, 2517 (1990).
16. M. A. Subramanian and A. K. Ganguli, *Mater. Res. Bull.* 26, 91 (1991).
17. T. Itoh and H. Uchikawa, *Jpn. J. Appl. Phys.* 28, L591 (1989).
18. M. F. Tai, W. N. Wang, and H. C. Ku, *Jpn. J. Appl. Phys.* 27, L2287 (1988).
19. T. Itoh, H. Uchida, and H. Uchikawa, *Jpn. J. Appl. Phys.* 27, L2052 (1988).
20. R. S. Liu, Y. T. Huang, W. H. Lee, S. F. Wu, and P. T. Wu, *Physica C* 156, 791 (1988).
21. Y. Torii, H. Takei, and K. Tada, *Jpn. J. Appl. Phys.* 28, L 2192 (1989).
22. M. A. Subramanian, C. C. Torardi, J. Gopalkrishnan, P. L. Gai, J. C. Calabrese, T. R. Askew, R. B. Flippen, and A. W. Sleight, *Science* 242, 249 (1988).
23. M. A. Subramanian, P. L. Gai, and A. W. Sleight, *Mater. Res. Bull.* 25, 101 (1990).
24. S. Li and M. Greenblatt, *Physica C* 157, 365 (1989).
25. A. K. Ganguli, K. S. Nanjundaswamy, and C. N. R. Rao, *Physica C* 156, 788 (1988).
26. J. C. Barry, Z. Iqbal, B. L. Ramakrishna, R. Sharma, H. Eckharat, and F. Reidinger, *J. Appl. Phys.* 65, 5207 (1989).
27. J. B. Parise, P. L. Gai, M. A. Subramanian, J. Gopalkrishnan, and A. W. Sleight, *Physica C* 159, 245 (1989).
28. C. Martin, J. Povost, D. Bourgault, B. Domenges, C. Michel, H. Hervieu, and B. Raveau, *Physica C* 157, 460 (1989).
29. S. Li, M. Greenblatt, Y. Geon, J. Chen, G. Liang, and M. Croft, *Physica C* 173, 279 (1991).

30. G. U. Kulkarni, G. Sankar, and C. N. R. Rao, *Appl. Phys. Lett.* 55, 388 (1989).
31. S. Li and M. Greenblatt, *Mater. Res. Bull.* 26, 229 (1991).
32. J. B. Parise, N. Herron, M. K. Crawford, and P. L. Gai, *Physica C* 159, 255 (1989).
33. R. A. Gunasekaran, G. M. Phatak, M. D. Sastry, U. R. K. Rao, and R. M. Iyer, *J. Mater. Sci. Lett.* 10, 760 (1991).
34. A. Sequeira, H. Rajagopal, I. K. Gopalakrishnan, P. V. P. S. S. Sastry, G. M. Phatak, J. V. Yakhmi, and R. M. Iyer, *Physica C* 156, 599 (1988).
35. H. C. Ku, M. F. Tai, J. B. Shi, M. J. Shieh, S. W. Hsu, G. H. Hwang, D. C. Ling, T. J. Watson-yang, and T. Y. Lin, *Jpn. J. Appl. Phys.* 28, L923 (1989).
36. T. Manako, Y. Shimakawa, Y. Kubo, T. Satoh, and H. Igarashi, *Physica C* 158, 143 (1989).
37. I. K. Gopalakrishnan, J. V. Yakhmi, and R. M. Iyer, *Physica C* 175, 183 (1991).
38. M. A. Subramanian, G. H. Kwei, J. B. Parise, J. A. Goldstone, and R. B. Von Dreele, *Physica C* 166, 19 (1990).
39. A. Sundaresan, A. K. Rajarajan, L. C. Gupta, M. Sharon, and R. Vijayaraghavan, *Physica C* 178, 193 (1991).
40. T. Nagashima, K. Watanabe, H. Saito, and Y. Fukai, *Jpn. J. Appl. Phys.* 27, L1077 (1988).
41. S. Matsuda, S. Takeuchi, A. Soeta, T. Suzuki, K. Aihara, and T. Kamo, *Jpn. J. Appl. Phys.* 27, 2062 (1988).
42. W. L. Lechter, M. S. Osofsky, R. J. Soulen, Jr., V. M. LeTourneau, E. F. Skelton, S. B. Qadri, W. T. Elam, H. A. Hoff, R. A. Hein, L. Humphreys, C. Skowronek, A. K. Singh, J. V. Gilfrich, L. E. Toth, and S. A. Wolf, *Solid State Commun.* 68, 519 (1988).
43. M. Kuroda and M. Araki, *Jpn. J. Appl. Phys.* 28, L1154 (1989).
44. I. K. Gopalakrishnan, J. V. Yakhmi, and R. M. Iyer, *Physica C* 172, 450 (1991).
45. T. Doi, K. Usami, and T. Kamo, *Jpn. J. Appl. Phys.* 29, L57 (1990).
46. Y. Kubo, Y. Shimakawa, T. Manako, T. Kondo, and H. Igarashi, *Physica C* 185–189, 1253 (1991), and references therein.
47. E. A. Hayri and M. Greenblatt, *Physica C* 156, 775 (1988).
48. A. Soeta, T. Suzuki, S. Takeuchi, T. Kamo, K. Usami, and S.-P. Matsuda, *Jpn. J. Appl. Phys.* 28, L1186 (1989).
49. D. Bourgault, C. Martin, C. Michel, M. Hervieu, and B. Raveau, *Physica C* 158, 511 (1989).
50. D. Bourgault, C. Martin, C. Michel, M. Hervieu, J. Provost, and B. Raveau, *J. Solid State Chem.* 78, 326 (1989).
51. M. A. Subramanian, *Mater. Res. Bull.* 25, 191 (1990).
52. T. Mochiku, T. Nagashima, M. Watahiki, Y. Fukai, and H. Asano, *Jpn. J. Appl. Phys.* 28, L1926 (1989).
53. J. B. Goodenough and A. Manthiram, *J. Solid State Chem.* 88, 155 (1990).
54. A. K. Ganguli, V. Manivannan, A. K. Sood, and C. N. R. Rao, *Appl. Phys. Lett.* 55, 2664 (1989).
55. M. H. Whangbo and M. A. Subramanian, *J. Solid State Chem.* 91, 403 (1991).
56. M. Huve, C. Michel, C. Martin, M. Hervieu, A. Maignan, J. Provost, and B. Raveau, *Physica C* 179, 214 (1991).
57. T. Itoh and H. Uchikawa, *Jpn. J. Appl. Phys.* 28, L200 (1989).
58. S. Adachi, O. Inoue, H. Hirano, Y. Takahashi, and S. Kawashima, *Jpn. J. Appl. Phys.* 28, L775 (1989).
59. T. Itoh and H. Uchikawa, *Phys. Rev. B* 39, 4690 (1989).
60. T. Itoh and H. Uchikawa, *Jpn. J. Appl. Phys.* 28, L1790 (1989).
61. T. Nagashima, M. Watahiki, and Y. Flaki, *Jpn. J. Appl. Phys.* 28, L930 (1989).

62. A. K. Ganguli, R. Nagarajan, K. S. Nanjundaswamy, and C. N. R. Rao, *Mater. Res. Bull.* 24, 103 (1989).
63. P. S. Nakajima, M. Kikuchi, Y. Syono, N. Kobayashi, and Y. Muto, *Physica C* 168, 57 (1990).
64. D. Shindo, K. Hiraga, S. Nakajima, M. Kikuchi, Y. Syono, N. Kobayashi, K. Hojou, T. Soga, S. Furuno, and H. Otsu, *Physica C* 159, 794 (1989).
65. T. Manako, Y. Shimkawa, Y. Kubo, T. Satoh, and H. Igarashi, *Physica C* 156, 315 (1988).
66. C. Michel, E. Suard, V. Caignaert, C. Martin, A. Maignan, M. Hervieu, and B. Raveau, *Physica C* 178, 29 (1991).
67. T. Manako, Y. Shimakawa, Y. Kubo, T. Satoh, and H. Igarashi, *Advances in Superconductivity, Proc. First International Symposium on Superconductivity* (ISS'88), Aug. 28–31, 1988, Nagoya, ed. K. Kitazawa and T. Ishiguro, Springer-Verlag, Tokyo, 1988, p. 819.
68. R. Vijayaraghavan, A. K. Ganguli, N. Y. Vasanthacharya, M. K. Rajumon, G. U. Kulkarni, G. Sankar, D. D. Sarma, A. K. Sood, N. Chandrabhas, and C. N. R. Rao, *Supercond. Sci. Technol.* 2, 195 (1989).
69. J. Mizuki, Y. Kubo, T. Manako, Y. Shimakawa, and H. Igarashi, *Physica C* 156, 781 (1988).
70. J. H. Wang, Z. Z. Sheng, C. Dong, X. Fei, L. Sheng, and A. M. Hermann, *Physica C* 158, 507 (1989).
71. C. N. R. Rao, A. K. Ganguli, and R. Vijayaraghavan, *Phys. Rev. B* 40, 2565 (1989).
72. O. Inoue, S. Adachi, and S. Kawashima, *Jpn. J. Appl. Phys.* 28, L1375 (1989).
73. J. M. Liang, R. S. Liu, Y. T. Huang, S. F. Wu, P. T. Wu, and L. J. Chen, *Physica C* 165, 347 (1990).
74. Y. T. Huang, R. S. Liu, W. N. Wang, and P. T. Wu, *Physica C* 162–164, 39 (1989).
75. Y. T. Huang, R. S. Liu, W. N. Wang, and P. T. Wu, *Jpn. J. Appl. Phys.* 28, L1514 (1989).
76. R. Vijayaraghavan, N. Rangavittal, G. U. Kulkarni, E. Grantscharova, T. N. Guru Row, and C. N. R. Rao, *Physica C* 179, 183 (1991).
77. J. K. Liang, Y. L. Zhang, G. H. Rao, X. R. Cheng, S. S. Xie, and Z. X. Zhao, *Solid State Commun.* 70, 661 (1989).
78. J. K. Liang, Y. L. Zhang, G. H. Rao, X. R. Cheng, S. S. Xie, and Z. X. Zhao, *Z. Phys. B Condensed Matter* 76, 277 (1989).
79. P. T. Wu, R. S. Liu, J. M. Liang, Y. T. Huang, S. F. Wu, and L. J. Chen, *Appl. Phys. Lett.* 54, 2464 (1989).
80. R. M. Iyer, G. M. Phatak, K. Gangadharan, M. D. Sastry, R. M. Kadam, P. V. P. S. S. Sastry, and J. V. Yakhmi, *Physica C* 160, 155 (1989).

19

Thermal Stability of Tl–Ba–Ca–Cu–O High-T_c Superconductors

Ken H. Sandhage and Patrick K. Gallagher
The Ohio State University
Columbus, Ohio

I. INTRODUCTION

In March 1988, Sheng and Hermann reported that compositions in the rare-earth-free Tl–Ba–Ca–Cu–O system [1] could be produced with zero resistance above 100 K. Subsequent research has revealed that a family of high-T_c superconducting compounds exist in this system. These compounds consist of perovskite-like blocks containing CuO_2 layers sandwiched between Tl–O monolayers or bilayers, with Ba cations positioned in planes adjacent to the Tl–O layers and Ca cations located between the CuO_2 layers within the perovskite blocks [2]. The ideal compositions of the monolayer and bilayer Tl–O compounds are $Tl_1Ba_2Ca_{n-1}Cu_nO_{2n+3}$ (where $1 \leq n \leq 5$) and $Tl_2Ba_2Ca_{n-1}Cu_nO_{2n+4}$ (where $1 \leq n \leq 4$), respectively. The value of n indicates the number of CuO_2 layers per unit cell (e.g., $Tl_1Ba_2Cu_1O_5$ and $Tl_2Ba_2Cu_1O_6$ possess one CuO_2 layer per unit cell, $Tl_1Ba_2Ca_1Cu_2O_7$ and $Tl_2Ba_2Ca_1Cu_2O_8$ possess two CuO_2 layers per unit cell, etc.).

The superconducting properties achieved with Tl–Ba–Ca–Cu–O compositions over the past 4 years have been impressive. The highest confirmed value of T_c ($R = 0$) reported for an oxide superconductor at ambient pressures is 127 K, which was obtained with the $Tl_2Ba_2Ca_2Cu_3O_y$-type compound synthesized in bulk form [3]. Thallium-bearing superconductors have also been produced with relatively high values of critical current density, J_c. Thin polycrystalline $Tl_2Ba_2Ca_1Cu_2O_y$ and $Tl_2Ba_2Ca_2Cu_3O_y$ films (< 1 μm thick), produced by electron beam evaporation onto single-crystal substrates, have exhibited transport critical current densities, $J_{c,t}$, on the order of 10^5 A/cm² at 77 K in the absence of an applied magnetic field [4,5]. Such thin films exhibited little drop ($\approx 20\%$) in $J_{c,t}$ (76 K) upon application of in-plane magnetic fields up to 800 Oe. Thick films (2 to 3 μm thick) of the $Tl_1Ba_2Ca_2Cu_3$ O_y-type compound on polycrystalline yttria-stabilized zirconia, produced by vapor-phase thallation of spray-pyrolyzed Ba–Ca–Cu-bearing nitrate solutions, have been shown to possess J_c,s of $>10^5$ and $>10^4$ A/cm^2 at 77 K in zero applied field and at 60 K in a 2-T field oriented parallel to the c axis, respectively [6]. Intragranular critical current densities (obtained from magnetization measurements) for bulk, polycrystalline $Tl_1Ba_2Ca_2Cu_3O_y$-, $(Tl,Pb)_1Sr_2Ca_2Cu_3O_y$-, and $(Tl,Pb)_1(Ba,Sr)_2Ca_2Cu_3O_y$-type compounds have also been found to be quite high (10^4 to 10^5 A/cm^2 in a 1-T field at 77 K) [7–9], indicating that the thermal activation of vortices is a less severe problem in these single-Tl–O-layer compounds than in double-Tl–O-layer compounds or in Bi(Pb)–Sr–Ca–Cu–O-based superconductors.

Despite these attractive properties, the processing of Tl–Ba–Ca–Cu–O superconductors is complicated by the volatility of thallium oxide at the temperatures required for superconducting oxide phase formation, the formation of low-melting-point liquids, and the multiplicity of solid phases known to exist in this system (not to mention the toxicity of thallium compounds). A thorough understanding of the thermal stability of thallium-based superconductors is required to optimize processing conditions to obtain high-J_c superconductors in a reproducible fashion. The purpose of this chapter is to provide a brief review of several important thermodynamic features of the Tl–Ba–Ca–Cu–O system, including the formation of volatile thallium-bearing oxides, the stability of thallium-bearing compounds, and cation and anion nonstoichiometry in thallium-bearing superconductors. For the rest of this chapter, the chemical formulas of thallium-bearing compounds will be abbreviated by the appropriate (ideal) Tl:Ba:Ca:Cu ratio (e.g., $TlBa_2Ca_1Cu_2O_7$ will be referred to as 1212, $Tl_2Ba_2Cu_1O_6$ as 2201, etc.).

II. VOLATILIZATION OF THALLIUM-BEARING OXIDES

The equilibria between condensed and vapor phases in the Tl–O binary system will be considered before the volatilization of thallium from Tl–Ba–Ca–Cu–O superconductors is discussed. The Tl–O binary system is relatively complicated. First, a number of solid compounds have been reported in this system, including Tl_2O_3, Tl_2O, Tl_2O_4, $2Tl_2O_3·3Tl_2O$, $Tl_2O_3·Tl_2O$, and $3Tl_2O·Tl_2O_3$ [10,11]. Nonstoichiometric oxides, TlO_x ($x = 1.52, 1.53, 1.54, 1.75$), have also been synthesized [10]. Second, a range of melting points has been observed for Tl–O compounds. For example, $Tl_2O_{(s)}$ has been reported to melt at 300°C [10] or at 579°C [11,12]. $Tl_2O_{3(s)}$ has been reported to melt at 717 to 725°C [13–15], between 750 and 775°C [16,17], or at 834°C [11,12]. These discrepancies have been attributed to the formation of solid and liquid Tl–O solutions containing thallium cations with multiple valences [16,18]. The liquidus temperatures for such solutions are likely to vary with oxygen partial pressure, as has been observed for the melting of CuO. This explanation is consistent with an observed decrease in the melting point of Tl_2O_3 at reduced oxygen partial pressure (the melting points of Tl_2O_3 in air, flowing oxygen, and >1 atm oxygen are 717 to 725°C [13–15], $>750°C$ [16,17], and $>825°C$ [16], respectively). Finally, several thallium-bearing vapor species have been reported in the Tl–O system, including $Tl(v)$, $Tl_2O(v)$, $TlO(v)$, $Tl^+(v)$, $Tl_2^+(v)$, and $Tl_2O^+(v)$ [10–12].

Thermodynamic data over a range of temperatures are available for $Tl_2O_3(s,l)$, $Tl_2O(s,l)$, $Tl_2O(v)$, and $Tl(v)$ [11,12]. Equilibrium reactions between these species can be written as follows:

$$2Tl(s,l) + \tfrac{1}{2}O_2(v) = Tl_2O(s,l) \tag{1}$$

$$Tl_2O(s,l) + O_2(v) = Tl_2O_3(s,l) \tag{2}$$

$$Tl_2O(v) + O_2(v) = Tl_2O_3(s,l) \tag{3}$$

$$2Tl(v) + \tfrac{3}{2}O_2(v) = Tl_2O_3(s,l) \tag{4}$$

$$Tl_2O(v) = Tl_2O(s,l) \tag{5}$$

$$2Tl(v) + \tfrac{1}{2}O_2(v) = Tl_2O(s,l) \tag{6}$$

$$Tl_2O(v) = 2Tl(s,l) + \tfrac{1}{2}O_2(v) \tag{7}$$

$$Tl(v) = Tl(s,l) \tag{8}$$

Equations describing the vapor pressures of $Tl_2O(v)$ and $Tl(v)$ in equilibrium with condensed Tl, Tl_2O, and Tl_2O_3 phases are listed below.

Gas-phase equilibria with $Tl_2O_3(s,l)$:

$$P_{Tl_2O} = [P_{O_2}]^{-1} a_{Tl_2O_3(s,l)} \exp \left[\frac{\Delta G°(3)}{RT} \right] \tag{9}$$

$$P_{Tl} = [P_{O_2}]^{-3/4} [a_{Tl_2O_3(s,l)}]^{1/2} \exp \left[\frac{\Delta G°(4)}{2RT} \right] \tag{10}$$

Gas-phase equilibria with $Tl_2O(s,l)$:

$$P_{Tl_2O} = a_{Tl_2O(s,l)} \exp \left[\frac{\Delta G_.°(5)}{RT} \right] \tag{11}$$

$$P_{Tl} = [P_{O_2}]^{-1/4} [a_{Tl_2O(s,l)}]^{1/2} \exp \left[\frac{\Delta G°(6)}{2RT} \right] \tag{12}$$

Gas-phase equilibria with $Tl(s,l)$:

$$P_{Tl_2O} = [P_{O_2}]^{1/2} [a_{Tl(s,l)}]^2 \exp \left[\frac{\Delta G°(7)}{RT} \right] \tag{13}$$

$$P_{Tl} = a_{Tl(s,l)} \exp \left[\frac{\Delta G°(8)}{RT} \right] \tag{14}$$

P_i and a_i refer to the partial pressure and activity, respectively, of species i. The activities of $Tl(s,l)$, $Tl_2O(s,l)$, and $Tl_2O_3(s,l)$ are also related by the following equations:

$$\frac{a_{Tl_2O(s,l)}}{[a_{Tl(s,l)}]^2} = [P_{O_2}]^{1/2} \exp \left[-\frac{\Delta G°(1)}{RT} \right] \tag{15}$$

$$\frac{a_{Tl_2O_3(s,l)}}{a_{Tl_2O(s,l)}} = P_{O_2} \exp \left[-\frac{\Delta G°(2)}{RT} \right] \tag{16}$$

With knowledge of the activities of the condensed phases at a given temperature, one can calculate the equilibrium vapor pressures of $Tl_2O(v)$ and $Tl(v)$. Unfortunately, the activity coefficients of thallium-bearing species in solid and liquid Tl–O solutions have not been reported, so that precise calculations of equilibrium vapor pressures over condensed thallium-bearing phases cannot be made. For purposes of illustration, however, a Kellogg-type diagram, shown in Fig. 1, has been constructed at 677°C (950 K) from the data in Refs. 11 and 12, using the zeroth-order assumption that the condensed phases do not exhibit mutual solid or liquid solubility. The diagram has been divided into three separate regions, corresponding to the stability fields of the pure condensed phases $Tl(l)$, $Tl_2O(l)$, and $Tl_2O_3(s)$. This simplified diagram suggests that $Tl_2O(v)$ is likely to be a more dominant vapor species than $Tl(v)$ at 677°C over a range of oxygen partial pressures (10^{-10} atm < P_{O_2} < 100 atm). Under highly reducing conditions, however, $Tl(v)$ becomes the dominant vapor species. The pressures of both $Tl(v)$ and $Tl_2O(v)$ are expected to decrease with increasing oxygen pressure under sufficiently oxidizing conditions, whereas the pressure of $Tl(v)$ becomes independent of P_{O_2} under sufficiently reducing conditions. An important consequence is that elevated oxygen pressures should be used to minimize the total vapor pressure resulting from $Tl(v)$ and $Tl_2O(v)$.

If some amount of mutual solubility is considered for the condensed phases, so that the activities of $Tl_2O_3(s,l)$, $Tl_2O(l)$, and $Tl(l)$ are less than unity, then the values of P_{Tl_2O} and P_{Tl} at any given value of oxygen pressure and temperature will be reduced, as indicated by Eqs. (9) to (14) [18]. Thus the values of P_{Tl_2O} and P_{Tl} shown in Fig. 1 may be overestimated. Cubicciotti and Keneshea [18] found, from transpiration experiments, that the dominant vapor species over $Tl_2O_3(s)$ at

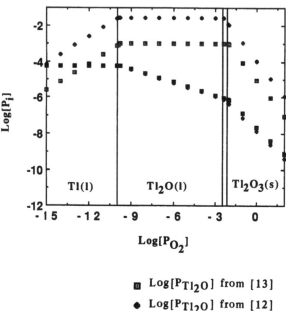

Figure 1   Kellogg-type diagram for thallium vapor species in equilibrium with $Tl(l)$, $Tl_2O(l)$, and $Tl_2O_3(s)$ at 950 K from the data of Refs. 12 and 13. Several simplifying assumptions have been made in creating this diagram, as discussed in the text. (Data from Ref. 11.)

530 to 675°C for oxygen partial pressures over the range 0.01 to 0.40 atm was $Tl_2O(v)$, consistent with Fig. 1. These authors determined the following equation for the equilibrium pressure of $Tl_2O(v)$ over $Tl_2O_3(s)$ in the temperature range 530 to 675°C:

$$\log[P_{Tl_2O}] = \left(-19.63 \times \frac{10^3}{T}\right) + 16.6 - \log[P_{O_2}] \tag{17}$$

The value of $\log[P_{Tl_2O}]$ obtained from this equation for the oxygen pressure of air ($\log[P_{O_2}] = -0.68$) at 677°C is -3.4, which is relatively close to the value predicted from the P_{Tl_2O} curve in Fig. 1 obtained from the data of [11]. However, recent thermogravimetric experiments by Wahlbeck et al. [19] indicate that the values of P_{Tl_2O} predicted by Eq. (17) are too high; that is, the P_{Tl_2O} data of Wahlbeck et al. [19] would fall below the P_{Tl_2O} curves shown in Fig. 1. Further work is required to resolve these discrepancies.

Because the activity of Tl_2O_3 in thallium-bearing superconductors is less than the activity of pure $Tl_2O_3(s)$, the vapor pressures of $Tl_2O(v)$ and $Tl(v)$ over thallium-bearing superconductors should be lower than over $Tl_2O_3(s)$. Indeed, from mass spectrometry measurements, Wahlbeck et al. [19] found that the $Tl_2O(v)$ pressure at 750 K over a specimen comprised of a mixture of 2223- and 2212-type compounds was about 230 times lower than that over pure $Tl_2O_3(s)$. Through the use of Eq. (9), these authors obtained a value of 0.0044 ± 0.001 for the activity of $Tl_2O_3(s)$ in these specimens at 750 K (for a pure, solid Tl_2O_3 reference state).

Bulk specimens of thallium-bearing superconductors are often produced by firing a mixture of oxides (e.g., Tl_2O_3, BaO_2, CaO, CuO) in flowing oxygen at peak temperatures in excess of 800°C. Owing to the high pressure of $Tl_2O(v)$ over unreacted Tl_2O_3 at such temperatures, a significant amount of thallium can be lost during such firing. For example, the weight loss of powders having the nominal 6223 composition during annealing in flowing oxygen at 850°C [20] is shown in Fig. 2. The powders were prepared by calcination of coprecipitated (Chem-Prep) precursors or by calcination of blended oxides (mixed-oxide) [20]. Both types of powders were found to lose > 80% of their original thallium within 30 min at 850°C. The rate of weight loss was particularly rapid during the first 5 min of firing, when a significant amount of unreacted Tl_2O_3 remained in the powders. After longer annealing times (>15 min in Fig. 2), the Tl_2O_3 reacted to form multicomponent oxides, so that the local Tl_2O_3 activity decreased (relative to pure, unreacted Tl_2O_3), which resulted in a decrease in the $Tl_2O(v)$ pressure and a lower rate of weight loss. While a reduction in the firing temperature can decrease the vapor pressure of $Tl_2O(v)$, prolonged annealing at modest temperatures can still lead to noticeable thallium loss from Tl–Ba–Ca–Cu–O compounds. For example, Martin et al. [21] found that 2212 pelleted specimens lost a detectible amount of thallium after annealing for 12 h at temperatures as low as 400°C.

A number of authors have prepared superconducting oxide phases by annealing a mixture of precursor oxides (e.g., Tl_2O_3, BaO_2, CaO, and CuO) in open systems lacking a source of excess $Tl_2O(v)$. Superconducting oxide phase formation in such open systems occurs simultaneously with thallium volatilization. Ruckenstein and Cheung [22,23] examined the evolution of superconducting oxide phases during the firing of mixed oxides (Tl_2O_3, BaO_2, CaO, CuO) of nominal compositions 2223,

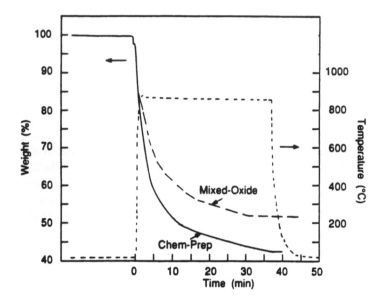

Figure 2 Weight loss as a function of calcination time at 850°C in flowing oxygen for coprecipitated (Chem-Prep) and blended-oxide (mixed-oxide) precursors of nominal 6223 composition. (From Ref. 21.)

1223, 1324, and 1122 in an alumina crucible capped with silver foil at 867°C in air for 2.5 to 150 min. Superconducting oxide phases were observed to form in the following order during annealing for up to 120 min:

For 1212 compositions:

1201 → 2201 → 2201 + 2212 + 1212 → 2212 + 1212 → 1212

For 1223 compositions:

1201 + 2201 → 2212 + 2223 → 2223 → 2223 + 1223 → 1223

For 1234 compositions:

2201 → 2212 + 2223 → 2223 → 2223 + 1234

For 2223 compositions:

2201 → 2201 + 2212 → 2212 + 2223

For the 2223 starting composition, the order of appearance of the superconducting oxide phases (2201, then 2212, then 2223) was consistent with gradual insertion of CaO and CuO layers into the single-CuO_2-layer compound 2201. The evolution of the 2212 from 2201, and of 2223 from 2212, have been reported by a number of authors [22,24–27]. A similar evolution of phases was not observed for the 1212, 1223, and 1234 compositions, however (e.g., the 1223 phase did not form by the sequence 1201 → 1212 → 1223). For these compositions, the double-Tl–O-layer compounds formed before the desired single-Tl–O-layer superconducting oxide phases were produced. The gradual formation of single-Tl–O-layer compounds from double-Tl–O-compounds, after annealing at 885 to 890°C in pure oxygen, has also been reported by other authors [8,16,22,24,26,28–31]. It is interesting to note that the three-CuO_2-layer compound, 2223, did not form from the 1212 composition, which was depleted in calcium and copper relative to the other compositions. Other authors have also observed that compositions rich in calcium or calcium and copper tend to promote the formation of the three-layer phases [32–39]. Unfortunately, residual nonsuperconducting calcium–cuprate phases tend to remain after the firing of such calcium- and copper-enriched compositions.

Ruckenstein and Cheung [22,23] also observed that single-Tl–O-layer compounds formed after annealing 2223 starting compositions for $\geq$150 min, as a result of thallium volatilization. A significant amount of thallium was also lost from the 1223 and 1212 compositions after annealing for 120 min. This resulted in the formation of significant amounts of nonsuperconducting $BaCuO_2$, CaO, and CuO phases.

A number of authors have enhanced the rate of superconducting oxide formation, so as to minimize the reaction time and resulting thallium volatilization, by reacting Tl_2O_3 [or $Tl_2O(v)$] with prereacted binary or ternary Ba–Ca–Cu oxides [1,30,36,37,41,42]. For example, Sugise et al. [30] observed that $Ba_2Ca_2Cu_3O_y$ thick films exposed to $Tl_2O(v)$ (evolving from Tl_2O_3 or $Tl_{1.5}Ba_2Ca_2Cu_3O_y$ powder) were largely transformed into the 2223-type phase after 10 min at 895°C in an oxygen atmosphere. Wu et al. [41,42] prepared phase-pure 2212-type compound (as determined by x-ray diffraction) by firing a mixture of Tl_2O_3, CuO, and $CaBa_2CuO_4$ (with an overall 2212 stoichiometry) for 6 to 14 h at 830°C in air. This low-temperature anneal resulted in relatively little thallium loss for specimens wrapped in gold foil (2.0% thallium loss after 6 h, 7.3% loss after 14 h). Phase-pure 2223-type compound (by x-ray diffraction) was also produced by firing a reactant mixture of the Tl_2O_3, CaO, CuO, and $CaBa_2CuO_4$ phases (with an overall 2223 composition) for 2 h in air at 870 or 880°C.

Thallium volatilization complicates the study of phase equilibria. However, even in the absence of significant thallium loss [e.g., in closed systems or in open systems with a buffered $Tl_2O(v)/O_2$ atmosphere], complex phase relations exist in the Tl–Ba–Ca–Cu–O system at various pressures of $O_2(v)$ and $Tl_2O(v)$ and at various temperatures, as discussed in the following section.

III. STABILITY OF Tl–Ba–Ca–Cu–O COMPOUNDS (IN CLOSED SYSTEMS OR OPEN, BUFFERED SYSTEMS)

Equilibration experiments in the Tl–Ba–Ca–Cu–O system are nontrivial. To establish a well-defined state of equilibrium, the oxygen partial pressure, the partial pressures of thallium vapor species, the total pressure, the nonthallium cation ratios, and the temperature must be controlled [16]. Fixing the $O_2(v)$ and $Tl_2O(v)$ pressures is usually accomplished by (1) heat treating, at an elevated total pressure, specimens of known composition inside ampoules that have been sealed at room temperature with a known gas mixture [16,43], or (2) annealing, at ambient pressure, specimens in an open system in which the $Tl_2O(v)$ pressure is established by heating at $Tl_2O(v)$ vapor source (e.g., Tl_2O_3 or Tl–Ba–Ca–Cu–O compositions) in a controlled oxygen atmosphere [17,44]. Because the *a*-axis lengths of the thallium-bearing superconducting oxides are similar, intergrowths of superconducting oxide phases are frequently observed, particularly for superconducting thallium compounds with two or more CuO_2 layers per unit cell. Intergrowths of compounds containing different numbers of CuO_2 layers (e.g., 2223 inside 1223 grains, 2212 inside 2223 grains, 1212 inside 1223 grains) or Tl–O layers (e.g., 1234 of 1256 inside 1245 grains) have been observed by electron microscopy [2,4,17,33,45–54]. Removal of metastable intergrowths can require prolonged annealing times (i.e., many days) [16,17]. Care must also be taken to minimize reaction of Tl–Ba–Ca–Cu–O compositions with crucible materials. Crucible materials that have exhibited relatively little reaction with Tl–Ba–Ca–Cu–O specimens include Au, MgO, or Tl–Ba–Ca–Cu–O powder of similar composition [16,17,43,44].

The stability of a given superconducting oxide phase is strongly related to the partial pressure of $Tl_2O(v)$ in the gas phase. Aselage et al. [16,43] have examined the equilibration of Tl–Ca–Ba–Cu–O specimens sealed inside silica ampoules at room temperature in pure oxygen at 610 torr. These authors independently controlled the temperature of a $Tl_2O(v)$ source (pure Tl_2O_3), and the temperature of the Tl–Ca–Ba–Cu–O specimens, so as to be able to vary the $Tl_2O(v)$ pressure at a given specimen temperature. The oxygen pressure within the sealed ampoules was estimated to be ≥ 3 atm for specimen temperatures in the range 825 to 925°C. In a first series of experiments [43], these authors kept the thallium vapor source and the specimens at the same temperature. Equilibration of 0212, 0223, and 0234 starting compositions with the thallium vapor source at 875 to 925°C yielded Tl–Ca–Ba oxides of approximate composition $(Ca,Ba)_{3-4}Tl_{10}O_y$, $CaBa_{0.8}Tl_2O_y$, and $Ca_2Tl_2O_y$, along with CuO, as the stable phases; that is, no superconducting phases were observed under these conditions. In a second set of experiments, Tl–Ba–Ca–Cu–O specimens were equilibrated at 890°C with a range of $Tl_2O(v)$ pressures. The $Tl_2O(v)$ pressure was changed by heating the thallium vapor source (pure Tl_2O_3) to various temperatures. For thallium source temperatures between 725 and 800°C, a mixture of 2212- and 2223-type compounds was found to be stable. As the calcium-to-copper ratio of the precursors increased from 1:2 to 2:3 to 3:4 under these conditions, the 2223-type phase became more prevalent than the 2212-type phase (although a mixture of these phases was still observed). With source temperatures below 700°C, less thallium was observed in the specimens, and $m = 1$ phases (particularly 1223) and barium and calcium cuprates were formed. Thus the 2212- and 2223-type phases were found to be stable only within a certain range of $Tl_2O(v)$ pressures at 890°C and elevated oxygen pressures; at sufficiently low or high $Tl_2O(v)$ pressures, the $m = 2$ phases became unstable.

Aselage et al. [17] also examined the equilibria of Tl–Ba–Ca–Cu–O specimens with $Tl_2O(v)$ at an oxygen pressure of 1 atm. Tl–Ba–Ca–Cu–O specimens with Ba/Ca/Cu ratios of 2:1:2, 2:2:3, and 2:3:4 were equilibrated at 880°C in flowing oxygen with $Tl_2O(v)$ produced from a thallium vapor source (pure Tl_2O_3) heated from 600°C to 840°C. Equilibration of specimens with a Ba/Ca/Cu ratio of 2:1:2 with $Tl_2O(v)$ produced from thallium source temperatures of 510 to 700°C yielded the 1212-type compound and a small amount of $BaCuO_2$. When the thallium source temperature was raised to 750°C, mixtures of 1212- and 2212-type compounds were observed in the 2:1:2 specimens. At 775 to 840°C, the thallium source melted, and 2212 and a Tl–Ca–Ba–O phase (of approximate composition $CaBa_{0.8}Tl_2O_y$) were the stable phases. For specimens with a Ba/Ca/Cu ratio of 2:2:3 and 2:3:4, equilibration at a thallium source temperature of 600°C yielded CaO, Ca_2CuO_3, and $Ba_2Cu_3O_5$ phases (superconducting oxide phases were not stable under these conditions). At 700°C, the 1223-type compound was the dominant oxide phase, with 1212- or 2223-type compounds present in small amounts for the 2:2:3 or 2:3:4 specimens, respectively. At 750°C, a mixture of 2223- and 2122-type compounds were observed (with more 2223 seen in the 2:3:4 specimen than in the 2:2:3 specimen). Between 775 and 800°C, the thallium source melted and 2212 became the dominant oxide phase, with 2223 or a Tl–Ca–O phase (of approximate composition Ca$_2$Tl$_2$O$_y$) as minor phases. At 840°C, the CaBa$_{0.8}$Tl$_2$O$_y$, Ca$_2$Tl$_2$O$_y$, and an additional Tl–Ba–Ca–O phase [of approximate composition (Ca,Ba)$_{3-4}$Tl$_{10}$O$_y$] were observed with CuO (no superconducting phase were stable). Thus the work of Aselage et al. [16,17,43] indicates that the $m = 2$ phases (2212, 2223) are stable at 880 to 890°C only over a certain range of Tl$_2$O(v) pressures for oxygen pressures in the range 1 to ≈3 atm. As the Tl$_2$O(v) pressure is reduced below the stability limit of the $m = 2$ phases at a given temperature and oxygen pressure, the $m = 1$ phases (1212 and 1223) begin to form. This transformation from $m = 1$ to $m = 2$ phases with increasing Tl$_2$O(v) pressure has also been observed by DeLuca et al. [44]. These authors annealed thick Ba–Ca–Cu–O films (Ba/Ca/Cu = 2:2:3) at 895°C for 30 min in flowing oxygen downstream from a thallium source (Tl$_2$O$_3$). As the thallium source temperature increased from ≈800–810°C to above 810°C, the thallium content of the film increased and the 2212-type compound replaced the 2223-type compound as the dominant oxide phase.

Several authors have used Tl–Ba–Ca–Cu–O bulk specimens as thallium vapor sources, instead of pure Tl$_2$O$_3$, to provide a desired Tl$_2$O(v) pressure during annealing of Tl–Ba–Ca–Cu–O films. Ginley et al. [4,5] observed that proper control of the thallium content in Tl–Ba–Ca–Cu–O films was essential to obtaining good superconducting properties. These authors prepared thin films of the 2212- and 2223-type compounds (0.2 to 0.7 μm thick) by electron-beam evaporation, followed by annealing at 750°C in 1 atm oxygen near polycrystalline Tl–Ba–Ca–Cu–O pellets of the target film composition. The 750°C anneals resulted in the desired thallium contents in the films, as evidenced by the formation of the 2212- or 2223-type compound as the predominant phase by x-ray diffraction analyses. Sugise et al. [30] found that exposure of Ba$_2$Ca$_2$Cu$_3$O$_y$ thick films (10 μm thick) to a Tl$_{1.5}$Ba$_2$Ca$_2$Cu$_3$O$_y$ vapor source at 895°C in a 1-atm oxygen atmosphere resulted in the formation of the 2223-type compound as the predominant phase. The 2223 phase was found to be stable for exposure times ranging from 10 to 60 min with this vapor source; and use of Tl$_2$O$_3$ was the vapor source under similar conditions resulted in transformation of the 2223 phase to the 2212 phase for exposure times ≥10 min. The results of Aselage et al., discussed above, indicate that prolonged exposure of Ba–Ca–Cu–O films to Tl$_2$O(v) evolving from molten Tl$_2$O$_3$ under these conditions would have resulted in complete decomposition of Tl–Ba–Ca–Cu–O superconducting phases.

The stability of a given superconducting oxide phase is also dependent on the oxygen partial pressure. Ahn et al. [57] annealed Tl–Ba–Ca–Cu–O pellets composed of the 2212-type compound and calcium–copper oxide phases (with a composition of 1.7:2:2:3) over a range of oxygen partial pressures for 1 to 12 hours at 700 to 900°C in a sealed quartz chamber. At any given temperature, annealing above a critical oxygen partial pressure resulted in the retention of 2212 as the dominant phase, whereas annealing below this oxygen pressure led to the formation of 2223 as the dominant phase. Because these authors did not observe a transformation of the 2223-type phase back to the 2212-type phase after annealing above the critical oxygen pressure (i.e., for 12 h in 0.5-atm oxygen at 850°C), they concluded that this critical oxygen partial pressure was a kinetic boundary corresponding to the stability limit of a liquid phase that assisted in the formation of the 2223-

type phase from the 2212-type phase [58]. However, Aselage et al. [17] have argued that more than 12 h is required at 850°C to transform the 2223 phase back to the 2212 phase, and as a result, the P_{O_2}–T boundary line described by Ahn et al. may still be considered a thermodynamic boundary between the 2223- and 2212-type phases. In any event, the results of Ahn et al. are consistent with the observation that the temperature required to form the 2223-type phase in a closed system (e.g., in a silica ampoule filled with pure oxygen gas) is generally higher than in an open system [57]. If the atmosphere inside an open and a closed system are the same at room temperature, the oxygen pressure in the closed system will be higher at elevated temperatures, so that the temperature required to form the 2223-type compound from the 2212-type compound will also be higher in the closed system. Further, because 2223 phase formation can occur at lower temperatures with reduced oxygen pressures, thin 2223-type films can be produced with minimal substrate interaction [57,59]. At sufficiently low oxygen pressures, however, thallium superconductors will become unstable, owing to the reduction of thallium and copper cations.

IV. SOLID–LIQUID EQUILIBRIA IN THE Tl–Ba–Ca–Cu–O SYSTEM

Another important thermodynamic feature of the Tl–Ba–Ca–Cu–O system is the presence of low-melting liquids [4,5,60]. DeLuca et al. [44] found that thick films (2 to 3 μm thick) of textured 2223-type phase could be produced with relatively high $J_{c,t}$ (77 K, zero field) values (up to 28,800 A/cm^2) when the films were annealed at temperatures just below those required for the formation of appreciable amounts of liquid. They suggested that small amounts of liquid may accelerate the formation of a highly crystallized, textured 2223 microstructure. However, excessive liquid formation was found to result in the formation of unoriented 2212-type grains and a general reduction in critical current density. Such work indicates that careful control of liquid formation is necessary during the processing of Tl–Ba–Ca–Cu–O superconductors to obtain good superconducting properties in a reproducible fashion.

Kotani et al. [60] examined the influence on melting behavior of the calcium and copper contents of Tl–Ba–Ca–Cu–O specimens. Mixtures of Tl_2O_3, CaO, BaO_2, and CuO with nominal starting compositions, $Tl_2Ba_2Ca_{1.5+x}Cu_{2.5+y}O_z$ (where x was varied from 0 to 6.5 and y was varied from 0 to 5), were partially melted in gold tubes at >950°C and cooled slowly to room temperature. As the nominal calcium and copper contents were both increased, the dominant superconducting phases observed after firing varied from 2201 to 2212 to 2223. Differential thermal analyses (presumably conducted in air) of multiphased compositions that contained mostly 2201-, 2212-, 2223-, and 2234-type phases revealed that melting events occurred at 941, 907, 896, and 893°C, respectively. Specimens comprised of a mixture of the 2223 phase and Ca_2CuO_3 were also found to exhibit small endothermic peaks at ≈885°C. These authors suggested that the 893, 896, 907, and 941°C melting events corresponded to peritectic melting of the 2234-, 2223-, 2212-, and 2201-type phases, respectively, while the 885°C melting event was assumed to be a eutectic melting point. Melting of mixed oxides of nominal 2212 and 2223 compositions at ≥870°C in air has also been reported by Wu et al. [41,42]. The compositions of the differential thermal analysis (DTA) specimens examined by Kotani et al. were not reported, so that exact correlation of a given melting temperature with a given composition could not be determined. Nonetheless, based on their DTA and x-ray diffraction results, Kotani et al. have proposed the approximate pseudo-binary-phase diagram shown in Fig. 3 [60]. This diagram is also consistent with reports that the peritectic melting of the 2223 compound results in the formation of the 2212 phase [35,44,61].

As discussed in Section II, the melting point of Tl$_2$O$_3$ is believed to increase as the oxygen partial pressure increases. The formation of liquids during the heat treatment of Tl–Ba–Ca–Cu–O compositions has also been found to be influenced by the oxygen partial pressure. Goretta et al. [56] observed that severe melting occurred when a nominal 2223 composition was annealed at 860°C in 0.1 atm of oxygen; less melting was observed for similar specimens annealed at 890°C and 0.3 atm of oxygen. Morris et al. [40] observed that the formation of 2212 or 2223 compounds from precursor oxides at elevated pressures (200 bar) was inhibited during reaction below 850°C, presumably because the higher pressures hindered the formation of a reactive, transient liquid phase.

The formation of liquids in the Tl–Ba–Ca–Cu–O system also appears to be influenced by the Tl$_2$O(v) pressure. DeLuca et al. [44] observed that a liquid phase formed during exposure of spray-pyrolyzed Ca$_2$Ba$_2$Cu$_3$O$_y$ thick films (2 to 3 μm thick) at 885°C to a Tl$_2$O(v) source (Tl$_2$O$_3$) held at 770 to 820°C in air. The amount of liquid produced in the films increased as the Tl$_2$O(v) pressure increased [i.e., as the temperature of the Tl$_2$O(v) source increased]. Exposure to the Tl$_2$O(v) source of similar films in flowing oxygen was observed to result in liquid formation, only if the film was heated to ≥890°C and if the Tl$_2$O(v) source was heated to ≥815°C. Thus an increase in the oxygen partial pressure and/or a decrease in the Tl$_2$O(v) pressure yielded a reduced amount of liquid.

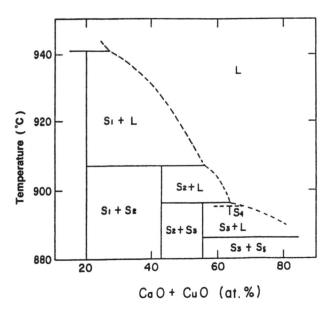

Figure 3   A speculated pseudo-binary-phase diagram for the Tl–Ba–Ca–Cu–O system. S$_1$, S$_2$, S$_3$, S$_4$, S$_5$, and L refer to the 2201, 2212, 2223, 2234, calcium copper oxide, and liquid phases, respectively. (From Ref. 60.)

The formation of intermediate, transient liquids is thought to play a role in the formation of Tl–Ba–Ca–Cu–O phases during the firing of precursor oxides. Ruckenstein and Cheung [22,23] observed the formation of a glassy-appearing phase after annealing mixed oxides (Tl_2O_3, BaO_2, CaO, CuO) of nominal composition 2223 at 867°C in air for 5 to 60 min. The glassy phase formed as a matrix around Tl–Ba–Ca–Cu–O phases and CaO particles. Careful examination revealed that the glassy phase was separated from the CaO particles by a layer of the 2212-type phase. These authors believed that at elevated temperatures, the glassy phase was a liquid that possessed a composition close to 2203. This approximate composition was assigned to the liquid from the observation that the amount of melting at 867°C increased as the nominal starting composition of calcium-free specimens changed from 2201 to 2202 to 2203. The amount of melting for 2204 and 2205 was also extensive, but slightly less than for 2203. After cooling, the liquid-like regions consisted of CuO, 2201, and some unknown phases, consistent with recrystallization of a liquid of composition 2203. Ruckenstein and Cheung [23] suggested that the 2203-type liquid forms as an intermediate phase that reacts directly with CaO particles to form 2212-type compound. The importance of liquids on superconducting oxide phase formation in this and other systems has also been discussed by several other authors [40,44,51,57,58,62].

V. NONSTOICHIOMETRY IN Tl–Ba–Ca–Cu–O SUPERCONDUCTORS

The superconducting oxide compositions shown in Fig. 3 are represented as line compounds. In fact, however, the Tl–Ba–Ca–Cu–O superconductors have been found to exhibit noticeable cation and oxygen nonstoichiometry. Since the electronic properties of ceramic materials are often strongly dependent on point defect concentrations, an understanding of cation and oxygen nonstoichiometry in thallium-bearing superconductors is important.

A number of authors have applied refinement methods to x-ray and neutron diffraction data in order to determine the cation site occupancies in single-crystal or polycrystalline specimens containing 2223-, 2212-, and 2201-type phases [35,38,46,54,63–76]. While such refinements have generally indicated that barium and copper sites are fully occupied, relatively low and high occupation factors have often been reported for the thallium and calcium sites, respectively, which has led several authors to suggest that thallium cations may be substituting on calcium sites (with vacancies present on the thallium sites), or that excess calcium cations may be substituting on thallium sites. Some authors have also predicted a calcium deficiency from refinements of diffraction data, which suggests that calcium vacancies may be present in some cases [37]. Calcium–thallium site substitution is not unreasonable, considering the similarity of the ionic radii of Ca^{2+} and Tl^{3+} [77]. Morosin et al. [67,71–73] observed a decrease in the *c*-axis length of 2212- and 2223-type single crystals as the calculated substitution of thallium on calcium sites increased, which is consistent with the slightly smaller size of the Tl^{3+} cation relative to the Ca^{2+} cation. These authors also reported that the maximum calculated amount of thallium substituted on calcium sites for the 2223-type compound was less than for the 2212-type compound. High-resolution electron microscopy of polycrystalline 2212- and 2223-type compounds (prepared by firing mixed oxides at 910°C in flowing oxygen) suggests that calcium substitution may be occurring on thallium sites [78]. Calcium–thallium site substitution and cation vacancies have also been predicted from diffraction data for single-Tl–O-layer compounds (e.g., 1212, 1223) [50,68,69,71,72].

Chemical analyses have also indicated a significant variation in the compositions of thallium-bearing superconductors from the ideal compositions. Electron microprobe analyses (EPMA) of the 2201-, 2212-, and 2223-type phases (prepared by firing mixed oxides enclosed in gold foil in flowing oxygen at 880 to 910°C from several minutes to several hours) have indicated significant thallium deficiency; 2212- and 2223-type phases have also exhibited both enrichment and depletion of calcium [33,35,76,79]. Energy-dispersive x-ray analyses (EDX) have also revealed thallium and calcium deficiencies, relative to the ideal compositions of 2212-, 2223-, 2234-, and 1245-type grains in polycrystalline specimens [54,80,81]. Interpretation of EPMA and EDX results can be complicated, however, by the presence of fine-scale intergrowths. For specimens with large intergrowth concentrations, electron microprobe analysis will represent an average composition of the intergrowth and matrix phases, not the lattice composition of the matrix phase. For example, EPMA data obtained by Parkin et al. [47] indicated that the 1223-type phase present in multiphased, polycrystalline specimens was thallium-rich and calcium-poor relative to the ideal stoichiometry. These authors attributed the relatively high thallium composition to 2223 intergrowths that were subsequently observed (by high-resolution electron microscopy) inside the 1223 grains. Hence careful interpretation of EPMA and EDX data requires electron microscopy.

Consistent with EPMA and EDX analyses, thallium deficiencies in 2201- and 2212-type compounds have also been detected from wet-chemical analyses by Paranthaman et al. [82,83]; these authors were able to synthesize phase-pure $Tl_{2-x}Ba_2Cu_1O_y$-type compound (according to x-ray diffraction analysis) with a thallium deficiency as high as $x = 0.5$. Phase-pure 2201 was found to be thermodynamically unstable in pure oxygen at modest temperatures (e.g., 500°C) for $Tl_{2-x}Ba_2Cu_1O_y$-type compositions with $x < 0.5$. (Tl_2O_3 was found to be "extruded" from specimen compositions with $x < 0.5$ at modest temperatures.)

Despite the general trends discussed above for cation nonstoichiometry, significant differences in the thallium and calcium contents for any given type of thallium-bearing compound have been reported by different authors. Discrepancies may be due, at least in part, to differences in composition resulting from variations in processing conditions from laboratory to laboratory. For example, Gopalakrishnan et al. [84] observed that the thallium site occupancy in phase-pure 2212 specimens decreased as the specimens were annealed for longer times at 677°C in argon or oxygen, presumably due to thallium volatilization. An increase in thallium deficiency in the 2201-type compound with increased annealing time at 900°C in air has also been reported by Paranthaman et al. [82,83]. Aselage et al. [17] observed that the thallium content in single-phase 1212 and 2212 specimens increased as the $Tl_2O(v)$ pressure, to which these specimens were exposed, increased. The calcium and thallium contents in the 2212-type phase were also observed to vary with the nominal calcium and copper concentrations in the starting precursors [17]. A reduction in the initial nominal thallium content of polycrystalline $Tl_{2-x}Ba_2Ca_1Cu_2O_y$ specimens (from $x = 0$ to 0.4) has also been reported to reduce the thallium–calcium site substitution in the 2212-type compound [63].

The oxygen content of Tl–Ba–Ca–Cu–O superconducting compounds can be varied with appropriate heat treatments, although this variation is considerably smaller than for $REBa_2Cu_3O_{7-y}$ compounds. The oxygen content (or oxygen non-stoichiometry) of certain Tl–Ba–Ca–Cu–O superconductors can have a strong effect on the critical temperature, T_c. This is particularly true for the 2201-type compound. Several authors have reported that exposure of the 2201-type compound to oxidizing conditions at modest temperatures (e.g., 1 atm O_2 at 350°C [85]) renders this compound nonsuperconducting, while exposure to inert or reducing conditions at modest temperatures (e.g., argon or H_2–Ar mixtures at $\approx$300 to 350°C [85,86] can yield T_c (onset) values as high as 92 K [21,85–91].

The increase in T_c is also accompanied by a small weight loss and an increase in the room-temperature c-axis length, with both phenomena resulting from oxygen evolution [21,85,87,90,91]. Such changes in weight, c-axis length, and T_c can be reversed by varying the oxygen content in the atmosphere during heat treatment at sufficiently low temperatures, which indicates that these effects are not due to irreversible thallium volatilization. The relationship between oxygen content and superconducting properties is opposite to that observed for the $REBa_2Cu_3O_{7-y}$ compounds, where the optimum superconductivity is associated with the maximum oxygen content. In this respect the Tl-based superconductors more closely resemble those of the Pb–Sr–Ca–Y–Cu–O system [92].

Paranthaman et al. [83], who used wet-chemical techniques to measure the thallium and oxygen contents directly, also found that the T_cs increased as the absolute oxygen content in $Tl_{1.92}Ba_2CuO_x$ specimens decreased from $x = 5.95$ (nonsuperconducting) to $x = 5.92$ [T_c (onset) = 26 K] to $x = 5.59$ [T_c (onset) = 63 K] [83]. While the reversible change in oxygen content of the 2201-type compound is considerably smaller than for the $REBa_2Cu_3O_{7-y}$ compounds, the rate of change in T_c with oxygen content is more dramatic for the 2201-type compound; for example, a change of 0.1 oxygen atom per formula unit in the 2201-type compound yields a change in T_c of $\approx$80 K. Such dramatic changes in T_c with oxygen content for 2201 indicate that variations in the cooling rate from the peak sintering temperature and the oxygen partial pressure during cooldown may alter the T_c observed, even for powders [21]. Since such processing parameters are often not reported, differences in T_c obtained from different laboratories are not surprising. In any event, it is important to note that the 2201 phase is not a stable superconducting phase in pure oxygen at modest temperatures (e.g., <500°C).

Unlike the other Tl–Ba–Ca–Cu–O compounds, which are reported to possess only tetragonal symmetry, the 2201-type compound has been found to exist in both orthorhombic and tetragonal forms [2,73,75,79,85,86,88–90]. The orthorhombic version can be produced by annealing the 2201-type compound in 1-atm oxygen at 300°C or by slow cooling in air or oxygen from peak sintering temperatures of $\geq$880°C. The tetragonal version can be produced by rapid quenching from temperatures above 880°C or by annealing the orthorhombic version at $\geq$300°C in a reducing atmosphere. Manthiram et al. [82] note that the transformation from tetragonal to orthorhombic 2201 can also be avoided by starting with a thallium-deficient specimen of composition $Tl_{1.5}Ba_2Cu_1O_y$. With proper adjustment of oxygen content, both orthorhombic and tetragonal 2201 phases have been found to superconduct, so that the crystal symmetry is not a limiting factor for superconductivity in this compound [82,83].

While refinement of neutron diffraction data has indicated that excess oxygen may be located on interstitial sites in the Tl–O layers of the 2201 compound [93], direct measurements of the oxygen contents (by iodometric titration) of fully oxygenated, phase-pure 2201- and 2212-type specimens (annealed in oxygen at ≥300°C) reveal deficiencies of oxygen in these compound(s) [82,83].

The T_c of the 2212-type compound (prepared by annealing mixed oxides at peak temperatures of 880 to 900°C, either wrapped in gold foil in flowing oxygen or sealed in ampoules at room temperature in an air or oxygen atmosphere) has been reported to increase (from as low as 85 K to as high as 118 K) as oxygen is removed from the lattice at modest temperatures (300 to 600°C) [4,5,21,85,87]. For the 2223-type compound, the influence of oxygen content on T_c is reported to be much smaller than for the 2201-type compound. Some authors have reported an increase in T_c after inert or vacuum-heat treatment to remove oxygen from the 2223-type compound [3,69,94], while others have observed either no effect or the opposite effect [21,30,42,85,87,95]. Such mixed reports for the 2223-type compound suggests that an optimum oxygen content may exist, with a corresponding maximum T_c, for this compound; that is, too little or to much oxygen may reduce T_c below the maximum value.

The single-Tl–O-layer compounds 1212, 1223, and 1234 (prepared by annealing mixed oxides at peak temperatures of 880 to 900°C in flowing oxygen or in stagnant oxygen present inside a sealed silica ampoule) have also exhibited an increase in T_c with oxygen loss (e.g., after annealing from 200 to 500°C in inert atmospheres) [21,96–98]. As for the double-Tl–O-layer compounds, the change in T_c per unit oxygen loss was observed to increase as the number of CuO_2 layers in the unit cell increased (i.e., ΔT_c was largest for 1212 and smallest for 1234) [96–98].

It should be noted that a variety of other interrelated factors could influence T_c in Tl–Ba–Ca–Cu–O compounds in addition to the oxygen nonstoichiometry. Such factors include cation nonstoichiometry/substitution, strain effects, and structural modulations or intergrowths [2,16,17,29,32,50,68,71,72,94].

VI. GENERAL COMMENTS

Owing to the complexities of the Tl–Ba–Ca–Cu–O system discussed in this chapter, a good deal more research remains to be conducted to obtain a more detailed understanding of several thermodynamic features of this system. Among the important questions that remain to be resolved are:

1.   What are the subsolidus compatibility fields for the five-component Tl–Ba–Ca–Cu–O system? How are such compatibility fields influenced by the $O_2(v)$ and $Tl_2O(v)$ pressures?

2.   What are the compositions of the solid and liquid decomposition products that form upon peritectic melting of thallium-bearing superconductors? Rapid quenching experiments conducted under well-controlled conditions, like those reported for bismuth-based superconductors [62], are needed.

3.   How are the limits of cation nonstoichiometry for the single- and double-layer-Tl–O superconducting compounds influenced by the $O_2(v)$ and $Tl_2O(v)$ pressures, the total pressure, and temperature?

ACKNOWLEDGMENT

The authors are grateful to the Midwest Superconducting Consortium for its continued support of their research and publication in this general field.

REFERENCES

1. Z. Z. Sheng, and A. M. Hermann, *Nature (London)* 332, 138 (1988).
2. S. S. P. Parkin, V. Y. Lee, A. I. Nazzal, R. Savoy, T. C. Huang, G. Gorman, and R. Beyers, *Phys. Rev. B* 38, 6531 (1988).
3. T. Kaneko, H. Yamauchi, and S. Tanaka, *Physica C* 178, 377 (1991).
4. D. S. Ginley, J. F. Kwak, E. L. Venturini, B. Morosin, and R. J. Baughman, *Physica C* 160, 42 (1989).
5. D. S. Ginley, J. F Kwak, R. P. Hellmer, R. J. Baughman, E. L. Venturini, M. A. Mitchell, and B. Morosin, *Physica C* 156, 592 (1988).
6. J. E. Tkaczyk, J. A. DeLuca, P. L. Karas, P. J. Bednarczyk, M. F. Garbauskas, R. H. Arendt, K. W. Lay, and J. S. Moodera, *Appl. Phys. Lett.* (1992).
7. R. S. Liu, D. N. Zheng, J. W. Loram, K. A. Mirza, A. M. Campbell, and P. P. Edwards, *Appl. Phys. Lett.* 60, 1019 (1992).
8. D. E. Peterson, P. G. Wahlbeck, M. P. Maley, J. O. Willis, P. J. Kung, J. Y. Coulter, K. V. Salazar, D. S. Phillips, J. F. Bingert, E. J. Peterson, and W. L. Hults, *Physica C*, in press.
9. T. Kamo, T. Doi, A. Soeta, T. Yuasa, N. Inoue, K. Aihara, and S. Matsuda, *Appl. Phys. Lett.* 59, 3186 (1991).
10. A. G. Lee, *The Chemistry of Thallium*, Elsevier, Amsterdam, 1971, pp. 92–95.
11. L. B. Pankratz, *Thermodynamic Properties of Elements and Oxides*, Bulletin 672, U.S. Bureau of Mines, Washington, D.C., 1982.
12. R. H. Lamoreaux, D. L. Hildebrand, and L. Brewer, *J. Phys. Chem. Ref. Data* 16, 419 (1987).
13. G. P. Wirtz and D. C. Siebert, *Mater. Res. Bull.* 6, 381 (1971).
14. S. A. Shchukarev, G. A. Semenov, and I. A. Rat'kovskii, *Russ. J. Inorg. Chem.* 6, 1423 (1961).
15. M. D. Karkhanavala and S. H. Daroowalla, *J. Indian Chem. Soc.* 46, 1112 (1969).
16. T. L. Aselage, J. A. Voigt, E. L. Venturini, W. F. Hammetter, S. B. VanDeusen, D. L. Lamppa, R. G. Tissot, and M. O. Eatough, in *Superconductivity and Ceramic Superconductors II*, Ceram. Trans. 18, ed. K. M. Nair, U. Balachandran, Y.-M. Chiang, A. S. Bhalla, 1991, pp. 9–22.
17. T. L. Aselage, E. L. Venturini, S. B. Van Deusen, T. J. Headley, M. O. Eatough, and J. A. Voigt, *Physica C* (1992), in press.
18. D. Cubicciotti and F. J. Keneshea, *J. Phys. Chem.* 71, 808 (1967).
19. P. G. Wahlbeck, R. R. Richards, and D. L. Meyers, *J. Chem. Phys.* 95, 9122 (1991).
20. J. A. Voigt, B. C. Bunker, W. F. Hammetter, D. S. Ginley, E. L. Venturini, J. F. Kwak, and D. L. Lamppa, in *High Temperature Superconducting Compounds: Processing and Related Properties*, ed. S. H. Whang and A. DasGupta, The Minerals, Metals, and Materials Society, Warrendale, Pa., 1989, pp. 291–301.
21. C. Martin, A. Maignan, J. Provost, C. Michel, M. Hervieu, R. Tournier, and B. Raveau, *Physica C* 168, 8 (1990).
22. E. Ruckenstein and C. T. Cheung, *J. Mater. Res.* 4, 1116 (1989).
23. C. T. Cheung and E. Ruckenstein, *J. Mater. Res.* 5, 245 (1990).
24. R. Sugise and H. Ihara, *Jpn. J. Appl. Phys.* 28, 334 (1989).
25. R. Sugise, M. Hirabayashi, N. Terada, M. Jo, M. Tokumoto, T. Shimomura, and H. Ihara, *Jpn. J. Appl. Phys.* 27, L2310 (1988).

26. J. R. Thompson, J. Brynestad, D. M. Kroeger, Y. C. Kim, S. T. Sekula, D. K. Christen, and E. D. Specht, *Phys. Rev. B* 39, 6652 (1989).
27. S. Narain and E. Ruckenstein, *Supercond. Sci. Technol.* 2, 236 (1989).
28. R. Sugise, M. Hirabayashi, N. Terada, M. Jo, T. Shimomura, and H. Ihara, *Physica C* 157, 131 (1989).
29. B. Morosin, E. L. Venturini, D. S. Ginley, B. C. Bunker, E. B. Stechel, K. F. McCarty, J. A. Voigt, J. F. Kwak, J. E. Schirber, D. Emin, R. J. Baughman, N. D. Shinn, W. F. Hammetter, D. Boehme, and D. R. Jennison, in *High Temperature Superconducting Compounds: Processing and Related Properties*, S. H. Whang and A. Das-Gupta, eds., The Minerals, Metals, and Materials Society, Warrendale, Pa., 1989, pp. 651–670.
30. R. Sugise, M. Hirabayashi, N. Terada, M. Jo, F. Kawashima, and H. Ihara, *Jpn. J. Appl. Phys.* 27, L2314 (1988).
31. R. Sugise, M. Hirabayashi, N. Terada, M. Jo, T. Shimomura, and H. Ihara, *Jpn. J. Appl. Phys.* 27, L1709 (1988).
32. R. M. Iyer, G. M. Phatak, K. Gangadharan, M. D. Sastry, R. M. Kadam, P. V. P. S. S. Sastry, and J. V. Yakhmi, *Physica C* 160, 155 (1989).
33. S. S. P. Parkin, V. Y. Lee, E. M. Engler, A. I. Nazzal, T. C. Huang, G. Gorman, R. Savoy, and R. Beyers, *Phys. Rev. Lett.* 60, 2539 (1988).
34. R. S. Liu and P. P. Edwards, *Physica C* 179, 353 (1991).
35. M. Kikuchi, T. Kajitani, T. Suzuki, S. Nakajima, K. Hiraga, N. Kobayashi, H. Iwasaki, Y. Syono, and Y. Muto, *Jpn. J. Appl. Phys.* 28, L382 (1989).
36. S. Adachi, K. Mizuno, K. Setsume, and K. Wasa, *Physica C* 171, 543 (1990).
37. A. Sequeira, H. Rajagopal, I. K. Gopalakrishnan, P. V. P. S. S. Sastry, G. M. Phatak, J. V. Yakhmi, and R. M. Iyer, *Physica C* 156, 599 (1988).
38. A. W. Hewat, E. A. Hewat, J. Brynestad, H. A. Mook, and E. D. Specht, *Physica C* 152, 438 (1988).
39. R. M. Hazen, L. W. Finger, R. J. Angel, C. T. Prewitt, N. L. Ross, C. G. Hadidiacos, P. J. Heaney, D. R. Veblen, Z. Z. Sheng, A. El Ali, and A. M. Hermann, *Phys. Rev. Lett.* 60, 1657 (1988).
40. D. E. Morris, M. R. Chandrachood, and A. P. B. Sinha, *Physica C* 175, 156 (1991).
41. N. Wu, Y. Der Yao, S. Lee, S. Wong, and E. Ruckenstein, *Physica C* 161, 302 (1989).
42. N.-L. Wu and Y. der Yao, *Mater. Res. Soc. Symp. Proc.* 169, 405 (1990).
43. T. L. Aselage, J. A. Voigt, and K. D. Keefer, *J. Am. Ceram. Soc.* 73, 3345 (1990).
44. J. A. DeLuca, M. F. Garbauskas, R. B. Bolon, J. G. McMullen, W. E. Balz, and P. L. Karas, *J. Mater. Res.* 6, 1415 (1991).
45. M. Verwerft, G. van Tendeloo, and S. Amelinckx, *Physica C* 156, 607 (1988).
46. C. C. Torardi, M. A. Subramanian, J. C. Calabrese, J. Gopalakrishnan, K. J. Morrissey, T. R. Askew, R. B. Flippen, U. Chowdry, and A. W. Sleight, *Science* 240, 631 (1988).
47. S. S. P. Parkin, V. Y. Lee, A. I. Nazzal, R. Savoy, R. Byers and S. J. Laplaca, *Phys. Rev. Lett.* 61, 750 (1988).
48. D. Cubicciotti and H. Edding, *J. Chem. Eng. Data* 12, 548 (1967).
49. M. Hervieu, C. Martin, J. Provost, and B. Raveau, *J. Solid State Chem.* 76, 419 (1988).
50. B. Morosin, D. S. Ginley, P. F. Hlava, M. J. Carr, R. J. Baughman, J. E. Schirber, E. L. Venturini, and J. F. Kwak, *Physica C* 152, 413 (1988).
51. P. E. D. Morgan, J. J. Ratto, R. M. Housley, and J. R. Porter, *Mater. Res. Soc. Symp. Proc.* 121, 421 (1988).
52. R. Beyers, S. S. P. Parkin, V. Y. Lee, A. I. Nazzal, R. Savoy, G. Gorman, T. C. Huang, and S. LaPlaca, *Appl. Phys. Lett.* 53 432 (1988).
53. S. Ijima, T. Ichihashi, Y. Shimakawa, T. Manako, and Y. Kubo, *Jpn. J. Appl. Phys.* 27, L1054 (1988).

54. M. A. Subramanian, *Ceram. Trans.* 13, 59 (1990).
55. M. Hansen and K. Anderko, *Constitution of Binary Alloys*, 2nd ed., McGraw-Hill, New York, 1958.
56. K. C. Goretta, J. L. Routbort, D. Shi, J. G. Chen, M. Xu, and M. C. Hash, in *High-Temperature Superconductors: Fundamental Properties and Novel Materials Processing*, ed. D. Christen, J. Narayan, L. Schneemeyer, *Mater. Res. Soc. Symp. Proc.* 169, 365 (1990).
57. B. T. Ahn, W. Y. Lee, and R. Beyers, *Appl. Phys. Lett.* 60, 2150 (1992).
58. J. J. Ratto, J. R. Porter, R. M. Housley, and P. E. D. Morgan, *Jpn. J. Appl. Phys.* 29, 244 (1990).
59. W. Y. Lee, S. M. Garrison, M. Kawasaki, E. Venturini, B. T. Ahn, R. Beyers, J. Salem, R. Savoy, and J. Vazquez, *Appl. Phys. Lett.* 60, 772 (1992).
60. T. Kotani, T. Kaneko, H. Takei, and K. Tada, *Jpn. J. Appl. Phys.* 28, L1378 (1989).
61. M. Okada, T. Yuasa, T. Matsumoto, K. Aihara, M. Seido, and S. Matsuda, *Jpn. J. Appl. Phys.* 29, 2732 (1990).
62. K. H. Sandhage, G. N. Riley, Jr., and W. L. Carter, *J. Met.* 43, 21 (1991).
63. C. Michel, C. Martin, M. Hervieu, A. Maignan, J. Provost, M. Huve, and B. Raveau, *J. Solid State Chem.* 96, 271 (1992).
64. M. A. Subramanian, J. C. Calabrese, C. C. Torardi, J. Gopalakrishnan, T. R. Askew, R. B. Flippen, K. J. Morrissey, U. Chowdry, and A. W. Sleight, *Nature* 332, 420 (1988).
65. Y. Shimakawa, Y. Kubo, T. Manako, Y. Nakabayashi, and H. Igarashi, *Physica C* 156, 97 (1988).
66. D. E. Cox, C. C. Torardi, M. A. Subramanian, J. Gopalakrishnan, and A. W. Sleight, *Phys. Rev. B* 38, 6624 (1988).
67. B. Morosin, D. S. Ginley, E. L. Venturini, R. J. Baughman, and C. P. Tigges, *Physica C* 172, 413 (1991).
68. B. Morosin, E. L. Venturini, and D. S. Ginley, *Physica C* 185–189, 657 (1991).
69. B. Morosin, E. L. Venturini, and D. S. Ginley, *Physica C* 183, 90 (1991).
70. M. Eibschutz, L. G. Van Uitert, W. H. Grodkiewicz, and D. E. Cox, *Physica C* 162–164, 530 (1989).
71. B. Morosin, R. J. Baughman, D. S. Ginley, J. E. Schirber, and E. L. Venturini, *Physica C* 161, 115 (1990).
72. B. Morosin, D. S. Ginley, E. L. Venturini, R. J. Baughman, J. E. Schirber, and C. P. Tiggs, *Mater. Res. Soc. Symp. Proc.* 169, 141 (1990).
73. T. Zetterer, H. H. Otto, G. Lugert, and K. F. Renk, *Z. Phys. B* 73, 321 (1988).
74. H. H. Otto, T. Zetterer, and R. F. Renk, *Naturwissenschaften* 75, 509 (1988).
75. A. W. Hewat, P. Bordet, J. J. Capponi, C. Chaillout, J. Chenavas, M. Godinho, E. A. Hewat, J. L. Hodeau, and M. Marezio, *Physica C* 156, 369 (1988).
76. Y. Syono, M. Kikuchi, S. Nakajima, T. Suzuki, T. Oku, K. Hiraga, N. Kobayashi, H. Iwasaki, and Y. Muto, *Mater. Res. Soc. Symp. Proc.* 156, 229 (1989).
77. R. D. Shannon, *Acta Crystallogr. A.* 32, 751 (1976).
78. K. Hiraga, D. Shindo, M. Hirabayashi, M. Kikuchi, N. Kobayashi, and Y. Syono, *Jpn. J. Appl. Phys.* 27, L1848 (1988).
79. T. C. Huang, V. Y. Lee, R. Karimi, R. Beyers, and S. S. P. Parkin, *Mater. Res. Bull.* 23, 1307 (1988).
80. S. J. Hibble, A. K. Cheetham, A. M. Chippindale, P. Day, and J. A. Hriljac, *Physica C* 156, 604 (1988).
81. P. L. Gai, M. A. Subramanian, J. Gopalakrishnan, and E. D. Boyes, *Physica C* 159, 801 (1989).
82. A. Manthriam, M. Paranthaman, and J. B. Goodenough, *Physica C* 171, 135 (1990).

83. M. Paranthaman, A. Manthiram, and J. B. Goodenough, *J. Solid State Chem.* 87, 479 (1990).
84. I. K. Gopalakrishnan, P. V. P. S. S. Sastry, H. Rajagopal, A. Sequeira, J. V. Yakhmi, and R. M. Iyer, *Physica C* 159, 811 (1989).
85. Y. Shimakawa, Y. Kubo, T. Manako, and H. Igarashi, *Mater. Res. Soc. Symp. Proc.* 169, 1061 (1990).
86. A. Maignan, C. Martin, M. Huve, J. Provost, M. Hervieu, C. Michel, and B. Raveau, *Physica C* 170, 350 (1990).
87. Y. Shimakawa, Y. Kubo, T. Manako, and H. Igarashi, *Phys. Rev. B* 40, 11400 (1989).
88. D. M. de Leeuw, W. A. Groen, J. C. Jol, H. B. Brom, and H. W. Zandbergen, *Physica C* 166, 349 (1990).
89. K. V. Ramanujachary, S. Li, and M. Greenblatt, *Physica C* 165, 377 (1990).
90. Y. Shimakawa, Y. Kubo, T. Manako, T. Satoh, S. Ijima, T. Ichihashi, and H. Igarashi, *Physica C* 157, 279 (1989).
91. M. Kikuchi, S. Nakajima, Y. Syono, K. Nagase, R. Suzuki, T. Kajitani, N. Kobayashi, and Y. Muto, *Physica C* 166, 497 (1990).
92. P. K. Gallagher, H. M. O'Bryan, R. J. Cava, A. C. W. P. James, D. W. Murphy, W. W. Rhodes, J. J. Krajewski, W. F. Peck, and J. J. Waszczak, *Chem. Mater.* 1, 277 (1989).
93. J. B. Parise, C. C. Torardi, M. A. Subramanian, J. Gopalakrishnan, A. W. Sleight, and E. Prince, *Physica C* 159, 239 (1989).
94. E. L. Venturini, B. Morosin, D. S. Ginley, J. F. Kwak, J. E. Schirber, R. J. Baughman, and R. A. Graham, *Mater. Res. Soc. Symp. Proc.* 156, 239 (1989).
95. A. Schilling, H. R. Ott, and F. Hulliger, *Physica C* 157, 144 (1989).
96. S. Nakajima, M. Kikuchi, Y. Syono, K. Nagase, T. Oku, N. Kobayashi, D. Shindo, and K. Hiraga, *Physica C* 170, 443 (1990).
97. S. Nakajima, M. Kikuchi, Y. Syono, and N. Kobayashi, *Physica C* 185–189, 673 (1991).
98. S. Nakajima, M. Kikuchi, Y. Syono, T. Oko, K. Nagase, N. Kobayashi, D. Shindo, and K. Hiraga, *Physica C* 182, 89 (1991).

20
Magnetic Properties of Thallium-Based Superconductors

Timir Datta
University of South Carolina
Columbia, South Carolina

I. INTRODUCTION

In this chapter we describe one of the most important characteristic behaviors of any superconductor: magnetism. Diamagnetism is the most distinguishing property exhibited by superconductors. Although dc conductivity approaches infinity, the total electromagnetic response of a superconductor is controlled principally by its zero inductance.

Perfect diamagnetism is proof positive of superconductivity. Later in this chapter we note that good-quality Tl single crystals exhibit ideal diamagnetic behavior. We also discuss how magnetism provides microscopic information about the superconducting (condensed) ground state.

The ground state of these materials appears to be anisotropic. That is, the physical properties are different along the different crystal axes. These superconductors are strongly two-dimensional with a non-s-wave symmetry. The magnetic penetration depth (λ) is a power-law function of temperature; λ is more than an order of magnitude greater along the c direction (λ_c) than in the ab plane (λ_{ab}).

The magnetic behavior at high temperature and field is similar to conventional (soft) type II superconductors. Magnetic estimates of the critical current density are high (e.g., $\sim 10^{10}$ A/m^2 at 20 K with 2-T field), but the overall bulk current is low. Even at modest temperatures (ca. 20 K), the flux line motion is observed to be considerable. Flux motion is an obstacle in large-scale power applications. We discuss the progress that has been made in understanding the magnetic properties of Tl materials; however, much work remains.

II. BACKGROUND

For almost two decades after the discovery of superconductivity by Onnes, no relationship was recognized between this remarkable phenomenon and magnetism. Onnes himself had noted that applying a magnetic field stronger than a critical value (H_c) destroys superconductivity as effectively as does raising temperature [1].

It was believed that superconductors were merely perfect electrical conductors [i.e., had infinite conductivity below the transition temperature (T_c)]. It was a surprise when Meissner and Ochsenfeld concluded that magnetic behavior is different from what is expected for zero resistance; in reality, superconductors are perfect diamagnets. The principal property of a perfect conductor is zero resistance, and that of a superconductor is zero inductance.

To understand the consequences of zero resistance and of zero inductance, let us consider the magnetic behavior under identical conditions. Due to infinite conductivity, or zero resistance, the magnetic induction, or flux B, is completely frozen ($dB/dt = 0$) in a perfect conductor. Any attempt to change the total flux is canceled instantly by the law of induction. As can be seen in Fig. 1, if the initial flux is zero (not zero), the flux after cooling is still zero (not zero). Hence for the same final values of field and temperature (H,T), the magnetization of a perfect conductor is not a thermodynamic state function [2]. The final state depends on the (B,T) path chosen. A plasma of totally ionized gas is a laboratory approximation to a perfect conductor.

Zero inductance of a superconductor leads to the same final state no matter what the ($\mathbf{B},T$) path is as shown in Fig. 2. Notice first that the superconductor prevents an external magnetic field from entering into it, a behavior termed *mag-*

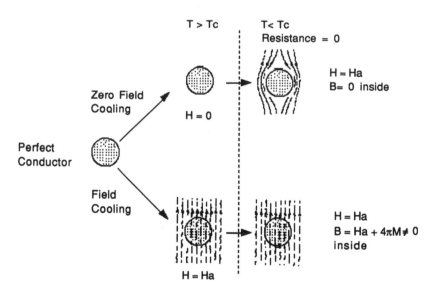

Figure 1 History-dependent magnetism of a zero-resistance or perfect conductor. Observe that the final flux in the solid is zero in the first case. In the second (lower) case, when the field was applied before it was cooled below the perfect-conduction transition temperature (T_c), the field is frozen into the conductor flux is not zero.

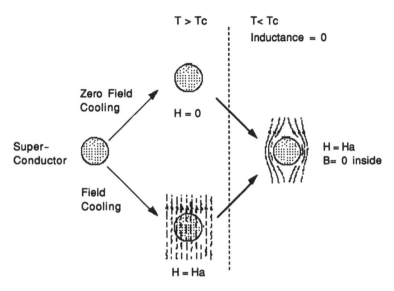

Figure 2 Diamagnetism is a thermodynamic property of a zero inductor or superconductor. No matter which path is followed to produce the final (H,T) condition, a superconductor attains the perfect diamagnetic state. The flux inside the superconductor is zero.

netic exclusion. Second, if the field is introduced into the medium before cooling, flux will be expelled after superconductivity sets in. For either situation, the magnetic induction inside is zero. Perfect diamagnetism is a state function of a superconductor.

Expulsion of the magnetic flux ($\mathbf{B} = 0$) is known as the *Meissner effect.* From Maxwell's equation for induction, the rate of change of flux is identically zero when $\mathbf{B} = 0$ because

$$-\frac{d\mathbf{B}}{dt} = 0 = \text{curl } \mathbf{E} \tag{1}$$

which automatically produces zero emf, so both $\mathbf{B}$ and $\mathbf{E}$ can be zero. Infinite conductivity is a necessary consequence of the Meissner effect, but the converse is not true.

It took over 30 years after the discovery of the Meissner effect to develop a microscopic theoretical model for superconductivity [3,4]. But the role of magnetism was central in this understanding, as indicated by the difference in the thermodynamic free energies between the normal and condensed states, gauge symmetry breaking, and spin correlation of Cooper pairs.

Magnetism has been, and continues to be, important in experimental studies. Diamagnetism and field expulsion (i.e., a large negative susceptibility and at least a partial Meissner effect) are among the most recognized indicators of superconductivity. At the same time, many of these measurements are the basic ingredients of "materials characterization." For example, the onset temperature of diamagnetism, as indicated by the susceptibility χ, or, more precisely, the position and width of the peak of its temperature derivative $(d\chi/dT)$, are convenient experimental definitions of T_c and a measure of the quality of the superconducting transitions [5].

Furthermore, magnetic investigations provide an immense range of scientific and technological information. Many fundamental microscopic quantities relating to the superconducting ground state, such as its symmetry and dimensionality, coherence length, effective mass, and energy gap, may be determined from the magnetic properties. Some of these are the topic of this chapter.

III.   IONIC AND NUCLEAR MAGNETISM

In most respects the magnetic properties of thallium-based superconductors are similar to those of other copper oxides [6,7]. In these materials, magnetism is exhibited in several forms, one of which is antiferromagnetism of the Cu^{2+} ions. The theoretical possibility of a magnetic route to pairing and superconducting condensation has contributed greatly to research into the behavior of the Cu ions. Quite generally, the interplay between superconductivity and Cu^{2+} moments is very complex [8].

Magnetism also arises due to the presence of magnetic ions such as Fe, Ni, or the magnetic rare earths. In addition, the ^{203}Tl and ^{205}Tl isotopes are magnetic, with one-half nuclear moment (i.e., spin $I = \frac{1}{2}$) [9].

Many of these phenomena are studied by scattering and resonant spectroscopic methods such as neutron diffraction, electron spin resonance (ESR) [10], nuclear magnetic resonance (NMR), and related techniques. In some respects, ionic and nuclear magnetism may be regarded as extrinsic to superconductivity; however, they may be very useful in understanding these superconducting materials.

IV.   NORMAL-STATE BEHAVIOR AND PAULI PARAMAGNETISM

A characteristic of any conductor, such as a normal metal or a superconducting solid above T_c, is Pauli paramagnetism. In the paramagnetic state there is no magnetic order; an external magnetic field can enforce order to produce a net magnetization in the direction of the field. The origin of this behavior in a metal is the spin (moment) of the conduction electrons. The density of these electrons is high, and they form a degenerate Fermi system. In the degenerate condition, not all the electrons can be influenced by the external field. Only those electrons in energy states that are thermally excitable above the Fermi surface can line up with the field. This is why (Pauli) conduction-electron paramagnetism is temperature independent. This paramagnetism is weak because each of these extended wave electrons contributes a small amount (μ_b), 1 Bohr magneton, of magnetic moment.

When the material contains localized moments, magnetization can be strong and temperature dependent. Localized moment behavior is well represented by Curie paramagnetism or Curie–Weiss (C-W) models [11]. The moments with ferromagnetic (antiferromagnetic) interaction tend to order parallel (antiparallel) to each other. In the paramagnetic state, or above the ordering temperature, the temperature-dependent C-W susceptibility $\chi_{C\text{-}W}$ is a two-parameter function. For ferromagnetic coupling, $\chi_{C\text{-}W}$ is often expressed as:

$$\chi_{C\text{-}W} = \frac{C}{T - \Theta} \tag{2}$$

where the Curie parameter C is proportional to the square of the magnetic moment, and Θ is a parameter with a dimension of temperature. Θ represents the strength of the coupling between the moments. The sign of Θ distinguishes between (positive) ferromagnetic and (negative) antiferromagnetic interactions.

In a real specimen of thallium superconductor, the degenerate behavior is accompanied by other related forms of magnetism, such as χ_{core} diamagnetism, due to the orbital moments of the core electrons, or χ_{L-P}, the Landau–Peierls contribution arising from the motion of the conduction electrons is a magnetic field. These two forms of magnetism are generally independent of temperature.

Quantitatively, the core contribution is often estimated by Wiedemann's law, the composition-weighted sum of Pascal's constants [12] for the ions. For typical high-T_c material, due to large values of effective mass and stoner factor, χ_{L-P} is only a few percent of χ_P.

A van Vleck type of term, χ_V, due to the population of electrons in the excited states, may also exist. It is generally small; however, in the case of anomalous thermal expansion, χ_V can be large and a sensitive function of temperature [13]. The total measured susceptibility, $\chi(T > T_c)$, is the sum of all the individual contributions, given by

$$\chi(T) = \chi_P + \chi_{L-P} + \chi_{core} + \chi_V + \chi(T)_{C-W} \tag{3}$$

Figure 3 shows the total magnetic response of two types of thallium materials [14] in their normal (i.e., nonsuperconducting) state. In both cases, temperature-dependent paramagnetism is obvious.

Equation (3) is really a multi (usually three)-parameter model for the temperature dependence of the experimental $\chi(T)$ value, where the temperature-dependent part is chosen to be a Curie (–Wiess) type of function. In most cases, this is adequate. Reliable measurements of $\chi(T)$ in the normal state, well above the superconducting transition and over a sufficiently large range of temperature, can be used to separate out such parameters as χ_P, C, and Θ.

Parameter determination is best done by nonlinear least-squares analysis. Care has to be taken to assure both that the error is minimized and that the residue is free of any systematic trend [15].

The localized moments and their interaction can be estimated from the values of C and Θ. The density of states at the Fermi surface $D(E_F)$ can be calculated from χ_P by the following expression:

$$D(E_F) = (3.1 \times 10^4/\text{f.u. of Cu})\chi_P \tag{4}$$

Here the density of states is expressed in states per electron volt per formula unit (f.u.) of copper ions, and χ_P is expressed in emu/mol. It is important to take into account the number of Cu^{2+} ions in a formula unit, which depends on the composition of the Tl material. For example, the f.u. value of Cu is 3 in $Tl_2Ba_2Ca_2Cu_3O_{10}$, and it is 2 in $TlSr_2CaCu_2O_9$. From χ_P, the value of the Sommerfeld constant (γ) can also be calculated as follows:

$$\gamma = \frac{\pi^2 k^2}{3\mu_B^2} D(E_F) \tag{5}$$

where γ is in mJ/mol Cu · K². In the absence of reliable calorimetric data, this is a rather important piece of information. Furthermore, compared with the calori-

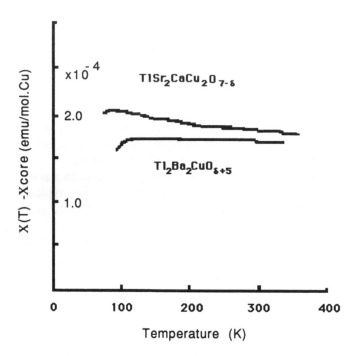

Figure 3 Temperature-dependent magnetism for two types of Tl copper oxides in their normal state. For the barium compound, a simple Curie term (C/T) was also subtracted from the measured susceptibility. (From Ref. 14.)

metric value, the magnetic γ can provide an estimate of the enhancement of the thermal effective mass. Typically, in Tl-2223 $D(E_F)$ is about 2/(eV Cu · f.u.), and γ is about 5 mJ/mol Cu · K^2 [16]. Tl, as well as many other high-T_c oxides, appears to have a small density of states (Sommerfeld constant) for such high values of T_c; this may indicate that a new mechanism for superconductivity is operative in these solids.

Another quantity that is determined by Pauli-paramagnetism is the field-induced decrease in the free energy of the electrons at the Fermi surface. If the paramagnetic response is strong, the superconducting (Cooper) pairs are broken. The resulting free spins align with the field, and the energy of the material is lowered. The value (H_P) of the field at which superconductivity is lost and the material becomes a magnetically aligned normal conductor is known as the Pauli paramagnetic limit of the critical field [17–19]. H_P can be related to the thermo-dynamic critical field [$H_c(0)$] as follows:

$$\chi_P H_P^2(0) = \frac{1}{4\pi} H_c^2(0) \tag{6}$$

In terms of T_c and the strong coupling parameter (α), one has

$$\mathbf{H_P} = 2.8\alpha T_c \tag{7}$$

Here α includes the differences in the renormalizations of the density of states for χ_P and $H_c(0)$.

V. LOW-FIELD BEHAVIOR AND PERFECT DIAMAGNETISM

Diamagnetism, the topic of this section, is a strongly temperature-dependent property associated directly with the superconducting transition. As we stated earlier, the superconducting state is perfectly diamagnetic. There are two ways of representing perfect diamagnetism, the state of zero magnetic flux density [17].

In the bulk diamagnetic description, a superconductor is regarded as a highly magnetic material, so that $\mathbf{B}_i$, the induction in the sample (in unrationalized cgs units), is given by

$$\mathbf{B}_i = \mathbf{H}_a + 4\pi(1 - N)\mathbf{M} \tag{8}$$

and

$$\mathbf{M} = \chi\mathbf{H}_a \tag{9}$$

where $\mathbf{H}_a$ is the applied field, $\mathbf{M}$ the magnetization or moment per unit volume, N the demagnetization factor, and χ the volume susceptibility. If the superconductor is shaped like a long cylinder so that demagnetization effect can be neglected, zero flux ($\mathbf{B} = 0$) requires that χ be equal to -1 in rationalized units, or $-\frac{1}{4}\pi \sim -0.0795$ emu in unrationalized units.

In another model, a superconductor is considered magnetically similar to any normal conductor. An external field produces the magnetization and flux density appropriate for a metal. In addition, persistent screening currents are created. These currents go around the boundary of the superconductor and produce an induction (flux) that exactly cancels the external flux.

These models are entirely equivalent and explain how zero induction is produced. For any given situation, one chooses the description that is convenient. Both perfect diamagnetism and persistent currents can be observed in the Tl–copper oxide superconductors [20,21].

Diamagnetism in a Tl-2212 single crystal is shown in Fig. 4. These measurements were made in a superconducting quantum interference device (SQUID) magnetometer; clearly, at temperatures below the transition, (105 K for this specimen) magnetic susceptibility is large and negative. Flux exclusion is represented by the curve marked "shielding." In the lowest-temperature region (1.6 K—not shown in this figure), susceptibility reaches a value very close to unity. Such perfect diamagnetism can be observed only when the specimen is cooled down to the lowest temperature in a zero or extremely small field before the small experimental field (ca. 0.5 mT) is applied. Typically, the background field is less than a 1 mG. This is the zero-field cooled (ZFC) procedure. Under this condition, almost the entire volume of the specimen is free of flux and is perfectly diamagnetic.

The percentage of the ideal value ($-\frac{1}{4}\pi$) that the measured volume susceptibility attains at low temperatures is called the superconducting volume fraction. For the crystal of Fig. 4, the volume fraction is 100%. This fraction indicates how much of the total volume, or bulk of the material, is superconducting. Estimates can be reliable only if the ZFC conditions are met. Corrections for the demagnetization effects due to the sample shape and field orientations are also to be included. In small flat samples, the demagnetizing effects can increase the response 10- to 20-fold [22,23]. A poor-quality sample with 5 to 10% superconducting components can produce a nominally perfect signal, leading to overestimation in quan-

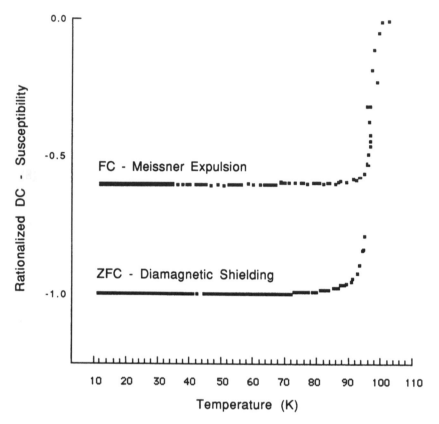

Figure 4 Perfect diamagnetism [i.e., volume susceptibility negative unity (or -0.08 emu)] of a TlBa(2223) single crystal. The lower curve (ZFC) represents shielding or flux exclusion. The upper curve (FC) represents the flux expulsion of Meissner signal. Partial flux retention due to flux trapping is also shown by the difference between the ZFC and FC signals.

titative measurements. Also, a demagnetization contribution can be used to increase the sensitivity of the magnetic techniques for detecting trace amounts of superconductive components in a solid.

The upper curve in Fig. 4 represents data that were obtained when the single crystal was field cooled (FC). In this procedure the applied field is turned on at a high temperature to permeate through the sample in its normal state; then the sample is cooled below T_c with the field held steady. These data correspond to field expulsion or the Meissner effect.

The FC response is smaller (less negative) than the ZFC response at the corresponding temperatures. The FC − ZFC difference is due to the amount of field retained inside the material by trapping. The trapped field is in the direction of the external field and can be observed if the external field is turned off. At that time the sample response changes sign and shows a positive moment equal to the difference between the FC and ZFC values. This sign reversal and flux retention are positive proof of superconductivity. In the absence of traps (i.e., in the type I or Meissner limit), as temperature is lowered, the field is totally expelled; the FC and ZFC values merge and produce a perfect (100%) value for the Meissner fraction. This fraction decreases with the increase in flux retention.

A spuriously large Meissner fraction may also be inferred if the background field in the (nominal) ZFC process is not small enough. Compared to the ZFC behavior, the FC curve shows a weaker temperature dependence at low temperatures and a stronger dependence near the superconducting transition. The transition width is broader when the field is perpendicular to the *c* axis of the crystal [23].

VI. MAGNETIC PENETRATION DEPTH

In Section V we described flux exclusion or diamagnetic shielding and have shown that in a good-quality thallium superconductor, diamagnetism is ideal. However, shielding does not take place discontinuously. The field *B* does not go to zero abruptly at the boundary of the superconductor, but falls off gradually. For an isotropic superconductor, occupying the positive half of space with *x* equal to zero plane as the left bounding surface (Fig. 5), *B* is given by

$$B(x) = B_a \exp\left(\frac{-x}{\lambda}\right) \tag{10}$$

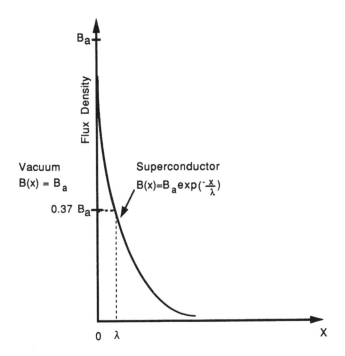

Figure 5 Pictorial description of the magnetic flux in a semi-infinite superconductor. The superconductor occupies the positive half of the figure. Outside the flux density is B_a; inside it falls off exponentially. Penetration depth is the distance at which the flux is decreased to 0.37, the value at the free surface.

where B_a is the externally applied field, x the distance into the material, and λ the penetration depth. Penetration depth is an important scale of length for a super-conductor. It is the distance over which a current or field may change. λ depends on temperature; its value at zero kelvin, $\lambda(0)$, is commonly termed the *London penetration depth*. Penetration depth contains information about certain micro-scopic parameters of the material (i.e., m^*, the effective mass, and n, the carrier density) as follows:

$$\lambda(0) = l\left(\frac{m'l}{4\pi r_0}\right)^{1/2} = \left(\frac{m^*c^2}{4\pi ne^2}\right)^{1/2} \tag{11}$$

where $m' = m^*/m_e$, $r_0 = e^2/mc^2$ is the classical radius, and $l = n^{-1/3}$ is the average carrier separation.

In the two-fluid description, $\lambda(T)$ is a direct measure of the superfluid density at low temperatures. Hence $\lambda(T)$ reflects the symmetry of the superconducting state and the corresponding energy gap; $\lambda(T)$ is an important property of the superconducting ground state [24,25].

Physically, field penetration reduces the measured susceptibility to less than -0.0795 emu because this process reduces the actual volume in which flux is zero. The actual volume V^*, in which diamagnetism is perfect, is equal to $f(T)V$, where $f(T)$ is a correction factor to account for field penetration and V is the total volume of the specimen. The demagnetization factor (N) may also be affected by temper-ature. As the flux enters, the effective aspect ratio (shape factor) of the diamagnetic region changes. The total moment $(\mathbf{m})$ of a specimen can be expressed as

$$\mathbf{m}(T) = \frac{-1}{4\pi}(\mathbf{H}_i)V^* = \left(\frac{-1}{4\pi}\right)Vf(T)\frac{\mathbf{H}_a}{1-N} \tag{12}$$

or

$$\chi(T) = \mathbf{m}\frac{(T)}{\mathbf{H}_aV} = \frac{\mathbf{M}(T)}{\mathbf{H}_a} = \frac{(-1/4\pi)f(T)}{1-N} \tag{13}$$

where $\mathbf{H}_i$ is the magnetic field inside the material.

Tl superconductors are anisotropic. In an anisotropic solid, when the external field is oriented arbitrarily with respect to the crystal axes, the magnetic response is quite complicated. In general, susceptibility is less than ideal at all temperatures. Let us consider two field orientations of high symmetry [23]. In both cases the single crystal will be assumed to be thin along the c axis and flat (planar) along either a or b axes. The dimensions along a and b are assumed to be comparable and equal to w, $d/w \ll 1$, where d is thickness.

When the field is parallel to the c axis, the sample may be approximated by a flat disk of radius R, $R = (\text{area}/\pi)^{1/2}$. In this orientation, $\chi(T)$ is given by [26]

$$\frac{\chi(T)}{\chi(0)} = \left[1 + \frac{3}{x(0)}\right][1 - 3Z(x) + 2Z^2(x)] \tag{14}$$

where $x = R/\lambda(T)$, $x(0) = R/\lambda(0)$, and $Z(x) = I_1(x)/xI_0(x)$. The zero- and first-order Bessel functions, $I_0(x)$ and $I_1(x)$, reflect the azimuthal symmetry of this configuration.

When the field is oriented parallel to the larger faces of the crystal (*ab* planes), or perpendicular to the *c* axis, field penetration from both the faces and the edges contributes to the temperature dependence of χ. Under this condition, χ(*T*) may be expressed by exponential basis functions as follows:

$$\frac{\chi(T)}{\chi(0)} = \left[1 + 2\left(\frac{\lambda_{ab}^{(0)}}{d} + \frac{\lambda_c^{(0)}}{w}\right)\right]$$
$$\left[1 - \frac{8}{\pi^2}\sum_{n=\text{odd}}\frac{1}{n^2}\left(\frac{\tanh y_n}{y_n} + \frac{\tanh z_n}{z_n}\right)\right] \quad (15)$$

where $y_n = (d/2\lambda_{ab})[(n\pi\lambda c/w)^2 + 1]^{1/2}$ and $z_n = (w/2\lambda c)[(n\pi\lambda_{ab}/d)^2 + 1]^{1/2}$.

In the low-temperature regime ($T \sim 0.5T_c$), Eqs. (14) and (15) provide a direct method for determining λ_c and λ_{ab}, the magnetic penetration depths along the *c* and *ab* directions, respectively [23]. In the BCS model [27], the relative difference in λ(*T*) and λ(0) is given as follows:

$$\frac{\lambda(0) - \lambda(T)}{\lambda(0)} = \frac{\Delta\lambda(T)}{\lambda(0)} = \frac{T_c}{T}\exp\left(\frac{-E_g}{2kT}\right) \quad (16)$$

The temperature dependencies of λ_{ab} and λ_c for a TlBa(2212) single crystal are shown in Fig. 6. The absolute values of λ_c are about 10 times larger than those

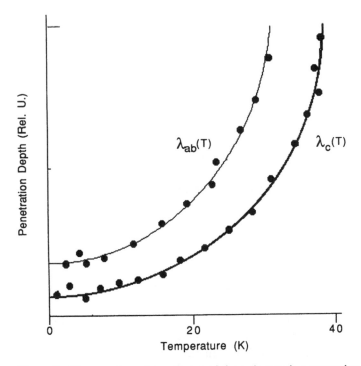

Figure 6 Temperature dependence of the anisotropic penetration depths [$\lambda_c(T)$ and $\lambda_{ab}(T)$] in TlBa(2212) single crystals for temperatures less than $0.4T_c$. For both directions the temperature dependence is described by quadratic equations: $\lambda = AT + BT^2$. The *y* intercepts for the two curves have been shifted to distinguish them clearly.

of λ_{ab}. Over the range of the y axis indicated in this figure, λ_c increases by about a factor of 6 and λ_{ab} increases by about a factor of 2. Proportionately, the *ab* plane penetration depth increases more strongly with temperature.

Both of these relationships show a nonexponential power-law dependence given by

$$\frac{\Delta\lambda(T)}{\lambda(0)} = \begin{cases} At + Bt^2 & t = \dfrac{T}{T_c} < 0.4 \\ At & t = \rightarrow 0.0 \end{cases} \tag{17}$$

As the temperature approaches zero, linear T dependence may be indicative of a non-BCS s-wave-type ground state [24,28] for these high-T_c Tl compounds. Typical values of $\lambda(0)$ and the ratio of the anisotropic effective mass are given below.

$\lambda_{ab(0)}$: 200 ± 20 nm
$\lambda_{c(0)}$: 2500 ± 200 nm
λ_c/λ_{ab}: 12
m_c^*/m_{ab}^*: 180

Anisotropy in the effective mass may also be estimated from torque magnetization measurements. For $Tl_2Ba_2CaCu_2O_x$ thin films, the torque magnetization effective mass anisotropy has been reported to be about 10^4 [29].

At temperatures close to T_c, χ as expressed by Eqs. (14) and (15) is not applicable because at high temperatures, shielding is overcome by the entry of flux and $\lambda(T)$ is no longer a faithful indicator of $\lambda(T)$.

VII. MIXED-STATE MAGNETISM AND TYPE II SUPERCONDUCTIVITY

From the list at the end of Section VI it will be clear the zero-temperature penetration depth is rather large in all directions. In fact, λ is much larger than the corresponding coherence length (ξ); hence the Ginsburg–Landau parameter $\kappa = \lambda/\xi \gg 0.707$. Large κ is a key to the vast potential for high-field, high-current applications of all the high-T_c materials, including the thallium copper oxide compounds. $\kappa > 0.707$ permits superconductivity to coexist with magnetic field; also, the value of the upper critical field (H_{c2}), the field required to quench superconductivity, is increased by a factor of 1.4κ over the thermodynamic critical field (H_c).

A large value of κ is one of the many differences between low-temperature superconductors (LTS) and copper oxides. In the LTS, coherence length is large and κ is small. Over the last few decades, considerable effort had to be focused on the task of increasing the κ value of commercial LTS materials. This was achieved principally by realization of the Anderson effect; that is, with T_c constant, the coherence length was reduced by defects and nonmagnetic impurities.

In high-T_c systems, ξ is short, about 2 nm in the *ab* (Cu–O) planes, and is approximately equal to the thickness of the adjacent Cu–O layers along the perpendicular direction. Furthermore, the charge carrier mean free path is much larger than the coherence length. These two conditions dictate that even pure single-crystal Tl materials will be type II superconductors in the clean limit. Polycrystalline or ceramics are naturally also type II.

For **H** above the lower critical value (H_{c1}), magnetic flux may enter as localized flux regions or vortices. Each vortex contains one quantum of flux Φ_0 ($= 2.07 \times 10^{-15}$ Wb). The spaces between and away from the vortices remain flux free, as in the Meissner state. This condition of flux intermingled with superconductivity is the well-known mixed state. In the (H,T) phase space of Fig. 7c, the mixed state is represented by the region bounded by the $H_{c2}(T)$ and $H_{c1}(T)$ curves. In this state, diamagnetism is less than ideal. Vortices carry flux in the same direction as the external field. If n vortices enter, the net flux or moment of the material is increased (diamagnetic moment decreased) by an amount equal to $n\phi_0$. In the Tl materials with H stronger than the lower critical field, mixed-state properties similar to those of other type II superconductors may be observed.

In the absence of defects, the vortices minimize their free energy at low temperatures by forming an Abrikosov lattice [19]. This vortex lattice may undergo transitions when H and T are varied. At high fields the vortex density is large, the average vortex separation is small, and the vortices can influence each other strongly. At intermediate temperatures, a strongly interacting vortex system may form non-Abrikosov phases, which are entangled and give rise to novel magnetic behaviors [30]. Near T_c, the flux lines have large thermal energy and may behave as a fluid.

In principle, from the features of **M** or **B** isothermals as a function of **H**, the values of H_{c1} and H_{c2} can be determined. These features are indicated by points A and C in Figs. 7a and b. Both H_{c2} and its slope at $T_c (dH_{c2}/dT_c)$ are important quantities. From them the values of electronic parameters such as Fermi velocity and others can be estimated [18].

For example, from the value of the upper critical field, the superconducting coherence length ξ can be obtained as follows:

$$H_{c2} = \frac{\Phi_0}{2\pi\xi^2} \quad \text{or} \quad \xi = \left(\frac{\phi_0}{2\pi}\right)^{1/2} \tag{18}$$

In practice, some operational definitions, such as assuming T_c to be the temperature at 10 or 50% of the transition curve, also become important. Moreover, parameters such as the external field, composition of oxygen [31], and other structural factors broaden the transition; these can introduce systematic errors. In a real specimen, vortices enter gradually and the isotherms are rounded out (Fig. 8), as indicated by point A. Under such conditions, it becomes difficult to decide exactly what H_{c1} or H_{c2} is. Care has to be taken to eliminate the extrinsic dynamical and hysteretic effects. Figure 8 also shows how point A is shifted during two measurements [32]. Flux motion is large near transition, particularly in soft superconductors like the Tl compounds. For these reasons, the exact values of some of the properties, such as $d(T_c)/dH$, H_{c1}, and H_{c2} for these materials, are still uncertain.

Hysteresis of the magnetic induction (**B**) or magnetization (**M**) with the field (**H**) is an important property of a type II superconductor. This property is best shown as hysteresis loops. A hysteresis loop is a closed magnetization versus field isotherm taken over at least one cycle of field reversal. Two examples of these loops for a Tl-ceramic specimen [33] is provided in Fig. 9.

The critical current (J_c) can also be estimated from the hysteresis loops. In the Bean approximation [19], J_c (A/cm^2) is given by

$$J_c = 30\left(\frac{\Delta M}{d}\right) \tag{19}$$

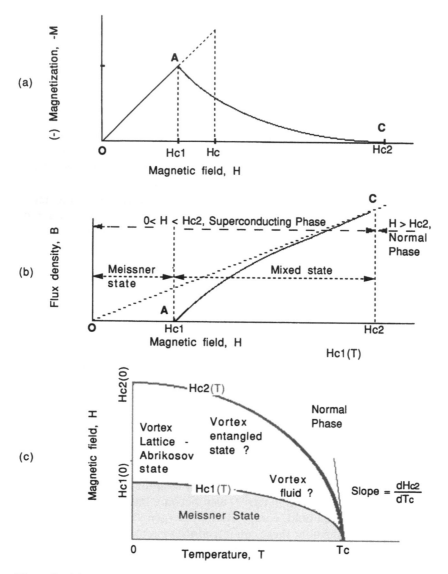

Figure 7   (a) Type II superconductivity. The values of the three fields (H_{c1}, H_c, and H_{c2}) are shown on the minus M versus H isotherm. For fields less than the lower critical value (H_{c1}), the superconducting medium is in the Meissner state; it is completely free of flux, and magnetization (in emu) is equal to -0.08 times the applied field. At H_{c1}, flux starts entering, and M becomes a sublinear function of H. This is reflected in the abrupt change in the $M(H)$ slope and is marked by point A. The dashed line indicates the magnetization for a type I material with the same value of T_c or H_c. The region between H_{c1} and H_{c2} represents the mixed state. At H_{c2}, bulk superconductivity is quenched and the entire volume is in the normal state. (b) Idealized behavior of a $B(H)$ isotherm. Notice that B is zero below H_{c1} and asymptotically reaches the normal-state value at H_{c2}. The field axis is not drawn to scale. In the Tl copper oxides, H_{c1} is about 10^2 Oe and H_{c2} can be 10,000 times higher than H_{c1}. (c) Sketch of the (H,T) phase space of the high-T_c oxides. The slope of the $H_{c2}(T)$ curve is an important materials parameter. Three flux configurations in the mixed state are marked. The exact boundary and the correct nature of the non-Abrikosov states are yet to be determined.

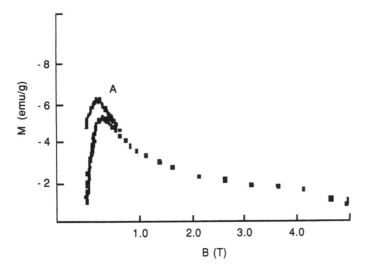

Figure 8 $M(H)$ for a ceramic specimen of TlBa(2223) at 4.2 K. Point A marks the peak of (negative) magnetization. The two curves represent the initial and subsequent data, respectively. The smoothing of the peak and the difficulty in a precise determination of H_{c1} in a real situation, in contrast to the ideal case of Fig. 7a, should be clear. Similar practical difficulties are also faced in determining H_{c2}.

where ΔM is the asymptotic difference of magnetization (emu) and d (cm) is a size transverse to the field direction. The largest source of error in applying the Bean equation is the uncertainty of determining the effective value of d because d corresponds to the length scale of the screening current. Only in the ideal simple situation is d equal to the macroscopic sample size. In a ceramic specimen, d may be the grain size. An estimate for ΔM is indicated on the Fig. 9. The larger curve in Fig. 9 is the data at a temperature of 5 K, and the small curve corresponds to a temperature of 40 K; clearly, at higher temperatures, $J_c(\Delta M)$ decreases significantly. This magnetic determination provides an upper limit of J_c and often results in values considerably higher than those obtained from transport measurements. In any case, the experimental J_c value is always smaller than the theoretical or depairing current density (J_c^*), given by

$$J_c^* = \frac{q_n E_g}{2P_f} = \frac{q_n E_g}{2m^* V_f} \tag{20}$$

where q ($= 2e$) is the charge, n the density of the carriers, E_g the energy gap, and $P_f (V_f)$ the Fermi momentum (velocity). Compared with the conventional (metallic) low-T_c superconductors in the Tl materials, the values of both E_g and m^* are large, but n [34] and V_f are small.

Figure 10 shows the ZFC and FC magnetization versus temperature for a $Tl_2Ba_2CaCu_2O_x$ ceramic in 1- and 10-mT fields. The most striking feature is how rapidly the temperature (T^*) at which ZFC and FC responses converge increases with the applied field. The $T^*(H)$ or $H(T^*)$ curve in the (H,T) space is often called the irreversibility line. For temperatures higher than T^*, ZFC and FC curves lie

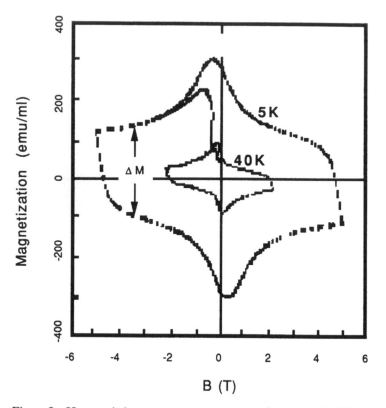

Figure 9 Hysteresis loops at two temperatures in a ceramic-TlBa sample. The asymptotic difference in the magnetization (ΔM) is also shown. The critical current density at the measurement temperature can be estimated from ΔM by Eq. (19).

on top of each other. That is, magnetization is the same irrespective of the (H,T) history of the specimen.

Typically, the reduced t^* $(=1 - T^*/T_c)$ is an algebraic function of the field (i.e., $T^* \sim H^x$). The inset of Fig. 10 shows this scaling behavior of T^*; for this material, $x = 0.67$ or $T^* \sim H^{2/3}$. The exact significance of T^*, such as its relation to critical current, deepening temperature, or whether it is a phase transition into a new phase (entangled, molten, or glassy flux lattice state), remains to be understood. However, T^* is one of the clearest metric parameters of these materials. Several other such quantities—namely, T_c, Meissner fraction, and percentage of perfect diamagnetism—were described before.

VIII. THERMODYNAMIC FLUCTUATIONS AND ABOVE-T_C DIAMAGNETISM

Due to thermodynamic fluctuations, there is a finite probability that superconductive condensation will take place, even above the transition temperature. The fraction of the regions that are in the condensed state show properties characteristic of superconductivity, such as a lowering of resistance and dM'. Here dM' is an increased diamagnetism in excess of what would be expected in the normal state

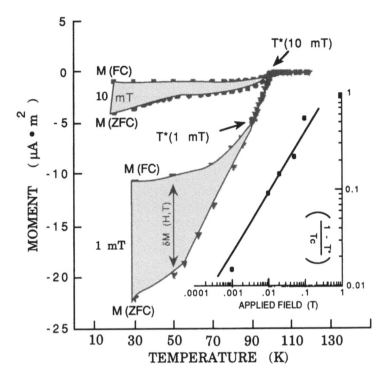

Figure 10 Flux trapping and the onset of ZFC-FC irreversibility in a TlBa-ceramic. The irreversibility temperature (T^*) for the two-thirds scaling of the $T^*(H)$ curve is shown in the inset.

that temperature [19]. The corresponding volume susceptibility ($d\chi'$) due to this fluctuation diamagnetism is given by

$$d\chi(T) = \frac{\pi\xi^2 <r^2> kT}{\Phi_0^2 V} \tag{21}$$

These fluctuations are most pronounced at temperatures in the critical region and are stronger in lower dimensions. Short coherence length and high T_c combine to increase the critical region in the copper oxide compounds. The temperature dependence of dM' or $d\chi'$ can be utilized to determine the effective dimensionality of the superconducting state. This is another example of a macroscopic measurement providing microscopic information about the superconductor.

The principal source of error in the experimental studies of excess diamagnetism is in the presumed magnetic behavior of the normal state with no superconductivity. To determine the excess diamagnetism, the value of the normal-state contribution has to be subtracted from the total measured magnetization. The question is: How much is this background? It is not convenient to suppress the transition and determine the background normal response because the field has to be large, and any nonlinearity can introduce errors in estimating the low-field limit of dM'. For this reason the background value is often extrapolated from the high-temperature limit of the experimental normal-state data.

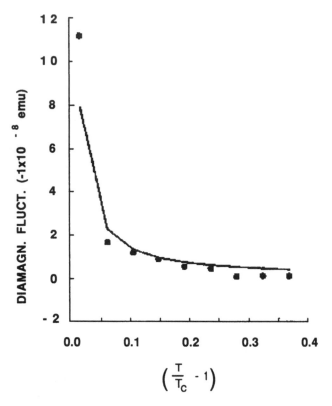

Figure 11 Superconductive fluctuation and excess diamagnetism above T_c. These D'_m data are from a ceramic-TlBa sample. The solid curve is the theoretical behavior of the fluctuation in a two-dimensional superconducting state.

Figure 11 shows excess diamagnetism [16] as a function of the reduced temperature t ($= T/T_c - 1$) in a polycrystalline sample of $TlSr_2GaCu_2O_9$. The solid line is the theoretical prediction from the Prange model for two dimensions. The resulting fluctuations are clearly consistent with the planar arrangement of the copper oxide planes. The principal source of superconductivity in these systems.

IX. TIME- AND FREQUENCY-DEPENDENT MAGNETISM

When the external magnetic field [$H(t)$] is harmonic in time, that is, when

$$H(t) = H_0 + h \cos \omega t \tag{22}$$

and the amplitude (h) is small enough, the induced moment of the specimen [$M(t)$] is also harmonic. The harmonic susceptibility components (χ'_n and χ''_n) can be obtained by cosine and sine transforms of $M(t)$:

$$M(t) = M(H_0) + h(\chi'_n \cos n\omega t + \chi''_n \sin n\omega t) \tag{23}$$

In general, both χ'_n and χ''_n are functions of temperature, H_0, and h. Furthermore, only the odd components are nonzero when the average field (H_0) is zero. The negative of the first quadrate or out-of-phase component (χ''_1) is often called the imaginary component of the ac susceptibility. The ac susceptibility (χ_{ac}) is defined to be

$$\chi(\omega) = \chi'(\omega) + i\chi''(\omega) \tag{24}$$

Typically, h is small (1 to 10^3 μT) and ω is between 1 and 10^3 Hz. The real component represents the dispersive response, or induction. The resistive response, or absorption, is described by the out-of-phase (imaginary) components.

In the Tl superconductors, as in the other copper oxides, the temperature dependence of χ_{ac} at low frequencies is similar to $\chi(T)$, the dc susceptibility [35], which was discussed earlier. The out-of-phase components $[\chi''(T)]$ show an asymmetric peak at T_{max} [32], somewhat lower than T_c, as indicated in Fig. 12. The temperature T_{max} depends on H_0, h, and ω. The higher harmonics also show this maximum behavior. The amplitude of these components decreases with the order n. Both the sharpness and the value of T_{max} are increased in the higher-order components [32]. In some cases there may be several other peaks in $\chi(T)$. These peaks correspond to loss terms representing the contributions of all the dissipative processes in the medium as it is subjected to the oscillatory external field.

Generally, due to the potential for complications from superficial effects, a diamagnetic $\chi'(T)$ is not as confirmatory a test of bulk superconductivity as the dc measurements. However, the ac technique is a powerful tool, particularly for studying the frequency-dependent electromagnetic properties of the thallium superconductors.

At a finite temperature, the vortices are not perfectly fixed in position, even if the applied field is held steady. Thermal agitations constantly move them away

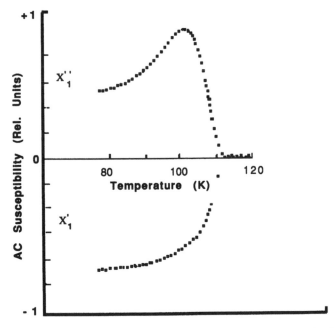

Figure 12 Ac susceptibility at 42 Hz. Note how similar is the temperature dependence of the dispersive response $[\chi'(T)]$ to the diamagnetism of Fig. 4. The peak in absorptive component $[\lambda''(T)]$ is shown by the top curve.

from the regions of high flux density. The vortices move in a sequence of diffusive steps. This equilibrating process proceeds against the grain of the random pinning sites (traps). Traps are areas where it is energetically favorable for the flux to thread through the medium or to form (condense) a flux vortex.

Clearly, small islands of normal regions, impurities, or defects in the crystal structure may act as traps. Usually, a region of (transverse) size comparable to the coherence length over which superconductivity is weak may act as a pinning center. Although the material as a whole is in the superconducting state, the superconducting order parameter is not zero. When the vortices are displaced, work is done and energy is dissipated.

The time dependence of the magnetic moment as flux enters or is expelled is related to how vortices move inside the superconducting medium. It was first investigated by Anderson and Kim [19,36] in low-temperature superconductors.

Microscopically, flux motion is very complicated because it can be influenced by a wide variety of factors, including field strength, anisotropy, trapping energy, temperature, current density, vortex density, and interaction between the flux lines. Phenomenologically, there are two classes of motion, slow (flux creep) or not slow (flux flow).

Anderson's model for flux creep is based on his key observation that it is energetically unfavorable to disturb the flux lattice by translating a single vortex at a time, but that motion can take place collectively in bundles. Each bundle is a group or domain of correlated vortices bounded by regions in which the vortices are mismatched. The energy cost per vortex for cooperative relaxation can be small. Even at temperatures well below T_c, flux reorganization can take place by thermal activation over a barrier. Anderson reduced the number of parameters to three. These represent the scale of the three basic physical quantities of length, energy, and time. L, the scale of length, is also the size of a typical bundle. The energy scale is kT. The time scale is $t = \Omega^{-1}$, where Ω, the attempt or flux-line vibration, is assumed to be in the range 10^5 to 10^{11} Hz. Other quantities are constructed out of these parameters. For example, the flux bundle jump rate (R) is expressed as

$$R = \Omega \exp\left(\frac{-F_0}{kT}\right) \tag{25}$$

where the activation free energy (F_0) is assumed to be only a small fraction of the condensation energy $(H_c^2 L^3/8\pi)$ but still much larger than kT. The flux flow velocity (V) is given by

$$V = V_0 \exp\left(\frac{-U}{kT}\right) \tag{26}$$

where

$$V_0 = \Omega L \tag{27}$$

The activation energy (U) is related to the net force density and is given by the gradient of the condensation energy. In this model the long-time dynamics of the trapped flux is logarithmic. That is, the proportional change in the magnetic moment $[M(t)]$ of a superconductor with trapped flux between the first and hundredth hour is the same as that between the first and hundredth year! Persistent currents also follow a log-time dependence. This logarithmic behavior is qualitatively different from the exponential dependence of the $L-C-R$ time constant observed in non-superconducting circuits. Both flux motion and ohmic resistance give rise to a potential drop along the direction of the current and result in the loss of energy; however, the dynamics of these two types of energy-dissipating processes are not similar. The log behavior is a reflection of the activated relaxation of the trapped field.

In a recent model [16,37], valid for a wide range of time, $M(t)$ is given by

$$M(t) = M(0)\left(1 - \frac{kT}{2F_0}\right)\ln\left(\frac{1+t}{t_0}\right) \qquad (28)$$

where t_0 is a characteristic time. Equation (28) indicates two important points. First, a linear $\ln t$ versus M value is observed only if the origin of time is correctly included. Second, a linear (not log-linear) time dependence may be expected initially. The slope (S) of M versus $\ln t$ is a measure of the competition between temperature and the intrinsic J_c (i.e., the critical current density if flux creep were zero). $S(T)$ is also related to the energy spectrum of the trap sites. It is not monotonic but shows one or more maxima as a function of temperature. The peaks are signatures of the dominant pinning process. The time-dependent magnetic moment of a Tl thin film is provided in Fig. 13. Linear $M(t)$ versus $\ln(t)$ values can be seen

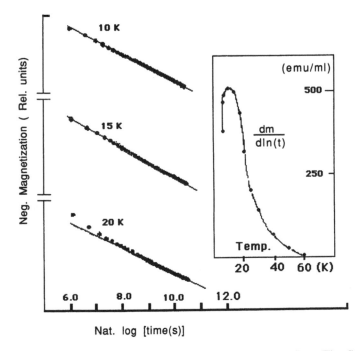

Figure 13 Evidence of flux motion in a TlBa thin film. The flux creep data, or the time dependence of the magnetic moment after a 500 Oe, was applied parallel to the c axis of the film. The time dependencies at three temperatures are shown. The inset shows the rate of flux creep as a function of temperature. (From Ref. 38.)

in all three sets of data. The slope of these straight lines are different, with that of the 15-K set being the sharpest. The temperature dependence of $S(T)$ has one broad peak at T_m, and is given in the inset of this figure. $S(T)$ depends on the field; strong fields tend to lower both T_m and the height of the peaks [$S(T_m)$]. Defects produced by irradiating thin films appear to increase the pinning potential [39]. Deeper pinning potential wells decrease $S(T_m)$ but increase the temperature at the peak.

X. SUMMARY AND CHALLENGE

Diamagnetic results of single crystals of Tl-copper oxides demonstrate perfect bulk superconductivity with clear, sharp transitions. T_c is dependent on the number of copper oxygen planar regions, material type, and quality. T_c can vary from about 20 K to well into the triple-digit numbers [6]. This system shows the widest T_c range and highest maximum transition temperature of all the superconductors.

We understand the high-T_c phenomenon much better today than we did when the Tl materials were first examined [40]. In contrast to the conventional conductors, the Sommerfeld constant of these compounds is rather small, especially for such a large T_c. Temperature dependence of the magnetic penetration depth indicates a low-symmetry non-s-wave ground state. Possibly, there are some regions on the Fermi surface with no-gap or pseudogap.

The London penetration depth is very large (ca. 200 nm). Effective mass anisotropy is also large ($m_c^*/m_{ab}^* \sim 10^2$). Consistent with this anisotropy, diamagnetic fluctuations indicate the existence of low (two)-dimensional superconductivity.

Most members in this highly anisotropic system show diverse types of magnetic phenomena rich in detail. We also know that to explain magnetization of these materials, glassy superconductivity models need not be invoked. Generally, most of the macroscopic behavior is very similar to that of soft type II conventional superconductors.

Softness is evidenced in the broadening of the transition and the reduced irreversibility at the high-field, high-temperature region of the phase space. The problem is compounded by the fact that the precise determination of many fundamental quantities, such as T_c, dH_{c2}/dT_c, flux lattice transition, and others required for theoretical understanding, is obscured by these experimental difficulties.

In many large-scale applications, the superconducting elements are required to carry large currents (10^9 to 10^{12} A/m^2) at high fields (1 to 5 T); under these conditions, even small amounts of flux motion are not acceptable. Over the last three decades, excellent progress has been made in developing metallurgical processes for hardening the low-temperature superconductors. Hardening creates sufficiently strong and numerous pinning centers, so that under operating conditions, the vortices remain practically stationary. Aside from energy-loss considerations in certain applications, such as those for persistent magnets, the field produced needs to be held steady to within a close tolerance. For example, in the magnetic resonance imaging (MRI) magnets, the relative field stability has to be better than 1 part in 10 million per hour ($d \ln B/dt < 10^{-7} h^{-1}$). Comparable reductions of flux motion in the Tl materials have to be achieved before their full potential can be realized. Short coherence length results in the weakening of the proximity effect and the overlap of the superconducting wave functions across any defects. The small value of ξ makes it easy for the vortex to leave the traps or pinning sites. Because of weak pinning, granularity, and strong anisotropy, the techniques of manufacturing conventional hard superconductors, such as the creation of collective pinning, is not directly transferable to the new materials. The more-than-tenfold increase in thermal energy in going from liquid helium to liquid air temperatures makes strong pinning a formidable obstacle.

Convincing answers to problems such as whether there is a magnetic mechanism to high-T_c superconductivity, and how to stabilize the flux vortex lattice at high temperatures and fields, are still lacking. Such challenges can be met only by reliable data and clear theoretical understanding.

Fortunately, there are several promising factors in these puzzles. Unlike the old materials, no effort is needed to reduce the coherence length. These compounds have such short coherence lengths that they are naturally in the type II (clean) limit. In this respect, we are halfway there! Most important, researchers in laboratories all over the world are active, so new theoretical understanding and materials development are continuing at a remarkable pace. It is likely that the magnetic puzzle of the Tl–copper oxides will be solved in the next few years.

REFERENCES

1. P. F. Dahl, *Superconductivity*, American Institute of Physics, New York, 1992.
2. H. E. Stanley, *Introduction to Phase Transitions and Critical Phenomena*, Oxford University Press, Oxford, 1971.
3. J. Bardeen, in *Concise Encyclopedia of Magnetic and Superconducting Materials*, ed. Jan Evettes, Pergamon Press, Oxford, 1992, p. 554.
4. R. D. Parks, ed., *Superconductivity*, Vols. 1 and 2, Marcel Dekker, New York, 1969.
5. T. Datta, C. P. Poole, H. A. Farach, C. Almasan, J. Estrada, D. U. Gubser, and S. A. Wolf, *Phys. Rev. B* 37, 7843 (1988).
6. C. P. Poole, T. Datta, and H. A. Farach, *Copper Oxide Superconductors*, Wiley, New York, 1988.
7. D. M. Ginsberg, ed., *Physical Properties of High Temperature Superconductors*, Vols. I, II, and III, World Scientific, Singapore, 1988, 1990, 1992.
8. T. Datta, in *Concise Encyclopedia of Magnetic and Superconducting Materials*, ed. (1992), Jan Evettes, Pergamon Press, Oxford, p. 408.
9. N. Winzek, F. Hentsch, M. Mehring, Hj. Mattausch, R. Kremer, and A. Simon, *Physica C* 168, 327 (1990).
10. N. H. Tea, M. B. Salamon, T. Datta, H. M. Duan, and A. M. Hermann, *Phys. Rev. B* 45, 5628 (1992).
11. C. Kittel, *Introduction to Solid State Physics*, 5th ed. Wiley, New York, 1975.
12. L. N. Mulay, *Magnetic Susceptibility*, Interscience, New York, 1963.
13. T. Datta, A. Barrientos, J. Amirzadeh, and E. R. Jones, Jr., *Appl. Phys.* 61, 3555 (1987).
14. Y. Kubo, Y. Shimakawa, T. Manako, T. Kondo, and H. Igarashi, *Physica C* 185–189, 1253 (1991).
15. J. A. Estrada, Ph.D. dissertation, University of South Carolina, 1989.
16. C. C. Almasan, Ph.D. dissertation, University of South Carolina, 1989.
17. A. C. Rose-Innes and E. H. Rhoderick, *Introduction to Superconductivity*, 2nd ed., Pergamon Press, Oxford, 1978.
18. T. P. Orlando, E. J. McNiff, S. Fonner, and M. R. Beasley, *Phys. Rev. B* 19, 4545 (1979).

19. M. Tinkham, *Introduction to Superconductivity*, McGraw-Hill, New York, 1975.
20. Ning Hong, T. Datta, H. Duan, and A. M. Hermann, *Bull. Am. Phys. Soc.* 35, R16-7, 786 (1990).
21. G. S. Grader, E. M. Gyorgy, L. G. Van Uitert, W. H. Grodkiewicz, T. R. Kyle, and M. Eibschultz, *Appl. Phys. Lett.* 53, 319 (1988).
22. C. C. Almasan, M. C. de Andrade, Y. Dalichaouch, J. J. Neumeier, C. L. Seaman, M. B. Maple, R. P. Guertin, M. V. Kuric, and J. C. Garland, *Phys. Rev. Lett.* 69 (1993), to appear.
23. Hong Ning, Ph.D. dissertation, University of South Carolina, 1991.
24. J. Annett, N. Goldenfeld, and S. R. Renn, *Phys. Rev. B* 43, 2778 (1991).
25. Hong Ning, H. Duan, P. D. Kirven, A. M. Hermann, and T. Datta, *J. Supercond.* 5, 503 (1992).
26. H. Ning et al., *Bull. Am. Phys. Soc.* 37, K14-10, 478 (1992).
27. J. R. Cooper, C. T. Chu, L. W. Zhou, B. Dunn, and G. Gruner, *Phys. Rev. B* 37, 638 (1988).
28. H. Ning and T. Datta, Unpublished (1991).
29. K. E. Gray, R. T. Kampwirth, and D. E. Farrell, *Phys. Rev. B* 41, 819 (1990).
30. F. Zuo, M. B. Salamon, T. Datta, and K. Ghiron, *J. Appl. Phys.* 69(8), Apr. 15, 1991.
31. N.-L. Wu and H. T. Chu, *Physica C* 167, 269 (1990).
32. R. Navaro, F. Lera, A. Badia, C. Rillo, J. Bartolome, W. L. Lechter, and L. E. Toth, *Physica C* 183, 73 (1991).
33. D. E. Morris, M. R. Chandrachood, and A. P. B. Sinha, *Physica C* 175, 156 (1991).
34. M. Paranthaman, M. Foldeski, and A. M. Hermann, *Physica C* 192, 161 (1992).
35. S. F. Wahid and N. K. Jaggi, *Physica C* 170, 395 (1990).
36. P. W. Anderson and Y. B. Kim, *Rev. Mod. Phys.* 36, 39 (1964).
37. R. M. Hagen, in *Physical Properties of High Temperature Superconductors*, ed., D. M. Ginsberg, Vol. II, World Scientific, Singapore, 1990. Vol. II, X. Ginsberg, ed. (1990), p. 121.
38. V. K. Chan and S. H. Liou, *Phys. Rev. B* 45, 5547 (1992).
39. E. L. Venturini, J. C. Barbour, D. S. Ginley, R. J. Baughman, and B. Morosin, *Appl. Phys. Lett.* 6, 2456 (1990).
40. R. Poole, *Science* 240, 146 (1988).
41. C. C. Almasan, C. L. Seaman, Y. Dalichaouch, and M. B. Maple, *Physica C* 174, 93 (1991).
42. T. Datta, C. P. Poole, C. Almasan, J. Estrada, S. A. Wolf, D. U. Gubser, Z. Z. Sheng, and A. M. Hermann, *MRS International Meeting on Advances in Materials*, Vol. 6, Materials Research Society, Pittsburgh, Pa., 1989, p. 951.
43. G. Deutscher, in *Concise Encyclopedia of Magnetic and Superconducting Materials*, ed. J. Evettes, Pergamon Press, Oxford, 1992, p. 82.
44. S. Kondoh, Y. Ando, M. Onoda, and M. Sato, *Solid State Commun.* 65, 1329 (1988).
45. V. Kovachev, *Energy Dissipation in Superconducting Materials*, Clarendon Press, Oxford, 1992.
46. R. Kubiac, K. Westerholt, G. Pelka, H. Bach, and Y. Khan, *Physica C* 166, 523 (1990).
47. Y. Kubo, T. Kondo, Y. Shimakawa, T. Manako, and H. Igarashi, *Phys. Rev. B* 45, 5553 (1932).
48. J. W. Lynn, ed., *High Temperature Superconductivity*, Springer-Verlag, New York, 1990.
49. Z. Z. Sheng, A. M. Hermann, A. El Ali, C. Almasan, J. Estrada, T. Datta, and R. J. Matson, *Phys. Rev. Lett.* 60, 937 (1988).
50. Y. Shimakawa, Y. Kubo, T. Manako, T. Satoh, S. Ijima, T. Ichihashi, and H. Igarashi, *Physica C* 157, 279 (1989).

51. R. Sugise, M. Hirabayashi, N. Terada, M. Jo, T. Shimomura, and H. Ihara, *Physica C* 157, 131 (1989).
52. J. R. Thompson, D. K. Christen, H. A. Deeds, Y. C. Kim, J. Brynestad, S. T. Sekula, and J. Budai, *Phys. Rev. B* 41, 7293 (1990).
53. E. L. Venturini, J. F. Kwak, D. S. Ginley, B. Morosin, and R. J. Bauhman, in *Proc. High-T$_c$ Superconductors: Magnetic Interaction*, ed. L. H. Benett, Y. Flom, and G. C. Vezzoli, World Scientific, Singapore, 1988, p. 169.

21

Effects of Pressure on Thallium-Based Superconductors

Norman E. Moulton and Earl F. Skelton
Naval Research Laboratory
Department of the Navy
Washington, D.C.

I. INTRODUCTION

Pressure has been an important research tool in understanding the nature of superconductivity and in the development of superconducting materials. This was so even before the discovery of the high-transition-temperature (T_c) superconductors. For example, pressure was vital in understanding the origin of superconductivity in certain organic charge transfer salts based on the cation molecule ditetramethyltetraselenofulvalenium (TMTSF), since all but one of these salts is superconducting only at elevated pressures.

The current revolution in the quest for higher-T_c superconductors began with the discovery of Bednorz and Müller of superconductivity in La–Ba–Cu–oxide [1]. An important experiment on this material was the measurement of dT_c/dP. Because of the unusually large pressure effect, 10^{-3} K/bar [2], workers were led to substitute smaller ions for La, in the hope of generating a permanent "internal pressure." This idea led to the discovery of the Y–Ba–Cu–oxide, $YBa_2Cu_3O_7$, with a T_c above 90 K [3].

Before discussing the application of high pressure to studies of the thallium-based systems, it may be helpful to review high-pressure experimental techniques. We then review the effects of pressure on the structure and bulk properties of thallium-based superconductors and compare these with other systems. Readers unfamiliar with high-pressure research who seek more information are referred to review articles [4, pp. 437–457; 5] and recent proceedings of two international high-pressure groups: International Association for the Advancement of High Pressure Science and Technology, AIRAPT, and the European High Pressure Research

Group, EHPRG.* The SI unit of pressure is the pascal (Pa) defined as 1 N-m^{-2}. In high-pressure experiments performed on condensed-matter phases, pressures are usually reported in GPa (1 bar $\equiv 10^6$ dyn-cm^{-2}, so 1 bar = 10^5 Pa and 10 kbar = 1 GPa).

The usual purpose of subjecting a material to high pressure is to cause a change in interatomic distances. This is accomplished while leaving its thermal energy invariant. Spatial changes resulting from applied pressure are usually an order of magnitude greater than those brought about by changes in temperature. For example, changing the temperature of most materials by 1000 K, which is readily accomplished in the laboratory, will produce a fractional change in the length of a few tenths of a percent; on the other hand, pressures of a few tens of GPa, which are also easily generated, will in most materials cause fractional length changes of several percent. Pressure adds a second dimension to phase space and can sometimes yield new compounds that would not be possible otherwise. A recent example is the single perovskite cell material, $LaCuO_3$, and 19 different alloys developed from it [7]. Pressure further enables one to utilize different thermodynamic paths to the same region of (P,T)-space.

It is usually the goal of static high-pressure experiments to control how changes in interatomic distances are affected. In hydrostatic experiments, the pressure is transmitted to the sample by way of a fluid. The presence of the fluid ensures the absence of shear stresses and hence a uniform pressure on all parts of the sample. The opposite of a hydrostatic pressure state would be a uniaxial stress experiment, in which pressure is applied along one axis of the sample only.

In a hydrostatic experiment, the pressure is increased through either the introduction of more fluid into a fixed volume or by reducing the volume of the system. Hydrostatic conditions are maintained until the fluid solidifies. The present-day limit of purely hydrostatic pressures is defined by the point at which helium freezes at 300 K; this is about 11.7 GPa [8]. When the fluid freezes, shear stresses are introduced by the pressure-transmitting material and the effect of pressure becomes less well defined. It is sometimes difficult to compare the results of such measurements with theory. If the degree of nonhydrostaticity is severe, odd effects are possible. For superconductors, this can result in excessive broadening of a transition [9] or even a change in the sign of dT_c/dP [10].

II. PRESSURE-GENERATING SYSTEMS

There are several standard systems for performing high-pressure experiments. The simplest is a large volume cell, sometimes called a pressure "bomb" [11, p. 395]. This is a hollow cell of a few cubic centimeters of experimental volume, usually fabricated of hardened, high-strength steel or beryllium copper alloys. The sample is immersed in a pressure-transmitting fluid inside the cell. Typical fluids used include oils, mixtures of organic liquids, and helium gas. Pressure is generated by isothermal volume reduction using a piston intensifier or by advancing a screw plug. Measurements of physical properties such as strain, electrical resistivity, and magnetic susceptibility are carried out via electrical feed-throughs. In general, these

---

*Proceedings of both EHPRG and AIRAPT are published in Ref. 6.

cells are used for experiments in the pressure regime below about 1 GPa, although some cells have reached 2 GPa, the limiting factor being the strength of the containment vessel. The large volume of pressure-transmitting fluid in the cell stores a significant amount of energy. If care is not taken to remain well within the operating limits of the material used in cell construction, an explosion can occur.

The best hydrostatic experiments employ helium as a pressure-transmitting medium. Other fluids, such as the 4:1 methanol/ethanol mixture can also be used, but care must be taken to work above its freezing temperature. If the pressure is changed before the fluid freezes, the nonhydrostatic stresses are much less than when the pressure is changed after the pressure medium has solidified. Some experimental setups do not facilitate the use of these fluids. In such systems a variety of other materials are used to transmit pressure. Kerosene, glycerine, Vaseline, various oils, and soft solids such as NaCl, crushed pyrophyllite, and various polymers have been used. In some experiments no fluid is used; the sample itself serves as the pressure medium, often resulting in large pressure gradients across the sample volume. These techniques offer varying degrees of quasi-hydrostaticity, but it should be understood that they offer good first-order measurements only. For high-precision measurements, hydrostatic conditions are required. A large volume cell used for $T_c(P)$ measurements is shown in Fig. 1.

Another type of high-pressure device commonly used is the Bridgman anvil (Fig. 2) [11, pp. 400–402]. In this apparatus, two anvils formed of high-strength steel or cemented tungsten carbide are arranged in opposition with the sample contained in a gasket between them. The anvils are supported by large, compression binding rings arranged with an interference fit to provide inward-acting radial stresses. The gasket also usually contains some type of pressure-transmitting medium and apparatus for measuring physical properties (e.g., susceptibility coils). Typical gasket materials are pyrophyllite and steel; typical fluids used are oils, resins, and viscous fluids such as Vaseline or glycerine. Pressure is generated by applying a force to one of the anvils while the other is held fixed. Pressures above 20 GPa have been achieved using this type of apparatus, again the limiting factor being the strength of the materials. In one instance, a pressure of 40 GPa was achieved using carboloy pistons tipped with sintered diamond [13].

In the hierarchy of pressure-producing apparatus, the multianvil systems follow the uniaxial devices. Four that have been developed are the tetrahedral press, the cubic press, the multianvil sliding system (MASS), and the split sphere [4, pp. 260–262]. In each case the anvils are advanced simultaneously toward the high-pressure region where the sample is contained by a suitable gasketing material. Information from the sample is obtained electronically using appropriate feed-throughs and from x-rays scattered from the sample and passing through the gasketing material.

An apparatus that is gaining wide use in the study of high-T_c materials and probably the most powerful instrument for basic high-pressure materials research is the diamond anvil cell (DAC) [14]. It offers three distinct advantages over older, larger systems:

1. It is compact. A typical DAC can be held in the hand, and for this reason can readily be heated or cooled.
2. It is comparatively inexpensive. A DAC can be constructed for a small fraction of the cost of the larger systems.

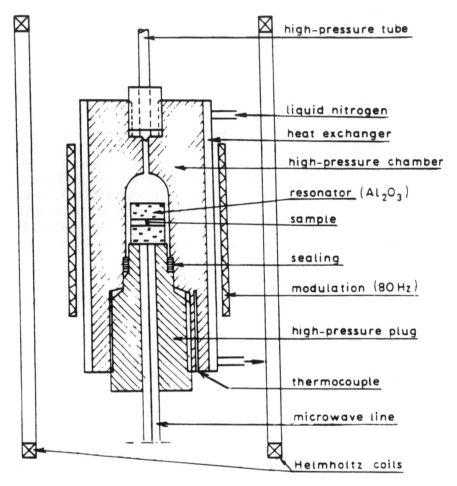

Figure 1 Large-volume high-pressure bomb used for $T_c(P)$ studies. The pressure is changed by adding or removing fluid through the high-pressure tube. (From Ref. 12.)

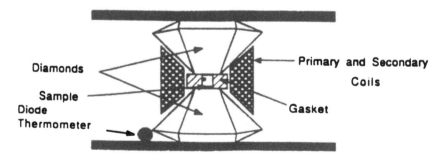

Figure 2 Schematic drawing of a DAC used for $T_c(P)$ studies.

3. The diamond anvils are transparent to a broad spectrum of electromagnetic radiation. The pressure chamber can readily be probed in a variety of ways and samples studied.

Since an early design of a DAC for low-temperature research [15], several improved models have been reported in the literature [16,17]. The heart of the DAC is a pair of gem-quality diamond anvils that are forced together over small contact areas. A sample is contained between the diamonds by a gasket usually also containing a pressure-transmitting fluid. The diamond anvils are flawless gem-cut diamonds with the tip polished to an optical flat, creating a culet. This serves as the pressure-generating surface; the extreme pressures are produced with modest forces by keeping the surface area small. Diamonds used range in size from over a carat to below 0.2 carat. In general, smaller diamonds are used to produce higher pressures. Typical sample sizes range from 10^{-4} to 10^{-6} mm^3. The largest static pressures (hundreds of GPa) are generated by using bevelled diamonds. Typical gasket materials are hardened steels. Inconel, and beryllium–copper. Pressure fluids used include oils, alcohol mixtures, viscous fluids, and helium gas.

The most attractive feature of the DAC is that it provides a window into the pressure chamber. Visible radiation can be used to measure the pressure in situ based on the calibrated shift of ruby fluorescence radiation [18]. Also, x-rays can be used to monitor the structure of the sample under pressure [19] and at extremes of temperature [15,20]. Physical properties of high-T_c samples have been measured using electrical resistivity [9], magnetic susceptibility [21], and Raman scattering [22].

Uniaxial experiments are extremely difficult to perform and are carried out at modest pressures, below 1 GPa. The principal difficulty lies in generating a uniform pressure over the surface of the sample, usually a single crystal. Stress gradients often result in destruction of the crystal. Usually, a crystal or thin film of material is placed between two parallel plates and compressed while a measurement is made of a physical property under consideration [23]. Tensile stresses can be generated by bending the substrate upon which a thin film is grown [24] or by pulling the crystal [25].

III. PRESSURE MEASUREMENT

Pressure measurement is a subject that deserves some attention. At room temperature, any number of techniques can be applied. A primary standard exists up to 2.6 GPa [26]; above this, pressures are often estimated from theoretical equations of state combined with measured temperatures and x-ray diffraction data. Physical properties of materials, such as structural or electrical phase transitions, have been calibrated and are useful secondary standards.

In experiments involving studies of samples at high pressures and low temperatures, care must be taken. There are three typical approaches to the measurement of pressure in such experiments. In the first, the pressure is measured at room temperature before the temperature of the cell is decreased. The obvious shortcoming of these measurements is that the pressure at which physical properties of the sample are measured is not known. Lacking low-temperature pressure measurement capabilities, researchers often assume that they cool along an isobar, but this is rarely the case. We have found through experience with DACs that the pressure increased by as much as 100% on cooling from room temperature to 77 K. Moreover, these changes usually vary from gasket to gasket, and from sample to sample, thus eliminating calibration as an alternative. It may be possible in some cases to perform legitimate pressure calibrations, but care must be taken to ensure that the proper pressure dependence is measured. One group performed over 100 experiments before a satisfactory calibration was reached for their system [16].

A second approach is to include a superconducting manometer within a sample chamber [16]. In this case the cell is cooled to the critical temperature of the manometer, the pressure dependence of which is carefully calibrated. In studies of high-T_c materials, Pb is generally used. This means that the experimental measurements are carried out in the approximate range 80 to 125 K and the pressure measurements are carried out below 10 K. Again, obvious problems may exist with this method and care must be used in calibration. In addition, there is disagreement between measurements of the critical temperature of lead when either helium or 4:1 methanol/ethanol is used as a hydrostatic fluid [27].

The third approach, and probably the most reliable, is to measure the pressure in situ at low temperatures. The shift in wavelength of the R_1 and R_2 fluorescence peaks from ruby has been calibrated at low temperatures using primary standards, and yields values accurate to a few percent [28]. This approach has only been tried with diamond anvil cells, although in principle, it could be used with high-volume cells if such cells provided an optical path to the high-pressure region. This can be achieved using any number of closed-cycle refrigerators (for smaller cells), helium transfer refrigerators, or other cryogenic systems in which a cold finger can be enclosed in a shroud with optical windows. Such systems can also be used to perform in situ x-ray measurements at low temperatures. In these experiments the pressure may be measured using an x-ray calibrant for which the equation of state is known, such as NaCl, whose equation-of-state calculations [29] have been extended to low temperatures [30].

IV. EFFECT OF PRESSURE ON STRUCTURE

The effect of pressure on the T_cs of the new classes of ceramic oxides is demonstrative, but its effect on their structure is minimal. The lower-T_c compounds, $La_{1.8}Sr_{0.2}CuO_4$ and $YBa_2Cu_3O_{7-\delta}$, have been studied at room temperature to 22.0 and 15.0 GPa, respectively [31]. In a separate experiment, $YBa_2Cu_3O_{7-\delta}$, was studied to 10.6 GPa [32]. Recent investigations were also carried out on $BiSrCaCu_2O_x$ to 50 GPa [33]. We have performed energy-dispersive x-ray diffraction measurements on $Tl_2Ba_2CaCu_2O_{8-\delta}$ to over 50 GPa [34]. In all cases, no evidence of any structural transformations have been reported.

At 10.6 GPa, the orthorhombic axial dimensions a, b, and c of $YBa_2Cu_3O_{7-\delta}$ were observed to change by only 2.0, 2.3, and 1.1%, respectively. The corresponding approximately linear, volume compression corresponds to an isothermal bulk modulus of 196 ± 17 GPa [32]. This is only slightly larger than the room-temperature bulk modulus for $La_{1.8}Sr_{0.2}CuO_4$ of 160 ± 13 GPa [31], but more than double the value of 62.5 GPa reported for $Tl_2Ba_2CaCu_2O_{8-\delta}$ [33].

Whereas the linear thermal expansivity along the c-axis for the three families of compounds, $La_{1.9}Ba_{0.1}CuO_4$, $YBa_2Cu_3O_{7-\delta}$, and $Tl_2Ba_2CaCu_2O_{8-\delta}$, is approximately the same [35], the linear compressibility in that direction more than doubles, from 1.6×10^{-3} GPa^{-1} for La$_2$CuO$_4$ to 3.8×10^{-3} GPa^{-1} for Tl$_2$Ba$_2$Ca$_2$Cu$_3$O$_{10-\delta}$. This has been interpreted as a stabilizing of the structure by the Cu–O octahedra upon intercalation of additional layers [36].

V. EFFECTS OF PRESSURE ON T_c

When a new superconductor is discovered, one of the first measurements is often that of $T_c(P)$. If T_c increases with applied pressure, there is reason to believe that the material can be transformed into a permanent higher T_c state. Experiments that have been conducted on $T_c(P)$ of thallium-based materials are divided into two classes, those performed at pressures at or below 2 GPa and those above. A representative set of the experiments from the first group is given in Table 1. The usual experimental technique is to use a clamp cell with a hydrostatic fluid and a lead manometer. Other experiments were performed in gas cells with helium or in piston-cylinder cells with various fluids and gauges. As stated previously, experiments conducted under these conditions are susceptible to large uncertainties in the pressure in the temperature region where the transition occurs.

In addition, samples are measured in a variety of stress states, ranging from ceramic sinters, to polycrystalline powders, to single crystals. Studies with sinters and polycrystalline samples may be susceptible to proximity effects, because as the pressure is increased, the separation distance between neighboring crystallites will decrease. An ideal experiment would be to use helium as a pressure medium, an optical standard for in situ pressure measurement, and high-quality single-crystal samples. Each of the experiments in Table 1 possesses some of these ideals, but none possess them all.

Despite the wide variance in sample quality and composition, and the possible uncertainties associated with the various experimental techniques employed, the results are surprisingly uniform. In the low-pressure range all samples exhibit an approximately linear pressure derivative, dT_c/dP (Fig. 3). For Tl$_2$Ba$_2$CaCuO$_{8-\delta}$, a value of about 2 K/GPa is observed, while values range between about 2.5 and 5 K/GPa for Tl$_2$Ba$_2$Ca$_2$Cu$_3$O$_{10-\delta}$. These results agree with the general trend pointed out for all hole-carrier-type high-T_c systems: namely, a positive pressure derivative at low pressures [48]. Hence the proposed theoretical explanations for this behavior were similar. A review by Griessen contains a compilation of the results of early high-pressure work on high-T_c systems and a synopsis of some of the prevalent theories to explain the effects observed [49].

The electron carrier systems L$_{2-x}$M$_x$CuO$_{4-\delta}$ (L = Pr, Nd, Sm, Eu; M = Ce, Th) initially seemed to be starkly different from the hole carriers in this respect, exhibiting a nearly zero or negative pressure dependence in the lower-pressure regime [48,50]. This suggested an asymmetry in the high-T_c mechanism. Results of experiments on the hole system Tl$_2$Ba$_2$CuO$_{6-\delta}$ [41,51], however, have changed that perception. Several papers have appeared indicating that the pressure derivative of this system is negative. To our knowledge, this is one of only two hole systems for which this is the case, the other being (Y$_{1-x}$Pr$_x$)Ba$_2$Cu$_3$O$_{7-\delta}$ in certain ranges of x [37]. An interesting observation for Tl$_2$Ba$_2$CuO$_{6-\delta}$ is that the effect depends on the temperature at which the pressure is changed [41]. When the pressure was changed at room temperature, dT_c/dP is negative, as in other ex-

Table 1 Compilation of Parameters of Several Experiments Carried Out on Thallium-Based Superconductors at pressures at or Below 2 GPa

Ref.	Sample[a]	Cell type	Pressure measurement	Hydrostatic fluid	Max. P (GPa)	T_c at 1 atm (K)	dT_c/dP (K/GPa)
37	2212, pc	Clamp	Pb manometer	Si oil	1.6	101	1.8
38	2212, sinter	Piston-cylinder	Manganin guage	Kerosene, diffusion pump oil	0.8	117.1	2.0
39	2212, bulk state unknown	Gas cell	Unknown	He gas	0.8	106	2.0
40	2223, pc	Clamp	Pb manometer	Isoamyl alcohol/ isopentane 1:1	1	99.5	5.0
37	2223, pc	Clamp	Pb manometer	Si oil	1.6	106	2.4
41	2201, pc	Gas cell	Pb manometer; He isochores	He gas	0.6	58	−0.64
37	2201, pc	Clamp	Pb manometer	Si oil	1.6	8	−1.4
						19.1	−2.7
						69.3	−3.9
42	2201, pc	Piston-cylinder	Phase-transition calibration	Fluorinert	2.0	16	−2.8
						20	−3.3
						49	−6.0
						51	−5.9
						61	−3.2
						82	−1.3
						87	−0.88
43	2212, sc	Clamp	Unknown	Silicon oil kerosene 1:1	1.3	83	3.3
							2.7
44	5526, sc	Gas cell	Unknown	He gas	0.6	106	0.23
45	Multiphase, ceramic	Clamp	Conventional manometer	Petroleum ether	0.5	113.0	2.2
46	TlBaCa$_3$Cu$_3$Oy	Clamp	Pb manometer	Kerosene, diffusion pump oil	1.3	119.4	3.2
47	Tl$_2$Ba$_{1.7}$Ca$_{1.3}$Cu$_3$O$_9$	Piston-cylinder	Pb manometer	Kerosene, diffusion pump oil	1.8	110	1.83

[a]2201, Tl$_2$Ba$_2$CuO$_{6+\delta}$; 2212, Tl$_2$Ba$_2$CaCu$_2$O$_{8-\delta}$; 2223, Tl$_2$Ba$_2$Ca$_2$Cu$_3$O$_{10-\delta}$; 5526, Tl$_5$Ba$_5$Ca$_2$Cu$_6$O$_x$; pc, polycrystalline sample; sc, single-crystal sample.

periments, but when the pressure was varied at low temperature (slightly above the freezing point of He), much weaker pressure dependencies were observed for samples with varying oxygen contents. It was proposed that the effect is due to the lack of thermal energy required for rearrangement of oxygen in the depleted samples, noting that the derivative is almost zero for the stoichiometric system when pressure is changed at room temperature. Future studies must take this possibility into account to avoid erroneous results.

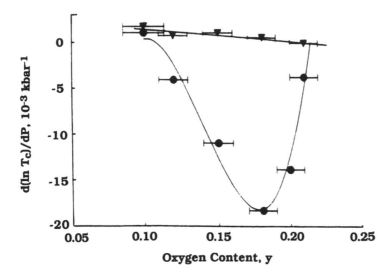

Figure 3 $d \ln(T_c)/dP$ for $Tl_2Ba_2CuO_{6+y}$ for various values of y for changes in pressure made at room temperature (circles) and at low temperatures (triangles). The absence of any significant effect or the low-temperature data is interpreted as possible evidence that oxygen ordering is an important component of dT_c/dP. (From Ref. 41.)

Several experiments have been conducted in DAC and Bridgman anvil type systems on $Tl_2Ba_2CaCuO_{8-\delta}$ and $Tl_2Ba_2Ca_2Cu_3O_{10-\delta}$ samples of varying quality. A summary of some of this work is given in Table 2. None of the experiments summarized in Table 2 has all the ideal characteristics of a hydrostatic experiment: He gas as a pressure medium, high-quality single-crystal samples, and in situ pressure measurement. Nevertheless, as in the lower-pressure work, varying experimental situations on samples of varying quality have produced similar results. In all cases, T_c increases with pressure, in reasonable agreement with low-pressure results, and then reaches a maximum before falling off at higher pressures (Fig. 4). There is little or no hysteresis in this pressure dependence.

Results on $Tl_2Ba_2CuO_{6+\delta}$ to 8 GPa in a cubic anvil cell have verified that T_c has a negative pressure dependence at lower pressures but a positive curvature at higher pressures, suggesting that it might eventually change sign if subjected to high enough pressure [21]. Data on two $Tl_2Ba_2Ca_2Cu_3O_{10-\delta}$ crystals are plotted in Fig. 5 [57]. In each case, T_c increased from about 116 K to 131.7 ± 0.5 K at the highest pressure. The initial slope of $T_c(P)$ is 4.8 ± 0.3 K/GPa, about twice the value reported for $Tl_2Ba_2CaCuO_{8-\delta}$ by Moulton et al. [21].

The pressure dependence of $Tl_2Ba_2CaCuO_{8-\delta}$ and $Tl_2Ba_2Ca_2Cu_3O_{10-\delta}$ is shared by several hole-conducting high-T_c superconductors (Fig. 6). Maxima have been observed in the yttrium-based systems $YBa_2Cu_3O_{7-\delta}$ [58], $YBa_2Cu_4O_{8-\delta}$ [56,59], $YBa_2Cu_3O_{7-\delta}$, $Y_{0.9}Ca_{0.1}Ba_2Cu_4O_{8-\delta}$, $La_{2-x}Sr_2CuO_{4-y}$, and $CaBaLaCu_3O_{7-\delta}$ [56], and in the bismuth-based system $Bi_2Sr_2CaCu_2O_{8-\delta}$ [60].

A theoretical explanation of the mechanism of high-T_c superconductivity in these systems must account for this behavior and there are several schools of thought

Table 2 Compilation of Parameters of Several Experiments Carried Out on Thallium-Based Superconductors at Pressures Above 2 GPa[a]

Ref.	Sample	Cell type	Pressure measurement	Hydrostatic fluid	Max. P (GPa)	T_c at 1 atm (K)	dT_c/dP (K/GPa)	$P(T_{c,max})$ (GPa)
52	2212, sinter	Metal gasket	Manganin gauge	Methanol/ ethanol	4	107	0.5	2
53	2212, pc	Metal gasket	Manganin gauge	Methanol/ ethanol	3.5	107 106.6	1.4 0.8	2.5 2.5
54	2223, ceramic	Bridgman anvil	Unknown	Unknown	19	~105	1.2	~10 ~15
42	2212, pc	Cubic anvil	Phase transition calibration	Fluorinert	8	117	1.7	3
21	2212, sc	DAC	Ruby scale in situ at low temp.	Si oil Methanol/ ethanol Methanol/ ethanol Methanol/ ethanol	6	112.1 114.1 106.7 107.2	3.6 2 2 2	3.8 ≈5 ≥5 ≥5
55	2212, sc	DAC	Ruby scale room temp.	Methanol/ ethanol	13.6	Unknown	112	3
56	2201, pc	Cubic anvil	Phase transition calibration	Fluorinert	8	~50	−7.3	>8

[a]The abbreviations used in Table 1 are employed here.

on the topic. The most prominent theory is that increased pressure causes an increase in carrier concentration in the Cu−O planes, where superconductivity is thought to occur. To explain the maximum T_c, there is presumed to be an optimal carrier concentration and, once exceeded, T_c will decrease with increasing pressure. Evidence supporting this idea has come from two observations. First, several experiments have been performed in which the planar carrier concentration is varied, by changing the oxygen stoichiometry for both Bi-based systems [61] and for Tl-based systems [62,63] and by changes in Sr content for $La_{2-x}Sr_xCuO_4$ [64]. In these systems there seems to be a carrier concentration at which T_c at ambient pressure is indeed maximized.

Second, there is significant experimental evidence for a pressure-induced increase in the hole density in the Cu−O planes (Fig. 7). Hall effect measurements on systems $CaBaLaCu_3O_{7-\delta}$, $Bi_2Sr_2CaCu_2O_{8-\delta}$, $YBa_2Cu_4O_{8-\delta}$, and $Tl_2Ba_2CuO_{6+\delta}$ all show decreases in the temperature-dependent Hall coefficient, R_H, with increasing pressure [56]. However, there are conflicting results for $YBa_2Cu_3O_{7-\delta}$: One group observed no effect of pressure on R_H [65], whereas another group measured a pressure-induced decrease similar to that seen in the aforementioned systems [56]. Others have reported a null effect in the $La_{2-x}Sr_xCuO_{4-y}$ system as well [66]. Neutron-scattering experiments support the idea of pressure-induced changes in the carrier concentration [67], as do upper critical field measurements

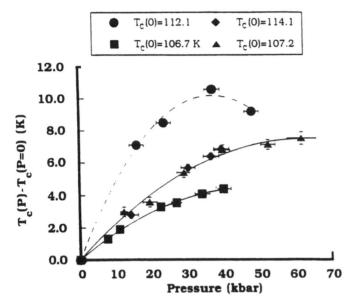

Figure 4 $T_c(P)$ for several single crystals of $Tl_2Ba_2CaCu_2O_{8-\delta}$. Note that $T_c(P)$ is nonlinear and exhibits a maximum for one of the samples. (From Ref. 21.)

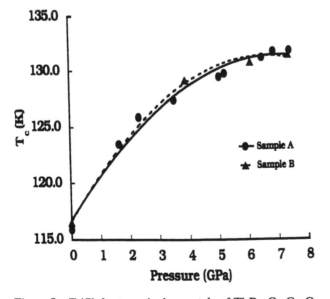

Figure 5 $T_c(P)$ for two single crystals of $Tl_2Ba_2Ca_2Cu_3O_{10-\delta}$.

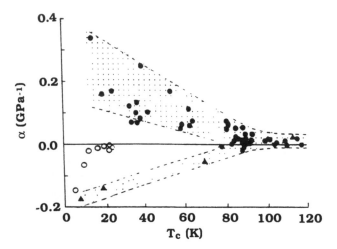

Figure 6 $\alpha(\equiv dT_c/T_c dP)$ versus T_c for all known high-temperature superconductors, including hole-carrier systems (filled symbols) and electron-carrier systems (open circles); triangles correspond to data from $Tl_2Ba_2CuO_{6+\delta}$. Note that almost all hole systems possess positive values for α. (From Ref. 51.)

[59]. In summary, the majority of the work reported in the literature supports this notion.

However, recent measurements are in conflict with the foregoing hypothesis. At low pressures, systems that clearly exhibit an optimal carrier concentration at ambient pressure possess positive pressure derivatives regardless of the ambient pressure carrier density. This phenomenon has been observed in both $Bi_2Sr_2CaCu_2O_{8-\delta}$ [61] and $(La_{1-x}Sr_x)_2CuO_4$ [66]. Similarly, the fully oxygenated $YBa_2Cu_3O_{7-\delta}$ is overdoped but exhibits a small positive pressure dependence. It was found that by adding holes to $YBa_2Cu_3O_{7-\delta}$, by doping with Ca while keeping

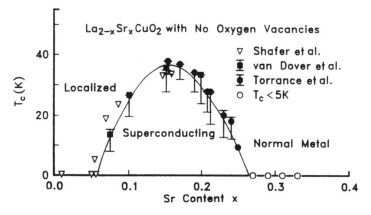

Figure 7 Dependence of T_c on hole concentration in $La_{2-x}Sr_xCuO_4$. A clear extremum exists for a hole concentrations in the range from about 0.13 to 0.22. The question mark in the figure indicates that oxygen vacancies may exist in that sample. (From Ref. 64.)

the oxygen stoichiometry constant, T_c decreased [68]. If increased carrier concentration is the sole mechanism responsible for the change in T_c, it should follow that samples with carrier concentrations below the optimal level should exhibit a positive dT_c/dP value and those with oxygen contents beyond optimum should have a negative dT_c/dP value. Calculations based on pressure-induced carrier increases predict that the pressure derivative for fully oxygenated $YBa_2Cu_3O_{7-\delta}$ should be about -0.5 K/GPa, opposite in sign to the value observed [69].

Wijngaarden et al. have suggested an explanation for this effect which is consistent with the idea that carrier concentration is the dominant parameter [59]. In their view, the increase in carrier concentration, δ, causes the T_c versus δ curve to shift to higher δ and increase in temperature. At high pressures, there is a different optimal carrier concentration and a different maximum T_c attainable. This approach described the tendencies observed in $(La_{1-x}Sr_x)_2CuO_4$ at low pressures ($P \sim 1$ GPa) [66]. However, it also has been observed that R_H was pressure independent, which conflicts with the estimate of carrier increase obtained by Wijngaarden et al. from upper critical field data.

Therefore, it is doubtful that fluctuations in carrier concentration are the sole reason for pressure-induced changes in T_c. An alternative explanation, which also involves the effect of pressure on the lattice, was offered in the paper by the present authors [21]. In this, high-pressure ac susceptibility measurements of T_c were performed on single crystals of $Tl_2Ba_2CaCu_2O_{8-\delta}$. The results were explained qualitatively by the theory of Kresin and Morawitz [70], developed initially to explain the carrier dependence in $La_{2-x}Sr_xCuO_4$. It utilizes the strong-coupling BCS theory and the highly anisotropic quasi-two-dimensional Fermiology that exists because of the layered structure of the Cu–O planes. Using the strong-coupling form of the Eliashberg form of the BCS theory, they find that T_c is nonmonotonic and a very sensitive function of the Fermi wavevector, k_F. This is due to the electron–phonon coupling parameter, λ, being proportional to $(k_F)^2$ when $2k_F$ is smaller than a limiting value of the phonon momentum, q_c, for the most important phonon mode, and proportional to $(k_F)^{-1}$ when $2k_F >> q_c$. Thus as $2k_F$ crosses through q_c, T_c is a maximum and decreases on either side of this optimum condition.

In a quasi-two-dimensional superconductor, it was shown [59,71] that $(k_F)^2 = 2\pi n c$, where n is the carrier concentration and c is the distance between conducting two-dimensional subsystems. Thus not only is T_c strongly affected by changes in n, it is clearly just a strongly a function of c. If, indeed, n is increased with increasing P, this effect will either compete with or complement the effect of reducing c, depending on whether the system is optimized with respect to these parameters. An increase in P will cause a decrease in c and an increase in n; thus the actual value of k_F will be a sensitive function of the compressibility and the functional form of the pressure dependence of n. A quantitative analysis based on changes in interplanar spacing has not yet been performed since compressibility data on $Tl_2Ba_2CaCu_2O_{8-\delta}$ are not available.

The overall picture presented by all the experimental data of the effects of pressure on T_c seems to be even more complicated than that suggested by the Kresin–Morawitz theory. Significant uniaxial pressure derivatives seem to exist in the Cu–O planes. These derivatives are of a magnitude comparable to the interplanar pressure derivatives. One class of experiments that indicate this are uniaxial stress measurements. These experiments are difficult to perform without causing distortion of the sample or without generating nonuniform stress fields, which can lead to improper comparisons with theory because of the loss of translational symmetry. For example, some experiments were performed in simple compression cells in which a free sample is uniaxially compressed between the faces of two metal disks. It has been noted that this configuration can produce inhomogeneous in-plane stresses [69]. Single crystals of $YBa_2Cu_3O_{7-\delta}$ have been compressed along the c-axis in this way, yielding a value of $dT_c/dP \approx 0.8$ KG/Pa [23]. This suggests that the entire contribution to the pressure derivative in $YBa_2Cu_3O_{7-\delta}$ ($dT_c/dP \approx 0.7$ K/GPa) comes from interplanar compression. Estimated strain derivatives have been extracted from hydrostatic experiments performed on thin films of $(RE)Ba_2Cu_3O_{7-\delta}$ (RE = Yb, Dy, Y, and Gd), using known elastic moduli of the film and substrate [72]. The total dT_c/dP was found to be about 0.77 K/GPa, in good agreement with the experimental value of 0.83 K/GPa. This is a combination uniaxial derivatives $dT_c/dP_a = -3.1$ K/GPa, $dT_c/dP_b = +3.8$ K/GPa, and $dT_c/dP_c = +3.4$ K/GPa. In other thin-film experiments, uniaxial stresses were generated by bending the substrate, although nonuniform stress fields must have resulted. In this it was observed that $dT_c/d\sigma_{ab} = 0.45$ to 0.50 K/GPa, where σ_{ab} is the in-plane stress. Since the derivative of T_c with respect to hydrostatic compression is given by $dT_c/dP = 2(dT_c/d\sigma_{ab}) + dT_c/d\sigma_c$, it is concluded that the derivative with respect to pressure along the c axis should be insignificant. As of this writing, uniaxial experiments have not provided any definitive results concerning the question of interplanar pressure derivatives but suggest that intraplanar contributions are significant.

In thermal expansion measurements, the dependence of the critical temperature on stress applied along each of the crystallographic axes is arrived at by using the Ehrenfest relation [69]: $dT_c/dP_i = T_c(\Delta\alpha_i/\Delta C_p)$, where $i = a, b, c$. Measurements of the discontinuity in the specific heat, ΔC_p, and in the thermal expansivity, $\Delta\alpha_i$, yield a value of the pressure derivative for that axis. In experiments performed on $YBa_2Cu_3O_{7-\delta}$ with oxygen stoichiometries of 6.5 and 6.9, it was observed that $\Delta\alpha_i$ was about equal in magnitude and opposite in sign for a and b, and about zero for the c direction in the 6.9 case and large and positive in the 6.5 case. The interpretation of this is that the observed small positive pressure derivative in the fully oxygenated material is an artifact of the near cancellation of the a and b components, while the large positive pressure derivative of the oxygen-deficient sample is the result of an additional positive discontinuity in the thermal expansion of the c axis. For other systems, the intraplanar pressure derivatives are comparable to those of the interplanar values [73].

The wide range of experimental values—positive, negative, and near zero— for dT_c/dP_i for high-T_c systems suggests that these systems are not necessarily optimized with respect to inter- and intraplanar configurations. Thus the pressure derivative of a system can have positive or negative contributions attributable to these structural modifications, depending on whether the system is driven toward the optimum configuration or away from it by pressure. Possible causes for intraplanar pressure derivative include changes in in-plane fields due to changes in atomic positions, changes in phonon frequencies from lattice stiffening, and changes in electronic band structure. For discussions of relevant theories, we refer readers to excellent reviews by Griessen [49], Wijngaarden and Griessen [74], and Schilling and Klotz [69], and references therein.

Another mechanism that might contribute significantly, as mentioned above, is ordering in the oxygen sublattice. When a material is subjected to increased pressure, it responds by a reduction in volume; therefore, ordering of the oxygen sublattice is a definite possibility. Neutron-scattering experiments on $YBa_2Cu_3O_{7-\delta}$ have shown that ordering of oxygen defects results in a reduced cell volume [67]. It has also been observed in $Tl_2Ba_2CuO_{6+\delta}$ that there is a definite pressure dependence of dT_c/dP upon the temperature at which the pressure was changed [41]. The interpretation of this is that when the pressure was changed at high temperatures, there was sufficient thermal energy to permit the oxygen atoms to reorder, whereas at lower temperatures this was not the case. Since this effect has not been seen in other systems, including $YBa_2Cu_3O_{7-\delta}$ [41] and $Tl_2Ba_2CaCu_2O_{8-\delta}$ [61], it may be possible that oxygen reordering occurs freely at all experimental temperatures in these systems. In the pressure-induced ordering scheme, the contribution to T_c would rise with P until the optimum configuration was reached. Samples with nearly stoichiometric oxygen configurations would therefore undergo less reordering and hence exhibit less of a rise in T_c at lower pressures. Additional structural experiments are needed to confirm this.

We note that in the work of Tissen [55], an irreversible phase separation was observed in $Tl_2Ba_2CaCu_2O_{8-\delta}$ single crystals at 6.9 GPa. It was proposed, based on the T_cs of the two phases, that they included an oxygen-rich and an oxygen-deficient phase, although this was not verified. This would suggest high-oxygen mobility in this system.

Thus a scenario to explain the possible observed pressure dependencies has been developed from a wide body of experimental evidence. There are at least four contributing factors, all of which have pressure dependencies that may be optimized at different pressures. Consider, as an example, a system in which the carrier concentration is beyond optimum, with several somewhat randomly oriented oxygen vacancies present. As the pressure is increased, ordering of the oxygen sublattice tends to increase T_c. Carrier injection into the planes would decrease T_c, and changes in interplanar separation and configuration could have either effect, depending on whether the initial configuration was driven toward or beyond the optimum by pressure. As the oxygen sublattice reaches the optimum state of ordering, the negative contributions due to overdoping, suboptimal interplanar spacing, and suboptimal intraplanar configuration become dominant, causing T_c to drop. In this picture, samples near their structural optimum (e.g., $YBa_2Cu_3O_{7-\delta}$ with T_c near 90 K or $Tl_2Ba_2Ca_2Cu_3O_{10-\delta}$ with T_c near 125 K) should exhibit very small increases in T_c at low pressures and then drop significantly at higher pressures. This was observed in $YBa_2Cu_3O_{7-\delta}$ [58], while experiments on fully stoichiometric $Tl_2Ba_2Ca_2Cu_3O_{10-\delta}$ have yet to be carried out. This picture also accounts for the observed behavior of substoichiometric systems, in which dT_c/dP is high at low pressures and decreases with increasing pressure.

Observations of the thallium-based systems have presented data vital to an understanding of the effects of pressure on high-T_c materials in general. The observed maximum in T_c in $Tl_2Ba_2CaCu_2O_{8-\delta}$ and $Tl_2Ba_2Ca_2Cu_3O_{10-\delta}$ and the peculiar behavior of $Tl_2Ba_2CuO_{6+\delta}$ have aided in the understanding that there may be complex-order effects that play a role in the behavior of high-T_c systems under pressure. Theoretical explanations of the observed pressure dependencies will have to take this process into account before a real understanding of high-T_c systems under pressure will be realized.

ACKNOWLEDGMENTS

The authors are grateful to S. A. Wolf, D. D. Berkley, and D. H. Liebenberg for helpful discussions. One of us (N.E.M.) would also like to thank the Office of Naval Technology for support.

REFERENCES

1. J. G. Bednorz and K. A. Müller, *Z. Phys. B* 64, 189 (1986).
2. C. W. Chu, P. H. Hor, R. L. Meng, L. Gao, Z. J. Huang, and Y. Q. Wang, *Phys. Rev. Lett.* 58, 405 (1987).
3. M. K. Wu, J. R. Ashburn, C. J. Torng, P. H. Hor, R. L. Meng, L. Gao, Z. J. Huang, Y. Q. Wang, and C. W. Chu, *Phys. Rev. Lett.* 58, 908 (1987).
4. E. F. Skelton and A. W. Webb, in *Encyclopedia of Physical Science and Technology*, Vol. 13, ed. R. A. Meyers, Academic Press, 1987, pp. 437–457.
5. A. Jayaraman, *Rev. Mod. Phys.* 55, 65 (1983).
6. M. Ross, ed., *High Pressure Research*, Gordon and Breach, New York, 1992.
7. A. W. Webb, E. F. Skelton, S. B. Qadri, V. Browning, and E. R. Carpenter, Jr., *Phys. Rev. B* 45, 2480 (1992).
8. M. G. E. van Hinsberg, W. L. Vos, and J. A. Schouten, *High Pressure Res.* 4, 592 (1990).
9. A. Driessen, R. Griessen, N. Koeman, E. Salomons, R. Brouwer, D. G. de Groot, K. Heeck, H. Hemmes, and J. Rector, *Phys. Rev. B* 36, 5602 (1987).
10. R. N. Shelton, A. R. Moodenbaugh, P. D. Dernier, and B. T. Matthias, *Mater. Res. Bull.* 10, 1111 (1975).
11. I. L. Spain and J. Paauwe, *High Pressure Technology*, Vol. I, Marcel Dekker, New York, 1977.
12. J. Stankowski et al., *Physica C* 160, 170 (1989).
13. F. P. Bundy, *Rev. Sci. Instrum.* 46, 1318 (1975).
14. A. Jayaraman, *Rev. Sci. Instrum.* 57, 1013 (1986).
15. E. F. Skelton, I. L. Spain, and F. J. Rachford, in *High Pressure and Low Temperature Physics*, ed. C. W. Chu and J. A. Woollam, Plenum Press, New York, 1978, p. 337.
16. B. Bireckoven and J. Wittig, *J. Phys. E Sci. Instrum.* 21, 841 (1988).
17. D. Braithwaite, G. Chouteau, and G. Martinez, *Meas. Sci. Technol.* 1, 1347 (1990).
18. H. K. Mao, J. Xu, and P. M. Bell, *J. Geophys. Res.* 91, 4673 (1986).
19. I. L. Spain, S. B. Qadri, C. S. Menoni, A. W. Webb, and E. F. Skelton, in *Physics of Solids Under High Pressure*, ed. J. S. Schilling and R. N. Shelton, North-Holland, Amsterdam, 1981, p. 73.
20. M. W. Schaefer, E. F. Skelton, and D. Schiferl, in *Synchrotron Radiation Applications in Mineralogy and Petrology*, ed. A. Barto-Kyriakidis, Theophrastus Publications, Athens, 1988, pp. 13–31.
21. N. E. Moulton, S. A. Wolf, E. F. Skelton, D. H. Liebenberg, T. A. Vanderah, A. M. Hermann, and H. M. Duan, *Phys. Rev. B* 44, 12,632 (1991).
22. L. V. Gasparov, O. V. Misochko, M. I. Eremets, A. V. Lomsadze, and V. V. Struzhkin, *Sov. Phys. JETP* 71, 945 (1990).
23. M. F. Crommie, A. Y. Liu, A. Zettl, M. L. Cohen, P. Parilla, M. F. Hundley, W. N. Creager, S. Hoen, and M. S. Sherwin, *Phys. Rev. B* 39, 4231 (1989).
24. G. L. Belenky, S. M. Green, A. Royburd, C. J. Lobb, S. J. Hagen, R. L. Green, M. G. Forrester, and J. Talvacchio, *Phys. Rev. B* 44, 10,117 (1991).

25. X. F. Chen, G. X. Tessema, and M. J. Skove, *Physica C* 181, 340 (1991).
26. D. P. Johnson and P. L. M. Heydemann, *Rev. Sci. Instrum.* 38, 1294 (1967).
27. J. Thomasson, C. Ayache, I. L. Spain, and M. Villedieu, *J. Appl. Phys.* 68, 5933 (1990).
28. R. A. Noack and W. B. Holtzapfel, in *High Pressure Science and Technology*, ed. K. D. Timmerhaus and M. S. Barber, Plenum Press, New York, 1979, Vol. 1, p. 748.
29. D. L. Decker, *J. Appl. Phys.* 42, 3239 (1971).
30. E. F. Skelton, A. W. Webb, S. B. Qadri, S. A. Wolf, R. C. Lacoe, J. L. Feldman, W. T. Elam, E. R. Carpenter, Jr., and C. Y. Huang, *Rev. Sci. Instrum.* 55, 849 (1984).
31. M. R. Dietrich, W. H. Fietz, J. Ecke, and C. Politis, *Jpn. J. Appl. Phys.* 26, 1113 (1987).
32. S. Block, G. J. Piermarini, R. G. Munro, and W. Wong-Ng, *Adv. Ceram. Mater.* 2, 601 (1987).
33. J. Staun-Olsen, S. Steenstrup, L. Gerward, and B. Sundqvist, *Phys. Scr.* 44, 211 (1991).
34. N. E. Moulton, S. B. Qadri, A. W. Webb, and E. F. Skelton, to be published.
35. E. F. Skelton, W. T. Elam, M. S. Osofsky, P. Lubitz, M. Z. Harford, and A. K. Singh, *Phys. Rev.* 39, 2779 (1989).
36. W. H. Fietz, C. A. Wassilew, H. A. Ludwig, B. Obst, C. Politis, M. R. Dietrich, and H. Wuhl, *High Pressure Res.* 4, 423 (1990).
37. J. G. Lin, K. Matsuishi, Y. Q. Wang, Y. Y. Xue, P. H. Hor, and C. W. Chu, *Physica C* 175, 627 (1991).
38. S. H. Han, Y. L. Zhang, H. Han, and Y. Liu, *Physica C* 156, 113 (1988).
39. J. E. Schirber, B. Morosin, and D. S. Ginley, *Physica C* 157, 237 (1989).
40. R. Kubiak, K. Westerholt, G. Pelka, H. Bach, and Y. Khan, *Physica C* 166, 523 (1990).
41. R. Sieburger and J. S. Schilling, *Physica C* 173, 403 (1991).
42. N. Môri, H. Takahashi, Y. Shimakawa, T. Manako, and Y. Kubo, *J. Phys. Soc. Jpn.* 59, 3839 (1990).
43. Z. Liu, D. Zheng, Y. Yan, J. Wang, L. Wang, and Z. Zhao, *Chin. Phys. Lett.* 8, 29 (1991).
44. B. Morosin, D. S. Ginley, E. L. Venturini, P. F. Hlava, R. J. Baughman, J. F. Kwak, and J. E. Schirber, *Physica C* 152, 223 (1988).
45. J. Stankowski, B. Czyzak, M. Krupski, J. Baszynski, T. Datta, C. Almasan, Z. Z. Sheng, and A. M. Hermann, *Physica C* 160, 170 (1989).
46. M. Lin, W. Tang, X. Meng, Z. Lin, W. Han, Q. Tu, Y. Ren, and Z. Liu, *Phys. Rev.* 41, 2517 (1990).
47. J. L. Zhang, C. G. Cui, S. L. Li, Y. L. Zhang, X. R. Cheng, and Q. S. Yang, *Mod. Phys. Lett.* 2, 879 (1988).
48. C. Murayama, N. Môri, S. Yomo, H. Takagi, S. Uchida, and Y. Tokura, *Nature* 339, 293 (1989).
49. R. Griessen, *Phys. Rev. B* 36, 5284 (1987).
50. J. T. Markert, J. Beille, J. J. Neumeier, E. A. Early, C. L. Seaman, T. Moran, and M. B. Maple, *Phys. Rev. Lett.* 64, 80 (1990).
51. J. G. Lin, K. Matsuishi, Y. Q. Wang, Y. Y. Xue, P. H. Hor, and C. W. Chu, *Physica C* 175, 627 (1991).
52. J. Diederichs, W. Reith, B. Sundqvist, J. Niska, K. E. Easterling, and J. S. Schilling, *Supercond. Sci. Technol.* 4, S97 (1991).
53. C. Allgeier, R. Sieburger, H. Gossner, J. Diederichs, H. Neumaier, W. Reith, P. Müller, and J. S. Schilling, *Physica C* 162–164, 741 (1989).
54. I. V. Berman, N. B. Brandt, Y. P. Kurkin, E. A. Naumova, I. L. Romashkina, V. I. Sidorov, A. I. Akimov, V. I. Gapal'skaya, and E. K. Stribut, *JETP Lett.* 49, 769 (1989).

55. V. G. Tissen, *Superconductivity* 4, 223 (1991).
56. N. Môri, C. Murayama, H. Takahashi, H. Kaneko, K. Kawabata, Y. Iye, S. Uchida, H. Takagi, Y. Tokura, Y. Kubo, H. Sasakura, and K. Yamaya, *Physica C* 185–189, 40 (1991).
57. D. D. Berkley, E. F. Skelton, N. E. Moulton, and D. H. Liebenberg, *Phys. Rev. B* (1993), to be published.
58. S. Klotz, W. Reith, and J. S. Schilling, *Physica C* 172, 423 (1991).
59. R. J. Wijngaarden, J. J. Scholtz, and E. N. Griessen, in *Frontiers of High Pressure Research*, ed. H. D. Hochheimer and E. D. Etters, Plenum Press, New York, 1991.
60. E. A. Alekseeva, I. V. Berman, N. B. Brandt, A. A. Zhukov, I. L. Romashkina, and V. I. Sidorov, *JETP Lett.* 51, 467 (1990).
61. R. Sieburger, P. Müller, and J. S. Schilling, *Physica C* 181, 335 (1991).
62. C. Martin, A. Maignan, J. Provost, C. Michel, M. Hervieu, R. Tournier, and B. Raveau, *Physica C* 168, 8 (1990).
63. Y. Shimakawa, Y. Kubo, T. Manako, and H. Igarashi, *Phys. Rev. B* 40, 11400 (1989).
64. J. B. Torrance, Y. Tokura, A. I. Nazzal, A. Bezinge, T. C. Huang, and S. S. P. Parkin, *Phys. Rev. Lett.* 61, 1127 (1988).
65. I. D. Parker and R. H. Friend, *J. Phys. C Solid State Phys.* 21, L345 (1988).
66. N. Tanahashi, Y. Iye, T. Tamegai, C. Murayama, N. Môri, S. Yomo, N. Okazaki, and K. Kitazawa, *Jpn. J. Appl. Phys.* 28, L762 (1989).
67. J. D. Jorgensen, S. Pei, P. Lightfoot, H. Shi, A. P. Parilla, and B. W. Veal, *Physica C* 167, 571 (1991).
68. J. J. Neumeier, T. Bjørnholm, M. B. Maple, and I. K. Schuller, *Phys. Rev. Lett.* 63, 2516 (1989).
69. J. S. Schilling and S. Klotz, in *Physical Properties of High Temperature Superconductors*, Vol. III, ed. D. M. Ginsberg, World Scientific, Singapore, 1992.
70. V. Kresin and H. Morawitz, *Solid State Commun.* 74, 1203 (1990).
71. V. Z. Kresin and S. A. Wolf, *J. Supercond.* 1, 143 (1988).
72. S. L. Bud'ko, O. Nakamura, J. Guimpel, M. B. Mapel, and I. L. Schuler, *Physica C* 185–189, 1947 (1991).
73. W. Schnelle, O. Hoffels, E. Braun, H. Broicher, and D. Wohlleneben, in *Proc. NATO Advanced Study Institute: Physics and Materials Science of High Temperature Superconductors II*, ed. R. Kossowsky, B. Raveau, and S. Patapis, Kluwer Academic Press, Dordrect, The Netherlands, 1992.
74. R. J. Wijngaarden and R. Griessen, in *Studies of High Temperature Superconductors*, ed. A. Narlikar, Nova Science, Commack, N.Y., 1988, p. 29.

22
Specific Heat of Thallium Oxide Superconductors

Robert A. Fisher and Norman E. Phillips
Lawrence Berkeley Laboratory
University of California at Berkeley
Berkeley, California

Joel E. Gordon
Amherst College
Amherst, Massachusetts
and Lawrence Berkeley Laboratory
University of California at Berkeley, Berkeley California

I. INTRODUCTION

The discovery of superconductivity in the Tl–Ba–Ca–Cu–O (TBCCO) system [1] generated intense interest and led to numerous structural studies and measurements of physical properties on these compounds. Interpretations of the measurements are in most cases complicated by the evident occurrence, or at least the possibility of the occurrence, of more than one phase in the samples. Furthermore, even ignoring that complication, the specific-heat measurements have been neither detailed nor extensive enough to establish the types of correlations among sample-dependent parameters that have been recognized from the much more extensive data on $YBa_2Cu_3O_{7-\delta}$ (YBCO) and which have provided a basis for the current understanding of the specific heat of that material. For these reasons, it is not useful to discuss specific-heat measurements on the Tl–O superconductors except in the context of what is known about other high-T_c superconductors (HTSCs). The measurements on YBCO are more complete and better understood than those for any other high-T_c superconductor, which suggests a discussion of the specific heat of the Tl–O superconductors based on a comparison with measurements on YBCO and their interpretation. Such an approach is not entirely satisfactory, both because the current understanding of the specific heat of YBCO is itself incomplete—and based to a significant degree on comparisons with the specific heat of conventional superconductors—and because the complications associated with oxygen stoichiometry ($\delta \neq 0$) which are attendant on the occurrence of CuO chains, as well as the CuO_2 planes, in YBCO may have no counterpart in the Tl–O superconductors. Nevertheless, this seems to be the most satisfactory approach, and it is the one taken here.

This chapter is organized in six sections. Section II is a brief description of the information obtained from specific-heat measurements on conventional superconductors, and Section III is a summary of the measurements on YBCO and their interpretation. Since there is a natural division of the specific-heat-derived results into "low-temperature" ($1 \leq T \leq 10$ K) results and results near T_c, which is reflected in the discussion in Section III, the low-temperature measurements and measurements near T_c in Tl–O superconductors are presented in Sections IV and V, respectively. Section VI includes additional discussion of the parameters derived from specific-heat data for the Tl–O superconductors and their relation to the corresponding parameters for other high-T_c superconductors, the electron density of states at the Fermi surface, the phonon spectrum, and the strength of the coupling.

II. SPECIFIC HEAT OF CONVENTIONAL SUPERCONDUCTORS

The two components of the specific heat, C, common to all materials containing conduction electrons are the lattice and the conduction electron contributions, C_l and C_e, respectively. C_l is generally taken to be independent of applied magnetic field, H, and the same in the normal and superconducting states. (In fact, for HTSCs there are indications, both theoretical and experimental, that differences in the phonon spectrum, and therefore in C_l, between the normal and superconducting states should be taken into account. However, the effects are small, still not well defined, and not relevant to the present state of the data for the Tl–O superconductors or their interpretation. This complication will not be considered in the following.) At low temperatures,

$$C_l = B_3 T^3 + B_5 T^5 + B_7 T^7 + \cdots \tag{1}$$

where

$$B_3 = \tfrac{12}{5}\pi^4 R \Theta_0^{-3} \tag{2}$$

and Θ_0 is the Debye characteristic temperature in the limit of zero temperature. The higher-order terms in Eq. (1) give some information about phonon dispersion, but they are not sensitive to the details of the phonon spectrum. They are frequently replaced by Einstein terms that are interpreted in terms of singularities in the phonon spectrum, but in fact the difference in C_l from that associated with the peaks in the phonon spectrum produced by more-or-less normal phonon dispersion is usually not meaningful.

In the normal state (i.e., for $H > H_{c2}$, the upper critical field) C_e is

$$C_{en} = \gamma_T \tag{3}$$

with

$$\gamma = \tfrac{1}{3}\pi^2 k_B^2 N(E_F) \tag{4}$$

where $N(E_F)$ is the density of electron states at the Fermi energy for both spin directions. Band-structure calculations give the band structure or "bare" density of states $N_{\text{bs}}(E_{\text{F}})$, which is related to $N(E_{\text{F}})$ by

$$N(E_{\text{F}}) = (1 + \lambda)N_{\text{bs}}(E_{\text{F}}) \tag{5}$$

where λ represents the electron–phonon enhancement. For $H = 0$ (i.e., in the superconducting state) the temperature dependence of C_e is changed dramatically by the development of the gap in the electron density of states at E_{F} and becomes

$$C_{\text{es}} \approx a\gamma T_c \exp\left(\frac{-bT_c}{T}\right) \tag{6}$$

At T_c there is a discontinuity in C_e, $\Delta C(T_c)$, that is directly related to the temperature dependence of the gap at T_c, and with the assumption of a model for the temperature dependence of the gap over the entire interval to 0 K, to the 0-K gap, $2\Delta_0$. The temperature dependencies of C_{en} and C_{es} are illustrated as C_e/T versus T in Fig. 1. The solid curve for C_{es} corresponds to the weak-coupling BCS case, in which $2\Delta_0 = 3.53k_{\text{B}}T_c$. The dashed curve corresponds to a strong-coupling case, represented here by the α model. In that model [2], which represents C_{es} quite well for a number of conventional superconductors, the BCS temperature dependence of the gap is preserved, but the ratio $\alpha = 2\Delta_0/k_{\text{B}}T_c$ is taken as an adjustable parameter. For the dashed curve $\alpha = 3.0$ and $\Delta C(T_c)/\gamma T_c = 4.13$: to be compared with the weak-coupling values, 1.764 and 1.43, respectively. For conventional superconductors, the highest known values are $\alpha \approx 2.4$ and $\Delta C(T_c)/\gamma T_c \approx 2.8$, but the still higher values represented by the dashed curve may be relevant for YBCO. For very strong coupling, the shape of the specific-heat anomaly at T_c is

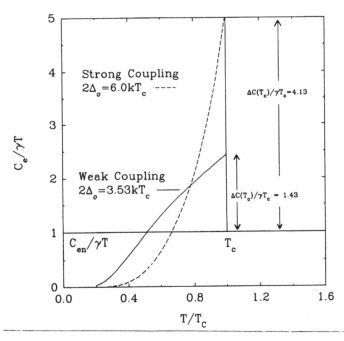

Figure 1 Electron contribution to the normal and superconducting-state specific heats for conventional superconductors. See the text for details.

qualitatively different: The ratio $(dC_{es}/dT)_{Tc}/\gamma$ is greater, and the curvature of C_{es}/T is positive.

For most conventional superconductors it is quite practical to measure C for $H > H_{c2}$ and to analyze the data with Eqs. (1) and (3) to obtain reliable values of the coefficients in those equations. The values of C_l so determined can be used to subtract that contribution from the C measured for $H = 0$, and thus to obtain C_{es}. In many cases C_l is small enough relative to C_e, even at T_c, that the features illustrated in Fig. 1, including the sharp (limited only by temperature resolution) discontinuity at T_c, have been well determined. Thus specific-heat measurements on conventional superconductors have given: for $H > H_{c2}$, the value of $N(E_F)$, which is of fundamental importance in understanding the properties of both the normal and superconducting states, the value of Θ_0, which defines the phonon spectrum in the long-wavelength limit, and the higher-order coefficients in Eq. (1), which give some (limited) information about phonon dispersion; for $H = 0$, the temperature dependence of C_{es} and the value of $\Delta C(T_c)$, which give information about the strength of the electron–electron coupling that produces the transition to the superconducting state.

III.   GENERAL FEATURES OF THE SPECIFIC HEAT OF HTSCs

In this description of the general features of the specific heat of HTSCs, most features are illustrated by results for YBCO. Only a few references to original work are given; other references, and more detail, can be found in a recent review [3].

A.   Overview

The total specific heat (including C_l) of a reasonably representative polycrystalline sample of YBCO, for $H = 0$, is shown in Fig. 2 as C/T versus T. In this case C_l is so large relative to C_e that the specific-heat "anomaly" at T_c is only about 3% of the total. This is typical of HTSCs and limits the accuracy with which data near T_c can be analyzed to obtain detailed information about C_{es}. Furthermore, because the values of H_{c2} are so high, on the order of 100 T, data for $H > H_{c2}$ and their analysis into C_{en} and C_l are not available. Thus the direct determination of $N(E_F)$ that has been made routinely for conventional superconductors has not been made for HTSCs. On the other hand, since C_l is the dominant contribution above a few Kelvin, it has been possible to obtain reasonably accurate estimates of C_l, and representative results for several HTSCs are shown in Fig. 3 as C_l/T^3 versus $\log T$.

B.   Low-Temperature Specific Heat

There are two contributions to the low-temperature specific heat that set HTSCs apart from conventional superconductors, and that are qualitatively apparent in the low-temperature inset to Fig. 2. One is a low-temperature "upturn" in C/T. As shown by its magnetic field dependence in fields of a few tesla—it becomes a Schottky anomaly with H and T dependencies expected for Cu^{2+} magnetic moments—it is associated with Cu^{2+} moments that for $H = 0$, order in the vicinity of 0.1 K under the influence of internal interactions. For $H = 0$ and $T \geq 0.4$ K, as in Fig. 2, it is only the high-temperature tail of the $H = 0$ anomaly that is observed. In fields of a few tesla, the applied field is large enough, relative to the internal fields,

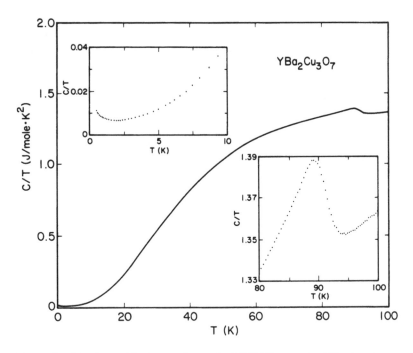

Figure 2 Total specific heat of a typical YBCO sample. The insets are expanded representations of the data at low temperatures and in the vicinity of T_c.

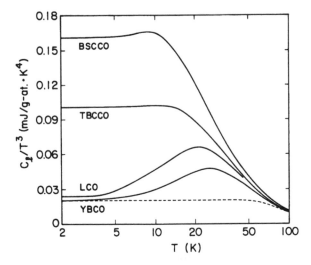

Figure 3 Lattice specific heats of YBCO, LCO, TBCCO, and BSCCO. The dashed curve represents the Debye specific heat function for $\Theta_0 = 450$ K, the value for a particular YBCO sample.

that a Schottky anomaly is a good approximation, and the concentration of Cu^{2+} moments associated with this feature, n_2, is determined by the in-field data (see below). Because the distribution of internal fields broadens the anomaly relative to a Schottky anomaly, the high-temperature "tail" is not well represented by the T^{-2} dependence characteristic of a magnetic Schottky anomaly and requires additional terms in negative powers of T. Since certain superconducting-state properties are correlated with n_2, it is clear that these moments are located, at least predominantly, in the YBCO phase. The other is a "linear" term $\gamma^*(H)T$. The field dependence of this term is well represented by

$$\gamma^*(H) = \gamma^*(0) + H\frac{d\gamma^*}{dH} \tag{7}$$

with $d\gamma^*/dH$ independent of H to within experimental uncertainty. The $H = 0$ component [4], $\gamma^*(0)$, received a great deal of attention in the first few years of research on HTSCs both because it has no counterpart in conventional superconductors and because it corresponded to a prediction of early versions of the RVB theory. For YBCO, there is no doubt that $\gamma^*(0)$ includes a contribution associated with impurity phases, $BaCuO_2$ in particular [5]. The total concentration of Cu^{2+} moments in a sample, n, which includes those in impurity phases such as $BaCuO_2$, n_1, as well as those in the YBCO lattice, n_2, can be determined from the Curie–Weiss term in the high-temperature magnetic susceptibility. Thus n_1 and n_2 are both determined. To within an accuracy of the order of ± 1 mJ/mol · K^2, the value of $\gamma^*(0)$ is well represented by a sum of two terms, one proportional to n_1 and the other proportional to n_2. This suggests that a part of $\gamma^*(0)$ is associated with impurity phases and the rest is associated with nonsuperconducting regions of the samples that are present in an amount proportional to n_2 and associated with the Cu^{2+} moments that produce the upturn [6]. Thus, at least in this case, any intrinsic contribution to $\gamma^*(0)$ associated with the superconducting state is ≤ 1 mJ/mol · K^2. The H-dependent part of $\gamma^*(H)$, $H\,d\gamma^*/dH$, however, is similar to a contribution that is well known in conventional type II superconductors and is associated with increasing flux penetration for $H_{c1} < H < H_{c2}$.

The analysis of low-temperature data for YBCO into the components C_l, $\gamma^*(H)T$, and C_m, the contribution associated with Cu^{2+} moments, is illustrated in Fig. 4 for $H = 0$ and 7 T. In this figure another contribution, not mentioned above, but well known in other materials, is also apparent: For $H = 7$ T there is an upturn in C/T associated with the interaction of nuclear magnetic moments with H. This contribution, C_h, is readily distinguishable from the $H = 0$ upturn associated with Cu^{2+} electronic moments by its magnitude and also by its H and T dependencies—for the temperatures of interest here, it is accurately proportional to $(H/T)^2$.

C. Specific-Heat Anomaly at T_c

The normal/superconducting transition is substantially broadened in HTSCs relative to that in many conventional superconductors. There are no doubt two contributions to the breadth of the transition as observed in the specific heat, both related to the short value of the coherence length, ξ, although otherwise quite different in origin. One is simply sample inhomogeneity—fluctuations in concentration in the case of solid solutions or in oxygen stoichiometry in YBCO or other atomic-

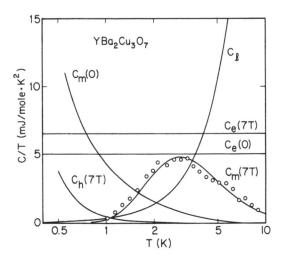

Figure 4 Analysis of the low-temperature specific heat of a YBCO sample into four contributions, for $H = 0$ and 7 T. The points represent experimental data from which three of the contributions have been subtracted to illustrate the accuracy of the determination of the fourth contribution—that associated with Cu^{2+} magnetic moments.

scale defects which are typical of oxides such as the HTSCs. In HTSCs with ξ on the order of the lattice parameter, these defects can produce regions differing in the superconducting properties, whereas the greater values of ξ in conventional superconductors have an averaging effect. The small values of ξ also produce an intrinsic broadening of the transition by fluctuation effects that is orders of magnitude larger than in conventional superconductors. In the latter case fluctuation broadening is too small to be observed, but in HTSCs, although the magnitude of the effect is sample dependent (at least in YBCO), there is good evidence for its existence. The example in the second inset in Fig. 2 shows a reasonably sharp transition for a polycrystalline YBCO sample; sharper transitions have been observed in a few polycrystalline samples; but generally the sharpest transitions have been observed in single crystals, for which analysis of the data for fluctuation effects is most convincing [7]. An example of fluctuations in a polycrystalline sample is shown in Fig. 5.

For YBCO it is clear that the shape of the specific-heat anomaly at T_c is influenced by sample inhomogeneities, by strong-coupling effects, and by fluctuations. Furthermore, analysis of the shape of the anomaly is always complicated by uncertainty in the subtraction of the "background" or lattice specific heat. In principle all of these effects can be represented by analytical expressions, but there are different possibilities, among which somewhat arbitrary choices have to be made, and the results of the analysis are not uniquely determined by the data. For example, positive curvature in C/T below T_c may be associated with strong-coupling effects or fluctuations; positive curvature just above T_c with fluctuations or sample inhomogeneities. Nevertheless, for a particular sample, quite consistent values of the mean-field contribution to the anomaly, the discontinuity $\Delta C(T_c)$, are obtained by different reasonable approaches. Figure 6 provides an example: The dashed straight-line representation is a simple entropy-conserving construction that gives

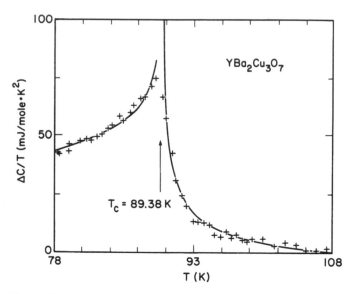

Figure 5 Specific heat of polycrystalline YBCO near T_c. $\Delta C = C(\text{obs}) - C(\text{fit})$ where $C(\text{fit})$ is the sum of a term proportional to temperature and a harmonic lattice contribution. The curves represent the sum of BCS and three-dimensional Gaussian fluctuation contributions.

one value of $\Delta C(T_c)$; the dash-dot curve is a different extrapolation of the high-temperature data based on a harmonic-lattice fit to the data extending to 280 K which gives a value of $\Delta C(T_c)$ that is only slightly different; the solid curve is a fit with an expression that takes strong-coupling effects into account, allows for a Gaussian distribution of T_cs, and gives essentially the same value for $\Delta C(T_c)$. Similar results have been obtained when fluctuation effects have been included.

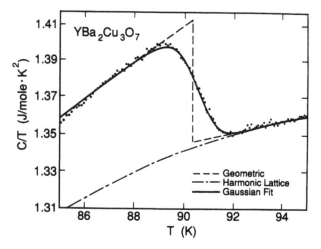

Figure 6 Specific-heat data for YBCO near T_c, with several constructions used to determine $\Delta C(T_c)$ as described in the text.

Measurements on series of different YBCO samples in two different laboratories [6,8] have shown, however, that the values of $\Delta C(T_c)$ do show a strong sample-to-sample dependence, varying by factors of more than 2. In both cases this variation was shown to be correlated with sample-to-sample variations in the low-temperature specific heat, specifically the low-temperature upturn in C/T and $\gamma^*(0)$ (although the actual parameters on which the correlations were based were different in the two cases).

D. Interpretation of Sample-to-Sample Variations of the Specific Heat of YBCO as Reflecting Varying Volume Fractions of Superconductivity

$\Delta C(T_c)$ is one of three sample-dependent parameters derived from specific-heat data that might be expected to reflect sample-to-sample variations in the volume fraction of superconductivity, f_s, in fact to be proportional to f_s. The other two are $d\gamma^*/dH$ [see Eq. (7)] and ΔS, an entropy change associated with the effect of magnetic field on the specific-heat anomaly at T_c that is defined in detail in Ref. 9. In fact, all three of these parameters are mutually proportional with respect to sample dependence, and they can be used to assign relative values of f_s [9]. The f_s scale has been made absolute on the basis of the observation that the same three parameters decrease approximately linearly with increasing n_2, the concentration of Cu^{2+} magnetic moments. This was taken as an indication that the Cu^{2+} moments are themselves, or are associated with, defects that suppress the superconducting transition and produce nonsuperconducting regions. The limit $n_2 = 0$ was taken as determining $f_s = 1$ and the values of $\Delta C(T_c)$, $d\gamma^*/dH$, and ΔS characteristic of the ideal, fully superconducting state. For example [6], for $f_s = 1$, $\Delta C(T_c)/T_c = 77$ mJ/mol · K^2. [Similar conclusions, in that case for $\Delta C(T_c)/T_c$ alone, were reached on the basis of a related criterion and reported in Ref. 8.]

IV. LOW-TEMPERATURE SPECIFIC HEAT OF TBCCO

Specific heats at low temperatures have been measured in the TBCCO system for the compositions Tl-2201, Tl-2212, and Tl-2223. (This notation is used to distinguish the Tl–O superconductors from the corresponding phases of the Bi–O superconductors, and to specify the mole numbers of Tl, Ba, Ca, and Cu, by the four digits, in that order.) Table 1 summarizes parameters characterizing these samples and their low-temperature specific heat, and gives references to the original work. The Tl-2201 and Tl-2212 materials were probably predominantly single phase. However, the Tl-2223 phases are best synthesized by starting with an off-stoichiometric mixture, and consequently, the Tl-2223 polycrystalline samples contained some or all of the possible additional phases Tl_2O_3, BaO, $BaCuO_2$, CuO, CaO, and other TBCCO compounds (e.g., Tl-1212, Tl-1223, Tl-2201, and Tl-2212).

Figure 7 is a plot of C/T versus T^2 for sample 7, a polycrystalline, mixed-phase sample (Tl-2212 + Tl-2223) for $H = 0$ and 7 T. These were possibly the first low-temperature specific-heat measurements on TBCCO. (Samples 5, 6, and 7 seem to be the only samples that have been measured in magnetic fields.) In Fig. 8, C/T versus T^2 is compared for a single crystal and a polycrystalline aggregate of Tl-2223, samples 1 and 2, respectively, while Fig. 9 is a similar plot for Tl-2212, and Tl-2201, samples 8 and 9, respectively.

Table 1 Parameters Characterizing the Low-Temperature Properties of TBCCO[a]

Sample	Starting composition	Superconducting phase	T_c (K)	$-4\pi\chi_v$	n	n_2	$\gamma^*(0)$	θ_0	Ref.
1	Tl-2112[b]	Tl-2223	105	—	—	0.0004	<1	270	10
2	Tl-2112[b]	Tl-2223	—	—	—	—	5	280	10
3	—[c]	Tl-2223	114	—	—	—	63	(290)	11
4	Tl-1133[d]	Tl-2223	122	0.24	0.03	—	—	290	12
5	Tl-3223[e]	Tl-2223	120	0.50	—	0.01	28	270	14
6	Tl-3223[e]	Tl-2223	112	0.40	—	0.01	~0	260	14
7	Tl-1112[f]	Tl-2223 and Tl-2212	113	0.30	—	0.025	16	268	13
8	Tl-2212[g]	Tl-2212	99	0.51	0.05	—	<6.8	254	12
9	Tl-2201[g]	Tl-2201	<12	0.08	0.03	—	0	238	12

[a]Dashes signify that either those parameters were not measured or were not available. n and n_2 are the ratio of the Cu^{2+} localized magnetic moments to the total Cu content of the starting composition except for sample 1, where the ratio is with respect to the Cu content of Tl-2223. n was derived from Curie–Weiss law fits to magnetic-susceptibility data above T_c, and n_2 was derived from an analysis of the low-temperature specific-heat measurements (see the text). $\gamma^*(0)$ has the units of mJ/mol·K^2, where the molecular weight and the number of atoms was calculated from the stoichiometry of the starting composition for samples 4 and 7, and for the superconducting phase stoichiometry for the others. Values of θ_0 and $\gamma^*(0)$ in parentheses were scaled from a graph of C/T versus T^2 for sample 3. Since there may be flux pinning, and because of the possible presence of nonsuperconducting phases, the Meissner fraction $(-4\pi\chi_v)$ is a lower limit on f_s.
[b]The starting composition of the melt was Tl-2112. For sample 1, a single crystal of Tl-2223 was separated from the solidified melt, and a polycrystalline mass of Tl-2223, sample 2, was broken from it. The authors state that a small amount of Tl-2212 was probably present in sample 2.
[c]Preparation details were not stated and no indication was given for the presence or absence of other phases.
[d]Since the starting composition was Tl-1113, the Tl-2223 sample was multiphase. The other phases could not be identified with certainty but were consistent with CuO, $BaCuO_2$, CaO, and Tl_2O_3. The phases Tl-1212, Tl-1223, Tl-2201, and Tl-2212 were not detected. All analysis was by x-ray.
[e]Since the starting composition was Tl-3223, the Tl-2223 samples were multiphase. The Tl-2223 phase concentration was given as 90% from x-ray analysis. The authors state that sample 6 was "structurally more perfect" than sample 5, and they speculate that the structural imperfection may be related to a finite $\gamma^*(0)$. Specific-heat measurements were made in the magnetic field range $0 \leq H \leq 16$ T.
[f]The starting composition of Tl-1112 resulted in a multiphase sample of approximately equal amounts of Tl-2223 and Tl-2212. CuO was also detected. The analysis was by x-ray.
[g]This sample was stated to be single phase on the basis of an x-ray analysis.

The specific heat displayed in Fig. 7 is similar to that observed for YBCO: a low-temperature upturn in C/T at $H = 0$ that becomes a Schottky anomaly in 7 T, and a finite $\gamma^*(0)$. All of the low-temperature specific-heat measurements on TBCCO show a low-temperature upturn in C/T, but the upturn for the single crystal (see Fig. 3) is much smaller than that for the polycrystalline samples. These results are similar to those observed for single-crystal [10] and polycrystalline [13] BSCCO samples, which have the same structure. In the TBCCO system $\gamma^*(0)$ ranges from 0 to 63 mJ/mol · K^2.

Since Cu is the only potentially magnetic element present in TBCCO, the C/T upturn can be assumed to be associated with localized Cu^{2+} electronic magnetic moments, as it is in YBCO. Furthermore, as in YBCO, in a magnetic field of 7 T (see Fig. 7) this upturn becomes a Schottky anomaly that can be fitted with $g = 2$ and $S = \frac{1}{2}$ (characteristic of Cu^{2+} at low temperatures). The amplitude of that

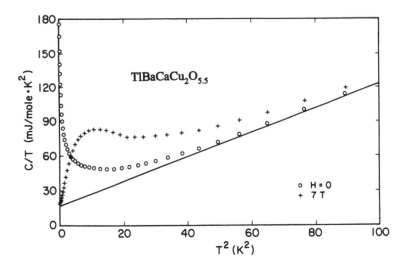

Figure 7 C/T versus T^2 for a polycrystalline sample of a mixture of Tl-2212 + Tl-2223 phases for $T \leq 10$ K and $H = 0$ and 7 T. The straight line represents the T and T^3 terms of the least-squares fit to the zero-field data. (From Ref. 13.)

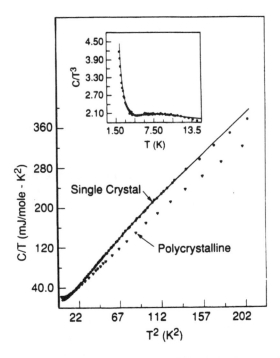

Figure 8 C/T versus T^2 for Tl-2223 single-crystal and polycrystalline samples. The curve represents the best fit of the single-crystal data in the range $3.5 \leq T \leq 11$ K—see the text for details. The inset shows C/T^3 versus T for the single crystal with the same fit. (From Ref. 10.)

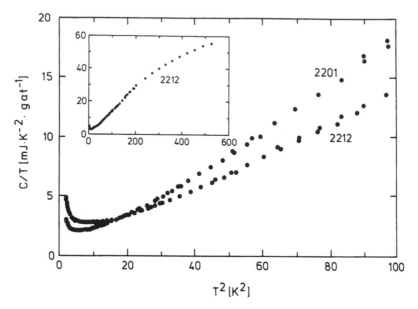

Figure 9 C/T versus T^2 for Tl-2201 and Tl-2212 polycrystalline samples for $T \leq 10$ K. The inset shows the same data for the Tl-2212 sample for $T \leq 25$ K. (From Ref. 12.)

anomaly gives a reliable estimate of the amount of Cu^{2+} present (see n_2 of Table 1). For that sample [13], the amount of Cu^{2+} was about an order of magnitude larger than that found for a typical YBCO sample, but comparable to that of BSSCO polycrystalline samples [13]. Junod et al. [12] have measured the magnetic susceptibility above T_c for TBCCO (samples 4, 8, and 9), and from fitting the data to a Curie–Weiss law they have derived the total Cu^{2+} concentration of magnetic moments (see Table 1). These concentrations are comparable to the value of n_2 for sample 7. The Cu^{2+} magnetic moments in TBCCO samples could be associated either with localized Cu^{2+} moments located in the CuO_2 planes of the TBCCO structure or with impurity phases.

For samples 3, 8, and 9, the $H = 0$ upturns in C/T were fitted with T^{-2} terms; for sample 7, additional terms were required because the data extended to lower temperatures. Urbach et al. [10] interpreted the upturn as being due to spin-glass ordering of Cu^{2+}. They used a term proportional to T^{-1}, which corresponds to the high-temperature limit for spin-glass ordering, to fit the upturn, and calculated n_2 from the coefficient of that term using a spin-glass model and an estimate of the exchange energy based on the antiferromagnetic interactions of La_2CuO_4 and $YBa_2Cu_3O_6$. However, neither the validity of the T^{-1} representation of the upturn nor the applicability of the spin-glass model is clear. The fitting expression was apparently not really adequate to represent the data; it was used only above 3.5 K, where the T^{-1} term contributes only 14% of the total C; and the T^{-1} term drops to 4% of the total at 5 K. It is clear from the magnitude of the upturn that n_2 is small, but the assumptions underlying the estimate leave room for doubt as to its quantitative significance.

The fact that to within experimental error $\gamma^*(0) = 0$ for nearly all BSCCO samples (single crystal or polycrystalline), while $\gamma^*(0)$ is finite for YBCO, LMCO (M = Ca, Sr, Ba) and for some, but not all, TBCCO samples is interesting and deserves further study. For YBCO it has been shown [6] that $\gamma^*(0)$ can be interpreted as arising from a combination of the "linear term" associated [5] with some forms of $BaCuO_2$, present as impurity phases, and the presence of normal metallic material interspersed with the superconducting material, and present in an amount proportional to n_2. For LSrCO, $\gamma^*(0)$ is associated entirely with normal material [15]. [Too few measurements have been made for LBaCO and LCaCO to allow any conclusions to be drawn about the origin of the observed finite $\gamma^*(0)$.] Structurally, the TBCCO and BSCCO compounds are alike; however, one significant difference may be that the former contain Ba, and hence there is the possibility of $BaCuO_2$ phases, which could contribute to $\gamma^*(0)$. This view is supported by the fact that for a single crystal of TBBCO (Tl-2223) [10], which might be expected to have relatively small amounts of $BaCuO_2$ present, $\gamma^*(0) \approx 0$. However, there remains an inconsistency in that for YBCO. A part of $\gamma^*(0)$ seems to be associated with Cu^{2+} moments (n_2), while for many BSCCO samples these moments are present in relatively high concentrations, but $\gamma^*(0)$ is at least small, if not zero. One possibility is that unlike YBCO, BSCCO and TBCCO have no CuO chains in their structure and none of the attendant problems with oxygen stoichiometry. Perhaps the Cu^{2+} moments in YBCO that give rise to a finite $\gamma^*(0)$ are associated with these chain sites and not the CuO_2 planes.

For YBCO [6] and LSrCO [15] there is a correlation between the size of $\Delta C(T_c)/T_c$ (which has been taken as a measure of the fraction of the sample that is superconducting) and that part of $\gamma^*(0)$ not due to impurity phases. At present there are insufficient data for either BSCCO or TBCCO to make such a comparison. (See Section V for a discussion of the specific heat near T_c for TBCCO.)

As shown in Table 1, the limiting low-temperature Debye temperature, Θ_0, decreases as the number of CuO_2 planes in the unit cell decreases. (The variation in Θ_0 for the Tl-2223 samples is undoubtedly due to the presence of other phases— except possibly for the single-crystal sample [10].) For comparable BSCCO and TBCCO structures Θ_0 is about the same. Figure 3 is a plot of C_l/T^3 versus log T for LCO, YBCO, BSCCO (Bi-2212), and TBCCO (Tl-2212 + Tl-2223). For both the YBCO and LCO [similar to LMCO (M = Ca, Sr, Ba)] the T^3 region lies below 5 K and there is a pronounced peak near 20 K that shows the presence of a peak in the phonon density of states near 50 K. For both BSCCO and TBCCO the T^3 region extends about a factor of 2 higher in temperature with only a small peak in C_l/T^3 that occurs just below the rapid decrease of C_l/T^3 at higher temperatures. The lattices for both BSCCO and TBCCO, with Θ_0 in the range 250 to 290 K, are much softer than those of LCO and YBCO, for which Θ_0 values range from 450 to 500 K.

A hyperfine specific heat has been observed only in the case of sample 7. C_h was much smaller than expected for the combined Cu and Tl contributions. The size of the measured C_h suggests that the hyperfine contribution is due to the Tl nuclear moments and therefore that the Cu nuclear moments are not relaxing sufficiently rapidly to contribute. These results are similar to those found for the BSCCO compounds [13].

V. SPECIFIC-HEAT ANOMALY AT T_c for TBCCO

The early specific-heat measurements on YBCO provided evidence of a bulk superconducting transition near 90 K, but the observed anomalies were relatively broad, with heights considerably lower than those observed in later experiments. As sample quality improved, the YBCO anomalies became higher, narrower, and sufficiently well characterized to provide clear evidence that a simple weak-coupling BCS picture is not adequate for a full understanding of the superconducting phase transition. That such a definitive view is not yet possible in the case of the superconducting Tl–O superconductors (or, for that matter, in the case of other HTSCs) is made clear in Fig. 10, where some of the recent measurements on TBCCO and BSCCO are compared with those on YBCO [16]. These specific-heat data, unlike some of the earlier Tl–O and Bi–O HTSC results (see, e.g., Refs. 17 and 18) do show departures from the "background" (lattice plus normal electron) specific heat. Nevertheless, the striking difference between the Tl-2212 and Tl-2223 anomalies, on the one hand, and that of YBCO, on the other, strongly suggests that much work on sample preparation remains to be done before interpretations of the Tl specific-heat data near T_c can be made with confidence. To be sure, band-structure calculations (see Section VI) suggest that the bare density of states, $N_{bs}(E_F)$, for the two Tl HTSCs are only about half as large as for YBCO. For that reason it is possible that the specific-heat anomalies in the Tl-2212 and Tl-2223 compounds will never be as large, and therefore not as sharply defined, as that in YBCO. Nevertheless, the experience with YBCO sample preparation, and the improvement already made in producing single-phase Tl superconductors, give hope that the next few years will help to clarify what is presently a confusing picture of super-

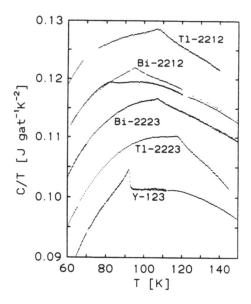

Figure 10 C/T versus T in the vicinity of T_c for polycrystalline samples of YBCO, Bi-2212, Bi-2223, Tl-2212, and Tl-2223. The data for the Bi-2223 and Y-123 are shifted by 0.005 J/g-at · K^2 upward and downward, respectively. The higher T_c for the Bi-2212 sample resulted after it was heat treated in an argon atmosphere. (From Ref. 23.)

conductivity in these materials. In fact, some very recent results [19] indicate that the anomaly in Tl-2223 may be at least as pronounced as that in YBCO, and that therefore the similarities between this HTSC and YBCO may be greater than Fig. 10 suggests.

The specific heats of $Tl_2Ba_2Ca_{n-1}Cu_nO_{2n+4}$ for $n = 1$, 2, and 3 have been measured [12]. Tl-2201 was found to show no calorimetric evidence of a superconducting transition, although there was evidence from ac susceptibility data of an incomplete superconducting transition that began just above 12 K. Figure 10 shows that the specific heats of both Tl-2212 and Tl-2223 provide evidence for superconducting transitions at temperatures above 100 K. A sample [13] that contained roughly equal parts of Tl-2212 and Tl-2223 showed a broad anomaly that began at ≈113 K. These data are shown in Fig. 11, together with data obtained in a 7-T magnetic field. The difference between the zero-field and 7-T data are also shown as the dashed line in Fig. 12, where they are compared with $\Delta C = C(0) - C(5\,T)$ as computed from reversible magnetization [20] data on Tl-2223. Fisher et al. [13] compare their results with similar data obtained on YBCO and estimate that the observed differences between the zero- and the in-field results are likely to be two to three times smaller than would be expected for a fully superconducting sample.

The specific heat of Tl-2212 has been measured from low temperatures up to 300 K by two groups [12,21,22], and in the region 60 to 140 K by several others [17–19,23]. Atake et al. [21,22] measured C for three different Tl-2212 samples in which the number of Tl atoms was 2.10, 1.94, and 1.82 (samples a, b, and c, respectively). They found values for $\Delta C(T_c)$ and T_c of 2.6, 2.8, and 2.2 J/mol · K and 104, 96, and 89 K, respectively. In the case of sample a, they used the technique described by Sharifi et al. [24] for analyzing the specific heat near T_c. This technique, similar to that employed by Gordon et al. [25], is based on the assumption that

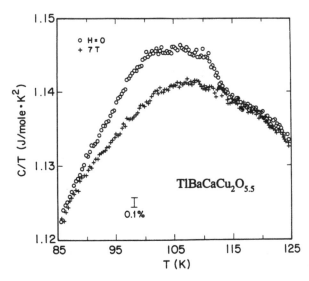

Figure 11 C/T versus T for a polycrystalline sample of a mixture of Tl-2212 + Tl-2223 phases in the region of T_c for $H = 0$ and 7 T. (From Ref. 13.)

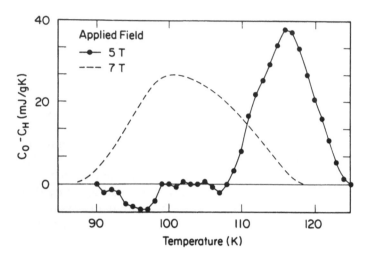

Figure 12 Comparison of the effects of magnetic field on the specific heat derived from magnetization measurements for polycrystalline Tl-2223 (solid circles) and from direct measurements [13] for a mixed phase polycrystalline sample of Tl-2212 + Tl-2223 (dashed curve). (From Ref. 20.)

the measured C is the sum of a lattice term and an electronic term that has a Gaussian distribution in T_c. To account for the possibility of strong coupling and for a fluctuation contribution, Atake et al. [21] approximate their data with an adjustable eight-parameter fit in which the fluctuation term is assumed to be due to three-dimensional Gaussian fluctuations, and in which the mean-field part of the anomaly is assumed to vary as $|1 - T/T_c|$. Figure 13 shows their data after the computed "background" (lattice plus normal electron) specific heat has been subtracted off. The smooth curve in the figure is their fit to the data. From this fit they find that the sample has a spread in T_c of ~ ± 7 K, a mean coherence length ξ ~ 5.4 Å, and a γ of 21 mJ/mol · K^2. The fit also indicates BCS weak coupling. The authors rightly caution, however, that a sample with a narrower distribution in T_c might yield different results.

The data of Wohlleben et al. [23] and Braun et al. [16] on Tl-2212 near T_c (see Fig. 10) are quite similar to those reported by Atake et al. [21,22] and by Junod et al. [12]. They differ, however, from earlier results [17,18] that showed no cusp in the specific-heat data but did have discontinuities in the slopes of the C/T versus T data. Seidler et al. [18] suggested that these results could imply the existence of a third-order phase transition in the Tl and Bi oxide superconductors; however, their own later data (as well as the results of the other groups) indicate that no such hypothesis need be entertained. Wohlleben et al. [23] and Braun et al. [16] obtain a fit to the "background" specific heat by fitting the data well above and below T_c, and then subtract the "best" background fit from the data in the vicinity of T_c. The authors represent the remaining ΔC as the sum of a BCS mean-field specific heat plus a fluctuation term. The former authors [23] assume that this term arises solely from critical fluctuations, $C_{fl} \propto \log |1 - T/T_c|$, whereas Braun et al. [16] assume the fluctuations to have this form only within ± 5 K of T_c, and to have the $|1 - T/T_c|^{-1}$ dependence characteristic of two-dimensional Gaussian

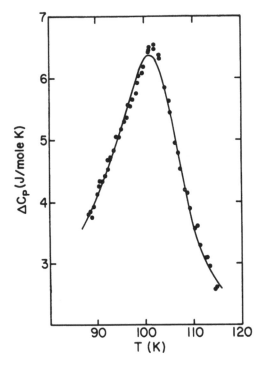

Figure 13 Electronic specific heat of Tl-2122 (Tl = 2.1). The solid curve is a result of a least-squares fit—see the text for details. (From Ref. 21.)

fluctuations for $|T - T_c| > 5$ (as opposed to the $|1 - T/T_c|^{-1/2}$ dependence of the three-dimensional Gaussian fluctuations assumed by Atake et al. [21]). Figure 14a shows the ΔC data near T_c as well as the fit corresponding to an anomaly made up of mean-field and two-dimensional Gaussian fluctuation components. These authors [16] obtain a value for γ of 3.7 mJ/mol · K², while for the same data, Wohlleben et al. [23] obtain 12 mJ/mol · K². Both values are considerably lower than the value of 21 mJ/mol · K² obtained by Atake et al. [21], a result that surely reflects the different assumptions regarding the fluctuation term as well as possible differences in the fitting procedures and in sample composition.

Several groups [12,14,16,19,23,26] have reported observations of an anomaly in the specific heat of Tl-2223. The techniques for making this Tl–O superconductor differ among the various groups (the data reported by Wohlleben et al. [23] and by Braun et al. [16] were taken on the same samples). Junod et al. [12] found x-ray evidence for nonsuperconducting phase impurities. Panova et al. [14] report that their samples were at least 90% Tl-2223, while Braun et al. [16] and Bandyopadhyay [19] report that x-ray analysis indicated a sample with the Tl-2223 structure. The specific-heat measurements of Gavrichev et al. [26] showed anomalies at 97 and 125 K and a break in the slope of C/T versus T at 110 K. This break indicates the presence of Tl-2122 in the sample, while the two anomalies are evidence for the existence of both Tl-1122 and Tl-2223 phases in this nominal Tl-2223 material.

Wohlleben et al. [23] and Braun et al. [16] analyze the Tl-2223 specific-heat data using the same techniques that they had used on Tl-2212. That is, they assume

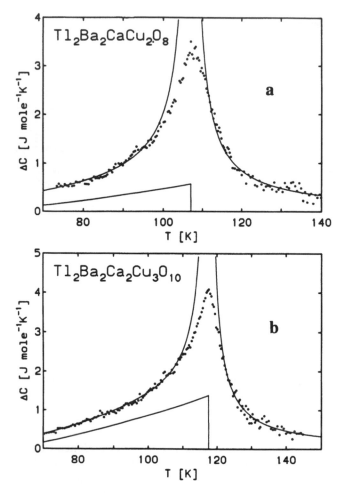

Figure 14   (a) Specific-heat anomaly for Tl-2212 after subtraction of a background poly-nominal. The curved line represents a fit made up of a mean-field BCS contribution and two-dimensional Gaussian fluctuations (the triangular ramp is the mean-field contribution). (b) A similar analysis was made for Tl-2223—however, with a crossover to critical fluctuations in a region near T_c. (From Refs. 16 and 23.)

the anomaly to be due to a BCS mean-field term plus a fluctuation term. The former group [23] assumes this fluctuation term to be due solely to critical fluctuations and obtain a value for γ of 15.4 mJ/mol · K^2. The latter group [16] reexamined the Tl-2223 data and concluded that they can be better fitted assuming two-dimensional fluctuations for $|1 - T/T_c| > 5$ K, with a crossover to critical fluctuations in the region nearer T_c. This analysis yields $\gamma \sim 8$ mJ/mol · K^2. These results are shown in Fig. 14b. The data of Bandayopadhyay et al. [19] are shown in Fig. 15. These authors do not attempt to include a fluctuation term in their analysis, although the curvature in the data above T_c (see the inset in Fig. 15) probably indicates that mean-field analysis of the type they carry out should be

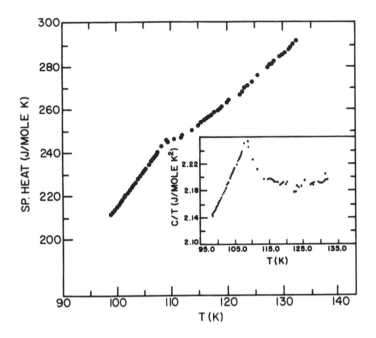

Figure 15 C versus T data for Tl-2223. The inset shows C/T versus T near T_c. (From Ref. 19.)

modified. The authors report a value of 50.9 mJ/mol $\cdot$ K^2 for $\Delta C(T_c)/T_c$, a result that gives $\gamma \sim 36$ mJ/mol $\cdot$ K^2 assuming the BCS weak-coupling relationship, $\Delta C(T_c)/T_c = 1.43\gamma$. The considerable difference between their value for γ and that calculated by either of the other two groups [12,16] reflects, in part, their failure to include a fluctuation term. However, it also appears that the data of Bandyopadhyay et al. [19] show a surprisingly pronounced anomaly. In fact, an "entropy-conserving" construction to estimate $\Delta C(T_c)/T_c$ from their data gives a discontinuity of almost 85 mJ/mol $\cdot$ K^2, a value comparable to that characteristic of a fully superconducting YBCO sample. Panova et al. [14] have made electrical-resistance, thermal-emf, magnetic-susceptibility, and specific-heat measurements on two samples of Tl-2223. They measured C in magnetic fields to 16 T and estimated that for $H = 0$ the samples had a 40 to 50% Meissner fraction. They also found the samples to have $\Delta C(T_c)/T_c \sim 30$ mJ/mol $\cdot$ K^2. From the BCS weak-coupling result, $\gamma = \Delta C(T_c)/1.43T_c$, they obtain $\gamma = 21$ mJ/mol $\cdot$ K^2, a value consistent with that which they calculate from their high-temperature magnetic-susceptibility data.

Braun et al. [16] point out that evidence for fluctuation effects in the Tl–O superconductors can also be found in their own thermal expansion data and in the values for $\Delta C = C(0) - C(H)$ calculated from reversible magnetization measurements [20]. The authors [16] note that their measured specific-heat anomalies have the same triangular shape as that obtained from the magnetization data [20]. However, the measured values are considerably larger than $C(0) - C(5\,T)$, thereby indicating that fields considerably larger than 5 T would be necessary to suppress completely the specific-heat anomaly near T_c. This conclusion is consistent with

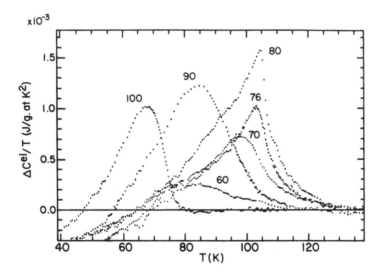

Figure 16 $\Delta C/T$ for $(Y_{1-x}Ca_x)Sr_2Cu_2(Tl_{0.5}Pb_{0.5})O_7$, where ΔC, the value of C relative to that for a reference sample, was measured using a differential calorimeter. The labels show $100\times$. (From Ref. 29.)

that of Fisher et al. [13]. Figure 12 is a graph of $C(0) - C(5\ T)$ for Tl-2223 as calculated from the reversible magnetization data [20]. The dashed curve in the figure shows the $C(0) - C(7\ T)$ data of Fisher et al. [13].

The general class of Tl compounds that we have discussed thus far can be regarded as the $m = 2$ subset of the series $Tl_mBa_2Ca_{n-1}Cu_nO_{2(n+1)+m}$. In the $m = 2$ case the CuO_2 planes responsible for superconductivity are separated by two insulating TlO sheets and T_c increases with $n = 1, 2,$ and 3. The $m = 1$ series of compounds has a single TlO layer separating the CuO_2 planes, and for a given n has a T_c 10 to 15 K lower than that for the corresponding $m = 2$ compound. Kim et al. [27] have suggested that the $m = 1$ series should have a higher critical current, J_c, than the $m = 2$ series because Josephson tunneling between CuO_2 layers is greater through a single TlO sheet than through the double TlO layer of the $m = 2$ series. This suggestion has been confirmed by Liu et al. [28], who have measured both J_c and C for the lead-stabilized Tl-1223 compound, $(Tl_{0.5}Pb_{0.5})Sr_2Ca_2Cu_3O_9$. They find that J_c is greater than for Tl-2223 and is, in fact, comparable to the J_c of YBCO. Their measurements of C show a broad anomaly that begins at $\sim$115 K and has $\Delta C(T_c)/T_c \approx 29$ mJ/mol $\cdot$ K^2. According to the authors, the high-temperature "tail" of the anomaly is consistent with a contribution to C from three-dimensional Gaussian fluctuations. They note that this fluctuation term is more like that present in YBCO [7] than the two-dimensional fluctuation observed in Tl-2223 [16]. They predict that this Tl-1223 HTSC, with a J_c comparable to that of YBCO and a T_c that is about 25 K higher, may prove to have considerable technological promise.

One other set of specific-heat measurements on an HTSC thallium-containing compound deserves mention. The system $(Y_{1-x}Ca_x)Sr_2Tl_{0.5}Pb_{0.5}O_7$ shows behavior [29] reminiscent of $La_{2-x}Sr_xCuO_4$. Here $Tl_{0.5}Pb_{0.5}$ replaces the Cu atom in the

Table 2 γ and Other Parameters Characteristic of TBCCO near and above T_c.[a]

Compound	T_c (K)	$\Delta C(T_c)/T_c$	γ_{mf}	γ_{fl}	f_s	$\chi_0 \times 10^4$	γ_χ	$N_{bs}(E_F)$	γ_{bs}	Refs.
Tl-2201	12<	—	—	—	0.08	0.425	3.1			12
				—				1.24	2.92	33
Tl-2212	104	25[b]	17[c]	21	—	—	—			21, 22
	100	35 ± 10[b]	24 ± 8	—	0.51	2.9	21			12
	109	32[d]	22[c]	3.7, 12	0.20	—	—			16, 23
								2.82	6.7	34
Tl-2223	117	34[d]	24[c]	8, 15.4	0.25	—	—			16, 23
	122	20 ± 10[b]	14 ± 7	—	0.24	4.0	29			12
	115	50.9[b]	36	—	0.90	—	—			19
	120	35[b]	24	—	0.50	3.0	22			14
	112	27[b]	19	—	0.40	3.0	22			14
	—	—	—	—	—	2.19	16			35
								3.8	9.0	34
(TlPb)-1223	115	29[b]	20	—	—	—	—			28
YBCO[e]	91	66[f]	46	—	0.86[g]	2.8	20			3
								6.78[h]	16	36

[a] $\Delta C/T$ and γ in mJ/mol·K^2; χ_0 in emu/mol; $N_{bs}(E_F)$ in states/eV/unit cell. A mole is the gram-formula weight of the compound listed. $\Delta C(T_c)/T_c$ is the mean-field height of the anomaly; $\gamma_{mf} = \Delta C(T_c)/1.43T_c$; γ_{fl} is the mean-field weak-coupled BCS value of γ when fluctuation contributions are used in the data analysis; γ_χ is calculated from χ_0, the temperature-independent part of the magnetic susceptibility above T_c, by assuming that it is equal to the Pauli susceptibility; γ_{bs} is calculated from $N_{bs}(E_F)$, the band-structure density of states at the Fermi energy; f_s is the fraction of superconducting material present in the sample (see Section V).

[b] $\Delta C(T_c)/T_c$ as given by the authors (see references for details).

[c] These authors also fit their data near T_c using a fluctuation contribution plus a mean-field BCS contribution. γ_{fl} is calculated from this mean-field contribution and is different from $\gamma_{mf} = \Delta C(T_c)/1.43T_c$. (See Section V and references for details).

[d] $\Delta C(T_c)/T_c$ was read from Fig. 14.

[e] Values of parameters listed for YBCO are given for the purpose of comparison with the TBCCO parameters.

[f] $\Delta C(T_c)/T_c$ is an average of the values obtained from the various methods used on the data shown in Fig. 6 (see Section III.C).

[g] f_s for YBCO was obtained using the correlations described in Section III.D.

[h] This value of $N_{bs}(E_F)$ is a corrected value of that listed in Ref. 36. (A. J. Freeman, private communication.)

CuO chains of YBCO while Sr replaces Ba. The substitution of Ca^{2+} for some of the Y^{3+} permits a variation of hole concentration in the material and allows the experimenter to achieve a wide variation in both T_c and $\Delta C(T_c)/T_c$. The results are shown in Fig. 16. The authors use a differential calorimeter that permits the electronic contribution to the total specific heat to be determined with high precision. As is evident from the figure, the system has an optimum composition near $x = 0.8$, with both T_c and $\Delta C_{el}(T_c)/T_c$ decreasing as x varies on either side of the optimal value.

Table 2 is a compilation of results on several of the Tl-2212 and Tl-2223 Tl–O superconductors. Included in the table are estimates of f_s, the volume fraction of superconductivity, predicted on the basis of Meissner effect and/or ac susceptibility measurements. Because of the uncertainties that arise in attempting to interpret these magnetic measurements, the listed values of f_s should be regarded as no more than a rough guide to the actual volume fractions. Also included in the table are values for (1) χ_o, the temperature-independent paramagnetic suscep-tibility; (2) the value of γ_χ inferred from χ_o by setting it equal to the Pauli suscep-tibility—assuming that the core diamagnetic contribution to χ_o is essentially offset by the Van Vleck paramagnetic contribution and that the Landau–Peierls dia-magnetic term is small enough to be ignored; (3) γ_{bs}, the value of γ calculated from $N_{bs}(E_F)$ without taking into account enhancement effects [i.e., $\gamma_{bs} = \gamma/(1 + \lambda)$; see Eq. (5)].

VI.   FURTHER DISCUSSION AND SUMMARY

A.   Lattice Specific Heat

Kulkarni et al. [30] have calculated the phonon spectra and lattice specific heats of the Tl–O HTSCs. They predict that $\Theta_{300} \sim 600$ K, in agreement with the experimental results reported in Ref. 14 on Tl-2223. However, their predictions regarding the range over which $C_l \propto T^3$ and the values of Θ_0 are not in accord with the experimental results (see Table 1 and Fig. 3).

B.   Low-Temperature Upturn C/T and $\gamma^*(0)$

Within the experimental uncertainty of ~ 1 mJ/mol $\cdot$ K^2, $\gamma^*(0)$ for YBCO can be represented as the sum of a contribution from impurity phase, particularly BaCuO$_2$, and a contribution from nonsuperconducting regions in the YBCO associated with Cu^{2+} moments present in a concentration n_2; but with the same uncertainty, $\gamma^*(0) \neq 0$ for all samples. On the other hand, $\gamma^*(0) = 0$ for most BSCCO samples and for some TBCCO samples. The fact that $\gamma^*(0) = 0$ for BSCCO samples, includ-ing some for which n_2 is large, suggests that the relation between Cu^{2+} moments and $\gamma^*(0)$ is different from that for YBCO. If there are localized Cu^{2+} moments in the CuO$_2$ planes of BSCCO and TBCCO, they may produce nonsuperconducting material that is not metallic (i.e., which does not have a γ associated with it).

C.   Fluctuations

The data on Tl-2212 [16,21,23], Tl-2223 [16,19,23], and (TlPb)-1223 [28] provide evidence for a fluctuation contribution to C near T_c. In Ref. 21 it is assumed that this contribution comes from three-dimensional Gaussian fluctuations, whereas in Refs. 16, 23, and 28 the contribution for $|T - T_c| > 5$ K is attributed to two-dimensional fluctuations. The authors in Ref. 16 argue that there is a crossover from two-dimensional Gaussian fluctuations to three-dimensional critical fluctua-tions as $|T - T_c|$ approaches zero. Sokolov [31] has examined the results from Refs. 16 and 23 and argues that they span the crossover region but do not adequately penetrate the temperature region close to T_c to be considered as evidence for true three-dimensional critical behavior.

Because strong-coupling effects can produce upward curvature in the C_{es}/T versus T data below T_c (see Fig. 2), the evidence for fluctuations for $T > T_c$ is somewhat clearer than that for $T < T_c$. This is especially true in the data on (TlPb)-1223 [28]. However, even above T_c the apparent evidence for a fluctuation con-tribution to C will depend on how the estimate is made for the background specific heat, and also on whether or not broadening of the anomaly due to sample inhomogeneities is included in the data analysis. Only in Ref. 21 is the possibility of such broadening included in the analysis.

D. Strong Coupling

In most of the results reported here, the authors assume BCS weak coupling to be applicable, that is, they assume that (in the absence of a fluctuation term) $\Delta C(T_c)/T_c = 1.43\gamma$. This assumption proves to be consistent with the fitting procedure used in Ref. 21, but as is noted in Refs. 16 and 21, the data do not point unambiguously to such a conclusion. The best-defined specific-heat anomaly for any of the Tl superconductors is that reported by Bandyopandhyay [19]. In their analysis the authors do not include the effects of fluctuations, strong coupling, or broadening due to inhomogeneities. Nevertheless, the data above T_c indicate the possible presence of a fluctuation contribution. However, data below T_c are like many of the results on YBCO in that there is little evidence of curvature just below T_c but give a very large value of the ratio $T_c (dC/dT)/\Delta C(T_c)$. A large value for this ratio is characteristic of strong rather than weak coupling (see Fig. 1). The results in Ref. 19 give hope that as sample quality improves, specific-heat data on the Tl–O compounds may provide clearer evidence of strong coupling (or its absence) and of two-dimensional versus three-dimensional character.

E. Electron Density of States Near the Fermi Energy and γ

Table 2 lists values of $N_{bs}(E_F)$ obtained from band-structure calculations and the values of the γ_{bs}, the value of γ calculated from $N_{bs}(E_F)$ without any allowance for enhancement effects [i.e., $\gamma_{bs} = \gamma/(1 + \lambda)$; γ_{bs} (mJ/mol $\cdot$ K^2) $= 0.424N_{bs}(E_F)$ (states/eV/unit cell)]. Also listed are the values of χ_0, the temperature-independent part of the magnetic susceptibility, and the associated gamma, γ_x; γ_x (mJ/mol $\cdot$ K^2) $= 7.3 \times 10^4\chi_0$ (emu/mol). Values for $\gamma_{mf} = \Delta C(T_c)/1.43T_c$, the value for γ to be expected for a weak-coupled BCS superconductor, and γ_{fl}, the mean-field weak-coupled BCS value of γ when fluctuations were used in the data analysis, are also tabulated. In Table 2 the relatively small values of γ_{fl} obtained from Refs. 16 and 23 reflect the fact that the authors subtract a sizable fluctuation contribution from the measured $\Delta C(T_c)$ before using the BCS relationship to calculate γ_{fl}. Except for these cases, γ_{mf} and γ_x are greater than γ_{bs}, a result that might be taken to indicate substantial values of λ, and that strong coupling is present in the Tl–O superconductors. However, such a conclusion is clearly inconsistent with the weak-coupling assumption that $\Delta C(T_c)/\gamma_{mf}T_c = 1.43$ (however, see Ref. 32). The fact that most of the values for γ_{mf} and γ_x for a given compound are comparable may be significant, but one should be wary of leaping to this conclusion too readily. The values of $N(E_F)$, and therefore of γ_x, obtained from the susceptibility data will depend on whether or not corrections for the core diamagnetism and Van Vleck paramagnetism are included, and also on the assumptions made concerning the possibility of a Stoner enhancement of χ_0.

The variation in the values of γ listed in Table 2 make it evident that as yet little can be said with confidence about the size of γ in Tl–O HTSCs, or about the possible renormalization of γ by phonon or other enhancement effects. In what is generally a confusing set of results, there does exist one bit of clarity: χ_0, expressed in units of emu/(g-at Cu), is essentially the same for the Tl–O HTSCs as for YBCO.

However, the $N_{bs}(E_F)$ calculated (in states/eV/Cu atom) for these materials [33–35] is substantially lower than that calculated for YBCO [36]. This situation for $N_{bs}(E_F)$ mirrors that for the $\Delta C(T_c)/T_c$s observed, thus indicating some measure of agreement between the specific-heat results and the band-structure calculations. The data reported in Ref. 19 are an exception to this generalization. Should subsequent measurements on Tl-2223 confirm these results, further band-structure calculations will be in order.

In summary, it is evident that better specific-heat measurements on better samples, especially single crystals, should help to remove some of the ambiguities in the data discussed in this chapter. It is to be hoped that such measurements, in combination with other work on the Tl–O superconductors, will clarify the role played by fluctuations, by possible strong-coupling effects, and by dimensionality in determining the characteristics of the only materials yet found to be superconducting at temperatures greater than 115 K.

ACKNOWLEDGMENTS

We thank the authors who supplied preprints of their work, particularly those who provided figures. We are especially grateful to Lorna Woelfel, who skillfully handled the many revisions of the manuscript, and to W. Slocombe, for his assistance and expert draftsmanship in preparing some of the figures. This work was supported by the Director of the Office of Energy Research, Office of Basic Energy Sciences, Materials Sciences Division of the U.S. Department of Energy under Contract DE-AC03-76SF00098. Additional support for J.E.G. was provided by the Research Corporation.

NOTE ADDED IN PROOF

Calculations of the phonon spectra and lattice specific heats [37,38] of the Tl–O HTSCs have recently been reported by Agrawal et al. These calculations have been discussed by de Wette and Kulkarni [39], who argue that the values of θ_0 calculated in Refs. 37 and 38 disagree with the experimental values far more than those of Kulkarni et al. [30].

Loram et al. [40] have recently reported specific-heat results on Tl-2223. They find evidence for 2-D fluctuations and a possible crossover to critical fluctuations near T_c. The authors do not report a value for $\Delta C(T_c)/T_c$, but an entropy-conserving construction to their data yields a value of $\sim$40 mJ/mol·K^2. In this paper the authors also report that $(Ca_{0.8}Y_{0.2})(Tl_{0.5}Pb_{0.5})Sr_2Cu_2O_7$ exhibits 2-D fluctuations, in contrast to the 3-D behavior of the lead-stabilized Tl-1223 compound reported in Ref. 28.

REFERENCES

1. Z. Z. Sheng and A. M. Hermann, *Nature (London)* 332, 55 (1988); Z. Z. Sheng and A. M. Hermann, *Nature (London)* 332, 138 (1988).
2. H. Padamsee, J. E. Neighbor, and C. Schiffman, *J. Low Temp. Phys.* 12, 387 (1973).
3. N. E. Phillips, R. A. Fisher, and J. E. Gordon, in *Progress in Low Temperature Physics*, Vol. 13, ed. D. F. Brewer, North-Holland, Amsterdam, 1992, p. 267.
4. This component was recognized in very early measurements; the first published reports were: M. E. Reeves, T. A. Friedmann, and D. M. Ginsberg, *Phys. Rev. B* 35, 7207

(1987); L. E. Wenger, J. T. Chen, G. W. Hunter, and E. M. Logothegis, *Phys. Rev. B* 35, 7213 (1987); B. D. Dunlap, M. V. Nevitt, M. Slaski, T. E. Klippert, Z. Sungaila, A. G. McKale, D. W. Capone, R. B. Poeppel, and B. K. Flandermeyer, *Phys. Rev. B* 35, 7210 (1987).

5. A. P. Ramirez, R. J. Cava, G. P. Espinosa, J. P. Remeika, B. Batlogg, S. Zahurak, and E. A. Rietman, *Mater. Res. Soc. Symp. Proc.* 99, 459 (1987); R. Kuentzler, Y. Dossmann, S. Vilminot, and S. el Hadiqui, *Solid State Commun.* 65, 1529 (1988); D. Eckert, A. Junod, T. Graf, and J. Muller, *Physica C* 153–155, 1038 (1988).

6. N. E. Phillips, R. A. Fisher, J. E. Gordon, S. Kim, A. M. Stacy, M. K. Crawford, and E. M. McCarron III, *Phys. Rev. Lett.* 65, 357 (1990).

7. S. E. Inderhees, M. B. Salamon, J. P. Rice, and D. M. Ginsberg, *Phys. Rev. Lett.* 66, 232 (1991).

8. A. Junod, D. Eckert, T. Graf, G. Triscone, and J. Muller, *Physica C* 162–164, 1401 (1989).

9. R. A. Fisher, J. E. Gordon, and N. E. Phillips, in *Lattice Effects in High-T$_c$ Superconductors*, eds. Y. Bar-Yam, T. Egami, J. Mustre-de Leon, and A. R. Bishop, World Scientific, Singapore, 1992, p. 317.

10. J. S. Urbach, D. B. Mitzi, A. Kapitulnik, J. Y. T. Wei, and D. E. Morris, *Phys. Rev. B* 39, 12391 (1989).

11. Y. Muto, N. Kabayashi, and T. Sasaki, in *The Science of Superconductivity and New Materials*, ed. S. Nakajima, World Scientific, Singapore, 1989, p. 98.

12. A. Junod, D. Eckert, G. Triscone, V. Y. Lee, and J. Muller, *Physica C* 159, 215 (1989); A. Junod, D. Eckert, G. Triscone, J. Muller and V. Y. Lee, *Physica C* 162–164, 476 (1989).

13. R. A. Fisher, S. Kim, S. E. Lacy, N. E. Phillips, D. E. Morris, A. G. Markelz, J. Y. T. Wei, and D. S. Ginley, *Phys. Rev. B* 38, 11942 (1988).

14. G. Kh. Panova, M. N. Khlopkin, N. A. Chernoplekov, A. V. Suetin, B. I. Savel'ev, A. I. Akimov, L. P. Poluchankma, and A. P. Chernyakova, *Supercond. Phys. Chem. Technol.* 4, 60 (1991).

15. A. Amato, R. A. Fisher, N. E. Phillips, and J. B. Torrance, *Physica B* 165–166, 1337 (1990).

16. E. Braun, W. Schnelle, H. Broicher, J. Harnischmacher, D. Wohlleben, C. Allgeier, W. Reith, J. S. Schilling, J. Bock, E. Preisler, and G. J. Vogt, *Z. Phys. B Condensed Matter* 84, 333 (1991).

17. E. Braun, W. Schnelle, F. Seidler, P. Böhm, W. Braunisch, Z. Drzazga, S. Ruppel, H. Broicher, H. Geus, M. Galffy, B. Roden, I. Felner, and D. Wohlleben, *Physica C* 162–164, 496 (1989).

18. F. Seidler, P. Böhm, H. Geus, W. Braunisch, E. Braun, W. Schnelle, Z. Drzazga, N. Wild, B. Roden, H. Schmidt, and D. Wohlleben, *Physica C* 175, 375 (1989).

19. A. K. Bandyopadhyay, P. Maruthikumar, G. L. Bhalla, S. K. Agarwal, and A. V. Narlikar, *Physica C* 165, 29 (1990).

20. J. Gohng and D. K. Finnemore, *Phys. Rev. B* 42, 7946 (1990); M. M. Fang, J. E. Ostenson, D. K. Finnemore, D. E. Farrell, and N. P. Barsal, *Phys. Rev. B* 39, 222 (1989).

21. T. Atake, H. Kawaji, M. Itoh, T. Nakamura, and Y. Saito, *Thermochem. Acta* 183, 143 (1991).

22. T. Atake, H. Kawaji, M. Itoh, T. Nakamura, and Y. Saito, *Physica C* 162–164, 488 (1989).

23. D. Wohlleben, E. Braun, W. Schnelle, J. Harnischmacher, S. Ruppel, and R. Dömel, in *International Conference on Superconductivity* (ICSC), ed. S. K. Joshi, C. N. R. Rao, and S. V. Subramanyam, World Scientific, Singapore, 1990, p. 194.

24. F. Sharifi, J. Giapintzakis, D. M. Ginsberg, and D. J. van Harlingen, *Physica C* 161, 555 (1989).
25. J. E. Gordon, M. L. Tan, R. A. Fisher, and N. E. Phillips, *Solid State Commun.* 69, 625 (1989).
26. K. S. Gavrichev, V. E. Gorbunov, I. A. Konovalova, V. B. Lazarev, E. A. Tishchenko, and I. S. Shaplygin, *Inorg. Mater.* 26, 943 (1990).
27. D. H. Kim, K. E. Gray, R. T. Kampwirth, J. C. Smith, D. S. Richeson, T. J. Marks, J. H. Kang, J. Talvecchio, and M. Eddy, *Physica C* 177, 431 (1991).
28. R. S. Liu, N. Zheng, J. W. Loram, K. A. Mirza, A. M. Campbell, and P. P. Edwards, *Appl. Phys. Lett*, 60 (1992), in press.
29. J. W. Loram, K. A. Mirza, and R. S. Liu, *Supercond. Sci. Technol.* 4, S286 (1991).
30. A. D. Kulkarni, F. W. de Wette, J. Prade, V. Schröder, and W. Kress, *Phys. Rev. B* 43, 5451 (1991).
31. A. I. Sokolov, *Physica C* 174, 208 (1991).
32. F. Marsiglio, R. Aiko, and J. P. Carbotte, *Phys. Rev. B* 36, 5245 (1987).
33. D. R. Hamann and L. F. Mattheis, *Phys. Rev. B* 38, 5138 (1988).
34. J. Yu, S. Massidda, and A. J. Freeman, *Physica C* 152, 273 (1988).
35. W. Reith, P. Müller, C. Allgeier, R. Hoben, J. Heise, J. S. Schilling, and K. Andres, *Physica C* 156, 319 (1988).
36. S. Massida, J. Yu, A. J. Freeman, and S. D. Koelling, *Phys. Lett. A* 122, 198 (1987).
37. B. K. Agrawal, J. S. Negri, S. Agrawal and P. S. Yadav, *Physica C* 192, 237 (1992).
38. B. K. Agrawal, J. S. Negri, P. S. Yadav and S. Agrawal, *Phys. Rev. B* 45, 3152 (1992).
39. F. W. de Wette and A. D. Kulkarni, *Physica C* 196, 399 (1992).
40. J. W. Loram, J. R. Cooper, J. M. Wheatley, K. A. Mirza, and R. S. Liu, *Phil. Mag. B* 65, 1405 (1992).

23
Infrared Studies of Thallium Cuprate Superconductors

Karl F. Renk

Institut für Angewandte Physik
Universität Regensburg
Regensburg, Germany

I. INTRODUCTION

Presently, the thallium cuprate superconductors [1–3] have the highest superconducting transition temperatures, up to 125 K, of all known high-temperature superconductors. It is therefore of great interest to study their properties. On one hand, it would be important to understand the microscopic mechanisms that can lead to the high transition temperatures, and on the other hand, it is interesting to think about possible applications that make use of the high transition temperatures.

Infrared studies can contribute to an understanding of basic properties. Because of the large transition temperatures, one also expects large energy gaps, with excitation energies in the infrared spectral region. Large energy gaps would make it possible to develop lossless reflectors for infrared radiation. Another aspect of application follows directly from the high transition temperature: At high temperatures relaxation processes are fast, and therefore high-speed detectors for radiation seem to be an interesting spectroscopic application of thallium cuprate superconductors.

From a basic point of view, thallium cuprate superconductors are especially suitable for systematic studies because of a large variety of different members of the thallium cuprate family. There is the homologous series $Tl_2Ba_2Ca_nCu_{n+1}O_{2n+6}$ ($n = 0,1,2,3$), with the highest T_c of 125 K for $n = 2$, for which TlO double layers between the CuO_2 layers are characteristic [1–3]. In this series the crystal structure of the members that are superconducting is tetragonal, while orthorhombic $Tl_2Ba_2CuO_6$ is a normal metal also at low temperatures. There is, furthermore, the $TlBa_2Ca_nCu_{n+1}O_{2n+5}$ ($n = 0,1,2,3$) series with TlO monolayers in the crystal structure and T_cs up to 116 K [4–7]. By replacing Tl partly by Pb, it is possible to realize a system $Tl_{1-x}Pb_xSr_2CaCu_2O_{7-y}$, with T_cs of 85 to 100 K [8,9], that allows one to vary T_c by changing the concentration x of Pb and thus to prepare material of similar crystal properties and composition that is normal conducting or super-conducting.

For infrared studies it is important to have well-defined material, a condition, of course, for all studies, and furthermore to have high-quality material near the surfaces within a range of the penetration depth of the infrared radiation which is about a London penetration depth (on the order of 100 nm). This condition of high-quality surface regions supposes strong requirements on sample preparation. During the preparation, usually performed at high temperatures well above room temperature, one must prevent thallium evaporation from the surface, which would lead to damage. Because of the difficulty in preparing samples of high surface quality and because of the toxicity of thallium, there have been performed up to now only a limited number of infrared investigations that allow conclusions with respect to basic properties and to applications. Nevertheless, the presently available results of infrared studies are interesting enough to be presented in a comprehensive article.

Most interesting is the observation of anomalous phonon behavior for the thallium compound with the highest transition temperature $Tl_2Ba_2Ca_2Cu_3O_{10}$ [10]. One of the infrared active phonons shows a strong decrease of its resonance frequency at T_c, and almost all phonons show strongly temperature-dependent oscillator strengths above T_c but most strongly at T_c [10]. A comparison of the far-infrared reflectivity of normal conducting and superconducting $Tl_{1-x}Pb_xSr_2CaCu_2O_{7-y}$ indicates [11] that phonon anomalies occur in the superconducting but not in the normal conducting compound. These studies have been performed using poly-crystalline pellets.

The present results allow us to determine characteristic properties of infrared active phonons [10–12]. Theoretical, lattice dynamical studies [13] have delivered a good understanding of infrared and Raman active modes. In a study of a single-crystal $Tl_2Ba_2CaCu_2O_8$ [14] it has been shown that the infrared reflectivity at room temperature corresponds to that of the CuO_2 planes in $YBa_2Cu_3O_{7-\delta}$ and that the TlO interlayers do not contribute noticeably to the infrared conductivity (i.e., that the TlO layers seem to be insulating layers); a theoretical analysis gives evidence that the experimental reflectivity curve can be described by Eliashberg's theory with strong electron–phonon interaction [14]. This study was performed with an infrared radiation beam focused to a small spot of a crystal at room temperature, thus selecting a high-quality region of a crystal.

The size of the presently available single-crystal thallium cuprate samples was not sufficient for studies at low sample temperatures. First informations on energy gap behavior have been obtained by measuring the optical response of current-biased thallium cuprate thin films [15]. The study performed with polycrystalline films gives evidence for an energy gap $2\Delta(0) \approx 2\,kT_c$ that may be attributed to excitations with the electric field parallel to the crystal c direction. These experiments also indicate that high-speed infrared detectors can be fabricated using thallium cuprate thin films [15].

In this chapter characteristic infrared properties of thallium cuprate superconductors are described. A survey of anomalous phonon properties is given and an analysis, performed by use of the lattice dynamical results [13], is presented that makes it possible to characterize infrared active phonon modes obtained by re-flectivity measurements [10–13]; it will be shown, by an analysis of results of the three members (n = 1,2,3) of the $Tl_2Ba_2Ca_nCu_{n+1}O_{2n+6}$ series and of $Tl_{1-x}Pb_xSr_2CaCu_2O_7$, that there are characteristic phonon modes with frequencies that are not much different for the different compounds. The discussion concen-trates on results obtained for samples prepared and studied in our laboratory in Regensburg. Special emphasis has been put on a well-defined characterization of the samples. This was most important because the similarity of different phases of thallium cuprate superconductors can easily lead to phase mixtures during sample preparation.

Samples studied by infrared techniques have been prepared as polycrystalline pellets, single crystals, and thin films. For characterization x-ray diffraction, resis-tivity, dc and ac magnetic susceptibility measurements, and scanning microscope studies have been performed. It should be mentioned that general surveys on infrared properties of high-temperature superconductors have been given about one year after the high-T_c research started [17] and also more recently [18].

II. SAMPLE PREPARATION AND CHARACTERIZATION

A. Polycrystalline Pellets

Polycrystalline samples of the n = 0,1,2 members of the homologous series $Tl_2Ba_2Ca_nCu_{n+1}O_{2n+6}$ have been prepared [12] by a solid-state reaction of the thoroughly mixed appropriate amounts of Tl_2O_3, $Ba(NO_3)_2$, CaO, and CuO. Pressed pellets (typically 13 mm in diameter and 10 mm in height) were placed into an aluminum oxide crucible together with a gold boat containing a small amount of Tl_2O_3 and covered with a fitting lid. Under these conditions a saturated Tl_2O and O_2 rich vapor existed during the annealing procedure and Tl loss due to the high volatility of thallium oxide was reduced in this way.

The samples have been heated with a rate of 200°C/h up to 800°C and held at this temperature for 8 h, cooled down to 600°C within 2 h, and then taken out of the furnace. The sintered material was reground and pressed again to pellets (10 tons/cm^2) with 2 to 3 mm thickness. The second heat treatment has been different for the different phases. Annealing in flowing oxygen at 800°C and cooling in the furnace to room temperature lead to the normal-conducting orthorhombic form of the 2201 (*o*-2201) phase. The superconducting, tetragonal 2201 (*t*-2201) phase was obtained if pellets with the appropriate stoichiometry were fired at 850°C in air for half an hour and then quenched to liquid nitrogen temperature. The 2212 phase was obtained by annealing a sample with the adequate stoichiometry at 820°C in flowing oxygen for 30 min, furnace cooling to 600°C and quenching to room tem-perature after 1 h. The 2223 phase could be formed by annealing a 2223 sample in air at 840°C for 5 min, furnace cooling to 600°C, holding there for 90 min, and then quenching to room temperature.

Figure 1a to d shows x-ray powder diffraction patterns obtained by use of Co-$K_{\alpha1}$ radiation (wavelength 1.788965 Å) for the (nonsuperconducting) orthorhombic $Tl_2Ba_2CuO_6$ (*o*-2201 phase), tetragonal $Tl_2Ba_2CuO_6$ (*t*-2201), $Tl_2Ba_2CaCu_2O_8$ (2212 phase), and $Tl_2Ba_2Ca_2Cu_3O_{10}$ (2223 phase) [19,20]. Only minor impurity phases (marked by crosses) were found (i.e., the samples had high purity). An impurity

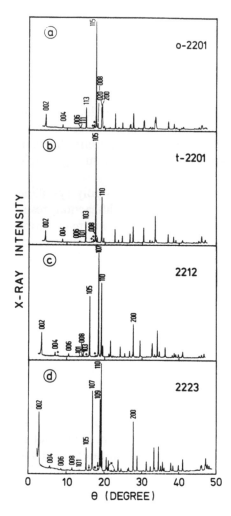

Figure 1 X-ray powder diffraction patterns of orthorhombic (a) $Tl_2Ba_2CuO_6$, (b) tetragonal $Tl_2Ba_2CuO_6$, (c) $Tl_2Ba_2CaCu_2O_8$, and (d) $Tl_2Ba_2Ca_2Cu_3O_{10}$. Impurity phases are marked with arrows (2212 phase), ○ (surface phase), × (CuO), or + (other phases).

content of the samples of few percent was estimated (i.e., the samples consisted of almost single-phase material).

A detailed analysis of the x-ray diffraction data gives evidence [19,20] that Tl vacancies occur and that these are responsible for the hole doping of the CuO_2 planes (i.e., for a deviation of the Cu oxidation state from a value of +2). Chemical formulas and apparent oxidation states of Cu are shown in Table 1. The oxidation values obtained from a bond strength analysis and the formula are indicated. Accordingly, there is a Tl deficiency (and a Ca deficiency) that determines mainly the oxidation state. While it is near +2.0 for the nonsuperconducting compound, it increases up to +2.31 for the compound 2223 with the highest T_c. In this compound there are three adjacent CuO_2 layers, one of them seems to contain Cu

Table 1 Chemical Formulas and Apparent Oxidation States of Cu for
$Tl_{2-x}Ba_2Ca_nCu_{n+1}O_{2n+6}$

		Oxidation state of Cu	
n	Formula	From bond strength	From formula
0	$Tl_{1.90}Ba_2CuO_{5.87}$ (*o*)	2.0+	(2.04+)
	$Tl_{1.93}Ba_2CuO_6$ (*t*)	2.17+	2.21+
1	$Tl_{1.90}Ba_2Ca_{0.90}Cu_2O_8$	2.24+	2.25+
2	$Tl_{1.87}Ba_2Ca_{1.86}Cu_3O_{10}$	2.23+ (mean)	2.22+ (mean)
		2.31+ (×2)	
		2.07+ (×1)	

atoms with an oxidation state near +2, while the two other layers contain Cu atoms with the high oxidation state.

Dc resistivity measurements were performed for the different phases by the standard low-frequency four-probe method with silver paint contacts. All samples showed, for room temperature, specific electrical resistances between 300 μΩ · cm and 800 μΩ · cm and a decrease of resistance with decreasing temperature (Fig. 2). For the *o*-2201 phase normal metallic behavior and no transition to superconductivity down to 4.2 K was observed, while the *t*-2201 phase became superconducting at 17 K, the 2212 phase at 105 K, and the 2223 phase at 119 K. The *o*-2201 phase shows a behavior of the resistivity (Fig. 2) that is characteristic for normal metals, with residual resistivity at low temperature and a strong increase at high temperatures. The superconducting phases show an almost linear dependence as it seems to be characteristic for high-temperature superconductors [21]. Deviations from the linear dependence near T_c may be due to fluctuation effects.

Table 2 characterizes samples that were used for a systematic study of infrared active phonons (Section IV). T_c denotes zero resistance temperature, T_{cm} the temperature of the transition midpoint, and ΔT_c the difference between $T_{90\%}$ and $T_{10\%}$. For the 2201 phase T_c was strongly dependent on the second annealing process, T_c was maximized by fast quenching from the highest temperature, at which the 2201 phase was stable. Table 2 also contains data on polycrystalline $Tl_{1-x}Pb_xSr_2CaCu_2O_7$ samples. The electrical resistivity of Pb-doped 1212 samples is shown in Fig. 3; ρ_0 was the resistivity at room temperature and decreased with increasing Pb concentration ($\rho_0 \approx 400$ μΩ · cm for $x = 0.45$).

X-ray studies of the TlO monolayer samples showed [22,23] that these had high purity and that the Tl atoms have an off-center position in the tetragonal elementary cell.

Measurements of the magnetic dc susceptibility have been performed with a modified SHE SQUID magnetometer (VTS 50), which allows measurements in the temperature range 1.5 K ≤ *T* ≤ 300 K and in the range of the magnetic field $0 \leq B \leq 5$ T. Bulk samples with typical dimensions $2 \times 2 \times 2$ mm^3 were cut from the pellets, and magnetic shielding and the Meissner effect have been measured in the following procedure [24,25]. The samples were first cooled in a zero magnetic

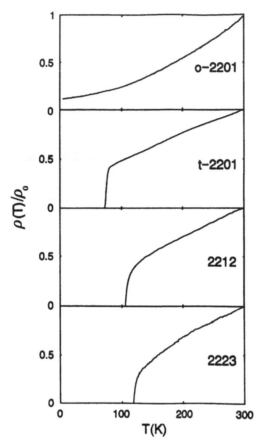

Figure 2 Resistivity of the $Tl_2Ba_2Ca_nCu_{n+1}O_{2n+6}$ ($n = 0,1,2$) phases.

field down to about 5 K. After applying a magnetic field (3 mT) a diamagnetic magnetization as a result of superconducting shielding currents appeared.

Figure 4 shows the results for samples of the (a) t-2201, (c) 2212, and (d) 2223 phases, respectively. Also shown is (b) the susceptibility for a $Tl_{0.55}Pb_{0.45}Sr_2CaCu_2O_7$ sample. The onset of the diamagnetic signal gives T_cs that agree with those obtained from resistivity measurements. For the t-2201 phase (Fig. 4) shielding and Meissner

Table 2 Characteristic Temperatures for Thallium Cuprates Investigated by Infrared Techniques

Phase	T_c (K)	$T_{c,\text{onset (K)}}$	ΔT_c (K)
o-2201	—	—	
t-2201	71	75	2
2212	105	112	4
2223	119	128	4
1212 (20% Pb)	—	—	—
1212 (45% Pb)	70	76	5

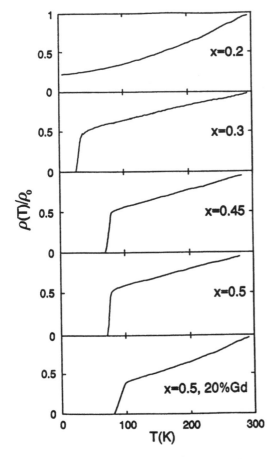

Figure 3 Resistivity of $Tl_{1-x}Pb_xSr_2CaCu_2O_7$.

signals are not much different, possibly because of weak pinning. The shielding signal for the 1212 phase (b) shows a step near 50 K that may be due to different pinning centers. The signals for the samples of the 2212 and 2223 phases indicate that these samples were of high phase purity.

Scanning electron microscope pictures [26] show that the surface of the pellets consisted of almost randomly oriented crystallites of typical size a few micrometers along the (a,b) plane and less than a micrometer along the c direction.

B. Single Crystals

For the preparation of crystals that are sufficiently large for infrared studies an elaborated technique was developed [14]. It was started with a Ba–Ca–Cu–O precursor from $Ba(NO_3)_2$, CaO, and CuO (with stoichiometric ratio 2:1.5:2.5) presintered in air at 890°C for 4 h. This material has been reground, mixed with Tl_2O_3 powder according to the stoichiometric ratio 2.5:2:1.5:2.5, and pressed to pellets. A special arrangement of gold foils and pressed pellets creating cavities of a few cubic millimeters was placed into an alumina crucible in order to obtain a local separation of growing crystals from solidified flux and original pellets during

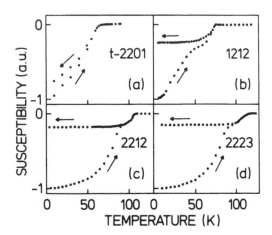

Figure 4 Shielding and Meissner signals of (a,c,d), the $Tl_2Ba_2Ca_nCu_{n+1}O_{2n+6}$ ($n = 0,1,2$) phases and (b) $Tl_{0.55}Pb_{0.45}Sr_2CaCu_2O_7$.

the growth procedure. The crucible was covered with a fitting lid and pasted up with an Al_2O_3-based high-temperature adhesive, heated to 930°C within 2 h, held at this temperature for 10 h, slowly cooled down to 855°C with 5°C/h, and then cooled to room temperature at a 100°C/h cooling rate. Free-standing crystals with mirrorlike surfaces were grown in the preformed gold foil cavities with sizes up to $2 \times 2 \times 0.2$ mm^3. By x-ray analysis using the Buerger precession method, no other phases besides the 2212 phase could be detected.

Results of dc resistivity measurements [14] performed with the four-point contact method after van der Pauw are shown in Fig. 5. The specific resistivity in the (a,b) plane decreases almost linearly from room temperature to 110 K and then drops to zero. The resistivity along the c direction is about 10^3 times larger than in the (a,b) plane and increases slightly with decreasing temperature.

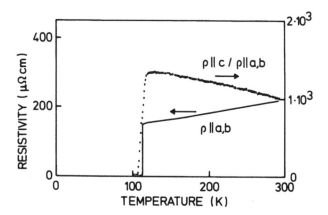

Figure 5 Dc resistivity $\rho\|(a,b)$ of a $Tl_2Ba_2CaCu_2O_8$ crystal parallel and (of another crystal) perpendicular to the CuO_2 layers ($\rho\|c$).

The result of a measurement of the dc susceptibility [27] is shown in Fig. 6. There is only a small difference between Meissner and shielding signal, giving evidence that the crystal contained a low concentration of pinning centers, or a very low activation energy, as the deviation between the two signals at low temperature may indicate. The superconducting transition was sharp, with T_c = 109 K. Samples of large size (up to $2 \times 2 \times 0.2$ mm^3) were obtained; however, these had on their surfaces regions of minor quality. Infrared studies have been performed by focusing the light to regions of high quality [14].

C. Thin Films

Polycrystalline high-T_c TlBaCaCuO thin films have been prepared with different deposition techniques, including laser deposition [28,29], sputtering [30], and chemical vapor deposition [31]. Films were prepared in two-stage processes. They were nonsuperconducting after the first stage. To establish high-T_c phases the films were annealed in furnaces, either in the presence of Tl_2O_3 [28] or together with Tl-containing ceramic pellets [29–32]. For the resulting films T_cs up to 120 K [32] and critical current densities up to 10^5 A/cm^2 at 100 K were reported [30]. The films consisted of crystallites of typical dimensions of few micrometers. The surface morphology varied from needle-shaped [29] to platelike crystallites [32]. A recently developed in situ preparation technique [33] consists of in situ preparation of TlBaCaCuO thin films by laser ablation combined with thermal evaporation; in the thallium-free films of appropriate composition, Tl is introduced in situ at high temperature by thermal evaporation of Tl_2O_3. For laser ablation an excimer laser (wavelength 308 nm, pulse repetition rate 5 Hz, pulse energy 0.5 J) was used. The arrangement is shown in Fig. 7. The laser beam was focused onto a rotating $Ba_2Ca_2Cu_3O_x$ ceramic pellet. The MgO <100> substrate was mounted close to a wire heater at a distance of about 4 cm from the target. In the corners of the substrate, silver paste contacts were prepared. The evaporation source consisted of a small heated Al_2O_3 crucible containing Tl_2O_3. The crucible was moved under the substrate by a motor after laser deposition.

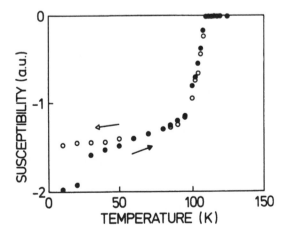

Figure 6   Meissner and shielding signal for a $Tl_2Ba_2CaCu_2O_8$ crystal.

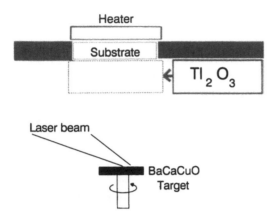

Figure 7   Apparatus for in situ preparation of TlBaCaCuO thin films.

During deposition the substrate temperature was at about 800°C. The oxygen pressure in the chamber was approximately 0.3 mbar. A typical deposition time was 10 min, resulting in film thicknesses of about 0.5 μm. The resistance of the deposited BaCaCuO film was about 10^5 Ω. After laser deposition of the BaCaCuO thin film the oxygen pressure was increased to 0.8 bar and then the thermal evaporation source of Tl_2O_3 was shifted under the substrate. The resistance dropped to a value of about 10^4 Ω. The temperature of the substrate was increased to 850 to 860°C and the temperature of the Tl source was increased to a value higher than 700°C within 5 min. The resistance of the film decreased continuously during the time of diffusion (typically, 15 min). The temperature of the Tl source was adjusted every second minute to minimize the resistance of the film, a minimum value of typically 20 Ω. Then the sample was cooled down to room temperature within a few minutes, causing a further drop in the resistance. A Tl_2O_3 vapor pressure was maintained during cooling to prevent loss of Tl in the film. X-ray diffraction measurements were performed with CuK_α radiation. The x-ray data (Fig. 8) of the film reveal a mixture of the 2212, 1234, and 2223 phases with partial orientation of the c axis perpendicular to the film plane for the different phases. There are also peaks of other orientation for the 2212 and 2223 phases. The unidentified peaks may represent insulating phases.

Superconducting properties of the films were studied by measuring at low temperatures both the resistance (standard 4-probe technique) and the ac susceptibility χ'. The susceptibility measurement was carried out by placing the film between two small coils. One of them excited magnetic fields of a strength of ~5 mOe and 900 Hz perpendicular to the plane of the film. The other was connected to a lock-in amplifier and received the field transmitted through the film. An experimental resistivity curve is shown in Fig. 9a. The resistivity had a value of about 400 μOhm·cm. Onset of transition was at ~120 K and the midpoint at ~115 K. Zero resistance was reached at 112 K. The normal state resistivity above 150 K was proportional to temperature. For the films critical currents of typically 10^4 to 10^5 A/cm² at liquid nitrogen temperature were found. The ac susceptibility (Fig. 9b) reveals a sharp drop below 112 K and shows that most of the film became

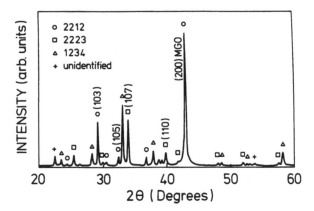

Figure 8 X-ray diffraction pattern of a TlBaCaCuO thin film. The marked peaks represent different phases, with the *c* axis perpendicular to the MgO <100> substrate.

superconducting near 105 K. A tail that extends to low temperatures was due to stray fields around the sample (size 8×8 mm^2).

Scanning electron microscope photographs indicated granular structure with an average grain size smaller than 1 μm. The grain size strongly depended on the substrate temperature during the diffusion of Tl; at higher substrate temperatures more platelike structures with a size of about 10 μm were found [33].

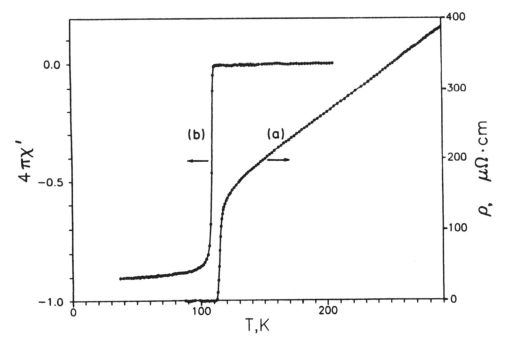

Figure 9 (a) Resistivity and (b) ac susceptibility of a TlBaCaCuO thin film.

III. BASIC THEORETICAL BACKGROUND AND METHODS OF INFRARED STUDIES

A. Dynamic Conductivity, Optical Constants, and Reflectivity

An electromagnetic wave of wave vector $\mathbf{q}$ and frequency ω

$$\mathbf{E} = \mathbf{E}_0 e^{i(qr - \omega t)} \tag{1}$$

in a medium is joint with a current density

$$\mathbf{j} = \varepsilon_0 \mathbf{E} + \varepsilon_0 \chi \mathbf{E} + \sigma^c \mathbf{E} \tag{2}$$

where the terms on the right are the densities of displacement, polarization, and conduction current, respectively; $\varepsilon_0 = 8.9 \times 10^{-12}$ ASV^{-1} m^{-1}; χ is the electric susceptibility, σ^c the conductivity. Writing

$$\mathbf{j} = (\sigma_1 + i\sigma_2)\mathbf{E} = (\varepsilon_1 + i\varepsilon_2)\mathbf{E} \tag{3}$$

where $\sigma_1 + i\sigma_2$ is the dynamic conductivity (including all types of current) and $\varepsilon_1 + i\varepsilon_2$ the dielectric constant, one finds that

$$\sigma_1 + i\sigma_2 = -i\omega\varepsilon_0(\varepsilon_1 + i\varepsilon_2) \tag{4}$$

and from the wave equation for electromagnetic waves,

$$q^2 = \frac{\omega^2}{c^2}(\varepsilon_1 + i\varepsilon_2) \tag{5}$$

and with $(n + ik)^2 = \varepsilon_1 + i\varepsilon_2$,

$$q = \frac{\omega}{c}(n + ik) \tag{6}$$

where $q = |\mathbf{q}|$ and n and k are real and imaginary part of the refractive index.

The reflection at a surface can be described by the Fresnel formula for perpendicular incidence,

$$R^{1/2}e^{i\Theta} = -\frac{n + ik - 1}{n + ik + 1} \tag{7}$$

where R is the reflectivity given by

$$R = \frac{(n - 1)^2 + k^2}{(n + 1)^2 + k^2} \tag{8}$$

and Θ the phase shift of the reflected wave. When the reflectivity is known for all frequencies, Θ can be determined by the Kramers–Kronig relation

$$\Theta(\omega) = -\frac{\omega}{\pi}\int_0^\infty \frac{\ln R(\omega') - \ln R(\omega)}{\omega'^2 - \omega^2} \, d\omega' \tag{9}$$

and the pairs n, k and ε_1, ε_2 and σ_1, σ_2 follow.

For samples that are not totally opaque, such as thin films on substrates, the measurement of both reflectivity and transmissivity at a frequency allows to determine n and k directly at this frequency. The Fresnel coefficient for transmission of radiation at a surface is given by

$$t = \frac{2}{n + ik + 1} \tag{10}$$

The high-T_c superconductors are highly anisotropic. The optical constants that follow from a Kramers–Kronig transformation of reflectivity data for polycrystalline material correspond to effective values that should be discussed by an effective medium theory [34]; first attempts at effective medium theories are described in Refs. 12 and 26.

B.   Free Carrier Conductivity

The dynamical conductivity of a normal metal can be described in the simplest way by the Drude conductivity

$$\sigma_1 + i\sigma_2 = \frac{Ne^2/m}{\Gamma - i\omega} \tag{11}$$

where N, e, m, and Γ are concentration, charge, effective mass, and scattering rate of the carriers. Taking into account that the effective mass and the scattering rate can be frequency dependent, a generalized Drude formula can be used,

$$\sigma_1 + i\sigma_2 = \frac{Ne^2/m(0)}{\gamma(\omega) - i\omega m(\omega)/m(0)} \tag{12}$$

where $\gamma(\omega)$ and $m(\omega)$ are frequency-dependent scattering rate and mass. This modification has first been introduced [34] to describe heavy fermion systems and has been used [35] to describe high-temperature superconductors.

In the superconducting state, for $T \ll T_c$, the carriers are undamped and the conductivity is, for a London-type superconductor, given by

$$\sigma = \frac{1}{\mu_0 \lambda_L^2} \left[\frac{\pi}{2} \delta(\omega) + \frac{i}{\omega} \right] \tag{13}$$

where the first term describes the loss less dc current, and the second loss less high-frequency currents. In the Bardeen–Cooper–Schrieffer (BCS) theory, Eq. (13) holds approximately for frequencies below the superconducting gap frequency defined by $h\nu_g = 2\Delta(0)$, where $2\Delta(0)$ is the energy gap. For $\nu > \nu_g$, Eq. (13) has to be modified according to ohmic currents; see Refs. 17 and 18 and for basic surveys, Refs. 36 and 37.

C.   Infrared Active Phonons

We describe the dielectric function by the sum of Lorentzian curves

$$\varepsilon_1 + i\varepsilon_2 = \sum_{j=1}^{n} \frac{S_j \omega_j^2}{\omega_j^2 - \omega^2 - i\omega\Gamma_j} \tag{14}$$

where the strength S_j is the contribution of the jth phonon to the static dielectric constant, ω_j the eigenfrequency, Γ_j the damping constant, and n the number of infrared active phonons. The oscillator strength of the jth phonon is

$$\int \sigma_1, j(\omega) \, d\omega \approx \frac{\pi}{2} \varepsilon_0 \omega_j^2 S_j$$

where the integration is taken over the Lorentzian line.

D. Interband and Midinfrared Absorption

Interband transitions well known from transitions in the visible spectral region may also occur in high-temperature superconductors in the infrared and far-infrared spectral regions, according to the complicated band structure of thallium cuprates. One type of interband transition can occur between subbands that arise because of coupling of the carrier motion in adjacent CuO_2 layers due to interchain charge transfer processes; the transfer energy is on the order of a few tens of meV.

Broadband midinfrared absorption that is not due to free carrier, phonon, and interband origin, for which evidence has been reported for La [38] and Y cuprates [17] (see also Ref. 39), may occur in Tl cuprates. The total dielectric function may then be described by

$$\varepsilon = \varepsilon^c + \varepsilon^p + \varepsilon^i + \varepsilon^m + 1 \tag{15}$$

taking account of free carriers (c), infrared active phonons (p), interband transitions (i), and midinfrared (m) absorption process. There is experimental evidence [i.e., asymmetric (Fano) rather than Lorentzian lineshapes for phonon resonances] that the infrared active phonons couple to the background described by the sum $\varepsilon^c + \varepsilon^i + \varepsilon^m$ [40].

E. Fourier Transform Spectroscopy

Infrared reflection and transmission spectra were obtained by use of an infrared Fourier spectrometer. By comparing the spectral intensity of radiation reflected from the sample or a gold mirror, respectively, the reflectivity of the sample was obtained. In addition to specular reflectivity, the sum of specular and diffuse reflectivity was measured, by collecting radiation in a large solid angle [41]. In this arrangement the focus of the incoming beam was small and it was possible to select a small area of a sample (typically, 0.5 mm diameter) for a reflection experiment; this arrangement has been used for studying single crystals.

It should be mentioned that far-infrared spectroscopy has been used successfully to study superconducting energy gaps in conventional superconductors [42–44] and for heavy fermion systems [45,46].

F. Optical Response of Polycrystalline Films

It has been found [47] that the optical response of current-carrying polycrystalline films can be used to determine energy gap values. While fast response was observed for $\nu < \nu_g$, bolometric response due to breaking up of Cooper pairs was found for $\nu > \nu_g$ (see Section V.C).

IV. INFRARED ACTIVE PHONONS

A. Reflectivity of the $Tl_2Ba_2Ca_nCu_{n+1}O_{2n+6}$ Series

For a survey Fig. 10 shows far-infrared reflectivity spectra of the homologous series $Tl_2Ba_2Ca_nCu_{n+1}O_{2n+6}$ ($n = 0,1,2$) for low temperature [12,10]. Common to all phases is a high reflectivity at small frequencies and a decrease toward large frequencies, and furthermore, a pronounced phonon structure. The reflectivity of the 2201 phases and the 2212 phase remains quite high at large frequencies, corresponding to a high surface quality that was obtained without further treatment of the samples after sintering. The smooth background is mainly due to electronic excitations in the (a,b) plane while the phonon structure is due to infrared active phonons with displacements in the c direction.

Common to all phases (Fig. 10) are two resonance-like reflection minima near 80 and 140 cm^{-1} that correspond mainly to Ba and Cu vibrations (Section IV.E) and to reststrahlen-like maxima near 580 cm^{-1} that can be attributed mainly to vibration of oxygen in the BaO plane against oxygen in the TlO plane (Section IV.E).

Reflectivity curves for the $n = 0, 1$ phases at different temperatures are shown in Figs. 11 to 13. The reflectivity of both the nonsuperconducting (Fig. 11) and the superconducting (Fig. 12) 2201 phases does not change much below 100 K, while strong changes are seen for the 2212 phase (Fig. 13). The sample has had a high surface quality; above about 700 cm^{-1} diffuse reflection occurred (inset of Fig. 13), indicating diffraction at the crystallites.

B. Anomalous Behavior of Phonons in $Tl_2Ba_2CaCu_2O_8$

For a discussion of characteristic properties we regard the dynamic conductivity. The result of a Kramers–Kronig analysis of the reflectivity curves is shown in Fig.

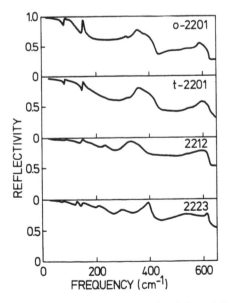

Figure 10 Far-infrared reflectivity of the $Tl_2Ba_2Ca_nCu_{n+1}O_{2n+6}$ ($n = 0,1,2$) series at low temperature ($T < 50$ K).

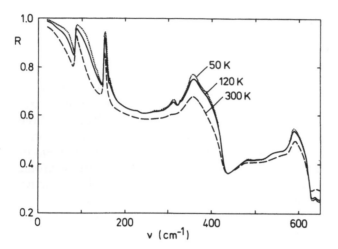

Figure 11   Reflectivity of (normal-conducting) orthorhombic $Tl_2Ba_2CuO_6$.

14. The spectra indicate six infrared active phonons. A description of the resonances by Lorentzian oscillators according to Eq. (14) delivers the data of Table 3. Damping of the modes increases with increasing resonance frequency; the $j = 2$ mode becomes narrower at low temperature.

One of the phonons ($j = 4$) shows a 3% decrease of frequency by the condensation, while the other phonon resonance frequencies are not noticeably dependent on temperature. The phonon shift is anomalously large; it can be compared to an anomalous shift of a phonon in $YBa_2Cu_3O_{7-\delta}$ (B_{1g} mode) observed by Raman spectroscopy [48,49]. The corresponding infrared active phonon in $YBa_2Cu_3O_{7-\delta}$ (at 310 cm^{-1}) also shows a softening, however only by about 1% [50]. The results (Table 3) indicate, furthermore, that the oscillator strength for all but the $j = 1$

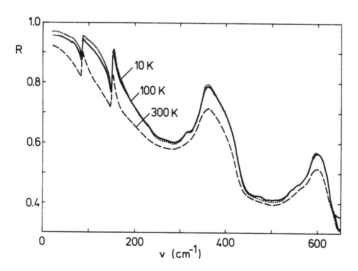

Figure 12   Reflectivity of tetragonal $Tl_2Ba_2CuO_6$ ($T_c \simeq 54$ K).

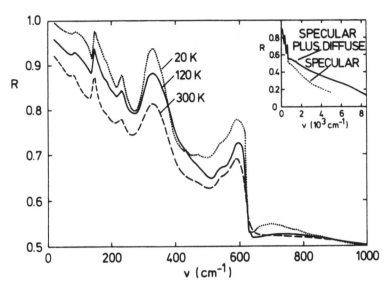

Figure 13 Far-infrared reflectivity of $Tl_2Ba_2CaCu_2O_8$ ($T_c \simeq 105$ K) and infrared reflectivity (inset).

mode are nearly twice as large in the superconducting state as in the normal state. This represents a second anomaly. The total strength (ΣS_j in Table 3) increases by about 50% by the cooling to the superconducting state.

It is still an open question whether the absolute values of the oscillator strengths are intrinsic properties of the material or dependent on the microstructure of the polycrystalline samples that should be described with an effective medium theory; for a discussion of this point, see Refs. 12 and 26. The main features with respect to the temperature dependence should, however, not be noticeably dependent on the polycrystalline structure.

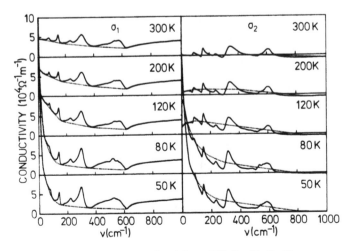

Figure 14 Dynamical conductivity of $Tl_2Ba_2CaCu_2O_8$.

Table 3   Infrared Active Phonons of $Tl_2Ba_2CaCu_2O_8$ for 300 K (and 20 K)

j	v_j (cm^{-1})	S_j	Γ_j (cm^{-1})
1	75	11 (11)	10
2	142	7 (13)	20 (10)
3	222	2 (5)	30
4	311 (302)	10 (16)	45
5	460	2 (4)	70
6	565	4 (7)	70
		$\Sigma S_j = 36$ (56)	

C.   Strong Phonon Anomalies for $Tl_2Ba_2Ca_2Cu_3O_{10}$

The 2223 phase presents anomalies even stronger than these observed for the 2212 phase. The condition for the observation was a careful sample preparation, with special emphasis on the surface region probed by the infrared radiation. Experimental results [10] obtained with samples of properties described in Section II are discussed next.

Figure 15 shows reflectivity curves, again with strong phonon resonances superimposed on a high background. Diffuse reflection (inset of Fig. 15) sets in above 600 cm^{-1}. Most remarkable is a shift in a reflection maximum (near 300 cm^{-1}) to higher frequencies by cooling from room temperature to 115 K (T_c = 114 K for this sample) and strong softening below T_c.

The conductivity curves (Fig. 16) indicate that there are eight infrared active phonons. There is a sudden change of the mode near 300 cm^{-1} just at T_c (Fig. 17). The phonon softens by 7% (Fig. 18)—the largest frequency change seen for phonons in high-temperature superconductors. Additionally, the oscillator strength and the phonon damping decrease with decreasing temperature (Fig. 18).

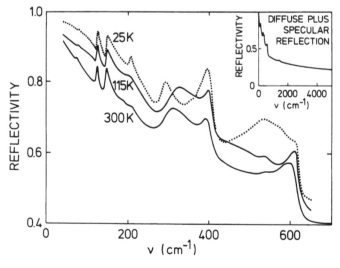

Figure 15   Reflectivity of polycrystalline $Tl_2Ba_2Ca_2Cu_3O_{10}$ ($T_c \approx$ 112 K).

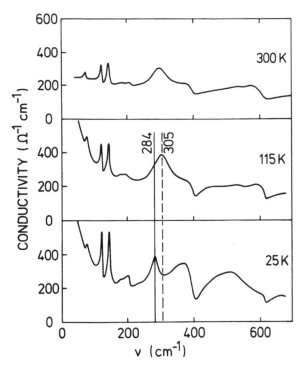

Figure 16   Conductivity of polycrystalline $Tl_2Ba_2Ca_2Cu_3O_{10}$.

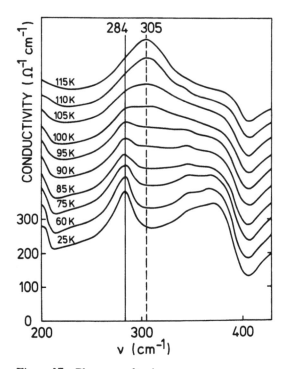

Figure 17   Phonon softening.

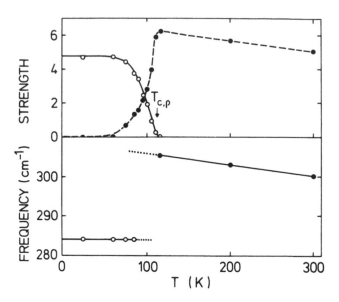

Figure 18   Anomalous phonon shift and strengths for $Tl_2Ba_2Ca_2Cu_3O_{10}$.

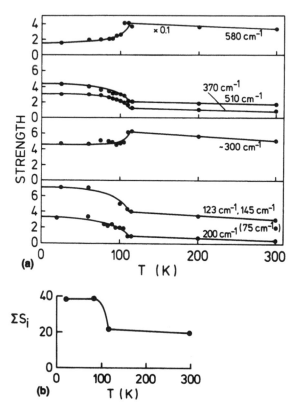

Figure 19   (a) Strengths of c-axis phonons of $Tl_2Ba_2Ca_2Cu_3O_{10}$ and (b) sum of the strengths.

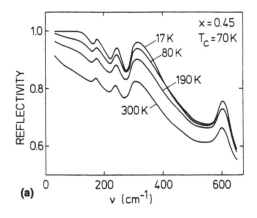

(a)

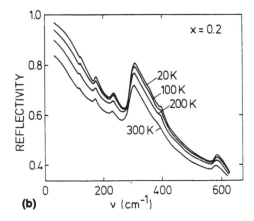

(b)

Figure 20   Reflectivity of (a) superconducting and (b) normal-conducting $Tl_{1-x}Pb_xSr_2CaCu_2O_7$.

The other phonons show only weak changes in their resonance frequencies and damping constants. However, there occurs a large change of oscillator strengths at T_c (Fig. 19a). Condensation of carriers has the consequence that oscillator strengths either decrease at T_c (for the anomalous, 300-cm^{-1} and 580-cm^{-1} phonon) or increase. The strong changes may be attributed to phonon-induced charge transfer processes. While for $Tl_2Ba_2CaCu_2O_8$, and also for $YBa_2Cu_3O_{7-\delta}$ [50], condensation of carriers leads to an increase in all oscillator strengths; for $Tl_2Ba_2Ca_2Cu_3O_{10}$, two of the modes show a decrease. This gives evidence that in the latter case Cooper pairs are involved in a phonon-induced charge transfer processes (Section IV.F).

The total oscillator strength (Fig. 19b) changes drastically at T_c, namely by a factor of about 2. Here, again, structural influence on the optical reflectivity can lead to too large absolute values of the oscillator strengths. However, the temperature dependencies are not expected to be strongly dependent on the structure, and should therefore not be regarded as artifacts.

D.   Superconducting and Normal Conducting $Tl_{1-x}Pb_xSr_2CaCu_2O_7$

In a study of $Tl_{1-x}Pb_xSr_2CaCu_2O_7$ compounds evidence was found [11] that phonon anomalies may be related with superconductivity. Figure 20a shows the reflectivity

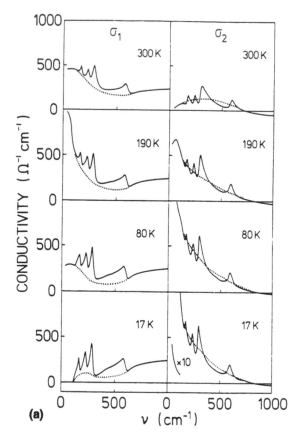

(a)

Figure 21 Dynamical conductivity of (a) superconducting and (b) normal-conducting $Tl_{1-x}Pb_xSr_2CaCu_2O_7$.

of superconducting phase ($x = 0.45$) and Fig. 20b of a normal conducting phase ($x = 0.2$). The conductivity curves (Fig. 21) show pronounced phonon structure. It is found from the conductivity curves that the oscillator strength increases strongly with decreasing temperature for all infrared active phonons in the superconducting phase (Fig. 22a), while it remains almost constant for the normal-conducting compound (Fig. 22b).

E.   Lattice Dynamics

Phonons in thallium cuprates have been studied on the basis of a shell model, taking account of short-range overlap potentials, long-range Coulomb potentials, and ionic polarizibilities (13); see also further lattice dynamical studies in Ref 13. In this study, Raman active phonon modes are also discussed.

Prominent displacements for infrared active phonon modes are shown in Fig. 23 for the four systems for which infrared data are available. In the figures the different signs are dots for Cu, circles for O, circles with crosses for Ba, large dots for Ca, and circled dots for Tl. In the figures calculated infrared active phonon

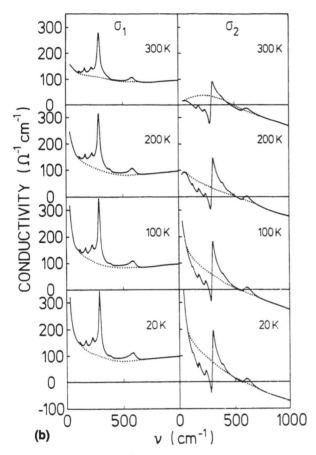

(b)

Figure 21 (continued)

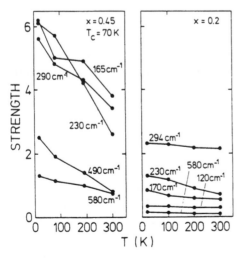

Figure 22 Phonon oscillator strengths for (left) superconducting and (right) normal-conducting $Tl_{1-x}Pb_xSr_2CaCu_2O_7$.

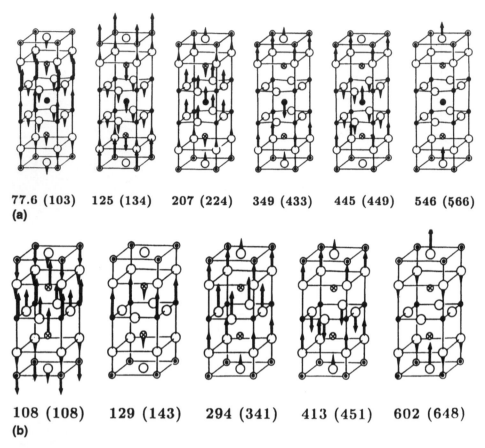

77.6 (103)   125 (134)   207 (224)   349 (433)   445 (449)   546 (566)
(a)

108 (108)   129 (143)   294 (341)   413 (451)   602 (648)
(b)

Figure 23 Infrared active c-axis phonons of (a) $Tl_{1-x}Pb_xSr_2CaCu_2O_7$, (b) $Tl_2Ba_2CuO_6$, (c) $Tl_2Ba_2CaCu_2O_8$, and (d) $Tl_2Ba_2Ca_2Cu_3O_{10}$.

frequencies and (in brackets) corresponding longitudinal frequencies in cm^{-1} are indicated.

A comparison of experimental phonon frequencies (Sections IV.A to IV.C) with the mode pictures and the calculated frequencies suggest an assignment given in Table 4. Common to all thallium cuprate listed in the table are the Ba–Tl vibration (with eigenfrequencies between 75 and 120 cm^{-1}), where the main displacements are Ba against Tl, but also with large displacements of Cu and O, then the Cu vibration (145 to 165 cm^{-1}), the O–Ba vibration (200 to 340 cm^{-1}) with vibration of all oxygen atoms mainly against Ba, the O_{Ba}–O_{Cu} vibration (460 to 520 cm^{-1}) and the O_{Ba}–O_{Tl} vibration (near 580 cm^{-1}). In the table O_{Ba} corresponds to the (bridging) oxygen atoms in the BaO layers, O_{Cu} to the oxygen atoms in the active CuO_2 layers, and O_{Cu_i} the oxygen atoms and Cu_i the copper atoms in the inactive CuO_2 layers (with $2+$ valencies of Cu_i, see Section II.A and Table 1) characteristic for the 2223 phase. Common to three of the compounds is the O_{Ba}–Ca vibration (near 290 cm^{-1}). Finally, there is the vibration Cu_i–Tl (vibration of Cu in the inactive CuO_2 layers against Tl) and the vibration O_{Cu}–O_{Cu_i} (vibration of Cu in the active CuO_2 layers against O in the inactive CuO_2 layer; note that for

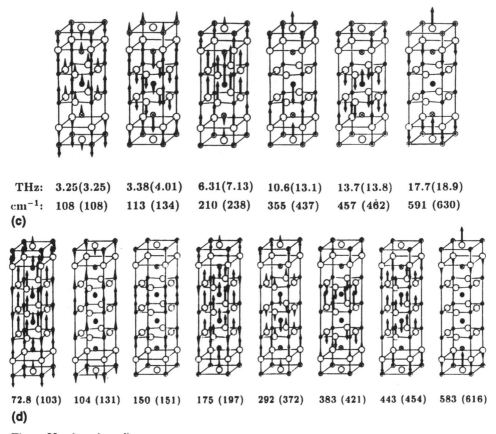

THz: 3.25(3.25)   3.38(4.01)   6.31(7.13)   10.6(13.1)   13.7(13.8)   17.7(18.9)
cm^{-1}: 108 (108)   113 (134)   210 (238)   355 (437)   457 (462)   591 (630)

(c)

72.8 (103)   104 (131)   150 (151)   175 (197)   292 (372)   383 (421)   443 (454)   583 (616)

(d)

Figure 23   (continued)

the 1212 phase the role of Ba is taken over by Sr. In the Cu vibration Cu vibrates against Ba and all O atoms (for the 1212 phase) or against Ba (2201), or against Ba and O_{Cu} (2212), or against Cu_i (2223). It is remarkable that most of the eigenmodes with similar mode pictures have similar eigenfrequencies in the different compounds, although these differ quite strongly with respect to atomic composition and microscopic structure.

Table 4   Experimental Frequencies at 10 K and (in Parentheses) Theoretical Frequencies of Infrared Active Phonons, and Assignment to Prominent Displacements

1212	2201	2212	2223	
120 (125)	85 (108)	75 (108)	75 (73)	Ba-Tl
			123 (104)	Cu_i-Tl
165 (78)	146 (129)	142 (113)	145 (150)	Cu
230 (207)	340 (294)	222 (210)	200 (175)	O-Ba
290 (349)		299 (355)	284 (292)	O_{Ba}-Ca
			370 (383)	O_{Cu}-O_{Cui}
490 (445)	~520 (413)	460 (457)	510 (443)	O_{Ba}-O_{Cu}
580 (546)	580 (602)	560 (591)	580 (583)	O_{Ba}-O_{Tl}

The O_{Ba}–Ca vibration shows the anomalous phonon softening at T_c for the 2212 phase (Section IV.B) and the 2223 phase (Section IV.C). This vibration corresponds according to the mode pictures (Fig. 23) to vibrations of atoms in the direct neighborhood of the CuO_2 layers and can therefore have a strong electrostatic interaction with the holes in the CuO_2 layers, possibly responsible for a strong electron–phonon interaction.

F. Theoretical Considerations to the Phonon Anomalies

The anomalous phonon shifts observed for the 2212 and 2223 phases and the strong temperature dependence of the oscillator strengths indicate strong electron–phonon interaction processes. In a theoretical study it has been shown [51] that the anomalous phonon shift of the phonon near 300 cm^{-1} can be explained if it is assumed that this phonon interacts with the charge carriers via low-frequency interband transitions. Condensation would shift the interband transitions because of opening of an energy gap; the interband transitions also correspond to excitations for $\mathbf{E} \| c$. The theoretical treatment is analogous to the theory of electron–phonon interaction of Raman active phonons that are, however, coupled to intraband transitions [51].

The temperature dependence of the oscillator strength may be attributed to phonon-induced charge fluctuations. Such fluctuations have been discussed for $YBa_2Cu_3O_{7-\delta}$ to explain an anomalously large oscillator strength of a phonon (with resonance frequency near 150 cm^{-1}) that corresponds mainly to a vibration of the Ba atoms against all oxygen atoms; charge transfer was suggested between oxygens in the chain and the bridging oxygen [52–54].

The experimental results for $Tl_2Ba_2CaCu_2O_8$ and $Tl_2Ba_2Ca_2Cu_3O_{10}$ suggest that charge transfer of holes occurs between adjacent CuO_2 layers. In this picture a phonon induces transfer of holes between the planes (e.g., by modulation of the charge transfer parameter that is responsible for the coupling between the planes). It follows from the experiments that such charge fluctuation processes become faster with decreasing temperature (up to T_c) possibly because of an increasing hole mobility. Below T_c the charge fluctuation processes are still faster for most of the phonons, including the anomalous phonon (near 300 cm^{-1}) of $Tl_2Ba_2CaCu_2O_8$, while two phonons of $Tl_2Ba_2Ca_2Cu_3O_{10}$ (at 300 cm^{-1} and 580 cm^{-1}) that show a decrease of the oscillator strength by condensation of charge carriers give evidence for a decrease in the charge transfer.

V. ELECTRONIC CONDUCTIVITY

A. Dynamical Conductivity of $Tl_2Ba_2CaCu_2O_8$ at Room Temperature

The dynamical conductivity in the (a, b) plane of $Tl_2Ba_2CaCu_2O_8$ has been measured [14] for the single crystal described in Section II.B. Figure 24 shows the (a, b) plane reflectivity of a $Tl_2Ba_2CaCu_2O_8$ single crystal at room temperature. The reflectivity is very similar to that of $YBa_2Cu_3O_{7-\delta}$ for $\mathbf{E} \| a$; it is large at small frequencies and decreases strongly with frequency and reaches small values near 10 000 cm^{-1}. There is, however, a difference: The reflectivity of $YBa_2Cu_3O_{7-\delta}$ is slightly larger in the range around 3000 cm^{-1} (Fig. 24). It follows from the reflectivity data for the $Tl_2Ba_2CaCu_2O_8$ crystal (Fig. 24), by a comparison with those of $YBa_2Cu_3O_{7-\delta}$,

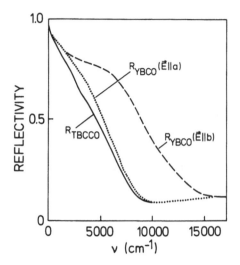

Figure 24   Reflectivity of a $Tl_2Ba_2CaCu_2O_8$ crystal and reflectivity of $YBa_2Cu_3O_{7-\delta}$.

that there is no contribution to the reflectivity from the TlO plane (as it is seen from the CuO chains). It is suggested that the TlO planes are almost nonconducting. This is consistent with large anisotropy of the dc resistivity (Fig. 5).

A Kramers–Kronig analysis delivers a dynamical conductivity (Fig. 25) that is Drude like at small frequencies and shows, as for $YBa_2Cu_3O_{7-\delta}$ in the CuO_2 plane, a high-frequency tail. For the Tl compound the effective scattering rate and the effective mass (inset of Fig. 25) show strong changes in the range of the phonon frequencies but no change at high frequencies. Possibly, interband transitions play a less important role than for $YBa_2Cu_3O_{7-\delta}$. The behavior of γ and m give evidence for strong electron–phonon interaction.

An analysis of the reflectivity data (Fig. 24) shows that the reflectivity behavior can be described by use of Eliashberg's theory [14]. The reflectivity curve can be

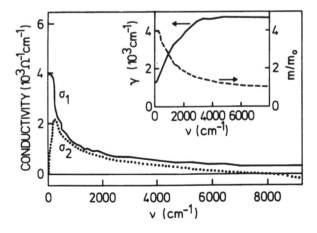

Figure 25   Dynamical conductivity, scattering rate, and effective mass (inset) for the (a,b) plane of $Tl_2Ba_2CaCu_2O_8$ at room temperature.

described with an Eliashberg function $\alpha^2 F(\omega)$ that has a strong maximum near 600 cm^{-1}. The analysis delivers evidence for electron–phonon interaction with a coupling constant $\lambda \approx 2.6$. The infrared results therefore give evidence for strong interaction of charge carriers with high-frequency phonons.

A theoretical study [54] has shown that the electron–phonon interaction can be strong in high-T_cs oxides, especially for long-range polar phonons. Influence of electron–phonon interaction on both the electronic and phononic excitations as reported in this and the preceding chapters are therefore supported by microscopic models.

It should be noted that a reflectivity curve similar to that of Fig. 24 for $Tl_2Ba_2CaCu_2O_8$ has been reported for a $Tl_2Ba_2Ca_2Cu_3O_{10}$ thin film [55]. The experimental results described in this chapter were obtained for a crystal at room temperature. As already mentioned, the high-quality areas of the crystal were not sufficient to measure with the presently available techniques the far-infrared reflectivity with sufficient accuracy to obtain conclusive results.

B. Reflectivity and Transmissivity of a Film at Low Temperature

For a polycrystalline thin film ($T_c \approx 112$ K) prepared in situ on MgO as described in Section II.C, transmissivity and reflectivity [56] have been measured (Fig. 26). Toward small frequencies the transmissivity strongly decreases and the reflectivity increases to a value near 1 for low temperature, as it is characteristic for superconductors. According to Section III.B, an analysis delivers a London penetration depth $\lambda_L \approx 400$ nm; this result is typical for polycrystalline material. The polycrystalline character of the sample (see Section II.C) is also seen in the phonon structure of both the transmission and reflection curves. There are characteristic

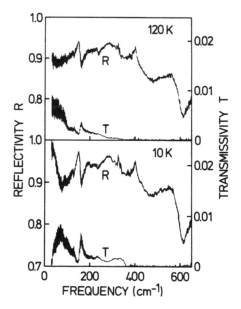

Figure 26 Transmissivity and reflectivity of a polycrystalline TlBaCaCuO thin film ($T_c \approx$ 112 K) at low temperature.

phonons appearing as minima in transmission and maxima in reflection already known from reflectivity studies (see Table 4); fine structure in the spectra is due to interference effects in the MgO plate (thickness 0.5 mm). Above T_c (Fig. 26) the phonon structure remains, but the reflectivity is, in comparison to low temperatures, smaller and the transmissivity at small frequencies is higher, which is characteristic for the normal state. The transmissivity above 200 cm^{-1} is strongly influenced by absorption of MgO that has a reststrahlen band at 400 cm^{-1}.

From the results one may guess a superconducting energy gap around 100 cm^{-1} [56] for the polycrystalline film material, consisting mainly of 2223 and 2212 phases (Section II.C).

It should be noted that further information on thallium cuprate superconductors is known with respect to phonons by Raman studies [57] and with respect to low-frequency, intragap absorption by microwave studies [58–62], and with respect to dynamical surface effects by electrochemical studies [63].

C. Energy Gap from Optical Response of Polycrystalline Films

In a recent experiment [15] it has been found that current-carrying polycrystalline Tl$_2$Ba$_2$CaCu$_2$O$_8$ films at temperatures below T_c showed fast optical response at far-infrared frequencies and that at a critical frequency the fast response disappeared and a slow, bolometric response appeared. Figure 27 shows the optical response of a film (size 5 mm × 1 mm, thickness ~ 1 μm, current ~ 10 mA) for radiation at 0.3 and 30 THz. At the smaller frequency, laser pulses of about 1 ns duration are resolved (inset), while at the higher frequency there is a bolometric response possibly due to a Cooper pair-breaking effect. The response time at the higher frequency corresponds to heat escape from the film to the substrate by thermal diffusion.

The response time (Fig. 28) changes drastically from less than 1 ns to 1 μs at a critical frequency v_c. The critical frequency may be attributed to an energy gap. It follows, with $v_c \approx 120$ cm^{-1}, that $hv_c \approx 2kT_c$ [15]. It is suggested that this gap

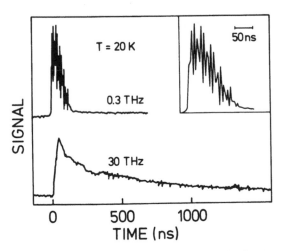

Figure 27 Response of a polycrystalline Tl$_2$Ba$_2$CaCu$_2$O$_8$ film to far-infrared (0.3 THz) and infrared (30 THz) radiation.

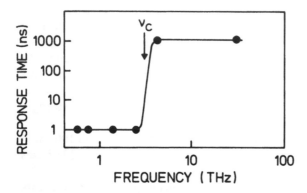

Figure 28   Response time of a $Tl_2Ba_2CaCu_2O_8$ film.

corresponds to the smallest gap occurring in randomly oriented material of $Tl_2Ba_2CaCu_2O_8$.

D.   High-Speed Far-Infrared Detectors

In the preceding section it has been shown that fast optical response up to a critical frequency of about 120 cm^{-1}. These detectors are useful for detection of short laser pulses. Cooling with liquid nitrogen is sufficient, as has been shown. In addition to the fast response, there is bolometric response due to heating of the film. The response time (about 1 μs in the experiment of Fig. 27) is given by the escape time of the heat from the film into the substrate. It should be mentioned that thallium cuprate films in magnet fields that are heated at their surfaces show large Nernst voltages, due to forced flux tube motion forced by the temperature gradient [64–67].

Fast response has also observed for $YBa_2Cu_3O_{7-\delta}$ thin films [68]. In these films fast response with 10^{-10} s response time has been observed; at weak excitation fast response has also been found for near-infrared and visible radiation [69]. Because of their higher T_cs, Tl–Ba–Ca–Cu–O superconductors should be most suitable as detector materials: at the higher temperatures relaxation processes occur faster than at lower temperatures.

VI.   CONCLUSIONS

Infrared studies of carefully prepared and characterized polycrystalline thallium cuprate high T_c samples that have high quality with respect to both bulk and surface properties indicate strong anomalies for phonons with atomic displacements along the c direction. An infrared active phonon that corresponds mainly to a vibration of the bridging oxygen atoms against the Ca atoms softens in $Tl_2Ba_2CaCu_2O_8$ by 3% of the resonance frequency (at 311 cm^{-1}) and in $Tl_2Ba_2Ca_2Cu_3O_{10}$ by 7% of the resonance frequency (at 305 cm^{-1}). The oscillator strengths of most of the phonons show strong temperature dependence, with a strong increase of oscillator strength below T_c for most of the phonons, including the phonon with the anomalous shift in the 2212 phase, while the phonon with the anomalous shift in the 2223 phase shows a strong decrease of strength; also, another phonon, at 580 cm^{-1}, of the 2223 phase that corresponds mainly to vibrations of the bridging oxygen ions against the oxygen ions in the TlO layers. These effects indicate strong electron–phonon interaction, although it remains unclear which role these phonons play for the condensation of charge carriers into the superconducting state. Anomalous behavior of the oscillator strength is also found for superconducting $Tl_{0.55}Pb_{0.45}Sr_2CaCu_2O_7$ but not for normal-conducting $Tl_{0.8}Pb_{0.2}Sr_2CaCu_2O_7$, giving evidence that phonon anomalies are related with superconductivity.

Taking account of lattice dynamical calculation of eigenfrequencies and eigenmode pictures, it was possible to relate observed infrared active phonons to well-defined modes. The analysis indicates that there are characteristic phonon modes that have nearly equal frequencies for different Tl cuprate superconductors.

The (a,b) plane conductivity of a $Tl_2Ba_2CaCu_2O_8$ single crystal at room temperature shows almost the same behavior as for the CuO_2 layers in $YBa_2Cu_3O_{7-\delta}$, indicating strong similarity in the conduction mechanism of these compounds that both have two adjacent CuO_2 layers (although of different distance from each other) in the unit cell. The experiments show that the TlO layers do not contribute significantly to the infrared conductivity and act as almost insulating sheets between the CuO_2 layers. The dynamical conductivity shows a complicated behavior, as for the CuO_2 layers in $YBa_2Cu_3O_{7-\delta}$: almost Drude-like behavior at small frequencies (up to about 300 cm^{-1} at room temperature) and a long tail at higher frequencies. The dynamical conductivity can be described by Eliashberg's theory.

With respect to the superconducting energy gap there is presently only a fragmentary knowledge. From results obtained by use of various experimental techniques energy gap values $2\Delta(0)/k_BT_c \sim 1$ to 2 have been suggested and attributed to the lowest gap values occurring in polycrystalline 2212 and 2223 thallium cuprates. Evidence for another gap with $2\Delta(0)/k_BT_c \sim 5$ to 6 may be concluded from reflectivity date for polycrystalline material of these two compounds; these values follow from the observation that condensation of charge carriers to Cooper pairs has a strong influence (for the 2212 and 2223 phases) on the reflectivity and the dynamical conductivity up to a frequency of about 400 cm^{-1}. The anomaly of the phonon near 300 cm^{-1} gives support to the belief that this phonon may have a resonance frequency near superconducting gap frequencies.

Finally, first experiments indicate that thallium cuprate films should be most suitable, because of their high T_cs, for fabricating ultrafast detectors with broadband response over a large frequency range, possibly from the submillimeter to the visible spectral range.

ACKNOWLEDGMENTS

Cooperation in the study of thallium cuprate superconductors with H. H. Otto, T. Zetterer, H. Lengfellner, J. Betz, J. Keller, and U. Schröder are gratefully acknowledged. The work was supported by the Bundesministerium für Forschung und Technologie and the Bayerischer Forschungsverbund Hochtemperatur-Supraleiter (FORSUPRA).

REFERENCES

1. Z. Z. Sheng and A. M. Hermann, *Nature* 332, 55 (1988).
2. Z. Z. Sheng and A. M. Hermann, *Nature* 332, 138 (1988).
3. S. S. P. Parkin, V. Y. Lee, E. M. Engler, A. I. Nazzal, T. C. Huang, G. Gorman, R. Savoy, and R. Beyers, *Phys. Rev. Lett.* 60, 2539 (1988).

4. M. A. Subramanian, J. C. Calabrese, C. C. Torardi, J. Gopalakrishnan, T. R. Ashkev, R. B. Flippen, K. J. Morissey, U. Chowdhry, and A. W. Sleight, *Nature* 332, 420 (1988).
5. T. Itoh, H. Uchida, and H. Uchikawa, *Jpn. J. Appl. Phys.* 27, L2052 (1988).
6. S. S. P. Parkin, V. Y. Lee, A. I. Nazzal, R. Savoy, and R. Beyers, *Phys. Rev. Lett.* 61, 750 (1988).
7. R. M. Hazen, L. W. Finger, R. J. Angel, C. T. Prewitt, N. L. Ross, C. D. Hadidiacos, P. J. Heaney, D. R. Veblen, Z. Z. Sheng, A. El Ali, and A. M. Hermann, *Phys. Rev. Lett.* 60, 1657 (1988).
8. M. Hervieu, C. Michel, and B. Raveau, *J. Less Common Met.* 150, 59 (1989).
9. T. Doi, K. Usami, and T. Kamo, *Jap. J. Appl. Phys.* 29, L57 (1990).
10. T. Zetterer, M. Franz, J. Schützmann, W. Ose, H. H. Otto, and K. F. Renk, *Phys. Rev. B* 41, 9499 (1990).
11. T. Zetterer, M. Franz, J. Schützmann, W. Ose, H. H. Otto, and K. F. Renk, *Solid State Commun.* 75, 325 (1990).
12. T. Zetterer, W. Ose, J. Schützmann, H. H. Otto, P. E. Obermayer, N. Tasler, H. Lengfellner, G. Lugert, J. Keller, and K. F. Renk, *J. Opt. Soc. Am. B* 6, 420 (1989).
13. A. D. Kulkarni, F. W. de Wette, J. Prade, U. Schröder, and W. Kress, *Phys. Rev. B* 41, 6409 (1990).
14. G. Jehl, T. Zetterer, H. H. Otto, J. Schützmann, S. Shulga, and K. F. Renk, *Europhys. Lett.* 17, (1992).
15. H. Lengfellner, Gi. Schneider, J. Betz, M. Hogan, W. Prettl, and K. F. Renk, *Europhys. Lett.* 15, 343 (1991).
16. T. Zetterer, Ph.D. thesis, Universität Regensburg, Regensburg, 1990.
17. T. Timusk and D. B. Tanner, in *Infrared Properties of High-T_c Superconductors*, ed. D. M. Ginsberg, World Scientific, Singapore, 1989, p. 339.
18. K. F. Renk, in *Studies of High Temperature Superconductors*, ed. A. Narlikar, Vol. 10, (Nova Science, Commack, N.Y.), 1992.
19. H. H. Otto, T. Zetterer, and K. F. Renk, *Naturwissenschaften* 75, 509 (1988).
20. T. Zetterer, H. H. Otto, G. Lugert, and K. F. Renk, *Z. Phys. B. Condensed Matter* 73, 321 (1988).
21. S. Martin, A. T. Fiory, R. M. Fleming, L. F. Schneemeyer, and J. V. Waszczak, *Phys. Rev. B* 41, 846 (1990).
22. H. H. Otto, T. Zetterer, and K. F. Renk, *Z. Phys. B Condensed Matter* 75, 433 (1989).
23. T. Zetterer, H. H. Otto, K. Meidenbauer, and K. F. Renk, *Physica C* 162–164, 514 (1989).
24. K. A. Müller, M. Takashige, J. G. Bednorz, *Phys. Rev. Lett.* 58, 1143 (1987).
25. B. Renker, I. Apfelstedt, H. Küpfer, C. Politis, H. Rietschel, W. Schauer, H. Wühl, U. Gottwick, H. Kneissel, U. Rauchschwalbe, H. Spille, and F. Steglich, *Z. Phys. B Condensed Matter* 67, 1 (1987).
26. K. F. Renk, W. Ose, T. Zetterer, J. Schützmann, H. Lengfellner, H. H. Otto, J. Keller, B. Roas, L. Schultz, and G. Saemann-Ischenko, *Infrared Phys.* 29, 791 (1989).
27. G. Jehl, Diplom thesis, Regensburg, 1990, unpublished.
28. B. Johs, D. Thomson, N. J. Iano, J. A. Woollam, S. H. Lui, A. M. Hermann, Z. Z. Sheng, W. Kiehl, Q. Shams, X. Fei, L. Sheng, and Y. H. Lui, *Appl. Phys. Lett.* 54, 1810 (1989).
29. H. Lengfellner, J. Betz, and K. F. Renk, *Appl. Phys. A* 48, 501 (1989).
30. M. Hong, S. H. Liou, D. D. Bacon, G. S. Grader, J. Kwo, and A. R. Kortan, *Appl. Phys. Lett.* 53, 2103 (1988).
31. G. Malandrino, D. S. Richeson, T. J. Marks, D. C. DeGroot, J. L. Schindler, and K. R. Kannewurf, *Appl. Phys. Lett.* 58, 182 (1991).
32. W. Y. Lee, V. Y. Lee, J. Salem, T. C. Huang, R. Savoy, D. C. Bullock, and S. S. P. Parkin, *Appl. Phys. Lett.* 53, 329 (1988).

33. J. Betz, A. Piehler, E. V. Pechen, and K. F. Renk, *J. Appl. Phys.* 71, 2478 (1992).
34. B. C. Webb and A. J. Sievers, *Phys. Rev. Lett.* 57, 1951 (1986).
35. G. A. Thomas, J. Orenstein, D. H. Rapkine, M. Capizzi, A. J. Millis, R. N. Bhatt, L. F. Schneemeyer, and J. Waszczak, *Phys. Rev. Lett.* 61, 1313 (1988).
36. M. Tinkham, *Introduction to Superconductivity*, McGraw-Hill, New York, 1975.
37. M. Tinkham, in *Far-Infrared Properties of Solids*, ed. S. S. Mitra and S. Nudelman, Plenum Press, New York, 1970, p. 223.
38. S. Tajima, T. Ido, S. Ishibashi, T. Itoh, H. Eisaki, Y. Mizuo, T. Arima, H. Takagi, and S. Uchida, *Phys. Rev. B* 43, 10496 (1991).
39. K. Kamarás, S. L. Herr, C. D. Porter, N. Tache, D. B. Tanner, S. Etemad, T. Venkatesan, E. Chase, A. Inham, X. D. Wu, M. S. Hegde, and B. Dutta, *Phys. Rev. Lett.* 64, 84 (1990); 64, 1962 (erratum) (1990).
40. V. M. Burlakov, S. V. Shulga, J. Keller, and K. F. Renk, *Physica C*, 190, 304 (1992).
41. W. Ose, P. E. Obermayer, H. H. Otto, T. Zetterer, H. Lengfellner, J. Keller, and K. F. Renk, *Z. Phys. B* 70, 307 (1988).
42. R. E. Glover and M. Tinkham, *Phys. Rev.* 104, 844 (1956); *Phys. Rev.* 108, 243 (1957).
43. D. M. Ginsberg and M. Tinkham, *Phys. Rev.* 118, 990 (1960); P. L. Richards and M. Tinkham, *Phys. Rev.* 119, 575 (1960).
44. L. H. Palmer and M. Tinkham, *Phys. Rev.* 165, 588 (1968).
45. F. E. Pinkerton, B. C. Webb, A. J. Sievers, J. W. Wilkins, and L. J. Sham, *Phys. Rev. B* 30, 3068 (1984).
46. P. E. Sulewski, A. J. Sievers, M. B. Maple, M. S. Torikachvili, J. L. Smith, and Z. Fisk, *Phys. Rev. B* 38, 5338 (1988).
47. H. Lengfellner, Gi. Schneider, J. Betz, M. Hogan, W. Prettl, and K. F. Renk, *Europhys. Lett.* 15, 343 (1991).
48. R. M. Macfarlane, H. Rosen, and H. J. Seki, *Solid State Commun.* 63, 831 (1987).
49. R. Thomsen, M. Cardona, B. Gegenheimer, R. Liu, and A. Simon, *Phys. Rev. B* 37, 9860 (1988).
50. G. Hastreiter, U. Hofmann, J. Keller, and K. F. Renk, *Solid State Commun.* 76, 1015 (1990).
51. R. Zeyher and G. Zwicknagl, *Solid State Commun.* 66, 617 (1988); R. Zeyher and G. Zwicknagl, *Physica C* 162–164, 1709 (1989).
52. L. Genzel, A. Wittlin, M. Bauer, M. Cardona, E. Schönherr, and A. Simon, *Phys. Rev. B* 40, 2170 (1989).
53. I. Batistić, A. R. Bishop, R. L. Martin, and Z. Tesanović, *Phys. Rev. B* 40, 6896 (1989).
54. R. Zeyher, *Z. Phys. B Condensed Matter* 80, 187 (1990).
55. I. Bozovic, J. H. Kim, J. S. Harris, Jr., and W. Y. Lee, *Phys. Rev. B* 43, 1169 (1991).
56. J. Betz, Ph.D. thesis, Universität Regensburg, Regensburg, 1993.
57. G. Burns, M. K. Crawford, F. H. Dacol, and N. Herron, *Physica C* 170, 80 (1990).
58. L. D. Chang, M. J. Moskowitz, R. B. Hammond, M. M. Eddy, W. L. Olson, D. D. Casavant, E. J. Smith, and M. Robinson, *Appl. Phys. Lett.* 55, 1357 (1989).
59. D. W. Cooke, P. N. Arendt, E. R. Gray, B. L. Bennett, D. R. Brown, N. E. Elliott, G. A. Reeves, A. D. Rollett, and K. M. Hubbard, *Appl. Phys. Lett.* 58, 1329 (1991).
60. D. W. Cooke, E. R. Gray, P. N. Arendt, G. A. Reeves, R. J. Houlton, N. E. Elliott, and D. R. Brown, *Appl. Phys. Lett.* 56, 2147 (1990).
61. Y. Hayashi, M. Fukui, T. Fujita, H. Shibayama, K. Iwahashi, and K. Adachi, *Jpn. J. Appl. Phys* 28, L-1746 (1989).
62. D. J. Keeble, D. S. Ginley, E. H. Poindexter, M. Harmatz, and H. G. Grimmeiss, *Supercond. Sci. Technol.* 3, 124 (1990).
63. A. Pinkowski, J. Doneit, K. Jüttner, W. Lorenz, G. Saemann-Ischenko, T. Zetterer, and M. Breiter, *Electrochim. Acta* 34, 1113 (1989).

64. H. Lengfellner, A. Schnellbögl, J. Betz, W. Prettl, and K. F. Renk, *Phys. Rev. B* 42, 6264 (1990).

65. H. Lengfellner, A. Schnellbögl, J. Betz, W. Prettl, and K. F. Renk, *Physica B* 165–166, 1219 (1990).

66. H. Lengfellner, A. Schnellbögl, J. Betz, K. F. Renk, and W. Prettl, *Int. J. Infrared Millimeter Waves* 11, 631 (1990).

67. H. Lengfellner and A. Schnellbögl, *Physica C* 174, 373 (1991).

68. A. D. Semenov, P. T. Lang, K. F. Renk, and I. G. Gogidze, *Solid State Common.* 80, 507 (1991).

69. A. D. Semenov, G. N. Gol'tsman, I. G. Gogidze, A. V. Sergeev, E. M. Gershenzon, P. T. Lang, and K. F. Renk, *Appl. Phys. Lett.*, 60, 903 (1992).

24
X-Ray Photoemission of Tl-Based Cuprate Superconductors

Yasuo Fukuda
Shizuoka University
Hamamatsu, Japan

Teruo Suzuki and Masayasu Nagoshi
NKK Corporation
Kanagawa, Japan

I. INTRODUCTION

Since the discovery of high-temperature (high-T_c) superconductors [1], various models have been proposed to explain the superconducting mechanism. However, the mechanism has not yet been elucidated, although much effort is being expended on the problem [2]. The physical properties in both the superconducting and normal states have been studied using many methods. The electronic structures are of special interest because (1) superconductivity appears around the boundary where insulator–metal transition occurs with carrier doping, holes, or electrons, depending on the systems, and (2) high-T_c superconductors show abnormal behavior compared to the normal metals: for example, in transport phenomena, magnetic properties, and so on [2].

The electronic structures of high-T_c superconductors have been studied using photoemission [3]; a clear Fermi edge was found by ultraviolet photoelectron spectroscopy (UPS) and inverse photoelectron spectroscopy (IPES) in the occupied and unoccupied valence bands, respectively, and the experimental results showed reasonable agreement with the band calculations. X-ray photoelectron spectroscopy (XPS) is a powerful method for identifying the valence of ions in the superconductors. It is very important to decide the valence of dopant ions because it strongly affects the superconducting properties that reflect the mechanism. For example, $YBa_2Cu_3O_7$ doped with Pr into Y sites does not show superconductivity [4], whereas the same material doped with the other rare earth metal does, which would relate to the electronic or magnetic properties of Pr. In this chapter we focus on the XPS studies of core levels for Tl-based cuprate superconductors. Valence-band studies by XPS are described briefly.

II. CORE LEVELS

It is very important to study the core levels of constituent atoms in the Tl-based cuprate superconductors in order not only to characterize the films and the bulks synthesized, but also to elucidate the electronic structures. We describe XPS core levels of Tl-4f, Cu-2p, O-1s, Ba-3d, and Ca-2p.

A. Tl-4f Core Level

The Tl-based cuprate superconductors are classified into two groups: $TlBa_2Ca_{n-1}Cu_nO_{2n+3}$ with a single Tl–O layer and $Tl_2Ba_2Ca_{n-1}Cu_nO_{2n+4}$ with double Tl–O layers. For the former, positive charge can be provided intrinsically for the ideal composition even though Tl has mono- or trivalence. However, the origin of holes for double-Tl–O-layered compounds would be questionable, assuming the trivalent Tl and the stoichiometric compositions.

Figure 1 shows the Tl-4f spectra of $TlBa_2CaCu_2O_7$ (Tl-1212), $Tl_2Ba_2Ca_1Cu_2O_8$ (Tl-2212), and $Tl_2Ba_2Ca_2Cu_3O_{10}$ (Tl-2223) along with Tl_2O_3 and Tl_2O as standard spectra for Tl^{3+} and Tl^+ [5,6], in which the first three samples were in situ cleaned by scraping under ultrahigh vacuum. The binding energies (E_b) of Tl-4$f_{7/2}$ are listed in Table 1. The E_b values of Tl^{3+} and Tl^+ are 117.4 and 118.6 eV, respectively. On the other hand, the E_b values of Tl-1212, Tl-2212, and Tl-2223 are 117.7, 118.1, and 118.2 eV, respectively. From these results it was concluded that Tl in Tl-1212 is trivalent, and that in Tl-2212 and Tl-2223 is between mono- and trivalent. To make it sure, we sputtered the samples above because it is expected that Tl(I)

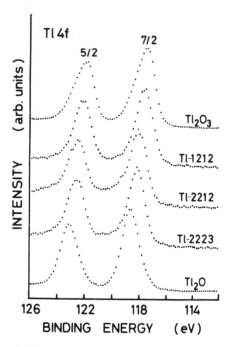

Figure 1   XPS Tl-4f spectra for Tl_2O_3, Tl-1212, Tl-2212, Tl-2223, and Tl_2O. $\frac{7}{2}$ and $\frac{5}{2}$ denote Tl-4$f_{7/2}$ and Tl-4$f_{5/2}$, respectively.

Table 1 Binding Energies of C-1s (Hydrocarbon) and Tl-4$f_{7/2}$ for the Tl-Based Cuprate Superconductors Tl_2O and Tl_2O_3[a]

Compound	Binding energy (eV)		Refs.
	C-1s	Tl-4$f_{7/2}$	
Tl_2O	284.9	118.6	5, 6
$Tl_2Ba_2Ca_2Cu_3O_{10}$			
Slightly contaminated	284.9	118.3	5, 6
Scraped in vacuum	Not observed	118.2	5, 6
$Tl_2Ba_{2-x}Ca_{2+x}Cu_3O_{10+y}$			
Cleaved single crystal	—	118.2	8
Cleaved polycrystalline	—	117.8	8
$Tl_2Ba_2CaCu_2O_8$			
Scraped in vacuum	—	118.1	8
$Tl_2Ba_2CuO_6$			
Scraped in vacuum	—	118.2	14
Tl_2O_3			
Scraped in vacuum	Not observed	117.4	5, 6
$TlBa_2CaCu_2O_7$			
Scraped in vacuum	—	117.7	5
$(Tl_{0.5}Pb_{0.5})Sr_2(Ca_{1-x}Th_x)Cu_2O_y$			
Scraped in vacuum	—	~118	7

[a]Energies for thin films are excluded because the composition is not clear.

would be reduced to Tl(0) or not and Tl(III) to Tl(I) or Tl(0) by sputtering. It was found that the E_b values of Tl_2O, Tl-2212, and Tl-2223 are kept constant and those of Tl_2O_3 and Tl-1212 are shifted to 118.0 and 118.1 eV, respectively, which are close to the E_b values of Tl-2212 and Tl-2223. Therefore, the results confirm the conclusion given above.

XPS study was also carried out on $(Tl_{0.5}Pb_{0.5})Sr2(Ca_{1-x}Th_x)Cu_2O_y$ bulk samples [7]. The E_b values of Tl-4$f_{7/2}$ and Tl-4$f_{5/2}$ were about 118 and 122.5 eV, which are not in agreement with our result. It was found that the Tl-4$f_{7/2}$ line consistently shifts slightly to the high-binding-energy side as Th increases. This was explained as follows: The oxidation state of Tl decreased as the hole concentration decreased, due to the substitution of Th^{4+} into Ca^{2+} sites.

XPS spectra were measured on both single- and polycrystalline $Tl_2Ba_{2-x}Ca_{2+x}Cu_3O_y$ [8]. For the single crystal, the E_b values of Tl-4$f_{7/2}$ and Tl-4$f_{5/2}$ were 118.2 and 122.6 eV, respectively, which are in good agreement with our result, although these shifted by 0.4 eV for polycrystalline to the low-binding-energy side. The origin of the binding energy shift was speculated to relate to oxygen stoichiometry.

To fabricate high-quality Tl–Ba–Ca–Cu–O superconducting thin films, the chemical state of Tl in the films was examined by XPS [9]. The change in Tl oxidation state with time was studied on a film deposited using an excimer laser (XeCl), in which Tl-4$f_{7/2}$ lines were deconvoluted into Tl^{3+} (117.4 for 4$f_{7/2}$ and 121.9 eV for 4$f_{5/2}$) and Tl^+ (118.6 and 123.1 eV). The percentage of Tl^{3+} was found to decrease as a function of time. The Tl-4$f_{7/2}$ states with E_b = 115.8 and 116.7

eV were also found for films deposited under no oxygen background pressure. They were attributed to negatively charged Tl because their binding energy was lower than that of elemental Tl [9].

The chemical state of Tl in thin films deposited by radio-frequency magnetron sputtering on $SrTiO_3$ substrates was characterized by XPS [10]. It was found that the E_b values of $Tl\text{-}4f_{7/2}$ were 118.0, 118.1, and 119.3 eV at the surface, in the bulk, and near the interface with the substrate, respectively. From the result, the valence of Tl in the films was suggested to be between $3+$ and $1+$, which is in good agreement with our results.

We should emphasize here that we have to be careful in explaining the data on thin films because Ar^+ ion sputtering was usually employed for cleaning the films. As pointed out earlier, Tl^{3+} is reduced to the lower valence by sputtering, although Tl with $1+$ and with $3+$ to $1+$ is not changed.

It is well known that T_c is largely dependent on the hole concentration in a $Cu\text{-}O_2$ layer, and the optimum value determining the highest T_c is about 0.2. For $Tl_2Ba_2Ca_2Cu_3O_{10}$ with a T_c higher than 100 K, the hole concentration in a $Cu\text{-}O_2$ layer was estimated experimentally to be 0.2 to 0.3 [11]. We have two possibilities for hole creation in Tl–Ba–Ca–Cu–O systems with double Tl–O layers. First, assuming a trivalence of Tl, partial substitution of Ca^{2+} for Tl^{3+} would provide holes. This was confirmed by electron probe microanalysis (EPMA) [12] and the analysis of high-resolution images obtained by transmission electron microscope (TEM) [13]: About 15% of Tl in $Tl_2Ba_2Ca_2Cu_3O_{10}$ is substituted for Ca. However, the hole concentration caused by the substitution would not be sufficient for T_cs above 100 K, because an average hole concentration in a $Cu\text{-}O_2$ layer is estimated to be 0.1. The second possibility is a charge transfer of $Tl^{3+} + (Cu\text{-}O_2)^0 \rightarrow Tl^{(3-\delta)+} + (Cu\text{-}O_2)^{\delta+}$, which leads to a formation of holes in the $Cu\text{-}O_2$ layers. The XPS results strongly support this possibility because the valence of Tl was found to be between $+3$ and $+1$, although it is difficult to estimate an exact valence. $Tl_2Ba_2CuO_6$ would have excess holes because of having only one $Cu\text{-}O_2$ layer, which led to an increase in T_cs with the substitution of La^{3+} for Ba^{2+} sites [14]. On the other hand, holes are automatically doped for $TlBa_2CaCu_2O_7$ even though Tl is trivalent.

B.   Cu-2p Core Level

In general, one of the characteristics of high-T_c superconductors is that insulator materials become metallic with hole or electron doping, with the result that the carrier-doped materials show superconductivity. It is well known that the $Cu\text{-}O_2$ layer in cuprate superconductors is a path of carriers that serve the superconductivity. Therefore, the electronic state of Cu that might be affected by the doping is important in a study of the mechanism.

The $Cu\text{-}2p_{3/2}$ core-level spectra of cuprate superconductors show a main line (M) at about 933 eV due to the well-screened core-hole final state of the $2p^5 3d^{10}$ L configuration, in which the L denotes a ligand hole, and a broad satellite (S) centered at about 942 eV due to the poorly screened final state of the $2p^5 3d^9$ configuration. The relative intensity of the satellite (I_S) with respect to the main peak (I_M) would essentially be related by the charge-transfer excitation energy between the Cu-3d and O-2p levels and the hybridization strength of the Cu–O bond, according to a simple CuO_4^{6-} cluster model in D_{4h} symmetry with a $3d_{x2-y2}$ orbital on Cu and the $2p\sigma_{xy}$ orbitals on the oxygens [15].

Figure 2 shows Cu-$2p$ spectra for $Tl_2Ba_2Ca_2Cu_3O_{10}$, in which the main lines ($\frac{3}{2}$ and $\frac{1}{2}$) and the satellites (S) are indicated. The binding energy of the main line is 933.8 eV and that of the satellite is centered at about 942 eV for $2p_{3/2}$. These binding energies were almost the same within experimental error for Tl-1212, Tl-2201, Tl-2212, and Tl-2223, which implies that most of copper in the superconductors is divalent [16]. For thin films prepared by radio-frequency sputtering [10], it was reported that the binding energies of the Cu-$2p$ main lines on the surface and in the bulk of the films lie near the Cu^+ state.

We find that the intensity ratios of the satellite to the main line, I_S/I_M, are 0.37, 0.45, 0.39, and 0.38 $\pm$ 0.04 for Tl-1212, Tl-2201, Tl-2212, and Tl-2223, respectively. The values for single- and polycrystalline Tl-2223 were compared [8]: The former (0.21) is less than the latter (0.38). It was found for $(Tl_{0.5}Pb_{0.5})Sr_2(Ca_{1-x}Th_x)Cu_2O_y$ [5] to be 0.26, 0.27, 0.23, 0.29, and 0.27 for $x = 0, 0.05, 0.10, 0.20$, and 0.30, respectively, which shows the almost constant value for x, although electrons would be doped by the substitution of Th. Furthermore, the ratio was obtained for the same family: 0.28 for $TlSr_2Ca_{0.5}Pr_{0.5}Cu_2O_y$ ($T_c \sim 90$ K) and 0.22 for $Tl_2Ba_2CaCu_2O_8$ ($T_c \sim 50$ K) [17]. On the other hand, it was between 04. and 0.5 for semiconductors, $Tl_{0.5}Pb_{0.5}Sr_2Ca_{0.5}Y_{0.5}Cu_2O_y$ and $Tl_{0.5}Pb_{0.5}Sr_2YCu_2O_y$ [17].

It was reported that the relative intensity of the satellite decreases with an increase in the T_cs of $Tl_2Sr_2Ca_{n-1}Cu_nO_{2n+4}$ and $TlSr_2Ca_{1-x}Nd_xCu_2O_7$ [18]. The relationship between the ratio and the T_c was discussed in terms of the charge-transfer energy and the hole concentration. Assuming that some parameters including the charge-transfer energy (Δ), the I_S/I_M value and the binding-energy separation between the main and the satellite lines (ΔE) were calculated. It was claimed that the result of the calculation is in good agreement with that of the

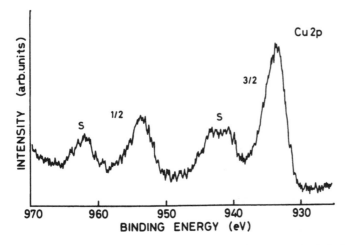

Figure 2 Typical Cu-$2p$ spectrum for Tl-2223. The labels $\frac{3}{2}$, $\frac{1}{2}$, and S denote Cu-$2p_{3/2}$, Cu-$2p_{1/2}$, and satellite, respectively.

experiment: the I_S/I_M value decreases and the ΔE increases with decreasing Δ. Furthermore, the I_S/I_M and the T_cs were related to x and the hole concentration (n_h) for $Bi_2Sr_2Ca_{1-x}R_xCu_2O_{8+\delta}$ (R = Y and Yb). From the discussion above it was pointed out that the intensity ratio is inversely related to n_h as well as the superconducting transition temperature.

Values of the ratios for Tl-based cuprate superconductors and T_c are listed in Table 2. As seen in the table, there seems to be no relationship between them. We emphasize that careful interpretation of XPS data should be required because XPS is a surface-sensitive technique. For example, the ratio was different between the single- (0.21) and polycrystalline (0.38) forms of $Tl_2Ba_2Ca_2Cu_3O_{10+\delta}$ despite the same T_c (118 K), which was ascribed to the nature of the exposed surface [8]. Another serious problem for measurement of the intensity ratio is a background subtraction that strongly affects the intensity of XPS core levels. We do not yet have the best subtraction method, although several methods have been proposed [19]. For resolution of the XPS spectrometer it would also be important to have the correct value of the ratio.

C. O–1s Core Level

The O–1s core-level lines are usually observed at about 529 eV and between 531 and 532 eV. It is well known that the former is attributed to oxygens in the crystal lattice of the cuprate superconductors and the others are due to contaminants containing oxygen: for example, hydroxide and carbonate groups [20,21]. Therefore, lines between 531 and 532 eV are not observed for the clean surfaces of

Table 2   Intensity Ratio of the Satellite to the Main Line (I_S/I_M) for Cu-$2p_{3/2}$ and T_c for Tl-Based Cuprate Superconductors

Compound	I_S/I_M	T_c (K)	Ref.
$Tl_2Ba_2Ca_2Cu_3O_{10}$	0.38	114	This work
$Tl_2Ba_{2-x}Ca_{2+x}Cu_3O_{10+y}$			
Polycrystalline	0.38	118	8
Single crystal	0.21	118	8
$Tl_2Ba_2Ca_2Cu_3O_{10}$	0.19	—	18
$Tl_2Ba_2CaCu_2O_8$	0.39	98	This work
$Tl_2Ba_2CaCu_2O_8$	0.34	—	18
$Tl_2Ba_2CaCu_2O_8$	0.22	—	17
$Tl_2Ba_2CuO_6$	0.45	non	This work
$TlBa_2CaCu_2O_7$	0.37	73	This work
$TlSr_2Ca_{0.75}Nd_{0.25}Cu_2O_7$	0.37	82	18
$TlSr_2Ca_{0.25}Pb_{0.75}Cu_2O_7$	0.43	78	18
$TlSr_2Ca_{0.5}Pr_{0.5}Cu_2O_y$	0.28	90	17
$Tl_{0.5}Pb_{0.5}Sr_2CaCu_2O_y$	0.26	77	7
$Tl_{0.5}Pb_{0.5}Sr_2Ca_{0.95}Th_{0.05}Cu_2O_y$	0.27	87	7
$Tl_{0.5}Pb_{0.5}Sr_2Ca_{0.9}Th_{0.1}Cu_2O_y$	0.23	89	7
$Tl_{0.5}Pb_{0.5}Sr_2Ca_{0.8}Th_{0.2}Cu_2O_y$	0.29	71	7
$Tl_{0.5}Pb_{0.5}Sr_2Ca_{0.7}Th_{0.3}Cu_2O_y$	0.27	54	7

single- and polycrystalline superconductors [22]. No lines higher in energy than 529 eV would become fingerprints on the clean surface.

Figure 3 shows the O–1s spectrum of $Tl_2Ba_2Ca_2Cu_3O_{10}$, in which two lines are observed at 528.8 and 531.0 eV, showing the existence of contaminants on the surface, even though the sample was in situ scraped. A 531-eV line was also found on the single crystal of $Tl_2Ba_{2-x}Ca_{2-x}Cu_3O_{10}$ [8]. It was attributed to impurities exposed during cleaning, a process that would preferentially occur at weak spots in the crystal. It was claimed that a feature at about 530 eV was also found, although the presence is unclear because the large feature at 531 eV overlaps with it. The 531-eV line has been discussed [8] in terms of a contribution from the d^9L grand state, an energy loss, and screening, but it is not clear so far. The 531-eV line was also observed for $(Tl_{0.5}Pb_{0.5})Sr_2(Ca_{1-x}Th_x)Cu_2O_y$ [7]. For Tl–Ba–Ca–Cu–O thin films [10], the binding energies of the main O–1s peaks were 530.8, 529.8, and 531.5 eV at the surface, in the bulk, and near the interface, respectively. It appeared to be very difficult to make a pure sample Tl–Ba–Ca–Cu–O system, although only clean O–1s spectra were obtained in Ref. 23.

D. Ba-3d and Ca-2p Core Levels

The chemical state of barium in cuprate superconductors seems to be complicated because the Ba-3d and Ba4d lines have several components [20,21]. It is well known that the binding energy of Ca-2$p_{3/2}$ for $Bi_2Sr_2Ca_{n-1}Cu_nO_{2n+4}$ is significantly lower than that for CaO [24]. In this section the binding energies of Ba-3d and Ca-2p for Tl-based cuprate superconductors are discussed in terms of the chemical states and compared with those for other cuprate superconductors.

Figure 4 shows the Ba-3d spectrum for $Tl_2Ba_2Ca_2Cu_3O_{10}$. The binding energies of Ba-3$d_{5/2}$ and Ba-3$d_{3/2}$ for various Tl-based cuprate superconductors are listed in Table 3. We observe the Ba-3$d_{5/2}$ and Ba-3$d_{3/2}$ lines at 779.4 and 794.6 eV, respectively, and weak and broad shoulders are also found at the lower- and higher-binding-energy sides of the main lines, respectively. The broad shoulder at higher energy was observed for all samples listed in the table except for $Tl_2Ba_2CaCu_2O_8$.

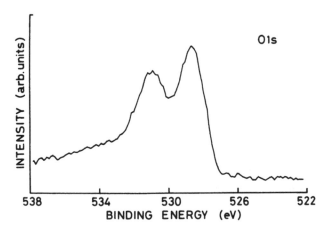

Figure 3 Typical O–1s spectrum for Tl-2223. The 531-eV peak is due to contaminants.

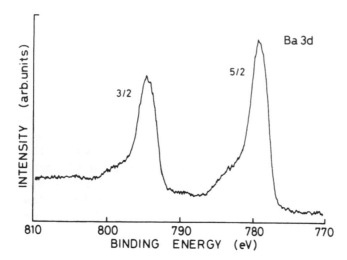

Figure 4 Typical Ba-3d spectrum for Tl-2223. The labels $\frac{5}{2}$ and $\frac{3}{2}$ denote Ba-3$d_{5/2}$ and Ba-3$d_{3/2}$, respectively.

The weak shoulder at lower energy was also found for all samples listed, but it became a main line for $Tl_2Ba_2CuO_6$.

The chemical state of barium was studied in detail for $YBa_2Cu_3O_y$ [20,21]. The high-resolution XPS spectra of Ba-3$d_{5/2}$ were divided into three components at about 780, 779, and 778 eV. The first was ascribed to barium carbonate and/or hydroxide and the second to barium in superconductors. The 778-eV line was observed previously, but the origin is not yet clear. The low-binding-energy line was also observed for Ba-4d on a cleaved single-crystal surface using synchrotron orbit radiation [25]. This peak was identified as a surface component since it is most intense at $E_F = 50$ eV in the constant final-state measurement. However, an opposite interpretation was reported [26,27]: The 778-eV line was attributed to the bulk state.

Table 3 Binding Energies of Ba-3$d_{5/2}$ and Ba-3$d_{3/2}$ for Various Tl-Based Cuprate Superconductors

Compound	Binding energy (eV)		Ref.
	$3d_{5/2}$	$3d_{3/2}$	
$TlBa_2CaCu_2O_7$	778.8	794.0	This work
$Tl_2Ba_2CuO_6$	777.9	793.2	This work
$Tl_2Ba_2CaCu_2O_8$	779.6	795.0	This work
$Tl_2Ba_2Ca_2Cu_3O_{10}$	779.4	794.6	This work
$Tl_2Ba_2Ca_2Cu_3O_{10}$			
Polycrystalline	779.2	—	8
Single crystal	778.3	—	8

Most of the Ba-$3d_{5/2}$ binding energy for the samples [Tl-1212, Tl-2223, and Tl-2223(poly)] listed in the table is at about 779 eV. However, it is higher for Tl-2212 than for the others. This would be due to the fact that the intensity of the O–1s line at 531 eV was larger than that for the others. The binding energies for Tl-2201 and Tl-2223 (single crystal) are close to 778 eV, which is due to the stronger intensity of the 778-eV line than that of the 779- and 980-eV lines. This might imply that barium was exposed to surfaces during the cleaning process.

Broad bands at about 784 and 799 eV can be seen in Fig. 4. The broad bands were also found at the higher-binding-energy side of the main lines [8], although the binding energies are a little different from that in Fig. 4. Since the energy separation of the bands is about 15 eV, which is equal to spin–orbit separation between Ba-$3d_{5/2}$ and Ba-$3d_{3/2}$ [27], they could be attributed to barium compound with no conductivity, leading to a charge-up shift to the higher-binding-energy side.

For the thin films [10], the binding energy of $3d_{5/2}$ was 779.7 eV for the surface, the bulk, and the interface. The value is close to that of barium carbonate and/or hydroxide [20,21]. Since the O–1s spectra consisted of lines with higher binding energy than 529 eV, the barium would be contaminated during the preparation process.

Figure 5 shows Ca-2p spectra of $Tl_2Ba_2Ca_2Cu_3O_{10}$, and binding energies of Ca-2$p_{3/2}$ and Ca-2$p_{1/2}$ for various Tl-based superconductors are listed in Table 4. They are close to about 345 eV, although values are a little higher for Tl-2212 than for the others. The value 345 eV is in good agreement with that for $Bi_2Sr_2Ca_{n-1}Cu_nO_y$ [28], but this is much lower than those for Ca^{2+} ions, which are usually between 346 and 348 eV [16]. Moreover, it is lower than that for Ca metal (345.7 eV) [24]. One reason might be the different Madelung potential for the superconductors from that for the calcium oxide.

It is indicated in Fig. 5 that the spectrum contains more than two spin–orbit pairs. For Tl-1212, the Ca-2$p_{3/2}$ and Ca-2$p_{1/2}$ lines could not be separated, due to overlap of the pairs. The two pairs were attributed to Ca–Ba disorder within the unit cell [8].

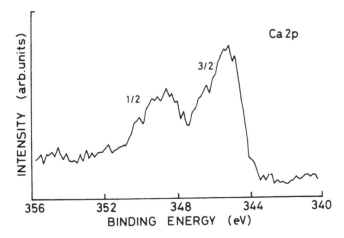

Figure 5 Typical Ca-2p spectrum for Tl-2223. The labels $\frac{3}{2}$ and $\frac{1}{2}$ denote Ca-2$p_{3/2}$ and Ca-2$p_{1/2}$, respectively.

Table 4   Binding Energies of Ca-$2p_{3/2}$ and Ca-$2p_{1/2}$ for
Various Tl-Based Cuprate Superconductors

Compound	Binding energy (eV)		Ref.
	$3d_{5/2}$	$3d_{3/2}$	
TlBa$_2$CaCu$_2$O$_7$	345.0	—	This work
Tl$_2$Ba$_2$CaCu$_2$O$_8$	345.8	349.6	This work
Tl$_2$Ba$_2$Ca$_2$Cu$_3$O$_{10}$	345.3	348.6	This work
	~345	~348.6	8

Ca-$2p_{3/2}$ lines for the thin films were observed at 345.1, 345.7, and 347.2 eV at the surface, in the bulk, and near the interface, respectively [10]. The latter is very close to the binding energy of CaO, although the first two are in agreement with the value of the bulk samples. This result suggests that calcium oxide was formed near the interface, which is of interest for study of the mechanism of thin-film formation.

III.   VALENCE BAND

Figure 6 shows the valence-band spectrum centered at about 4 eV below the Fermi level, which is in good agreement with previous work [8]. Shallow core-level peaks due to Tl-5d and Ba-5p are observed between 12 and 16 eV. A clear Fermi edge is not seen in the figure as well as in YBa$_2$Cu$_3$O$_y$ [22]. Moreover, it was not even observed for the single crystal [8].

According to a calculated photoemission spectrum at $h\nu$ = 1486.6 eV [29], the valence band consists primarily of Cu-d orbitals and weak Tl-s and Tl-d orbitals. The calculation predicted that the main band centers at about 3 eV and the density

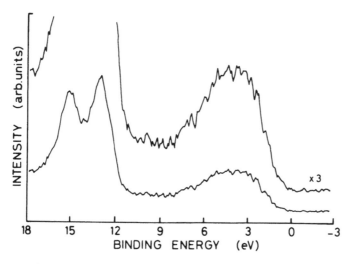

Figure 6   XPS valence-band spectra for Tl-2223. A clear Fermi edge is not seen in the spectra. Peaks between 12 and 16 eV are due to Tl-5d superimposed with Ba-5p.

of states at the Fermi level is large enough to detect, which is not in agreement with the experimental results described above. One reason for the invisible density of states at the Fermi level might be alternation of surface exposed to ultrahigh vacuum. We would need to keep the sample at a low temperature to detect the Fermi edge as well as in $YBa_2Cu_3O_y$ [30].

IV. SUMMARY

We described the electronic states of core levels of Tl-4f, Cu-2p, O–1s, Ba-3d, and Ca-2p and valence bands for various Tl-based cuprate superconductors. It was shown that XPS spectra not only gave us chemical information for constituent elements in the superconductors but was also used as a fingerprint. Although this technique is very useful, it has limitations. Since XPS is very sensitive to the surface layers, we have to pay attention to cleaning sample surfaces. Ar^+ ion sputtering would change the chemical states as described in Section II.A. Scraping and fracturing would be the best methods for cleaning the samples, although the former altered the surface electronic states for $Nd_{2-x}Ce_xCuO_y$ [31,32] and $Ba_{1-x}K_xBiO_3$ [32,33] and the grain boundary was exposed by the latter for $YBa_2Cu_3O_y$ [20]. XPS might not detect the electronic states of one unit cell with a large c axis, for example Bi- and Tl-based superconductors, but information on the $c/2$ unit cell in the c direction would be obtained [34]. Therefore, XPS could be used to examine the bulk electronic structures of superconductors.

Detailed band structures of Tl-based cuprate superconductors could be studied for single crystals using ultraviolet photoelectron spectroscopy (UPS) and inverse photoelectron spectroscopy (IPES), among which the later is an important technique for the study of unoccupied electronic states because the carriers are holes for many high-T_c cuprate superconductors. Although an IPES study was reported [35], it is not described in this chapter. Systematic study using XPS, UPS, and IPES on clean samples is required to elucidate the electronic structures of high-T_c superconductors.

ACKNOWLEDGMENTS

The authors thank Y. Syono, M. Kikuchi, M. Tachiki, and S. Nakajima for useful discussions. One of them (Y.F.) thanks the Ministry of Education, Science and Culture of Japan for support through a grant in aid for scientific research.

REFERENCES

1. J. G. Bednorz and K. A. Müller, *Z. Phys. B* 64, 189 (1986).
2. For example, H. Kamimura and A. Oshiyama, eds., *Mechanics of High-Temperature Superconductivity*, Springer-Verlag, Berlin, 1989.
3. P. A. P. Lindberg, Z. X. Shen, W. E. Spicer, and I. Lindau, *Surf. Sci. Rep.* 11, 1 (1990).
4. L. Soderholm, K. Zhang, D. G. Hinks, M. A. Beno, J. D. Jorgensen, C. U. Segre, and I. K. Schuller, *Nature* 328, 604 (1987).
5. T. Suzuki, M. Nagoshi, Y. Fukuda, S. Nakajima, M. Kikuchi, Y. Syono, and M. Tachiki, *Physica C* 162–164, 1387 (1989).
6. T. Suzuki, M. Nagoshi, Y. Fukuda, Y. Syono, M. Kikuchi, N. Kobayashi, and M. Tachiki, *Phys. Rev. B* 40, 5184 (1989).
7. W. M. Hurng, S. R. Horng, S. F. Wu, C. Y. Shei, W. H. Lee, and P. T. Wu, *Solid State Commun.* 77, 269 (1991).

8. H. M. Meyer III, T. J. Wagener, and J. H. Weaver, *Phys. Rev. B* 39, 7343 (1989).
9. K. H. Young, E. J. Smith, M. M. Eddy, and T. W. James, *Appl. Surf. Sci.* 52, 85 (1991).
10. G. Subramanyam, F. Radpour, V. K. Kapoor, and G. H. Lemon, *J. Appl. Phys.* 68, 1157 (1990).
11. J. Clayhold, N. P. Ong, P. H. Hor, and C. W. Chu, *Phys. Rev. B* 38, 7016 (1988).
12. M. Kikuchi, T. Kajitani, T. Suzuki, S. Nakajima, K. Hiraga, N. Kobayasi, H. Iwasaki, Y. Syono, and Y. Muto. *Jpn. J. Appl. Phys.* 28, L477 (1989).
13. K. Hiraga, D. Shindo, M. Hirabayashi, M. Kikuchi, N. Kobayashi, and Y. Syono, *Jpn. J. Appl. Phys.* 27, L1848 (1988).
14. S. Nakajima, M. Kikuchi, T. Oku, N. Kobayashi, T. Suzuki, Y. Syono, and M. Tachiki, *Physica C*, 458 (1989).
15. D. D. Sarma and S. G. Ovchinnikov, *Phys. Rev. B* 42, 6817 (1990).
16. C. D. Wagner, W. M. Riggs, L. E. Davis, J. F. Moulder, and G. E. Muilenberg, in *Handbook of X-ray Photoelectron Spectroscopy*, Perkin-Elmer, Eden Prairie, Minn., 1978.
17. C. N. R. Rao, A. K. Ganguli, and R. Vijayaraghavan, *Phys. Rev. B* 40, 2565 (1989).
18. C. N. R. Rao, G. R. Rao, M. K. Rajumon, and D. D. Sarma, *Phys. Rev. B* 42, 1026 (1990).
19. P. M. A. Sherwood, in *Practical Surface Analysis*, Vol. 1, *Auger and X-ray Photoelectron Spectroscopy*, ed. D. Briggs and M. P. Seah, Wiley, New York, 1990, p. 581.
20. Y. Fukuda, T. Suzuki, M. Nagoshi, Y. Syono, and M. Tachiki. *Mol. Cryst. Liq. Cryst.* 184, 389 (1990).
21. Y. Fukuda, M. Nagoshi, T. Suzuki, Y. Namba, Y. Syono, and M. Tachiki, *Phys. Rev. B* 39, 11494 (1989).
22. J. H. Weaver, H. M. Meyer III, T. J. Wagener, D. M. Hill, Y. Gao, D. Peterson, Z. Fisk, and A. J. Arko, *Phys. Rev. B* 38, 4668 (1988).
23. P. Steiner, S. Hufner, A. Jungmann, V. Kinsinger, and I. Sander, in *High T_c Superconductors: Electronic Structure*, ed. A. Bianconi and A. Marcell, Pergamon Press, Oxford, 1989, p. 169.
24. H. V. Doveren and J. A. T. Verhoeven, *J. Electron Spectrosc. Relat. Phenom.* 21, 265 (1980).
25. R. Liu, C. G. Olson, A. B. Yang, C. Gu, D. W. Lyuch, A. J. Arko, R. S. List, R. J. Bartlett, B. W. Veal, J. Z. Liu, A. P. Paulikas, and K. Vandervoot, *Phys. Rev. B* 40, 2650 (1989).
26. C. C. Chang, M. S. Hedge, X. D. Wu, B. Dutta, A. Inam, T. Venkatesan, B. J. Wilkens, and J. B. Wachtman, Jr., *J. Appl. Phys.* 67, 7483 (1990).
27. G. Frank, Ch. Zieler, and W. Gopel, *Phys. Rev. B* 43, 2828 (1991).
28. D. M. Hill, H. M. Meyer III, J. H. Weaver, C. F. Galls, and K. C. Goretta, *Phys. Rev. B* 38, 11331 (1988).
29. P. Marksteiner, J. Yu, S. Massidda, A. J. Freeman, J. Rediuger, and P. Weinberger, *Phys. Rev. B* 39, 2894 (1989).
30. R. S. List, A. J. Arko, Z. Fisk, S. W. Cheong, S. D. Conradson, J. D. Thompson, C. B. Pierce, D. E. Peterson, R. J. Bartlett, N. D. Shinn, J. E. Schirber, B. W. Veal, A. P. Paulikas, and J. C. Campuzano, *Phys. Rev. B* 38, 11966 (1988).
31. T. Suzuki, M. Nagoshi, Y. Fukuda, K. Oh-ishi, Y. Syono, and M. Tachiki, *Phys. Rev. B* 42, 4263 (1990).
32. M. Nagoshi, T. Suzuki, Y. Fukuda, K. Oh-ishi, K. Ueki, A. Tokiwa, M. Kikuchi, Y. Syono, and M. Tachiki, proceedings of the *4th International Symposium on Superconductivity* (ISS'91), PYP-19, Tokyo, Oct. 14–17, 1991, Springer-Verlag, Tokyo, in press.
33. M. Nagoshi, Y. Fukuda, T. Suzuki, K. Ueki, A. Tokiwa, M. Kikuchi, Y. Syono, and M. Tachiki, *J. Phys. Condensed Matter*, to be published.
34. M. P. Seah and W. A. Dench, *Surf. Interface Anal.* 1, 2 (1979).
35. P. Steiner, S. Hufner, A. Jungmann, V. Kinsinger, and I. Sander, *Z. Phys. B* 74, 173 (1989).

25

Nuclear Resonance Studies of Tl-Based High-T_c Superconductors

Hans B. Brom
Leiden University
Leiden, The Netherlands

I. INTRODUCTION

For high-T_c superconductors, nuclear resonance has been of crucial importance in the present understanding of the underlying mechanisms for the normal (and superconducting) properties of these materials [1]. Part of this can be traced back to the reliability of the results obtained: nuclear magnetic and nuclear electric quadrupolar resonance (NMR and NQR, respectively) are not that surface sensitive, and data on $YBa_2Cu_3O_{7-\delta}$, which is the best studied example among the high-T_c compounds, for various groups are in good agreement. In addition, nuclear resonance is an excellent probe for antiferromagnetic excitations in the material, which are hard to access otherwise and are probably linked to the superconducting mechanism.

In Section III we discuss the most important data obtained in $YBa_2Cu_3O_{7-\delta}$ by ^{89}Y, ^{63}Cu, and ^{17}O NMR. The normal-state NMR properties can be described satisfactorily by a one-spin model. Differences in the Cu and Y/O nuclear relaxation rates show the importance of magnetic correlations, the basis of the phenomenological model of Millis et al. [2]. The interpretation of recent measurements in the superconducting state [3–5] is still open. Many of the data can be explained by d-state pairing. Although NMR and neutron scattering results are closely connected, the existing data cannot be reconciled completely.

In Section IV we turn to the Tl-based high-T_c compounds $Tl_1Ba_2Ca_{n-1}Cu_nO_{2n+3}$ with $n = 1, 2, 3$ and $Tl_2Ba_2Ca_{n-1}Cu_nO_{2n+4}$ with $n = 1, 2, 3$. NMR studies have been concerned primarily with the Tl double-layer compounds. Both ^{203}Tl and ^{205}Tl are excellent NMR nuclei with nuclear spin $I = \frac{1}{2}$. In $Tl_2Ba_2CuO_6$ a slight increase of 0.02 in oxygen content strongly reduces the superconducting transition temperature from 80 K to below 4 K, while increasing the metallicity of the compound. This doping dependence is reflected in the NMR Knight shift and relaxation data. Due to the Ca^{2+} substitution by Tl^{3+}, $Tl_2Ba_2CaCu_2O_8$ bears a strong analogy with the oxygen-deficient 123 compounds. The relatively narrow normal-state width of the Tl resonance line allows an accurate study of properties of the flux pattern in a highly two-dimensional material. Finally, study of hole concentration in the various CuO_2 layers in $Tl_2Ba_2Ca_2Cu_3O_{10}$ by, for example, Cu NMR and indirectly by Tl NMR is feasible and might reveal screening of the inner layer by the outer ones [6].

As summarized in Section V, the high anisotropy of the material, Tl/Ca substitution, and "overdoping" in $Tl_2Ba_2CuO_6$ distinguish Tl compounds from other high-T_c superconductors, such as $YBa_2Cu_3O_{7-\delta}$. Further exploration of these differences will be rewarding.

II. NMR

Since the 1950s NMR has been a valuable tool in the study of static and dynamic properties of materials and at present has found a wide range of applications in physics, chemistry, biology, and medicine. Its still-growing influence stems from sophisticated pulse programs, efficient algorithms, combined with powerful computers (e.g., in two-dimensional NMR or NMR imaging) and line-narrowing procedures (e.g., magic angle spinning) [7,8]. In solid-state physics such as that of the high-T_c materials, these new developments, although certainly important, have a limited application. The nuclei under study often have linewidths that are too broad to be reduced. Line positions, relaxation rates, and linewidths measured by standard NMR pulse techniques are then the main parameters that can be obtained [9,10].

A standard technique in pulse NMR is the echo method. By application of a $\pi/2$ pulse the nuclear magnetization component along the direction of the static field B_0 (the z axis) is rotated to the plane $\perp B_0$. After a delay time τ a second (π) pulse, twice the duration of the $\pi/2$ pulse, is applied that refocuses the magnetization in the $\perp$ plane after a time 2τ. The height of the echo is then proportional to the magnetization originally along the z axis, M_z. By slightly different procedures two important relaxation rates can be measured. The measuring cycle of the nuclear Zeeman relaxation rate (T_1^{-1}) usually starts by a comb of $\pi/2$ pulses, by which M_z is destroyed. Thereafter one measures the recovery of M_z via the echo method. In a determination of the transverse or spin–spin relaxation rate T_2^{-1}, the memory loss in the refocusing process is measured by varying the delay time τ in the echo method. Every cycle starts with the same rotation of M_z to the $\perp$ plane. T_2 is a measure for the time dependence of the local field, which is often caused by the interaction between the spins. If the echo in the time domain is Fourier transformed to the frequency domain, the position and width of that curve can be analyzed. The shift of the resonance line with respect to its reference (the position of the same nucleus in a "neutral" surrounding) gives valuable information about the chemical surrounding (chemical shift). In metals unpaired spins of the electrons in the conduction band give an additional shift (Knight shift) that is sensitive to the character of the bands at the Fermi level (e.g., s- or d-like). One contribution to the linewidth stems from the T_2 processes mentioned above and is called homogeneous broadening. A distribution of static (on the time scale of the experiment) magnetic fields gives an inhomogeneous contribution.

In simple metals only the delocalized s-electrons are strongly hyperfine coupled to the nucleus and responsible for the Knight shift and the relaxation rates [10,11]. The relative shift of the line with respect to its atomic position, the Knight shift $K = \Delta B/B$, is expressed by,

$$K = \frac{\Delta B}{B} \approx \frac{8\pi}{3} < |u_k(0)|^2 > <\|>_{E_F} \chi_0 \tag{1}$$

The density of electrons at the Fermi energy level at the nuclear site is given by $< |u_k(0)|^2 > <\|>_{E_F}$, the spin susceptibility by χ_0. For transition metals such as Pt, the inner-core s-electrons can be polarized due to electrons in other states (e.g., d electrons) at the Fermi level. In Pt metal this core polarization effect is very strong and is the reason for the unusual large negative Knight shift [12]. In such a complicated case the Knight shift is the sum of the separate contributions of the s and d electrons:

$$K_{\text{tot}} = K_s + K_d \tag{2}$$

If there is a certain spectral density at the frequency of the nuclear Zeeman splittings (in an applied magnetic field B) there can be an exchange of energy that will equilibrate the populations of the Zeeman levels involved. If the thermal reservoir responsible for the spectral density is formed by electronic spins ($S = \frac{1}{2}$), the transition probabilities W that determine the Zeeman relaxation rates of the various nuclei are functions of the electronic susceptibility. For nuclear spin $I = \frac{1}{2}$ and magnetic field B oriented along the x axis the relaxation rate $T_{1n}^{-1} = 2W$ is exponential and described by

$$\frac{1}{T_{1n}} = \lim_{\omega \to 0} \frac{1}{2\hbar\mu_B^2} \sum_{q,\bar{x}} F_{\bar{x}}^2(q) S(q,\omega) = \lim_{\omega \to 0} \frac{k_B T}{2\hbar^2\mu_B^2} \sum_{q,\bar{x}} F_{\bar{x}}^2(q) \frac{\chi''(q,\omega)}{\omega} \tag{3}$$

where the direction perpendicular to the x axis is indicated by $\bar{x}$, q denotes the wave vector, and μ_B is the Bohr magneton. The real and imaginary parts of the electronic susceptibility are given by χ' and χ'' and are assumed to be isotropic. $S(q,\omega)$ is the dynamic structure factor or scattering function and is the same function that determines the scattered intensity in neutron data. The direct hyperfine interaction between a nuclear spin, I, and an electronic spin, S, at a particular site n can be written as $\mathcal{H}_{hf} = A_x I_n^x S_n^x + A_y I_n^y S_n^y + A_z I_n^z S_n^z$. The form factors $F_x(q)$ in Eq. (3) denote the components of the spatial Fourier transforms of the hyperfine tensor A. Using the same notation the expression for the Knight shift becomes

$$K_x = \frac{1}{2\mu_B\gamma_n\hbar} F_x(0)\chi_0 \tag{4}$$

In simple metals, where the Fermi contact term of the s electrons dominates, Eq. (4) reduces to Eq. (1). Equations (3) and (4) then lead to the Korringa relation [11]:

$$T_1 T K^2 = \frac{\hbar^2}{4\pi k_B} \frac{\gamma_e^2}{\gamma_n^2} \tag{5}$$

This equation links the relative Knight shift (K) to the nuclear-spin lattice relaxation rate (T_1^{-1}). The gyromagnetic ratio of the electrons, $\gamma_e = 2\mu_B/\hbar$ is typically a factor 10^3 larger than that of the nuclei γ_n. Although different independent electronic contributions to the Knight shift might cancel each other, they always add in the relaxation rate

$$\frac{1}{T_{\text{tot}}} = \frac{1}{T_s} + \frac{1}{T_d} \tag{6}$$

Modifications of the simple Korringa picture are needed if electron–electron interactions are important [13].

If the spin I of the nuclei is higher than $\frac{1}{2}$—^{63}Cu and ^{65}Cu both have $I = \frac{3}{2}$ and ^{17}O has a spin $I = \frac{5}{2}$—electric field gradients at the nucleus might give a splitting in zero magnetic field. If the zero-field splittings are large enough (for ^{63}Cu in the high-T_c compounds they are on the order of 20 to 30 MHz) the relaxation rates and Knight shifts can be determined not only by NMR but also by zero-field NQR. The NQR and NMR relaxation rates are related to the same transition probability W.

III. $YBa_2Cu_3O_{7-\delta}$

An early controversy in the theoretical description of the high-T_c materials is how far one can go in simplification without losing essential properties. In the high-T_c materials it is now accepted that the CuO_2 plane is the plane of interest. Furthermore, in these transition metal oxides [14], both Cu and O sites participate in the charge transfer. This means that the simplified picture presented by Anderson [15] of layers with only one relevant site needs additional justification. Indeed, starting from a layer with Cu and O sites, a one-site representation of the CuO_2 plane could be obtained [16] for the dilute doped case, although not under all circumstances [17]. If we apply such a picture to $YBa_2Cu_3O_{7-\delta}$, it means that the undoped material (O_6) has only Cu^{2+} spins. After doping, the extra spin of the doped hole is compensated by the existing Cu spins [i.e., only the Cu spin survives (one-spin model)]. If the additional hole, which resides mainly on the O sites, is not compensated, Cu and O spins are present (two-spin model). The example given is only (an important) one out of a variety of microscopic and phenomenological approaches that are developed to describe the static and dynamic NMR properties. It illustrates how the discussion about one or two spins has it consequences for the microscopic picture. In the following we pay special attention to the phenomenological one-spin approach of Millis et al. [2], referred to as the MMP model, which is notable for its simplicity and success.

NMR and NQR data in $YBa_2Cu_3O_{7-\delta}$ [5,18–23] show that a one-spin model works quite satisfactorily. In the MMP model these spins can be shown to be strongly antiferromagnetically coupled. How to make the connection to the underlying microscopic mechanism is not yet elucidated, although progress has been made. For example, recent calculations in the $t - J$ model [24] (one of the models based on a microscopic picture) seem to converge to the experimental relaxation data.

A major role in the NMR interpretation of the Knight shift data is played by the quantum chemical analysis by Mila and Rice [25]. They showed that the electron–nuclear interaction for the various nuclei in $YBa_2Cu_3O_{7-\delta}$ could be represented in the following way:

$$\mathcal{H}(^{63}Cu(2)) = A_{\parallel}{}^{63}I_n^z \cdot S_n^z + A_{\perp}{}^{63}I_n^x \cdot S_n^x + B \sum_k {}^{63}\mathbf{I}_n \cdot \mathbf{S}_{n+k} \tag{7}$$

where k labels the four unit cells nearest n and $A_{\parallel}$ and $A_{\perp}$ are the two relevant components of the hyperfine tensor.

$$\mathcal{H}(^{17}O) = \sum_l C_{\parallel}{}^{17}I_n^z \cdot S_{n+l}^z + C_{\perp}{}^{17}I_n^x \cdot S_{n+l}^x \tag{8}$$

with l labeling the two unit cells sharing n:

$$\mathcal{H}(^{89}Y) = \sum_m D_{\parallel}{}^{89}I_n^z \cdot S_{n+m}^z + D_{\perp}{}^{89}I_n^x \cdot S_{n+m}^x \tag{9}$$

In Eq. (9), m labels the eight unit cells nearest n.

As the expression for the Cu(2) Knight shift contains $A_{\parallel}$, $A_{\perp}$, and B as three free parameters and the experiment gives only the Knight shift in the $\parallel$ and $\perp$ directions, $K_{\parallel}$ and $K_{\perp}$, respectively,

$$K_{\parallel} = \frac{(A_{\parallel} + 4B)\chi_0}{\gamma_e^{63}\gamma_n\hbar^2} \tag{10}$$

$$K_{\perp} = \frac{4B(1 + \alpha)\chi_0}{\gamma_e^{63}\gamma_n\hbar^2} \tag{11}$$

one parameter remains undetermined. The unknown parameter is often chosen to be the ratio α between $A_{\perp}$ and $4B$ [26].

The description of the Knight shift in $YBa_2Cu_3O_{7-\delta}$ for various values of δ with the same parameter set (the only free parameter is α, although the spin susceptibility χ_0 is often also derived from the Knight shift data; see Fig. 1) is a strong point for the one-spin model [19,26,27].

The pronounced differences in the T dependencies of the relaxation rates between the Cu(2)—the copper sites in the CuO_2 plane—on the one hand, and O and Y, on the other (Fig. 2), can be traced back to the q dependence of the form factor F in Eq. (3) [20]. Formally,

$$^{63}W_{\parallel} \propto \lim_{\omega \to 0} \sum_q [A_{\perp} - 2B(\cos q_x a + \cos q_y a)]^2 S(q,\omega) \tag{12}$$

$$^{63}W_{\perp} \propto \lim_{\omega \to 0} \sum_q ([A_{\parallel} - 2B(\cos q_x a + \cos q_y a)]^2 + [A_{\perp} - 2B(\cos q_x a + \cos q_y a)^2]^2) S(q,\omega) \tag{13}$$

For isotropic hyperfine interactions between the electronic spin and the oxygen and yttrium nuclear spins, one gets

$$^{17}W \propto \lim_{\omega \to 0} \sum_q [2C^2(1 - \cos q_x a)]S(q,\omega) \tag{14}$$

$$^{89}W \propto \lim_{\omega \to 0} \sum_q \left[16D^2 \frac{\cos^2 q_z a}{2} (1 - \cos q_x a)(1 - \cos q_y a)\right]S(q,\omega) \tag{15}$$

The wave vectors q are measured from the zone corner $(Q,Q) = (\pi/a, \pi/a)$. Strong enhancements in $S(q,\omega)$ or $\chi''(q,\omega)$ for certain $\mathbf{q}$ values [e.g., due to antiferromagnetic (AF) correlations between the Cu spins] might be felt by the Cu spins, but for symmetry reasons will cancel at the oxygen or yttrium sites. The AF cor-

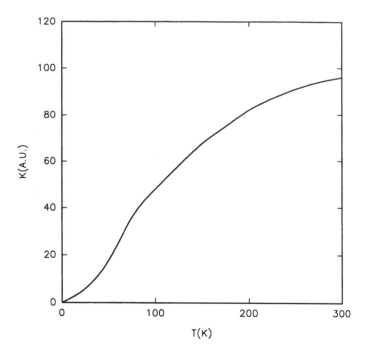

Figure 1 Knight shifts in arbitrary units for Cu, Y, and O in YBa$_2$Cu$_3$O$_{6.63}$. The data scale with the susceptibility. (From Refs. 19 and 27.)

relations have $(q_x, q_y) = (0,0)$ and drop out the expressions for ^{17}W and ^{89}W. Only ^{63}W is enhanced by $S(0,\omega)$.

This idea is the basis of the MMP model. A normal Fermi liquid and an antiferromagnetic part make up the susceptibility. The antiferromagnetic part has been assumed to have a Lorentzian shape with a correlation length ξ:

$$\chi''_{AF}(q,\omega) = \frac{\pi \chi_0 \hbar \omega}{\Gamma} \frac{(\xi/a)^4}{[1 + (\mathbf{q} - \mathbf{Q})^2 \xi^2]^2} \tag{16}$$

The real and imaginary part of the susceptibilities are linked via the Kramers–Kronig relations. Application leads to

$$\chi'_{AF}(q,\omega) = \chi_0 \frac{(\xi/a)^2}{[1 + (\mathbf{q} - \mathbf{Q})^2 \xi^2]} \tag{17}$$

The real and imaginary part of the normal Fermi liquid contributions of the quasi-particles to the susceptibility are

$$\chi''_{QP} = \frac{\omega \pi \chi_0 \hbar}{\Gamma} \tag{18}$$

and

$$\chi'_{QP} = \chi_0 \tag{19}$$

Apart from $\xi(T)$ and the broadening Γ, the model has as an additional parameter β, which gives the relative strength of the AF fluctuations to the zone center

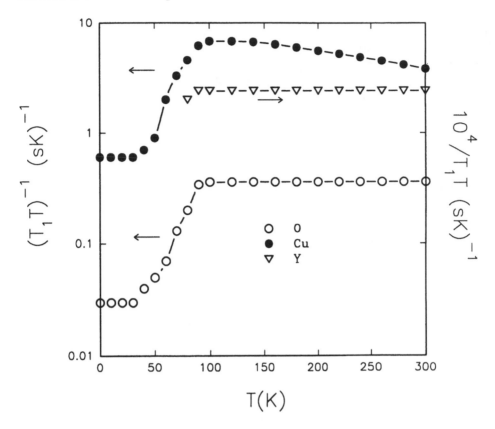

Figure 2 Korringa product $(T_1 T)^{-1}$ for different nuclei in $YBa_2Cu_3O_7$ [19,20]. The Y-relaxation rate (scale at the right) is a factor 10^3 to 10^4 lower than that of O or Cu (scale at the left).

fluctuations.

$$\chi''(q,\omega) = \chi''_{QP} + \beta\chi''_{AF} \qquad (20)$$

The proposed form of the susceptibility gives a satisfactory description of the T_1 data—probing the imaginary part—in $YBa_2Cu_3O_{7-\delta}$ with various oxygen deficiencies [4]. Also the real part of the susceptibility, which is accessible via T_2 [5], is in fair agreement with the computed value.

As shown above, with the three free parameters α, β, and $\xi(T)$ the MMP model reproduces the NMR data pretty well [26]. If the essential features are correct, it points to marginal Fermi liquid behavior: Strong antiferromagnetic correlations with a coherence length of a few lattice spacings coexist with free electron behavior.

The scattering function $S(q,\omega)$, which after summing in q-space, determines the relaxation rate of the various nuclei, is measured directly by neutron scattering. The agreement between NMR [28] and neutron data [29] on the same single crystals is not perfect. An even more troublesome question is raised by the pseudogap observed in the neutron spectra, which cannot be reconciled with the present NMR data.

Most of the discussion had dealt with properties in the normal state. Knight shift data in the superconducting state show the disappearance of the spin susceptibility [4,23], as expected for BCS-like spin pairing. Compared to the normal state, the relaxation rate below T_c is strongly diminished [18]. Absence of the coherence peak, expected for simple BCS singlet-state pairing [30], can be explained in various ways, one of which is d-state pairing. Also, the anisotropy and magnetic field dependence of the planar Cu(2) relaxation rate [4,31–33] below T_c might point to a BCS spin–singlet orbital d-wave pairing state [3,34].

IV. $Tl_2Ba_2CuO_{6+\delta}$, $Tl_2Ba_2CaCu_2O_8$, and $Tl_2Ba_2Ca_2Cu_3O_{10+\delta}$

As in $YBa_2Cu_3O_{7-\delta}$, most NMR studies are performed on oriented powders. Due to the anisotropy in the orbital diamagnetism, small crystallites can be oriented in a large magnetic field. If mixed with a slowly hardening epoxy, well-oriented samples can be obtained. For $YBa_2Cu_3O_{7-\delta}$ this technique has been proven to be very useful. Samples obtained in this way have the advantage over single crystals that the dimensions of the grains can be kept smaller than the penetration depth of the radio-frequency (RF) field, so that no RF signal is lost. Only grains that are single crystallites can be completely oriented. A small fraction of nonoriented powder might therefore be present and complicate the analysis.

The ^{205}Tl, ^{63}Cu, and ^{17}O NMR studies on the high-T_c Tl compounds have been inspired largely by the unusual normal and superconducting state properties of these materials [35–56]. The variation in the number of CuO_2 planes between 1 and 5 and the extremely high electronic anisotropy, as reflected in, for example, transport measurements, allow the study of the influence of these parameters on the T_c and other electronic properties. Furthermore, ^{63}Cu NQR spectra and the chemical shift tensors of the various nuclei give information about the crystal structure and ionicities. We discuss the NMR data per compound. At the end we make a comparison and discuss the outlook.

By ^{205}Tl NMR also, the peculiar properties of the flux "lattice" due to the low irreversibility or freezing temperature [42,47,51,52] can be studied. These studies are dealt with in a separate subsection.

A. CuO_2 Monolayers: $Tl_2Ba_2CuO_{6+\delta}$

$Tl_2Ba_2CuO_{6+\delta}$ shows a strong depression of T_c under a slight increase in oxygen content [57]. By an increase in δ of 0.023, T_c drops from 80 K to less than 4 K. The extra oxygen uptake is accompanied by a decrease in length of the c axis. If self doping (i.e., the transfer of electrons from the CuO_2 planes to the TlO layers) is possible, the number of holes in the CuO_2 plane can also be influenced by c-axis contraction. Measurements of the Tl Knight shift and relaxation might reveal possible changes in the Tl valence state or the number of conduction electrons. Figure 3 shows a typical (nonaligned) powder spectrum of $Tl_2Ba_2CuO_{6+\delta}$ [48,49].

The resonance line is located close to that of Tl^{3+} from pure $K_4Tl_2O_5$. The asymmetric line shape is due to the axial symmetry surrounding of the Tl site. Modulations in the structure will give additional line broadening [58–60]. The line shapes and positions of the ^{205}Tl resonances of samples with different oxygen contents are slightly different. At room temperature the relaxation rates [35,49] almost coincide. The data of the nonsuperconducting samples are almost Korringa-like, while those of the samples with the highest T_cs are reminiscent of that of Cu(2) in $YBa_2Cu_3O_{7-\delta}$ (Fig. 4).

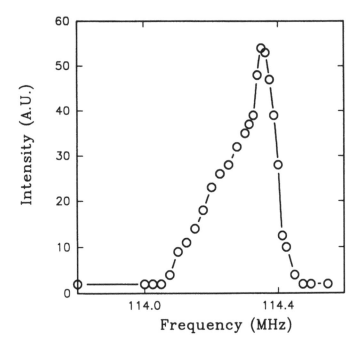

Figure 3 ^{205}Tl lineshape in a magnetic field of 4.7 T for $Tl_2Ba_2CuO_{6+\delta}$ [48,49]. The reference line of Tl^{1+} from a solution of Tl_2SO_4 in dilute nitric acid is at 114.08 MHz and of Tl^{3+} from pure $K_4Tl_2O_5$ at 114.28 MHz, which is close to the peak in the powder spectrum.

The proportionality of the Cu(2) and Tl–N relaxation rates suggests an admixture of the Cu spins in the Tl-6s orbitals. The location on top of the Cu(2) nuclei allows the antiferromagnetic correlations that enhance the Cu relaxation rate to have a similar effect on the thallium T_1. If we consider only the isotropic hyperfine interaction, the ratio between the Tl and Cu relaxation rates is determined by the Knight shifts (K) and nuclear gyromagnetic ratios (γ):

$$\frac{T_1(\text{Tl})^{-1}}{T_1(\text{Cu})^{-1}} = \frac{K^2(\text{Tl})\gamma^2(\text{Tl})}{K^2(\text{Cu})\gamma^2(\text{Cu})} \tag{21}$$

The hyperfine constant for the Tl-6s orbital is about 16 times larger than that of that for Cu-4s. The gyromagnetic ratio of the Tl nucleus is about 2.2 times larger than that of Cu. Due to these figures only 0.2% of Cu spins in the Tl-6s orbital are needed to explain the measurements. In a simple model calculation analogous to that of Mila and Rice [25], such an admixture is shown to be plausible [50].

If the crystallites in the powders measured are not completely oriented, it is difficult to draw precise conclusions about the anisotropy of the Knight shift. Alignment of single crystals, on the other hand, is relatively easy. For the Cu-monolayer compound, no data on aligned single crystals are available, but single-crystal data of $Tl_2Ba_2CaCu_2O_8$, where the same processes determine the NMR properties, show an almost isotropic Knight shift. The absence of a large anisotropy supports the 6s character of the relevant Tl orbital and a complete filling of the 5d orbitals. Therefore, the Tl valence is indeed 3 +, as was also concluded from the marker position [48]. As far as these Tl NMR data are concerned, there is no

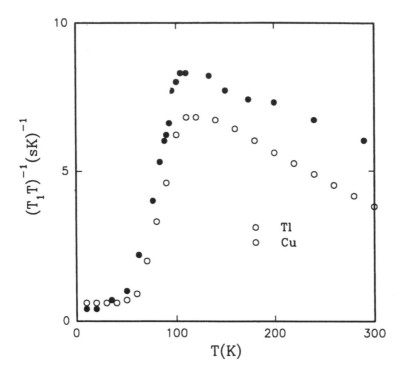

Figure 4 ^{205}Tl and ^{63}Cu relaxation rates in $Tl_2Ba_2CuO_{6+\delta}$. The ^{63}Cu relaxation rates are taken from $YBa_2Cu_3O_7$ (see Fig. 2). (From Refs. 20 and 50.)

evidence of self-doping: In the TlO layers the Tl valency remains $3+$ over the entire doping range.

Based on the NMR data the concentration dependence of the number of holes in this Tl compound is sketched in Fig. 5. Figure 5 is similar to that of Torrance [61] for $La_{2-x}Sr_xCuO_4$; only the maximum of T_c is adapted. It shows that by a small excess of oxygen, the hole concentrations can be made higher than is possible in $YBa_2Cu_3O_{7-\delta}$. As pointed out by De Castro et al. [62], the suppression of T_c might be correlated with the excess of holes with local d_z^2 symmetry in the context of the Zhang–Rice approach [16].

If in $Tl_2Ba_2CuO_{6+\delta}$ [35] the Cu and Tl relaxation rates for samples with different transition temperatures are compared, one finds that the $^{63}(1/T_1T)$, which is largely enhanced by the AF spin correlations in $La_{2-x}Sr_xCuO_4$ or $YBa_2Cu_3O_{7-\delta}$ and has the Curie–Weiss T-dependence of $C/(T + \Theta)$, is markedly depressed with increasing hole density. In the lightly doped material (the highest T_c) Curie–Weiss-like behavior is present (see Fig. 4). Also, the anisotropy $^{63}R = (1/T_1)_{ab}/(1/T_1)_c$ reveals the presence of AF correlations between nearest-neighbor Cu sites. In the heavier doped nonsuperconducting samples, a Korringa behavior is found. The anisotropy has disappeared [36,36]. These results point out that there may be an intimate interrelation between the occurrence of superconductivity and the presence of AF correlations, although the magnetic correlation length, ξ_M, is supposed to be as small as the lattice spacing a. The T_1 relaxation rates in the superconducting state compare well with those of $YBa_2Cu_3O_{7-\delta}$. This can be seen as evidence for d-state

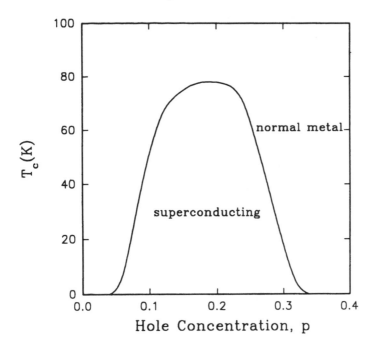

Figure 5 T_c as a function of the number of holes in $Tl_2Ba_2CuO_{6+\delta}$. This curve is obtained from the T_c dependence on hole concentration in $La_{2-x}Sr_xCuO_4$ [61] by multiplying with the ratio between the maximum transition temperatures. Only the part above $p = 0.2$ seems to be accessible for the Tl compound.

pairing [63], but other mechanisms, such as impurities, motion of the flux lattice (for the in-field data), and so on, can give additional relaxation paths.

B. CuO_2 Double Layer: $Tl_2Ba_2CaCu_2O_8$

NMR spectra show two Tl sites [39,44,51]. The most intense lines stems from the TlO layers and is referred to as the normal (N) line. The second, less intense line arises from Tl at a Ca site [44] and will be called the impurity or I line. In Fig. 6 we show the typical line shape for $H\|c$. The Tl–I position in the crystal is analogous to that of Y in $YBa_2Cu_3O_{7-\delta}$. The Tl substitution on the Ca site removes holes from the CuO_2 layer. A similar undoping is brought about in $YBa_2Cu_3O_{7-\delta}$ by oxygen vacancies in the CuO chains, where the depletion effects on the NMR properties are well studied. By the study of single crystals with the same transition temperature, samples with a narrow distribution in Tl substitution are possible [51], and the chemical and Knight shift can be studied accurately. For samples with a transition temperature T_c of 105 K, the angular dependence of the Tl shift of the N line follows the simple equation

$$\nu = -A \cos 2\theta + C \tag{22}$$

This equation shows the axial symmetry for the Tl site, as expected from the crystal structure. The values at room temperature in a field of 4.7 T are $A = 130$ kHz (or 1144 ppm) and $C = 114.765$ MHz (or 1232 ppm). The shifts are related

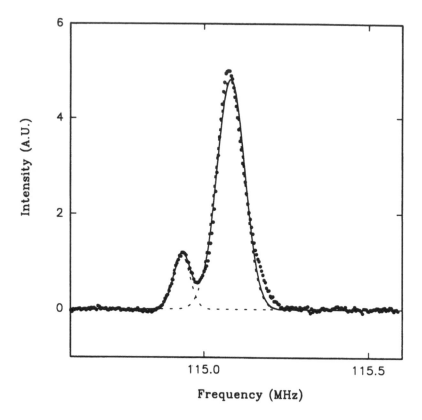

Figure 6 Example of the lineshape for $B\|c$, $B = 4.7$ T, and $T = 75$ K [70]. The drawn line is the best fit to the data (closed circles) with two Gaussians (dashed lines). The intensity ratio of the two lines (20:1, taking the T_2 effects into account) shows that about 10% of the Ca sites are occupied by Tl atoms.

to Tl^+ in dilute nitric acid. For the I line the difference in total Tl shift for a field oriented along the c axis and in the (a,b) plane is 838 ppm ($= 2A$ if the local symmetry of the I site has a similar axial symmetry as the N site) with an isotropic part of -2296 ppm [51]. The T dependence of the position of the lines is given in Fig. 7a.

The isotropy of the Knight shift is corroborated by isotropy in the relaxation rates. As in $Tl_2Ba_2CuO_{6+\delta}$, the positive Knight shift of the N line can be understood from the admixing of Cu-$3d$ character in the Tl-$6s$ orbital. The negative shift of the I line is similar to that of Y in $YBa_2Cu_3O_{7-\delta}$ and arises from indirect hyperfine coupling [64]. The T dependence of the two lines is analogous to that observed for oxygen/yttrium or copper in $YBa_2Cu_3O_{6.6}$ (see Fig. 1). In the latter the shifts in the normal state decrease with decreasing T and continue smoothly below T_c. There is no abrupt change at T_c. Details about the analysis of the linewidth (see Fig. 7b), especially in the superconducting state, are given in Section IV.D.

C. CuO$_2$ Triple Layer: $Tl_2Ba_2Ca_2Cu_3O_{10+\delta}$

The NMR data in $Tl_2Ba_2Ca_2Cu_3O_{10+\delta}$ [41,46,48] compare well with those obtained in $Tl_2Ba_2CaCu_2O_8$. A study that might be possible (but has not yet been performed) is to measure the number of holes in the three Cu layers. In that way one can check if the saturation of the transition temperature as a function of number of Tl layers is a consequence of the nonequal distribution of the number of holes between the planes [6].

D. Vortex Structure

In a type II superconductor an applied magnetic field, if above H_{c1} but below H_{c2}, penetrates the superconducting material in quantized units: the flux lines. In "normal" superconductors the flux lines are arranged in the triangular Abrikosov lattice. The field distribution due to the flux pattern is reflected in the second moment of the NMR line M_{2f} via

$$M_{2f} \propto \left[\frac{\lambda(0)}{\lambda(T)} \right]^2 \tag{23}$$

The T dependence of the penetration depth λ can often be described by the empirical two-fluid expression. If so, the T dependence of M_{2f} is given by

$$M_{2f} \propto 1 - \left(\frac{T}{T_c} \right)^4 \tag{24}$$

This contribution of the vortex solid to the linewidth ($\sim M_2^{1/2}$) is superposed on the normal-state width according to $M_2 = M_{2f} + M_{2n}$ and gradually disappears as T_c is approached. Because the T dependence of λ might reflect the microscopic pairing mechanism [30,65], its study in the high-T_c compounds is of special interest. For the Tl compounds several studies have been reported [42,44,47, 51–53].

Already in an early stage of the high-T_c superconductors, the internal field profile below T_c in $YBa_2Cu_3O_7$ was determined as a function of T by muon spectroscopy. As in NMR the muon lines are broadened by the presence of an inhomogeneous field. The main conclusion for $YBa_2Cu_3O_7$ is that the T dependence of both the muon [65,66] and the Y-NMR linewidths [67] are indeed well described by the empirical two-fluid formula.

In the presence of moving vortices, motional narrowing will prevent the line broadening to occur. In two-dimensional or very anisotropic three-dimensional superconductors, the flux lattice is molten above the Kosterlitz–Thouless critical temperature T_{KT} [68], which can be considerably less than T_c, and line narrowing occurs. $Tl_2Ba_2CaCu_2O_8$ is a good example of such a very anisotropic superconductor [69] and the Tl nucleus is an excellent probe for such a study. The normal-state linewidths are small enough to see the vortex broadening. [In a field of 4.7 T the linewidths of both lines just above T_c (see Fig. 7b) are on the order of 60 kHz (I) to 80 kHz (N), compared to a few hundred at the lowest temperature.]

To see the effect of the flux lattice in the superconducting state, one has to subtract the normal-state broadening mechanisms. A comparison of Y and Tl-I data shows that a spread in Knight shifts is the origin for the inhomogeneous linewidth of Tl–I [70]. For the N line, although the changes in width and shift are less pronounced, the same reasoning applies. The inhomogeneous relative width $\Delta K/K$ ($K = 550$ ppm and $\Delta K = 70$ kHz at 115 MHz) is about 1.1. This value is about three times larger than that for the I line, possibly caused by an extra spread in electron density due to the modulations in the Tl plane [59,60].

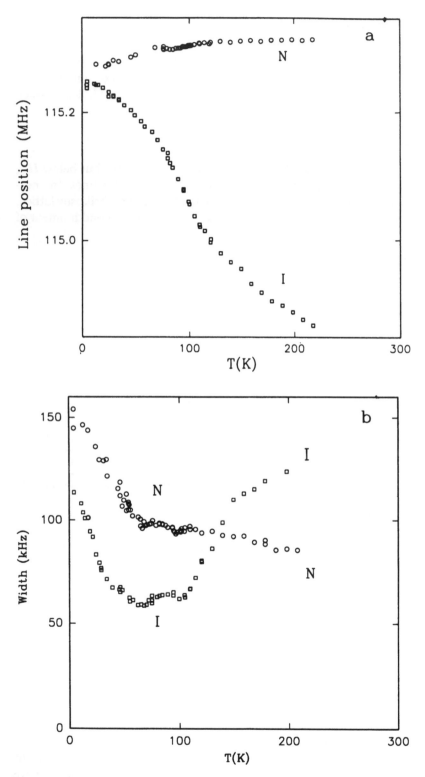

Figure 7 *T* dependence of the (a) position [51] and (b) width [70] of the N and I lines in Tl$_2$Ba$_2$CaCu$_2$O$_8$ for *B*||*c*, *B* = 4.7 T. Compared to the N line, the I line has the larger width at room temperature but the smaller at low temperatures.

The penetration of the magnetic field via flux lines and the shielding of the field from the interior (demagnetizing factor) give an inhomogeneous field distribution in the sample that is reflected in the width of the resonance line [71]. For platelet crystals the effect of the sample shape on the linewidth will be small compared to the broadening due to the flux pattern. The penetration depth in materials with strongly anisotropic electronic properties will be angular dependent. In case of axial symmetry, the anisotropy in λ can be accounted for by the introduction of an anisotropic mass in the London equations [72]

$$\frac{\lambda_\parallel}{\lambda_\perp} = \left(\frac{m_\parallel}{m_\perp}\right)^{1/2} = \gamma \tag{25}$$

where $\lambda_\perp$ and $\lambda_\parallel$ are the penetration depths when the screening currents are flowing in the plane perpendicular to the symmetry ($\parallel$) axis, respectively along that direction. For the two extreme orientations the ratio of the contributions of the flux lattices to the second moment of the resonance line is given by [73]

$$\frac{M_{2\parallel}}{M_{2\perp}} = \frac{\lambda_\parallel^2}{\lambda_\perp^2} \tag{26}$$

Measurements on nonaligned powders will give a mixture of the various angular contributions. In the case of axial symmetry and sufficient anisotropy, lineshape analysis of powders allows the derivation of $\lambda_\perp$ [73]. In aligned powders or single crystals the anisotropy in λ can be measured directly. In extreme anisotropic materials (the Tl compounds are almost two-dimensional) such a direct measurement of the anisotropy requires a very precise orientation of the magnetic field with respect to the symmetry axis [74], which is hard to realize experimentally. For that reason only results for $H\parallel c$ will be quoted here.

As shown in Fig. 7b, the linewidth in $Tl_2Ba_2CaCu_2O_8$ below T_c is almost T independent, down to about 50 K. Only below 50 K does it start to increase. In $Tl_2Ba_2CuO_6$ a similar situation exist [45,47]. This suggests that the vortices are moving above 50 K and start to freeze in a lattice below this temperature. Zero-field cooled and in field-cooled results in the O_6 compound became markedly different only below 20 K (which was conjectured to be the irreversibility temperature).

For better understanding of the linewidth in the superconducting state, it will be useful to consider the extreme limit of decoupled CuO_2 double layers. In that limit the melting behavior is well described by the Kosterlitz–Thouless theory of lattice melting [68]. At low temperatures only dislocation pairs exist, which can dissociate in single pairs at more elevated temperatures. This process finally leads to melting at the Kosterlitz–Thouless temperature.

Numerical calculations [52] show that the diffusion constant D for the vortices increases almost linearly with T up to roughly 50 K. Above this temperature D increases rapidly. At a temperature slightly below the upturn, motional narrowing prevents the NMR line to broaden above its 100-K value.

Figure 8 shows the results for the linewidth for the [205]Tl resonance in $Tl_2Ba_2CaCu_2O_8$ under the same conditions as in Fig. 7b. To make a comparison of this numerical simulation and the experimental data points, the experimental data have to be corrected for the contributions for nonvortex sources already

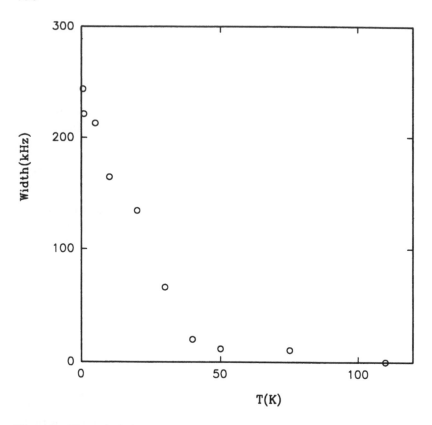

Figure 8 Numerical simulation of the T dependence of the ^{205}Tl NMR linewidth due to the presence of vortices, $B = 5$ T. (From Ref. 52.)

present in the normal state. Taking the 100-K width as the reference level, it can be seen that simulation and experiment agree rather well. The temperatures where drastic line broadening sets in are 50 K in the experiment and 40 K in the simulation.

V. SUMMARY AND OUTLOOK

The variety in number of CuO_2 and TlO planes in the Tl compounds, and not in the least the continuously changeable T_c of $Tl_2Ba_2CuO_{6+\delta}$, make them ideal for a systematic study of the influence of these parameters on the superconducting pairing mechanism. NMR and NQR in $Tl_2Ba_2CuO_{6+\delta}$ have shown the connection between magnetic correlations and the superconducting mechanism. By ^{205}Tl NMR in $Tl_2Ba_2CaCu_2O_8$ the intrinsic static and dynamic properties of the flux lattice in the very anisotropic limit have appeared to be accessible.

Although much resonance work has been done already, there are still challenging problems to solve. For a better understanding of the pairing mechanism, additional NMR/NQR experiments in the superconducting state of the highly anisotropic Tl compounds will be needed. Because of the unusual results obtained so far in $YBa_2Cu_3O_7$, not only the field direction but also its magnitude have to be varied. Also, saturation of the transition temperature as a function of CuO_2 layers might be clarified by a NMR/NQR study of the hole concentration in the various Cu layers in the CuO_2 triple-layer compound $Tl_2Ba_2Ca_2Cu_3O_{10}$. For the immediate future, NMR and NQR on the Tl compounds will remain an exciting subject.

ACKNOWLEDGMENTS

It is a pleasure to acknowledge Jaco T. Moonen, Derk Reefman, and Daniël van der Putten for useful comments and critical reading of the manuscript.

REFERENCES

1. H. Fukuyama, *Physica C* 185–189, xxv–xxxiv (1991)
2. A. J. Millis, H. Monien, and D. Pines, *Phys. Rev. B* 42, 167 (1990).
3. P. Monthoux, A. V. Balatsky, and D. Pines, *Phys. Rev. Lett.* 67, 3448 (1991).
4. M. Takigawa, J. L. Smith, and W. L. Hults, *Phys. Rev. B* 44, 7764 (1991).
5. C. H. Pennington and C. P. Slichter, *Phys. Rev. Lett.* 66, 381 (1991).
6. M. Di Stasio, K. A. Müller, and L. Pietronero, *Phys. Rev. Lett.* 64, 2827 (1990).
7. R. R. Ernst, G. Bodenhausen, and A. Wokaun, *Principles of Nuclear Magnetic Resonance in One and Two Dimensions*, Oxford University Press, 1987.
8. J. K. M. Sanders and B. K. Hunter, *Modern NMR Spectroscopy: A Guide for Chemists*, Oxford University Press, Oxford, 1988.
9. A. Abragam, *Principles of Nuclear Magnetism*, Clarendon Press Oxford, 1961/1986.
10. C. P. Slichter, *Principles of Magnetic Resonance*, Springer-Verlag, Berlin, 1990.
11. J. Korringa, *Physica* 16, 601 (1950).
12. M. Weinert and A. J. Freeman, *Phys. Rev. B* 28, 2626 (1983).
13. T. Moriya, *J. Phys. Soc. Jpn.* 18, 516 (1963).
14. J. Zaanen, G. A. Sawatzky, and J. W. Allen, *Phys. Rev. Lett.* 55, 418 (1985).
15. P. W. Anderson, *Science* 235, 1196 (1987).
16. F. C. Zhang and T. M. Rice, *Phys. Rev. B* 37, 3759 (1988).
17. V. J. Emery and G. Reiter, *Phys. Rev. B* 38, 4547 (1988); *Phys. Rev. B* 38, 11938 (1988).
18. T. Imai, T. Shimizu, H. Yasuoka, Y. Ueda, and K. Kosuge, *J. Phys. Soc. Jpn.* 57, 2280 (1988).
19. H. Alloul, P. Mendels, and P. Monod, *Phys. Rev. Lett.* 61, 746 (1988); H. Alloul, T. Ohno, and P. Mendels, *Phys. Rev. Lett.* 63, 1700 (1989).
20. P. C. Hammel, M. Takigawa, R. H. Heffner, Z. Fisk, and K. C. Ott, *Phys. Rev. Lett.* 63, 1992 (1989).
21. M. Takigawa, P. C. Hammel, R. H. Heffner, and Z. Fisk, *Phys. Rev. B* 39, 7371 (1989).
22. R. E. Walstedt, W. W. Warren, Jr., R. F. Bell, G. F. Brennert, G. P. Espinosa, R. J. Cava, L. F. Schneemeyer, and J. V. Waszczak, *Phys. Rev. B* 38, 9299 (1988).
23. S. E. Barret, D. J. Durand, C. H. Pennington, C. P. Slichter, T. A. Friedmann, J. P. Rice, and D. M. Ginsberg, *Phys. Rev. B* 41, 6283 (1990).
24. T. Tanamoto, K. Kuboki, and H. Fukuyama, *J. Phys. Soc. Jpn.* 60, 3072 (1991); H. Fukuyama, *ISSP A* 2473, 1 (1991).
25. F. Mila and T. M. Rice, *Physica C* 157, 561 (1989).
26. A. J. Millis and H. Monien, *Phys. Rev. B* 45, 3059 (1992).
27. H. Monien, D. Pines, and M. Takigawa, *Phys. Rev. B* 43, 258 (1991).

28. C. Berthier, Y. Berthier, P. Butaud, W. G. Clark, J. A. Gillet, M. Horvatic, P. Segransan, and J. Y. Henry, *Physica C* 185–189, 1141 (1991).
29. J. Rossat-Mignod, L. P. Regnault, C. Vettier, P. Burlet, J. Y. Henry, G. Lapertod, *Physica B* 169, 58 (1991); J. Rossat-Mignod, L. P. Regnault, C. Vettier, P. Bourges, P. Burlet, J. Bossy, J. Y. Henry, G. Lapertod, *Physica C* 185–189, 86 (1991).
30. M. Tinkham, *Introduction to Superconductivity*, R. E. Krieger, Malabar, Fla., 1975.
31. S. E. Barret, J. A. Martindale, D. J. Durand, C. H. Pennington, C. P. Slichter, T. A. Friedmann, J. P. Rice, and D. M. Ginsberg, *Phys. Rev. Lett.* 66, 108 (1991).
32. J. A. Martindale, S. E. Barret, C. A. Klug, K. E. O'Hara, S. M. DeSoto, C. P. Slichter, T. A. Friedman, and D. M. Ginsberg, *Phys. Rev. Lett.* 68, 702 (1992).
33. F. Borsa, A. Rigamonti, M. Corti, J. Ziolo, Ok-Bae Hyun, and D. R. Torgeson, *Phys. Rev. Lett.* 68, 698 (1992).
34. N. Bulut and D. J. Scalapino, *Phys. Rev. Lett.* 68, 706 (1992).
35. K. Fujiwara, Y. Kitaoka, K. Ishida, K. Asayama, Y. Shimakawa, T. Manako, and Y. Kubo, *Physica C* 184, 207 (1991).
36. Y. Kitaoka, K. Fujiwara, K. Ishida, K. Asayama, Y. Shimakawa, T. Manako, and Y. Kubo, *Physica C* 179, 107 (1991).
37. K. Asayama, Y. Kitaoka, and Y. Kohori, *Physica B* 171, 226 (1991).
38. K. Fujiwara, Y. Kitaoka, K. Asayama, Y. Shimakawa, T. Manako, and Y. Kubo, *J. Phys. Soc. Jpn.* 59, 3459 (1991).
39. K. Fujiwara, Y. Kitaoka, K. Asayama, H. Katayama-Yoshida, Y. Okabe, and T. Takahashi, *J. Phys. Soc. Jpn.* 57, 2893 (1988).
40. S. Kambe, Y. Yoshinari, H. Yashuoka, A. Hayashi, and Y. Ueda, *Physica C* 185–189, 1181 (1991).
41. Iu, I. Zhadanov, B. A. Aleksashin, K. N. Mikhalyov, V. V. Lavrentjev, S. V. Verhovskii, A. Yu. Yakubovskii, V. I. Ozhogin, L. D. Shustov, and A. B. Myasoedov, *Physica C* 183, 247 (1991).
42. Y.-Q. Song, Moohee Lee, W. P. Halperin, L. M. Tonge, and T. J. Marks, *Phys. Rev. B* 44, 914 (1991); Id, *Phys. Rev. B* 45, 4945 (1992).
43. E. Oldfield, C. Coretsopoulos, S. Yang, L. Reven, H. C. Lee, J. Shore, O. H. Han, E. Ramli, and D. Hinks, *Phys. Rev. B* 40, 6832 (1989).
44. F. Hentsch, N. Winzek, M. Mehring, Hj. Mattausch, and A. Simon, *Physica C* 158, 137 (1989).
45. M. Mehring, F. Hentsch, Hj. Mattausch, and A. Simon, *Z. Phys. B* 77, 355 (1989).
46. F. Hentsch, N. Winzek, M. Mehring, Hj. Mattausch, A. Simon, and R. Kremer, *Physica C* 165, 485 (1990).
47. M. Mehring, F. Hentsch, Hj. Mattausch, and A. Simon, *Solid State Commun.* 75, 753 (1990).
48. N. Winzek, F. Hentsch, M. Mehring, Hj. Mattausch, R. Kremer, and A. Simon, *Physica C* 168, 327 (1990).
49. D. M. de Leeuw, W. A. Groen, J. C. Jol, H. B. Brom, and H. Zandbergen, *Physica C* 166, 349 (1991).
50. H. B. Brom, D. Reefman, J. C. Jol, D. M. de Leeuw, and W. A. Groen, *Phys. Rev. B* 41, 7261 (1990).
51. J. C. Jol, D. Reefman, H. B. Brom, T. Zetterer, D. Hahn, H. H. Otto, and K. F. Renk, *Physica C* 175, 12 (1991).
52. D. Reefman and H. B. Brom, *Physica C* 183, 212 (1991); *Physica C* 185–189, 1459 (1991).
53. J. T. Moonen, D. Reefman, J. C. Jol, H. B. Brom, T. Zetterer, D. Hahn, H. H. Otto, and K. F. Renk, *Physica C* 185–189, 1891 (1991).
54. K. Tompa, I. Bakonyi, P. Bánki, I. Furó, S. Pekker, J. Vandlik, G. Oszlányi, and L. Mihály, *Physica C* 152, 486 (1988).

55. A. K. Rajarajan, K. V. Gopalkrishnan, R. Vijayaraghavan, and L. C. Gupta, *Solid State Commun.* 69, 213 (1989).
56. A. Poddar, P. Mandal, K. Ghosh Ray, A. N. Das, B. Ghosh, P. Choudhury, and S. H. Lahiri, *Physica C* 159, 226 (1989).
57. Y. Shimakawa, Y. Kubo, T. Manako, T. Satoh, S. Iijima, T. Ischihashi, and H. Igarashi, *Physica C* 157, 279 (1989); Y. Shimakawa, Y. Kubo, T. Manako, H. Igarashi, F. Izumi, H. Asano, *Phys. Rev. B* 42, 10165 (1990).
58. A. W. Hewat, P. Bordet, J. J. Capponi, C. Chaillout, J. Chenavas, M. Godinho, E. A. Hewat, J. L. Hodeau, and M. Marezio, *Physica C* 156, 369 (1988).
59. T. Zetterer, H. H. Otto, G. Lugert, and K. F. Renk, *Z. Phys. B* 73, 321 (1988); T. Zetterer, H. H. Otto, K. Meidenbauer, and K. F. Renk, *Physica C* 162–164, 514 (1989).
60. W. Dmowksi, B. H. Toby, T. Egami, M. A. Subramanian, J. Gopalakrishnan, and A. W. Sleight, *Phys. Rev. Lett.* 61, 2608 (1988); B. H. Toby, T. Egami, J. D. Jorgensen, and M. A. Subramanian, *Phys. Rev. Lett.* 64, 2415 (1990).
61. J. Torrance, Y. Tokura, A. I. Nazzal, A. Bezinge, T. C. Huang, and S. S. P. Parkin, *Phys. Rev. Lett.* 61, 1127 (1988).
62. C. Di Castro, L. F. Feiner, and M. Grillim, *Phys. Rev. Lett.* 66, 3209 (1991).
63. K. Ishida, Y. Kitaoka, T. Yoshimoto, N. Ogata, T. Kamino, and K. Asayama, *Physica C* 179, 29 (1991).
64. F. J. Adrian, *Phys. Rev. Lett.* 61, 2148 (1988).
65. B. Pümpin, H. Keller, W. Kündig, W. Odermatt, I. M. Savić, J. W. Schneider, H. Simmler, P. Zimmermann, E. Kaldis, S. Rusiecki, Y. Maeno, and C. Rossel, *Phys. Rev. B* 42, 8019 (1990).
66. T. M. Riseman, J. H. Brewer, B. R. Cyca, J. F. Carolan, W. N. Hardy, R. F. Kiefl, M. Celio, W. L. Kossler, Y. J. Uemura, G. M. Like, B. J. Sternlieb, H. Hart, K. W. Lay, H. Kojima, I. Tanaka, K. Kakurai, *Physica C* 162–164, 1555 (1989).
67. H. B. Brom and H. Alloul, *Physica C* 177, 297 (1991).
68. J. M. Kosterlitz and D. J. Thouless, *J. Phys. C* 6, 1181 (1973).
69. D. E. Farrell, R. G. Beck, M. F. Booth, C. J. Allen, E. D. Bukowski, and D. M. Ginsberg, *Phys. Rev. B* 42, 6758 (1990).
70. H. B. Brom, J. T. Moonen, and D. Reefman, *Appl. Magn. Res.* 3, 597 (1992).
71. E. H. Brandt, *J. Low Temp. Phys.* 73, 355 (1988); *Phys. Rev. B* 37, 2349 (1988).
72. V. G. Kogan, *Phys. Rev. B* 24, 1572 (1981).
73. W. Barford and J. M. F. Gunn, *Physica C* 156, 515 (1988).
74. S. L. Thiemann, Z. Radović, V. G. Kogan, *Phys. Rev. B* 39, 11406 (1989).

26
Normal-State Transport Properties of TBCCO

D. G. Naugle
Texas A&M University
College Station, Texas

A. B. Kaiser
Victoria University of Wellington
Wellington, New Zealand

I. INTRODUCTION

The electronic and thermal transport properties give information regarding the nature of the charge and thermal carriers and how they are scattered. To formulate an understanding of superconductivity in the high-T_c superconductors will require a good understanding of their normal-state transport properties. We consider here resistivity, Hall effect, thermoelectric power, and thermal conductivity, which have been investigated extensively for the high-temperature superconductors in general, both experimentally and theoretically. There is particular interest in the behavior of these coefficients in the normal state since (apart from thermal conductivity) they are zero in the superconducting state in the absence of a magnetic field.

The family of Tl-based high-T_c superconductors is an extremely rich family that encompasses a very large number of layered compounds of the type $Tl_m Ba_2 Ca_{n-1} Cu_n O_{2n+m+2+\delta}$ (TBCCO), where m is the number of Tl–O planes and n is the number of Cu–O planes. There are two distinct subgroups, one with one Tl–O plane ($m = 1$) [1,2] and a second with two Tl–O planes ($m = 2$), each with up to four Cu–O planes ($n = 1, 2, 3, 4$) [1]. For $m = 2$, intergrowths with $n = 5, 6$ have been observed in high-resolution electron microscopy studies, but single-phase samples of this structure have not been reported [3]. Also, a single Tl-layer compound with up to six Cu–O layers has been prepared [4], $(Tl,Pb)Sr_2 Ca_{n-1} Cu_n O_{2n+3}$ ($n = 2, 3, 4, 5, 6$), with the addition of Pb at the Tl site. The family of Bi-based cuprate high-T_c superconductors is structured in a manner similar to the Tl-based family, but only the phases $n = 2$, $m = 1, 2, 3$, have been reported for the Bi-based cuprates. Consequently, we should expect that the family of Tl-based superconductors with its large variation in Tl–O planes and Cu–O planes would offer a unique opportunity for testing the ideas to explain both superconductivity and normal-state transport. Similarly, the general trends in superconductivity and normal-state transport for the corresponding phases of the Bi- and Tl-based cuprates should be similar.

There is, however, a scarcity of reliable studies of transport in the Tl-based systems compared to Y–Ba–Cu–O (YBCO) and the Bi-based cuprates. We believe that several reasons have contributed to this: (1) the toxicity of Tl and the safety precautions required in production and handling of the materials; (2) the difficulties encountered in the production of high-quality, single-phase thin film or pressed powder samples; and (3) the unavailability until very recently of single-crystal samples big enough for high-precision transport measurements. A major factor in the last two items has been the high vapor pressure of Tl and Tl oxides at the temperatures required for the chemical reaction and growth of the crystalline phases.

Transport measurements of single-phase $Tl_2Ba_2Ca_2Cu_3O_{10+\delta}$ 2223 samples, the phase that exhibits the highest-reproducible superconducting transition temperature yet reported (127 K, $R = 0$) [5], are rare. This phase appears to be very difficult to prepare, and most of the samples purported to be of this phase exhibit evidence of an admixture of varying amounts of other phases, particularly the (2212) phase. There are fewer measurements for phases with $n > 3$ for either single or double Tl–O layers. Single-phase 2212 materials of reasonable quality have been used in several transport studies. Surprisingly, some of the more interesting transport measurements have been performed on 2201 phase samples, which were, at least initially, among the most difficult samples to prepare with reproducible properties. The key observation for this phase was that the properties were *very* sensitive to oxygen doping [6,7]. The T_c for this system can be varied from 85 K to 0 K with an increase in oxygen per unit formula of the order of $\delta = 0.1$. The variation of T_c with δ for the 2223 and 2212 phases is smaller, but still significant [8]. Similar oxygen-doping effects for the Bi-based cuprates have been reported [9,10]. In terms of normal-state transport properties the most systematic studies, however, have been for the Tl 2201 phase, where Hall effect [11], resistivity [11–13], and thermopower [12,13] measurements for comparable samples have been made as a function of oxygen doping and temperature and where measurements of the anisotropic conductivity [14,15] and Hall effect [15] for single-crystal samples have been reported.

The most common way of studying the change in normal-state transport properties with variation in the concentration of charge carriers is by site-selective substitution. Appropriate doping may occur for the Tl, Ba, Ca, or Cu sites. Although there are many reports of measurements of one or another transport property after selective doping at a particular site, there are no systematic studies of all three normal-state transport properties (resistivity, thermopower, and Hall effect) for a particular substitution. Consequently, a coherent interpretation of these diverse results is difficult. Rather complete measurements for undoped single-crystal and/or highly oriented film 2212 phase samples have been reported for resistivity [16,17], thermopower [18], and Hall effect [19]. Fewer transport measurements have been reported for the phases with single layers of Tl–O. Limited transport measurements have been reported [20,21] for some exotic Tl-based phases containing rare earth elements, such as Pr–Tl–Sr–Ca–Cu–O, which exhibit T_cs in the range 80 to 90 K. Generally, these have not been single-phase samples and thus have provided only qualitative information.

Transport measurements on the high T_c superconductors have revealed several unusual features that must be accounted for by theoretical models proposed to describe the superconductivity. One of the primary aims of these measurements has been to find clues to help determine the nature of the superconducting pairing mechanism that gives rise to such extraordinarily high values of the superconducting transition temperature. In the following sections we consider in turn the resistivity, Hall effect, thermopower, and thermal conductivity, giving in each section some background on theoretical concepts and their application to data for other high-T_c superconductors before discussing data on TBCCO.

II. RESISTIVITY

A. Background Summary

The resistivity, since it is linked so intimately to the occurrence of superconductivity, is probably the most widely measured normal-state property of the superconductors. An early review of normal-state transport in high-temperature superconductors was given by Allen et al. [22].

The dominant and striking characteristic of the normal-state resistivity in the *ab* plane is that in superconductors with high values of T_c, it is linear in temperature from T_c up to high temperatures. This behavior is surprisingly universal, although in some lightly doped samples the resistivity flattens out or increases as temperature is decreased and T_c is approached. Of course, a resistivity linear in temperature that results from scattering of the charge carriers by phonons is the standard behavior in normal metals, but in general there is a change to a higher power law at temperatures on the order of $T_D/5$, where T_D is the Debye temperature, sometimes a T^5 law in good crystalline metals, or a T^2 law in disordered metals. In some cuprate superconductors with relatively low values of T_c, the resistivity fails to show the expected flattening at low temperatures [22]. However, in others, a transition to a T^2 behavior is observed [23].

A variety of explanations for the linear in-plane resistivity has been advanced. Recent calculations for a strong-coupling treatment of electron–phonon resistivity using a screened ionic model [24] yield both the linear temperature dependence down to T_c and the magnitude of the resistivity; a value for the electron–phonon coupling constant of $\lambda \sim 3$ gave $T_c \sim 90$ K for $YBa_2Cu_3O_7$.

Scattering of the hole carriers by antiferromagnetic spin fluctuations has also been calculated [25,26] to give a linear law above about 100 K, with a T^2 behavior (superseded by the superconductivity) at low temperatures. This linear law is expected to be a rather general feature of scattering of electron or hole carriers by bosons of any type [27], and there are examples of alloys with low-energy spin fluctuations showing measured linear laws down to below 10 K.

Mott [28] has discussed the spin–bipolaron theory for high-temperature superconductors, in which a nondegenerate gas of bosons (the bipolarons) is the dominant carrier of the electrical current above T_c. With a constant diffusion constant for these bipolarons, the resistivity is predicted to increase with a linear T law, and the magnitude of ρ is of the order observed experimentally.

In the resonating valence bond (RVB) model of Anderson and Zou [29], the current is also carried by bosons above T_c (in this case, spinless "holons" of charge

2e); the linear resistivity is ascribed to scattering of these holons off chargeless "spinon" excitations, although this result is controversial [22]. Nagaosa and Lee [30], investigating the experimental consequences of gauge-field fluctuations on the uniform RVB state, also obtained a resistivity linear in T, as did Ioffe and Kotliar [31] in related work, and Varma et al. [32] in their phenomenological marginal Fermi liquid model.

The behavior of the resistivity ρ_c in the c direction is much less well defined, but it is orders of magnitude larger than that for in-plane directions. In some cases an approximately T^{-1} behavior is observed for ρ_c (as predicted by Anderson and Zou [29] in the holon–spinon model), but in more recent work $d\rho_c/dT$ is usually found to be positive, as for the in-plane directions. This suggests a metallic type of conduction could also be present in the c direction.

B. Experimental Survey for TBCCO

The number of papers reporting the temperature dependence of resistivity or resistance for Tl-based compounds is large. We discuss here only some of those papers that may illustrate a general trend and some of the papers that provide systematic studies of T_c and ρ as a function of doping. There are many studies of fluctuation conductivity just above T_c and flux flow resistance below T_c, but only the normal-state resistivity is included in this chapter. Most of the experimental references in subsequent sections also report resistivity, but not all will be referenced here.

One of the most interesting questions is how the in-plane resistivity component ρ_{ab} compares with the c-axis component ρ_c and the resistivity ρ of the corresponding polycrystalline sample for these highly anisotropic compounds. There are now measurements on single crystals of both the 2201 phase [14,15] and the 2212 phase [16]. Measurements with these small, highly anisotropic crystals normally use the Montgomery techniques [33]. Measurements of ρ_{ab} and ρ_c for single crystals of these two phases from Duan et al. [14,16] are shown in Fig. 1. The temperature dependence of ρ_{ab} and ρ_c are qualitatively similar for a given material, but the magnitude of ρ_c is larger by a factor of 10^2. A comparison of the resistivity of sintered powder samples of the (2201) phase can be obtained from Fig. 2. The different curves correspond to different values δ of excess oxygen [11]. The temperature dependence of the resistivity of the powder sample with a T_c similar to that of the single-crystal sample in Fig. 1 is similar but somewhat weaker than the temperature dependence of ρ_{ab} or ρ_c for the single crystal. The resistivity for this powder sample is greater than, but of the same order of magnitude as, ρ_{ab} for the single crystal. Manako et al. [15] have also reported resistivity measurements on single-crystal 2201 samples with different oxygen doping. Their data are for samples with T_cs corresponding roughly to curves for $\delta = 0.03, 0.08$, and 0.10 in Fig. 2 for the sintered powder samples. They find that the temperature dependence of ρ_{ab} fits very closely

$$\rho = \rho_0 + \beta T^n \tag{1}$$

with values n close to those for the corresponding curves in Fig. 2 (n varies from 0.95 for $\delta = 0.01$ to 1.88 for $\delta = 0.1$ in Fig. 2) and that ρ_{ab} is smaller but of the same order of magnitude as ρ for the sintered samples. They also find that the temperature dependence of ρ_c does not fit a simple power law as well, but there

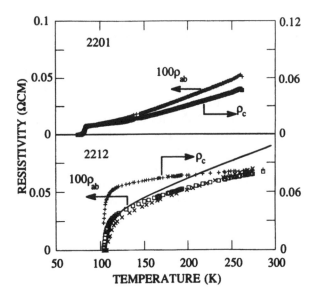

Figure 1 Anisotropic resistivity ρ_{ab} and ρ_c for a single-crystal 2201 sample from Duan et al. [14] and ρ_{ab} and ρ_c for a single-crystal 2212 sample from Duan et al. [16] and ρ_{ab} data from an oriented 2212 polycrystalline thin film from Wang et al. [19] (solid line).

are qualitative similarities between the shapes of the $\rho_c(T)$ curves and $\rho_{ab}(T)$ curves. In contrast to the measurements of Duan et al. [14,16], who find an anisotropy ratio closer to 10^2, Manako et al. [15] find a resistivity ratio close to 10^3 for single-crystal samples of the same phase.

For 2201 samples with maximum T_c, the normal-state resistivity is a linear function of temperature, but the dependence approaches quadratic as the super-

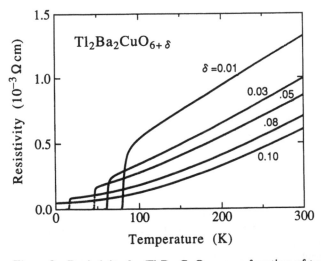

Figure 2 Resistivity for $Tl_2Ba_2CuO_{6+\delta}$ as a function of temperature for different values of δ from pressed powder samples of Kubo et al. [11]. Hall effect data for the same samples are displayed in Fig. 6.

conductor–normal metal (curve for $\delta = 0.1$) transition is approached for both single-crystal [15] and polycrystalline [11] samples. Resistivity and other transport measurements discussed below indicate that the addition of oxygen to this phase adds charge carriers. The resistivity curves for the single-crystal 2212 sample in Fig. 1 exhibit a somewhat weaker temperature dependence than usually observed for a typical polycrystalline sample of this phase (data in the *ab* plane from a single phase, highly oriented, thin-film sample [19] are included for comparison), and the temperature dependencies for ρ_{ab} and ρ_c for this single crystal exhibit more difference than for the 2201 crystals. The flattening of resistivity for this single-crystal sample as a function of temperature and the difference between ρ_c and ρ_{ab} may result from an oxygen deficiency that reduces the charge carrier density.

An example of underdoping form the optimal charge carrier concentration for maximum T_c is provided by the experiments of Poddar et al. [34], who studied the properties of $Tl_2Ba_2Ca_{1-x}Y_xCu_2O_{8+\delta}$. The resistivity data for selected samples are shown in Fig. 3. Resistivity and Hall effect measurements indicate that replacement of Ca by Y reduces the number of charge carriers and the value of T_c. Near $x \approx 0.5$ a metal insulator transition occurs. Underdoping "flattens" out the temperature dependence and near the metal–insulator transition leads to a negative $d\rho/dT$ value. Similar results have been observed by Wang et al. [35] for $Tl_2Ba_2(Ca_{2-x}Y_x)Cu_3O_y$.

A large number of experiments have studied the effects of doping on the resistivity and superconducting transition temperature. For most of these the follow-up experiments to measure thermopower and Hall effect have not been reported. The study by Presland et al. [36] for $Tl_{0.5+x}Pb_{0.5-x}Ca_{1-y}Y_ySr_2Cu_2O_{7+\delta}$ clearly

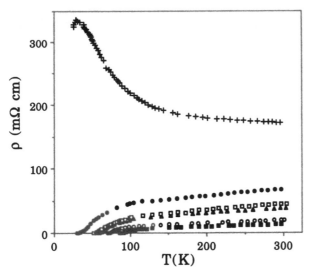

Figure 3 Temperature dependence of the resistivity of pressed powder samples of $Tl_2Ba_2Ca_{1-x}Y_xCu_2O_{8+\delta}$ for different values of x taken from Poddar et al. [34]. Hall effect data for some of the same samples are shown in Fig. 5; $x = 0$ (solid squares), $x = 0.20$ (open circle), $x = 0.25$ (solid triangles), $x = 0.30$ (open squares), $x = 0.35$ (solid circles), and $x = 0.48$ (crosses).

illustrates the fact that there is an optimal doping level for the maximum T_c in Tl-based superconductors. Replacement of Ca by Y appears to decrease the carrier number, while replacement of Tl with Pb in this single-layer-Tl compound increases the number. The total suppression of T_c at $y \approx 0.6$ is presumably a super-conductor–insulator transition ($T = 0$) while that at $x \approx 0.2$ (based on resistance data for a similar sample by Otto et al. [37]) is the superconductor–normal metal transition ($T = 0$), but only susceptibility measurements of T_c were reported in Ref 36. Kubo et al. [38] report a similar superconductor–normal metal transition in oxygen-doped $TlSr_2CaCu_2O_{7-\delta}$ discussed in more detail in the following section. Other studies of resistivity and T_c with doping for single-Tl–O-layer compounds include $TlBaSrCuO_{5+\delta}$ [39], $TlBa_{2-x}La_xCuO_{5+\delta}$ [40], $TlBa_{1+x}La_{1-x}CuO_{5+\delta}$ [41], and $TlCa_{1-x}Y(RE)_xSr_2Cu_2O_y$ [42]. It is interesting to note that for the $(Tl,Pb)Sr_2Ca_{n-1}Cu_nO_{2n+3}$ compounds with five and six Cu–O layers ($n = 5, 6$) the linear resistivity extrapolates to a negative value at $T = 0$ [4].

The effect of Cr doping in several Tl-based compounds has been reported [43], as has the effect of Ce doping [44]. Inclusion of different rare earths has also been reported by several groups [20,21,45–47].

III. HALL EFFECT

A. Background Summary

The primary objects of Hall effect measurements are to determine the sign of the charge carriers, and in some cases to estimate their density. The Hall coefficient is given by $R_H = E_H/j_x B_z$, where j_x is the current density in the x direction, E_H the Hall voltage in the y direction, and the applied magnetic field is B_z. For a single band of free carriers,

$$R_H = \pm (n_H e)^{-1} \tag{2}$$

where the Hall number n_H equals the density n of carriers of charge $\pm e$. The sign of R_H is positive for holes and negative for electrons, but great caution should be exercised in equating the Hall number n_H from this simple expression, with the actual carrier density n in cases where the carriers are not homogeneous, since cancellation effects may occur.

A review of the Hall effect in high-temperature superconductors has been given by Ong [48], in addition to a brief discussion in the earlier review of Allen et al. [22]. The relatively small size of the Hall number n_H in many high-T_c superconductors indicates that the carrier density is considerably lower than that in traditional metals, as expected. It is found, (e.g., Ref 49) for Pb- and Y-doped Bi 2212 compounds that n_H follows the same trend as the hole concentration determined by iodometric titration, but is considerably larger. A more puzzling aspect of the data is that in contrast to the usual metallic behavior, R_H generally has a substantial temperature dependence, and this behavior appears to correlate with the appearance of superconductivity [48].

Several possible causes of this temperature dependence have been put forward, Initially, it was shown that two-band models could account for the temperature dependence if the bands were of very different character and would lead to carrier densities n smaller than n_H [50]. In many cases it was found that the temperature dependence can be described approximately by

$$R_H^{-1} = A + BT \tag{3}$$

where A and B are constants. For $YBa_2Cu_3O_7$ the parameter A is often near zero, but for other superconductors the temperature dependence is weaker. A temperature dependence of this general type is often seen in conventional metals containing magnetic ions and is ascribed to skew scattering of carriers due to the spin–orbit interaction. Such magnetic skew scattering has been suggested as the cause of the temperature dependence of R_H in high-T_c superconductors [51]. However, Ong [48] finds a lack of the expected correlation with the magnetic properties of the superconductors, although the situation in these materials is rather complex.

Mott [52] has obtained an expression for R_H in the spin–bipolaron model which gives a T^{-1} dependence. A spin flip on the outer surface of the bipolaron is supposed to activate its movement through a distance of about 1 Å, with the jumps in the $+y$ and $-y$ directions having different probability in the applied field B_z.

Trugman [53] considered a weakly interacting Fermi liquid of quasi-particles that form when holes are added to the antiferromagnetic Mott insulator with one electron per site. A temperature-dependent Hall number was obtained, but agreement with the experimental temperature dependence was qualitative only and a detailed comparison was not made. In their investigations of the transport properties of a strongly antiferromagnetically correlated electron system in the temperature regime where Fermi liquid coherence ceases to exist, Ioffe and Kotliar [31] found an anomalous temperature dependence of the Hall coefficient, and Ioffe et al. [31] calculated that R_H decreased as the temperature increased. Nagaosa and Lee [30] for their uniform RVB model also obtained a temperature dependence for the Hall coefficient that could possibly be a decrease as temperature increased.

Recently, Anderson [54] has given an expression for the Hall angle θ_H in the context of his holon–spinon model by distinguishing the scattering rate $1/\tau_{tr}$, which appears in the linear resistivity, from the transverse Hall relaxation rate $1/\tau_H$, which is determined by scattering of the spin excitations (spinons) alone. He predicts the temperature dependence

$$\cot \theta_H \equiv \frac{\rho_{xx}}{R_H B_z} = \alpha T^2 + C \tag{4}$$

where α and C are constants.

The T^2 term arises from spinon–spinon scattering, and the constant term arises from spinon scattering from magnetic impurities in the Cu–O plane. Measurements of the Hall angle for $YBa_2Cu_{3-x}Zn_xO_{7-\delta}$ single crystals [55] are in good agreement with Eq. (4). The coefficient α, which is determined by the bandwidth of the spin excitations, was found to be essentially independent of x, and the constant C, which should be proportional to the concentration of magnetic scatterers, was found to increase linearly with x in this experiment. The value of the normal-state resistivity ρ_0 at $T = 0$ as determined by extrapolation from above T_c was also observed to increase linearly with x while T_c itself decreased linearly with x. Chien et al. [55] suggested that this simple dependence of the Hall angle on T and x would appear to rule out the usual multiband models which assume that the transport relaxation rate and the transverse (Hall) relaxation rate are the same within a multiplicative constant. However, Eq. (4) is also the behavior of the Hall angle expected for a conventional metal with R_H constant and $\rho = \rho_0 + \beta T^2$, where ρ_0 arises from impurity scattering and βT^2 from electron–electron, electron–phonon (in disordered metals), or electron–spin fluctuation scattering. Thus the form of Eq. (4) for the temperature dependence of the Hall angle is likely to be rather general, but aside from Anderson's model, there is no reason to expect such behavior when R_H is strongly temperature dependent.

We reemphasize that for no theory able to account even qualitatively for the transport coefficients can $n_H = (eR_H)^{-1}$ be identified with the charge carrier density, even though much of the data in the literature is presented in terms of the number of charge carriers per unit volume or per copper ion.

B. Experimental Survey for TBCCO

Experimental Hall effect data for Tl-based cuprates is very limited. As illustrated in Fig. 4, data for undoped samples with different numbers of Cu–O planes can be represented adequately by Eq. (3). For 2223 phase samples the early data by Clayhold et al. [57], although perhaps single phase, exhibits appreciable scatter, while that from Mandal et al. [58] is most likely a mixed-phase sample. Wang et al. [19] have measured R_H for single-phase 2212 thin films whose c axis is oriented perpendicular to the plane of the film and thus provide a good measurement of the in-plane coefficient, R_{Hab}. Earlier measurements by Woo et al. [59] were for films with no preferential orientation. Poddar et al. [34] have reported values of R_H for this phase using a single-phase pressed powder sample. The superconducting transition temperature, temperature dependence of the resistance, magnitude of the resistance, and temperature dependence of R_H below 110 K suggest that the sample may have been inhomogeneous. The only measurements for the 2201 phase that have been reported are those by Kubo et al. [11] with pressed powder samples

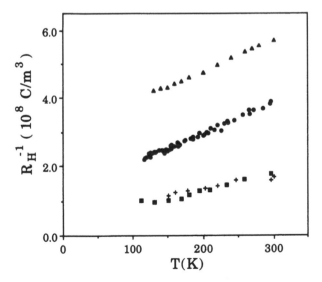

Figure 4 R_H^{-1} as a function of temperature for various Tl-based phases: 2223 from Mandal et al. [58] (crosses); 2212 from Wang et al. [19] (solid circles) and Poddar et al. [34] (solid squares); 2201 from Kubo et al. [11] (solid triangles).

containing an estimated small amount of excess oxygen. Data from the sample with $T_c = 81$ K which contained the smallest amount of excess oxygen ($\delta \sim 0.01$) are shown in Fig. 4.

Poddar et al. [34] have studied the system $Tl_2Ba_2Ca_{1-x}Cu_2O_{8+\delta}$ over the range $0 \le x \le 0.6$ using pressed powder samples. Values of R_H for different doping levels are illustrated in Fig. 5. Above $x = 0.1$, R_H increases as a function of x, and the temperature dependence is greatly reduced. The resistivity is roughly linear in temperature for $x \le 0.4$, with the absolute magnitude increasing with increasing x. The relative temperature dependence $\rho^{-1}(d\rho/dT)$ decreases with x as expected in approaching a metal–semiconductor transition. The superconducting transition temperature decreased with increasing x, and above $x = 0.4$ the temperature dependence of the normal-state resistivity changes from metallic-like to semiconducting-like. In the neighborhood of $x = 0.5$ a superconductor–insulator transition ($T = 0$) appears. The data are consistent with a reduction in the number of hole-like charge carriers in the Cu–O plane with increasing x.

In contrast to doping experiments by substitution at the Ca site, Kubo et al. [7] have doped 2201 samples by changing the oxygen content in the range $0 < \delta \le 0.1$. They have shown that the excess oxygen atoms are incorporated at interstitial sites between the double Tl–O layers. There is a dramatic decrease in T_c and normal-state resistivity with the increase in δ as shown in Fig. 2. The nature of the temperature dependence of resistivity changes and superconductivity vanishes for $\delta \approx 0.1$. As shown in Fig. 10 in the following section, the thermopower for a sample with $T_c < 4$ K [12] (presumably $\delta \approx 0.1$, based on the T_c versus δ scale in Ref. 7) is approximately proportional to T over the entire temperature range, consistent with the ideal diffusion thermopower of a metal. The variation of R_H^{-1} with tem-

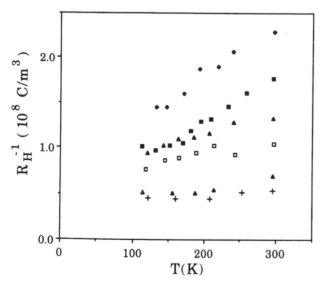

Figure 5 R_H^{-1} as a function of temperature for $Tl_2Ba_2Ca_{1-x}Y_xCu_2O_{8+\delta}$ with different values of x from Poddar et al. [34]. Resistivity curves for the same samples are shown in Fig. 3; $x = 0$ (solid squares), $x = 0.10$ (solid diamonds), $x = 0.25$ (solid triangles), $x = 0.30$ (open squares), $x = 0.40$ (open triangles), $x = 0.48$ (crosses).

perature and δ is shown in Fig. 6. The minimum at about 120 K is not understood, particularly in the "metallic" sample. Recent measurements on a single-crystal, oxygen-doped, metallic sample [15] show a similar magnitude and temperature dependence except that the minimum is shifted to a somewhat lower temperature. The data are consistent with an increase in the number of holelike charge carriers in the Cu–O plane with increased δ, which, for δ = 0.1, leads to a superconductor–normal metal transition ($T = 0$). ·

A similar superconductor–normal metal transition has been reported [38] for oxygen doping of the $TlSr_2CaCu_2O_{7-\delta}$ system. In this system the Tl–O layer is normally oxygen deficient. The addition of oxygen to change δ from 0.2 to 0.12 reduces T_c from 68 K to 0 K and changes the temperature dependence of the normal-state resistivity from linear in T to a T^2 dependence. The behavior of the Hall coefficient with increasing oxygen content (increasing number of hole-like excitations) is similar to that shown in Fig. 6 for the 2201 samples. The temperature dependence of R_H^{-1} for the two systems is similar and shows a minimum also. The value of n_H^{-1} for the metallic $TlSr_2CaCu_2O_{7-\delta}$ (δ ≈ 0.12), however, is a factor of 2 to 3 smaller than that of the metallic 2201 sample (δ = 0.1).

Chien et al. [55] have recently pointed out that the temperature dependence of the Hall angle in single-crystal $YBa_2Cu_{3-x}Zn_xO_{7-\delta}$ samples is consistent with the relation given by Anderson [54], Eq. (4), based on the holon–spinon model. They find that the coefficient α is of the expected magnitude and that the superconducting transition temperature decreases linearly with x, while C and the "residual resistivity" increase linearly with x, in qualitative agreement with this model.

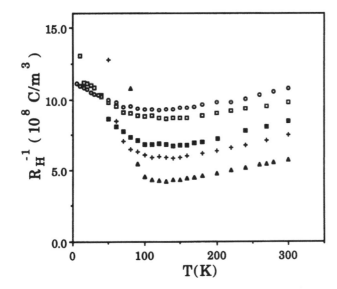

Figure 6 R_H^{-1} for $Tl_2Ba_2Cu_1O_{6+\delta}$ as a function of temperature for different values of δ from pressed powder samples of Kubo et al. [11]. Resistivity data for the same samples are displayed in Fig. 2, and thermopower data for similarly treated samples are displayed in Fig. 10; δ = 0.01 (solid triangles), δ = 0.03 (crosses), δ = 0.05 (solid squares), δ = 0.08 (open squares), and δ = 0.10 (open circles).

Values of cot θ_H in a field of 8 T for different Tl-based samples, including one of the metallic 2201 oxygen-doped samples, are plotted as a function of T^2 in Fig. 7.

The agreement between the experimental data and Eq. (4) is quite good. Of particular interest are the data for $Tl_2Ba_2CuO_{6+\delta}$ with different levels of oxygen doping δ. For some of the samples with small values of δ, R_H^{-1} is linear in T above T_c (see Fig. 6), but for the "metallic" sample ($\delta = 0.10$), which shows a quadratic temperature dependence of resistivity (see Fig. 2) and no superconducting transition, R_H^{-1} exhibits a minimum. The approximate linearity of cot θ_H as a function of T^2 of this sample reflects the T^2 variation of the resistivity in the "metallic samples," with R_H^{-1} showing only a small T dependence. The minimum of R_H^{-1} for $\delta = 0.10$ corresponds to a change in slope seen in cot θ_H below about 100 K, which is less noticeable than a minimum and is also compressed into a corner of the plot by the T^2 scale. The Hall angle, however, is described well by Eq. (4) for all reported values of δ, including the "metallic" sample ($\delta = 0.10$). In this system C decreases with doping, in contrast to the Zn-doped YBCO [55]; consequently, higher values of C are associated with higher values of T_c for the oxygen-doped 2201 samples. The value of α is the same for all of the 2201 samples, even the "metallic" one, and would imply a smaller bandwidth in this model for spinons in the 2201 system, 625 K compared to 830 K for the YBCO crystals [55]. The association of C with magnetic scattering in the Cu−O plane is not clear, though, for oxygen doping of the 2201 phase. The most interesting fact, however, is that although the resistivity and thermopower (as discussed in the next section) for the nonsuperconducting sample with $\delta = 0.10$ have a temperature dependence that appears metallic and cot θ_H approximately satisfies Eq. (4) in the "metallic limit"

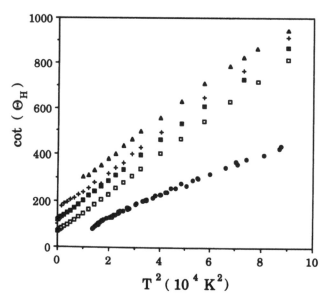

Figure 7 cot $\theta_H = \rho_{xx}/R_H B_z$ as a function of T^2 for selected high-T_c samples calculated from measured values of ρ_{xx}, R_H, and $B = 8$ T. $\delta = 0.01$ (solid triangles), $\delta = 0.05$ (crosses), $\delta = 0.08$ (solid squares), $\delta = 0.10$ (open squares) for $Tl_2Ba_2Cu_1O_{6+\delta}$ from Kubo et al. [11]; for an oriented thin-film 2212 sample from Wang et al. [19] (solid circles).

(as discussed in the preceding subsection), the slope of cot θ_H against T^2 is essentially the same for both the nonmetallic and metallic 2201 samples.

Also shown in Fig. 7 are data for an oriented 2212 thin film [19]. The temperature dependence of the Hall angle is again in reasonable agreement with Eq. (4), but there is a change in slope below 150 K, the temperature where deviations from a linear dependence of $\rho(T)$ due to inhomogeneities or fluctuations begin (see Fig. 1). In this model the slope above 150 K would imply a spinon bandwidth identical to that for the YBCO samples [55] for the 2212 film (625 K) and relatively little magnetic impurity scattering of the spinons. The success of Eq. (4) in describing the temperature dependence of the Hall angle for a large number of high-quality high-T_c samples suggests a generality for this expression for both the high-T_c and "overdoped" limits.

IV. THERMOPOWER

A. Background Summary

Thermoelectric power S is given by $S = \Delta V / \Delta T$, where ΔV is the voltage difference between the ends of the sample when a temperature difference of ΔT is applied. The diffusion of charge carriers down the temperature gradient occurs at a different rate to that up the gradient, the difference depending on the energy dependence of the density of states, velocity, and scattering rate of the carriers. The resulting thermopower can indicate the energy (relative to the Fermi level) of the carriers that make the dominant contribution to the conductivity. For a constant scattering rate, the thermopower is negative for electrons and positive for holes, although highly energy-dependent scattering can reverse the sign of S. A fuller discussion is given in a recent review of thermoelectricity in high-temperature superconductors [56].

In a metal the diffusion thermopower is small, because conduction occurs in the immediate vicinity of the Fermi level and is traditionally expected to be proportional to temperature, according to the Mott formula

$$S = \frac{\pi^2 k^2 T}{3e} \left[\frac{\partial \ln \sigma(\varepsilon)}{\partial \varepsilon} \right]_{\varepsilon_F} \tag{5}$$

where k is Boltzmann's constant and $\sigma(\varepsilon)$ is a conductivity-like function of electron energy ε. In practice, however, such simple behavior is virtually never observed. In crystalline metals, phonon drag causes large low-temperature peaks in S, and even when this term is suppressed by disorder, the measured thermopower is still nonlinear, owing to renormalization of the electronic mass and related electron–phonon effects at low temperature [60,61].

The thermopower of high-temperature superconductors shows a somewhat unusual behavior. The in-plane thermopower S_{ab} is small, usually just a few $\mu V/K$, as expected for metals, but it usually decreases in magnitude as temperature increases, even around room temperature, where the phonon drag term is sufficiently reduced to cause S to increase in magnitude as T increases in typical crystalline metals. In the La-bases superconductors, there is often a broad peak around 150 to 200 K. In $YBa_2Cu_3O_{7-\delta}$ the normal-state thermopower tends to be approximately constant (except sometimes near T_c), but its magnitude and even sign varies

with oxygen content; for δ close to zero, S is negative. In the Bi-based superconductors, S is usually positive just above T_c but then decreases linearly with temperature, sometimes crossing over to negative values at higher temperature. The
primary feature to be explained in each case is an approximately constant shift
of the experimental S versus T plot from the Mott formula prediction $S \propto T$,
the magnitude of this shift varying with the oxygen stoichiometry and number of
carriers [56].

Phonon drag seems unlikely as a cause of these constant shifts, since it should
reduce at higher temperatures. In Bi-based 2212 superconductors doped with Pb,
the size of this shift decreases as the Pb concentration increases and T_c decreases,
the thermopower continuing to show a linear decrease with temperature with little
change in slope [62]. Thus the thermopower becomes much closer to the standard
metallic behavior $S \propto T$. It is difficult to obtain this behavior pattern if the thermopower shift away from $S \propto T$ is due to a phonon drag term decreasing at high
temperatures.

A second conventional mechanism is electron–phonon renormalization of electronic properties [63], which has been used extensively to explain the offset (i.e.,
the shift away from $S \propto T$) observed in the thermopower of disordered materials
(often described as a "knee" in the thermopower seen around 50 K). This model
has been used previously to explain peaks and a change in thermopower sign seen
in amorphous Ag–Sn superconductors, behavior that is qualitatively similar to that
in high-temperature superconductors. This mechanisms gives a good account of
the high-T_c thermopowers [64] but requires a huge electron–phonon coupling and
partial cancellation of thermopower terms of different sign.

Normal band-structure effects, while giving rise to curvature in thermopower
as a function of temperature, do not produce sufficient nonlinearity to account for
the high-T_c data. However, an extremely narrow conduction band of width less
than kT_c very close to the Fermi surface could give thermopower above T_c with
little temperature dependence, and models have been proposed involving this possibility [65].

Another source of a constant thermopower is correlated hopping in a narrow
band in the Hubbard model, which has been applied to account for the decrease
in the thermopower of La_2CuO_4 from very large values as it is doped with small
amounts of Sr or Ba [66]. Trugman's model [53] for holes, added to the antiferromagnetic Mott insulator, gives similar behavior, but fails to provide thermopower
decreasing with temperature as seen at higher temperatures as the doping level
increases. This model is more appropriate for the low-doping limit.

In the spin–bipolaron model, Mott [28] also used the Heikes equation to derive
an approximately constant thermopower, which is small if $z \sim \frac{1}{2}$, as expected if the
carriers are bipolarons and conduction is in the impurity band (z is the ratio of
sites to carriers). A decrease in thermopower above T_c could result from a temperature dependence in z, although the linear decrease and change in sign seen in
some of the Bi-based superconductors would need to be accounted for in detail.

Unfortunately, very little has been predicted regarding thermopower in other
unconventional models for the superconductivity (e.g., the marginal Fermi liquid
or RVB picture, although Anderson et al. [67] cite the near vanishing of thermopower in $YBa_2Cu_3O_{7-\delta}$ as evidence for holons.

Curiously, the out-of-plane thermopower S_c resembles traditional metallic thermopower more closely than S_{ab}, but tends to be rather larger. S_c is generally positive for the Y- and Bi-based cuprate superconductors and usually extrapolates to positive values at $T = 0$, resembling the behavior usually seen in metallic diffusion thermopower with electron–phonon enhancement [64].

We also mention here that ceramic samples of $Ba_{1-x}K_xBiO_3$, which contain no magnetic ions, no Cu, and no planes, appear to show the same "anomalous" thermopower pattern as that in the Cu–O plane superconductors, which suggests that the origin of this thermopower behavior is not magnetic and does not require Cu–O planes [68].

B.   Experimental Survey for TBCCO

Some trends in the thermopower of polycrystalline Tl-based high-T_c superconductors are illustrated in Fig. 8. Much of the earlier work was focused on sintered samples of the 2223 phase materials [21,69–73]. The quality of the samples varied appreciably in these experiments, and many of the samples contained an admixture of other phases. Data from three different groups [21,69,70] for sintered 2223 phase material is displayed in the figure. Since the conductivity along the c-axis appears to be 100 to 1000 times smaller than in the ab plane, the data are expected to reflect S_{ab}. For comparison, thermopower measurements on a polycrystalline, single-phase 2223 thin-film sample with a high degree of orientation of the c-axis

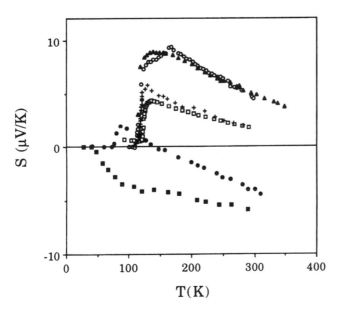

Figure 8   Thermopower as a function of temperature for sintered sample data: for 2223 phase Mitra et al. [69] (open circles), Bhatnagar et al. [21] (crosses) and Alcacer et al. [70] (open squares); $Tl_yPb_xCuSr_2Cu_2O_6$ from Rao et al., $x = 0$, $y = 0.9$ (solid circles) and $x = 0.25$, $y = 0.75$ (solid squares); for an oriented thin-film 2223 sample, Zhang et al. [74] (solid triangles).

normal to the plane of the film [74] are shown. Indeed, the results for the sintered samples and the oriented film are quite similar. Also shown are data for two doped single-Tl–O-layered high-T_c superconductors [75]. Universal behavior for the thermopower of Tl-based materials can be distinguished and can be described above T_c by

$$S = A' - B'T \tag{6}$$

The trend to this universal behavior can be distinguished in nearly all of the high-T_c superconductor families [56], but it is especially clear in the better Tl-based samples. The magnitude of the negative linear slope is consistent with metallic behavior, but the constant term is frequently large, so that the thermopower often extrapolates to a large positive value. An exception to that where the thermopower extrapolates to a negative value is also shown in Fig. 8. Measurements of thermopower for doped 2223 phase samples include replacement of Ca by Y [35] and Mg [76].

Measurements of thermopower for single-crystal [18] and single-phase sintered samples [77] of the 2212 phase compound have been reported. As may be seen in Fig. 9, the thermopower for this phase, both in the ab plane and along the c axis, is well described by Eq. (6). For the single-crystal samples [18] the magnitude of A' is quite large, about 34 μV/K [18] for both S_{ab} and S_c. The two samples of the sintered phase [77] had much smaller values, 16 and 8 μV/K. The values of A'

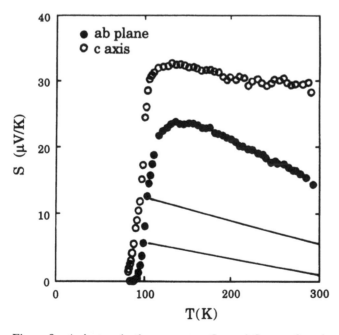

Figure 9   Anisotropic thermopower S_{ab} and S_c as a function of temperature for a single-crystal 2212 sample from Lin et al. [18]. Also shown are solid lines representing the best fit to the thermopower data above T_c for two sintered 2212 phase samples from Radhakrishnan et al. [77].

appear to correlate with the superconducting transition temperature in both the 2223 phase discussed above and the 2212 phase data with the smaller values of A' being associated with the higher-T_c samples for a given phase. In most measurements for the 2223 phase [21,69–72,74] the slope of the thermopower is approximately of the same size, $B' \approx 2$ to 4×10^{-2} μV/K, regardless of the variation between samples. The two polycrystalline 2212 samples have a similar slope [77], $B' \approx 2.5 \times 10^{-2}$ μV/K and 3.3×10^{-2} μV/K, with the larger value of B' correlated to the larger value of A' and lower value of T_c. For the single-crystal samples, as represented in Fig. 9, the value of B' for measurements along the c axis is approximately 1.8×10^{-2} μV/K, while for measurements in the ab plane the value is 6.5×10^{-2} μV/K, more than a factor of 3 greater in magnitude.

Some of the most interesting thermopower measurements [12,13] have been made on the 2201 samples. As discussed earlier, the properties of these samples can be changed dramatically by increasing the oxygen doping, δ. The effect of doping with Ce has also been reported [44]. The change in T_c and the nature of $\rho(T)$ as a function of δ for 2201 samples is shown in Fig. 2, while the change in the Hall coefficient for the identical samples [11] is illustrated in Fig. 6. The behavior of the thermopower for comparable samples [12] is shown in Fig. 10. Again the thermopower can be described by the linear dependence of Eq. (5) except for a small peak extending 10 to 30 K above T_c. This peak is not as evident in the data by Weeks et al. [13]. In the 2201 system, addition of oxygen (increasing δ) increases the carrier concentration, decreases T_c and ρ, changes the temperature dependence of ρ from a linear dependence toward a T^2 dependence, and alters the thermopower dramatically. The main change in the thermopower is the decease in A', but also the magnitude of the slope B' increases with increasing δ. Near the $(T = 0)$ superconductor–normal metal transition A' appears to approach zero, and the

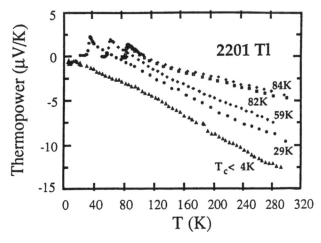

Figure 10 Thermopower as a function of temperature for $Tl_2Ba_2Cu_1O_{6+\delta}$ pressed powder samples from Obertelli et al. [12]. Doping levels δ can be estimated from Kubo et al. [7] based on the T_cs. Resistivity and Hall effect data for similarly treated samples from Kubo et al. [11] are shown in Figs. 2 and 6, respectively.

thermopower follows reasonably well the traditional metallic variation $S \propto T$. Thus the metallic character of these "overdoped" samples with higher carrier densities and T_cs far below the maximum value is reinforced by these thermopower data.

Some thermopower measurements have been made on compounds with only a single Tl-O plane [43,46,47,75]. These compounds show the same pattern of linear thermopowers with negative slopes. In most cases A' is positive, but in the case of Pb-substituted $Tl_{1-x}Pb_xSr_2CaCu_2O_{6+\delta}$, A' can be negative, as shown by the lower curve in Fig. 8. The shape of this curve where the extrapolated offset has the same sign as the slope of the linear component is similar to that seen in disordered metallic alloys in which the phonon drag is suppressed. In those alloys the apparent offset results from electron–phonon renormalization effects [60,61]. These results again emphasize the similarity to conventional metallic behavior in the absence of phonon drag.

V. THERMAL CONDUCTIVITY

A. Background Summary

In metals, both electrons and phonons contribute to the transport of heat. In pure metals the electronic contribution is usually dominant, but in alloys the phonon term is often significant, and in amorphous materials the phonon term is dominant. An upper limit for the electronic part κ_e of the thermal conductivity κ can be estimated for conventional metallic systems from the Wiedemann–Franz law, which relates κ_e and the electrical resistivity. If this law is also applicable to the high-temperature superconductors, it leads to the conclusion that the largest contribution to heat conduction comes from the phonon term, κ_{ph} [78].

Elementary kinetic theory gives the following expression for the thermal conductivity κ_{ph} due to phonons:

$$\kappa_{ph} = \tfrac{1}{3}C_{ph}vl_{ph} \tag{7}$$

where C_{ph} is the phonon contribution to the specific heat, v the phonon velocity (often taken as the speed of sound), and l_{ph} the mean free path of the phonons (assumed to be the same for all phonons in this simplified treatment). The specific-heat factor C_{ph} leads to an increase in κ_{ph} with temperature, whereas a decrease in the mean free path as temperature increases leads to a decrease in κ_{ph}. The former behavior dominates in amorphous materials in which l_{ph} shows little temperature variation, while the latter behavior dominates in good crystals at higher temperatures as increasing phonon–phonon scattering reduces l_{ph}. Measurements on high-T_c superconductors show a relatively small temperature dependence (of either sign) for κ above T_c (for details, see the review by Uher [79]).

Since the thermal conductivity is a property that is nonzero in the superconducting state, any change in κ near T_c is of particular interest. Data taken as temperature is lowered through T_c provide the most striking results on the thermal conductivity of the high-T_c superconductors: There is a sharp increase in the thermal conductivity κ as T decreases below T_c and a peak at a temperature around 50 K. This effect is ascribed to an increase in the phonon mean free path l_{ph} in Eq. (7) as the carriers condense into superconducting pairs which no longer scatter the phonons [78]. Single-crystal data show a pronounced peak below T_c for the in-plane thermal conductivity κ_{ab}, but until very recently no peak has been reported in the very limited data available for the out-of-plane thermal conductivity κ_c [79]. Theoretically, Tewordt and Wölkhausen have calculated the phonon thermal conductivity for high-T_c superconductors for the simple Bardeen–Cooper–Schrieffer (BCS) theory [80] and for strong-coupling s- and d-wave pairing [81], and obtain peaks below T_c qualitatively similar to those observed [79], although quantitative analyses are rather uncertain.

Presenting their marginal Fermi liquid model for the normal state of the Cu–O high-temperature superconductors, Varma et al. [32] state that the Wiedemann–Franz law is obeyed if it is assumed that the observed thermal conductivity κ is electronic. In fact, if this assumption is correct, the Wiedemann–Franz law is strongly violated, since the measured value of the ratio $\kappa\rho/T$ is too large by at least 100%, and often much more [79]. Further, no explanation is offered for the increase in κ below T_c. The thermal conductivity data therefore provide no support for their hypothesis. In general, theoretical predictions involving only the electronic thermal conductivity κ_e [30–32] have little relevance for the experimental data if the evidence mentioned above that phonons carry most of the heat current is correct.

B. Experimental Survey for TBCCO

Since thermal conductivity is a more difficult measurement to make than electrical conductivity or thermopower, data are correspondingly fewer. Figure 11 summarizes data for sintered samples of nominally 2223 phase material [82–84], although other minor phases are thought to be present in these samples; for the samples of Uher et al. [82], the fraction of other phases was estimated at 5% and the onset of the superconducting transition occurred at about 121 K. Measurements by Jezowski et al. [85] not shown in the figure are in general agreement with those shown in Fig. 11 that display a peak in κ. The absolute magnitude of the thermal

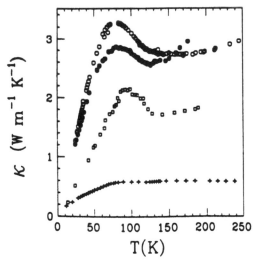

Figure 11 Thermal conductivity of sintered samples of the Tl-based 2223 phase, from Uher et al. [82] (top two data sets), Aliev et al. [83] (open squares), and Castello et al. [84] (crosses).

conductivity is comparable to that for other sintered superconductor samples [79]. Application of the Wiedemann–Franz law suggests that no more than 5% of the thermal conductivity is due to the flow of free charge carriers.

Except for one sample of Uher et al. [82] between T_c and 170 K, the thermal conductivity for these sintered 2223 samples increases as temperature increases above T_c. This behavior could be associated with large defect scattering, as seen in disordered materials, since the overall magnitude of κ is closer to that in disordered materials than in good insulating crystals (although increasing radiation losses can lead to an apparent increase in κ above 105 K [79]).

A dramatic feature of all but one of the data sets shown in Fig. 11 is the presence of a rather sharp increase in κ below T_c, with a pronounced peak at lower temperatures. Thus these measurements on the 2223 Tl-based materials confirm the enhanced thermal conductivity below T_c, consistent with an increase in phonon current as superconducting pairs condense. The exception is the sample of Castello et al. [84] with a small thermal conductivity, in which strong scattering of phonons by defects could mask the effect of the change in scattering by electrons.

For the sake of completeness, we mention that the thermal conductivity at very low temperatures (0.1 to 3 K) in these materials is well described by a power law T^x with $x = 2.4$ or 2.5 [82], which is similar to data for sintered Bi-based superconductors (as might be expected) but different from the linear behavior often seen for $YBa_2Cu_3O_7$.

Very recently, the first measurements of thermal conductivity on single-crystal TBCCO have been made on small 2212 crystals by Cao et al. [86]. The temperature dependence of κ_{ab} and κ_c is shown in Fig. 12. As for the 2223 sintered samples, above T_c κ increases as temperature increases, and a sharp increase in κ_{ab} is seen below T_c. The room-temperature value of the in-plane conductivity κ_{ab} was 1.5 ± 0.2 W·m^{-1} K^{-1}, which is smaller than most of the measured values for the 2223

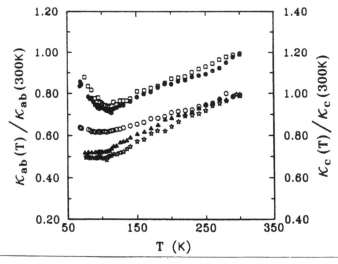

Figure 12   Reduced thermal conductivity from Cao et al. [86] for Tl 2212 phase single crystal: κ_{ab} (top two data sets) and κ_c (bottom three data sets).

sintered samples. The out-of-plane conductivity κ_c was approximately an order of magnitude smaller than κ_{ab}.

The most intriguing aspect of the new single-crystal data is that the anomaly across T_c is seen for the out-of-plane as well as for the in-plane thermal conductivity, although it is less pronounced for κ_c in the Tl-based crystals (the same authors, incidentally, also observed a sharp increase in κ_c below T_c in $YBa_2Cu_3O_7$ single crystals [83]). These results support the interpretation of the anomaly in terms of an increase in phonon current as scattering by electrons disappears, since this effect is expected to occur in both the in-plane and out-of-plane directions [81]. Greater defect scattering in the c direction would in general be expected to give a less pronounced effect in this direction, which could possibly account for the effect for the Tl-based crystals being more like a flattening in κ_c below T_c rather than a clearly defined peak, although other factors are likely to play a role [83].

VI. SUMMARY AND CONCLUSIONS

Selected experimental data can apparently be used to lend support to many different theories of transport in the normal state for Tl-based superconductors. The experimental data are sometimes confusing and often incomplete. Some of the confusion may be removed by more complete sets of measurements of a specific system and by systematic variations of sample doping to determine which differences between measurements on nominally similar samples can readily be understood in terms of small compositional differences (the variation with oxygen doping in the 2201 system is a good example). The techniques for sample preparation have advanced rapidly over the past few years to permit measurements on good-quality crystals. To be most useful, transport measurements on the same sample (or at least equivalent samples) should include $R_H(T)$, $\rho(T)$, and $S(T)$. We will try to indicate what characteristics of transport in Tl-based systems are well established by experiment, where further experiments are needed, and what limitations on the applicability of different theoretical models are indicated by these experiments.

Experimentally, the resistivity for optimally doped and "overdoped" samples can be described by a power law [Eq. (1)]. For the optimally doped samples a linear dependence is characteristic. The temperature dependence becomes quadratic in the overdoped case. For "underdoped" samples the resistivity becomes semiconductor-like. For single-crystal samples ρ_{ab} and ρ_c show strong similarities in temperature dependence, but ρ_c is two to three orders of magnitude greater than ρ_{ab}. The Hall coefficient is positive for all of the single-phase samples that have been reported for the Tl-based family of high-T_c samples. The underdoped and optimally doped samples generally exhibit a linear temperature dependence for R_H^{-1} which extrapolates to a positive intercept at zero temperature [Eq. (3)]. The temperature dependence for overdoped samples is much weaker (but it is not temperature independent as expected for good metallic samples) and appears to show a minimum. The thermopower above T_c can be described approximately as the sum of a constant and a linear term [Eq. (6)] for high-quality single-phase samples. The constant is nearly always positive and the linear term is always negative. The constant term is approximately zero for overdoped samples but increases dramatically for the underdoped samples. The temperature dependence appears to increase with doping. In many instances the transport measurements on sintered and polycrystalline thin-film single-phase samples agree reasonably well with those along the *ab* plane for the corresponding single-crystal materials.

Measurements with single-crystal samples and high-quality sintered or thin-film samples are still sufficiently rare that further transport measurements, particularly Hall effect, thermopower, and thermal conductivity, could be helpful, but it is very important that it be well established that the samples are single phase and that, particularly for the 2201 phase, the oxygen content is established. Doping of the 2201 phase by changing the oxygen content has been especially rewarding. Further measurements are now needed for samples in which the underdoping and overdoping is accomplished by substitutional doping in the Cu–O or the Tl–O layers. Unfortunately, it has proven much harder to make high-quality single-crystal or polycrystalline single-phase substitutionally doped samples. To exploit the diversity of the Tl-based family of high-T_c superconductors, measurements on the single- as well as double-Tl–O-layer members and on members with three and more Cu–O layers are needed. These efforts have really only begun. Lack of techniques for the preparation of high-quality samples is a major obstacle here, but one we expect will be overcome. We again stress that measurements of as many of the transport coefficients as possible, preferably all, should be made on each high-quality sample.

It is not easy to distinguish experimentally between the various mechanisms proposed to account for the approximately linear increase of resistivity with temperature. Apart form the data for the 2212 single crystal shown in Fig. 1b, there is no evidence of a reduction in the slope (i.e., "saturation") of the resistivity as a function of temperature, as seen, for example, in A15 compounds, in which scattering of the carriers at high temperatures is thought to be very strong. The Tl-based cuprate superconductors, particularly the 2201 compounds, do allow a test of whether the resistivity remains linear below 100 K in samples in which T_c is depressed to lower values by increasing the carrier concentration. The answer is clearly that a higher power law (T^2 for nonsuperconducting samples) is observed. Consequently, at least in these cases, there is more similarity to models involving scattering of conventional carriers by phonons or spin fluctuations than to theoretical models predicting a continuation of the high-temperature linear law [29], but it must also be noted that there are few conventional metals that exhibit such a large T^2 contribution over such a large temperature range. Further, the change in the extrapolated zero-temperature conductivity from positive to negative as n increases (up to five or six) for single-layer Tl compounds [4] argues against an intrinsic linear conductivity that extrapolates through the origin for "good" samples.

The observation of out-of-plane resistivities ρ_c that increase as temperature increases fails to provide any support for the expression proposed by Anderson and Zou [29], in which ρ_c should show the opposite behavior. The qualitative similarity of ρ_c and ρ_{ab}, as in some other cuprate superconductors, suggests metal-like conduction in the out-of-plane direction as well as in the Cu–O plane.

The positive sign of the Hall coefficient R_H indicates that unless some unusual effect reverses its sign, the dominant carriers are hole-like rather than electron-like. Like other cuprate superconductors, but unlike most metals, R_H is temperature dependent in Tl-based superconductors, although the temperature dependence appears to be less than in $Yba_2Cu_3O_7$. Several mechanisms producing a temperature-dependent Hall coefficient have been suggested, but there is a lack of specific prediction. However, the limited data for Tl-based compounds appear to be in reasonable agreement with the expression for the temperature-dependent Hall angle proposed by Anderson [54]. One of the problems interpreting Hall effect data is that the same expression also applies for conventional metals with a T^2 resistivity term and a temperature-independent Hall coefficient. Even the behavior of R_H in some conventional metals, in particular the effect of the spin−orbit interaction, is not fully understood. Nevertheless, Hall measurements on doped single-crystal and high-quality polycrystalline samples with three and four Cu−O layers may be particularly helpful in testing models for normal-state transport.

It seems from the linearity of thermopower that phonon drag does not play a major role in producing the shift of thermopower away from $S \propto T$ that characterizes the superconducting samples, in agreement with the conclusion from the data for Bi-based superconductors [62]. Contributions to this thermopower behavior based on conventional metallic models could arise from strong electron−phonon interactions or very sharp band features at the Fermi level. Other possibilities are bosonic conduction of different types in unconventional models, but there is a lack of specific predictions. In the underdoped limit where T_c is decreased below its maximum value, the thermopower becomes large and resembles a hopping-type thermopower, consistent with the larger resistivity that acquires a semiconductor-like temperature dependence in this limit. In the overdoped limit the thermopower approaches the traditional metallic variation given by the Mott formula [Eq. (5)], which is rarely ever observed in real metallic systems.

Regarding thermal conductivity data, there is a reduction in κ in the normal state relative to the superconducting state as in other cuprate superconductors, which is ascribed to an increased scattering of phonons by the unpaired electrons above T_c. Most interesting is the observation of this T_c anomaly in the out-of-plane thermal conductivity κ_c as well as for the in-plane direction. A lack of any T_c anomaly in the smaller c-axis thermal conductivity in earlier data was cited by Anderson [87] as a factor arguing for the electronic nature of the ab-plane thermal conductivity, so the evidence for the anomaly in κ_c favors the conventional explanation of the T_c anomaly in terms of the phonon current.

In general, one of the clear conclusions that emerges from the data is the similarity to conventional metallic behavior in the overdoped samples with relatively high carrier densities and low T_cs. This similarity is particularly clear in the data for the 2201 material for the resistivity [11−15], Hall coefficient [11,15], and thermopower [12,13]. Thus viable theories need to reproduce conventional metallic behavior smoothly as this regime is approached.

ACKNOWLEDGMENTS

We wish to thank J. L. Tallon for copies of data prior to publication. A.B.K. acknowledges a travel grant from the N.Z./U.S.A. Cooperative Science Programme and thanks Ctirad Uher for helpful discussions. D.G.N. acknowledges support from the Texas Advanced Technology Program (project 215) and the Robert A. Welch Foundation, Houston, Texas.

REFERENCES

1. R. M. Hazen, in *Physical Properties of High Temperature Superconductors II*, ed. D. M. Ginsberg, World Scientific, Singapore, 1990, pp. 175–188 and references therein.
2. R. Sugise, M. Hirabayashi, N. Terada, M. Jo, T. Shimomura, and H. Ihara, *Jpn, J. Appl. Phys.* 27, L1709 (1988).
3. M. Verwerft, G. Van Teudeloo, and S. Amelinckx, *Physica C* 156, 607 (1988).
4. H. Kushuhara, T. Kotani, H. Takei, and K. Tada, *Jpn. J. Appl. Phys.* 28, L1772 (1989).
5. T. Kaneko, H. Yamanchi, and S. Tanaka, *Physica C* 178, 377 (1991).
6. Y. Shimakawa, Y. Kubo, T Manako, T. Satoh, S. Iijima, T. Ichihashi, and H. Igarashi, *Physica C* 157, 279 (1989).
7. Y. Kubo, Y. Shimakawa, T. Manako, T. Satoh, S. Iijima, T. Ichihashi, and H. Igarashi, *Physica C* 162–164, 991 (1989).
8. Y. Shimakawa, Y. Kubo, T. Manako, and H. Igarashi, *Phys. Rev. B* 40, 11400 (1989).
9. R. G. Buckley, J. L. Tallon, I. W. M. Brown, M. R. Presland, N. E. Flower, P. W. Gilberd, M. Bowden, and N. B. Milestone, *Physica C* 156, 629 (1988).
10. V. P. Awana, S. B. Samanta, P. K. Dutta, E. Gmelin, and A. V. Narlikar, *J. Phys. Condensed Matter* 3, 8893 (1991).
11. Y. Kubo, Y. Shimakawa, T. Manako, and H. Igarashi, *Phys. Rev. B* 43, 7875 (1991).
12. S. D. Obertelli, J. R. Cooper, and J. L. Tallon, *Phys. Rev. B* 46, 14928 (1992).
13. D. E. Weeks, W. Kiehl, C. Dong, and A. M. Hermann, *Physica C* 176, 368 (1991).
14. H. M. Duan, R. M. Yandrofski, T. S. Kaplan, B. Dlugosch, J. H. Wang, and A. M. Hermann, *Physica C* 185–189, 1283 (1991).
15. T. Manako, Y. Shimakawa, Y. Kubo, and H. Igarashi, *Physica C* 190, 62 (1991).
16. H. M. Duan, W. Kiehl, C. Doug, A. W. Cordes, M. J. Saeed, D. L. Viar, and A. M. Hermann, *Phys. Rev. B* 43, 12925 (1991).
17. D. H. Kim, A. M. Goldman, J. H. Kang, and R. T. Kampwirth, *Phys. Rev. B* 40, 8834 (1989).
18. S. Y. Lin, L. Lu, D. L. Zhang, H. M. Duan, and A. M. Hermann, *Europhys. Lett.* 12, 641 (1990); *Physica B* 165–166, 1207 (1990).
19. P. S. Wang, J. C. Williams, D. Rathanyaka, B. Hennings, D. G. Naugle, and A. B. Kaiser, *Phys. Rev. B* 47, 1119 (1993).
20. Z. Z. Sheng, L. Sheng, X. Fei, and A. M. Hermann, *Phys. Rev. B* 39, 2918 (1989).
21. A. K. Bhatnagar, R. Pan, D. G. Naugle, P. J. Squattrito, A. Clearfield, Z. Z. Sheng, Q. A. Shams, and A. M. Hermann, *Solid State Commun.* 73, 53 (1990).
22. P. B. Allen, Z. Fisk, and A. Migliori, in *Physical Properties of High Temperature Superconductors I*, ed. D. M. Ginsberg, World Scientific Singapore, 1989, p. 213.
23. C. C. Tsuei, *Physica A* 168, 238 (1990).
24. R. Zeyher, *Phys. Rev. B* 44, 10404 (1991).
25. T. Moriya, Y. Takahashi, and K. Ueda, *J. Phys, Soc. Jpn.* 59, 2905 (1990).
26. D. Ihle, M. Kasner, and N. M. Plakida, *Z. Phys. B* 82, 193 (1991).
27. A. B. Kaiser and S. Doniach, *Int. J. Magn.* 1, 11 (1970).
28. N. F. Mott, *Adv. Phys.* 39, 55 (1990).
29. P. W. Anderson and Z. Zou, *Phys. Rev. Lett.* 60, 132 (1988).
30. N. Nagaosa and P. Lee, *Phys. Rev. B* 43, 1233 (1991).
31. L. B. Ioffe and G. Kotliar, *Phys. Rev. B* 42, 10348 (1990); L. B. Ioffe, V. Kalmeyer, and P. B. Wiegmann, *Phys. Rev. B* 43, 1219 (1991).
32. C. M. Varma, P. B. Littlewood, S. Schmitt-Rank, E. Abrahams, and A. E. Ruckenstein, *Phys. Rev. Lett.* 63, 1996 (1989).
33. H. C. Montgomery, *J. Appl. Phys.* 42, 2971 (1971).

34. A Poddar, P. Mandal, A. N. Das, B. Ghosh, and P. Choudhury, *Phys. Rev. B* 44, 2757 (1991).
35. J. H. Wang, D. Weeks, X. Fei, C. Dong, W. Kiehl, J. Trefny, Z. Z. Zhao, and A. M. Hermann, *Appl. Phys. Commun.* 10, 293 (1990).
36. M. R. Presland, J. L. Tallon, R. G. Buckley, R. S. Liu, and N. E. Flower, *Physica C* 176, 95 (1991).
37. H. H. Otto, T. Zetterer, and H. F. Renk, *Z. Phys. B* 75, 433 (1989).
38. Y. Kubo, Y. Shimakawa, T. Manako, T. Kondo, and H. Igarashi, *Physica C* 185–189, 1253 (1991); and in *Proc. 3rd ISTEC Workshop on Superconductivity*, May 13–15, 1991 Kumamoto, Japan, pp. 75–78; Y. Kubo, T. Kondo, Y. Shimakawa, T. Manako, and H. Igarashi, *Phys. Rev. B* 45, 5553 (1992).
39. I. K. Gopalakrishnan, J. V. Yakhmi, and R. M. Iyer, *Physica C* 172, 183 and 450 (1991).
40. H. C. Ku, M. F. Tai, J. B. Shi, M. J. Shieh, S. W. Hsu, G. H. Hwang, D. C. Ling, T. J. Watson-Yang, and T. Y. Lin, *Jpn. J. Appl. Phys.* 28, L923 (1989).
41. T. Manako, Y. Shimakawa, Y. Kubo, T. Satoh, and H. Igarashi, *Physica C* 158, 143 (1989).
42. C. N. R. Rao, A. K. Ganguli, and R. Vijayaraghavan, *Phys. Rev. B* 40, 2565 (1989).
43. Z. Z. Sheng, D. X. Gu, Y. Xin, D. O. Pederson, L. W. Finger, C. G. Hadidacios, and R. M. Hazen, *Mod. Phys. Lett. B* 5, 635 (1991).
44. J. H. Wang, D. E. Weeks, W. Kiehl, J. Trefny, and A. M. Hermann, *Appl. Phys. Commun.* 10, 1 (1990).
45. T. Itoh and H. Uchikawa, *Phys. Rev. B* 39, 4690 (1989).
46. A. K. Ganguli, V. Manivannan, A. K. Sood, and C. N. R. Rao, *Appl. Phys. Lett.* 55, 2664 (1989).
47. A. K. Ganguli, R. Vijayaraghavan, and C. N. R. Rao, *Physica C* 162–164, 867 (1989).
48. N. P. Ong, in *Physical Properties of High-Temperature Superconductors II*, ed. D. M. Ginsberg, World Scientific, Singapore, 1990, p. 459.
49. A. Maeda, M. Hase, I. Tsukada, K. Noda, S. Takebayashi, and K. Uchinokura, *Phys. Rev. B* 41, 6418 (1990).
50. C. Uher and A. B. Kaiser, *Phys. Lett. A* 125, 421 (1987).
51. A. T. Fiory and G. S. Grader, *Phys. Rev, B* 38, 9198 (1988).
52. N. F. Mott, *Phil. Mag. Lett.* 62, 273 (1990).
53. S. A. Trugman, *Phys. Rev. Lett.* 65, 500 (1990).
54. P. W. Anderson, *Phys. Rev. Lett.* 67, 2092 (1991).
55. T. R. Chien, Z. Z. Wang, and N. P. Ong, *Phys. Rev. Lett.* 67, 2088 (1991).
56. A. B. Kaiser and C. Uher in *High Temperature Superconductors*, Vol. 7, ed. A. V. Narlikar, Nova Science, Commack, N.Y., 1991, pp. 353–392.
57. J. Clayhold, N. P. Ong, P. H. Hor, and C. W. Chu, *Phys. Rev. B* 38, 7016 (1988).
58. P. Mandal, A. Poddar, A. N. Das, B. Ghosh, and P. Choudhury, *Phys. Rev. B* 40, 730 (1989).
59. K. C. Woo, K. E. Gray, R. T. Kampwirth, and J. H. Kang, *Physica C* 162–164, 1011 (1989).
60. D. G. Naugle, *J. Phys. Chem. Solids* 45, 367 (1984).
61. M. A. Howson and B. L. Gallagher, *Phys. Rep.* 170, 265 (1988).
62. C. Varoy, H. J. Trodahl, R. G. Buckley, and A. B. Kaiser, *Phys. Rev. B* 46, 463 (1992).
63. A. B. Kaiser, *Phys. Rev. B* 29, 7088 (1984).
64. A. B. Kaiser and G. Mountjoy, *Phys. Rev. B* 43, 6266 (1991).
65. S. Bar-Ad, B. Fisher, J. Ashkenazi, and J. Genossar, *Physica C* 156, 741 (1988).
66. J. R. Cooper, B. Alavi, L.-W. Zhou, W. P. Beyermann, and G. Grüner, *Phys. Rev. B* 35, 8794 (1987).

67. P. W. Anderson, G. Baskaran, Z. Zou, J. Wheatley, T. Hsu, B. S. Shastry, B. Doucot, and S. Liang, *Physica C* 153–155, 527 (1988).
68. C. Uher, S. D. Peacor, and A. B. Kaiser, *Phys. Rev. B* 43, 7955 (1991).
69. N. Mitra, J. Trefny, B. Yarar, G. Pine, Z. Z. Sheng, and A. M. Hermann, *Phys. Rev. B* 38, 7064 (1988).
70. L. Alcácer, M. Almeida, R. Braun, A. P. Goncalves, S. M. Green, E. B. Lopes, H. L. Luo, and C. Politis, *Mod. Phys. Lett. B* 2, 923 (1988).
71. S. Yan, T. Chen, H. Zhang, J. Peng, Z. Shen, C. Wei, Q. Wen, K. Wu, L. Tong, and H. Zhang, *Mod. Phys. Lett. B* 2, 1005 (1988).
72. Y. S. Song, Y. S. Choi, Y. W. Park, M. S. Jang, and S. K. Han, *J. Moscow Phys. Soc.* 1, 293 (1991).
73. I. S. Shchetkin, T. Sh. Osmanov, N. V. Abramov, and V. Z. Evseev, *Sov. J. Low Temp. Phys.* 17, 102 (1991); *Fiz. Nizk. Temp.* 17, 197 (1991).
74. H. Zhang, H. Ma, S. Yan, S. Liang, L. Wang, and S. Yan, *Physical B* 165–166, 1209 (1990).
75. C. N. R. Rao, T. V. Ramakrishnan, and N. Kumar, *Physica C* 165, 183 (1990).
76. S. S. Yan, T. Chen, H. B. Zhang, Z. H. Shen, and L. T. Tong, and H. Zhang, *Mod. Phys. Lett. B* 2, 1263 (1988).
77. V. Radhakrishnan, C. K. Subramaniam, R. Srinivasan, I. K. Gopalakrishnan, P. V. P. S. S. Sastry, J. V. Yakhmi, and R. M. Iyer, *Solid State Commun.* 73, 637 (1990).
78. C. Uher and A. B. Kaiser, *Phys. Rev. B* 36, 5680 (1987).
79. C. Uher, *J. Supercond.* 3, 337 (1990).
80. L. Tewordt and Th. Wölkhausen, *Solid State Commun.* 70, 839 (1989).
81. L. Tewordt and Th. Wölkhausen, *Solid State Commun.* 75, 515 (1990).
82. C. Uher, S. D. Peacor, and J. Shewchun, *Physica C* 177, 23 (1991).
83. F. G. Aliev, F. V. Moschalkov, and V. V. Pryadun, *Physica C* 162–164, 572 (1989).
84. D. Castello, M. Jaime, M. Nunez Regueiro, and C. Fainstein, in *Progress in High Temperature Superconductivity*, Vol. 25, ed. R. Nicolsky, World Scientific, Singapore, 1990, p. 452.
85. A. Jezowski, A. J. Zaleski, H. Misiorek, E. P. Khlybov, and V. V. Evdokimova, *Phys. Lett. A* 139, 265 (1989).
86. S.-C. Cao, D.-M. Zhang, D.-L. Zhang, H. M. Duan, and A. M. Hermann, *Phys. Rev. B* 44, 12571 (1991).
87. P. W. Anderson, in *Strong Correlations and Superconductivity*, ed. H. Fukuyama, S. Maekawa, and A. P. Malozemoff, Springer-Verlag, Heidelberg, 1989, p. 2.

27
Tunneling Spectroscopy of Thallium-Based High-Temperature Superconductors

John Moreland
National Institute of Standards and Technology
U.S. Department of Commerce
Boulder, Colorado

I. INTRODUCTION

There are several articles that review the field of electron tunneling spectroscopy (ETS) [1–4]. In addition, there have been several articles regarding ETS of high-temperature superconducting (HTS) materials summarizing much of the work to date [5–10]. Researchers have fabricated many kinds of tunnel junction configurations having HTS electrodes. The collected data show general trends in current–voltage (I–V) characteristics, but there are still inconsistencies between measurements on presumably similar tunneling configurations. Reported values for the ratio $2\Delta/k_b T_c$, where Δ is the superconducting energy gap derived from tunneling data, k_b is Boltzmann's constant, and T_c is the superconducting critical temperature, range from as low as 3 to as high as 20 for various HTS compounds.

In this chapter we focus on ETS of thallium-based HTS materials. A summary of published results is included with comments regarding interpretations of each measurement. Other phenomena that may have a bearing on the interpretation of tunneling spectra are addressed briefly, including a linear background conductance, Coulomb blockade effects, and bound states at superconductor/normal metal (SN) interfaces. We do not address many of the Josephson-like devices based on weak links, point contacts, microbridges, or step-edge junctions.

II. IDEAL JUNCTION GEOMETRIES

The large range of tunneling gaps may be due to local variations in stoichiometry and crystallographic orientation of the electrodes not reflected in the T_c measurement of the sample. It is clear from the transport properties of HTS materials that they are highly anisotropic and that the grain boundaries in polycrystalline HTS materials act as Josephson junctions. For this reason, it is important to perform tunneling measurements on materials with good stoichiometry throughout the sample. Large crystals and large-grained epitaxial films, for example, should have sufficient uniformity and size for fabricating tunneling structures required for explainable spectroscopic features. Measurements with HTS polycrystalline electrodes are otherwise open to various interpretations, and although quite often intriguing, are generally considered unconvincing.

The logical conclusion is that, ideally, junction configurations for tunneling spectroscopy should have single-crystal HTS electrodes. These can be fabricated using deposited or pressed-contact counter electrodes with native oxide barriers on an HTS electrode, or with a vacuum-gap tunnel barriers most often achieved with a scanning tunneling microscope (STM). The advantage of the STM is that the contact area is small enough (less than 1 nm^2) that it is possible to limit tunneling to the smallest single grain in a polycrystalline HTS sample. In addition, the capability of scanning the surface of the sample allows for spatially resolved spectroscopy, commonly referred to as scanning tunneling spectroscopy (STS) [11]. Deposited electrodes can be patterned to achieve this goal, but the size of the junction cannot be reduced below the fundamental limits of the patterning process. Minimizing the area of a pressed-contact junction, on the other hand, is very difficult to do in a controlled manner. Therefore, this type of junction requires that the sample be a large single crystal or a polycrystal with large enough grains to encompass the contact area. By definition, pressed contacts are different from point contacts because the electrodes are separated by a tunneling barrier in the former case, whereas the electrodes are shorted together in the latter case.

It is also possible to incorporate single-crystal HTS electrodes into static (nonscanning) mechanically adjustable tunnel barrier junctions such as break junctions [12]. A break junction in an HTS crystal affords the possibility of tunneling between natural surfaces of the crystal. There are other relatively new configurations that rely on epitaxial structures having thin-film HTS electrodes. An epitaxial trilayer junction can be made with an insulating film separating two single-crystal HTS films [13]. Here tunneling occurs predominately along the c-axis direction perpendicular to the film substrate. It is also possible to fabricate single-crystal, thin-film structures where tunneling occurs primarily perpendicular to the c axis and parallel the substrate [14]. Thus far such epitaxial configurations have not been achieved with thallium-based HTS materials.

III. TUNNELING SPECTROMETERS

There are several low-noise electronic circuits that have been developed for measuring the $I-V$ and $dI/dV-V$ characteristics of junctions [15–17]. For mechanically adjustable junctions or junctions where the resistances are large and change considerably as a function of voltage, it is important to take precautions to avoid measurement errors. STS or ETS with an STM operated in a static mode are techniques with the additional complication of incorporating a procedure that allows for $I-V$ measurements at a given point on the sample at a fixed tip–sample distance. Various sample-and-hold circuits have been developed [18,19]. In addition, difficulty arises for low-temperature measurements that require long leads to the tunneling tip which are susceptible to capacitive and inductive noise pickup. One solution is to use a cold preamplifier near the tunneling tip. Alternatively, a driven shield around the tunnel-current lead can be used to minimize capacitive pickup effects.

IV. ETS OF THALLIUM-BASED HTS MATERIALS

The following excerpt regarding break junction measurements of a thallium-based HTS crystal was taken from Ref. 20. It is included to illustrate some of the factors that influence interpretation of tunneling energy gap measurements in HTS materials.

The Meissner effect data for the crystal is shown in Fig. 1. The flux expulsion was measured using a SQUID magnetometer by cooling from above the superconducting transition in a field of 2.5 mT applied long the c axis of the crystal. The Meissner data was taken before the crystal is mounted onto the break junction substrate assembly. The data is normalized by the weight of the crystal (110 μg). The Meissner fraction, however, is not meaningful since the demagnetization factor is unknown and substantial when the field is normal to an irregular thin plate. The relatively small size of these crystals prevents measurements with the field parallel to the plate (the signal to noise was much improved when the field is along the c axis). The data should accurately reflect the Meissner onsets $(T_c$'s) and general nature of the transition, neglecting the unknown demagnetization factor. Notice that there appear to be two Meissner onsets in Fig. 1. A gradual transition begins at 105 K rising slowly and reaching only 25% of the low temperature value over the first 15 K below T_c. It appears that about 75% of this crystal has a T_c near 90 K as shown by the rapid increase in the signal at this point. We also measure the low frequency ac resistance of the sample as it is being cooled in the break

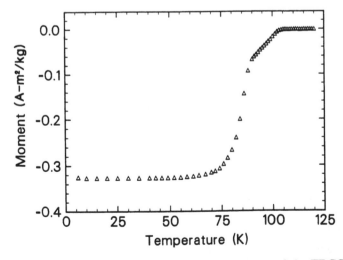

Figure 1 Magnetic moment versus temperature of the TBCCO crystal prior to breaking. Notice the two distinct onsets near 105 and 90 K. (From Ref. 21.)

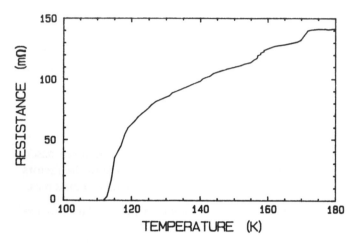

Figure 2 Resistance versus temperature of the TBCCO crystal prior to breaking. The midpoint of the transition occurs at 115 K with the resistance falling below 0.001 Ω (the detection limit) at 108 K. (From Ref. 21.)

junction apparatus. The results are shown in Fig. 2. The resistive midpoint occurs at 115 K with the resistance falling from about 0.08 Ω at 108 K. The results imply that there may be a thin layer of higher T_c phase at the surface of the crystal that effectively shorts out subsurface conduction.

Some of the tunneling data obtained for this crystal are shown in Fig. 3. Here we show the current and conductance versus voltage for the original setting of the junction in liquid helium at 4 K. These data were taken just after the crystal was fractured and break junction relaxed so that tunneling could be detected within the fracture. Notice the broad gap evident between the well-

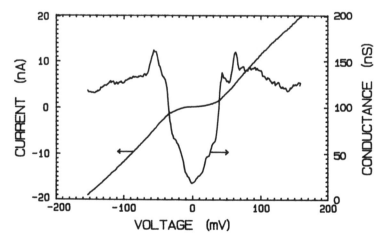

Figure 3 Current and conductance versus voltage at 4 K for the TBCCO break junction. A broad gap about zero bias is apparent in the curves. Equating the voltage difference between the peak conductance voltages to 4Δ implies that Δ = 30 meV. (From Ref. 21.)

formed conductance peaks symmetric about zero bias and the relatively flat conductance outside of the gap voltages. We are assuming that the electrode configuration of the fracture was of the superconductor–insulator–superconductor (SIS) type. Therefore the BCS energy gap, Δ, can be estimated from the voltage difference between voltages of the conductance peaks of 120 meV $\equiv 4\Delta$. This implies $\Delta = 30$ meV. After the initial setting, the junction was subsequently adjusted back to an open contact. The above procedure was repeated several times with the same result for Δ. However, once the fracture was relaxed so that a supercurrent could be observed in the tunneling $I-V$ characteristic (presumably forming a point contact in the fracture), subsequent settings of the fracture had lower tunneling Δ's. It seems that the physical contact between tunneling electrodes destroyed the superconducting properties of the crystal's fractured surfaces.

The temperature dependence of the crystal break junction $I-V$ curves was measured by raising the apparatus to various levels above the surface of the liquid helium. We found that it was possible to detect a supercurrent at temperatures as high as 95 K. It was not possible to accurately track Δ as a function of temperature due to radical shifting of the junction electrodes upon warming. However, we could detect a tunneling gap at temperatures as high as 40 K. At higher temperatures, the gap, even after repeated resetting of the junction, was obliterated due to broadening of the conductance peaks at the tunneling gap edges. We could only speculate about the causes of this phenomena. Spectral broadening may have been due to significant thermal smearing of the two dimensional fermi surface characteristic of the layered compounds or the possible activation of surface states within the gap with increasing temperature.

From the data we conclude that the ratio of $2\Delta/k_B T_c$ is between 6.1 and 7.7 if one chooses a range of T_c's between 115 and 90 K. Ideally, the crystal should be a single phase, thus eliminating uncertainty in the break junction gap measurement due local stoichiometric variations. However, the sample tested for this paper was a polycrystal. Thallium based high temperature superconductors are capable of syntactic growth as well as intergrowth within the crystal lattice [21]. Also, their is evidence for Tl substitution into Ca sites. X-ray diffraction measurements on the crystal used for this paper showed a predominate 2122 phase with traces of 2223 phase. This observation combined with the two onsets apparent in the Meissner data and the higher T_c from the resistance data indicate that, indeed, the crystal was a polycrystal.

The interpretation of $I-V$ curves of thallium-based HTS materials has thus far been a matter of locating relatively gross shifts in the tunneling current as a function of voltage as discussed above. There is some consensus regarding the "gap" values determined from tunneling $I-V$ curves. Typically, gap values are determined from derivatives of the $I-V$ curve by measuring the voltage differences between peak locations in the $dI/dV-V$ curves. Alternatively, the $dI/dV-V$ curve can be fitted in order to take into account a spread in the gap values, to including thermal smearing, or to include a damping factor that usually results in a lower gap value than the peak-to-peak method. Fitting $I-V$ curves is required at temperatures approaching T_c or when $dI/dV-V$ curves are sufficiently broadened at the gap edges. Table 1 summarizes some of the other tunneling measurements on the HTS

Table 1 Summary of ETS Measurements of Thallium-Based HTS Materials

T_c (K)	$2\Delta_{pp}$ (meV)	$2\Delta_{pp}/k_B T_c$	$2\Delta_{fit}$ (meV)	$2\Delta_{fit}/k_B T_c$	Method[a]	Comment[b]	Refs.
90–115	60 ± 2	6.1–7.7			BJ	SIS (+J)	20
94.5	50 ± 6	6.1 ± 0.7			PC/Pb	SIS' (−J)	22
120	46–68	4.4–6.5			PC/Pb	SIS' (−J)	23
114	60 ± 5	6.1–1			PC/Pb	SIS' (−J)	23
105	40	4.4	26–30[c]	2.9–3.3[c]	PC/PtIr	SIN	24, 25
	40	4.4			GB	SIS (−J)	24, 25
112	40	4.4	32–56[d]	3.3–5.8[d]	PC/Au	SIN	26
116	46	4.6			PC/W	SIN	27
106	98	10.6			PC/Ta	SIS' (−J)	28
	48	5.3			GB	SIS (−J)	28

[a]BJ, break junction; PC, pressed contact; GB, grain boundary.
[b]S, superconductor; I, insulator; N, normal metal.
[c]Gaussian distribution fit [29].
[d]Dynes fit [30].

thallium compounds performed at liquid helium temperatures well below T_c. It follows the format presented in a recent review article written by Hasagawa et al. [10] and includes a summary of the types of techniques used to do ETS and some comments on each measurement. This review is hopefully comprehensive to date. Note that under the comments column, use of the term $\pm J$ has been used to denote the presence or absence of a Josephson current flowing in the junctions at zero bias. If a Josephson current is not observed in what is thought to be an SIS junction, there is probably a normal layer on the surface of one or both of the superconducting electrodes.

V. POSSIBLE ORIGIN OF TUNNELING GAPS

The tunneling gaps that have been observed for HTS junctions bear remarkable resemblance to those observed in conventional superconducting junctions with some notable exceptions. The ratio of $2\Delta/k_b T_c$ is within the strong-coupling limits for BCS superconductivity, increasing with T_c from a value of about 4 for LSCO, 5 for YBCO, and 6 for TBCCO compounds. However, the subgap leakage current for HTS junctions is unusually high, and generally the normal-state conductance is linearly increasing out to several hundred millivolts in contrast to very low subgap currents and flat (constant) normal-state conductance as a function of voltage for conventional superconductors. In addition, the temperature dependence of the gap seems to follow the general behavior, which is not Bardeen–Cooper–Schrieffer (BCS)-like, where the conductance peaks at the gap edges do not shift in voltage but rather become thermally broadened, disappearing from the tunneling spectrum near T_c. The linear conductance background persists even above T_c.

Table 2 summarizes some of the explanations offered for the tunneling gaps and other phenomena in I–V curves of HTS junctions. The table is organized according to experimental observable, theoretical explanation, and pertinent references. It does not specifically apply to thallium-based HTS materials but to all

Table 2 Theoretical Explanations for Tunneling Spectra of HTS Materials

Experimental observable	Theoretical explanations	Ref.
Conductance peaks symmetric about zero bias	Superconducting energy gap	29
	SN bound state	31
	Coulomb blockade	32
Multiple conductance peaks symmetric about zero bias at constant intervals ("staircase $I-V$ curve")	Coulomb blockade	32
	Strong-coupling density of states	33
Multiple conductance dips and peaks symmetric about zero bias	Switching of microscopic currents	34
	Strong-coupling density of states	35
Zero-bias anomalies	Supercurrent	36
	Andreev reflection at SN interface	4
	Magnetic scattering	37
Linearly increasing background conductance with voltage	Resonant tunneling	38
	Inelastic scattering processes	39
	RVB density of states	40
	Coulomb blockade	41
	Two-dimensional anisotropy	42
	Density of states	43
	Geometrical effects	26

HTS materials generally. The references are selected because they give arguments for the application of the theoretical explanation to the experimental observable. In all cases there exist several references that might be used for this purpose.

There are so many plausible explanations for HTS $I-V$ curves that it is difficult to be conclusive. The fact that the size of the tunneling gaps generally scales with T_c for different HTS compounds and the fact that the tunneling gap disappears above T_c for a given junction provide strong evidence that the tunneling gaps observed for HTS materials are quite often due to the formation of a superconducting energy gap for quasi-particle pair formation. Even though comparatively few tunneling measurements have been made on thallium-based HTS materials compared to other HTS materials (especially measurements as a function of temperature), the $I-V$ characteristics appear to have the same general gap spectrum and therefore are probably caused by the same physical phenomena.

REFERENCES

1. P. K. Hansma, *Electron Tunneling Spectroscopy*, Plenum Press, New York, 1982.
2. E. Burstein and S. Lundquist, eds., *Tunneling Phenomena in Solids*, Plenum Press, New York, 1969.
3. E. L. Wolf, *Principles of Electron Tunneling Spectroscopy*, Oxford Science, New York, 1989.
4. W. L. McMillan and J. M. Rowell, in *Superconductivity*, Vol. 1, R. D. Parks, Marcel Dekker, New York, 1969, p. 598.
5. J. C. Phillips, *Physics of High T_c Superconductors*, Academic Press, New York, 1989, pp. 238–250.

6. J. Moreland, L. F. Goodrich, J. W. Ekin, T. E. Capobianco, and A. F. Clark, in *Advances in Cryogenic Materials*, Vol. 34, ed. A. F. Clark and R. P. Reed, Plenum Press, New York, 1988, pp. 625–632.

7. K. E. Gray, in *Novel Mechanisms of Superconductivity*, eds. S. A. Wolf and V. Z. Kresin, Plenum Press, New York, 1988, p. 611.

8. A. Barone, *Physica C* 153–155, 1712 (1988).

9. J. R. Kirtley, *Int. J. Mod. Phys. B* 4, 201 (1990).

10. T. Hasegawa, H. Ikuta, and K. Kitizawa, in *Physical Properties of High Temperature Superconductors III*, ed. D. M. Ginsberg, World Scientific, Singapore, 1992.

11. H. F. Hess, R. B. Robinson, R. C. Dynes, J. M. Valles, Jr., and V. Wasczak, *J. Vac. Sci. Technol. A* 8, 450 (1990).

12. J. Moreland and J. W. Ekin, *J. Appl. Phys.* 58, 3888 (1985).

13. I. Bozovic, J. N. Eckstein, M. Klausmeier-Brown, and G. Virshup, presented at the *MRS Spring Meeting*, San Francisco, Apr. 27–May 1, 1992; M. E. Klausmeier-Brown, G. F. Virshup, I. Bozovic, and J. N. Eckstein, *Appl. Phys. Lett.* 60, 2806 (1992).

14. R. P. Robertazzi, R. H. Koch, R. B. Laibowitz, and W. J. Gallagher, *Appl. Phys. Lett.* 61, 710 (1992).

15. J. G. Adler and J. Strauss, *Rev. Sci. Instrum.* 46, 158 (1975).

16. J. Lambe and R. C. Jacklevic, *Phys. Rev.* 165, 821 (1968).

17. A. F. Hebard and P. W. Shumate, *Rev. Sci. Instrum.* 45, 529 (1974).

18. A. P. Fein, J. R. Kirtley, and R. M. Feenstra, *Rev. Sci. Instrum.* 58, 1806 (1987).

19. H. G. Le Duc, W. J. Kaiser, and J. A. Stern, *Appl. Phys. Lett.* 50, 1921 (1987).

20. J. Moreland, D. S. Ginley, E. L. Venturini, and B. Morosin, *Appl. Phys. Lett.* 55, 1463 (1989).

21. D. S. Ginley, B. Morosin, R. J. Baughman E. L. Venturini, J. E. Schirber, and J. F. Kwak, *J. Crystal Growth* 91, 456 (1988).

22. I. Takeuchi, J. S. Tsai, Y. Shimakawa, T. Manako, and Y. Kubo, *Physica C* 158, 83 (1989).

23. J. S. Tsai, I. Takeuchi, Y. Shimakawa, T. Manako, and Y. Kubo, *Physica C* 162–164, 113 (1989).

24. S. Vieira, J. G. Rodrigo, M. A. Ramos, K. V. Rao, and Y. Makino, *Phys. Rev. B* 40, 11403 (1989).

25. S. Vieira, J. G. Rodrigo, M. A. Ramos, N. Agrait, K. V. Rao, Y. Makino, and J. L. Costa, *J. Appl. Phys.* 67, 5026 (1990).

26. Q. Huang, J. F. Zasadzinski, K. E. Gray, E. D. Bukowski, and D. M. Ginsberg, *Physica C* 161, 141 (1989).

27. W. Lamping, H. Jian, and W. Guowen, *Phys. Rev. B* 40, 10954 (1989).

28. S. J. Adler and C. J. Adkins, *Supercond. Sci. Technol.* 4, S190 (1991).

29. A. P. Fein, J. R. Kirtley, and M. W. Shafer, *Phys. Rev. B* 37, 9738 (1988).

30. R. C. Dynes, V. Narayanmurti, and J. P. Garno, *Phys. Rev. Lett.* 41, 1509 (1987).

31. P. J. M. van Bentum, H. F. C. Hoevers, H. van Kempen, L. E. C. van de Leemput, M. J. M. F. de Nivelle, L. W. M. Schreurs, R. J. M. Smokers, and P. A. A. Teunissen, *Physica C* 153–155, 1718 (1988).

32. P. J. M. van Bentum, R. T. M. Smokers, and H. van Kempen, *Phys. Rev. Lett.* 60, 2543 (1988).

33. J. Moreland, A. F. Clark, L. F. Goodrich, H. C. Ku, and R. N. Shelton, *Phys. Rev. B* 35, 8711 (1987).

34. T. Walsh, J. Moreland, R. H. Ono, and T. S. Kalkur, *Phys. Rev. B* 43, 11492 (1991).

35. L. N. Bulaevskii, O. V. Dolgov, I. P. Kazakov, S. N. Maksimovskii, M. O. Ptitsyn, V. A. Stepanov, and S. I. Vedeneev, *Supercond. Sci. Technol.* 1, 205 (1988).

36. T. Walsh, J. Moreland, R. H. Ono, and T. S. Kalkur, *Phys. Rev. Lett.* 66, 516 (1991).

37. L. Y. L. Shen and J. M. Rowell, *Phys. Rev.* 165, 566 (1968).

38. R. Berthe and J. Halbritter, *Phys. Rev. B* 43, 6880 (1991).
39. J. R. Kirtley and D. J. Scalapino, *Phys. Rev. Lett.* 65, 798 (1990).
40. P. W. Anderson and Z. Zou, *Phys. Rev. Lett.* 60, 132 (1988).
41. P. J. M. van Bemtum, H. van Kempen, L. E. C. van de Leemput, and P. A. A. Teunissen, *Phys. Rev. Lett.* 60, 369 (1988).
42. D. Mandrus, L. Forro, D. Koller, and L. Mihaly, *Nature* 351, (1991).
43. F. Sharifi, A. Pargellis, and R. C. Dynes, *Phys. Rev. Lett.* 67, 509 (1991).

28
Band Structures of Tl–Ba–Ca–Cu–O Materials

Roland E. Allen
Texas A&M University
College Station, Texas

Brent A. Richert
U.S. Air Force Academy
Colorado Springs, Colorado

I. INTRODUCTION

In this chapter we discuss the band structures of the Tl-based high-T_c superconductors, emphasizing a simple chemical model [1–3] that is in surprisingly good agreement with the more sophisticated calculations [4–14]. We mention that experiment indicates the basic premise of this model to be correct: High-temperature superconductors do exhibit quasi-particle energy bands and Fermi-liquid-like behavior [15–19]. Other experiments, specifically on the Tl-based materials [20–33], are consistent with the general features of the band structures presented below.

The thallium-based high-temperature superconductors share many features with their lower-T_c cousins (e.g., $La_{2-x}Sr_xCuO_4$, $YBa_2Cu_3O_7$, and the Bi-based materials). However, there are some differences: For example, a nearly unoccupied $Tl(6s)$–$O(2p)$ band is predicted to extend down to an energy below or very near the Fermi energy E_F [2,4,6]. This may affect transport in the Tl–O layers.

II. SIMPLE CHEMICAL MODEL

Even when many-body effects are included, the electron (quasi-particle) wavefunction satisfies a Schrödinger-like equation:

$$H\psi = E\psi \qquad H = H_1 + \Sigma \tag{1}$$

The one-electron Hamiltonian H_1 consists of three terms: (1) the kinetic energy operator $-(\hbar^2/2m)\nabla^2$, (2) the interaction of an electron with the positive ion cores, and (3) the interaction of this electron with the charge density $\rho(\mathbf{r}) = -en(\mathbf{r})$ of all the other electrons. The self-energy Σ is, in general, frequency-dependent, nonlocal, and non-Hermitian. There is also no known prescription for treating it exactly, so it must be approximated.

In quantitative calculations for real materials, it is customary to use the local density approximation (LDA): $\Sigma(\mathbf{r},\mathbf{r}';\omega)$ is replaced by an effective potential $v(n(\mathbf{r}))$. The most famous deficiency of the LDA is that it yields semiconductor band gaps that are too small. Similarly, La_2CuO_4 and $YBa_2Cu_3O_6$ are predicted to be metallic. The LDA results are, however, thought to be more meaningful for the actual high-T_c superconductors $La_{1.85}Sr_{0.15}CuO_4$, $YBa_2Cu_3O_7$, $Tl_2Ca_2Ba_2Cu_3O_{10}$, and so on.

Detailed discussions of LDA calculations can be found elsewhere [34,35]. Here we will emphasize a simpler description—a semiempirical tight-binding model in which the basis functions are regarded as being atomic orbitals (rather than, e.g., thousands of plane waves). There are several reasons why this description is more appropriate in the present context: (1) It makes direct contact with the basic notions of chemical bonding which are so important in the high-T_c materials; (2) it reproduces the results of the more complicated calculations surprisingly well; (3) the model given below is "universal" in that it applies to all the known cuprate and bismuthate high-temperature superconductors; and (4) this model can be easily used by anyone who wants to study the electronic properties of these materials.

In this simple approach, Eq. (1) is replaced by the matrix eigenvalue equation [36]

$$\overset{\leftrightarrow}{H}(\mathbf{k}) \cdot \mathbf{u}(\mathbf{k}n) = E(\mathbf{k}n)\mathbf{u}(\mathbf{k}n) \tag{2}$$

The component $u_{i\alpha}$ corresponds to the α orbital on the ith atom in the unit cell, and $E(\mathbf{k}n)$ is the energy for the nth band at wave vector $\mathbf{k}$. The Hamiltonian matrix $\overset{\leftrightarrow}{H}(\mathbf{k})$ contains two types of elements: the diagonal elements $\varepsilon_{i\alpha}$, which are regarded as atomic energies, and the off-diagonal elements $H_{i\alpha,i'\alpha}(\mathbf{ll}')$, which can be written in terms of overlap integrals $V_{ll'm}$ [37]. We will follow Harrison's parametrization [36]

$$V_{ll'm} = \eta_{ll'm} \frac{\hbar^2}{m_e d^2} \qquad (l,l' = s \text{ or } p) \tag{3a}$$

$$V_{ldm} = \eta_{ldm} \frac{\hbar^2 r_d^{3/2}}{m_e d^{7/2}} \qquad (l = s \text{ or } p) \tag{3b}$$

where d is the distance between the atomic nuclei, m_e is the electron mass, and $\eta_{ll'm}$ and r_d are parameters. Here m corresponds to a particular orientation of the orbitals; for example, $V_{pp\sigma}$ corresponds to the overlap of two p orbitals that are pointing directly toward each other.

In principle, the parameters $\varepsilon_{i\alpha}$ and $\eta_{ll'm}$ can be made to include many-body effects by fitting to experimental data—for example, the measured band gaps of semiconductors and the observed energy bands of high-T_c superconductors. In the latter case, however, there is still only a modest body of reliable data, so we have simply fitted to first-principles LDA calculations for two materials: $La_{2-x}Sr_xCuO_4$ [38,39] and $BaPb_{1-y}Bi_yO_3$ [40]. The parameters are then carried over without change to all the other cuprate and bismuthate high-T_c superconductors.

Our model also includes the following features:

1. The atomic energies $\varepsilon_{i\alpha}$ for Tl were extrapolated from those of Bi and Pb using Harrison's table [36]. We believe that this table treats chemical trends correctly.

2. The atomic energies for O, La, Sr, Ba, Y, and so on, were simply taken from Harrison's table, since the results appeared to be satisfactory without modification.

3. The valence s and p orbitals were included for O, and s and d orbitals for all the metal atoms. We originally included the excited d states for La, Sr, Ba, Y, and so on, merely to obtain the correct charge transfer, but these states were later seen experimentally, in inverse photoemission [41–44], at about the positions that we had predicted.

4. We included p orbitals for Bi, Pb, Tl, and Hg, since the p states for these atoms can lie near the Fermi energy. Also, whenever two of these large atoms were within 3.9 Å of each other, we included second-neighbor interactions, using the same interatomic parameters η as for nearest-neighbor interactions. (Some would argue that second-neighbor interactions are also important for the smaller atom O, but we did not find them to be necessary in the present context.)

5. We ignore the small structural distortions in $La_{2-x}Sr_xCuO_4$, $YBa_2Cu_3O_7$, and $Ba_{1-x}K_xBiO_3$, which have only a tiny effect on the band structure. We similarly ignore the long-range structural modulations in the Tl and Bi cuprates.

The motivation behind this model should be clear: It is meant to be as simple as possible without losing any of the essential physics. More sophisticated tight-binding models have been constructed, but they are also inevitably more complicated [45].

The parameters determined through the procedure above are given in Tables 1 and 2. It is now straightforward to calculate the electronic energies $E(\mathbf{k}n)$ and the associated state vectors $\mathbf{u}(\mathbf{k}n)$ using Eq. (2).

III. RESULTS

Our calculated electronic energy bands for $Tl_2Ba_2CuO_6$, along the symmetry lines of the Brillouin zone for the body-centered tetragonal (bct) crystal structure [46]

Table 1 Atomic Parameters for High-Temperature Superconductors[a]

	ε_s (eV)	ε_p (eV)	ε_d (eV)	r_d (Å)
Tl	−14.8	−8.3	−23.0	1.0
Pb[b]	−18.0	−9.4	−29.0	1.0
Bi[b]	−21.2	−10.5	−35.0	1.0
K	−4.2	—	−3.2	1.2
Ca	−5.4	—	−3.2	1.2
Sr	−5.0	—	−6.8	1.6
Y	−5.5	—	−6.8	1.6
Ba	−4.5	—	−6.6	1.6
La	−4.9	—	−6.6	1.6
Cu[c]	−12.0	—	−14.0	0.95
O	−29.0	−14.0	—	—

[a]The parameters for Tl are extrapolated from those of Pb and Bi. All unfitted parameters are taken from a standard table [36]. The atomic energies are referenced to the nominal vacuum level.
[b]Parameters fitted to $BaPb_{1-y}Bi_yO_3$.
[c]Parameters fitted to $La_{2-x}Sr_xCuO_4$.

Table 2   Interatomic
Parameters

$\eta_{ss\sigma}{}^{a}$	-1.1
$\eta_{sp\sigma}{}^{b}$	1.4
$\eta_{pp\sigma}{}^{b}$	1.5
$\eta_{pp\pi}{}^{b}$	-0.6
$\eta_{sd\sigma}{}^{a}$	-1.6
$\eta_{pd\sigma}{}^{a}$	-2.5
$\eta_{pd\pi}{}^{a}$	1.4

[a]Parameters fitted to
$La_{2-x}Sr_{x}CuO_{4}$.
[b]Parameters fitted to
$BaPb_{1-y}Bi_{y}O_{3}$.

are shown in Fig. 1, with the zero of energy shifted to E_F. (The atomic positions for all materials were taken from experiment [47–51], but those for $TlCa_3Ba_2Cu_4O_{11}$ were extrapolated from those of the analogous three-layer copper-oxide superconductor.) The bands are virtually dispersionless along ΓZ, perpendicular to the CuO_2 plane, and are dominated as usual by the $pd\sigma$ antibonding state of $Cu(d)$–$O(p)$ near the Fermi energy. This hole conduction band strongly resembles the previous results for the planar CuO_2 regions of $La_{1.85}Sr_{0.15}CuO_4$ [1,38,39,52–54], $YBa_2Cu_3O_7$ [1,55–58], and $Bi_2CaSr_2Cu_2O_8$ [2,59–63]. The Tl–Tl in-plane interactions cause the Tl-p bands to disperse by about 3 eV from Γ to D along the [100] direction in Fig. 1. These bands do not dip below the Fermi energy as do the Bi-p bands in $Bi_2CaSr_2Cu_2O_8$, but the $Tl(s)$–$O(p)$ hybrid states do cross below E_F to about -0.5 eV at Γ, forming occupied electron pockets. The valences for $Tl_2Ba_2CuO_6$ are shown in Table 3. (In determining the valences and local densities of states, we average over 24 sample wave vectors $\mathbf{k}$ within the irreducible part of the Brillouin zone for the bct and simple tetragonal structures.) Our notation for the atomic sites places oxygen site O(1) within the CuO_2 plane, site O(2) in the BaO region separating the Cu and Tl layers, and site O(3) in the TlO layer. Notice that the valence for Cu is close to those of the planar Cu sites in $La_{1.85}Sr_{0.15}CuO_4$ and $YBa_2Cu_3O_7$ [1]. In the present model, the total density of states at the Fermi energy, $\rho(E_F)$, is 2.0 states/eV cell.

The energy bands of $Tl_2CaBa_2Cu_2O_8$ are shown in Fig. 2. Two antibonding $Cu(d)$–$O(p)$ states protrude above the Fermi energy, corresponding to the two adjacent CuO_2 planes in this material. The $Tl(s)$–$O(p)$ antibonding bands disperse below the Fermi energy to -0.3 eV at the symmetry points Γ and Z, as in earlier calculations. The local densities of states for Tl, Ca, Ba, and Cu are shown in Fig. 3, with those for oxygen presented in Fig. 4. The Tl-d bands lie far below the Fermi energy, while the Tl-p bands, centered about 3 eV above E_F, interact somewhat weakly with the neighboring O(3) oxygens. Both the Ca and the Ba-d states are above E_F and are quite ionic. The Cu and O(1) sites show strong $pd\sigma$ interactions, with the antibonding bands protruding above E_F at X in Fig. 2. The total density of states at the Fermi energy is 2.7 states/eV cell. The valences for $Tl_2CaBa_2Cu_2O_8$ are shown in Table 3. The valence of Cu in this material is similar to that in $Tl_2Ba_2CuO_6$.

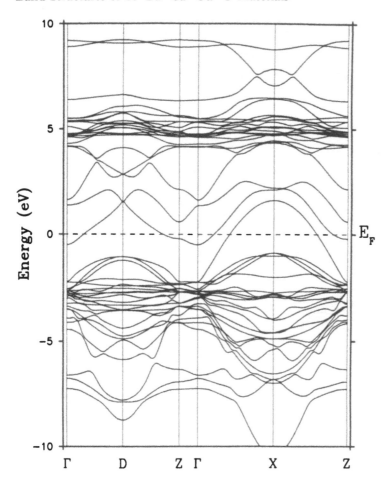

Figure 1 Electronic energy bands for $Tl_2Ba_2CuO_6$. The zero of energy has been shifted to the Fermi energy E_F. The $Cu(d)$–$O(p)$ antibonding band, with hole conduction states, crosses E_F along the ΓX symmetry line. A $Tl(s)$–$O(p)$ band dips below E_F, making both the CuO_2 planes and TlO layers metallic.

Table 3 Valences Δn

	Tl	Ca[a]	Ba	Cu[b]	Cu(2)	O(1)	O(2)	O(3)	O(4)
$Tl_2Ba_2CuO_6$	1.05	—	1.36	0.89	—	−1.10	−1.04	−0.72	—
$Tl_2CaBa_2Cu_2O_8$	1.05	1.45	1.37	0.86	—	−1.13	−1.03	−0.71	—
$Tl_2Ca_2Ba_2Cu_3O_{10}$	1.06	1.45	1.38	0.54	1.02	−1.27	−1.09	−1.08	−0.65
$TlCa_3Ba_2Cu_4O_{11}$	1.07	1.45	1.36	0.90	1.07	−1.12	−1.06	−1.10	−1.17

[a]Averaged over central Ca(1) and outer Ca(2) sites for $TlCa_3Ba_2Cu_4O_{11}$.
[b]Central Cu(1) site for $Tl_2Ca_2Ba_2Cu_3O_{10}$ and $TlCa_3Ba_2Cu_4O_{11}$.

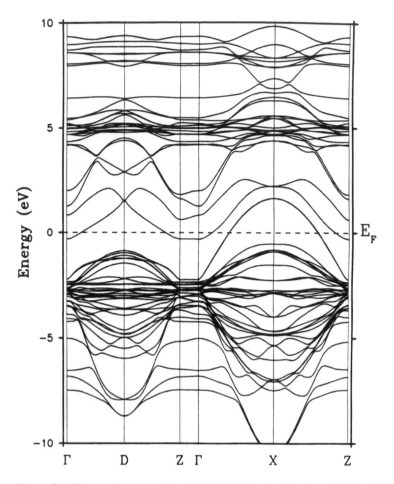

Figure 2 Electronic energy bands for $Tl_2CaBa_2Cu_2O_8$. Two $Cu(d)-O(p)$ antibonding bands, corresponding to the two adjacent CuO_2 planes, protrude above E_F.

The electronic energy bands of $Tl_2Ca_2Ba_2Cu_3O_{10}$ are shown in Fig. 5. This 125-K superconductor has three antibonding $Cu(d)-O(p)$ states that cross E_F, corresponding to the triple layer of CuO_2 planes stacked with Ca ions. Small electron pockets are again formed by $Tl(s)-O(p)$ antibonding states which dip below E_F to -0.6 eV at the symmetry points Γ and Z. The atomic valences in Table 3 show that $Cu(2)$, in the outer CuO_2 planes adjacent to the BaO layers, has a higher valence than in the preceding Tl materials. However, the central copper site $Cu(1)$ shows a decreased number of holes, with a valence of only 0.54. The notation for the oxygen atoms places $O(1)$ in the central CuO_2 plane, $O(2)$ in the outer CuO_2 planes, $O(3)$ in the BaO layer, and $O(4)$ in the TlO layer. The total density of states at E_F for $Tl_2Ca_2Ba_2Cu_3O_{10}$ is 5.3 states/eV cell.

The bands of the single-Tl-layer superconductor $TlCa_3Ba_2Cu_4O_{11}$ are shown in Fig. 6. This simple tetragonal crystal structure contains four CuO_2 layers separated by Ca ions. These layers contribute the four $Cu(d)-O(p)$ antibonding bands which form the hole conduction bands, while the TlO layer provides a single $Tl(s)-$

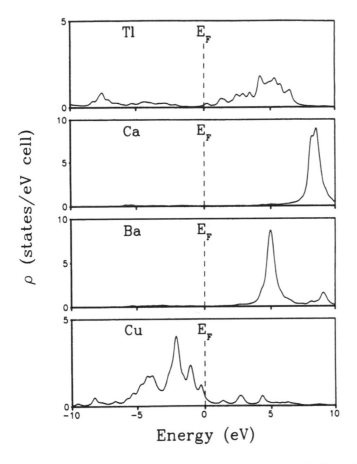

Figure 3 Local density of states for the metal atoms in $Tl_2CaBa_2Cu_2O_8$. Tl interacts strongly with the neighboring O(3) atoms, but Ca and Ba remain relatively ionic.

O(p) band which just crosses below E_F at Γ and Z. In this material, there are two inner copper oxide planes, labeled Cu(1) and O(1), which are relatively isolated from the remaining structure by the two outer Cu(2)–O(2) layers. The results of the valence calculation for this material in Table 3 show that the inner CuO_2 layers have fewer holes than the outer layers. The outer Cu(2) sites have the largest valence of all the Tl-containing superconductors considered here.

IV. EXPERIMENT

The Tl-based materials have received much less experimental attention than several other high-temperature superconductors, despite the fact that they have the highest T_cs and other desirable properties for application. In this section we mention several representative studies of the electronic structure.

Figure 7 shows the spectra of both occupied and unoccupied states for bulk ceramic $Tl_2Ba_{2-x}Ca_{2+x}Cu_3O_{10+y}$ as measured, respectively, with photoemission and inverse photoemission by Meyer et al. [20]. There is a low density of states within about 1 eV of the Fermi energy E_F, and there are structures at ~3, ~7,

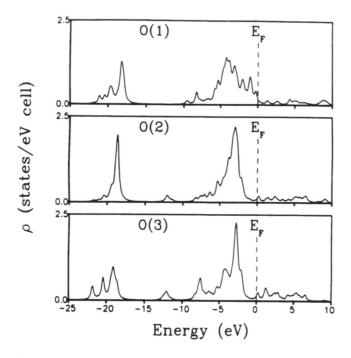

Figure 4 Local density of states for the oxygen atoms in $Tl_2CaBa_2Cu_2O_8$. The Cu–O(1) in-plane $pd\sigma$ interactions give metallic character at the Fermi energy, as in the other high-T_c copper oxide superconductors.

9.6, and 14 eV above E_F that are attributed to Tl-$6p$, Ba-$5d$, Ca-$3d$, and Ba-$4f$ empty states. As for other materials, the dominant peak below E_F is attributed to Cu–O states.

Figure 8 shows the spectrum of occupied states for $Tl_2Ba_2CaCu_2O_{8+y}$ measured in photoemission by Vasquez and Olson [21]. The material was an epitaxial c-axis-oriented thin film on a single-crystal $LaAlO_3$ (100) substrate, deposited by laser ablation. The surface was improved by etching with Br_2 in ethanol. A clear Fermi edge can be seen, particularly after etching.

Figure 9 represents x-ray absorption measurements by Krol et al. [22]. In this spectrum, for Cu-$2p_{3/2}$ excitations in the $O_{9.95}$ material, there are two clear peaks. The larger one is attributed to an excitation from a $3d^9$ initial state (formally Cu^{2+}) to $\underline{2p}3d^{10}$ final states ($2p$ hole plus excited d state). This peak is found to persist in the oxygen-depleted material $Tl_2Ba_2Ca_2Cu_3O_{9.7}$. The smaller peak is attributed to an excitation from a $3d^9\underline{L}$ initial state (formally Cu^{3+}, with L denoting an oxygen ligand hole) to a $\underline{2p}3d^{10}\underline{L}$ final state. This second peak is drastically reduced in the oxygen-depleted $\overline{O}_{9.7}$ material. Similar results for O 1s excitations also indicate that the O is removed almost entirely from the CuO_2 planes, as well as confirming that the three distinct oxygen sites have chemically different environments.

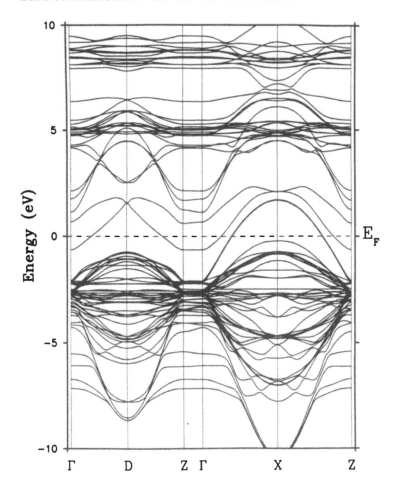

Figure 5 Electronic energy bands for $Tl_2Ca_2Ba_2Cu_3O_{10}$. The Cu–Ca–Cu–Ca–Cu layer structure contributes three Cu(d)–O(p) hole conduction bands in this material.

V. CONCLUSIONS

The calculated band structures of Section III are representative of those for all the thallium-based superconductors. They exhibit several features.

1. The antibonding Cu($3d$)–O($2p$) band that protrudes above the Fermi surface, and which is responsible for the conductivity and superconductivity in all the high-T_c cuprate superconductors. This band is obtained in a simple three-orbital picture [34,38], which retains only the Cu $d_{x^2-y^2}$ orbital and two O-p_σ orbitals in the CuO$_2$ planes. The model 3×3 Hamiltonian matrix gives a nonbonding state with $E(\mathbf{k}) = \varepsilon$, and bonding and antibonding states with

$$E(\mathbf{k}) = \varepsilon \pm 2V\left(1 - \frac{\cos k_x a + \cos k_y a}{2}\right)^{1/2} \tag{4}$$

where the Cu($3d$) and O($2p$) atomic energies are assumed to be nearly equal: $\varepsilon_d = \varepsilon_p = \varepsilon$. V is the interatomic matrix element and a is twice the Cu–O bond length.

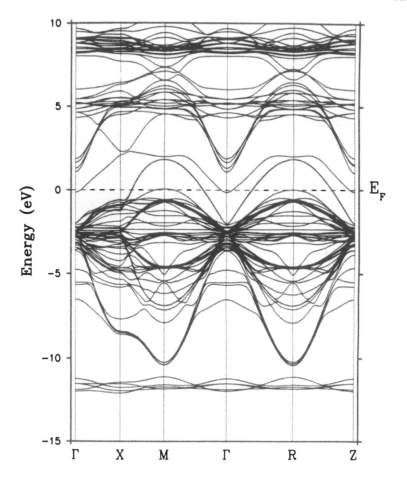

Figure 6 Electronic energy bands for $TlCa_3Ba_2Cu_4O_{11}$.

2. An antibonding $T(6s)-O(2p)$ band that comes down just below the Fermi energy, E_F, and which makes the Tl–O layers slightly metallic.

3. Various bands below E_F that can be seen in photoemission.

4. Various bands above E_F that can be seen in inverse photoemission.

More experimental studies of the electronic structure of the Tl-based high-temperature superconductors are needed. For example, the unoccupied Tl-s band that extends slightly below E_F in the calculations [2,4,6] has not yet been definitively seen in inverse photoemission [20]. If this band were to extend significantly below E_F, it could make the Tl layers metallic and influence transport. Also, even though a Fermi edge has been seen in photoemission [20,21], there is not yet the evidence for Fermi liquid behavior that has been observed in $YBa_2Cu_3O_7$ and $B_2Sr_2CaCu_3O_8$. The importance of the Tl-based materials, and the availability of good polycrystalline and single-crystal samples, makes further experimental observations highly desirable.

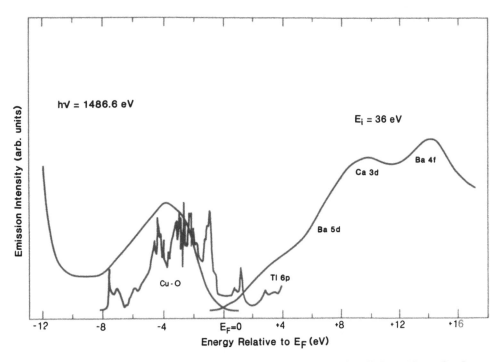

Figure 7 Photoemission and inverse photoemission results for $Tl_2Ba_{2-x}Ca_{2+x}Cu_3O_{10+y}$, compared with the density of states of Ref. 7. (From Ref. 20.)

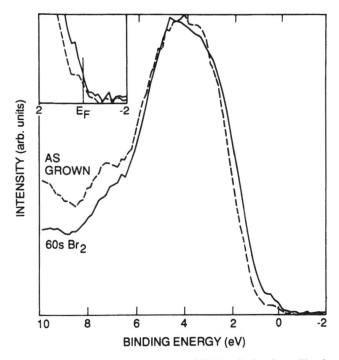

Figure 8 Valence band spectra of $Tl_2Ba_2CaCu_2O_{8+y}$ film from x-ray photoemission, before and after etching to improve the surface. The inset shows the region near the Fermi energy. (From Ref. 21.)

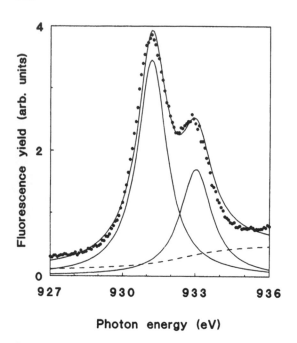

Figure 9   Cu-$2p_{3/2}$ x-ray absorption spectrum of $Tl_2Ba_2Ca_2Cu_3O_{9.95}$ with $T_c = 114$ K. The spectrum is deconvoluted with two Lorentzians (solid curves) and an inelastic background (dashed curve). (From Ref. 22.)

REFERENCES

1.   B. A. Richert and R. E. Allen, *Phys. Rev. B* 37, 7869 (1988).
2.   B. A. Richert and R. E. Allen, *J. Phys. Condensed Matter* 1, 9443 (1989).
3.   The model of Refs. 1 and 2 was applied by S. V. Meshkov, S. N. Molotkov, S. S. Nazin, I. S. Smirnova, and V. V. Tatarskii, *Physica C* 161, 497 (1989) and 166, 476 (1990); by Z.-H. Huang and R. E. Allen, *Physica C* 173, 173 (1991); and by B. A. Richert and R. E. Allen, *Phys. Rev. B* 37, 7496 (1988) and *J. Phys. Condensed Matter* 1, 9451 (1989). It also motivated the work of B. A. Richert, E. Schachinger, and R. E. Allen, *Physica C* 162–164, 1443 (1989).
4.   D. R. Hamann and L. F. Mattheiss, *Phys. Rev. B* 38, 5138 (1988).
5.   L. F. Mattheiss, *Phys. Rev. B* 42, 10108 (1990).
6.   J. Yu, S. Massidda, and A. J. Freeman, *Physica C* 152, 273 (1988).
7.   P. Marksteiner, J. Yu, S. Massidda, A. J. Freeman, J. Redinger, and P. Weinberger, *Phys. Rev. B* 39, 2894 (1989).
8.   R. V. Kasowski, W. Y. Hsu, and F. Herman, *Phys. Rev. B* 38, 6470 (1988). The prediction of $T_c = 135$ to 165 K for the 2234 material was, of course, not confirmed.
9.   F. Herman and R. V. Kasowski, *Physica C* 162–164, 1355 (1989).
10. S. T. Chui, R. V. Kasowski, and W. Y. Hsu, *Phys. Rev. B* 41, 273 (1990).
11. G.-L. Zhao, W. Y. Ching, and K. W. Wong, *J. Opt. Soc. Am. B* 6, 505 (1989).
12. P. Strange and J. M. F. Gunn, *J. Phys. Condensed Matter* 1, 6843 (1989).
13. H. Lee and R. R. Sharma, *Phys. Rev. B* 43, 7756 (1991).
14. M. Breza and R. Boča, *Solid State Commun.* 74, 801 (1990).
15. A. J. Arko et al., *Phys. Rev. B* 40, 2268 (1989).
16. C. G. Olson et al., *Phys. Rev. B* 42, 381 (1990).

17. J. C. Campuzano et al., *Phys. Rev. Lett.* 64, 2308 (1990).
18. B. O. Wells et al., *Phys. Rev. Lett.* 65, 3056 (1990).
19. Additional experimental evidence for Fermi liquid behavior is discussed by W. E. Pickett, H. Krakauer, R. E. Cohen, and D. J. Singh, *Science* 255, 46 (1992).
20. H. M. Meyer III, T. J. Wagener, J. H. Weaver, and D. S. Ginley, *Phys. Rev. B* 39, 7343 (1989).
21. R. P. Vasquez and W. L. Olson, *Physica C* 177, 223 (1991).
22. A. Krol, C. S. Lin, Y. L. Soo, Z. H. Ming, Y. H. Kao, J. H. Wang, M. Qi, and G. C. Smith, *Phys. Rev. B* 45, 10051 (1992).
23. G. Margaritondo, *J. Am. Ceram. Soc.* 73, 3161 (1990).
24. T. Suzuki, M. Nagoshi, Y. Fukuda, Y. Syono, M. Kikuchi, N. Kobayashi, and M. Tachiki, *Phys. Rev. B* 40, 5184 (1989); T. Suzuki, M. Nagoshi, Y. Fukuda, S. Nakajima, M. Kikuchi, Y. Syono, and M. Tachiki, *Physica C* 162–164, 1387 (1989).
25. W.-M. Hurng, S. R. Horng, S. F. Wu, C. Y. Shei, W. H. Lee, and P. T. Wu, *Solid State Commun.* 77, 269 (1991).
26. H. Romberg, N. Nücker, M. Alexander, J. Fink, D. Hahn, T. Zetterer, H. H. Otto, and K. F. Renk, *Phys. Rev. B* 41, 2609 (1990).
27. J. Yuan, L. M. Brown, W. Y. Liang, R. S. Liu, and P. P. Edwards, *Phys. Rev. B* 43, 8030 (1991).
28. D. Shindo, K. Hiraga, S. Nakajima, M. Kikuchi, Y. Syono, K. Hojou, T. Soga, S. Furuno, and H. Otsu, *Physica C* 165, 321 (1990).
29. V. Jayaram, G. U. Kulkarni, and C. N. R. Rao, *Solid State Commun.* 72, 101 (1989).
30. J. Tanaka, K. Kamiya, M. Shimizu, C. Tanaka, H. Ozeki, and S. Miyamoto, *Physica C* 162–164, 1117 (1989).
31. A. A. Maksimov, I. I. Tartakovskii, and V. B. Timofeev, *Physica C* 162–164, 1243 (1989).
32. V. B. Timofeev, A. A. Maksimov, O. V. Misochko, and I. I. Tartakovskii, *Physica C* 162–164, 1409 (1989).
33. C. F. Hague, V. Barnole, J.-M. Mariot, and M. Ohno, *Physica C* 162–164, 1321 (1989).
34. For a comprehensive early review of the electronic structure of copper-oxide superconductors, see K. C. Hass, *Solid State Phys.* 42, 213 (1989).
35. See, e.g., W. E. Pickett, *Comp. Phys. Rep.* 9, 115 (1989), and the papers cited in Refs. 4 and 6.
36. W. A. Harrison, *Electronic Structure and the Properties of Solids*, W. H. Freeman, San Francisco, 1980.
37. J. C. Slater and G. F. Koster, *Phys. Rev.* 94, 1498 (1954).
38. L. F. Mattheiss, *Phys. Rev. Lett.* 58, 1028 (1987).
39. J. Yu, A. J. Freeman, and J.-H. Xu, *Phys. Rev. Lett.* 58, 1035 (1987).
40. L. F. Mattheiss and D. R. Hamann, *Phys. Rev. B* 28, 4227 (1983).
41. T. J. Wagener, Y. Gao, J. H. Weaver, A. J. Arko, B. Flandermeyer, and D. W. Capone II, *Phys. Rev. B* 36, 3899 (1987).
42. Y. Gao, T. J. Wagener, J. H. Weaver, A. J. Arko, B. Flandermeyer, and D. W. Capone II, *Phys. Rev. B* 36, 3971 (1987).
43. J. A. Yarmoff, D. R. Clarke, W. Drube, U. O. Karlsson, A. Taleb-Ibrahimi, and F. J. Himpsel, *Phys. Rev. B* 36, 3967 (1987).
44. T. J. Wagener, H. M. Meyer III, D. M. Hill, Y. Hu, M. B. Jost, J. H. Weaver, D. G. Hinks, B. Dabrowski, and D. R. Richards, *Phys. Rev. B* 40, 4532 (1989).
45. Other tight-binding models are cited in Ref. 20 of the present Ref. 2.
46. A. W. Luehrmann, *Adv. Phys.* 17, 1 (1968).
47. $Tl_2Ba_2CuO_6$: C. C. Torardi, M. A. Subramanian, J. C. Calabrese, J. Gopalakrishnan, E. M. McCarron, K. J. Morrissey, T. R. Askew, R. B. Flippen, U. Chowdhry, and A. W. Sleight, *Phys. Rev. B* 38, 225 (1988).

48. $Tl_2CaBa_2Cu_2O_8$: M. A. Subramanian, J. C. Calabrese, C. C. Torardi, J. Gopalakrish-nan, T. R. Askew, R. B. Flippen, K. J. Morrissey, U. Chowdhry, and A. W. Sleight, *Nature* 332, 420 (1988).

49. $Tl_2Ca_2Ba_2Cu_3O_{10}$: C. C. Torardi, M. A. Subramanian, J. C. Calabrese, J. Gopala-krishnan, K. J. Morrissey, T. R. Askew, R. B. Flippen, U. Chowdhry, and A. W. Sleight, *Science* 240, 631 (1988).

50. $TlCa_2Ba_2Cu_3O_9$: S. S. P. Parkin, V. Y. Lee, A. I. Nazzal, R. Savoy, R. Beyers, and S. J. La Placa, *Phys. Rev. Lett.* 61, 750 (1988).

51. For $TlCa_3Ba_2Cu_4O_{11}$ the structure was extrapolated from that of Ref. 50, with the correct tetragonal unit cell parameters. All metal atom placements were correct to within 2% of the final experimental values, and oxygen atom placements were correct to within 1.5%, except for O in the BaO layer, which was displaced in the c direction by 8% (0.19 Å) from the experimental position. See R. M. Hazen, in *Physical Properties of High Temperature Superconductors II*, ed. by Donald M. Ginsberg, World Scientific, Singapore, 1990, p. 185.

52. W. E. Pickett, H. Krakauer, D. A. Papaconstantopoulos, and L. L. Boyer, *Phys. Rev. B* 35, 7252 (1987).

53. T. Oguchi, *Jpn. J. Appl. Phys.* 26, L417 (1987).

54. K. Takegahara, H. Harima, and A. Yanase, *Jpn. J. Appl. Phys.* 26, L352 (1987).

55. L. F. Mattheiss and D. R. Hamann, *Solid State Commun.* 63, 395 (1987).

56. S. Massidda, J. Yu, A. J. Freeman, and D. D. Koelling, *Phys. Lett. A* 122, 198 (1987).

57. T. Fujiwara and Y. Hatsugai, *Jpn. J. Appl. Phys.* 26, L716 (1987).

58. F. Herman, R. V. Kasowski, and W. Y. Hsu, *Phys. Rev. B* 36, 6904 (1987).

59. M. S. Hybertsen and L. F. Mattheiss, *Phys. Rev. Lett.* 60, 1661 (1988); L. F. Mattheiss and D. R. Hamann, *Phys. Rev. B* 38, 5012 (1988).

60. S. Massidda, J. Yu, and A. J. Freeman, *Physica C* 152, 251 (1988).

61. H. Krakauer and W. E. Pickett, *Phys. Rev. Lett.* 60, 1665 (1988).

62. F. Herman, R. V. Kasowski, and W. Y. Hsu, *Phys. Rev. B* 38, 204 (1988).

63. P. A. Sterne and C. S. Wang, *J. Phys. C* 21, L949 (1988).

29

Thallium Cuprate Research by the U.S. Government

M. S. Davis
Strategic Analysis Inc.
Arlington, Virginia

S. A. Wolf
Naval Research Laboratory
Department of the Navy
Washington, D.C.

I. INTRODUCTION

The thallium-based superconducting copper oxides discovered in 1988 [1] hold a unique position among the cuprates in that they collectively have the highest reported transition temperatures. Several of the various compounds that fall in this class have T_cs that are well above 100 K. It is in this regard that there is a considerable amount of government-sponsored research into developing these materials into useful forms so that they can form the basis of a robust superconducting technology. In fact, it is the diversity of polytypes of these compounds that make these materials (or a particular polytype of these materials) suitable for a large number of applications. In the following sections we outline many of the government programs supporting research on developing the thallium cuprates into useful forms and devices. Even though the list of work is quite extensive, it is by no means complete. We surveyed the various agencies and requested a detailed list of their sponsored research. Unfortunately, all of the agencies did not supply a complete list, so we have included only the research efforts that were reported to us.

II. DEPARTMENT OF COMMERCE

According to their report, the Department of Commerce has only one sponsored program to develop thallium cuprate superconductors. This program, at the Du Pont Experimental Research Station, is to develop advanced thallium/lead thin-film technology and demonstrate their performance by fabricating and testing key electronic components and devices. This program has a level of effort between 2 and 3 person-years. Neither the NIST laboratories at Gaithersburg or at Boulder reported any thallium research.

III. DEPARTMENT OF DEFENSE

The Department of Defense (DoD) has an extensive program in superconductivity, including both conventional and cuprate superconductors. Since many of the agencies in the DoD act independently, we treat them separately here.

A. DARPA

Among the DoD agencies, DARPA has an extensive program to develop technology based on the high-transition-temperature cuprate superconductors. One of DARPA's biggest and most ambitious programs is the development of a multichip interconnect module for high-performance semiconductor digital ICs. This program is managed by the Air Force Wright Labs and the primary contractor is E-Systems. This program has many subcontractors. One of the important subcontractors is Superconducting Technologies, Inc. (STI) of Santa Barbara, California. They have a 7-person-year effort to demonstrate the feasibility of using Tl-based high-transition-temperature superconductors (HTSCs) for thin-film interconnects for large-scale multichip modules (MCMs). Recently, Du Pont has also become a subcontractor to E-Systems to provide large-area thallium cuprate films and lithography expertise in this high-visibility, high-risk program. In a separate, unrelated program, DARPA, through the Office of Naval Research (ONR), is funding STI to develop a millimeter-wave downconvertor. This subassembly will downconvert a 35-GHz input signal to a 5-GHz IF output. This will be an integrated subassembly consisting of several thallium HTSC devices. This is a 4-person-year effort.

DARPA also has several contracts managed by ONR with Advanced Technology Materials, Inc. (ATM). One of these programs is to develop in situ thallium cuprate films by metalloorganic chemical vapor deposition (MOCVD). This program also aims at developing patterning and etching technology capable of producing passive microwave components. Also being supported are two phase I small business innovative research (SBIR) projects, one to develop lumped-element X-band microwave filters and the other to develop a microstrip low-loss switch. All three contracts involve 4-person-years of effort.

B. Strategic Defense Initiative Organization

The Strategic Defense Initiative Organization (SDIO) has a reasonably large program to look at superconducting materials and develop devices from the most promising materials. It has a very large SBIR program and through its agents has several contracts with STI to develop devices based on thallium cuprate thin films. The phase I contracts involve both microwave and Josephson junction–based devices. STI is developing a spatial filter at 60 GHz and is also developing an innovative microwave-based infrared detector that uses a microwave readout of the detected signal. In the weak-link junction area, STI is developing devices based on bicrystal junctions. These programs support 1.5 person-years.

C. Department of the Navy

As far as the individual services go, the Department of the Navy (DoN) has the largest program in superconductivity. Much of the fundamental science and materials research effort is carried out in the contract research program of the Office of Naval Research (ONR) and at the Naval Research Laboratory (NRL). Of course, these programs encompass all aspects of superconductivity, including work in conventional low-transition-temperature metallic superconductors. NRL in their basic materials program has discovered one polytype of the thallium-based materials, $Tl_1Sr_2Ca_1Cu_2O_x$, with a transition temperature of over 100 K. Currently, a program is starting to understand the pinning properties of the one thallium oxide layer cuprates and subsequently try to develop a superconducting conductor technology based on these materials. This has the potential of a several-person-year effort. Currently, it is at the 2-person-year level.

In addition to managing many of the DARPA contracts, ONR is supporting work at the Naval Air Warfare Center, China Lake Laboratory to develop the single-crystal growth of the thallium cuprates. This is a small effort and involves less than 0.5 of a person-year. The Naval Air Development Center is administering a phase I SBIR contract with STI to demonstrate a three-channel multiplexer at 10 GHz.

D. U.S. Air Force

The U.S. Air Force (USAF), in addition to managing the large DARPA MCM program and several SDIO SBIR contracts, has several of its own SBIR programs to develop thallium cuprate technology. STI is developing a 1-bit 90° phase shifter under one phase I SBIR and is investigating methods for Q-switching a resonator, including photoconductive switches under a second phase I SBIR. STI has a phase II SBIR to characterize the noise figure in thallium-based weak-link Josephson junctions and compare with Schottky diodes. These programs involve 4-person-years of effort at STI.

E. U.S. Army

The U.S. Army (USA), through its Electronic Technology and Devices Laboratory at Fort Monmouth, has two phase II SBIR contracts. One contract with STI is to design and test a 35-GHz microwave filter. This program involves 2-person-years of effort. The second SBIR is with ATM and involves the development of scale-up of a MOCVD reactor for producing large-area thallium cuprate films. This contract involves 1 person-year of effort.

IV. DEPARTMENT OF ENERGY

The Department of Energy (DOE) has the largest government agency program on superconductivity. Its program is divided between the more conventional superconductors and the cuprates. There are superconductivity programs spread through many of DOE organizations, including Basic Energy Sciences and their Renewable Energy Office. I was able to get information on the thallium efforts only from the various DOE National Laboratories, so the single-investigator efforts are not included in this report.

A. Argonne National Laboratory

The Argonne National Laboratory (ANL) has an extensive program to study and develop technology based on the thallium cuprates. Using the 1223 compound,

ANL is pursuing wire development using the powder-in-tube method. They are looking into various tube materials as well as optimizing the processing to get the best critical current characteristics. Research, aimed more closely at the fundamental transport and magnetic properties, has been performed on various thin and thick films of the thallium cuprates. This work has included the 2212, 1212 and 1223 polytypes of the thallium cuprates. Recent work has been a comparison of the critical current performance of the one-layer Tl–O and the two-layer Tl–O polytypes. These materials are characterized by their transport properties, the nature of the vortex dynamics, and correlation of these characteristics with various microstructural features. The ultimate goal of the ANL program is to develop a wire for high-current and high-magnetic-field applications. Overall, this program consumes 3.5 person-years of effort.

B.   National Renewable Energy Laboratory

The National Renewable Energy Laboratory (NREL) pioneered the electrodeposition of the copper oxide superconductors with early work in 1989 on Y-123 and the Bi compounds. The NREL effort was redirected in 1991 to emphasize the electrodeposition of the Tl-based compounds. This work was part of the DOE/CE program to develop processing of HTS materials in a wire or tape configuration suitable for power-related applications. NREL has recently fabricated tapes with Tl–Ba–Ca–Cu–O coatings on Ag foil with J_c of 3.2×10^4 A/cm^2 at 77 K in zero field. The field dependence for J_c showed no evidence for weak-link behavior with -10^4 A/cm^2 observed at 77 K in a 5.0-kG field. Higher values for J_c were measured on rigid ceramic substrates (MgO and SrTiO$_3$) with -6×10^4 A/cm^2 observed at 77 K in zero field. The goal for the NREL program is the demonstration of long lengths of "practical" flexible tape with J_c properties greater than 10^5 A/cm^2 at 77 K in zero field.

C.   Oak Ridge National Laboratory

The Oak Ridge National Laboratory (ORNL) is working jointly with industry to develop Tl-2212-based wire. One effort with General Electric (GE) addresses thick-film deposition on polycrystalline metallic and ceramic substrates using spin deposition (GE) and aerosol pyrolysis (ORNL) followed by controlled thallination. A second joint effort with the National Renewable Energy Laboratory (NREL) and GE is focused on the thallination and characterization of NREL's electrodeposited wire. In another program with IBM, the fabrication of epitaxial films using controlled thallination is being pursued. These films are also being heavy-ion irradiated to improve their flux pinning performance. The total effort in this area at ORNL is about 3 person-years and the primary focus is on the development of a wire technology.

D.   Los Alamos National Laboratory

The Los Alamos National Laboratory (LANL) has several programs to develop thallium-based cuprates. In the area of microwave properties, they are investigating the surface resistance of Tl-based films on metallic substrates. The goal is to develop three-dimensional HTS cavities for stabilizing oscillators for an X-band (8 to 12

GHz) radar receiver. For power applications, LANL is investigating the transport critical current of Tl-based HTS films and bulk materials. For a metal-clad tape, the goal is to demonstrate 10,000 A/cm^2 in fields of 2 T at 35 K. The overall effort is at the 3- to 4-person-year level.

E. Sandia National Laboratory

The Sandia National Laboratory (SNL) has an extensive program to develop thallium-based materials. This program has three main program areas. There is a fundamental science program whose main goal is to understand the relation between processing, structure, and pinning in the thallium cuprate materials. This program involves thin films, bulk materials, and single crystals of a large number of the polytypes in this system. Another program involves a wire development program. This work includes thin film, thick film, tape, and powder-in-tube work on materials developed at Sandia and elsewhere.

Finally, the third program involves developing a new microelectronics based on the superconducting flux-flow transistor and step-edge Josephson junctions. Included in this program is the development of advanced photolithography techniques and a novel confocal resonator for characterizing the surface resistance of large-area films. This program takes advantage of the weaker pinning in the Tl-based materials and produces better flux-flow electronics. Many different devices have been demonstrated using this technology. The step-edge junction technology has been utilized in long-junction-based amplifiers and single flux quantum logic. Overall this program is at the level of 6-person-years. The main goals are to develop applications based on optimized materials, particularly in the thallium-based cuprates.

V. NATIONAL AERONAUTICS AND SPACE ADMINISTRATION

A. Jet Propulsion Laboratory

The Jet Propulsion Laboratory (JPL) has a small contract with STI to develop in situ thallium-based films by laser ablation.

VI. NATIONAL SCIENCE FOUNDATION

The only specific work that we could identify with the National Science Foundation (NSF) was in the Stanford and University of Illinois Material Research Laboratories (MRLs). At Stanford there is some work on the molecular beam epitaxy (MBE) of thallium-based materials, and at the University of Illinois there is some work on the characterization of Tl-based cuprates. The estimate was about 1 person-year of effort spread through these two MRLs.

VII. SUMMARY

It is clear that there is a large effort on thallium-based cuprates sponsored by the major government agencies. The list is certainly not complete and represents only the results of a survey of the agencies. In any case, the breadth and depth of the efforts mentioned here are substantial.

ACKNOWLEDGMENT

We want to thank all the people at the various agencies who took the time to respond to our questions. Without them, this chapter would not have been possible.

REFERENCE

1. Z. Z. Sheng and A. M. Hermann, *Nature*, 332, 55 and 138 (1988).

30
Thallium Safety

Gary E. Myers
Industrial Hygienist
Naperville, Illinois

Thallium and most (perhaps all) of its compounds are potentially hazardous. Hundreds of deaths have resulted from accidental, as well as homicidal and suicidal, ingestion of Tl compounds. Soluble Tl compounds possess the somewhat unusual ability to penetrate the unbroken skin, adding another dimension to their hazard potential. In this chapter we examine hazardous properties of thallium and its compounds, and detail practices for storing, handling, and using them safely.

I. TOXICITY AND INDUSTRIAL EXPERIENCE

Illness and injury can result from exposure to toxic materials by inhalation, ingestion, or skin contact. The toxicity of an ingested material can be quantified in several ways, the most common of which are TD_{LO}, the lowest dose that will produce a toxic effect, such as nausea; LD_{LO}, the lowest dose that will cause death; and LD_{50}, the dose that will result in the death of half of a population of experimental animals. The doses normally are measured in milligrams of toxicant per kilogram of body weight (mg/kg). Materials with an LD_{50} of 50 mg/kg or less are considered to be highly toxic.

The LD_{50} values for Tl compounds, regardless of solubility or route of administration, are in the range 15 to 50 mg Tl/kg, for several species of laboratory animals. For humans, estimates [1, p. 1920] for oral LD_{LO} are in the neighborhood of 1 g of Tl in the form of soluble salts; this corresponds to about 15 mg Tl/kg. By comparison, reported [2] oral LD_{LO} arsenic trioxide doses for humans range from 1.5 to 29 mg/kg. Thus it appears that Tl compounds are about as lethal, orally, as As_2O_3.

Although lethal dose information is useful for comparing the relative toxicities of various materials, in the occupational setting the objective is the prevention of *all* toxic effects. The human oral TD_{LO} value for elemental Tl is about 5 mg/kg, and that value [2] for thallous sulfate is 2 mg/kg (about 100 to 140 mg for an average-size person). Again, there is quantitative similarity to arsenic trioxide.

Occupational exposures to hazardous materials are by inhalation more often than by ingestion, although the latter route is not unimportant. The American Conference of Governmental Industrial Hygienists (ACGIH) has developed a highly regarded set of exposure limits called threshold limit values (TLVs) [3]. The TLV for a chemical substance is the airborne concentration of that material in milligrams per cubic meter (mg/m^3) to which it is believed that the average worker can be exposed 8 hours per day, 5 days per week, for a working lifetime, with no ill effects. In other words, such workplace exposure will result in a cumulative inhalation dose below the TD_{LO} for that material. The TLV for soluble Tl compounds is 0.1 mg/ m^3 as Tl, as is the OSHA permissible exposure limit (PEL). At this writing, no TLV has been assigned to elemental Tl or insoluble Tl compounds, but it is believed [1, pp. 1929–1930] that $0.1 mg/m^3$ is appropriate for these as well. An active person breathes about 10 m^3 of air during a workday, so continuous workplace exposure to Tl and its compounds at the TLV concentration will result in the inhalation of about 1 mg/day Tl.

ACGIH specifically cautions that TLVs are not fine lines between safe and dangerous concentrations, nor are they to be used as a relative index of toxicity; however, using them to compare materials having similar toxicological properties is valid for our purposes. The TLV for arsenic and its soluble compounds is 0.2 mg/m^3; for lead and its inorganic compounds it is 0.15 mg/m^3 (the OSHA PEL for Pb is 0.05 mg/m^3); and for mercury it is 0.05 mg/m^3. Again, we see quantitative similarity to As, as well as to Hg and Pb. Discussions (private communication, 1988) with industrial hygienists employed by a Tl producer indicate that they consider the inhalation hazard of Tl to be roughly equivalent to that of Pb.

Soluble Tl compounds can be absorbed through the unbroken skin. Many of the common Tl salts, as well as thallous oxide (Tl_2O_3), are water soluble to some degree. Elemental Tl is insoluble in water, but it oxidizes readily in moist air [4], quickly becoming covered with a gray film of Tl_2O_3. It has also been reported [5] that vacuum-deposited thin films of Tl convert to the soluble carbonate (Tl_2CO_3) after 37 days' exposure to air. Skin contact with the corroded metal or with thallous oxide itself thus provides the potential for absorption. It is believed by some that insoluble Tl compounds can be converted to the soluble chloride in contact with perspiration, so skin contact with even insoluble compounds may result in Tl absorption.

There is little quantitative information about human Tl poisoning by skin absorption. A case study [6] describes a woman who regularly applied a cold cream containing 7.18% thallium acetate to her face, allowing it to stay on throughout the night; after about 6 months of use, she began to experience symptoms of Tl poisoning, but continued to use the cream for 3 years. She suffered severe visual impairment during the continued use, and there was residual damage to the central and peripheral nervous systems. In an animal study [1, p. 1921] thallium formomalonate, which was the most toxic Tl compound studied, was found to have a cutaneous LD_{50} of 57.7 mg/kg for an unspecified species.

Thallium is a systemic poison with multisystem toxicities. Mild forms of Tl poisoning cause joint pain, vague digestive complaints, numbing or "pins and needles" in the hands and feet, irritability, and sleep disorders. The signs and symptoms of severe Tl poisoning include nausea and vomiting, diarrhea that may alternate with constipation, skin and nail abnormalities, fever, leg pain, extreme tenderness in the soles of the feet, confusion and disorientation, and hair loss. Death may result from respiratory paralysis.

Thallium and its compounds have not been reported to be carcinogenic; indeed, some Tl compounds have shown antitumor activity [1, p. 1921]. There have been numerous reports of severe Tl poisoning of pregnant women (usually from attempted suicide or abortion), but in contrast to the well-known fetotoxic effects of lead, the fetuses seemed not to be seriously affected [7].

Thallium is excreted primarily in the urine, but also in the feces. The biological half-life (i.e., the time required to excrete half of the remaining body burden of Tl) is not well established. Published values range from approximately 3 to 8 days [8], to 23 days [9], with the most recent being "about 14 days" [10]. Prussian blue (ferriferrocyanide) appears to facilitate removal of Tl from tissues and subsequent elimination, and successful treatment with Prussian blue has been reported.

Thallium poisoning in the work environment is rare, but it is not unheard of. The following examples are typical of those reported in the literature. Workers in a Swedish plant that produced Tl-based rodenticide showed symptoms of Tl poisoning [1, p. 1928]. In 1958, six cases of industrial Tl intoxication, the first in the United States, were reported [11] among men using organic Tl compounds in the separation of industrial diamonds; since no Tl could be found in the workroom air, it is presumed that the intoxication resulted from skin contact. Workers in a plant in England that produces an alloy containing 7% Tl were exposed to Tl dust and fume,* and had skin contact. Fourteen of 60 were found to be excreting excessive amounts of Tl, although none showed any clinical evidence of Tl intoxication. Excretion dropped to normal levels when an education program and good industrial hygiene procedures, which included limiting the workroom airborne Tl concentration to 0.1 mg/m^3, were instituted [1, p. 1929]. There appear to be no known work-related fatalities from Tl poisoning.

Unfortunately, at this time little is known about the hazards of Tl-containing superconductor materials. In the absence of information to the contrary, it is prudent to assume that they are as hazardous as other thallium compounds.

II. CONTROLS AND PRACTICES FOR WORKING SAFELY WITH THALLIUM

Handling, weighing, mixing, and processing of Tl-containing powders can produce airborne Tl-containing dust, as can drilling, cutting, abrading, or grinding Tl-containing solids. Mists of solutions or suspensions of Tl materials, such as might be produced by a high-speed wetted cutoff wheel, can result in an inhalation exposure, as well as skin contact. Heated material can vaporize or produce Tl fume. The first line of defense against airborne Tl is the adoption of work practices that prevent or minimize dust generation. Airborne dust at the TLV concentration for Tl compounds is *not* visible.

---

*"Fume" is an aerosol of metal oxide particles.

Exposure to airborne Tl is best controlled by the use of effective local exhaust ventilation, or by containment. Local exhaust ventilation is the removal of airborne contaminants at the point of generation, by a properly designed exhaust ventilation pickup that is positioned to capture all airborne material before it reaches the operator's breathing zone or enters the general room air. Dilution ventilation, such as that provided by increasing the air exchange rate in the room, is not appropriate for the primary control of highly toxic materials because the air contaminants must pass through the breathing zone before removal. Containment is the isolation of a contaminant-generating process within a ventilated enclosure. Containment can be as sophisticated as a glove box, or as simple as a laboratory fume hood.

A fume hood should have an average face velocity (the velocity at which air enters the working opening) of 100 to 125 ft/min when used for highly toxic materials [12]. Higher face velocities are not recommended because high-velocity air may cause disruption of processes or spillage of materials, and the resulting turbulence may in fact degrade the protection. High-velocity particles that travel toward the hood face, such as those that might be produced by grinding, can overcome the inward flow of air and reach the operator's breathing zone. Operations that might produce such energetic particles should be evaluated to determine whether high-velocity local exhaust ventilation, or a glove box or similar type of containment, might be more appropriate. All apparatus and work within the hood should be kept at least 6 inches back from the face of the hood. If cooling fans are needed within the hood, they should be positioned and directed so as not to blow toward the face or to create turbulence in the vicinity of the face. External influences, such as fans, room-air supply grills, or even foot traffic past the hood, can produce turbulence that will compromise a hood's effectiveness. Careful attention should be given to sources of air movement near the hood. A properly operating laboratory fume hood will provide adequate control of airborne Tl for common laboratory-scale operations if the cautions above are observed. Canopy-type hood (i.e., hoods that do not provide complete enclosure on all sides but the working face) generally are ineffective for the control of airborne toxic materials.

Ventilation systems for the control of hazardous materials should be designed, installed, and tested by qualified and experienced personnel. Criteria, engineering data, and design examples have been published for most common operations [12]. Existing ventilation systems should be evaluated by an industrial hygienist prior to use with highly hazardous materials, or whenever an operation or process changes significantly. All engineering controls, including exhaust ventilation systems, should be tested periodically to ensure continued protection.

Respiratory protection is not acceptable as the sole or primary means of personnel protection unless engineering control of the hazard (e.g., local exhaust ventilation) is infeasible. Respirators may be worn during the interim, while engineering controls are being instituted and tested, or as backup protection. Respirators for use with Tl and its compounds in particulate form should be equipped with high-efficiency "dust, mist, and fume" filters. If volatile Tl compounds are in use, the respirator manufacturer should be consulted for chemical cartridge recommendations. If respirators are used, a respiratory protection program that meets OSH Act requirements in 29 CFR 1910.134 must be in place. If effective engineering controls are used, the need for respirators will be greatly reduced, or even eliminated.

Housekeeping is always important, but it is vital when highly toxic materials are in use. Thallium poisoning of workers that apparently was due in part to poor housekeeping has been reported [13]. Thallium-containing waste should be cleaned up promptly and placed in an airtight container, such as a taped plastic bag. Dry sweeping or brushing can be a source of copious amounts of airborne contamination and should be forbidden. Wet wiping or mopping is acceptable if performed carefully, but all cleaning materials should be sealed in an airtight container while still wet: Soiled cleaning wipes or mops, when dry, can be a source of airborne contamination if disturbed. Thallium waste should never be cleaned up with an ordinary vacuum cleaner. The filter on a shop or household vacuum is very inefficient for small particles, and using such a vacuum on toxic materials can produce a serious airborne hazard. Only HEPA (high-efficiency particulate air)-filtered vacuum cleaners are appropriate for use with hazardous solids. They are readily available from major industrial vacuum cleaner suppliers, as well as from many scientific supply houses.

Hand contamination can result in the ingestion of toxic materials. Even if toxic materials are not handled directly with bare hands, contaminated tools and controls, such as valves and switches, can cross-contaminate the hands. If workers then do not conscientiously wash their hands before taking a break, toxic materials can be transferred directly to foodstuffs or to a handkerchief, chewing gum, or smoking materials. In addition, toxic materials can be mistaken for foodstuffs or additives such as coffee creamer, and be ingested. Eating, drinking, and smoking should be prohibited in areas where Tl and its compounds are stored or used.

Skin absorption can be the result of direct skin contact or the wearing of contaminated clothing. All toxic materials should be handled only while wearing impervious gloves. For dry materials, any waterproof glove should be adequate; however, the glove manufacturer's permeation data should be consulted when gloves are used as protection from wet materials or solutions. Disposable gloves are preferred, and they should be disposed of as Tl waste after each use. If gloves are reused, they can be a source of contamination unless properly stored and carefully donned. For example, the placing of contaminated gloves in a pocket can contaminate the inside of the pocket, and subsequently the worker's hands. Tools and controls that are handled or manipulated while wearing contaminated gloves should be marked as potentially contaminated, and all personnel should understand and observe procedures that prevent these items from being handled or used with bare hands. Thorough hand washing should be performed immediately prior to leaving areas where toxic materials are used.

It is prudent to avoid skin contact with all Tl compounds. The conscientious observance of the foregoing hand contamination prevention practices will prevent most skin contact. If skin contamination does occur, the affected area should be washed thoroughly immediately. Medical advice about proper first-aid procedures should be obtained. Although it is highly recommended that reasonable precautions be taken to prevent skin contact, at the same time it also appears that occasional incidental contact (e.g., from a torn glove) is unlikely to result in a serious health hazard.

If clothing contamination is possible, disposable protective garments should be worn. It is important to realize that the act of removing contaminated clothing can generate airborne contamination, so protective clothing should be either vacuumed thoroughly with a HEPA-filtered vacuum cleaner prior to removal, or the worker should wear a respirator while disrobing.

III. EMPLOYEE EDUCATION AND HAZARD COMMUNICATION

The Hazard Communication section of the OSH Act, 29 CFR 1910.1200, requires some very specific actions on the part of employers. These requirements are quite detailed, but some general guidelines follow.

All hazardous material containers must be properly labeled. Portable containers into which hazardous chemicals are transferred from labeled containers, and which are intended only for the immediate use of the employee who performs the transfer, need not be labeled, although it is good practice to do so.

Employees must be informed about the hazards of the materials with which they work (more important than satisfying the law, a good employee education program is invaluable for preventing accidents and overexposure to hazardous materials). A written hazard communication program must be prepared, implemented, and maintained.

A material safety data sheet (MSDS) on any reagent or other material used must be provided to any employee who requests it. Manufacturers are required to provide MSDSs for their materials, and these must be kept in a location that is known to all employees and is readily accessible to them. Moreover, an employee who requests an MSDS must be provided with it within the work shift. In a large organization, this can present considerable logistic problems, particularly if shift work is involved, and attempts to satisfy this requirement range from the maintenance of separate MSDS files in each work area to the on-line provision of MSDSs by a networked computer system. The MSDS file(s) must be updated as new or revised MSDSs are provided by manufacturers.

It is not necessary to generate an MSDS for materials that are made and used in-house, but if those materials are sent outside the parent organization for any purpose, they must be accompanied by an MSDS. An MSDS must also be provided with any material that is offered for sale, but not with articles. The distinction between a "material" and an "article" is subtle but important. In general, an article is a finished product that does not release, or otherwise result in exposure to, a hazardous chemical under normal conditions of use (e.g., a magnet wound with superconducting wire), while a material is a chemical or other substance that would be used in research or the production of articles (e.g., any quantity of powdered superconductor material). The generation of an MSDS is a complex procedure that is best performed by knowledgeable and experienced personnel. Consultants are available to assist in MSDS generation if the necessary expertise is not available in-house.

A complete discussion of the requirements of 29 CFR 1910.1200 is beyond the scope of this chapter, but management and supervisory personnel should take the necessary actions to implement it. The OSH Act Occupational Exposure to Hazardous Chemicals in Laboratories standard, 29 CFR 1910.1450, also contains labeling and hazard communication requirements, as well as requirements for a chemical hygiene plan and other measures. If in-house expertise is not available, development of a hazard communication plan, education and training for employees, and advice on methods for compliance with all parts of the OSH Act can be obtained from consultants.

The Emergency Planning and Community Right to Know Act of 1986, Section 313 of Title III of the Superfund Amendments and Reauthorization Act, requires specific actions when a research or manufacturing organization uses hazardous materials. A discussion of these requirements also is beyond the scope of this chapter, and consultation is advised.

IV. WASTE HANDLING AND DISPOSAL

Employees who collect, process, package, or otherwise handle thallium wastes should be included in the hazard communication and education program, and practices and procedures should be developed to assure that they will not receive significant Tl exposure. The primary packaging of all Tl-containing wastes in airtight containers, as described above, is an example of such a practice. Janitorial personnel should not be allowed to clean a Tl-contaminated area unless they have been properly trained and equipped. If protective clothing is to be laundered, laundry personnel should be included in the training program. Clothing that is contaminated with hazardous materials should not be taken home to be laundered.

Various thallium compounds are defined as acute hazardous wastes or toxic wastes under the Resource Conservation and Recovery Act (RCRA), 40 CFR 261. RCRA is administered by the U.S. EPA, and there may be additional state regulations as well. All applicable regulations should be consulted for specific details about shipping and disposal requirements.

REFERENCES

1. H. E. Stokinger, *Patty's Industrial Hygiene and Toxicology*, 3rd rev. ed., Vol. 2A, ed. George D. and Florence E. Clayton, Wiley, New York, 1981.
2. *NIOSH Registry of Toxic Effects of Chemical Substances*, National Institute for Occupational Safety and Health, Cincinnati, Ohio, annual (paper edition); quarterly (microfiche). Availability: U.S. Government Printing Office. Also an on-line service of the National Library of Medicine's Toxicology Information Program.
3. *1991–92 Threshold Limit Values for Chemical Substances and Physical Agents and Biological Exposure Indices*, American Conference of Governmental Industrial Hygienists, Cincinnati, Ohio, 1991.
4. K. Wade and A. J. Banister, in *Comprehensive Inorganic Chemistry*, Vol. 1, ed. by A. F. Trotman-Dickenson, Pergamon Press, Oxford, 1973, p. 1123.
5. C. H. Champness, *Phosphorus Sulfur* 38, 399 (1987).
6. J. B. Rudolphy, *Arch. Ophthalmol.* 13, 1108 (1935).
7. W. Johnson, *Med. J. Aust.* 47, 540 (1960).
8. V. Zitko, *Sci. Total Environ.* 4, 185 (1975).
9. R. C. Baselt, *Biological Monitoring Methods for Industrial Chemicals*, Biomedical Publications, Davis, Calif., 1980, p. 246.
10. N. Blackman, E. London, B. Zeleke, and R. St. John, *Morbid. Mortal. Weekly Rep.* 36, 487 (1987).
11. E. M. Richeson, *Ind. Med. Surg.* 27, 607 (1958).

12. *Industrial Ventilation: A Manual of Recommended Practice*, Committee on Industrial Ventilation, American Conference of Governmental Industrial Hygienists, Lansing, Mich., published biennially, 20th ed., 1988, p. 17.
13. M. J. Coye, A. Burdick, J. Cone, and D. Lewis, *Report HETA-83-385-1469*, Hazard Evaluations and Technical Assistance Branch, NIOSH, U.S. Department of Health and Human Services, Cincinnati, Ohio, 1984.

31
Prospects and Potential for Thallium Cuprate Superconductors

J. V. Yakhmi
Bhabha Atomic Research Centre
Bombay, India

Allen M. Hermann
University of Colorado at Boulder
Boulder, Colorado

Among the high-T_c superconducting cuprates, the Tl-based high-temperature superconductors retain extraordinary superconducting features: the highest T_cs (127 to 128 K) and very low thin-film surface resistance, $R_S < 50$ μΩ at 77 K and 10 GHz. Improvements have also been made recently in flux-pinning characteristics of single-layer Tl-1223. Tl-based oxides have the further advantage that superconductivity can be obtained after annealing for comparatively short periods, making them easier to process and fabricate than Y-123. Toxicity problems related to Tl-based materials deterred some workers initially, but the confidence gained in handling them safely for nearly 5 years in dozens of laboratories, with some elementary precautions, coupled with the recent successes in fabrication of a number of passive/active devices based on Tl–Ba–Ca–Cu–O (TBCCO) superconductors, have encouraged a number of new laboratories to undertake programs aimed at growth and study of Tl-based superconducting films and their applications in technology. Tl oxide superconductors have held the interest of academic researchers through the discovery of many new superconducting or potentially superconducting families of compounds. These discoveries have been prompted by variation of the structure or chemistry of the existing Tl–Ba–Ca–Cu–O phases, such as Tl-1212. The process of sustained research and investigations worldwide in TBCCO family of superconductors has also enriched our state of knowledge in many important aspects related to an understanding of the evolution of superconductivity in high-temperature superconducting oxides, such as the influence of hole doping, internal redox chemistry, and so on.

In this chapter we take stock of the current achievements in this exciting field, point out certain important areas of investigation requiring more attention to fill the gaps in the level of understanding of Tl-based high-T_c oxides, and make an attempt to envisage the future directions in the research and development of these materials. The analysis of a vast amount of data published in the literature on the structure–property relationships on both the double- and single-layer thallium series of superconductors and the related substituted compounds has led not only to the generalization of many features that help in understanding their superconducting behavior, but has also brought forth certain as yet unexplained, anomalous features pertaining to some of these phases. Superconductivity in Tl cuprates is thought to arise, as in most high-T_c oxides, from the presence of holes in the CuO_2 sheets. According to the usual valence considerations of the phases with double Tl–O layers, $Tl_2Ba_2Ca_{n-1}Cu_nO_{2n+4}$ with ideal stoichiometry would have all the copper in the $2+$ state, making it insulating. The holes required for superconductivity can be created in the CuO_2 layers through the overlap of Tl-$6s$ bands with the CuO_2 layer x^2-y^2 bands. The possible mechanisms for creation of holes in the double-layer Tl series are suggested to be the substitution of Tl^{3+} with Ca^{2+}, vacancies at cation sites, excess oxygen in the structure, or charge transfer such as $Tl^{3-t}(Cu-O)^p$. Electron energy loss spectroscopy (EELS) data on $Tl_2Ba_2CaCu_2O_8$ clearly indicate a metallic character of the Tl_2O_2 layers and support the picture of self-doping due to a charge transfer from the CuO_2 layers to the Tl_2O_2 layers [1]. On the other hand, the single-layer Tl series require an additional hole per unit cell for charge neutrality under ideal stoichiometric conditions. The Tl-$6s$ bands for the single-layer thallium series lie above the Fermi level, making the overlap difficult [2]. However, it is thought that the x^2-y^2 bands perhaps rise enough to raise the Fermi level above the bottom of the Tl-$6s$ bands due to the shortening of the Cu–O bonds in the superconducting state. Tl vacancies appear to be the main source of holes in a Tl-1212 system, since the T_c for $Tl_{1-x}CaBa_2Cu_2O_7$ has been shown to increase with an increase in x [3]. On the other hand, Michel et al. [4] have suggested recently that the thallium nonstoichiometry does not affect the superconducting properties of a $Tl_2Ba_2CaCu_2O_8$ system noticeably, but it is instead the oxygen nonstoichiometry that plays a major role since it allows the hole carrier density to be optimized in order to reach a critical temperature as high as 120 K for a $Tl_{1.6}$ sample (by annealing it in an Ar/H_2 atmosphere). Recent phase stability studies [5], however, indicate that both Tl-1212 and Tl-2212 can accommodate variable stoichiometry while maintaining an almost constant T_c (i.e., an unchanged carrier density). One expects the concentration of Tl vacancies and the extent of Ca/Tl substitutions to be strongly dependent on the Tl_2O partial pressure. It is hoped that the role of cation vacancies, oxygen content, and antisite defects in influencing the carrier density, and therefore the T_c of Tl-based superconductors, will be understood more clearly as more and more systematic studies are performed.

The Tl-based cuprate family of superconductors includes structurally simple systems such as Tl-1201 and Tl-2201, which are overdoped with holes and become superconducting when the formal valence of copper is lowered to an optimal level. $Tl_2Ba_2CuO_{6+\delta}$, for which the T_c can increase above 90 K by decreasing the oxygen content, is a unique system for studying the hole-doping effects since the oxygen uptake/release process is reversible. With the application of pressure, too, the hole content in the CuO_2 sheets of Tl-2201 increases, causing a drop in T_c [6]. The coefficient dT_c/dp is quite large (ca. -2 K GPa^{-1}), signifying the ease with which the electrons can be moved from the CuO_2 conduction sheet to the BaO–TlO– TlO–BaO charge reservoir sheet. Tl-2201 system also happens to be the ideal test case to search for singularities at the superconductor-to-normal metal phase boundary, a study that is difficult to carry out with the vast majority of high-temperature superconductor systems since they belong to the insulator-to-superconductor phase crossover category. Tl-2201 goes from a 90-K superconductor state to a normal metallic state by a simple annealing process, and preliminary studies have shown that it exhibits only a gradual change in both its Hall coefficient data and spin susceptibility at this phase boundary, pointing to a Fermi-liquid-like behavior [7]. ^{205}Tl NMR presents itself as an additional tool for exploring such behavior in Tl-based high-T_c oxides. If one has to extend the antiferro-magnetic Fermi liquid theory to Tl-cuprate superconductors, the Cu^{2+} moments should reside on the Cu–O planes. Since the coupling between these Cu^{2+} moments and Tl ions, via oxygen, is weak, their contribution to the spin-lattice relaxation time, T_1, of thallium should be negligibly small. Geometrical considerations, too, should force the cancellation of the antiferromagnetic fluctuations of Cu^{2+} at Tl sites leading to a Korringa behavior. In the case of $Tl_2Ba_2Ca_2Cu_3O_{10+\delta}$, such measurements have been made on aligned samples in the normal state [8] and have shown that the relevant electron–nuclear interactions are isotropic at the Tl site, and their weak temperature dependence is associated with spin fluctuations in the Tl–O plane. We suggest that it is now possible to obtain more definitive information of this nature with the availability of better-characterized samples of different phases in the thallium family.

Extended x-ray absorption fine structure (EXAFS) data for $TlBa_2Ca_3Cu_4O_{11}$ have shown evidence for a structural change in the fluctuation region around T_c ($= 118$ K), involving the presence of two unique axial oxygen sites whose separation decreases around T_c, presumably due to a phonon-mediated charge transfer [9]. Similar evidence relating the elastic degrees of freedom and superconductivity has also been shown to occur for $YBa_2Cu_3O_7$. The true significance of such measurements can be appreciated only if detailed and coordinated series of optical and structural studies are made on well-characterized samples of different TBCCO phases.

Raman spectra on single crystals of $TlBa_2CaCu_2O_7$ and $Tl_2Ba_2CaCu_2O_8$ have shown significant softening and broadening of apical oxygen vibrational modes above 300 K, characteristic of anharmonicity of high-frequency O vibrations, a behavior observed for all superconducting cuprates with a pyramidal oxygen-coordination surrounding planar Cu–O sheets [10]. Structural instability aspects of high-T_c cuprates related to the apical oxygen anharmonicity are expected to receive more attention in view of the evidence for structural anomalies preceding T_c. Whereas the Cu–O chains in Y-123 show clearly a contribution to the infrared conductivity (for $\mathbf{E} \parallel \mathbf{b}$) due to mobile carriers, no such contribution is observable for TBCCO, suggesting the absence of mobile charge carriers in TlO layers [11]. Considering the large dc resistivity along the c direction, it appears as if the TlO rock salt block acts as an insulating barrier between the oxygen-depleted superconducting $BaOCa_nCu_{n+1}O_{2n+2}$ perovskite blocks.

Until recently, there have been very few systematic attempts devoted to phase stability studies to establish the preparative conditions of the Tl-2223 phase with the highest reproducible T_c of about 125 K [12], with the result that most researchers have been making do with Tl-2223 samples with T_cs in the range of 115 to 120 K, thus losing the singular advantage of the highest superconducting transition temperature characteristic of this phase. Of late there has been a spurt of activity in this direction, and detailed analyses of the conditions of synthesizing Tl-2223 samples with zero-resistance temperatures of 127 to 128 K have been reported by a number of groups [13–17]. In the case of samples annealed in vacuum-sealed quartz tubes, some unidentified diffusion process such as ordering was suggested to be responsible for T_c enhancement to 127 K [13,17]. Oxygen loading has been suggested to be the main factor responsible for maximization of T_c of Tl-2223 by Presland et al. [18], albeit for $T_c < 120$ K only, whereas Ahn et al. [16] report that the T_c for the Tl-2223 phase rises gradually to 127 K as the initial oxygen pressure, used for sealing the sample before reacting in sealed quartz tubes, is reduced to 0.03 atm. Perhaps there are other factors, in addition to the oxygen stoichiometry, that play an important role in this process and that still require to be established, such as thallium vacancies [12,14], an optimum cation ratio, as well as an appropriate microstructure. The influence of Tl deficiency and Ca substitution on vacant Tl sites in the enhancement of T_cs beyond 125 K also deserves to be analyzed in view of the sporadic claims of superhigh T_cs of Tl-deficient compositions in superconductors of filamentary nature [19,20].

An interesting and exciting offshoot of research in Tl-based cuprate superconductors has been the discovery and development of many new cuprates, some of them superconducting, obtained through substitutions or modification of the existing structures. In this respect the $TlBa_2CaCu_2O_7$ structure has truly been a fountainhead. Some of the new cuprate phases that have been discovered in recent times are (1) the 40-K 1222 superconductor $(Tl_{0.5}Pb_{0.5})Sr_2(Eu_{2-x}Ce_x)Cu_2O_9$ [21], (2) $(Pb,Cu)(Sr,R)_2(R',Ce)_2Cu_2O_z$ superconductor [22], (3) $(Pb,Cu)Sr_2(Ho,Ce)_3Cu_2O_{11+z}$ [23], (4) $(Tl,Cu)Sr_2(R,Ce)_3Cu_2O_{11}$ [24], and (5) $(Tl_{1-z}Pb_z)(Sr_{1-x}La_x)_2(Ln_{1-y}Ce_y)_2Cu_2O_9$ [25]. The last three may be superconducting systems. Of course, all these new phases (1 to 5) can be taken as belonging to a generalized homologous series $(Pb,Tl,Cu)(Sr,Ba,R)_2(R,Ce,Ca)_nCu_2O_{5+2n+z}$ (R = Ln; $1 \leq n \leq 3$), constructed by alternating the 1212 block and double MO_2-unit fluorite blocks. Incidentally, the Tl-1212 structure is perhaps the only high-T_c cuprate structure to accept full occupancy of either Ba or Sr and still retain superconductivity. The rapidly accumulating number of new cuprate phases based on Tl cuprates raises interesting questions regarding their electronic structure and crystal chemistry.

Tl–Ba–Ca–Cu–O superconducting films, especially Tl-2223 films, are of considerable interest for practical applications since they have T_cs above 120 K, good compositional tolerance, and lack of weak links at the grain boundaries. In some of the preceding chapters, different methods of obtaining TBCCO superconducting thin films have been covered. In all these methods, the high volatility of Tl hinders the in situ deposition process, which means that the fabrication of heterostructures remains a problem until alternative techniques are developed. The liquid gas solidification process [26] for growth of Tl-1223 films with $T_c = 103$ K ($R = 0$) on Mg⟨100⟩ holds promise because it does not require any postanneal or compensation for loss of Tl after initial deposition.

Over the last two years, there has been an increased use of $LaGaO_3$, $LaAlO_3$, and Al_2O_3 as substrates for growth of TBCCO thin films due to the small dielectric constants of these substrates, their low loss at high frequencies, and their close lattice match to the TBCCO films [27]. The structural stability and chemical compatibility of CeO_2 with both Al_2O_3 and Tl-2212 has been utilized to exploit CeO_2 as a buffer layer for preparation of superconducting epitaxial films of Tl-2212

(T_c = 97.5 K, transport J_c = 2.8 × 10⁵ A cm⁻² at 75 K) on low-dielectric-value sapphire, in view of the excellent dielectric properties of Al_2O_3 and its corresponding use in electronic applications [28]. The experience gained so far in different laboratories in the growth of TBCCO thin films is proving to be useful in the development of simple, new, low-cost techniques in the growth of Tl-cuprate films and/or in the achievement of improved superconducting characteristics. For example, superconducting thin films of Tl-2223 or Tl-2212 phases with controlled grain morphology have been deposited on different substrates by using simple thermal evaporation combined with a two-step annealing process [29]. A strong bolometric peak was observed at 100 K for patterned 2223 films prepared by this process. Holstein et al. [30] have succeeded in preparing very-low-surface-resistance, *c*-axis-oriented Tl-2212 films on (100) LaAlO3 (T_c = 107 K), exhibiting R_S values about 80 times lower than oxygen-free high-conductivity copper at 77 K and 10 GHz, demonstrating the potential of TBCCO superconductors to provide superior performance for passive microwave applications at 77 K. In fact, the exceptionally low R_S values of these films continue up to 95 K, pointing to their expanded usefulness. These low R_S films are also obtained through postannealing processes, which are amenable to scale-up to large-area films.

In addition to the high values of T_c, Tl-2223 thin films have shown high J_cs, up to 1 × 10⁷ A/cm² at 87 K for H = 10 Oe [31]. Unfortunately, the magnetic phase diagram of the vortex state of ceramic Tl-2223 shows evidence of a depinning line $B^*(T)$, where J_c becomes zero well below H_{c2}, leading to flux flow. Thermal depinning of centers with low activation barriers results in the values of J_c remaining high only below 35 K. Weak flux pinning in Tl-2223 or Tl-2212 is a problem, causing high dissipation and a sharp fall in J_c at relatively high magnetic fields, above about 40 K [32]. Recently, it has been demonstrated that Tl-1223 phase offers the advantage of improved coupling, between the perovskite blocks, in the *c* direction, which provides improved transport critical current density [($J_{CT}(H)$ behavior] (i.e., 105,000 A/cm² at 77 K and H = 0 and >10,000 A/cm² at 60 K and $H \perp c$ = 2 T). $Tl_xBa_2Ca_2Cu_3O_y$: 0.37 Ag (0.65 < x < 1.00) polycrystalline films on YSZ substrates prepared by reacting Ag-containing Ca–Ba–Cu–oxide precursor with thallium oxide vapor indeed show very encouraging results [33]. The addition of silver helps in the formation of textured interconnected films (T_c = 104 to 107 K) with the *c*-axis oriented perpendicular to the substrate surface. It is expected that these successes would provide an impetus to develop low-cost processes to exploit high-J_c Tl-1223 films for many applications.

Although the $J_{CT}(H)$ behavior of Y-123 material is decidedly superior to that of Bi- or Tl-based superconductors, the achievement of intergranular connectivity is a formidable problem to be solved before one can obtain useful lengths of superconducting wires of Y-123 material for use in magnets and other power devices. Despite recent reports of the improved $J_{CT}(H)$ behavior of Ag-sheathed (Bi, Pb)-2223 tapes, their utility may remain restricted to temperatures <20 K. The location of an irreversibility line for Tl-1223 at higher temperatures due to improved coupling for the Cu–O conducting planes, makes this material compare favorably with Y-123 for performance at 77 K. Additional pinning centers have been introduced successfully by taking the composition $(Tl_{0.5}Pb_{0.5})(Sr_{1.6}Ba_{0.4})Ca_2Cu_3O_y$, leading to improved $J_{CT}(H)$ behavior at 77 K for tapes fabricated from this material [34]. It is expected that Ag-sheathed single-phase Tl-1223 wires obtained through hot extrusion and subsequent cold rolling and annealing should also provide high J_{CT} values at 77 K, although currently they are low due to the mixed-phase nature of the superconductor in such efforts.

Among the "active" TBCCO devices fabricated to date are bolometric infrared (IR) detectors and superconducting quantum interference devices (SQUIDs). Another exciting development has been that of a flux-flow transistor (SFFT), a three-terminal device that exploits the weak pinning in TBCCO films. It can be switched between the flux-flow state and a "lossless" state by a combination of the field in the device and bias across the weak-link array [35]. It has been shown that the Tl-based SFFT device is faster than the comparable Y-123 device by virtue of lower flux pinning and hence higher flux velocities [36]. Martens et al. have used SFFT as a microwave amplifier in phase shifters for eventual use in space communications and in amplifiers for FIR detectors [37]. Tl step-edge junction technology is being developed at the Sandia National Laboratory so as to attain capability to produce high-quality junctions with relatively high yield and good uniformity within a junction and across a wafer. A reflective microwave switch made of TBCCO thin film has also been developed for signal control applications [38]. An area that deserves more attention is the study of nonbolometric response of TBCCO films. The potentially subpicosecond response times of HTS materials make them attractive for the development of ultrafast optical and electrical switches and devices for use in optical and microwave signal processing. Kwok et al. [39] have identified nonbolometric as well as bolometric responses in TBCCO films, both of which can be separated.

A wide range of passive devices have already been demonstrated by using TBCCO superconductors (e.g., interconnects, stripline, waveguides, etc.). A 1-ns microstrip delay line using Tl-2212 thin films was fabricated in 1990 [40]. It has also been demonstrated that Tl-2212 films can make efficient Josephson mixers at 77 K. When integrated with an antenna structure, these mixers have broad applications at (sub)millimeter wave and FIR frequencies [41]. It is natural to expect that as the processing technology for TBCCO thin films advances, it would become feasible to explore them for utilization in many more devices. Of course, the patterning and contacting technologies for these films, too, have to make progress simultaneously to move toward a strong device-based input.

Despite a large volume of research conducted on Tl-cuprate superconductors, information pertaining to some key areas is still very sketchy or just not available. For instance, a complete description of the phase diagram of the Tl–Ba–Ca–Cu–O system is not yet available, due primarily to the complexity arising from the large number of components involved, the oxygen pressure, Tl-vapor pressure, and even to the lack of information on some of the simple binary systems, such as BaO–$TlO_{1.5}$. Although the roles of parameters such as phase control, incongruent melting, and so on, are understood to a limited extent, the overall thermodynamic stability of the main TBCCO phases is yet to be established. Only thereafter can some important questions such as the nature of the phase transition from the double-layer to the corresponding single-layer thallium cuprates, the decomposition route of Tl-2223, and so on be answered. An appropriate knowledge of the phase stability of the major superconducting Tl phases would also help in the preparation of well-characterized single-phase ceramic samples, as well as growth of single crystals of high quality, both of which are the essential inputs for the evaluation of superconducting behavior of Tl-based superconductors toward a better understanding of the underlying physics, and for feasible efforts in device technology. Crystals of

TBCCO grown generally by the self-flux method contain a variety of structural defects whose nature and density are a function of the starting composition of the flux and the temperature profile. The synthesis reaction involves metastable phases that make it difficult to control the defect formation during single-crystal growth. Increased efforts in the growth of TBCCO single crystals alone can shed light on the role of exact stoichiometry of the melt employed. This aspect is well illustrated by the recent STM and STS studies that have shown that Tl-2223 single crystals grown from different stoichiometry melts (i.e., 4:1:3:6 and 4:1:3:10), although possessing the same average structure and composition can exhibit substantially different local-structure and electronic and superconducting properties [42]. It is hoped that in the very near future, as high-quality single crystals of the different TBCCO phases with suitable dimensions become available, it will become possible to conduct a number of fundamental investigations to gain an improved understanding of the evolution of superconductivity and its characteristics in Tl-based cuprates. Among these, we mention the following:

1. Reliable data on the anisotropy of the coherence length and the magnetic penetration depth are needed from measurement on high-quality Tl-2223 single crystals. If the anisotropy turns out to be high, as has been suggested, continuum (three-dimensional) models are not valid, and pancake vortices could be expected for $H \parallel C$.

2. The nonavailability of data on c-axis thermal conductivity coupled with the strong phonon-defect scattering in this direction prevent evaluation of the anisotropy of the coupling constant. Measurements of anisotropy of thermal conductivity of Tl-2212 single crystals [43] suggest that while phonon–electron scattering may be important, phonon–phonon scattering may also have to be taken into account for complete understanding of the out-of-plane anomaly. Availability of good-quality single crystals of Tl-2223 and other Tl-cuprate phases would help in covering this important area.

3. Since the superconductivity is believed to take place in the Cu–O planes, any phonons involved in its origin would involve displacements in the Cu–O plane. Characteristic deviations (viz., phonon anomalies such as softening) calculated from the dispersion curves of the "normal" compounds, as revealed by inelastic neutron-scattering measurements could possibly be interpreted as being related to prospective mechanisms associated with superconductivity.

4. Study of the effects of oxygen stoichiometry on the phonon spectra of TBCCO superconductors appears attractive in order to evaluate how the phonon spectra change with T_c because of removal/addition of oxygen and to determine the influence of the successive number of CuO_2 sheets on the phonon frequencies. The apical oxygen seems to be crucial to the transfer of charge between the superconducting CuO_2 sheets and the adjacent charge-reservoir Tl–O planes. It has been proposed that anharmonic vibrations of this atom can give rise to a strong enough coupling to explain the high value of T_c. It has also been suggested that the apical-oxygen p_z orbital stabilizes a local singlet state formed by CuO_2 plane $Cu_p d_{x^2-y^2}$ and $O_p p_\sigma$ orbitals favorable for superconductivity [44]. The dynamics of the apical oxygen can be studied by Raman spectroscopy since its A_{1g} c-axis stretching vibration is Raman active.

5. ^{89}Y NMR measurements on three different double-layered superconductors, all having $T_c = 45$ K, conducted recently [45] have shown that the NMR shifts for all three of them (Pb-1212, Co-doped Y-123 and oxygen-depleted Y-123) show

similar values and temperature dependence, indicating the similarity in spin susceptibilities in their Cu–O planes and probably a similarity in the hole densities, too, in their Cu–O planes. On the other hand, the normal state ^{63}Cu NMR shifts for three $Tl_2Ba_2CuO_{6+y}$ samples with $T_c = 0$, 40, and 72 K were found to be nearly temperature independent, with a remarkable resemblance to the case of the 90-K Y-123 [46]. With the availability of better-characterized samples, it would be possible to conduct NMR studies aimed at providing systematic information on the local magnetic behavior for samples belonging to underdoped or overdoped regimes.

6. Cu L_{III}-edge measurements on single crystals or oriented thin films of Tl cuprates are of interest in arriving at an accurate analysis of the relation between anisotropy, hole density, and superconductivity.

REFERENCES

1. H. Romberg, N. Nucker, M. Alexander, J. Fink, D. Hahn, T. Zetterer, H. H. Otto, and K. F. Renk, *Phys. Rev. B* 41, 2609 (1990).
2. D. Jung, M. H. Whangbo, N. Herron, and C. C. Torardi, *Physica C* 160, 381 (1989).
3. R. Vijayaraghavan, J. Gopalakrishnan, and C. N. R. Rao, *J. Mater. Chem.* 2, 327 (1992).
4. C. Michel, C. Martin, M. Hervieu, A. Maignan, J. Provost, M. Huve, and B. Raveau, *J. Solid State Chem.* 96, 271 (1992).
5. T. L. Aselage, E. L. Venturini, S. B. van Deusen, T. J. Headley, M. O. Eatough, and J. A. Voigt, *Physica C*, in press.
6. F. Izumi, J. D. Jorgensen, Y. Shimakawa, Y. Kubo, T. Manako, S. Pei, T. Matsumoto, R. L. Hitterman, and Y. Kanke, *Physica C* 193, 426 (1992).
7. Y. Kubo, Y. Shimakawa, T. Manako, and H. Igarashi, *Phys. Rev. B* 43, 7875 (1991).
8. Y.-Q. Song, M. Lee, W. P. Halperin, L. M. Tonge, and T. J. Marks, *Phys. Rev. B* 45, 4945 (1992).
9. P. G. Allen, J. M. de Leon, S. D. Conradson, and A. R. Bishop, *Phys. Rev. B* 44, 9480 (1991).
10. D. Mihalovic, K. F. McCarty, and D. S. Ginley, *Phys. Rev. B* 44, 237 (1991).
11. G. Jehl, T. Zetterer, H. H. Otto, J. Schützmann, S. Shulga, and K. F. Renk, *Europhys. Lett.* 17, 255 (1992).
12. R. M. Iyer, G. M. Phatak, K. Gangadharan, M. D. Sastry, T. M. Kadam, P. V. P. S. S. Sastry, and J. V. Yakhmi, *Physica C* 160, 155 (1989).
13. T. Kaneko, H. Yamauchi, and S. Tanaka, *Physica C* 178, 377 (1991).
14. R. S. Liu, J. L. Tallon, and P. P. Edwards, *Physica C* 182, 119 (1991).
15. M. J. Tsai, S. F. Wu, Y. T. Huang, and S. W. Lu, *Physica C* 198, 125 (1992).
16. B. T. Ahn, W. Y. Lee, and R. Beyers, *Appl. Phys. Lett.* 60, 2150 (1992).
17. T. Kaneko, K. Hamada, S. Adachi, and H. Yamauchi, *Physica C* 197, 385 (1992).
18. M. R. Presland, J. L. Tallon, R. G. Buckley, R. S. Liu, and N. E. Flower, *Physica C* 176, 95 (1991).
19. R. S. Liu, P. T. Wu, J. M. Ling, and L. J. Chen, *Phys. Rev. B* 39, 2792 (1989).
20. K. Jyodoi, K. Fukushima, A. Sonoda, S. Yasuyama, and A. Ohyoshi, *Jpn. J. Appl. Phys.* 30, L2025 (1991).
21. Z. Iqbal, A. P. B. Sinha, D. E. Morris, J. C. Barry, G. J. Auchterlonie, and B. L. Ramakrishna, *Appl. Phys. Lett.* 70, 2234 (1991).
22. T. Maeda, K. Sakuyama, N. Sakai, H. Yamauchi, and S. Tanaka, *Physica C* 177, 337 (1991).
23. T. Wada, A. Ichinose, F. Izumi, A. Nara, H. Yamauchi, H. Asano, and S. Tanaka, *Physica C* 179, 455 (1991).

24. T. Wada, K. Hamada, A. Ichinose, T. Kaneko, H. Yamauchi, and S. Tanaka, *Physica C* 175, 529 (1991).

25. T. Mochiku, T. Nagashima, Y. Saito, M. Watahiki, H. Asano, and Y. Fukai, *Jpn. J. Appl. Phys.* 29, L588 (1990).

26. H. Chou, H. S. Chen, A. K. Kortan, L. C. Kimerling, F. A. Thiel, and M. K. Wu, *Appl. Phys. Lett.* 58, 2836 (1991).

27. D. J. Werder and S. H. Liou, *Physica C* 179, 430 (1991).

28. W. L. Holstein, L. A. Parisi, D. W. Face, X. D. Wu, S. R. Foltyn, and R. E. Muenchausen, *Appl. Phys. Lett.*, in press.

29. H. C. Lai, K. D. Vernon-Parry, J. D. Chern, and C. M. Grovenor, *Supercond. Sci. Technol.* 4, 306 (1991).

30. W. L. Holstein, L. A. Parisi, C. Wilker, and R. B. Flippen, *Appl. Phys. Lett.* 60, 2014, (1992).

31. M. L. Chu, H. L. Chang, C. Wang, J. Y. Juang, T. M. Uen, and Y. S. Gou, *Appl. Phys. Lett.* 59, 1123 (1991).

32. W. Y. Lee, B. T. Ahn, R. Beyers, J. Salem, R. Savoy, J. Vazquez, S. M. Garrison, M. Kawasaki, and E. L. Venturini, 1992, to be published.

33. J. A. DeLuca, P. L. Karas, J. E. Tkaczyk, C. L. Briant, M. F. Garbauskas, and P. J. Bednarczyk, *Proc. MRS Spring Meeting*, 1992.

34. T. Kamo, T. Doi, A. Soeta, T. Yuasa, N. Inoue, K. Aihara, and S. P. Matsuda, *Appl. Phys. Lett.* 59, 3186 (1991).

35. J. S. Martens, G. K. G. Hohenwarter, J. B. Beyer, J. E. Nordman, and D. S. Ginley, *J. Appl. Phys.* 65, 4057 (1989).

36. J. S. Martens, D. S. Ginley, J. B. Beyer, J. E. Nordman, and G. K. G. Hohenwarter, *IEEE Trans. Appl. Supercond.* 1, 95 (1991).

37. J. S. Martens V. M. Hietala, T. E. Zipperian, S. R. Kurtz, D. S. Ginley, C. P. Tigges, J. M. Phillips, and N. Newman, *IEEE Trans. Appl. Supercond.* (June 1992).

38. J. S. Martens, V. M. Hietala, T. E. Zipperian, D. S. Ginley, C. P. Tigges, and G. K. G. Hohenwater, *IEEE Microwave Guided Wave Lett.* 1, 291 (1991).

39. H. S. Kwok, J. P. Zheng, and S. H. Liou, *Physica C* 175, 573 (1991).

40. L. C. Bourne, R. B. Hammond, M. Robinson, M. M. Eddy, W. L. Olson, and T. W. James. *Appl. Phys. Lett.* 56, 2333 (1990).

41. J. P. Hong, T. W. Kim, H. R. Fetterman, A. H. Cardona, and L. C. Bourne, *Appl. Phys. Lett.* 59, 991 (1991).

42. Z. Zhang, C. C. Chen, C. M. Lieber, B. Morosin, D. S. Ginley, and E. L. Venturini, *Phys. Rev. B* 45, 987 (1992).

43. S. C. Cao, D. M. Zhang, D. L. Zhang, H. M. Duan, and A. M. Hermann, *Phys. Rev. B* 44, 12571 (1991).

44. Y. Ohta, T. Tohayama, and S. Maekawa, *Phys. Rev. B* 43, 2968 (1991).

45. Z. P. Han, R. Dupree, A. Gencten, R. S. Liu, and P. P. Edwards, *Phys. Rev. Lett.* 69, 1256 (1992).

46. Y. Kitaoka et al., *Physica C* 179, 107 (1991).

Index

9 780367 402273